MERCRUISER | Stern Drive
1992-00 REPAIR MANUAL
GASOLINE ENGINES

SELOC

Managing Partners	Dean F. Morgantini
	Barry L. Beck
Executive Editor	Kevin M. G. Maher, A.S.E.
Production Managers	Melinda Possinger
	Ronald Webb

Manufactured in USA
© 2002 Seloc Publishing, Inc.
104 Willowbrook Lane
West Chester, PA 19382
ISBN: 13 978-0-89330-053-1
ISBN 10: 0-89330-053-5
17th Printing 2109876543

www.selocmarine.com
1-866-SELOC55

CONTENTS

CONTENTS

IGNITION AND ELECTRICAL SYSTEMS — 10

DRIVE SYSTEMS ALPHA — 11

DRIVE SYSTEMS BRAVO/BLACKHAWK — 12

DRIVE SYSTEMS TRANSMISSIONS — 13

TRIM AND TILT — 14

STEERING — 15

GLOSSARY

MASTER INDEX

SAFETY NOTICE

Proper service and repair procedures are vital to the safe, reliable operation of all marine engines, as well as the personal safety of those performing repairs. This manual outlines procedures for servicing and repairing engines and drive systems using safe, effective methods. The procedures contain many NOTES, CAUTIONS and WARNINGS which should be followed, along with standard procedures, to minimize the possibility of personal injury or improper service which could damage the vehicle or compromise its safety.

It is important to note that repair procedures and techniques, tools and parts for servicing these engines, as well as the skill and experience of the individual performing the work, vary widely. It is not possible to anticipate all of the conceivable ways or conditions under which the engine may be serviced, or to provide cautions as to all possible hazards that may result. Standard and accepted safety precautions and equipment should be used during cutting, grinding, chiseling, prying, or any other process that can cause material removal or projectiles.

Some procedures require the use of tools specially designed for a specific task. Before substituting another tool or procedure, you must be completely satisfied that neither your personal safety, nor the performance of the vessel, will be endangered. All procedures covered in this manual requiring the use of special tools will be noted at the beginning of the procedure by means of an **OEM symbol**

Additionally, any procedure requiring the use of an electronic tester or scan tool will be noted at the beginning of the procedure by means of a **DVOM symbol**

Although information in this manual is based on industry sources and is complete as possible at the time of publication, the possibility exists that some manufacturers made later changes which could not be included here. While striving for total accuracy, Seloc Publishing cannot assume responsibility for any errors, changes or omissions that may occur in the compilation of this data. We must therefore warn you to follow instructions carefully, using common sense. If you are uncertain of a procedure, seek help by inquiring with someone in your area who is familiar with these motors before proceeding.

PART NUMBERS

Part numbers listed in this reference are not recommendations by Seloc Publishing for any particular product brand name, simply iterations of the manufacturer's suggestions. They are also references that can be used with interchange manuals and aftermarket supplier catalogs to locate each brand supplier's discrete part number.

SPECIAL TOOLS

Special tools are recommended by the manufacturers to perform a specific job. Use has been kept to a minimum, but, where absolutely necessary, they are referred to in the text by the part number of the manufacturer if at all possible; and also noted at the beginning of each procedure with one of the following symbols: **OEM** or **DVOM**.

The **OEM** symbol usually denotes the need for a unique tool purposely designed to accomplish a specific task, it will also be used, less frequently, to notify the reader of the need for a tool that is not commonly found in the average tool box.

The **DVOM** symbol is used to denote the need for an electronic test tool like an ohmmeter, multi-meter or, on cetain later engines, a scan tool.

These tools can be purchased, under the appropriate part number, from your local dealer or regional distributor, or an equivalent tool can be purchased locally from a tool supplier or parts outlet. Before substituting any tool for the one recommended, read the SAFETY NOTICE at the top of this page.

Providing the correct mix of service and repair procedures is an endless battle for any publisher of "How-To" information. Users range from first time do-it yourselfers to professionally trained marine technicians, and information important to one is frequently irrelevant to the other. The editors at Seloc Publishing strive to provide accurate and articulate information on all facets of marine engine repair, from the simplest procedure to the most complex. In doing this, we understand that certain procedures may be outside the capabilities of the average DIYer. Conversely we are aware that many procedures are unnecessary for a trained technician.

SKILL LEVELS

In order to provide all of our users, particularly the DIYers, with a feeling for the scope of a given procedure or task before tackling it we have included a rating system denoting the suggested skill level needed when performing a particular procedure. One of the following icons will be included at the beginning of most procedures:

EASY. These procedures are aimed primarily at the DIYer and can be classified, for the most part, as basic maintenance procedures; battery, fluids, filters, plugs, etc. Although certainly valuable to any experience level, they will generally be of little importance to a technician.

MODERATE. These procedures are suited for a DIYer with experience and a working knowledge of mechanical procedures. Even an advanced DIYer or professional technician will occasionally refer to these procedures. They will generally consist of component repair and service procedures, adjustments and minor rebuilds.

DIFFICULT. These procedures are aimed at the advanced DIYer and professional technician. They will deal with diagnostics, rebuilds and internal engine/drive components and will frequently require special tools.

SKILLED. These procedures are aimed at highly skilled technicians and should not be attempted without previous experience. They will usually consist of machine work, internal engine work and gear case rebuilds.

Please remember one thing when considering the above ratings—they are a guide for judging the complexity of a given procedure and are subjective in nature. Only you will know what your experience level is, and only you will know when a procedure may be outside the realm of your capability. First time DIYer, or life-long marine technician, we all approach repair and service differently so an easy procedure for one person may be a difficult procedure for another, regardless of experience level. All skill level ratings are meant to be used as a guide only! Use them to help make a judgement before undertaking a particular procedure, but by all means read through the procedure first and make your own decision—after all, our mission at Seloc is to make boat maintenance and repair easier for everyone whether you are changing the oil or rebuilding an engine. Enjoy boating!

ACKNOWLEDGMENTS

Seloc Publishing expresses appreciation to the following companies who supported the production of this book:

- Mercury Marine—Fond du Lac, WI
- Belks Marine—Holmes, PA
- Marine Mechanics Institute—Orlando, FL

Thanks to John Hartung and Judy Belk of Belk's Marine for allowing us full access to their dealership for a portion of our photo-shoot.

Seloc Publishing would like to express thanks to the fine companies who participate in the production of all our books:

- Hand tools supplied by Craftsman are used during all phases of our vehicle teardown and photography.
- Many of the fine specialty tools used in our procedures were provided courtesy of Lisle Corporation.
- Much of our shop's electronic testing equipment was supplied by Universal Enterprises Inc. (UEI).

1

GENERAL INFORMATION AND BOATING SAFETY

HOW TO USE THIS MANUAL

This manual is designed to be a handy reference guide to maintaining and repairing your Mercruiser Stern Drive. We strongly believe that regardless of how many or how few years experience you may have, there is something new waiting here for you.

This manual covers the topics that a factory service manual (designed for factory trained mechanics) and a manufacturer owner's manual covers. It will take you through the basics of maintaining and repairing your engine/drive, step-by-step, to help you understand what the factory trained mechanics already know by heart. By using the information in this manual, any boat owner should be able to make better-informed decisions about what they need to do to maintain and enjoy their craft.

Even if you never plan on touching a wrench (and if so, we hope that you will change your mind), this manual will still help you understand what a mechanic needs to do in order to maintain your engine.

Can You Do It?

If you are not the type who is prone to taking a wrench to something, NEVER FEAR. The procedures in this manual cover topics at a level virtually anyone will be able to handle. And just the fact that you purchased this manual shows your interest in better understanding your engine and drive unit.

You may find that maintaining your engine/drive unit yourself is preferable in most cases. From a monetary standpoint, it could also be beneficial. The money spent on hauling your boat to a marina and paying a tech to service the engine could buy you fuel for a whole weekend's boating. If you are unsure of your own mechanical abilities, at the very least you should fully understand what a marine mechanic does to your boat. You may decide that anything other than maintenance and adjustments should be performed by a mechanic (and that's your call), but know that every time you board your boat, you are placing faith in the mechanic's work and trusting him or her with your well-being, and maybe your life.

It should also be noted that in most areas a factory trained mechanic will command a hefty hourly rate for off site service. This hourly rate is charged from the time they leave their shop to the time they return home. The cost savings in doing the job yourself should be readily apparent at this point.

Where To Begin

Before spending any money on parts, and before removing any nuts or bolts, read through the entire procedure or topic. This will give you the overall view of what tools and supplies will be required to perform the procedure or what questions need to be answered before purchasing parts. So read ahead and plan ahead. Each operation should be approached logically and all procedures thoroughly understood before attempting any work.

Avoiding Trouble

Some procedures in this manual may require you to "label and disconnect . . ." a group of lines, hoses or wires. Don't be lulled into thinking you can remember where everything goes—you won't. If you reconnect or install a part incorrectly, things may operate poorly, if at all. If you hook up electrical wiring incorrectly, you may instantly learn a very, very expensive lesson.

A piece of masking tape, for example, placed on a hose and another on its fitting will allow you to assign your own label such as the letter "A", or a short name. As long as you remember your own code the lines can be reconnected by matching letters or names. Do remember that tape will dissolve when saturated in fluids. If a component is to be washed or cleaned, use another method of identification. A permanent felt-tipped marker can be very handy for marking metal parts; but remember that fluids will remove permanent marker.

SAFETY is the most important thing to remember when performing maintenance or repairs. Be sure to read the information on safety in this manual.

Maintenance or Repair

Proper maintenance is the key to long and trouble-free engine life, and the work can yield its own rewards. A properly maintained engine performs better than one that is neglected. As a conscientious boat owner, set aside a Saturday morning, at least once a month, to perform a thorough check of items that could cause problems. Keep your own personal log to jot down which services you performed, how much the parts cost you, the date, and the amount of hours on the engine at the time. Keep all receipts for parts

purchased, so that they may be referred to in case of related problems or to determine operating expenses. As a do-it-yourselfer, these receipts are the only proof you have that the required maintenance was performed. In the event of a warranty problem, these receipts will be invaluable.

It's necessary to mention the difference between maintenance and repair. Maintenance includes routine inspections, adjustments, and replacement of parts that show signs of normal wear. Maintenance compensates for wear or deterioration. Repair implies that something has broken or is not working. A need for repair is often caused by lack of maintenance.

For example: draining and refilling the engine oil is maintenance recommended by all manufacturers at specific intervals. Failure to do this can allow internal corrosion or damage and impair the operation of the engine, requiring expensive repairs. While no maintenance program can prevent items from breaking or wearing out, a general rule can be stated: MAINTENANCE IS CHEAPER THAN REPAIR.

Directions and Locations

◆ See Figure 1

Two basic rules should be mentioned here. First, whenever the Port side of the engine (or boat) is referred to, it is meant to specify the left side of the engine when you are sitting at the helm. Conversely, the Starboard means your right side. The Bow is the front of the boat and the Stern is the rear.

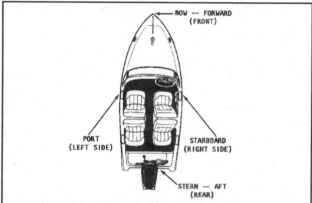

Fig. 1 Common terminology used for reference designation on boats of all size. These terms are used through out the manual

Most screws and bolts are removed by turning counterclockwise, and tightened by turning clockwise. An easy way to remember this is: righty-tighty; lefty-loosey; corny, but effective. And if you are really dense (and we have all been so at one time or another), buy a ratchet that is marked ON and OFF, or mark your own.

Professional Help

Occasionally, there are some things when working on an engine or drive unit that are beyond the capabilities or tools of the average Do-It-Yourselfer (DIYer). This shouldn't include most of the topics of this manual, but you will have to be the judge. Some engines require special tools or a selection of special parts, even for basic maintenance.

Talk to other boaters who use the same model of engine or drive unit and speak with a trusted marina to find if there is a particular system or component on your engine that is difficult to maintain. For example, although the technique of valve adjustment on some engines may be easily understood and even performed by a DIYer, it might require a handy assortment of shims in various sizes and a few hours of disassembly to get to that point. Not having the assortment of shims handy might mean multiple trips back and forth to the parts store, and this might not be worth your time.

You will have to decide for yourself where basic maintenance ends and where professional service should begin. Take your time and do your research first (starting with the information in this manual) and then make your own decision. If you really don't feel comfortable with attempting a procedure, DON'T DO IT! If you've gotten into something that may be over your head, don't panic. Tuck your tail between your legs and call a marine mechanic. Marinas and independent shops will be able to finish a job for you. Your ego may be damaged, but your boat will be properly restored to its full running order. So, as long as you approach jobs slowly and carefully, you really have nothing to lose and everything to gain by doing it yourself.

Purchasing Parts

◆ **See Figures 2 and 3**

When purchasing parts there are two things to consider. The first is quality and the second is to be sure to get the correct part for your engine. To get quality parts, always deal directly with a reputable retailer. To get the proper parts always refer to the information tag on your engine prior to calling the parts counter. An incorrect part can adversely affect your engine performance and fuel economy, and will cost you more money and aggravation in the end.

Fig. 2 By far the most important asset in purchasing parts is a knowledgeable and enthusiastic parts person

Just remember, a tow back to shore will cost plenty. That charge is per hour from the time the towboat leaves its home port, to the time it returns. Get the picture....$$$?

So who should you call for parts? Well, there are many sources for the parts you will need. Where you shop for parts will be determined by what kind of parts you need, how much you want to pay, and the types of stores in your neighborhood.

Your marina can supply you with many of the common parts you require. Using a marina as your parts supplier may be handy because of location (just walk right down the dock) or because the marina specializes in your particular brand of engine. In addition, it is always a good idea to get to know the marina staff (especially the marine mechanic).

The marine parts jobber, who is usually listed in the yellow pages or whose name can be obtained from the marina, is another excellent source for parts. In addition to supplying local marinas, they also do a sizeable business in over-the-counter parts sales for the do-it-yourselfer.

Almost every community has one or more convenient marine chain store. These stores often offer the best retail prices and the convenience of one-stop shopping for all your needs. Since they cater to the do-it-yourselfer, these stores are almost always open weeknights, Saturdays, and Sundays, when many jobbers are usually closed.

The lowest prices for parts are most often found in discount stores or the auto department of mass merchandisers. Parts sold here are name and private brand parts bought in huge quantities, so they can offer a

Fig. 3 Parts catalogs, giving application and part number information, are provided by manufacturers for most replacement parts

competitive price. Private brand parts are made by major manufacturers and sold to large chains under a store label.

Avoiding the Most Common Mistakes

There are 3 common mistakes in mechanical work:

1. Incorrect order of assembly, disassembly or adjustment. When taking something apart or putting it together, performing steps in the wrong order usually just costs you extra time; however, it CAN break something. Read the entire procedure before beginning disassembly. Perform everything in the order in which the instructions say you should, even if you can't immediately see a reason for it. When you're taking apart something that is very intricate, you might want to draw a picture of how it looks when assembled at one point in order to make sure you get everything back in its proper position. When making adjustments, perform them in the proper order; often, one adjustment affects another, and you cannot expect satisfactory results unless each adjustment is made only when it cannot be changed by another.

2. Overtorquing (or undertorquing). While it is more common for overtorquing to cause damage, undertorquing may allow a fastener to vibrate loose causing serious damage. Especially when dealing with aluminum parts, pay attention to torque specifications and utilize a torque wrench in assembly. If a torque figure is not available, remember that if you are using the right tool to perform the job, you will probably not have to strain yourself to get a fastener tight enough. The pitch of most threads is so slight that the tension you put on the wrench will be multiplied many times in actual force on what you are tightening.

3. Cross-threading. This occurs when a part such as a bolt is screwed into a nut or casting at the wrong angle and forced. Cross-threading is more likely to occur if access is difficult. It helps to clean and lubricate fasteners, then to start threading with the part to be installed positioned straight in. Always start a fastener, etc. with your fingers. If you encounter resistance, unscrew the part and start over again at a different angle until it can be inserted and turned several times without much effort. Keep in mind that some parts may have tapered threads, so that gentle turning will automatically bring the part you're threading to the proper angle, but only if you don't force it or resist a change in angle. Don't put a wrench on the part until it has been tightened a couple of turns by hand. If you suddenly encounter resistance, and the part has not seated fully, don't force it. Pull it back out to make sure it's clean and threading properly.

BOATING SAFETY

In 1971 Congress ordered the U.S. Coast Guard to improve recreational boating safety. In response, the Coast Guard drew up a set of regulations.

In addition to these federal regulations, there are state and local laws you must follow. These sometimes exceed the Coast Guard requirements. This section discusses only the federal laws. State and local laws are available from your local Coast Guard. As with other laws, "Ignorance of the

boating laws is no excuse." The rules fall into two groups: regulations for your boat and required safety equipment on your boat.

Most boats on waters within Federal jurisdiction must be registered or documented. These waters are those that provide a means of transportation between two or more states or to the sea. They also include the territorial waters of the United States.

Regulations For Your Boat

DOCUMENTING OF VESSELS

A vessel of five or more net tons may be documented as a yacht. In this process, papers are issued by the U.S. Coast Guard as they are for large ships. Documentation is a form of national registration. The boat must be used solely for pleasure. Its owner must be a U.S. citizen, a partnership of U.S. citizens, or a corporation controlled by U.S. citizens. The captain and other officers must also be U.S. citizens. The crew need not be.

If you document your yacht, you have the legal authority to fly the yacht ensign. You also may record bills of sale, mortgages, and other papers of title with federal authorities. Doing so gives legal notice that such instruments exist. Documentation also permits preferred status for mortgages. This gives you additional security and aids financing and transfer of title. You must carry the original documentation papers aboard your vessel. Copies will not suffice.

REGISTRATION OF VESSELS

If your boat is not documented, registration in the state of its principal use is probably required. If you use it mainly on an ocean, a gulf, or other similar water, register in the state where you moor it.

If you use your boat solely for racing, it may be exempt from the requirement in your state. States may also exclude dinghies. Some require registration of documented vessels and non-power driven boats.

All states, except Alaska, register boats. In Alaska, the U.S. Coast Guard issues the registration numbers. If you move your vessel to a new state of principal use, a valid registration certificate is good for 60 days. You must have the registration certificate (certificate of number) aboard your vessel when it is in use. A copy will not suffice. You may be cited if you do not have the original on board.

NUMBERING OF VESSELS

A registration number is on your registration certificate. You must paint or permanently attach this number to both sides of the forward half of your boat. Do not display any other number there.

The registration number must be clearly visible. It must not be placed on the obscured underside of a flared bow. If you can't place the number on the bow, place it on the forward half of the hull. If that doesn't work, put it on the superstructure. Put the number for an inflatable boat on a bracket or fixture. Then, firmly attach it to the forward half of the boat. The letters and numbers must be plain block characters and must read from left to right. Use a space or a hyphen to separate the prefix and suffix letters from the numerals. The color of the characters must contrast with that of the background, and they must be at least three inches high.

In some states your registration is good for only one year. In others, it is good for as long as three years. Renew your registration before it expires. At that time you will receive a new decal or decals. Place them as required by state law. You should remove old decals before putting on the new ones. Some states require that you show only the current decal or decals. If your vessel is moored, it must have a current decal even if it is not in use.

If your vessel is lost, destroyed, abandoned, stolen, or transferred, you must inform the issuing authority. If you lose your certificate of number or your address changes, notify the issuing authority as soon as possible.

SALES AND TRANSFERS

Your registration number is not transferable to another boat. The number stays with the boat unless its state of principal use is changed.

HULL IDENTIFICATION NUMBER

A Hull Identification Number (HIN) is like the Vehicle Identification Number (VIN) on your car. Boats built between November 1, 1972 and July 31, 1984 have old format HINs. Since August 1, 1984 a new format has been used.

Your boat's HIN must appear in two places. If it has a transom, the primary number is on its starboard side within two inches of its top. If it does not have a transom or if it was not practical to use the transom, the number is on the starboard side. In this case, it must be within one foot of the stern and within two inches of the top of the hull side. On pontoon boats, it is on the aft crossbeam within one foot of the starboard hull attachment. Your boat

also has a duplicate number in an unexposed location. This is on the boat's interior or under a fitting or item of hardware.

LENGTH OF BOATS

For some purposes, boats are classed by length. Required equipment, for example, differs with boat size. Manufacturers may measure a boat's length in several ways. Officially, though, your boat is measured along a straight line from its bow to its stern. This line is parallel to its keel.

The length does not include bowsprits, boomkins, or pulpits. Nor does it include rudders, brackets, outboard motors, outdrives, diving platforms, or other attachments.

CAPACITY INFORMATION

◆ See Figure 4

Manufacturers must put capacity plates on most recreational boats less than 20 feet long. Sailboats, canoes, kayaks, and inflatable boats are usually exempt. Outboard boats must display the maximum permitted horsepower of their engines. The plates must also show the allowable maximum weights of the people on board. And they must show the allowable maximum combined weights of people, engines, and gear. Inboards and stern drives need not show the weight of their engines on their capacity plates. The capacity plate must appear where it is clearly visible to the operator when underway. This information serves to remind you of the capacity of your boat under normal circumstances. You should ask yourself, "Is my boat loaded above its recommended capacity" and, "Is my boat overloaded for the present sea and wind conditions?" If you are stopped by a legal authority, you may be cited if you are overloaded.

Fig. 4 A U.S. Coast Guard certification plate indicates the amount of occupants and gear appropriate for safe operation of the vessel

CERTIFICATE OF COMPLIANCE

Manufacturers are required to put compliance plates on motorboats greater than 20 feet in length. The plates must say, "This boat," or "This equipment complies with the U. S. Coast Guard Safety Standards in effect on the date of certification." Letters and numbers can be no less than one-eighth of an inch high. At the manufacturer's option, the capacity and compliance plates may be combined.

VENTILATION

A cup of gasoline spilled in the bilge has the potential explosive power of 15 sticks of dynamite. This statement, commonly quoted over 20 years ago, may be an exaggeration, however, it illustrates a fact. Gasoline fumes in the bilge of a boat are highly explosive and a serious danger. They are heavier than air and will stay in the bilge until they are vented out.

Because of this danger, Coast Guard regulations require ventilation on many power boats. There are several ways to supply fresh air to engine and gasoline tank compartments and to remove dangerous vapors. Whatever the choice, it must meet Coast Guard standards.

■ **The following is not intended to be a complete discussion of the regulations. It is limited to the majority of recreational vessels. Contact your local Coast Guard office for further information.**

General Precautions

Ventilation systems will not remove raw gasoline that leaks from tanks or fuel lines. If you smell gasoline fumes, you need immediate repairs. The best device for sensing gasoline fumes is your nose. Use it! If you smell gasoline in an engine compartment or elsewhere, don't start your engine. The smaller the compartment, the less gasoline it takes to make an explosive mixture.

Ventilation for Open Boats

In open boats, gasoline vapors are dispersed by the air that moves through them. So they are exempt from ventilation requirements.

To be "open," a boat must meet certain conditions. Engine and fuel tank compartments and long narrow compartments that join them must be open to the atmosphere." This means they must have at least 15 square inches of open area for each cubic foot of net compartment volume. The open area must be in direct contact with the atmosphere. There must also be no long, unventilated spaces open to engine and fuel tank compartments into which flames could extend.

Ventilation for All Other Boats

Powered and natural ventilation are required in an enclosed compartment with a permanently installed gasoline engine that has a cranking motor. A compartment is exempt if its engine is open to the atmosphere. Diesel powered boats are also exempt.

VENTILATION SYSTEMS

There are two types of ventilation systems. One is "natural ventilation." In it, air circulates through closed spaces due to the boat's motion. The other type is "powered ventilation." In it, air is circulated by a motor driven fan or fans.

Natural Ventilation System Requirements

A natural ventilation system has an air supply from outside the boat. The air supply may also be from a ventilated compartment or a compartment open to the atmosphere. Intake openings are required. In addition, intake ducts may be required to direct the air to appropriate compartments.

The system must also have an exhaust duct that starts in the lower third of the compartment. The exhaust opening must be into another ventilated compartment or into the atmosphere. Each supply opening and supply duct, if there is one, must be above the usual level of water in the bilge. Exhaust openings and ducts must also be above the bilge water. Openings and ducts must be at least three square inches in area or two inches in diameter. Openings should be placed so exhaust gasses do not enter the fresh air intake. Exhaust fumes must not enter cabins or other enclosed, non-ventilated spaces. The carbon monoxide gas in them is deadly.

Intake and exhaust openings must be covered by cowls or similar devices. These registers keep out rain water and water from breaking seas. Most often, intake registers face forward and exhaust openings aft. This aids the flow of air when the boat is moving or at anchor since most boats face into the wind when anchored.

Power Ventilation System Requirements

◆ See Figure 5

Powered ventilation systems must meet the standards of a natural system. They must also have one or more exhaust blowers. The blower duct can serve as the exhaust duct for natural ventilation if fan blades do not obstruct the air flow when not powered. Openings in engine compartment, for carburetion are in addition to ventilation system requirements.

Required Safety Equipment

Coast Guard regulations require that your boat have certain equipment aboard. These requirements are minimums. Exceed them whenever you can.

TYPES OF FIRES

There are four common classes of fires:
- Class A—fires are in ordinary combustible materials such as paper or wood.
- Class B—fires involve gasoline, oil and grease.
- Class C—fires are electrical.

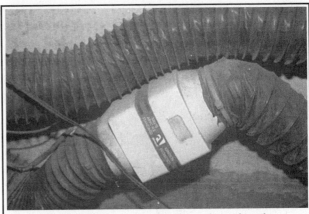

Fig. 5 Typical blower and duct system to vent fumes from the engine compartment

- Class D—fires involve ferrous metals

One of the greatest risks to boaters is fire. This is why it is so important to carry the correct number and type of extinguishers onboard.

The best fire extinguisher for most boats is a Class B extinguisher. Never use water on Class B or Class C fires, as water spreads these types of fires. You should never use water on a Class C fire as it may cause you to be electrocuted.

FIRE EXTINGUISHERS

◆ See Figure 6

If your boat meets one or more of the following conditions, you must have at least one fire extinguisher aboard. The conditions are:
- Inboard or stern drive engines
- Closed compartments under seats where portable fuel tanks can be stored
- Double bottoms not sealed together or not completely filled with flotation materials
- Closed living spaces
- Closed stowage compartments in which combustible or flammable materials are stored
- Permanently installed fuel tanks
- Boat is 26 feet or more in length.

Contents of Extinguishers

Fire extinguishers use a variety of materials. Those used on boats usually contain dry chemicals, Halon, or Carbon Dioxide (CO_2). Dry chemical extinguishers contain chemical powders such as Sodium Bicarbonate—baking soda.

Carbon dioxide is a colorless and odorless gas when released from an extinguisher. It is not poisonous but caution must be used in entering compartments filled with it. It will not support life and keeps oxygen from reaching your lungs. A fire-killing concentration of Carbon Dioxide is lethal. If you are in a compartment with a high concentration of CO_2, you will have no difficulty breathing. But the air does not contain enough oxygen to support life. Unconsciousness or death can result.

HALON EXTINGUISHERS

Some fire extinguishers and 'built-in' or 'fixed' automatic fire extinguishing systems contain a gas called Halon. Like carbon dioxide it is colorless and odorless and will not support life. Some Halons may be toxic if inhaled.

To be accepted to the Coast Guard, a fixed Halon system must have an indicator light at the vessel's helm. A green light shows the system is ready. Red means it is being discharged or has been discharged. Warning horns are available to let you know the system has been activated. If your fixed Halon system discharges, ventilate the space thoroughly before you enter it. There are no residues from Halon but it will not support life.

Although Halon has excellent fire fighting properties, it is thought to deplete the earth's ozone layer and has not been manufactured since January 1, 1994. Halon extinguishers can be refilled from existing stocks of the gas until they are used up, but high federal excise taxes are being

Fig. 6 An approved fire extinguisher should be mounted close to the operator for emergency use

charged for the service. If you discontinue using your Halon extinguisher, take it to a recovery station rather than releasing the gas into the atmosphere. Compounds such as FE 241, designed to replace Halon, are now available.

Fire Extinguisher Approval

Fire extinguishers must be Coast Guard approved. Look for the approval number on the nameplate. Approved extinguishers have the following on their labels: "Marine Type USCG Approved, Size..., Type..., 162.208/," etc. In addition, to be acceptable by the Coast Guard, an extinguisher must be in serviceable condition and mounted in its bracket. An extinguisher not properly mounted in its bracket will not be considered serviceable during a Coast Guard inspection.

Care and Treatment

Make certain your extinguishers are in their stowage brackets and are not damaged. Replace cracked or broken hoses. Nozzles should be free of obstructions. Sometimes, wasps and other insects nest inside nozzles and make them inoperable. Check your extinguishers frequently. If they have pressure gauges, is the pressure within acceptable limits? Do the locking pins and sealing wires show they have not been used since recharging?

Don't try an extinguisher to test it. Its valves will not reseat properly and the remaining gas will leak out. When this happens, the extinguisher is useless.

Weigh and tag carbon dioxide and Halon extinguishers twice a year. If their weight loss exceeds 10 percent of the weight of the charge, recharge them. Check to see that they have not been used. They should have been inspected by a qualified person within the past six months, and they should have tags showing all inspection and service dates. The problem is that they can be partially discharged while appearing to be fully charged.

Some Halon extinguishers have pressure gauges the same as dry chemical extinguishers. Don't rely too heavily on the gauge. The extinguisher can be partially discharged and still show a good gauge reading. Weighing a Halon extinguisher is the only accurate way to assess its contents.

If your dry chemical extinguisher has a pressure indicator, check it frequently. Check the nozzle to see if there is powder in it. If there is, recharge it. Occasionally invert your dry chemical extinguisher and hit the base with the palm of your hand. The chemical in these extinguishers packs and cakes due to the boat's vibration and pounding. There is a difference of opinion about whether hitting the base helps, but it can't hurt. It is known that caking of the chemical powder is a major cause of failure of dry chemical extinguishers. Carry spares in excess of the minimum requirement. If you have guests aboard, make certain they know where the extinguishers are and how to use them.

Using a Fire Extinguisher

A fire extinguisher usually has a device to keep it from being discharged accidentally. This is a metal or plastic pin or loop. If you need to use your extinguisher, take it from its bracket. Remove the pin or the loop and point the nozzle at the base of the flames. Now, squeeze the handle, and discharge the extinguisher's contents while sweeping from side to side. Recharge a used extinguisher as soon as possible.

If you are using a Halon or carbon dioxide extinguisher, keep your hands away from the discharge. The rapidly expanding gas will freeze them. If your fire extinguisher has a horn, hold it by its handle.

Legal Requirements for Extinguishers

You must carry fire extinguishers as defined by Coast Guard regulations. They must be firmly mounted in their brackets and immediately accessible.

A motorboat less than 26 feet long must have at least one approved hand-portable, Type B-1 extinguisher. If the boat has an approved fixed fire extinguishing system, you are not required to have the Type B-1 extinguisher. Also, if your boat is less than 26 feet long, is propelled by an outboard motor, or motors, and does not have any of the first six conditions described at the beginning of this section, it is not required to have an extinguisher. Even so, it's a good idea to have one, especially if a nearby boat catches fire, or if a fire occurs at a fuel dock.

A motorboat 26 feet to under 40 feet long, must have at least two Type B-1 approved hand-portable extinguishers. It can, instead, have at least one Coast Guard approved Type B-2. If you have an approved fire extinguishing system, only one Type B-1 is required.

A motorboat 40 to 65 feet long must have at least three Type B-1 approved portable extinguishers . It may have, instead, at least one Type B-1 plus a Type B-2. If there is an approved fixed fire extinguishing system, two Type B-1 or one Type B-2 is required.

WARNING SYSTEM

Various devices are available to alert you to danger. These include fire, smoke, gasoline fumes, and carbon monoxide detectors. If your boat has a galley, it should have a smoke detector. Where possible, use wired detectors. Household batteries often corrode rapidly on a boat.

You can't see, smell, nor taste carbon monoxide gas, but it is lethal. As little as one part in 10,000 parts of air can bring on a headache. The symptoms of carbon monoxide poisoning—headaches, dizziness, and nausea—are like sea sickness. By the time you realize what is happening to you, it may be too late to take action. If you have enclosed living spaces on your boat, protect yourself with a detector. There are many ways in which carbon monoxide can enter your boat.

PERSONAL FLOTATION DEVICES

Personal Flotation Devices (PFDs) are commonly called life preservers or life jackets. You can get them in a variety of types and sizes. They vary with their intended uses. To be acceptable, they must be Coast Guard approved.

Type I PFDs

A Type I life jacket is also called an offshore life jacket. Type I life jackets will turn most unconscious people from facedown to a vertical or slightly backward position. The adult size gives a minimum of 22 pounds of buoyancy. The child size has at least 11 pounds. Type I jackets provide more protection to their wearers than any other type of life jacket. Type I life jackets are bulkier and less comfortable than other types. Furthermore, there are only two sizes, one for children and one for adults.

Type I life jackets will keep their wearers afloat for extended periods in rough water. They are recommended for offshore cruising where a delayed rescue is probable.

Type II PFDs

◆ See Figure 7

A Type II life jacket is also called a near-shore buoyant vest. It is an approved, wearable device. Type II life jackets will turn some unconscious people from facedown to vertical or slightly backward positions. The adult

Fig. 7 Type II approved flotation devices are recommended for inshore and island cruising on calm water. Use them were there is a good chance of fast rescue

Fig. 8 Type IV buoyant cushions are made to be thrown to people in the water. If you can squeeze air out of the cushion, it is faulty and should be replaced

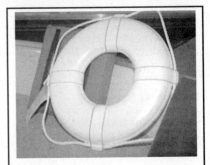

Fig 9. Type IV throwables, such as this ring life buoy, are not designed as personal flotation devices for unconscious people, non-swimmers, or children

size gives at least 15.5 pounds of buoyancy. The medium child size has a minimum of 11 pounds. And the small child and infant sizes give seven pounds. A Type II life jacket is more comfortable than a Type I but it does not have as much buoyancy. It is not recommended for long hours in rough water. Because of this, Type IIs are recommended for inshore and inland cruising on calm water. Use them where there is a good chance of fast rescue.

Type III PFDs

Type III life jackets or marine buoyant devices are also known as flotation aids. Like Type IIs, they are designed for calm inland or close offshore water where there is a good chance of fast rescue. Their minimum buoyancy is 15.5 pounds. They will not turn their wearers face up.

Type III devices are usually worn where freedom of movement is necessary. Thus, they are used for water skiing, small boat sailing, and fishing among other activities. They are available as vests and flotation coats. Flotation coats are useful in cold weather. Type IIIs come in many sizes from small child through large adult.

Life jackets come in a variety of colors and patterns—red, blue, green, camouflage, and cartoon characters. From a safety standpoint, the best color is bright orange. It is easier to see in the water, especially if the water is rough.

Type IV PFDs

Type IV ring life buoys, buoyant cushions and horseshoe buoys are Coast Guard approved devices called throwables. They are made to be thrown to people in the water, and should not be worn. Type IV cushions are often used as seat cushions. Cushions are hard to hold onto in the water. Thus, they do not afford as much protection as wearable life jackets.

The straps on buoyant cushions are for you to hold onto either in the water or when throwing them. A cushion should never be worn on your back. It will turn you face down in the water.

Type IV throwables are not designed as personal flotation devices for unconscious people, non-swimmers, or children. Use them only in emergencies. They should not be used for, long periods in rough water.

Ring life buoys come in 18, 20, 24, and 30 inch diameter sizes. They have grab lines. You should attach about 60 feet of polypropylene line to the grab rope to aid in retrieving someone in the water. If you throw a ring, be careful not to hit the person. Ring buoys can knock people unconscious

Type V PFDs

Type V PFDs are of two kinds, special use devices and hybrids. Special use devices include boardsailing vests, deck suits, work vests, and others. They are approved only for the special uses or conditions indicated on their labels. Each is designed and intended for the particular application shown on its label. They do not meet legal requirements for general use aboard recreational boats.

Hybrid life jackets are inflatable devices with some built-in buoyancy provided by plastic foam or kapok. They can be inflated orally or by cylinders of compressed gas to give additional buoyancy. In some hybrids the gas is released manually. In others it is released automatically when the life jacket is immersed in water.

The inherent buoyancy of a hybrid may be insufficient to float a person unless it is inflated. The only way to find this out is for the user to try it in the water. Because of its limited buoyancy when deflated, a hybrid is recommended for use by anon-swimmer only if it is worn with enough inflation to float the wearer.

If they are to count against the legal requirement for the number of life jackets you must carry on your vessel, hybrids manufactured before February 8, 1995 must be worn whenever a boat is underway and the wearer is not below decks or in an enclosed space. To find out if your Type V hybrid must be worn to satisfy the legal requirement, read its label. If its use is restricted it will say, "REQUIRED TO BE WORN" in capital letters.

Hybrids cost more than other life jackets, but this factor must be weighed against the fact that they are more comfortable than Type I, II or III life jackets. Because of their greater comfort, their owners are more likely to wear them than are the owners of Type I, II or III life jackets.

The Coast Guard has determined that improved, less costly hybrids can save lives since they will be bought and used more frequently. For these reasons a new federal regulation was adopted effective February 8, 1995. The regulation increases both the deflated and inflated buoyancys of hybrids, makes them available in a greater variety of sizes and types, and reduces their costs by reducing production costs.

Even though it may not be required, the wearing of a hybrid or a life jacket is encouraged whenever a vessel is underway. Like life jackets, hybrids are now available in three types. To meet legal requirements, a Type I hybrid can be substituted for a Type I life jacket. Similarly Type II and III hybrids can be substituted for Type II and Type III life jackets. A Type I hybrid, when inflated, will turn most unconscious people from facedown to vertical or slightly backward positions just like a Type I life jacket. Type I and III hybrids function like Type II and III life jackets. If you purchase a new hybrid, it should have an owner's manual attached which describes its life jacket type and its deflated and inflated buoyancys. It warns you that it may have to be inflated to float you. The manual also tells you how to don the life jacket and how to inflate it. It also tells you how to change its inflation mechanism, recommended testing exercises, and inspection and maintenance procedures. The manual also tells you why you need a life jacket and why you should wear it. A new hybrid must be packaged with at least three gas cartridges. One of these may already be loaded into the inflation mechanism. Likewise, if it has an automatic inflation mechanism, it must be packaged with at least three of these water sensitive elements. One of these elements may be installed.

Legal Requirements

A Coast Guard approved life jacket must show the manufacturer's name and approval number. Most are marked as Type I, II, III, IV or V. All of the newer hybrids are marked for type.

You are required to carry at least one wearable life jacket or hybrid for each person on board your recreational vessel. If your vessel is 16 feet or more in length and is not a canoe or a kayak, you must also have at least one Type IV on board. These requirements apply to all recreational vessels that are propelled or controlled by machinery, sails, oars, paddles, poles, or another vessel. Sailboards are not required to carry life jackets.

You can substitute an older Type V hybrid for any required Type I, II or III life jacket provided that its approval label shows it is approved for the activity the vessel is engaged in, approved as a substitute for a life jacket of the type required on the vessel, used as required on the labels, and used in accordance with any requirements in its owner's manual, if the approval label makes reference to such a manual.

A water skier being towed is considered to be on board the vessel when judging compliance with legal requirements.

You are required to keep your Type I, II or III life jackets or equivalent hybrids readily accessible, which means you must be able to reach out and get them when needed. All life jackets must be in good, serviceable condition.

General Considerations

The proper use of a life jacket requires the wearer to know how it will perform. You can gain this knowledge only through experience. Each person on your boat should be assigned a life jacket. Next, it should be fitted to the person who will wear it. Only then can you be sure that it will be ready for use in an emergency.

Boats can sink fast. There may be no time to look around for a life jacket. Fitting one on you in the water is almost impossible. This advice is good even if the water is calm, and you intend to boat near shore. Most drownings occur in inland waters within a few feet of safety. Most victims had life jackets, but they weren't wearing them.

Keeping life jackets in the plastic covers they came wrapped in and in a cabin assures that they will stay clean and unfaded. But this is no way to keep them when you are on the water. When you need a life jacket it must be readily accessible and adjusted to fit you. You can't spend time hunting for it or learning how to fit it.

There is no substitute for the experience of entering the water while wearing a life jacket. Children, especially, need practice. If possible, give your guests this experience. Tell them they should keep their arms to their sides when jumping in to keep the life jacket from riding up. Let them jump in and see how the life jacket responds. Is it adjusted so it does not ride up? Is it the proper size? Are all straps snug? Are children's life jackets the right sizes for them? Are they adjusted properly? If a child's life jacket fits correctly, you can lift the child by the jacket's shoulder straps and the child's chin and ears will not slip through. Non-swimmers, children, handicapped persons, elderly persons and even pets should always wear life jackets when they are aboard. Many states require that everyone aboard wear them in hazardous waters.

Inspect your lifesaving equipment from time to time. Leave any questionable or unsatisfactory equipment on shore. An emergency is no time for you to conduct an inspection.

Indelibly mark your life jackets with your vessel's name, number, and calling port. This can be important in a search and rescue effort. It could help concentrate effort where it will do the most good.

Care of Life Jackets

Given reasonable care, life jackets last many years. Thoroughly dry them before putting them away. Stow them in dry, well-ventilated places. Avoid the bottoms of lockers and deck storage boxes where moisture may collect. Air and dry them frequently.

Life jackets should not be tossed about or used as fenders or cushions. Many contain kapok or fibrous glass material enclosed in plastic bags. The bags can rupture and are then unserviceable. Squeeze your life jacket gently. Does air leak out? If so, water can leak in and it will no longer be safe to use. Cut it up so no one will use it, and throw it away. The covers of some life jackets are made of nylon or polyester. These materials are plastics. Like many plastics, they break down after extended exposure to the ultraviolet light in sunlight. This process may be more rapid when the materials are dyed with bright dyes such as "neon" shades.

Ripped and badly faded fabric are clues that the covering of your life jacket is deteriorating. A simple test is to pinch the fabric between your thumbs and forefingers. Now try to tear the fabric. If it can be torn, it should definitely be destroyed and discarded. Compare the colors in protected places to those exposed to the sun. If the colors have faded, the materials have been weakened. A fabric-covered life jacket should ordinarily last several boating seasons with normal use. A life jacket used every day in direct sunlight should probably be replaced more often.

SOUND PRODUCING DEVICES

All boats are required to carry some means of making an efficient sound signal. Devices for making the whistle or horn noises required by the Navigation Rules must be capable of a four second blast. The blast should be audible for at least one-half mile. Athletic whistles are not acceptable on boats 12 meters or longer. Use caution with athletic whistles. When wet, some of them come apart and loose their "pea." When this happens, they are useless.

If your vessel is 12 meters long and less than 20 meters, you must have a power whistle (or power horn) and a bell on board. The bell must be in operating condition and have a minimum diameter of at least 200 mm (7.9 inches) at its mouth.

VISUAL DISTRESS SIGNALS

◆ See Figure 10

Visual Distress Signals (VDS) attract attention to your vessel if you need help. They also help to guide searchers in search and rescue situations. Be sure you have the right types, and learn how to use them properly.

It is illegal to fire flares improperly. In addition, they cost the Coast Guard and its Auxiliary many wasted hours in fruitless searches. If you signal a distress with flares and then someone helps you, please let the Coast Guard or the appropriate Search And Rescue Agency (SAR) know so the distress report will be canceled.

Recreational boats less than 16 feet long must carry visual distress signals on coastal waters at night. Coastal waters are:

- The ocean (territorial sea)
- The Great Lakes
- Bays or sounds that empty into oceans
- Rivers over two miles across at their mouths upstream to where they narrow to two miles.

Recreational boats 16 feet or longer must carry VDS at all times on coastal waters. The same requirement applies to boats carrying six or fewer passengers for hire. Open sailboats less than 26 feet long without engines are exempt in the daytime as are manually propelled boats. Also exempt are boats in organized races, regattas, parades, etc. Boats owned in the United States and operating on the high seas must be equipped with VDS.

A wide variety of signaling devices meet Coast Guard regulations. For pyrotechnic devices, a minimum of three must be carried. Any combination can be carried as long as it adds up to at least three signals for day use and at least three signals for night use. Three day/night signals meet both requirements. If possible, carry more than the legal requirement.

■ **The American flag flying upside down is a commonly recognized distress signal. It is not recognized in the Coast Guard regulations, though. In an emergency, your efforts would probably be better used in more effective signaling methods**.

Types of VDS

VDS are divided into two groups; daytime and nighttime use. Each of these groups is subdivided into pyrotechnic and non-pyrotechnic devices.

Fig. 10 Internationally accepted distress signals

DAYTIME NON-PYROTECHNIC SIGNALS

A bright orange flag with a black square over a black circle is the simplest VDS. It is usable, of course, only in daylight. It has the advantage of being a continuous signal. A mirror can be used to good advantage on sunny days. It can attract the attention of other boaters and of aircraft from great distances. Mirrors are available with holes in their centers to aid in "aiming." In the absence of a mirror, any shiny object can be used. When another boat is in sight, an effective VDS is to extend your arms from your sides and move them up and down. Do it slowly. If you do it too fast the other people may think you are just being friendly. This simple gesture is seldom misunderstood, and requires no equipment.

DAYTIME PYROTECHNIC DEVICES

Orange smoke is a useful daytime signal. Hand-held or floating smoke flares are very effective in attracting attention from aircraft. Smoke flares don't last long, and are not very effective in high wind or poor visibility. As with other pyrotechnic devices, use them only when you know there is a possibility that someone will see the display.

To be usable, smoke flares must be kept dry. Keep them in airtight containers and store them in dry places. If the "striker" is damp, dry it out before trying to ignite the device. Some pyrotechnic devices require a forceful "strike" to ignite them.

All hand-held pyrotechnic devices may produce hot ashes or slag when burning. Hold them over the side of your boat in such a way that they do not burn your hand or drip into your boat.

Nighttime Non-Pyrotechnic Signals

An electric distress light is available. This light automatically flashes the international morse code SOS distress signal (••• —- •••). Flashed four to six times a minute, it is an unmistakable distress signal. It must show that it is approved by the Coast Guard. Be sure the batteries are fresh. Dated batteries give assurance that they are current.

Under the Inland Navigation Rules, a high intensity white light flashing 50-70 times per minute is a distress signal. Therefore, use strobe lights on inland waters only for distress signals.

Nighttime Pyrotechnic Devices

◆ See Figure 11

Aerial and hand-held flares can be used at night or in the daytime. Obviously, they are more effective at night.

Currently, the serviceable life of a pyrotechnic device is rated at 42 months from its date of manufacture. Pyrotechnic devices are expensive. Look at their dates before you buy them. Buy them with as much time remaining as possible.

Like smoke flares, aerial and hand-held flares may fail to work if they have been damaged or abused. They will not function if they are or have been wet. Store them in dry, airtight containers in dry places. But store them where they are readily accessible.

Aerial VDSs, depending on their type and the conditions they are used in, may not go very high. Again, use them only when there is a good chance they will be seen.

A serious disadvantage of aerial flares is that they burn for only a short time. Most burn for less than 10 seconds. Most parachute flares burn for less than 45 seconds. If you use a VDS in an emergency, do so carefully. Hold hand-held flares over the side of the boat when in use. Never use a road hazard flare on a boat, it can easily start a fire. Marine type flares are carefully designed to lessen risk, but they still must be used carefully.

Aerial flares should be given the same respect as firearms since they are firearms! Never point them at another person. Don't allow children to play with them or around them. When you fire one, face away from the wind. Aim it downwind and upward at an angle of about 60 degrees to the horizon. If there is a strong wind, aim it somewhat more vertically. Never fire it straight up. Before you discharge a flare pistol, check for overhead obstructions. These might be damaged by the flare. They might deflect the flare to where it will cause damage.

Disposal of VDS

Keep outdated flares when you get new ones. They do not meet legal requirements, but you might need them sometime, and they may work. It is

Fig. 11 Moisture protected flares should be carried onboard any vessel for use as a distress signal

illegal to fire a VDS on federal navigable waters unless an emergency exists. Many states have similar laws.

Emergency Position Indicating Radio Beacon (EPIRB)

There is no requirement for recreational boats to have EPIRBs. Some commercial and fishing vessels, though, must have them if they operate beyond the three mile limit. Vessels carrying six or fewer passengers for hire must have EPIRBs under some circumstances when operating beyond the three mile limit. If you boat in a remote area or offshore, you should have an EPIRB. An EPIRB is a small (about 6 to 20 inches high), battery-powered, radio transmitting buoy-like device. It is a radio transmitter and requires a license or an endorsement on your radio station license by the Federal Communications Commission (FCC). EPIRBs are activated by being immersed in water or by a manual switch.

Equipment Not Required But Recommended

Although not required by law, there are other pieces of equipment that are good to have onboard.

SECOND MEANS OF PROPULSION

All boats less than 16 feet long should carry a second means of propulsion. A paddle or oar can come in handy at times. For most small boats, a spare trolling or outboard motor is an excellent idea. If you carry a spare motor, it should have its own fuel tank and starting power. If you use an electric trolling motor, it should have its own battery.

BAILING DEVICES

All boats should carry at least one effective manual bailing device in addition to any installed electric bilge pump. This can be a bucket, can, scoop, hand operated pump, etc. If your battery "goes dead" it will not operate your electric pump.

FIRST AID KIT

◆ See Figure 12

All boats should carry a first aid kit. It should contain adhesive bandages, gauze, adhesive tape, antiseptic, aspirin, etc. Check your first aid kit from time to time. Replace anything that is outdated. It is to your advantage to know how to use your first aid kit. Another good idea would be to take a Red Cross first aid course.

Fig.12 Always carry an adequately stocked first aid kit on board for the safety of the crew and guests

Fig. 13 Choose and anchor of sufficient weight to secure the boat without dragging. In some cases separate anchors may be needed for different situations

Fig. 14 Carry a few tools and some spare parts, and learn how to make minor repairs.

ANCHORS

◆ See Figure 13

All boats should have anchors. Choose one of suitable size for your boat. Better still, have two anchors of different sizes. Use the smaller one in calm water or when anchoring for a short time to fish or eat. Use the larger one when the water is rougher or for overnight anchoring.

Carry enough anchor line of suitable size for your boat and the waters in which you will operate. If your engine fails you, the first thing you usually should do is lower your anchor. This is good advice in shallow water where you may be driven aground by the wind or water. It is also good advice in windy weather or rough water. The anchor will usually hold your bow into the waves.

VHF-FM RADIO

Your best means of summoning help in an emergency or in case of a breakdown is a VHF-FM radio. You can use it to get advice or assistance from the Coast Guard. In the event of a serious illness or injury aboard your boat, the Coast Guard can have emergency medical equipment meet you ashore.

TOOLS AND SPARE PARTS

◆ See Figure 14

Carry a few tools and some spare parts, and learn how to make minor repairs. Many search and rescue cases are caused by minor breakdowns that boat operators could have repaired. If your engine is an inboard or stern drive, carry spare belts and water pump impellers and the tools to change them.

Courtesy Marine Examination

One of the roles of the Coast Guard Auxiliary is to promote recreational boating safety. This is why they conduct thousands of Courtesy Marine Examinations each year. The auxiliarists who do these examinations are well-trained and knowledgeable in the field.

These examinations are free and done only at the consent of boat owners. To pass the examination, a vessel must satisfy federal equipment requirements and certain additional requirements of the coast guard auxiliary. If your vessel does not pass the Courtesy Marine Examination, no report of the failure is made. Instead, you will be told what you need to correct the deficiencies. The examiner will return at your convenience to redo the examination.

If your vessel qualifies, you will be awarded a safety decal. The decal does not carry any special privileges, it simply attests to your interest in safe boating.

SAFETY IN SERVICE

It is virtually impossible to anticipate all of the hazards involved with maintenance and service, but care and common sense will prevent most accidents.

Use common sense whenever you work on your boat or motor. If a situation arises that doesn't seem right, sit back and have a second look. It may save an embarrassing moment or potential damage to your boat.

Do's

• Do keep a fire extinguisher and first aid kit handy.
• Do wear safety glasses or goggles when cutting, drilling, grinding or prying, even if you have 20-20 vision. If you wear glasses for the sake of vision, wear safety goggles over your regular glasses.
• Do shield your eyes whenever you work around the battery. Batteries contain sulfuric acid. In case of contact with the eyes or skin, flush the area with water or a mixture of water and baking soda, then seek immediate medical attention.
• Do use adequate ventilation when working with any chemicals or hazardous materials.
• Do disconnect the negative battery cable when working on the electrical system. The secondary ignition system contains EXTREMELY HIGH VOLTAGE. In some cases it can even exceed 50,000 volts.
• Do follow manufacturer's directions whenever working with potentially hazardous materials.
• Do properly maintain your tools. Loose hammerheads, mushroomed punches and chisels, frayed or poorly grounded electrical cords, excessively worn screwdrivers, spread wrenches (open end), cracked sockets, or slipping ratchets can cause accidents.
• Do be sure that adjustable wrenches are tightly closed on the nut or bolt and pulled so that the force is on the side of the fixed jaw. Better yet, avoid the use of an adjustable if you have a fixed wrench that will fit.

Don'ts

• Don't run the engine in an enclosed area or anywhere else without proper ventilation—EVER! Carbon monoxide is poisonous; it takes a long time to leave the human body and you can build up a deadly supply of it in your system by simply breathing in a little every day.
• Don't work around moving parts while wearing loose clothing. Short sleeves are much safer than long, loose sleeves. Hard-toed shoes with neoprene soles protect your toes and give a better grip on slippery surfaces. Jewelry, watches, large belt buckles, or body adornment of any kind is not safe working around any vehicle. Long hair should be tied back under a hat.
• Don't use pockets for toolboxes. A fall or bump can drive a screwdriver deep into your body. Even a rag hanging from your back pocket can wrap around a spinning shaft.
• Don't smoke when working around gasoline, cleaning solvent or other flammable material. Don't smoke when working around the battery. When the battery is being charged, it gives off explosive hydrogen gas.
• Don't use gasoline to wash your hands; there are excellent soaps available. Gasoline contains dangerous additives which can enter the body through a cut or through your pores. Gasoline also removes all the natural oils from the skin so that bone-dry hands will suck up oil and grease.
• Don't use screwdrivers for anything other than driving screws! A screwdriver used as an prying tool can snap when you least expect it, causing injuries. At the very least, you'll ruin a good screwdriver.

2

TOOLS AND EQUIPMENT

TOOLS AND EQUIPMENT

Safety Tools

WORK GLOVES

◆ **See Figures 1 and 2**

Unless you think scars on your hands are cool, enjoy pain and like wearing bandages, get a good pair of work gloves. Canvas or leather is the best. And yes, we realize that there are some jobs involving small parts that can't be done while wearing work gloves. These jobs are not the ones usually associated with hand injuries.

A good pair of rubber gloves (such as those usually associated with dish washing) or vinyl gloves is also a great idea. There are some liquids such as solvents and penetrants that don't belong on your skin. Avoid burns and rashes. Wear these gloves.

And lastly, an option. If you're tired of being greasy and dirty all the time, go to the drug store and buy a box of disposable latex gloves like medical professionals wear. You can handle greasy parts, perform small tasks, wash parts, etc. all without getting dirty! These gloves take a surprising amount of abuse without tearing and aren't expensive. Note however, that it has been reported that some people are allergic to latex or the powder used inside some gloves, so pay attention to what you buy.

EYE AND EAR PROTECTION

◆ **See Figures 3 and 4**

Don't begin any job without a good pair of work goggles or impact resistant glasses! When doing any kind of work, it's all too easy to avoid eye injury through this simple precaution. And don't just buy eye protection and leave it on the shelf. Wear it all the time! Things have a habit of breaking, chipping, splashing, spraying, splintering and flying around. And, for some reason, your eye is always in the way!

If you wear vision correcting glasses as a matter of routine, get a pair made with polycarbonate lenses. These lenses are impact resistant and are available at any optometrist.

Often overlooked is hearing protection. Power equipment is noisy! Loud noises damage your ears. It's as simple as that! The simplest and cheapest form of ear protection is a pair of noise-reducing ear plugs; cheap insurance for your ears. And, they may even come with their own, cute little carrying case.

More substantial, more protection and more money is a good pair of noise-reducing earmuffs. They protect from all but the loudest sounds. Hopefully those are sounds that you'll never encounter since they're usually associated with disasters.

WORK CLOTHES

Everyone has "work clothes." Usually these consist of old jeans and a shirt that has seen better days. That's fine. In addition, a denim work apron is a nice accessory. It's rugged, can hold some spare bolts, and you don't feel bad wiping your hands or tools on it. That's what it's for.

When working in cold weather, a one-piece, thermal work outfit is invaluable. Most are rated to below zero (Fahrenheit) temperatures and are ruggedly constructed. Just look at what the marine mechanics are wearing and that should give you a clue as to what type of clothing is good.

Chemicals

There is a whole range of chemicals that you'll find handy for maintenance work. The most common types are, lubricants, penetrants and sealers. Keep these handy onboard. There are also many chemicals that are used for detailing or cleaning.

When a particular chemical is not being used, keep it capped, upright and in a safe place. These substances may be flammable, may be irritants or might even be caustic and should always be stored properly, used properly and handled with care. Always read and follow all label directions and be sure to wear hand and eye protection!

LUBRICANTS & PENETRANTS

◆ **See Figure 5**

Anti-seize is used to coat certain fasteners prior to installation. This can be especially helpful when two dissimilar metals are in contact (to help prevent corrosion that might lock the fastener in place). This is a good practice on a lot of different fasteners, BUT, NOT on any fastener which might vibrate loose causing a problem. If anti-seize is used on a fastener, it should be checked periodically for proper tightness.

Lithium grease, chassis lube, silicone grease or synthetic brake caliper grease can all be used pretty much interchangeably. All can be used for coating rust-prone fasteners and for facilitating the assembly of parts that are a tight fit. Silicone and synthetic greases are the most versatile.

■ **Silicone dielectric grease is a non-conductor that is often used to coat the terminals of wiring connectors before fastening them. It may sound odd to coat metal portions of a terminal with something that won't conduct electricity, but here is it how it works. When the connector is fastened the metal-to-metal contact between the terminals will displace the grease (allowing the circuit to be completed). The grease that is displaced will then coat the non-contacted surface and the cavity around the terminals, SEALING them from atmospheric moisture that could cause corrosion.**

Silicone spray is a good lubricant for hard-to-reach places and parts that shouldn't be gooped up with grease.

Penetrating oil may turn out to be one of your best friends when taking something apart that has corroded fasteners. Not only can they make a job easier, they can really help to avoid broken and stripped fasteners. The most familiar penetrating oils are Liquid Wrench® and WD-40®. A newer penetrant, PB Blaster® also works well. These products have hundreds of uses. For your purposes, they are vital!

Before disassembling any part (especially on an exhaust system), check the fasteners. If any appear rusted, soak them thoroughly with the penetrant and let them stand while you do something else (for particularly rusted or

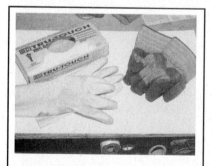

Fig. 1 Three different types of work gloves. The box contains latex gloves

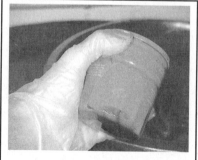

Fig. 2 Latex gloves come in handy when you are doing messy jobs

Fig. 3 Don't begin any job without a good pair of work goggles or impact resistant glasses. Also good noise reducing earmuffs are cheap insurance to protect your hearing

Fig. 4 Things have a habit of breaking, chipping, splashing, spraying, splintering and flying around. And, for some reason, your eye is always in the way

Fig. 5 Anti-seize, penetrating oil, lithium grease, electronic cleaner and silicone spray. These products have hundreds of uses and should be a part of your chemical tool collection

Fig. 6 Sealants are essential for preventing leaks

frozen parts you may need to soak them a few days in advance). This simple act can save you hours of tedious work trying to extract a broken bolt or stud.

SEALANTS

◆ See Figures 6 and 7

Sealants are an indispensable part for certain tasks, especially if you are trying to avoid leaks. The purpose of sealants is to establish a leak-proof bond between or around assembled parts. Most sealers are used in conjunction with gaskets, but some are used instead of conventional gasket material.

The most common sealers are the non-hardening types such as Permatex® No.2 or its equivalents. These sealers are applied to the mating surfaces of each part to be joined, then a gasket is put in place and the parts are assembled.

■ **A sometimes overlooked use for sealants like RTV is on the threads of vibration prone fasteners.**

One very helpful type of non-hardening sealer is the "high tack" type. This type is a very sticky material that holds the gasket in place while the parts are being assembled. This stuff is really a good idea when you don't have enough hands or fingers to keep everything where it should be.

The stand-alone sealers are the Room Temperature Vulcanizing (RTV) silicone gasket makers. On some engines, this material is used instead of a gasket. In those instances, a gasket may not be available or, because of the shape of the mating surfaces, a gasket shouldn't be used. This stuff, when used in conjunction with a conventional gasket, produces the surest bonds.

RTV does have its limitations though. When using this material, you will have a time limit. It starts to set-up within 15 minutes or so, so you have to assemble the parts without delay. In addition, when squeezing the material out of the tube, don't drop any glops into the engine. The stuff will form and set and travel around the oil gallery, possibly plugging up a passage. Also, most types are not fuel-proof. Check the tube for all cautions.

CLEANERS

◆ See Figures 8 and 9

There are two types of cleaners on the market today: parts cleaners and hand cleaners. The parts cleaners are for the parts; the hand cleaners are for you. They are not interchangeable.

There are many good, non-flammable, biodegradable parts cleaners on the market. These cleaning agents are safe for you, the parts and the environment. Therefore, there is no reason to use flammable, caustic or toxic substances to clean your parts or tools.

As far as hand cleaners go, the waterless types are the best. They have always been efficient at cleaning, but leave a pretty smelly odor. Recently though, just about all of them have eliminated the odor and added stuff that actually smells good. Make sure that you pick one that contains lanolin or some other moisture-replenishing additive. Cleaners not only remove grease and oil but also skin oil.

■ **Most women will tell you to use a hand lotion when you're all cleaned up. It's okay. Real men DO use hand lotion! Believe it or not, using hand lotion BEFORE your hands are dirty will actually make them easier to clean when you're finished with a dirty job. Lotion seals your hands, and keeps dirty and grease from sticking to your skin.**

Fig. 7 On some engines, RTV is used instead of gasket material to seal components

Fig. 8 The new citrus hand cleaners not only work well, but they smell pretty good too. Choose one with pumice for added cleaning power

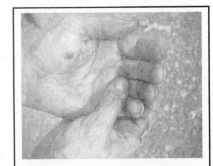

Fig. 9 The use of hand lotion seals your hands and keeps grease from sticking to your skin

TOOLS

◆ See Figure10

Tools; this subject could fill a completely separate manual. The first thing you need to ask yourself, is just how involved do you plan to get. If you are serious about your maintenance you will want to gather a quality set of tools

to make the job easier, and more enjoyable. BESIDES, TOOLS ARE FUN!!!

Almost every do-it-yourselfer loves to accumulate tools. Though most find a way to perform jobs with only a few common tools, they tend to buy more over time, as money allows. So gathering the tools necessary for maintenance does not have to be an expensive, overnight proposition.

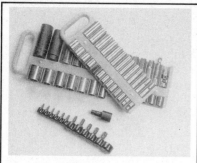

Fig. 10 Socket holders, especially the magnetic type, are handy items to keep tools in order

Fig. 11 A 3/8 inch socket set is probably the most versatile tool in any mechanic's tool box

Fig. 12 A swivel (U-joint) adapter (left), a 1/4 inch-to-3/8 inch adapter (center) and a 3/8 inch-to1/4 inch adapter (right)

When buying tools, the saying "You get what you pay for …" is absolutely true! Don't go cheap! Any hand tool that you buy should be drop forged and/or chrome vanadium. These two qualities tell you that the tool is strong enough for the job. With any tool, go with a name that you've heard of before, or, that is recommended buy your local professional retailer. Let's go over a list of tools that you'll need.

Most of the world uses the metric system. However, some American-built engines and aftermarket accessories use standard fasteners. So, accumulate your tools accordingly. Any good DIYer should have a decent set of both U.S. and metric measure tools.

■ Don't be confused by terminology. Most advertising refers to "SAE and metric", or "standard and metric." Both are misnomers. The Society of Automotive Engineers (SAE) did not invent the English system of measurement; the English did. The SAE likes metrics just fine. Both English (U.S.) and metric measurements are SAE approved. Also, the current "standard" measurement IS metric. So, if it's not metric, it's U.S. measurement.

Hand Tools

SOCKET SETS

◆ See Figures 11, 12 , 13, 14, 15, 16 and 17

Socket sets are the most basic hand tools necessary for repair and maintenance work. For our purposes, socket sets come in three drive sizes: 1/4 inch, 3/8 inch and 1/2 inch. Drive size refers to the size of the drive lug on the ratchet, breaker bar or speed handle.

A 3/8 inch set is probably the most versatile set in any mechanic's toolbox. It allows you to get into tight places that the larger drive ratchets can't and gives you a range of larger sockets that are still strong enough for heavy-duty work. The socket set that you'll need should range in sizes from 3/8 inch through 1 inch for standard fasteners, and a 6mm through 19mm for metric fasteners.

You'll need a good 1/2 inch set since this size drive lug assures that you won't break a ratchet or socket on large or heavy fasteners. Also, torque wrenches with a torque scale high enough for larger fasteners are usually

1/2 inch drive.

1/4 inch drive sets can be very handy in tight places. Though they usually duplicate functions of the 3/8 inch set, 1/4 inch drive sets are easier to use for smaller bolts and nuts.

As for the sockets themselves, they come in standard and deep lengths as well as 6 or 12 point.
6 and 12 points refers to how many sides are in the socket itself. Each has advantages. The 6 point socket is stronger and less prone to slipping which would strip a bolt head or nut. 12 point sockets are more common, usually less expensive and can operate better in tight places where the ratchet handle can't swing far.

Standard length sockets are good for just about all jobs, however, some stud-head bolts, hard-to-reach bolts, nuts on long studs, etc., require the deep sockets.

Most manufacturers use recessed hex-head fasteners to retain many of the engine parts. These fasteners require a socket with a hex shaped driver or a large sturdy hex key. To help prevent torn knuckles, we would recommend that you stick to the sockets on any tight fastener and leave the hex keys for lighter applications. Hex driver sockets are available individually or in sets just like conventional sockets.

More and more, manufacturers are using Torx ® head fasteners, which were once known as tamper resistant fasteners (because many people did not have tools with the necessary odd driver shape). They are still used where the manufacturer would prefer only knowledgeable mechanics or advanced Do-It-Yourselfers (DIYers) to work.

Torque Wrenches

◆ See Figure 18

In most applications, a torque wrench can be used to assure proper installation of a fastener. Torque wrenches come in various designs and most stores will carry a variety to suit your needs. A torque wrench should be used any time you have a specific torque value for a fastener. Keep in mind that because there is no worldwide standardization of fasteners, the charts at the end of this section are a general guideline and should be used with caution. If you are using the right tool for the job, you should not have to strain to tighten a fastener.

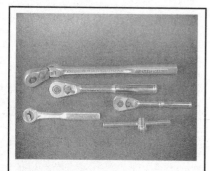

Fig. 13 Ratchets come in all sizes and configurations from rigid to swivel-headed

Fig. 14 Standard length sockets (top) are good for just about all jobs. However, some bolts may require deep sockets (bottom)

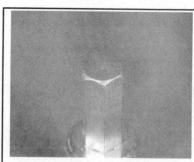

Fig. 15 Hex-head fasteners retain many components on modern powerheads. These fasteners require a socket with a hex shaped driver

Fig. 16 Torx ® drivers.......

Fig. 17 . . . and tamper resistant drivers are required to remove special fasteners installed by the manufacturers

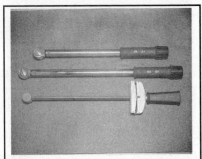

Fig. 18 Three types of torque wrenches. Top to bottom: a 3/8 inch drive beam type that reads in inch lbs., a 1/2 inch drive clicker type and a 1/2 inch drive beam type

BEAM TYPE

◆ See Figures 19 and 20

The beam type torque wrench is one of the most popular styles in use. If used properly, it can be the most accurate also. It consists of a pointer attached to the head that runs the length of the flexible beam (shaft) to a scale located near the handle. As the wrench is pulled, the beam bends and the pointer indicates the torque using the scale.

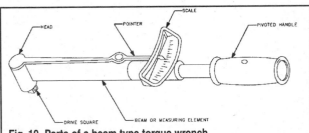

Fig. 19 Parts of a beam type torque wrench

CLICK (BREAKAWAY) TYPE

◆ See Figures 21 and 22

Another popular torque wrench design is the click type. The clicking mechanism makes achieving the proper torque easy and most use ratcheting head for ease of bolt installation. To use the click type wrench you pre-adjust it to a torque setting. Once the torque is reached, the wrench has a reflex signaling feature that causes a momentary breakaway of the torque wrench body, sending an impulse to the operator's hand.

Breaker Bars

◆ See Figure 23

Breaker bars are long handles with a drive lug. Their main purpose is to

provide extra turning force when breaking loose tight bolts or nuts. They come in all drive sizes and lengths. Always take extra precautions and use proper technique when using a breaker bar.

WRENCHES

◆ See Figures 24, 25, 26, 27 and 28

Basically, there are 3 kinds of fixed wrenches: open end, box end, and combination.

Open end wrenches have 2-jawed openings at each end of the wrench. These wrenches are able to fit onto just about any nut or bolt. They are extremely versatile but have one major drawback. They can slip on a worn or rounded bolt head or nut, causing bleeding knuckles and a useless fastener.

Box-end wrenches have a 360 deg circular jaw at each end of the wrench. They come in both 6 and 12 point versions just like sockets and each type has the same advantages and disadvantages as sockets.

Combination wrenches have the best of both. They have a 2-jawed open end and a box end. These wrenches are probably the most versatile.

As for sizes, you'll probably need a range similar to that of the sockets, about ¼ inch through 1 inch for standard fasteners, or 6mm through 19mm for metric fasteners. As for numbers, you'll need 2 of each size, since, in many instances, one wrench holds the nut while the other turns the bolt. On most fasteners, the nut and bolt are the same size so having two wrenches of the same size comes in handy.

■ **Although you will typically just need the sizes we specified, there are some exceptions. Occasionally you will find a nut that is larger. For these, you will need to buy ONE expensive wrench or a very large adjustable. Or you can always just convince the spouse that we are talking about SAFETY here and buy a whole (read expensive) large wrench set.**

One extremely valuable type of wrench is the adjustable wrench. An adjustable wrench has a fixed upper jaw and a moveable lower jaw; the lower jaw is moved by turning a threaded drum. The advantage of an adjustable wrench is its ability to be adjusted to just about any size fastener.

Fig. 20 A beam type torque wrench consists of a pointer attached to the head that runs the length of the flexible beam (shaft) to a scale located near the handle

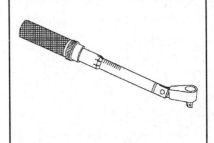

Fig. 21 A click type or breakaway torque wrench—note this one has a pivoting head

Fig. 22 Setting the proper torque on a click type torque wrench involves turning the handle until the proper torque specification appears on the dial

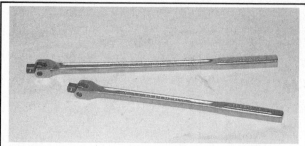

Fig. 23 Breaker bars are great for loosening large or stuck fasteners

The main drawback of an adjustable wrench is the lower jaw's tendency to move slightly under heavy pressure. This can cause the wrench to slip if it is not facing the right way. Pulling on an adjustable wrench in the proper direction will cause the jaws to lock in place. Adjustable wrenches come in a large range of sizes, measured by the wrench length.

PLIERS

◆ **See Figure 29**

Pliers are simply mechanical fingers. They are, more than anything, an extension of your hand. At least 3 pair of pliers are an absolute necessity—standard, needle nose and channel lock.

In addition to standard pliers there are the slip-joint, multi-position pliers such as ChannelLock® pliers and locking pliers, such as Vise Grips®.

Slip joint pliers are extremely valuable in grasping oddly sized parts and fasteners. Just make sure that you don't use them instead of a wrench too often since they can easily round off a bolt head or nut.

Locking pliers are usually used for gripping bolts or studs that can't be removed conventionally. You can get locking pliers in square jawed, needle-nosed and pipe-jawed. Locking pliers can rank right up behind duct tape as the handyman's best friend.

SCREWDRIVERS

You can't have too many screwdrivers. They come in 2 basic flavors, either standard or Phillips. Standard blades come in various sizes and thicknesses for all types of slotted fasteners. Phillips screwdrivers come in sizes with number designations from 1 on up, with the lower number designating the smaller size. Screwdrivers can be purchased separately or in sets.

HAMMERS

◆ **See Figure 30**

You always need a hammer for just about any kind of work. You need a ball-peen hammer for most metal work when using drivers and other like tools. A plastic hammer comes in handy for hitting things safely. A soft-faced dead-blow hammer is used for hitting things safely and hard. Hammers are also VERY useful with non air-powered impact drivers.

Other Common Tools

There are a lot of other tools that every DIYer will eventually need (though not all for basic maintenance). They include:

- Funnels (for adding fluid)
- Chisels
- Punches
- Files
- Hacksaw

INCHES	DECIMAL		DECIMAL	MILLIMETERS
1/8″	.125		.118	3mm
3/16″	.187		.157	4mm
1/4″	.250		.236	6mm
5/16″	.312		.354	9mm
3/8″	.375		.394	10mm
7/16″	.437		.472	12mm
1/2″	.500		.512	13mm
9/16″	.562		.590	15mm
5/8″	.625		.630	16mm
11/16″	.687		.709	18mm
3/4″	.750		.748	19mm
13/16″	.812		.787	20mm
7/8″	.875		.866	22mm
15/16″	.937		.945	24mm
1″	1.00		.984	25mm

Fig. 24 Comparsion of US measurement and metric wrench sizes

Fig. 25 Always use a backup wrench to prevent rounding flare nut fittings

Fig. 26 Note how the flare wrench sides are extended to grip the fitting tighter and prevent rounding

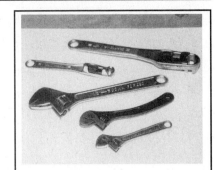

Fig. 27 Several types and sizes of adjustable wrenches

- Portable Bench Vise
- Tap and Die Set
- Flashlight
- Magnetic Bolt Retriever
- Gasket scraper
- Putty Knife
- Screw/Bolt Extractors
- Prybar

Hacksaws have just one use—cutting things off. You may wonder why you'd need one for something as simple as maintenance, but you never know. Among other things, guide studs to ease parts installation can be made from old bolts with their heads cut off.

A tap and die set might be something you've never needed, but you will eventually. It's a good rule, when everything is apart, to clean-up all threads, on bolts, screws and threaded holes. Also, you'll likely run across a situation in which stripped threads will be encountered. The tap and die set will handle that for you.

Gasket scrapers are just what you'd think, tools made for scraping old gasket material off of parts. You don't absolutely need one. Old gasket material can be removed with a putty knife or single edge razor blade. However, putty knives may not be sharp enough for some really stubborn gaskets and razor blades have a knack of breaking just when you don't want them to, inevitably slicing the nearest body part! As the old saying goes, "always use the proper tool for the job". If you're going to use a razor to scrape a gasket, be sure to always use a blade holder.

Putty knives really do have a use in a repair shop. Just because you remove all the bolts from a component sealed with a gasket doesn't mean it's going to come off. Most of the time, the gasket and sealer will hold it tightly. Lightly driving a putty knife at various points between the two parts will break the seal without damage to the parts.

A small—10 inches (20–25 centimeters) long—prybar is extremely useful for removing stuck parts.

■ Never use a screwdriver as a prybar! Screwdrivers are not meant for prying. Screwdrivers, used for prying, can break, sending the broken shaft flying!

Screw/bolt extractors are used for removing broken bolts or studs that have broke off flush with the surface of the part.

Special Tools

◆ See Figure 31

Almost every marine engine around today requires at least one special tool to perform a certain task. In most cases, these tools are specially designed to overcome some unique problem or to fit on some oddly sized component.

When manufacturers go through the trouble of making a special tool, it is usually necessary to use it to assure that the job will be done right. A special tool might be designed to make a job easier, or it might be used to keep you from damaging or breaking a part.

Don't worry, MOST basic maintenance procedures can either be performed without any special tools OR, because the tools must be used for such basic things, they are commonly available for a reasonable price. It is usually just the low production, highly specialized tools (like a super thin 7-point star-shaped socket capable of 150 ft. lbs. (203 Nm) of torque that is used only on the crankshaft nut of the limited production what-dya-callit engine) that tend to be outrageously expensive and hard to find. Luckily, you will probably never need such a tool.

Special tools can be as inexpensive and simple as an adjustable strap wrench or as complicated as an ignition tester. A few common specialty tools are listed here, but check with your dealer or with other boaters for help in determining if there are any special tools for YOUR particular engine. There is an added advantage in seeking advice from others, chances are they may have already found the special tool you will need, and know how to get it cheaper.

Electronic Tools

BATTERY TESTERS

The best way to test a non-sealed battery is using a hydrometer to check the specific gravity of the acid. Luckily, these are usually inexpensive and are available at most parts stores. Just be careful because the larger testers are usually designed for larger batteries and may require more acid than you will be able to draw from the battery cell. Smaller testers (usually a short, squeeze bulb type) will require less acid and should work on most batteries.

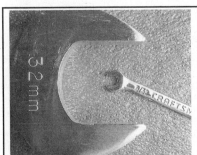

Fig. 28 Occasionally you will find a nut that requires a particularly large, or particularly small wrench. Rest assured that the proper wrench to fit is available at your local tool store

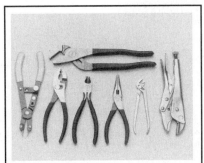

Fig. 29 Pliers and cutters come in many shapes and sizes. You should have an assortment on hand

Fig. 30 Three types of hammers. Top to bottom: ball peen, rubber dead-blow, and plastic

Fig. 31 Almost every marine engine around today requires at least one special tool to perform a certain task

Fig. 32 The Battery Tender ® is more than just a battery charger, when left connected, it keeps your battery fully charged

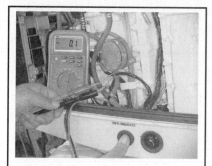

Fig. 33 Multimeters such as this one from UEI, are an extremely useful tool for troubleshooting electrical problems

Electronic testers are available and are often necessary to tell if a sealed battery is usable. Luckily, many parts stores have them on hand and are willing to test your battery for you.

BATTERY CHARGERS

◆ See Figure 32

If you are a weekend boater and take your boat out every week, then you will most likely want to buy a battery charger to keep your battery fresh. There are many types available, from low amperage trickle chargers to electronically controlled battery maintenance tools that monitor the battery voltage to prevent over or undercharging. This last type is especially useful if you store your boat for any length of time (such as during the severe winter months found in many Northern climates).

Even if you use your boat on a regular basis, you will eventually need a battery charger. Remember that most batteries are shipped dry and in a partial charged state. Before a new battery can be put into service it must be filled and properly charged. Failure to properly charge a battery (which was shipped dry) before it is put into service will prevent it from ever reaching a fully charged state.

MULTIMETERS

◆ See Figure 33

Multimeters are an extremely useful tool for troubleshooting electrical problems. They can be purchased in either analog or digital form and have a price range to suit any budget. A multimeter is a voltmeter, ammeter and ohmmeter (along with other features) combined into one instrument. It is often used when testing solid-state circuits because of its high input impedance (usually 10 megaohms or more). A brief description of the multimeter main test functions follows:

• Voltmeter—the voltmeter is used to measure voltage at any point in a circuit, or to measure the voltage drop across any part of a circuit. Voltmeters usually have various scales and a selector switch to allow the reading of different voltage ranges. The voltmeter has a positive and a negative lead. To avoid damage to the meter, always connect the negative lead to the negative (–) side of the circuit (to ground or nearest the ground side of the circuit) and connect the positive lead to the positive (+) side of the circuit (to the power source or the nearest power source). Note that the negative voltmeter lead will always be black and that the positive voltmeter will always be some color other than black (usually red).

• Ohmmeter—the ohmmeter is designed to read resistance (measured in ohms) in a circuit or component. Most ohmmeters will have a selector switch which permits the measurement of different ranges of resistance (usually the selector switch allows the multiplication of the meter reading by 10, 100, 1,000 and 10,000). Some ohmmeters are "auto-ranging" which means the meter itself will determine which scale to use. Since most meters are powered by an internal battery, the ohmmeter can be used like a self-powered test light. When the ohmmeter is connected, current from the ohmmeter flows through the circuit or component being tested. Since the ohmmeter's internal resistance and voltage are known values, the amount of current flow through the meter depends on the resistance of the circuit or component being tested. The ohmmeter can also be used to perform a continuity test for suspected open circuits. In using the meter for making continuity checks, do

not be concerned with the actual resistance readings. Zero resistance, or any ohm reading, indicates continuity in the circuit. Infinite resistance indicates an opening in the circuit. A high resistance reading where there should be none indicates a problem in the circuit. Checks for short circuits are made in the same manner as checks for open circuits, except that the circuit must be isolated from both power and normal ground. Infinite resistance indicates no continuity, while zero resistance indicates a dead short.

✳✳ WARNING

Never use an ohmmeter to check the resistance of a component or wire while there is voltage applied to the circuit.

• Ammeter—an ammeter measures the amount of current flowing through a circuit in units called amperes or amps. At normal operating voltage, most circuits have a characteristic amount of amperes, called "current draw" which can be measured using an ammeter. By referring to a specified current draw rating, measuring the amperes and then comparing the two values, one can determine what is happening within the circuit to aid in diagnosis. An open circuit, for example, will not allow any current to flow, so the ammeter reading will be zero. A damaged component or circuit will have an increased current draw, so the reading will be high. The ammeter is always connected in series with the circuit being tested. All of the current that normally flows through the circuit must also flow through the ammeter; if there is any other path for the current to follow, the ammeter reading will not be accurate. The ammeter itself has very little resistance to current flow and, therefore, will not affect the circuit, but it will measure current draw only when the circuit is closed and electricity is flowing. Excessive current draw can blow fuses and drain the battery, while a reduced current draw can cause motors to run slowly, lights to dim and other components to not operate properly.

GAUGES

Compression Gauge

◆ See Figure 34

An important element in checking the overall condition of your engine is to check compression. This becomes increasingly more important on outboards with high hours. Compression gauges are available as screw-in types and hold-in types. The screw-in type is slower to use, but eliminates the possibility of a faulty reading due to escaping pressure. A compression reading will uncover many problems that can cause rough running. Normally, these are not the sort of problems that can be cured by a tune-up.

Vacuum Gauge

◆ See Figures 35 and 36

Vacuum gauges are handy for discovering air leaks, late ignition or valve timing, and a number of other problems.

Eventually, you are going to have to measure something. To do this, you will need at least a few precision tools in addition to the special tools mentioned earlier.

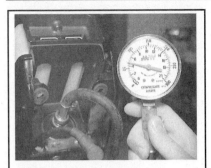

Fig. 34 Cylinder compression test results are extremely valuable indicators of internal engine condition

Fig. 35 Vacuum gauges are useful for many diagnostic tasks including testing of some fuel pumps

Fig. 36 In a pinch, you can also use the vacuum gauge on a hand-operated vacuum pump

Measuring Tools

MICROMETERS & CALIPERS

Micrometers and calipers are devices used to make extremely precise measurements. The simple truth is that you really won't have the need for many of these items just for simple maintenance. You will probably want to have at least one precision tool such as an outside caliper, but that should be sufficient to most basic maintenance procedures.

Should you decide on becoming more involved in boat engine mechanics, such as repair or rebuilding, then these tools will become very important. The success of any rebuild is dependent, to a great extent on the ability to check the size and fit of components as specified by the manufacturer. These measurements are made in thousandths and ten-thousandths of an inch.

Micrometers

◆ See Figure 37

A micrometer is an instrument made up of a precisely machined spindle that is rotated in a fixed nut, opening and closing the distance between the end of the spindle and a fixed anvil.

Outside micrometers can be used to check the thickness parts such shims or the outside diameter of components like the crankshaft journals. They are also used during many rebuild and repair procedures to measure the diameter of components such as the pistons. The most common type of micrometer reads in 1/1000 of an inch. Micrometers that use a vernier scale can estimate to 1/10 of an inch.

Inside micrometers are used to measure the distance between two parallel surfaces. For example, in engine rebuilding work, the inside mike measures cylinder bore wear and taper. Inside mikes are graduated the same way as outside mikes and are read the same way as well.

Remember that an inside mike must be absolutely perpendicular to the work being measured. When you measure with an inside mike, rock the mike gently from side to side and tip it back and forth slightly so that you span the widest part of the bore. Just to be on the safe side, take several

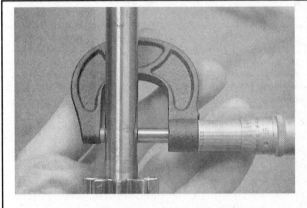

Fig. 37 Outside micrometers can be used to measure the thickness of shims or the outside diameter of a shaft

readings. It takes a certain amount of experience to work any mike with confidence.

Metric micrometers are read in the same way as inch micrometers, except that the measurements are in millimeters. Each line on the main scale equals 1mm. Each fifth line is stamped 5, 10, 15, and so on. Each line on the thimble scale equals 0.01mm. It will take a little practice, but if you can read an inch mike, you can read a metric mike.

Calipers

◆ See Figures 38, 39 and 40

Inside and outside calipers are useful devices to have if you need to measure something quickly and precise measurement is not necessary. Simply take the reading and then hold the calipers on an accurate steel rule.

Fig. 38 Calipers, such as this dial caliper, are the fast and easy way to make precise measurements

Fig. 39 Calipers can also be used to measure depth . . .

Fig. 40 . . . and inside diameter measurements, usually to 0.001 inch accuracy

Fig. 41 Here, a dial indicator is used to measure the axial clearance (end play) of a crankshaft during a powerhead rebuilding procedure

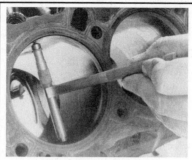

Fig. 42 Telescoping gauges are used during powerhead rebuilding procedures to measure the inside diameter of bores

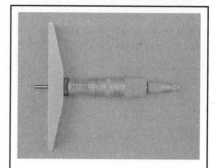

Fig. 43 Depth gauges are used to measure the depth of the bore or other small holes

DIAL INDICATORS

◆ See Figure 41

A dial indicator is a gauge that utilizes a dial face and a needle to register measurements. There is a movable contact arm on the dial indicator, when the arms moves the needle rotates on the dial. Dial indicators are calibrated to show readings in thousandths of an inch and typically are used to measure end-play and runout on various parts.

Dial indicators are quite easy to use, although they are relatively expensive. A variety of mounting devices are available so that the indicator can be used in a number of situations. Make certain that the contact arm is always parallel to the movement of the work being measured.

TELESCOPING GAUGES

◆ See Figure 42

A telescope gauge is used during rebuilding procedures (NOT usually basic maintenance) to measure the inside of bores. It can take the place of an inside mike for some of these jobs. Simply insert the gauge in the hole to be measured and lock the plungers after they have contacted the walls. Remove the tool and measure across the plungers with an outside micrometer.

DEPTH GAUGES

◆ See Figure 43

A depth gauge can be inserted into a bore or other small hole to determine exactly how deep it is. One common use for a depth gauge is measuring the distance the piston sits below the deck of the block at top dead center. Some outside calipers contain a built-in depth gauge so money can be saved just buying one tool.

FASTENERS, MEASUREMENTS AND CONVERSATIONS

Bolts, Nuts and Other Threaded Retainers

◆ See Figures 44 and 45

Although there are a great variety of fasteners found in the modern boat engine, the most commonly used retainer is the threaded fastener (nuts, bolts, screws, studs, etc). Most threaded retainers may be reused, provided that they are not damaged in use or during the repair.

■ Some retainers (such as stretch bolts or torque prevailing nuts) are designed to deform when tightened or in use and should not be reused.

Whenever possible, we will note any special retainers which should be replaced during a procedure. But you should always inspect the condition of a retainer when it is removed and you should replace any that show signs of damage. Check all threads for rust or corrosion that can increase the torque necessary to achieve the desired clamp load for which that fastener was originally selected. Additionally, be sure that the driver surface of the fastener has not been compromised by rounding or other damage. In some cases a driver surface may become only partially rounded, allowing the driver to catch in only one direction. In many of these occurrences, a fastener may be installed and tightened, but the driver would not be able to grip and loosen the fastener again. (This could lead to frustration down the line should that component ever need to be disassembled again).

A - Length
B - Diameter (major diameter)
C - Threads per inch or mm
D - Thread length
E - Size of the wrench required
F - Root diameter (minor diameter)

Fig. 44 Threaded retainer sizes are determined using these measurements

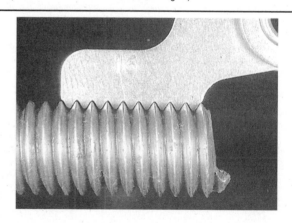

Fig. 45 Thread gauges measure the threads-per-inch and the pitch of a bolt or stud's threads

If you must replace a fastener, whether due to design or damage, you must always be sure to use the proper replacement. In all cases, a retainer of the same design, material and strength should be used. Markings on the heads of most bolts will help determine the proper strength of the fastener. The same material, thread and pitch must be selected to assure proper installation and safe operation of the vehicle afterwards.

Thread gauges are available to help measure a bolt or stud's thread. Most part or hardware stores keep gauges available to help you select the proper size. In a pinch, you can use another nut or bolt for a thread gauge. If the bolt you are replacing is not too badly damaged, you can select a match by finding another bolt that will thread in its place. If you find a nut which threads properly onto the damaged bolt, then use that nut to help select the replacement bolt. If however, the bolt you are replacing is so badly damaged (broken or drilled out) that its threads cannot be used as a gauge, you might start by looking for another bolt (from the same assembly or a similar location) which will thread into the damaged bolt's mounting. If so, the other bolt can be used to select a nut; the nut can then be used to select the replacement bolt.

In all cases, be absolutely sure you have selected the proper replacement. Don't be shy, you can always ask the store clerk for help.

✳✳ WARNING

Be aware that when you find a bolt with damaged threads, you may also find the nut or drilled hole it was threaded into has also been damaged. If this is the case, you may have to drill and tap the hole, replace the nut or otherwise repair the threads. Never try to force a replacement bolt to fit into the damaged threads.

Torque

Torque is defined as the measurement of resistance to turning or rotating. It tends to twist a body about an axis of rotation. A common example of this would be tightening a threaded retainer such as a nut, bolt or screw.

Measuring torque is one of the most common ways to help assure that a threaded retainer has been properly fastened.

When tightening a threaded fastener, torque is applied in three distinct areas, the head, the bearing surface and the clamp load. About 50 percent of the measured torque is used in overcoming bearing friction. This is the friction between the bearing surface of the bolt head, screw head or nut face and the base material or washer (the surface on which the fastener is rotating). Approximately 40 percent of the applied torque is used in overcoming thread friction. This leaves only about 10 percent of the applied torque to develop a useful clamp load (the force which holds a joint together). This means that friction can account for as much as 90 percent of the applied torque on a fastener.

Standard and Metric Measurements

Specifications are often used to help you determine the condition of various components, or to assist you in their installation. Some of the most common measurements include length (in. or cm/mm), torque (ft. lbs., inch lbs. or Nm) and pressure (psi, in. Hg, kPa or mm Hg).

In some cases, that value may not be conveniently measured with what is available in your toolbox. Luckily, many of the measuring devices which are available today will have two scales so Standard or Metric measurements may easily be taken. If any of the various measuring tools that are available to you do not contain the same scale as listed in your specifications, use the accompanying conversion factors to determine the proper value.

The conversion factor chart should be used by taking the given specification and multiplying it by the necessary conversion factor. For instance, looking at the first line, if you have a measurement in inches such as "free-play should be 2 in." but your ruler reads only in millimeters, multiply 2 in. by the conversion factor of 25.4 to get the metric equivalent of 50.8mm. Likewise, if the specification was given only in a Metric measurement, for example in Newton Meters (Nm), then look at the center column first. If the measurement is 100 Nm, multiply it by the conversion factor of 0.738 to get 73.8 ft. lbs.

CONVERSION FACTORS

LENGTH–DISTANCE

Inches (in.)	x 25.4	= Millimeters (mm)	x .0394	= Inches
Feet (ft.)	x .305	= Meters (m)	x 3.281	– Feet
Miles	x 1.609	= Kilometers (km)	x .0621	= Miles

VOLUME

Cubic Inches (in3)	x 16.387	= Cubic Centimeters	x .061	= in3
IMP Pints (IMP pt.)	x .568	= Liters (L)	x 1.76	= IMP pt.
IMP Quarts (IMP qt.)	x 1.137	= Liters (L)	x .88	= IMP qt.
IMP Gallons (IMP gal.)	x 4.546	= Liters (L)	x .22	= IMP gal.
IMP Quarts (IMP qt.)	x 1.201	– US Quarts (US qt.)	x .833	= IMP qt.
IMP Gallons (IMP gal.)	x 1.201	= US Gallons (US gal.)	x .833	= IMP gal.
Fl. Ounces	x 29.573	= Milliliters	x .034	= Ounces
US Pints (US pt.)	x .473	= Liters (L)	x 2.113	= Pints
US Quarts (US qt.)	x .946	= Liters (L)	x 1.057	= Quarts
US Gallons (US gal.)	x 3.785	= Liters (L)	x .264	= Gallons

MASS–WEIGHT

Ounces (oz.)	x 28.35	= Grams (g)	x .035	= Ounces
Pounds (lb.)	x .454	= Kilograms (kg)	x 2.205	= Pounds

PRESSURE

Pounds Per Sq. In. (psi)	x 6.895	– Kilopascals (kPa)	x .145	= psi
Inches of Mercury (Hg)	x .4912	– psi	x 2.036	= Hg
Inches of Mercury (Hg)	x 3.377	= Kilopascals (kPa)	x .2961	= Hg
Inches of Water (H2O)	x .07355	= Inches of Mercury	x 13.783	= H2O
Inches of Water (H2O)	x .03613	= psi	x 27.684	= H2O
Inches of Water (H2O)	x .248	= Kilopascals (kPa)	x 4.026	= H2O

TORQUE

Pounds–Force Inches (in–lb)	x .113	= Newton Meters (N·m)	x 8.85	– in–lb
Pounds–Force Feet (ft–lb)	x 1.356	= Newton Meters (N·m)	x .738	= ft–lb

Metric Bolts

Relative Strength Marking	4.6, 4.8			8.8		
Bolt Markings						
Usage	Frequent			Infrequent		
Bolt Size	Maximum Torque			Maximum Torque		
Thread Size x Pitch (mm)	Ft-Lb	Kgm	Nm	Ft-Lb	Kgm	Nm
6 x 1.0	2–3	.2–.4	3–4	3–6	.4–.8	5–8
8 x 1.25	6–8	.8–1	8–12	9–14	1.2–1.9	13–19
10 x 1.25	12–17	1.5–2.3	16–23	20–29	2.7–4.0	27–39
12 x 1.25	21–32	2.9–4.4	29–43	35–53	4.8–7.3	47–72
14 x 1.5	35–52	4.8–7.1	48–70	57–85	7.8–11.7	77–110
16 x 1.5	51–77	7.0–10.6	67–100	90–120	12.4–16.5	130–160
18 x 1.5	74–110	10.2–15.1	100–150	130–170	17.9–23.4	180–230
20 x 1.5	110–140	15.1–19.3	150–190	190–240	26.2–46.9	160–320
22 x 1.5	150–190	22.0–26.2	200–260	250–320	34.5–44.1	340–430
24 x 1.5	190–240	26.2–46.9	260–320	310–410	42.7–56.5	420–550

SAE Bolts

SAE Grade Number	1 or 2			5			6 or 7		
Bolt Markings — Manufacturers' marks may vary—number of lines always two less than the grade number.									
Usage	Frequent			Frequent			Infrequent		
Bolt Size (inches)—(Thread)	Maximum Torque			Maximum Torque			Maximum Torque		
	Ft-Lb	kgm	Nm	Ft-Lb	kgm	Nm	Ft-Lb	kgm	Nm
1/4 —20	5	0.7	6.8	8	1.1	10.8	10	1.4	13.5
—28	6	0.8	8.1	10	1.4	13.6			
5/16 —18	11	1.5	14.9	17	2.3	23.0	19	2.6	25.8
—24	13	1.8	17.6	19	2.6	25.7			
3/8 —16	18	2.5	24.4	31	4.3	42.0	34	4.7	46.0
—24	20	2.75	27.1	35	4.8	47.5			
7/16 —14	28	3.8	37.0	49	6.8	66.4	55	7.6	74.5
—20	30	4.2	40.7	55	7.6	74.5			
1/2 —13	39	5.4	52.8	75	10.4	101.7	85	11.75	115.2
—20	41	5.7	55.6	85	11.7	115.2			
9/16 —12	51	7.0	69.2	110	15.2	149.1	120	16.6	162.7
—18	55	7.6	74.5	120	16.6	162.7			
5/8 —11	83	11.5	112.5	150	20.7	203.3	167	23.0	226.5
—18	95	13.1	128.8	170	23.5	230.5			
3/4 —10	105	14.5	142.3	270	37.3	366.0	280	38.7	379.6
—16	115	15.9	155.9	295	40.8	400.0			
7/8 — 9	160	22.1	216.9	395	54.6	535.5	440	60.9	596.5
—14	175	24.2	237.2	435	60.1	589.7			
1— 8	236	32.5	318.6	590	81.6	799.9	660	91.3	894.8
—14	250	34.6	338.9	660	91.3	849.8			

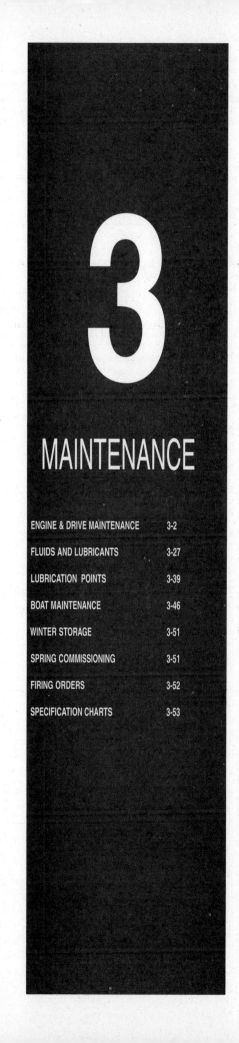

3

MAINTENANCE

ENGINE AND DRIVE MAINTENANCE

Serial Number Identification

An engine specifications decal can generally be found on top of the flame arrestor, or the side of the rocker arm cover, on most models – all pertinent serial number information can be found here, unfortunately this decal is not always legible on older boats so please refer to the following procedures for each individuals unit's serial number location.

ENGINE

◆ See Figures 1, 2 and 3

▪ ■ Serial numbers tags are frequently difficult to see when the engine is installed in the boat; a mirror can be a handy way to read all the numbers.

All Engines

The engine serial number is stamped on the starboard rear side of the engine; above and behind the starter motor for all stern drive applications. On inboards, it can be found in essentially the same place, stamped on a flange at the of the engine, but the starter is actually slightly above and behind it.

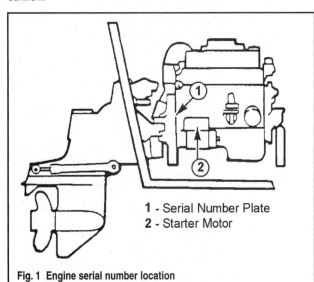

1 - Serial Number Plate
2 - Starter Motor

Fig. 1 Engine serial number location
3.0L engine shown, 4.3L similar

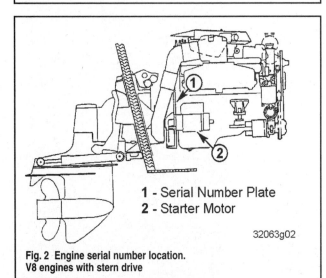

1 - Serial Number Plate
2 - Starter Motor

32063g02

Fig. 2 Engine serial number location.
V8 engines with stern drive

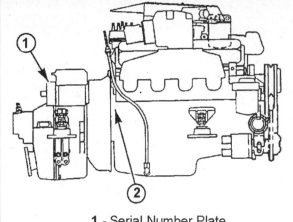

1 - Serial Number Plate
2 - Starter Motor

Fig. 3 Engine serial number location.
V8 engines w/inboard

STERN DRIVE

Alpha and Bravo

◆ See Figures 4 and 5

The stern drive serial number decal can be found on the upper port side of the unit. The serial number will be on the right side of the decal, while the gear ratio will be on the left side. Make sure you don't confuse the two!

Blackhawk

◆ See Figure 6

The stern drive serial number decal can be found on the upper starboard side of the unit. The serial number will be on the upper left side of the decal, while the gear ratio will be on the lower left side. Make sure you don't confuse the two!

TRANSOM ASSEMBLY

◆ See Figures 7 and 8

The transom assembly serial number decal can be found on the upper end of the unit, facing away from the stern.

1- Stern Drive Unit Serial Number
2- Stern Drive Unit Gear Ratio
Fig 4 Stern drive unit serial number location. Alpha One

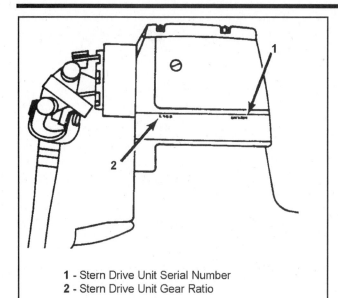

1 - Stern Drive Unit Serial Number
2 - Stern Drive Unit Gear Ratio

**Fig. 5 Stern drive unit serial number location
Alpha One Generation II and Bravo**

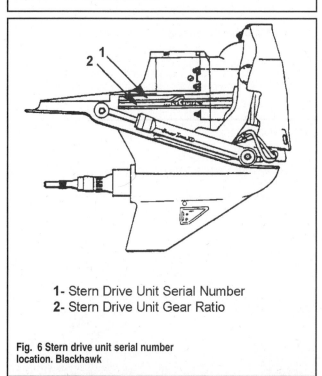

**1- Stern Drive Unit Serial Number
2- Stern Drive Unit Gear Ratio**

**Fig. 6 Stern drive unit serial number
location. Blackhawk**

1- Transom Assembly Serial Number

Fig. 7 Transom assembly serial number location. Alpha One

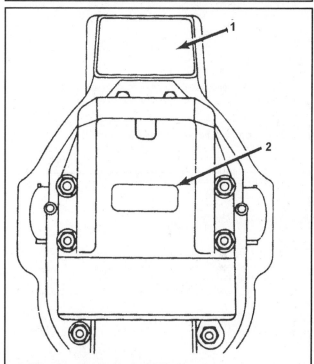

1 - Transom Assembly Serial Number
2 - Engine Designation Decal

**Fig. 8 Transom assembly serial number location. Alpha One
Generation II and Bravo. Note also the engine designation sticker**

Flame Arrestor

In a marine engine compartment, the minimal amount of dust and dirt in the air mean that a marine air filter requires less maintenance than its counterpart in the automotive world. However, the maintenance of a marine air filter is equally important.

The marine filter prevents dirt from entering the engine and scoring the cylinder walls. This lessens oil consumption and extends the engine's life. The air filter on some engines is also used as an intake silencer to quiet the intake air sound as it rushes into the cylinder head from the intake ports.

Over time, the air filter element will become clogged with dirt and oil, decreasing the amount of air entering the engine and lowering engine output. If an excessive amount of oil is clogging the filter, this could be an indication of worn cylinders or piston ring failure causing high pressure in the crankcase.

The maintenance interval for the flame arrestor cleaning is at the end of the first boating season, and then every 100 hours of engine operation or once a year, whichever comes first. On Horizon models, the interval is increased to every 300 hours or three years, whichever comes first.

REMOVAL & INSTALLATION

All Models W/Carburetor Or TBI (Exc. 7.4L and 8.2L w/EFI)

◆ **See Figures 9, 10, 11, 12, 13 and 14**

1. Remove or open the engine compartment cover.
2. Tag and disconnect the crankcase ventilation hoses (hose on the 1990-97 3.0L and 1992 4.3L) from the arrestor and the rocker arm covers.
3. Remove the nut and washer securing the flame arrestor to the carburetor or TBI unit. Many models have a decorative cover over the arrestor.

4. Lift off the flame arrestor.

5. Clean the arrestor in solvent and dry with compressed air if possible; otherwise make sure that it dries completely by air. Clean the hoses and then inspect them for cracks or deterioration. Replace if necessary.

6. Position the arrestor over the stud and reconnect the ventilation hoses.

7. Install the washer and nut. Tighten securely. Close the engine compartment.

7.4L, 8.1L And 8.2L Engines w/EFI

METAL SIDE MOUNTED

◆ See Figure 15

1. Remove or open the engine compartment cover.

2. Remove the four nuts and washers securing the flame arrestor to the throttle body. Many models have a decorative cover over the arrestor.

3. Lift off the flame arrestor.

4. Clean the arrestor in solvent and dry with compressed air if possible; otherwise make sure that it dries completely by air.

5. Position the arrestor and install the cap. Install the washers and nuts. Tighten to 50 inch lbs. (6 Nm). Close the engine compartment.

PLASTIC SIDE MOUNTED

◆ See Figure 16

1. Remove or open the engine compartment cover.

2. Carefully wiggle the crankcase ventilation hose off of its neck on the side of the arrestor.

3. Remove the two screws securing the flame arrestor to the throttle body. They can be found on the bottom of the casing.

4. Lift off the flame arrestor.

5. Clean the arrestor in solvent and dry with compressed air if possible; otherwise make sure that it dries completely by air.

6. Position the arrestor and install the cap. Install the two screws and tighten securely. Close the engine compartment.

PLASTIC FRONT MOUNTED

◆ See Figure 17

1. Remove or open the engine compartment cover.

2. Remove the starboard side engine cover.

3. Loosen the retaining clamp at the throttle body and slide off the arrestor.

4. Clean the arrestor in solvent and dry with compressed air if possible;

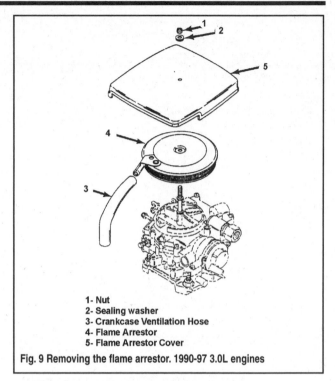

1- Nut
2- Sealing washer
3- Crankcase Ventilation Hose
4- Flame Arrestor
5- Flame Arrestor Cover

Fig. 9 Removing the flame arrestor. 1990-97 3.0L engines

otherwise make sure that it dries completely by air.

5. Position the arrestor and tighten the retaining clamp.

6. Install the engine cover and close the engine compartment.

350 Mag MPI

◆ See Figures 18, 19, 20, 21, 22 and 23

1. Remove or open the engine compartment hatch.

2. Loosen the mounting nut on the flame arrestor cover and remove the cover. Some models may not have a cover.

3. If equipped with an arrestor bracket, remove the mounting nuts on the fuel rail and lift off the bracket. Make sure you don't pull out the fuel rail stud.

4. The flame arrestor is mounted in one of two ways—a clamp or a nut. Either way, loosen the nut or bolt and lift off the arrestor.

5. Clean the arrestor in solvent and dry with compressed air if possible; otherwise make sure that it dries completely by air.

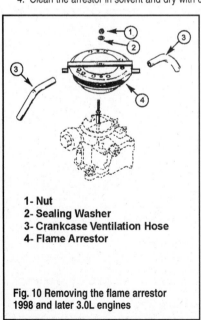

1- Nut
2- Sealing Washer
3- Crankcase Ventilation Hose
4- Flame Arrestor

Fig. 10 Removing the flame arrestor 1998 and later 3.0L engines

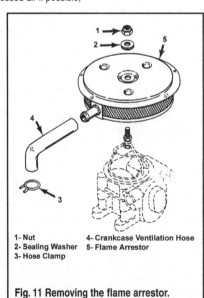

1- Nut   4- Crankcase Ventilation Hose
2- Sealing Washer   5- Flame Arrestor
3- Hose Clamp

Fig. 11 Removing the flame arrestor. 1992 4.3L engines

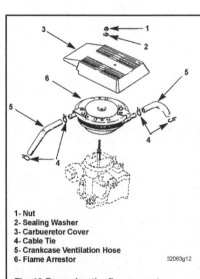

1- Nut
2- Sealing Washer
3- Carbueretor Cover
4- Cable Tie
5- Crankcase Ventilation Hose
6- Flame Arrestor

32063g12

Fig. 12 Removing the flame arrestor. 1993-97 4.3L engines and 1992-97 V8 engines

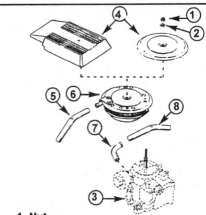

1- Nut
2- Sealing Washer
3- Carburetor
4- Cover (Depending on Model)
5- Crankcase Ventilation Hose
6- Flame Arrestor
7- Postive Crankcase Vent (PVC) Hose (on Carburetor)
8- Postive Crankcase Vent (PVC) Hose

Fig. 13 Removing the flame arrestor. 1998 and later
V6 and V8 engines w/carburetor

6. Position the arrestor on the throttle body and tighten the clamp bolt or mounting nut securely.

7. Install the bracket, if equipped, over the arrestor and onto the fuel rail. Install the thin washer, then the thick one and finally the nut.

8. Install the cover.

Fuel Filter

Regular replacement of the fuel filter will decrease the risk of blocking the flow of fuel to the engine, which could leave you stranded on the water. Fuel filters are usually inexpensive and replacement is a simple task. Change your fuel filter on a regular basis to avoid fuel delivery problems to the engine. All filters should be replaced no less than once a year or every 100 hours of operation, although halving this interval is cheap insurance!

REMOVAL & INSTALLATION

MODERATE

3.0L And 1992 4.3L Engines

These engines have at least two filters. One can be found inline at the carburetor, while the other will be in the mechanical fuel pump housing. Many engines will also have a water separating filter.

** CAUTION

Observe all applicable safety precautions when working around fuel. Whenever servicing the fuel system, always work in a well-ventilated area. Do not allow fuel spray or vapors to come in contact with a spark or open flame. Do not smoke while working around gasoline. Keep a dry chemical fire extinguisher near the work area. Always keep fuel in a container specifically designed for fuel storage; also, always properly seal fuel containers to avoid the possibility of fire or explosion.

CARBURETOR FUEL INLET FILTER

◆ See Figure 24

1. Remove or open the engine compartment cover.
2. Disconnect the negative battery cable.
3. Remove the flame arrestor as detailed in this section.
4. Position an open end wrench on the fuel inlet filter nut at the carburetor. Position another wrench on the fuel line nut. Loosen the fuel line nut while holding the inlet nut with the other wrench and pull out the fuel line. Make sure you plug the line to prevent any fuel from spilling.
5. Loosen and then carefully remove the inlet nut from the carburetor body. Pull out the two gaskets, the filter element and the spring.
6. Although the element can be cleaned and reused, we recommend replacement with a new one. Insert the spring into the carburetor and then slide in the filter. Be sure that the open end of the filter faces out (toward the inlet nut).
7. Position the large gasket over the inlet nut threads and the small one inside the nut. Screw the nut into the carburetor and tighten it to 18 ft. lbs. (24 Nm).
8. Position the fuel line nut into the inlet, seat it a few turns with your fingers and then tighten it to 18 ft. lbs. (24 Nm).

FUEL PUMP FILTER

◆ See Figures 25 and 26

■ Fuel pumps are equipped with a sight tube—if there is any evidence of fuel in the tube, the pump must be replace immediately.

1. Remove or open the engine compartment hatch.
2. Disconnect the negative battery cable and then remove the flame arrestor.
3. Remove the safety wire from the screw at the bottom of the pump on 3.0L engines, or at the top of the pump on early 4.3L engines. Loosen the screw and release the filter bowl bail from the housing. There will be fuel in the bowl so have plenty of rags available.
4. Carefully pry the bowl from the pump housing and remove the spring, filter element and gasket.
5. Position the spring, new filter and gasket into the bowl. Make sure that the open end faces the pump. Hold the bowl in position on the pump and snap the retaining bail into place.
6. Install the screw and then the safety wire.
7. Reconnect the battery cable and start the engine. Check that there are no fuel leaks and then install or close the engine compartment hatch.

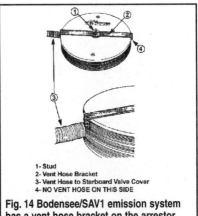

1- Stud
2- Vent Hose Bracket
3- Vent Hose to Starboard Valve Cover
4- NO VENT HOSE ON THIS SIDE

Fig. 14 Bodensee/SAV1 emission system has a vent hose bracket on the arrestor

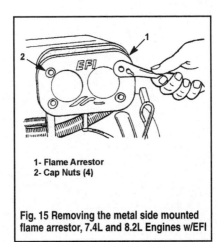

1- Flame Arrestor
2- Cap Nuts (4)

Fig. 15 Removing the metal side mounted flame arrestor, 7.4L and 8.2L Engines w/EFI

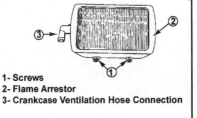

1- Screws
2- Flame Arrestor
3- Crankcase Ventilation Hose Connection

Fig. 16 Removing the plastic side mounted flame arrestor
7.4L and 8.2L Engines w/EFI

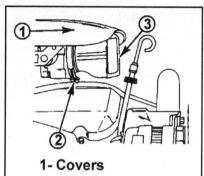

1- Covers
2- Clamp
3- Flame Arrestor

Fig. 17 Removing the plastic front mounted flame arrestor 7.4L and 8.2L Engines w/EFI

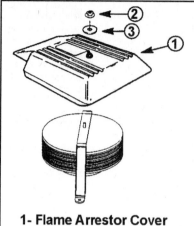

1- Flame Arrestor Cover
2- Nut
3- Washer

Fig. 18 Removing the flame arrestor cover on the 350 Mag MPI

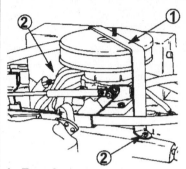

1- Bracket
2- Nuts and fuel Rail Stud

Fig. 19 Some models have a bracket over the flame arrestor

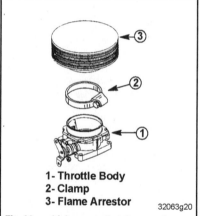

1- Throttle Body
2- Clamp
3- Flame Arrestor

32063g20

Fig. 20 ...which means that the arrestor is secured by a clamp

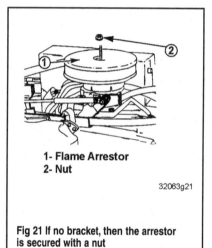

1- Flame Arrestor
2- Nut

32063g21

Fig 21 If no bracket, then the arrestor is secured with a nut

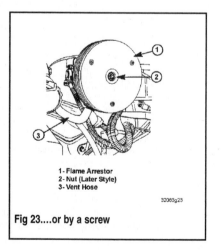

1- Flame Arrestor
2- Clamp (Earlier Style)
3- Vent Hose

Fig. 22 On Black Scorpion models, the arrestor is either secured with a clamp...

1- Flame Arrestor
2- Nut (Later Style)
3- Vent Hose

32063g23

Fig 23....or by a screw

1- Fuel Line
2- Fuel Inlet Filter Nut
3- Gasket (Large)
4- Gasket (Small)
5- Filter
6- Spring

32063g24

Fig. 24 Fuel inlet filter 3.0L and 1992 4.3L engines

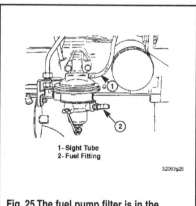

1- Sight Tube
2- Fuel Fitting

32063g25

Fig. 25 The fuel pump filter is in the bottom half of the pump. 3.0L engines

1993-2001 4.3L—Carburetor
V8 Engines—Carburetor

These engines have a fuel inlet filter, found inline at the carburetor. Many engines will also have a water separating filter.

✳✳ CAUTION

Observe all applicable safety precautions when working around fuel. Whenever servicing the fuel system, always work in a well-ventilated area. Do not allow fuel spray or vapors to come in contact with a spark or open flame. Do not smoke while working around gasoline. Keep a dry chemical fire extinguisher near the work area. Always keep fuel in a container specifically designed for fuel storage; also, always properly seal fuel containers to avoid the possibility of fire or explosion.

2BBL CARBURETORS
4BBL ROCHESTER CARBURETORS

◆ See Figure 25

1. Remove or open the engine compartment cover.
2. Disconnect the negative battery cable.
3. Remove the flame arrestor as detailed in this section.
4. Position an open end wrench on the fuel inlet filter nut at the carburetor. Position another wrench on the fuel line nut. Loosen the fuel line nut while holding the inlet nut with the other wrench and pull out the fuel line. Make sure you plug the line to prevent any fuel from spilling. Loosen and then carefully remove the inlet nut from the carburetor body. Pull out the two gaskets, the filter element and the spring.
5. Although the element can be cleaned and reused, we recommend replacement with a new one. Insert the spring into the carburetor and then slide in the filter. Be sure that the open end of the filter faces out (toward the inlet nut).
6. Position the large gasket over the inlet nut threads and the small one inside the nut. Screw the nut into the carburetor and tighten it to 18 ft. lbs. (24 Nm).
7. Position the fuel line nut into the inlet, seat it a few turns with your fingers and then tighten it to 18 ft. lbs. (24 Nm).

4 BBL WEBER CARBURETORS

◆ See Figure 27

Models that are equipped with a Weber 4bbl carburetor have a fuel inlet filter incorporated into the carburetor body. Disassembly of a substantial portion of the carburetor is required—please refer to the Fuel System section found later in this manual for detailed procedures.

V6 and V8 Engines—Fuel Injected

Models with the fuel injection system do not use a fuel filter other than the water separating filter.

Fuel/Water Separator

Most new engines have a factory-installed water separating fuel filter. This type filter is also available as an accessory for all other engines.

A water separating filter, as its name suggests, removes water and other fuel system contaminants before they reach the carburetor and helps minimize potential problems. The presence of water in the fuel will alter the proportion of air/fuel mixture to the "lean" side, resulting in a higher operating temperature and possible damage to pistons, if not corrected.

The filter consists of a mounting plate and disposable canister filter (much like an oil filter). The manufacturer recommends the disposable canister be changed at least once each year or every 100 hours of operation, whichever comes first.

The filter is installed between the fuel tank and the fuel pump.

Most models come equipped with a water separating fuel filter; although some early models may not be equipped with this type filter unit and should be, at the first opportunity. Such a kit is not expensive, and contains instructions for correct installation. If retrofitting one, notice there are two inlet fittings and two outlet fittings on the mounting plate. There are many ways to connect the fuel line incorrectly, but only one way to connect the lines correctly!

1- Gasket	4- Bowl
2- Filter element	5- Bail
3- Spring	6- Bail screw

Fig. 26 ...but on 4.3L engines with a mechanical fuel pump, its in the top half of the pump

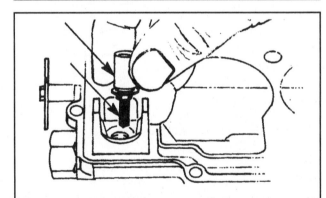

1- Fuel Inlet Seat (with Gasket)
2- Fuel Inlet Filter

Fig. 27 Access to the inlet filter on engines with a 4bbl Weber requires carburetor disassembly

If an installation kit has been purchased, then simply follow the instructions supplied with the kit for the engine being serviced.

SERVICE

◆ See Figures 28, 29 and 30

✳✳ CAUTION

Observe all applicable safety precautions when working around fuel. Whenever servicing the fuel system, always work in a well-ventilated area. Do not allow fuel spray or vapors to come in contact with a spark or open flame. Do not smoke while working around gasoline. Keep a dry chemical fire extinguisher near the work area. Always keep fuel in a container specifically designed for fuel storage; also, always properly seal fuel containers to avoid the possibility of fire or explosion.

1. Remove or open the engine compartment hatch. Disconnect the negative battery cables.

2. On models equipped with a filter cover, unsnap the latch and remove the upper and lower covers from the filter.

3. Remove the canister filter from the mounting plate by rotating the canister counterclockwise. An oil filter wrench may be necessary to break the filter free. Keep the filter upright to avoid spilling fuel. Properly dispose of the fuel and fuel saturated canister. Make sure you have plenty of rags handy just in case! The old canister filter cannot be cleaned and used a second time. Never attempt to reuse the filter!

4. Coat the sealing ring(s) of a new canister filter with clean engine oil (there may be two rings so make sure that the old one comes out and the new one goes in!). Install the filter onto the mounting plate and tighten it securely by hand. Never use an oil filter wrench to tighten the canister.

5. Install the filter covers if so equipped.

6. Reconnect the battery cables and start the engine. Check that there are no fuel leaks and then close the engine hatch.

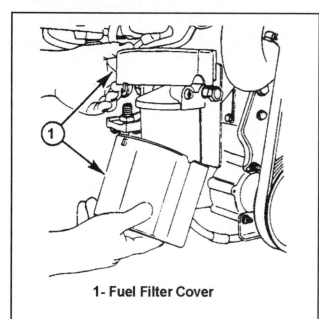

1- Fuel Filter Cover

Fig. 28 Many models utilize a cover over the water separating fuel filter

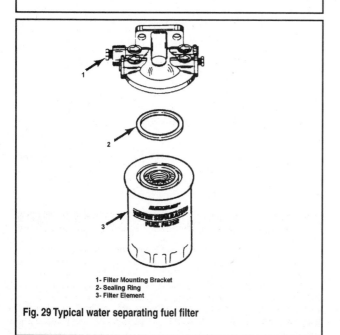

1- Filter Mounting Bracket
2- Sealing Ring
3- Filter Element

Fig. 29 Typical water separating fuel filter

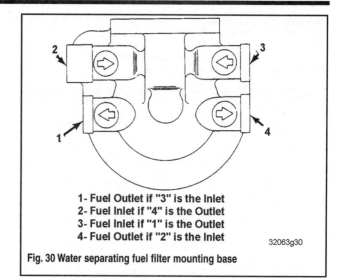

1- Fuel Outlet if "3" is the Inlet
2- Fuel Inlet if "4" is the Outlet
3- Fuel Inlet if "1" is the Outlet
4- Fuel Outlet if "2" is the Inlet

32063g30

Fig. 30 Water separating fuel filter mounting base

Belts

INSPECTION

◆ See Figures 31, 32, 33 and 34

V-belts and serpentine belts should be inspected on a regular basis for signs of glazing or cracking. A glazed belt will be perfectly smooth from slippage, while a good belt will have a slight texture of fabric visible. Cracks will usually start at the inner edge of the belt and run outward. All worn or damaged drive belts should be replaced immediately. It is best to replace all drive belts at one time, as a preventive maintenance measure, during this service operation.

Inspect the alternator, power steering and water pump V-belts or the serpentine belt every 100 hours or 6 months (whichever comes first) for evidence of wear such as cracking, fraying, and incorrect tension.

Determine the V-belt tension at a point halfway between the pulleys by pressing on the belt with moderate thumb pressure. The belt should deflect 1/4 in. (6mm). If the defection is found to be too much or too little, make adjustments as necessary.

Determine serpentine belt tension at the halfway point of the longest span between pulleys Press on the belt with moderate thumb pressure. The belt should deflect 1/2 in. (13mm) for most components, 1/4 in. (6mm) for the power steering pump and the seawater pump. If the defection is found to be too much or too little, loosen the mounting bolts and make adjustments as necessary.

■ When replacing belts, we recommend cleaning the inside of the belt pulleys to extend the service life of the belts. Never use automotive belts, marine belts used on your engine are heavy duty and not interchangeable.

ADJUSTMENT

V-Belts

◆ See Figures 35, 36, 37 and 38

Before you attempt to adjust any of your engine's belts, apply penetrating oil to the bracket fasteners to make them easier to loosen.

1. Remove or open the engine compartment hatch.

2. Disconnect the negative battery cable.

3. Loosen the component pivot bolt.

4. Loosen the component pump pulley adjustment bracket bolt.

5. Using a wooden lever, pry the component toward or away from the engine until the proper tension is achieved.

Do not overtighten the drive belts; the pump and/or alternator bearings can be damaged.

6. Tighten the component pulley adjustment bracket bolt to 16 ft. lbs. (28 Nm).
7. Tighten the component pivot bolt to 35 ft. lbs. (48 Nm).
8. Reconnect the negative battery cable.

Fig. 31 An example of a healthy drive belt

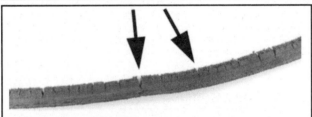

Fig. 32 Deep cracks in this belt will cause flex, building up heat that will eventually lead to belt failure

Fig. 33 The cover of this belt is worn, exposing the critical reinforcing cords to excessive wear

Fig. 34 Installing too wide a belt can result in serious belt wear and/or breakage

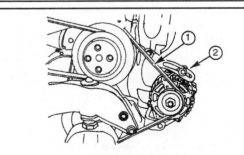

1- Check Point
2- Alternator Pivot Bolt

Fig. 35 Adjusting the alternator belt. 3.0L engines, 4.3L similar

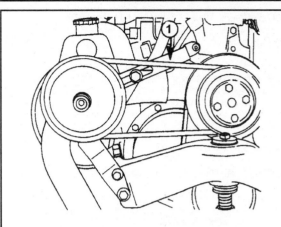

1- Measure Belt Deflection here

Fig. 36 Adjusting the power steering belt. 3.0L engines

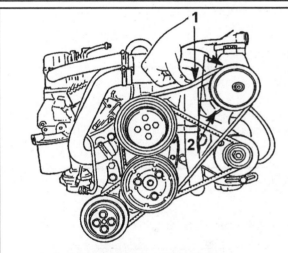

1- Belt should depress 1/4 in. (6 mm)
2- Screws and Nuts

Fig. 37 Adjusting the power steering belt 4.3L and 1992-97 V8 engines

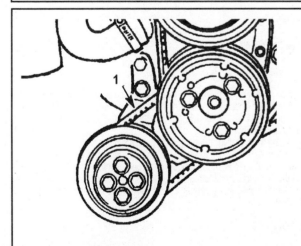

1- Depress here

Fig. 38 Adjusting the seawater pump belt 4.3L and 1992-97 5.0L, 5.7L engines

SEAWATER PUMP—1992-97 7.4L AND 8.2L ENGINES

◆ **See Figure 39**

These engines utilize a combination fuel pump and seawater pump that is adjustable by means of an idler pulley on the lower side of the belt. Loosen the outer locknut on the pulley and then turn the inner adjusting bolt (end of threaded shaft) until the proper belt tension is achieved. Tighten the locknut to 30 ft. lbs. (41 Nm) while holding the adjusting nut so as not to change the tension adjustment.

Serpentine Belts

◆ **See Figures 40, 41, 42, 43, 44, 45, 46 and 47**

1. Remove or open the engine hatch and disconnect the battery cables.
2. Loosen the tensioner pulley adjustment stud locknut with a 5/8 in. wrench. Leave the wrench on the nut.
3. Put on 5/16 in. socket over the adjustment stud and turn it until belt deflection equals 1/4 in. at the longest span between two pulleys. Hold the socket on the stud to maintain tension and tighten the locknut securely.
4. Connect the battery cables and run the engine for a few minutes. Recheck the belt tension.

REMOVAL & INSTALLATION

V-Belts

The replacement of the inner belt on multi-belted engines may require the removal of the outer belts.

To replace a drive belt, loosen the pivot and mounting bolts of the component which the belt is driving, then, using a wooden lever or equivalent, pry the component inward to relieve the tension on the drive belt; always be careful where you locate the prybar, or damage to components may result. Slip the belt off the component pulley, and match the new belt with the old belt for length and width.

1. 1992-97 7.4L and 8.2L engines utilize and idler pulley on the seawater pump. Loosen the locknut and then loosen the adjusting bolt so that the belt can be slipped off.
2. These measurements must be equal. It is normal for an old belt to be slightly longer than a new one. After a new belt is installed correctly, properly adjust the tension.

■ **When removing more than one belt, be sure to mark them for identification. This will help avoid confusion when replacing the belts.**

Serpentine Belts

◆ **See Figure 47**

ALL ENGINES EXC. 8.1L

1. Remove or open the engine hatch and disconnect the battery cables.
2. Loosen the tensioner pulley adjustment stud locknut with a 5/8 in. wrench.
3. Put on 5/16 in. socket over the adjustment stud and turn it until the belt is loose. Remove the belt.
4. Install the belt over the pulleys as detailed in the routing diagrams.
5. Put on 5/16 in. socket over the adjustment stud and turn it until belt deflection equals 1/4 in. at the longest span between two pulleys. Hold the socket on the stud to maintain tension and tighten the locknut securely.
6. Connect the battery cables and run the engine for a few minutes. Recheck the belt tension.

8.1L ENGINES

1. Remove or open the engine hatch and disconnect the battery cables.
2. Loosen the alternator mounting bolt.
3. Loosen the tensioner pulley adjustment stud locknut with a 5/8 in. wrench.

4. Put on 5/16 in. socket over the adjustment stud and turn it until the belt is loose.
5. Move the alternator to further relieve tension on the belt and then remove the belt.
6. Install the belt over the pulleys as detailed in the routing diagrams.
7. Swivel the alternator until the belt is tensioned and tighten the mounting bolt to 35 ft. lbs. (48 Nm).
8. Put on 5/16 in. socket over the adjustment stud and turn it until belt deflection equals 1/4 in. at the longest span between two pulleys. Hold the socket on the stud to maintain tension and tighten the locknut securely.
9. Connect the battery cables and run the engine for a few minutes. Recheck the belt tension.

Thermostat

The thermostat is a simple temperature sensitive valve that opens and closes to control cooling water flow through the engine. In operation the thermostat hovers somewhere between open and closed. As engine load and temperature increase, the thermostat opens to allow more cooling water into the engine. As temperature and load decrease the thermostat closes.

A sticking thermostat will either allow the temperature to rise well above the normal operating temperature before it opens, or, if stuck in the open position, will never allow the engine to reach operating temperature.

REMOVAL & INSTALLATION

Seawater Cooling System

3.0L AND 4.3L ENGINES

◆ **See Figures 48, 49 and 50**

1. Open or remove the engine hatch cover and disconnect the negative battery cables.
2. Drain all water from the cylinder block and exhaust manifold(s) as detailed in the Cooling System section later in this manual.
3. Loosen the hose clamps and then wiggle the coolant hoses off of the thermostat housing.
4. Remove the mounting bolts (2) with their lock washers and then remove the thermostat housing cover. Some models may have a lifting eye incorporated in the housing. Take note of its positioning.
5. Lift out the thermostat and discard it. If you are not sure that it is inoperable, perform the testing procedures outlined in this section.
6. Make sure that you carefully scrape off all remaining gasket material from the thermostat housing and cover and then insert a new thermostat (143 degrees, 1990-97 3.0L, 1992 4.3L engines and 1998-2001 3.0L w/closed cooling system) into the housing. The element must be pointing into the housing and the flange on the unit should be seated into the recess in the housing.

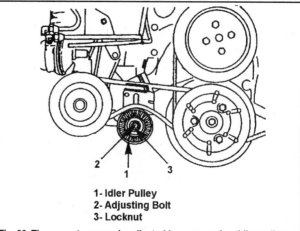

1- Idler Pulley
2- Adjusting Bolt
3- Locknut

Fig. 39 The seawater pump is adjusted by means of an idler pulley

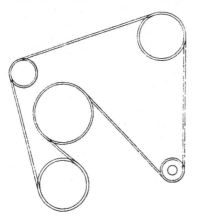

Alpha With Power Steering

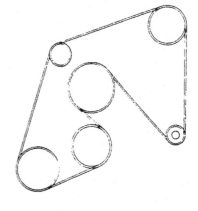

Bravo With Power Steering

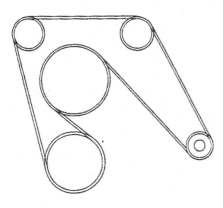

Alpha With Closed Cooling Without Power Steering

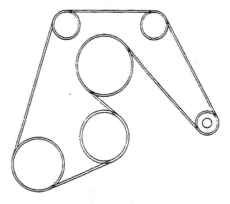

Bravo Without Power Steering

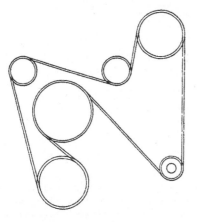

Alpha With Closed Cooling and Power Steering

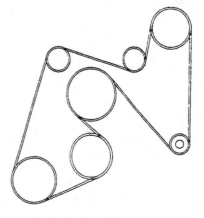

Bravo With Closed Cooling and Power Steering

32063g40

Fig. 40 Serpentine belt routing 4.3L engines below serial number OL619083; V8 (exc. 1998 -2001 7.4L/8.2L) engines below OL618999

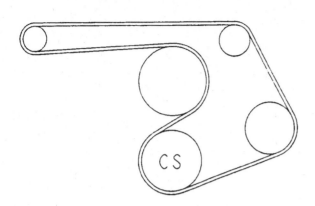

Alpha With Power Steering

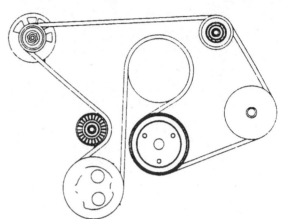

Bravo With Power Steering

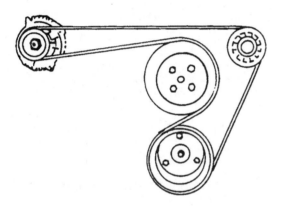

Alpha Without Power Steering

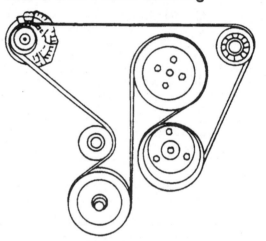

Bravo Without Power Steering

32063g41

Fig. 41 Serpentine belt routing 4.3L engines above serial number OL619084; V8 (exc. 1998-2001 7.4L/8.2L) engines above OL619000

■ 1998 and later 3.0L models came equipped with one of two thermostats. Models up to and including serial number 0L34099 use a brass thermostat rated at 140 degrees. Models with serial number 0L341000 and above come equipped with a stainless steel thermostat rated at 160 degrees. 1993 and later 4.3L models also come equipped with either of the two.

7. On 1998 and later 4.3L engines, position the O-ring in the housing, install the thermostat and then slide in the sleeve so it aligns with the groove in the housing.

8. Coat both sides of a new housing gasket with Quicksilver Perfect Seal (or similar) and position it onto the housing so that the holes line up. If your engine is equipped with an audio warning temperature switch, the thermostat will have continuity rivets, do not use a sealer or the alarm will not work properly.

9. Install the housing cover and mounting bolts/washers and tighten to 30 ft. lbs. (41 Nm).

10. Reconnect the hoses and tighten the clamps being careful not to pinch the hose. This is a good time to inspect the hoses! Connect the batteries and then start the engine and check for leaks.

V8 ENGINES

◆ See Figures 49 and 51

1. Open or remove the engine hatch cover and disconnect the negative battery cables.

2. Drain all water from the cylinder block and exhaust manifolds as detailed in the Cooling System section later in this manual.

3. On models with stainless steel thermostat housings, loosen the hose clamps and then wiggle the coolant hoses off of the thermostat housing.

4. Remove the mounting bolts (2) with their lock washers and then remove the thermostat housing cover. Some models may have a lifting eye incorporated in the housing—take note of its positioning.

5. Lift out the thermostat, spacers and gaskets. Discard the thermostat. If you are not sure that it is inoperable, perform the testing procedures outlined in this section.

6. Make sure that you carefully scrape off all remaining gasket material from the thermostat housing and cover.

7. On cast iron housings, position the O-ring in the housing, install the thermostat (143 or 160 degrees) so the element is pointing into the housing

and the flange on the unit should be seated into the recess in the housing and then slide in the sleeve so it aligns with the groove in the housing.

8. Coat both sides of a new housing gasket with Quicksilver Perfect Seal (or similar) and position it onto the housing so that the holes line up. If your engine is equipped with an audio warning temperature switch, the thermostat/gasket will have continuity rivets, do not use a sealer or the alarm will not work properly.

9. Install the housing cover and mounting bolts/washers and tighten to 30 ft. lbs. (41 Nm).

10. Reconnect the hoses and tighten the clamps being careful not to pinch the hose. This is a good time to inspect the hoses! Connect the batteries and then start the engine and check for leaks.

11. On stainless steel housings, coat both sides of a new gasket with Quicksilver Perfect Seal (or similar) and position it onto the manifold housing so that the holes line up. If your engine is equipped with an audio warning temperature switch, the thermostat/gasket will have continuity rivets, do not use a sealer or the alarm will not work properly

12. Slide the sleeve in and position the thermostat so that it seats properly.

13. Position the O-ring over the thermostat and install the housing cover. Tighten the bolts to 30 ft. lbs. (41 Nm).

14. On all models, reconnect the hoses and tighten the clamps being careful not to pinch the hose. This is a good time to inspect the hoses! Connect the batteries and then start the engine and check for leaks.

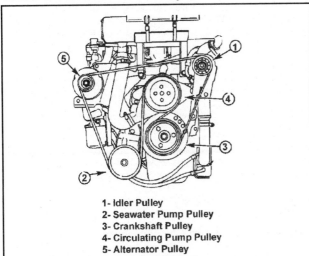

1- Idler Pulley
2- Seawater Pump Pulley
3- Crankshaft Pulley
4- Circulating Pump Pulley
5- Alternator Pulley

Fig. 42 Serpentine belt routing. 1998-2001 7.4L/8.2L inboard engines

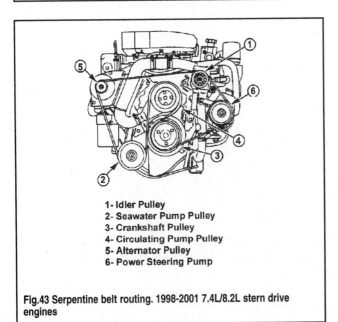

1- Idler Pulley
2- Seawater Pump Pulley
3- Crankshaft Pulley
4- Circulating Pump Pulley
5- Alternator Pulley
6- Power Steering Pump

Fig.43 Serpentine belt routing. 1998-2001 7.4L/8.2L stern drive engines

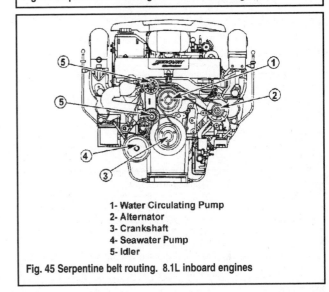

1- Water Circulating Pump
2- Alternator
3- Crankshaft
4- Seawater Pump
5- Idler
6- Power Steering

32063g44

Fig. 44 Serpentine belt routing. 8.1L stern drive engines

1- Water Circulating Pump
2- Alternator
3- Crankshaft
4- Seawater Pump
5- Idler

Fig. 45 Serpentine belt routing. 8.1L inboard engines

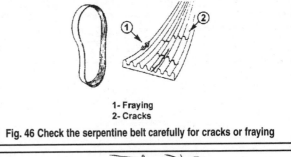

1- Fraying
2- Cracks

Fig. 46 Check the serpentine belt carefully for cracks or fraying

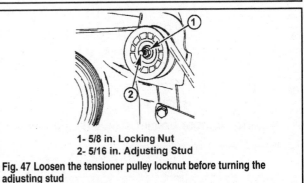

1- 5/8 in. Locking Nut
2- 5/16 in. Adjusting Stud

Fig. 47 Loosen the tensioner pulley locknut before turning the adjusting stud

Fig. 48 Installing the thermostat

Closed Cooling System

3.0L ENGINES

◆ **See Figure 52**

1. Open or remove the engine hatch cover and disconnect the negative battery cables.
2. Drain all water from the cylinder block and exhaust manifold(s) as detailed in the Cooling System section of this manual.
3. Loosen the hose clamps and then wiggle the coolant hoses off of the thermostat housing.
4. Remove the mounting bolts (2) with their lock washers and then remove the thermostat housing cover.
5. Lift out the thermostat and discard it. If you are not sure that it is inoperable, perform the testing procedures outlined in this section.
6. Make sure that you carefully scrape off all remaining gasket material from the thermostat housing and cover and then insert a new thermostat (143 degrees) into the housing. The element must be pointing into the housing (this means the pointed side is facing up) and the flange on the unit should be seated into the recess in the housing.
7. Coat both sides of a new housing gasket with Quicksilver Perfect Seal (or similar) and position it onto the housing so that the holes line up.
8. Install the housing cover and mounting bolts/washers and tighten to 30 ft. lbs. (41 Nm).
9. Reconnect the hoses and tighten the clamps being careful not to pinch the hose. This is a good time to inspect the hoses! Fill the system with a 50/50 mixture of ethylene glycol antifreeze and water. Connect the batteries and then start the engine and check for leaks.

1993-97 4.3L ENGINES

◆ **See Figure 53**

1. Open or remove the engine hatch cover and disconnect the negative battery cables.
2. Drain all water from the cylinder block and exhaust manifold(s) as detailed in the Cooling System section of this manual.
3. Loosen the hose clamps and then wiggle the coolant hoses off of the thermostat housing.
4. Remove the mounting bolts (2) with their lock washers and then remove the thermostat housing cover. Certain engines may be equipped with a lifting eye incorporated into the housing - be sure to note the position of the eye before separating the two.
5. Lift out the thermostat and discard it. If you are not sure that it is inoperable, perform the testing procedures outlined in this section.
6. Remove the cork gasket and housing sleeve on models so equipped.
7. Remove the thermostat housing and all gaskets. Discard the gaskets
8. Make sure that you carefully scrape off all remaining gasket material from the thermostat housing, cover and manifold, and then position the lower gasket on the manifold. This gasket should have continuity rivets for the audio warning temperature switch, so do not use Quicksilver Perfect Seal or any other sealer. Position the housing on the gasket.
9. Insert a new thermostat into the housing. The element must be pointing into the housing (this means the pointed side is facing up) and the flange on the unit should be seated into the recess in the housing. On models with a housing sleeve, make sure that the side with a turned in lip is UP, or facing the thermostat and that the cork gasket is positioned under the thermostat..
10. Coat both sides of a new housing gasket with Quicksilver Perfect Seal (or similar) and position it onto the housing so that the holes line up with those in the cover.
11. Install the housing cover and mounting bolts/washers and tighten to 30 ft. lbs. (41 Nm).
12. Reconnect the hoses and tighten the clamps being careful not to pinch the hose. This is a good time to inspect the hoses! Fill the system with a 50/50 mixture of ethylene glycol antifreeze and water. Connect the batteries and then start the engine and check for leaks.

1998-2001 4.3L ENGINES
V8 ENGINES EXC. 8.1L

◆ **See Figures 53, 54 and 55**

1. Open or remove the engine hatch cover and disconnect the negative battery cables.
2. Drain all water from the cylinder block and exhaust manifold(s) as detailed in the Cooling System section of this manual.
3. Loosen the hose clamps and then wiggle the coolant hoses off of the thermostat housing.

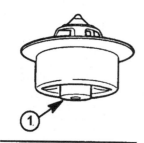

Brass

1- Serial number break:OL340999 and below

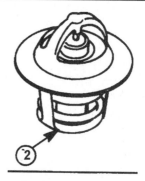

Stainless Steel

2-- Serial Number break:OL341000 and above

Fig. 49 Two different types of thermostats

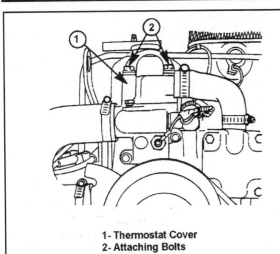

1- Thermostat Cover
2- Attaching Bolts

Fig. 50 Removing the thermostat housing cover on models with the brass thermostat 3.0L engines

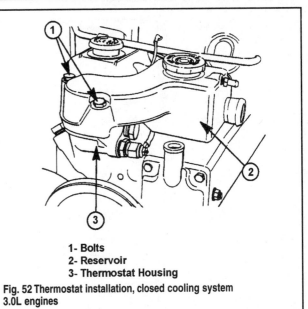

1- Bolts
2- Reservoir
3- Thermostat Housing

Fig. 52 Thermostat installation, closed cooling system 3.0L engines

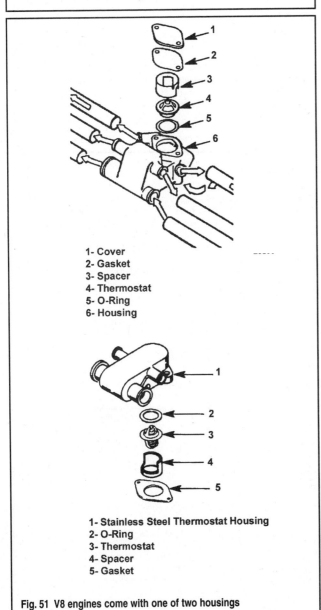

1- Cover
2- Gasket
3- Spacer
4- Thermostat
5- O-Ring
6- Housing

1- Stainless Steel Thermostat Housing
2- O-Ring
3- Thermostat
4- Spacer
5- Gasket

Fig. 51 V8 engines come with one of two housings

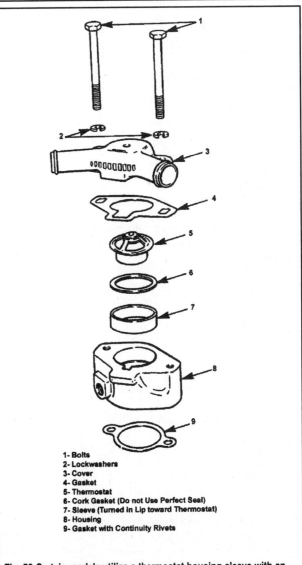

1- Bolts
2- Lockwashers
3- Cover
4- Gasket
5- Thermostat
6- Cork Gasket (Do not Use Perfect Seal)
7- Sleeve (Turned in Lip toward Thermostat)
8- Housing
9- Gasket with Continuity Rivets

Fig. 53 Certain models utilize a thermostat housing sleeve with an additional cork gasket

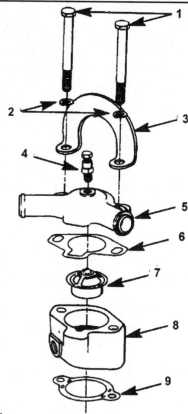

1- Bolts
2- Lockwashers
3- Lifting Eye (MCM only)
4- Hex head Bleeder
5- Cover
6- Gasket
7- Thermostat
8- Housing
9- Gasket with continuity rivets

Fig. 54 Exploded view of the thermostat and housing 1992-97 V8 engines

4. Remove the mounting bolts (2) with their lock washers and then remove the thermostat housing cover.

5. Lift out the thermostat and discard it. If you are not sure that it is inoperable, perform the testing procedures outlined in this section.

6. Remove the thermostat housing and all gaskets. Discard the gaskets

7. Make sure that you carefully scrape off all remaining gasket material from the thermostat housing, cover and manifold, and then position the lower gasket on the manifold. This gasket should have continuity rivets for the audio warning temperature switch, so do not use Quicksilver Perfect Seal or any other sealer if it does. Position the housing on the gasket.

8. Insert a new thermostat into the housing. The element must be pointing into the housing (this means the pointed side is facing up) and the flange on the unit should be seated into the recess in the housing.

9. Coat both sides of a new housing gasket with Quicksilver Perfect Seal (or similar) and position it onto the housing so that the holes line up with those in the cover.

10. Install the housing cover and mounting bolts/washers and tighten to 30 ft. lbs. (41 Nm).

11. Reconnect the hoses and tighten the clamps being careful not to pinch the hose. This is a good time to inspect the hoses! Fill the system with a 50/50 mixture of ethylene glycol antifreeze and water. Connect the batteries and then start the engine and check for leaks.

8.1L ENGINES

1. Open or remove the engine hatch cover and disconnect the negative battery cables.

2. Drain all water from the cylinder block and exhaust manifold(s) as detailed in the Cooling System section of this manual.

3. Remove the heat exchanger.

4. Remove the mounting bolts (2) with their lock washers and then remove the thermostat retainer.

5. Lift out the thermostat and discard it. If you are not sure that it is inoperable, perform the testing procedures outlined in this section.

6. Make sure that you carefully scrape off all remaining gasket material from the thermostat retainer and crossover, and then position a new O-ring into the crossover.

7. Insert a new thermostat into the crossover. The element must be pointing into the housing (this means the pointed side is facing up) and the flange on the unit should be seated into the recess in the housing.

8. Install the retainer and mounting bolts/washers and tighten them finger-tight.

9. Install the heat exchanger and then tighten the retainer mounting bolts to 12 ft. lbs. (16 Nm).

10. Fill the system with a 50/50 mixture of ethylene glycol antifreeze and water. Connect the batteries and then start the engine and check for leaks.

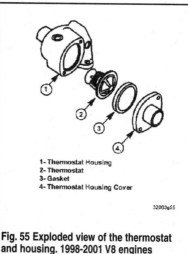

1- Thermostat Housing
2- Thermostat
3- Gasket
4- Thermostat Housing Cover

32063g55

Fig. 55 Exploded view of the thermostat and housing. 1998-2001 V8 engines

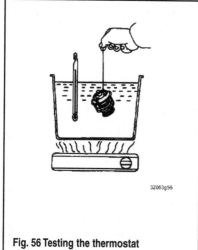

32063g56

Fig. 56 Testing the thermostat

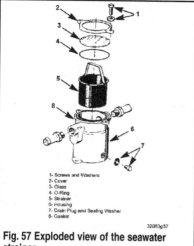

1- Screws and Washers
2- Cover
3- Glass
4- O-Ring
5- Strainer
6- Housing
7- Drain Plug and Sealing Washer
8- Gasket

32063g57

Fig. 57 Exploded view of the seawater strainer

TESTING

See Figure 56

1. Inspect the thermostat at room temperature. If the thermostat is fully open, it is defective and must be replaced. Hold the thermostat up to the light and check it for leaks. A light leak around the perimeter indicates the thermostat is not closing, and therefore, it must be replaced.

2. Attach a length of thread to the thermostat. Now, suspend the thermostat and a thermometer inside a container filled with water (do not use distilled water or ethylene glycol!). Take care to be sure neither the thermostat or the thermometer touches the container. If either one does touch the container, the test will be unreliable

3. Heat the water until the thermostat just begins to open – when this happens confirm that the temperature is the same as the thermostat rating. The thermometer reading must agree with the rating stamped on the thermostat. If the unit fails the test, it must be replaced.

4. Continue to heat the water until a temperature 25 degrees above the rating is reached. At this time the thermostat should be completely open; if not, replace it.

5. Turn the heat off and allow the water to cool to a temperature 10 degrees below the rating. The thermostat should now be completely closed; if not, replace it.

Seawater Strainer

Many models will utilize a seawater strainer, either factory-installed or aftermarket. It is a good idea to check the unit with regularity, particularly if the boat is operated in dirty waters.

CLEANING & INSPECTION

■ **For Removal & Installation procedures, please refer to Section 9, Cooling Systems.**

V6 And V8 Engines

◆ **See Figure 57**

1. If your vessel is equipped with a seacock, close it. If not equipped, disconnect the seawater inlet line at the strainer and plug it securely to prevent water overflow.

2. Loosen the two retaining screws and then remove them and the strainer cover.

3. Carefully lift off the glass plate and the O-ring.

4. Remove the drain plug and washer from the bottom of the housing. You may want to have a container handy to collect the water from the housing.

5. Lift out the strainer and clear the mesh of any debris. Rinse the strainer and the housing with water..

6. Screw the drain plug in securely and then insert the strainer.

7. Check the condition of the O-ring and install it and the glass plate. We suggest replacing the O-ring regardless of its condition.

8. Install the cover and tighten the screws securely, but not so tight as to warp the cover.

9. Open the seacock or unplug and reconnect the inlet line.

10. Start the engine and check for leaks.

Cylinder Compression

Cylinder compression test results are extremely valuable indicators of internal engine condition. The best marine mechanics automatically check an engine's compression as the first step in a comprehensive tune-up. A compression test will uncover many mechanical problems that can cause rough running or poor performance.

CHECKING COMPRESSION

1. Make sure that the proper amount and viscosity of engine oil is in the crankcase, then ensure the battery is fully charged.

2. Warm-up the engine to normal operating temperature, then shut the engine **OFF**. If the boat is out of the water, make sure to install a flush test kit.

3. Remove the flame arrestor and open the choke or throttle fully.

4. Disable the ignition system by removing and grounding the coil wire at the distributor (see the Ignition section for models with DDIS or EST ignition systems).

5. Tag and disconnect all spark plug wires and then remove the plugs themselves..

6. Install a screw-in type compression gauge into the No. 1 cylinder spark plug hole until the fitting is snug. Please refer to the firing order illustrations for location of the No. 1 cylinder. When fitting the compression gauge adapter to the cylinder head, make sure the bleeder of the gauge (if equipped) is closed.

7. According to the tool manufacturer's instructions, connect a remote starting switch to the starting circuit.

8. With the ignition switch in the **OFF** position, use the remote starting switch to crank the engine through at least five compression strokes (approximately 5 seconds of cranking) and record the highest reading on the gauge.

9. Repeat the test on each cylinder, cranking the engine approximately the same number of compression strokes and/or time as the first.

10. Compare the highest readings from each cylinder to that of the others. The indicated compression pressures are considered within specifications if the lowest reading cylinder is within 75 percent of the pressure recorded for the highest reading cylinder. For example, if your highest reading cylinder pressure was 150 psi (1034 kPa), then 75 percent of that would be 113 psi (779 kPa). So the lowest reading cylinder should be no less than 113 psi (779 kPa).

11. Compression readings that are generally low indicate worn, broken, or sticking piston rings, scored pistons or worn cylinders.

12. If a cylinder exhibits an unusually low compression reading, squirt a tablespoon of clean engine oil into the cylinder through the injector hole and repeat the compression test. If the compression rises after adding oil, it means that the cylinder's piston rings and/or cylinder bore are damaged or worn. If the pressure remains low, the valves may not be seating properly (a valve job is needed), or the head gasket may be blown near that cylinder.

13. If compression in any two adjacent cylinders is low (with normal compression in the other cylinders), and if the addition of oil doesn't help raise compression, there is leakage past the head gasket. Oil and coolant in the combustion chamber, combined with blue or constant white smoke from the tailpipe, are symptoms of this problem. However, don't be alarmed by the normal white smoke emitted from the tailpipe during engine warm-up during cold weather. There may be evidence of water droplets on the engine oil dipstick and/or oil droplets in the cooling system if a head gasket is blown.

Spark Plugs

The spark plug performs four main functions:
- It fills a hole in the cylinder head.
- It acts as a dielectric insulator for the ignition system.
- It provides spark for the combustion process to occur.
- It removes heat from the combustion chamber.

It is important to remember that spark plugs do not create heat, they help remove it. Anything that prevents a spark plug from removing the proper amount of heat can lead to pre-ignition, detonation, premature spark plug failure and even internal engine damage.

In the simplest of terms, the spark plug acts as the thermometer of the engine. Much like a doctor examining a patient, this "thermometer" can be used to effectively diagnose the amount of heat present in each combustion chamber.

Spark plugs are valuable tuning tools, when interpreted correctly. They will show symptoms of other problems and can reveal a great deal about the engine's overall condition. By evaluating the appearance of the spark plug's firing tip, visual cues can be seen to accurately determine the engine's overall operating condition, get a feel for air/fuel ratios and even diagnose driveability problems.

As spark plugs grow older, they lose their sharp edges and material from the center and ground electrodes is slowly eroded away. As the gap between these two points grows, the voltage required to bridge this gap increases proportionately. The ignition system must work harder to compensate for this higher voltage requirement and hence there is a greater rate of misfires or incomplete combustion cycles. Each misfire means lost horsepower, reduced fuel economy and higher emissions. Replacing worn out spark plugs with new ones (with sharp new edges) effectively restores the ignition system's efficiency and reduces the percentage of misfires, restoring power, economy and reducing emissions.

How long spark plugs last will depend on a variety of factors, including engine compression, fuel used, gap, center/ground electrode material and the conditions in which the engine is operated.

SPARK PLUG HEAT RANGE

◆ See Figure 58

Spark plug heat range is the ability of the plug to dissipate heat from the combustion chamber. The longer the insulator (or the farther it extends into the engine), the hotter the plug will operate; the shorter the insulator (the closer the electrode is to the block's cooling passages) the cooler it will operate.

Selecting a spark plug with the proper heat range will ensure that the tip will maintain a temperature high enough to prevent fouling, yet be cool enough to prevent pre-ignition. A plug that absorbs little heat and remains too cool will quickly accumulate deposits of oil and carbon since it is not hot enough to burn them off. This leads to plug fouling and consequently to misfiring. A plug that absorbs too much heat will have no deposits but, due to the excessive heat, the electrodes will burn away quickly and might possibly lead to pre-ignition or other ignition problems.

Pre-ignition takes place when plug tips get so hot that they glow sufficiently to ignite the air/fuel mixture before the actual spark occurs. This early ignition will usually cause a pinging during heavy loads and if not corrected will result in severe engine damage. While there are many other things that can cause pre-ignition, selecting the proper heat range spark plug will ensure that the spark plug itself is not a hot-spot source.

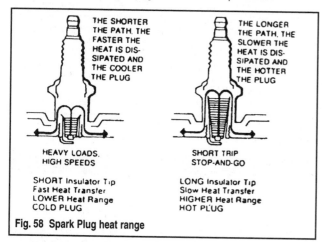

THE SHORTER THE PATH, THE FASTER THE HEAT IS DISSIPATED AND THE COOLER THE PLUG

THE LONGER THE PATH, THE SLOWER THE HEAT IS DISSIPATED AND THE HOTTER THE PLUG

HEAVY LOADS. HIGH SPEEDS

SHORT Insulator Tip
Fast Heat Transfer
LOWER Heat Range
COLD PLUG

SHORT TRIP STOP-AND-GO

LONG Insulator Tip
Slow Heat Transfer
HIGHER Heat Range
HOT PLUG

Fig. 58 Spark Plug heat range

SPARK PLUG SERVICE

■ **New technologies in spark plug and ignition system design have pushed the recommended replacement interval higher and higher. However, this depends on usage and conditions.**

Spark plugs should only require replacement once a season. The electrode on a new spark plug has a sharp edge but with use, this edge becomes rounded by wear, causing the plug gap to increase. As the gap increases, the plug's voltage requirement also increases. It requires a greater voltage to jump the wider gap and about two to three times as much voltage to fire a plug at high speeds than at idle.

Tools needed for spark plug replacement include: a ratchet, short extension, spark plug socket (there are two types; either 13/16 inch or 5/8 inch, depending upon the type of plug), a combination spark plug gauge and gapping tool and a can of anti-seize type compound.

REMOVAL & INSTALLATION

1. When removing spark plugs, work on one at a time. Don't start by removing the plug wires all at once, because unless you number them, they may become mixed up. Take a minute before you begin and number the wires with tape.
2. Disconnect the negative battery cable or turn the battery switch **OFF**.
3. If the engine has been run recently, allow the engine to thoroughly cool. Attempting to remove plugs from a hot cylinder head could cause the plugs to seize and damage the threads in the cylinder head, especially on aluminum heads!
4. Carefully twist the spark plug wire boot to loosen it, then pull the boot using a twisting motion to remove it from the plug. Be sure to pull on the boot and not on the wire, otherwise the connector located inside the boot may become separated from the high-tension wire.

■ **A spark plug wire removal tool is recommended as it will make removal easier and help prevent damage to the boot and wire assembly.**

5. Using compressed air (and safety glasses), blow debris from the spark plug area to assure that no harmful contaminants are allowed to enter the combustion chamber when the spark plug is removed. If compressed air is not available, use a rag or a brush to clean the area. Compressed air is available from both an air compressor or from compressed air in cans available at photography stores.

■ **Remove the spark plugs when the engine is cold, if possible, to prevent damage to the threads. If plug removal is difficult, apply a few drops of penetrating oil to the area around the base of the plug and allow it a few minutes to work.**

6. Using a spark plug socket that is equipped with a rubber insert to properly hold the plug, turn the spark plug counterclockwise to loosen and remove the spark plug from the bore.

✸✸ WARNING

Avoid the use of a flexible extension on the socket. Use of a flexible extension may allow a shear force to be applied to the plug. A shear force could break the plug off in the cylinder head, leading to costly and frustrating repairs. In addition, be sure to support the ratchet with your other hand—this will also help prevent the socket from damaging the plug.

7. Evaluate each cylinder's performance by comparing the spark plug condition. Check each spark plug to be sure they are all of the same manufacturer and have the same heat range rating. Inspect the threads in the spark plug opening of the block and clean the threads before installing the plug.
8. When purchasing new spark plugs, always ask the dealer if there has been a spark plug change for the engine being serviced. Many times manufacturers will update the type of spark plug used in an engine to offer better efficiency or performance.
9. Crank the engine through several revolutions to blow out any material that might have become dislodged during cleaning. Always use a new gasket (if applicable), but never use gaskets on taper seat plugs. The gasket must be fully compressed on clean seats to complete the heat transfer process and to provide a gas tight seal in the cylinder.
10. Inspect the spark plug boot for tears or damage. If a damaged boot is found, the spark plug boot and possible the entire wire will need replacement.
11. Check the spark plug gap prior to installing the plug. Most spark plugs do not come gapped to the proper specification.
12. Apply a thin coating of anti-seize on the thread of the plug.

This is extremely important on aluminum head engines.

13. Carefully thread the plug into the bore by hand. If resistance is felt before the plug completely bottomed, back the plug out and begin threading again.

✳✳ WARNING

Do not use the spark plug socket to thread the plugs. Always carefully thread the plug by hand or using an old plug wire to prevent the possibility of cross-threading and damaging the cylinder head bore.

14. Carefully tighten the spark plug. If the plug you are installing is equipped with a crush washer, tighten the plug until the washer seats, then turn it 1/4 turn to crush the washer. Whenever possible, spark plugs should be tightened to the factory torque specification:
  - 1992-97: 3.0L—15 ft. lbs. (20 Nm)
  - 1998-01: 3.0L—22 ft. lbs. (30 Nm)
  - 4.3L—15 ft. lbs. (20 Nm)
  - 1992-97: V8s—15 ft. lbs. (20 Nm)
  - 1998-01: V8s—15 ft. lbs. (20 Nm); except new heads and the 8.1L engine which should have an initial torque of 22 ft. lbs. (30 Nm)

15. Apply a small amount of silicone dielectric grease to the end of the spark plug lead or inside the spark plug boot to prevent sticking, then install the boot to the spark plug and push until it clicks into place. The click may be felt or heard. Gently pull back on the boot to assure proper contact.

16. Connect the negative battery cable or turn the battery switch **ON**.

17. Start the engine and insure proper operation.

READING SPARK PLUGS

◆ **See Figures 59, 60, 61, 62, 63, 64 and 65**

Reading spark plugs can be a valuable tuning aid. By examining the insulator firing nose color, you can determine much about the engine's overall operating condition.

In general, a light tan/gray color tells you that the spark plug is at the optimum temperature and that the engine is in good operating condition.

Dark coloring, such as heavy black wet or dry deposits usually indicate a fouling problem. Heavy, dry deposits can indicate an overly rich condition, too cold a heat range spark plug, possible vacuum leak, low compression, overly retarded timing or too large a plug gap.

If the deposits are wet, it can be an indication of a breached head gasket, oil control from ring problems or an extremely rich condition, depending on what liquid is present at the firing tip.

Look for signs of detonation, such as silver specs, black specs or melting or breakage at the firing tip.

Compare your plugs to the illustrations shown to identify the most common plug conditions.

Fouled Spark Plugs

A spark plug is fouled when the insulator nose at the firing tip becomes coated with a foreign substance, such as fuel, oil or carbon. This coating makes it easier for the voltage to follow along the insulator nose and leach

Fig. 59 A normally worn spark plug should have light tan or gray deposits on the firing tip (electrode)

Fig. 60 A carbon-fouled plug, identified by soft, sooty black deposits, may indicate an improperly tuned engine

Fig. 61 A physically damaged spark plug may be evidence of severe detonation in that cylinder. Watch the cylinder carefully between services, as a continued detonation will not only damage the plug but will most likely damage the engine

Fig. 62 An oil-fouled spark plug indicates an engine with worn piston rings

Fig. 63 This spark plug has been left in the engine too long, as evidenced by the extreme gap. Plugs with such an extreme gap can cause misfiring and stumbling accompanied by a noticeable lack of power

Fig. 64 A bridged or almost bridged spark plug, identified by the build-up between the electrodes caused by excessive carbon or oil build-up on the plug

Tracking Arc
High voltage arcs between a fouling deposit on the insulator tip and spark plug shell. This ignites the fuel/air mixture at some point along the insulator tip, retarding the ignition timing which causes a power and fuel loss.

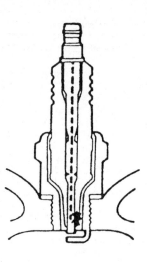

Wide Gap
Spark plug electrodes are worn so that the high voltage charge cannot arc across the electrodes. Improper gapping of electrodes on new or "cleaned" spark plugs could cause a similar condition. Fuel remains unburned and a power loss results.

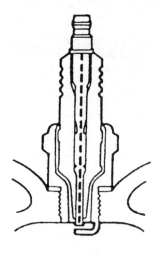

Flashover
A damaged spark plug boot, along with dirt and moisture, could permit the high voltage charge to short over the insulator to the spark plug shell or the engine. A buttress insulator design helps prevent high voltage flashover.

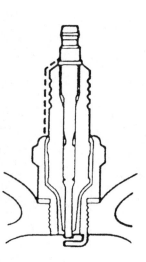

Fouled Spark Plug
Deposits that have formed on the insulator tip may become conductive and provide a "shunt" path to the shell. This prevents the high voltage from arcing between the electrodes. A power and fuel loss is the result.

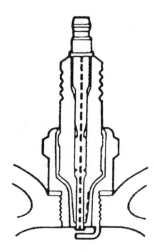

Bridged Electrodes
Fouling deposits between the electrodes "ground out" the high voltage needed to fire the spark plug. The arc between the electrodes does not occur and the fuel air mixture is not ignited. This causes a power loss and exhausting of raw fuel.

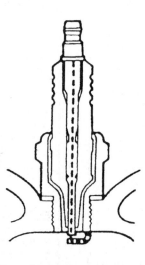

Cracked Insulator
A crack in the spark plug insulator could cause the high voltage charge to "ground out." Here, the spark does not jump the electrode gap and the fuel air mixture is not ignited. This causes a power loss and raw fuel is exhausted.

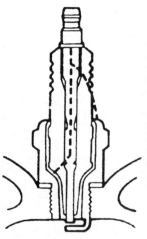

Fig. 65 Typical spark plug problems showing damage that may indicate engine problems

back down into the metal shell, grounding out, rather than bridging the gap normally.

Fuel, oil and carbon fouling can all be caused by different things but in any case, once a spark plug is fouled, it will not provide voltage to the firing tip and that cylinder will not fire properly. In many cases, the spark plug cannot be cleaned sufficiently to restore normal operation. It is therefore recommended that fouled plugs be replaced.

Signs of fouling or excessive heat must be traced quickly to prevent further deterioration of performance and to prevent possible engine damage.

Overheated Spark Plugs

When a spark plug tip shows signs of melting or is broken, it usually means that excessive heat and/or detonation was present in that particular combustion chamber or that the spark plug was suffering from thermal shock.

Since spark plugs do not create heat by themselves, one must use this visual clue to track down the root cause of the problem. In any case, damaged firing tips most often indicate that cylinder pressures or temperatures were too high. Left unresolved, this condition usually results in more serious engine damage.

Detonation refers to a type of abnormal combustion that is usually preceded by pre-ignition. It is most often caused by a hot spot formed in the combustion chamber.   As air and fuel is drawn into the combustion chamber during the intake stroke, this hot spot will "pre-ignite" the air fuel mixture without any spark from the spark plugs.

Detonation

Detonation exerts a great deal of downward force on the pistons as they are being forced upward by the mechanical action of the connecting rods. When this occurs, the resulting concussion, shock waves and heat can be severe. Spark plug tips can be broken or melted and other internal engine components such as the pistons or connecting rods themselves can be damaged.

Left unresolved, engine damage is almost certain to occur, with the spark plug usually suffering the first signs of damage.

■ **When signs of detonation or pre-ignition are observed, they are symptom of another problem. You must determine and correct the situation that caused the hot spot to form in the first place.**

INSPECTION & GAPPING

 MODERATE

◆ **See Figures 66 and 67**

A particular spark plug might fit hundreds of engines and although the factory will typically set the gap to a pre-selected setting, this gap may not be the right one for your particular engine.

Insufficient spark plug gap can cause pre-ignition, detonation, even engine damage. Too much gap can result in a higher rate of misfires, noticeable loss of power, plug fouling and poor economy.

Check the spark plug gap before installation. The ground electrode (the L-shaped one connected to the body of the plug) must be parallel to the center electrode and the specified size wire gauge must pass between the electrodes with a slight drag.

Do not use a flat feeler gauge when measuring the gap on a used plug, because the reading may be inaccurate. A round wire-type gapping tool is the best way to check the gap. The correct gauge should pass through the electrode gap with a slight drag. If you're in doubt, try a wire that is one size smaller or larger. The smaller gauge should go through easily, while the larger one shouldn't go through at all.

Wire gapping tools usually have a bending tool attached. Use this tool to adjust the side electrode until the proper distance is obtained. Never attempt to bend the center electrode. Also, be careful not to bend the side electrode too far or too often as it may weaken and break off within the engine, requiring removal of the cylinder head to retrieve it.

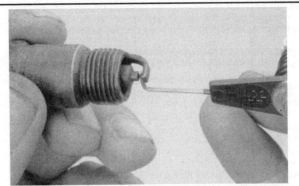

Fig. 66 Using a wire-type spark plug gapping tool to check the distance between center and ground electrodes

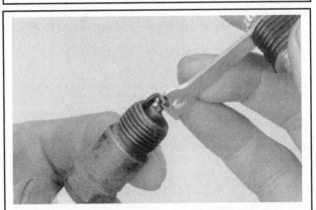

Fig. 67 Most spark plug gapping tools have an adjusting tool used to bend the ground electrode. USE IT! This tool greatly reduces the chance of breaking off the electrode and is much more accurate

Spark Plug Wires

TESTING

 MODERATE

Each time you remove the engine cover, visually inspect the spark plug wires for burns, cuts or breaks in the insulation. Check the boots on the coil and at the spark plug end. Replace any wire that is damaged.

Once a year, usually when you change your spark plugs, check the resistance of the spark plug wires with an ohmmeter. Wires with excessive resistance will cause misfiring and may make the engine difficult to start. In addition worn wires will allow arcing and misfiring in humid conditions.

Remove the spark plug wire from the engine. Test the wires by connecting one lead of an ohmmeter to the coil end of the wire and the other lead to the spark plug end of the wire. Resistance should measure approximately 7000 ohms per foot of wire. If a spark plug wire is found to have excessive (high) resistance, the entire set should be replaced.

REMOVAL & INSTALLATION

 MODERATE

When installing a new set of spark plug wires, replace the wires one at a time so there will be no confusion. Coat the inside of the boots with dielectric grease to prevent sticking. Install the boot firmly over the spark plug until it clicks into place. The click may be felt or heard. Gently pull back on the boot to assure proper contact. Repeat the process for each wire.

■ It is important to route the new spark plug wire the same as the original and install it in a similar manner on the engine. Improper routing of spark plug wires may cause engine performance problems.

Breaker Points and Condesner

Early 3.0L models were equipped with a standard breaker points ignition system. There are two ways to check breaker point gap: with a feeler gauge or with a dwell meter. Either way you choose, you are adjusting the amount of time (in degrees of distributor rotation) that the points will remain open. If you adjust the points with a feeler gauge, you are setting the maximum amount the points will open when the rubbing block on the points is on one of the high points of the distributor cam. When you adjust the points with a dwell meter, you are measuring the number of degrees (of distributor cam rotation) that the points will remain closed before they start to open as a high point of the distributor cam approaches the rubbing block of the points.

Although using a feeler gauge is reasonably accurate when setting new point sets, this method can be unreliable when checking used points due to the rough surface caused by pitting associated with wear and tear. Adjusting the dwell should always be considered the more accurate method.

There are two rules that should always be followed when adjusting or replacing points:

• The points and condenser are a matched set; NEVER replace one without replacing the other.

• When you change the point gap or dwell of the engine, you also change the ignition timing. Always adjust the timing after a point or dwell adjustment.

■ Marine distributors have a corrosion-resistant coating applied to the return spring on top of the breaker plate and on the two small springs under the plate—NEVER use automotive parts as a replacement!

REMOVAL AND INSTALLATION

◆ See Figures 68, 69, 70 and 71

1. Remove or open the engine compartment hatch cover. Disconnect the negative battery cable or turn the battery switch to OFF.

2. Loosen the distributor cap retaining screws (two) and carefully lift off the cap. Although it is not necessary remove the spark plug wires from the cap, we recommend that you first tag all the wires just to be safe.

3. Note the position of the rotor and pull the rotor straight up and remove it. Check the rotor carefully for a burned or corroded center contact, cracks or carbon tracks.

4. Loosen the primary terminal nut and then disconnect the lead wire. Do the same for the condenser lead wire.

5. Loosen the condenser/breaker point mounting screws and then lift them up and off the breaker plate. Clean any dirt or oil left on the plate.

6. Coat the distributor cam with a small amount of distributor Cam Lubricant (NEVER use grease or oil), wipe the new point set clean and position it on the breaker plate. Tighten the mounting screws, leaving the lock screw slightly loose..

7. Reconnect the lead wires for the condenser and primary.

8. Check that the points are in alignment. If not, carefully bend the stationary arm until they align properly. If you are still not satisfied, get a new set of points. Never adjust alignment on used points.

9. Adjust the point gap to 0.016 in. (0.5mm). Install the rotor in the same position is was removed. Install the cap, connect the battery cables and check for proper operation. Adjust the ignition timing

ADJUSTMENT WITH FEELER GAUGE

◆ See Figures 72 and 73

1. Perform the first three steps of the removal procedure above.

2. Connect a remote starter switch as detailed in the manufacturers' instructions. Have a friend bump the engine over until you see that the breaker point rubbing block is resting on the high point of the distributor cam; the points should open to their fullest extent.

3. Insert an 0.016 in. flat feeler gauge between the points. The gauge should be snug but not tight. If adjustment is required, loosen the lock screw and insert a screwdriver in to the adjustment slot on the breaker plate; move the point set until a slight drag can be felt on the feeler gauge and then tighten the lock screw. Always check the gap a final time after tightening the screw as the points sometimes move when tightening the screw.

4. Install the rotor and distributor cap. Connect the battery cables and adjust the ignition timing.

ADJUSTMENT WITH DWELL METER

1. Perform the first three steps of the removal procedure above.

2. Connect a dwell meter as per the manufacturer's instructions–usually the positive lead of the meter to the negative side of the coil and the negative lead of the meter to ground.

3. Connect a remote starter switch as detailed in the manufacturer's instructions and then crank the engine. Observe the dwell reading on the meter, it should be 30 to 45 degrees. If not in range, loosen the lock screw slightly and then adjust the point opening by means of the adjustment slot. Increasing the point gap lowers the dwell reading, while decreasing the gap raises it. When the reading is within specifications, tighten the lock screw and then recheck the dwell one final time.

4. Install the rotor and distributor cap. Connect the battery cables and adjust the ignition timing.

✳✳ CAUTION

Dwell should be checked between idle and 1750 rpm. Any dwell variations of more than 3 degrees from idle to 1750 rpm indicate possible wear in the distributor.

Ignition Timing

As the engine must be running while performing this operation we recommend that it is undertaken with the boat in the water. If not, make certain that an engine flushing kit has been installed.

ADJUSTMENT

3.0L Engines w/Breaker Point or DDIS Ignition Systems

◆ See Figure 68

■ The dwell must be correctly adjusted and within specifications before performing this procedure.

1. Connect a suitable timing light to the No. 1 spark plug lead (see firing order illustrations for location of the No. 1 cylinder). Connect the power supply lead to the battery as detailed in the light manufacturer's instructions.

2. Connect a tachometer to the engine as detailed by the manufacturer. Do not use the tachometer on the instrument panel as it will not provide the necessary accuracy.

3. Locate the timing marks on the engine timing cover (just above the crankshaft pulley) and place a bit of white paint where the proper mark should be (8 degrees BTDC). Timing marks are generally shown in 2 degree increments from TDC.

4. Start the engine and allow it to reach normal operating temperature at idle. While still idling, point the light at the timing marks. The strobe will make it appear that the mark on the tab and the mark on the pulley stand still in alignment.

5. If the timing requires adjustment, loosen the clamp bolt at the base of the distributor (DDIS models have a motion sensor in place of the distributor) and then carefully rotate the distributor or sensor until the correct marks line up.

6. Tighten the clamp bolt to 20 ft. lbs. (27 Nm) and check the timing one last time. If still correct, disconnect the light and tachometer.

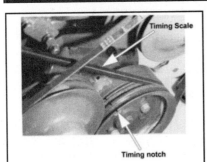

Fig. 68 Timing marks on the 3.0L engine

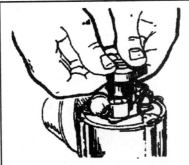

Fig. 69 Pull the rotor straight up to remove it

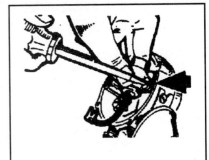

Fig. 70 The condenser is held in place by a screw and clamp

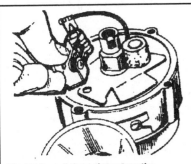

Fig. 71 Install the point set on the breaker plate, then attach the wires

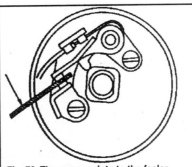

Fig. 72 The arrow points to the feeler gauge used to measure point gap

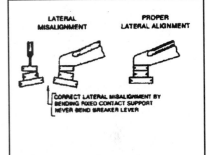

Fig. 73 Check the points for proper alignment after installation

3.0L Engines w/EST Ignition Systems

◆ See Figures 74, 75 and 76

■ Failure to follow the timing procedure instructions exactly will result in improper timing and cause performance problems at the least and possibly severe engine damage.

1. Connect a suitable timing light to the No. 1 spark plug lead (see firing order illustrations for location of the No. 1 cylinder). Connect the power supply lead to the battery as detailed in the light manufacturer's instructions.

2. Connect a tachometer to the engine as detailed by the manufacturer. Do not use the tachometer on the instrument panel as it will not provide the necessary accuracy.

3. Locate the timing marks on the engine timing cover (just above the crankshaft pulley) and place a bit of white paint where the proper mark should be (1990-97: 1 degree BTDC; 1998-2001: 0L096999 and below—1 degree BTDC, 0L097000-0L0340999-1degree ATDC, 0LO341000 and above—2 degrees ATDC). Timing marks are generally shown in 2 degree increments from TDC.

4. Start the engine and allow it to reach normal operating temperature at idle.

5. Disable the timing advance system by installing a jumper wire across the two white leads on the distributor 3-pin connector. This jumper wire is available from your local dealer (p/n 91-818812A1), or you can easily fabricate on with a 6 in. length of 16 gauge wire and two bullet terminal ends.

6. Bypass the shift interrupt switch by disconnecting the two wires at the switch and connecting them together. DO NOT FORGET TO RECONNECT THESE TWO WIRES!

7. While still idling, point the light at the timing marks. The strobe will make it appear that the mark on the tab and the mark on the pulley stand still in alignment.

8. If the timing requires adjustment, loosen the clamp bolt at the base of the distributor (DDIS models have a motion sensor in place of the distributor) and then carefully rotate the distributor or sensor until the correct marks line up.

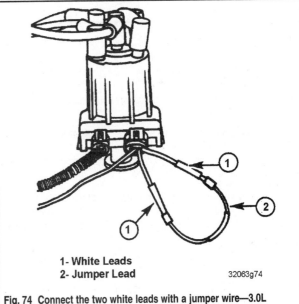

1- White Leads
2- Jumper Lead   32063g74

Fig. 74 Connect the two white leads with a jumper wire—3.0L w/ESRT

9. Tighten the clamp bolt to 20 ft. lbs. (27 Nm) and check the timing one last time.

10. Reconnect the two shift interrupt switch leads to the switch and then remove the jumper wire at the base of the distributor.

11. With the timing light still connected and the engine at idle, check that the timing advanced to 10-14 degrees BTDC after the jumper was removed on models with initial timing of 1 degree BTDC; 12-16 degrees BTDC on models with initial timing of 1 degree ATDC; or, 13-17 degrees on models with initial timing of 2 degrees ATDC.

12. Run the engine up to 2400-2800 rpm and check the timing again.

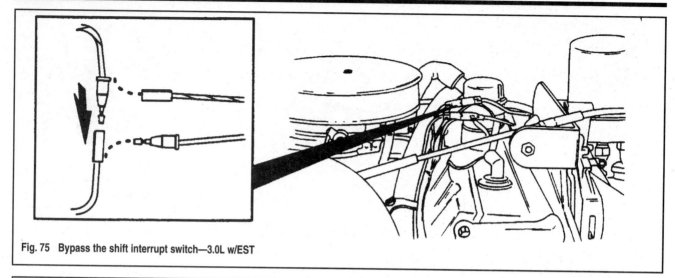

Fig. 75 Bypass the shift interrupt switch—3.0L w/EST

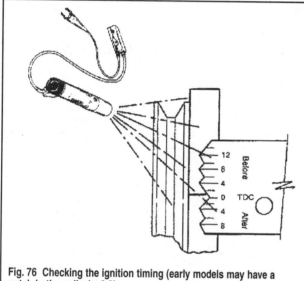

Fig. 76 Checking the ignition timing (early models may have a notch in the pulley)—3.0L

Total advance should be:
- 25-29 degrees on 1990-97 models
- 21-25 degrees on 1998-01 models w/initial timing of 1 degree BTDC
- 23-27 degrees on 1998-2001 models w/initial timing of 1 degree ATDC
- 24-28 degrees on 1998-2001 models w/initial timing of 2 degrees ATDC

If not check the ignition module as detailed in the Ignition section.

13. Disconnect the timing light and the tachometer.

V6 And V8 Engines

CARBURETORS W/THUNDERBOLT IV

◆ See Figure 77

This system is used on 1992-97 V6 models and certain 1992-97 V8 models.

■ **The dwell must be correctly adjusted and within specifications before performing this procedure.**

1. Connect a suitable timing light to the No. 1 spark plug lead (see firing order illustrations for location of the No. 1 cylinder). Connect the power supply lead to the battery as detailed in the light manufacturer's instructions.

2. Connect a tachometer to the engine as detailed by the manufacturer. Do not use the tachometer on the instrument panel, as it will not provide the necessary accuracy.

3. Locate the timing marks on the engine timing cover (just above the crankshaft pulley) and place a bit of white paint where the proper mark should be (8 degrees BTDC). Timing marks are generally shown in 2 degree increments from TDC.

4. Start the engine and allow it to reach normal operating temperature at idle. While still idling, point the light at the timing marks. The strobe will make it appear that the mark on the tab and the mark on the pulley stand still in alignment.

5. If the timing requires adjustment, loosen the clamp bolt at the base of the distributor (DDIS models have a motion sensor in place of the distributor) and then carefully rotate the distributor or sensor until the correct marks line up.

6. Tighten the clamp bolt to 20 ft. lbs. (27 Nm) and check the timing one last time. If still correct, disconnect the light and tachometer.

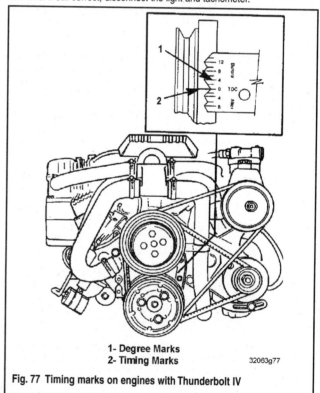

1- Degree Marks
2- Timing Marks

32063g77

Fig. 77 Timing marks on engines with Thunderbolt IV

CARBURETORS W/THUNDERBOLT V

This system is used on 1998-01 V6 models and certain V8 models.

■ **The idle speed and dwell must be correctly adjusted and within specifications before performing this procedure.**

1. Connect a suitable timing light to the No. 1 spark plug lead (see firing order illustrations for location of the No. 1 cylinder). Connect the power supply lead to the battery as detailed in the light manufacturer's instructions.
2. Connect a tachometer to the engine as detailed by the manufacturer. Do not use the tachometer on the instrument panel, as it will not provide the necessary accuracy.
3. Locate the timing marks on the engine timing cover (just above the crankshaft pulley/damper) and place a bit of white paint where the proper mark should be (10 degrees BTDC—carbureted; 8 degrees BTDC—EFI models). Timing marks are generally shown in 2 degree increments from TDC.
4. Locate the ignition system timing lead and connect it to a good ground (-) to lock the ignition module into base timing mode. The lead should be a purple and white wire near the distributor or in front of the engine near the fuel line.
5. Start the engine and allow it to reach normal operating temperature at idle.
6. Disconnect the throttle cable at the carburetor on 1998 and later V8 models.
7. While still idling, point the light at the timing marks. The strobe will make it appear that the mark on the tab and the mark on the damper/pulley stand still in alignment.
8. If the timing requires adjustment, loosen the clamp bolt at the base of the distributor and then carefully rotate the distributor until the correct marks line up.
9. Tighten the clamp bolt to 20 ft. lbs. (27 Nm) on all engines except the 1998 and later V8, which should be tightened to 18 ft. lbs. (25 Nm).
10. Disconnect the jumper wire from the timing lead and ground.
11. On 1998 and later V8s, reinstall and adjust the throttle cable.
12. Check the timing one last time. If still correct, disconnect the light and tachometer.

WITH TBI/MPI

Although ignition timing is adjustable on these models, it is generally controlled by the EFI electronic control module. In order to adjust the timing, the ECM must be forced to enter into its service mode by using a scan tool. This done, the ECM will stabilize the base timing to allow for adjustment by conventional means of rotating the distributor.

Mercruiser does not offer ignition timing adjustments for the 8.1L engines.

■ **The idle speed and dwell must be correctly adjusted and within specifications before performing this procedure.**

1. Open or remove the engine compartment hatch.
2. Connect a suitable timing light to the No. 1 spark plug lead (see firing order illustrations for location of the No. 1 cylinder). Connect the power supply lead to the battery as detailed in the light manufacturer's instructions.
3. Connect a tachometer to the engine as detailed by the manufacturer. Do not use the tachometer on the instrument panel, as it will not provide the necessary accuracy.
4. Locate the timing marks on the engine timing cover (just above the crankshaft pulley/damper) and place a bit of white paint where the proper mark should be as detailed in the Tune-Up Specifications chart. Timing marks are generally shown in 2 degree increments from TDC.
5. Start the engine and allow it to reach normal operating temperature at idle.
6. Turn off the engine, locate the data link connector (DLC) on the EFI/MPI main harness and plug in a scan or timing tool.
7. Restart the engine and allow the idle to stabilize. On TBI or MPI/MEFI-1 engines, adjust the throttle lever so that the idle moves to about 1200 rpm (1800 rpm on 7.4L/8.2L). On MPI/MEFI-2 and -3 engines the computer will automatically adjust the idle to 1200 rpm when the scan tool is set to the service mode.
8. While still idling, point the light at the timing marks. The strobe will make it appear that the mark on the tab and the mark on the damper/pulley stand still in alignment.

9. If the timing requires adjustment, loosen the clamp bolt at the base of the distributor and then carefully rotate the distributor until the correct marks line up.
10. Tighten the clamp bolt to 18 ft. lbs. (25 Nm) and then recheck the timing. On 7.4L/8.2L engines tighten the bolt to 30 ft. lbs. (40 Nm).
11. Disconnect the scan/timing tool. On engines where the throttle lever was adjusted up to 1200 rpm, adjust it back to the normal idle position.
12. Run the engine up to 1300 rpm and then slowly allow it to settle back down to idle. Recheck the idle and adjust the throttle cable if necessary.
13. Check the timing one last time. If still correct, disconnect the light and tachometer.

Idle Speed and Mixture

ADJUSTMENT

Preliminary Idle

ALL ENGINES W/2BBL

◆ **See Figures 78 and 79**

■ **The following adjustments will provide sufficient idle speed and mixture settings to get your engine started. Final adjustments must always be made with the engine running.**

1. Locate the idle speed screw on your carburetor and turn it in or out until it is just resting on the idle cam, but is not moving it. The idle screw is threaded into the throttle linkage on the side of the carburetor and has a spring between the screw head and the linkage.
2. Turn the screw inward (clockwise) two additional turns.
3. Locate the idle mixture screw on the base of the carburetor and turn it in (clockwise) until it just lightly seats itself and then back it out 1-1/4 turns.

✳✳ CAUTION

Be careful not to turn the idle mixture screw in past the seated position or you risk damaging the seat or needle.

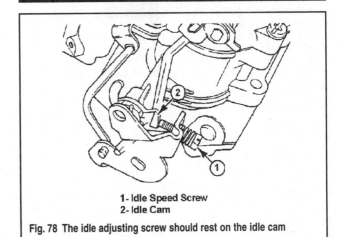

1- Idle Speed Screw
2- Idle Cam

Fig. 78 The idle adjusting screw should rest on the idle cam

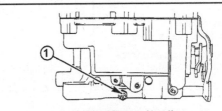

1- Idle Mixture Needle

Fig. 79 Never tighten the idle mixture screw against its seat

■ On 1998-01 models equipped with emissions carburetors, the idle mixture screw is sealed. Fuel mixture is not adjustable on these models.

ALL ENGINES W/4BBL

◆ See Figures 80, 81, 82 and 83

■ The following adjustments will provide sufficient idle speed and mixture settings to get your engine started. Final adjustments must always be made with the engine running.

1. Locate the idle speed screw on your carburetor and turn it in until it is just resting on the throttle lever cam, but is not moving it. The idle screw is threaded into a block on the side of the carburetor and has a spring between the screw head and the linkage.

2. Locate the idle mixture screws on the base of the carburetor and turn them in (clockwise) until they just lightly seat themselves and then back each one out 2–3 turns on 1993-97 4.3L engines and V8 engines with a Rochester; 1-1/4 turns on 1998-2001 4.3L engines and V8 engines with a Weber.

✳✳ CAUTION

Be careful not to turn the idle mixture screw in past the seated position or you risk damaging the seat or needle.

■ On models equipped with emissions carburetors, the idle mixture screw is sealed. Fuel mixture is not adjustable on these models.

Final Idle

3.0L ENGINES
4.3L ENGINES (EXC. 1998-2001 4BBL)
V8 ENGINES (EXC. THUNDERBOLT V IGNITION)

◆ See Figures 80, 81, 82 and 83

1. With the boat in the water and the engine running, connect a tachometer as per the manufacturer's instructions.
2. With the drive in Neutral, run the engine at approximately 1500 rpm until it reaches normal operating temperature.
3. Shift the drive into Forward gear, at idle. Make sure there's someone at the helm and you're in open water when doing this!
4. Being careful not to lose the spacer or anchor stud, disconnect the throttle cable barrel at the anchor stud.
5. Check the idle speed on the tachometer. If not within specifications (see Tune-Up Specifications chart) adjust the idle speed screw until the proper rpm is achieved.

6. When the engine is idling at the proper speed, turn the idle mixture screw, on non-emissions models, clockwise until the engine speed begins to drop due to a lean mixture. Remember the setting.
7. Now turn the idle mixture screw counterclockwise until the idle speed begins to drop due to a rich mixture. Remember the setting.
8. Turn the mixture screw to a point midway between the settings from the two previous steps until the engine runs smoothly. Check the idle speed once again and readjust if not within specifications.
9. Turn off the engine, remove the tachometer and reconnect the throttle cable at the anchor stud.

4.3L ENGINES (1998-2001 4BBL)
V8 ENGINES (THUNDERBOLT V IGNITION)

1. Connect a tachometer as per the manufacturer's instructions.
2. Locate the ignition timing lead near the distributor and connect a jumper wire between it and a good ground (-). This lead should be a purple and white wire; grounding it will force the ignition module into base timing mode.
3. Start the engine. With the drive in Neutral, run the engine at approximately 1500 rpm until it reaches normal operating temperature.
4. Disconnect the throttle cable from the carburetor.
5. Shift the drive into Forward gear, at idle. Make sure you're in open water when doing this!
6. Check the idle speed on the tachometer. If not within specifications (see Tune-Up Specifications chart) adjust the idle speed screw until the proper rpm is achieved.
7. When the engine is idling at the proper speed, check that the idle mixture screws are backed out 1-1/4 turns, on non-emissions models.
8. Install and adjust the throttle cable.
9. Turn off the engine, remove the tachometer and disconnect the jumper wire from the timing lead and ground.

4.3L ENGINES (1998-2001 SAV1 EMISSION)

Idle speed and mixture are set at the factory on these models. Mixture screws are sealed. Idle is adjustable only via the propane-enrichment method—we recommend taking the boat to an authorized service facility with the proper tools in order to complete this procedure.

V8 ENGINES (EFI)

Idle speed is constantly monitored by the electronic control module (ECM) and controlled by the idle air control valve (IAC). Idle speed and mixture are not adjustable. Please refer to Section 8 for further information on the fuel injection system.

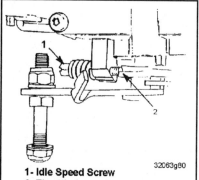

32063g80

1- Idle Speed Screw
2- Throttle Lever Contact Point

Fig. 80 The idle speed adjusting screw should rest on the throttle lever Weber 4bbl

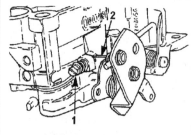

1- Idle Stop Screw
2- Throttle Lever

Fig. 81 The idle speed adjusting screw should rest on the throttle lever Rochester 4bbl

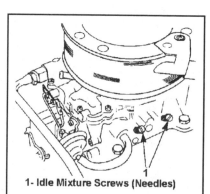

1- Idle Mixture Screws (Needles)

Fig. 82 Never tighten the idle mixture screws against their seats–Weber 4bbl

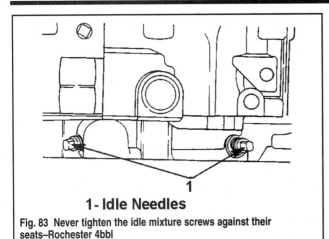

1- Idle Needles

Fig. 83 Never tighten the idle mixture screws against their seats–Rochester 4bbl

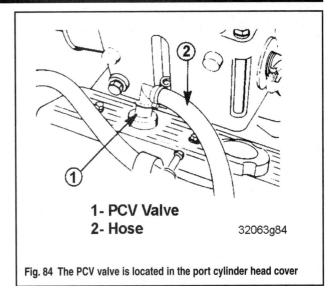

1- PCV Valve
2- Hose   32063g84

Fig. 84 The PCV valve is located in the port cylinder head cover

PCV Valve

4.3L and 5.7L engines with the Bodensee/SAV1 emissions system are equipped with a positive crankcase ventilation (PCV) circuit that utilizes a PCV valve in the rocker cover in order to ventilate unburned crankcase gases back into the engine via the intake manifold in order that they can be re-burned. The PCV valve should be replaced every boating season.

REMOVAL & INSTALLATION

◆ **See Figure 84**

1. Locate the PCV valve in the Port cylinder head cover.
2. Carefully wiggle it back and forth while pulling upward on the valve itself until it pops out of the cover.
3. Loosen the clamp and disconnect the breather hose from the valve.
4. Reconnect the hose and press the valve back into the cylinder head.

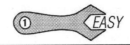

INSPECTION

◆ **See Figure 85**

Start the engine and allow it to reach normal operating temperature. Pop out the PCV valve as detailed above and cover the opening with your thumb. You should be able to feel significant vacuum; if not replace the valve. With the valve still in your fingers, shake it back and forth a few times; you should be able to hear the inside components moving around.

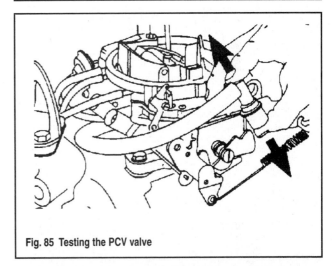

Fig. 85 Testing the PCV valve

Valve Adjustment

All engines covered in this manual are equipped with hydraulic valve lifters and do not require periodic valve adjustment. Adjustment to zero lash is maintained automatically by the hydraulic pressure in the lifters. For initial adjustment procedures after cylinder head work, please refer to Section 6.

FLUIDS AND LUBRICANTS

Fluid Disposal

Used fluids such as engine oil, gear oil, antifreeze and power steering fluid are hazardous waste and must be disposed of properly and responsibly. Before draining any fluids it is always a good idea to check with your local authorities; in many areas there are recycling programs available for easy disposal. Service stations, Parts stores and Marinas also often will accept waste fluids for recycling.

Be sure of your local recyclers policies before draining any fluids, as many will not accept fluids that have been mixed together.

Fuel And Oil Recommendations

FUEL

All engine covered in this manual are designed to run on unleaded fuel. Never use leaded fuel in your boat's engine. The minimum octane rating of fuel being used for your engine must be at least 87, which means regular unleaded, but some engines may require higher octane ratings. Fuel should

be selected for the brand and octane that performs best with your engine.

The use of a fuel too low in octane (a measure of anti-knock quality) will result in spark knock. Newer systems have the capability to adjust the engine's ignition timing to compensate to some extent, but if persistent knocking occurs, it may be necessary to switch to a higher grade of fuel. Continuous or heavy knocking may result in engine damage.

ENGINE OIL

◆ **See Figure 86**

Nothing affects the performance and durability of an engine more than the engine oil. If inferior oil is used, or if your engine oil is not changed regularly, the risk of piston seizure, piston ring sticking, accelerated wear of the cylinder walls or liners, bearings and other moving components increases significantly.

Maintaining the correct engine oil level is one of the most basic (and essential) forms of engine maintenance. Get into the habit of checking your oil on a regular basis; all engines naturally consume small amounts of oil, and if left neglected, can consume enough oil to damage the internal components of the engine. Assuming the oil level is correct because you "checked it the last time" can be a costly mistake.

If your engine has not been operated for more than 6 months it should be primed prior to starting.

When shutting the engine down, always let the engine idle a few minutes to bring engine temperature down to a normal level. Since the engine is, at least in part, cooled by engine oil, it is necessary to allow the engine oil temperature to stabilize prior to shutdown. Not allowing the temperature to stabilize can damage vital engine components.

Every container of engine oil for sale in the U.S. should have a label describing what standards it meets. Engine oil service classifications are designated by the American Petroleum Institute (API), based on the chemical composition of a given type of oil and testing of samples. The ratings include "S" (normal gasoline engine use) and "C" (commercial, fleet and diesel) applications. Over the years, the S rating has been supplemented with various letters, each one representing the latest and greatest rating available at the time of its introduction. During recent years these ratings have changed and most recently (at the time of this manual's publication), the rating is SH or CH-4. Each successive rating usually meets all of the standards of the previous alpha designation, but also meets some new criteria, meets higher standards and/or contains newer or different additives. Since oil is so important to the life of your engine, you should obviously NEVER use an oil of questionable quality. Oils that are labeled with modern API ratings, including the "energy conserving" donut symbol, have been proven to meet the API quality standards. Always use the highest grade of oil available. The better quality of the oil, the better it will lubricate the internals of your engine.

In addition to meeting the classification of the API, your oil should be of a viscosity suitable for the outside temperature in which your engine will be operating. Oil must be thin enough to get between the close-tolerance moving parts it must lubricate. Once there, it must be thick enough to separate them with a slippery oil film. If the oil is too thin, it won't separate the parts; if it's too thick, it can't squeeze between them in the first place—either way, excess friction and wear takes place. To complicate matters, cold-morning starts require a thin oil to reduce engine resistance, while high speed boating requires a thick oil which can lubricate vital engine parts at temperatures.

According to the Society of Automotive Engineers' (SAE) viscosity classification system, an oil with a high viscosity number (such as SAE 40 or SAE 50) will be thicker than one with a lower number (SAE 10W). The "W" in 10W indicates that the oil is desirable for use in winter operation, and does not stand for "weight". Through the use of special additives, multiple-viscosity oils are available to combine easy starting at cold temperatures with engine protection at high speeds. For example, a 10W40 oil is said to have the viscosity of a 10W oil when the engine is cold and that of a 40 oil when the engine is warm. The use of such an oil will decrease engine resistance and improve efficiency.

Mercury Marine has run through various recommendations with regard to multi-viscosity oils over the years so be certain to check you owner's manual before purchasing your oil.

1990-97 models come with the recommendation that you use only

Quicksilver Marine oil as detailed in the accompanying chart; single grade only, in a pinch allowing for 20W-40 or 20W-50, but strongly discouraging the use of any other multi-weight oils.

1998 and later models come with the recommendation to us Quicksilver Marine oil in a special 25W-40 viscosity, but continue to discourage anything but temporary use of other multi-weight oils (20W-40 or 20W-50 only).

Priming

■ **Anytime your boat's engine has not been run for more than 6 months it is a good idea to prime the engine prior to starting it.**

1. Check that the proper amount of oil is in the crankcase (see the Capacities chart).
2. Remove the spark plugs.
3. With the ignition key in the OFF position, connect a remote starter according to the manufacturer's instructions. If a remote starter is unavailable, disconnect the Purple wire from the ignition coil and then use the ignition key to turn the engine over. Make sure that you tape the terminal on the Purple wire to prevent it from touching ground.
4. Crank the engine for 15 seconds and then allow the starter to cool for 1 minute. Repeat this sequence 2 more times until a total cranking time of 45 seconds has been reached.
5. Remove the remote starter or reconnect the Purple wire to the coil.
6. Install the spark plugs and start the engine.

OIL LEVEL CHECK

✳✳ CAUTION

The EPA warns that prolonged contact with used engine oil may cause a number of skin disorders, including cancer! You should make every effort to minimize your exposure to used engine oil. Protective gloves should be worn when changing the oil. Wash your hands and any other exposed skin areas as soon as possible after exposure to used engine oil. Soap and water, or waterless hand cleaner should be used.

When checking the oil level, it is best that the boat be level and the oil be at operating temperature. Checking the level immediately after stopping the engine will give a false reading; always wait about 5 minutes before checking.

It is normal for an engine to naturally consume oil during the course of operation, particularly during break-in on a new engine. You should not be alarmed if the oil level in your engine drops slightly between inspections. In fact certain of MerCruiser's high performance engines (7.4L/8.2L) may use up to a quart of oil every 5 hours when operated at full throttle.

Also the color of the oil is usually a pitch black color. Smelling the oil is a better indicator of oil condition than the color. If the oil smells burned, it should be replaced immediately.

Over-filled crankcases can cause a fluctuation or drop in oil pressure, and, particularly on MerCruiser engines, clattering from the rocker arms. Take great care in checking and filling your engine with oil. Always maintain it between the ADD and FULL or OP RANGE markings on the dipstick. Beware of false readings by checking the level too soon after adding oil.

■ **It takes a little while for fresh oil poured into the engine to reach the crankcase. Wait for about 3 minutes and then check the oil level again.**

Oil level should be checked each day the engine is operated and it is best to check it while in the water.

1. Run the engine until it reaches normal operating temperature. Shut it off and allow the oil to settle for at least 5 minutes.
2. Locate the engine oil dipstick. The 8.1L engine has three, one on each side of the rear of the engine and one at the front.
3. Clean the area around the dipstick to prevent dirt from entering the engine.
4. Remove the dipstick and note the color of the oil. Wipe the dipstick clean with a rag.

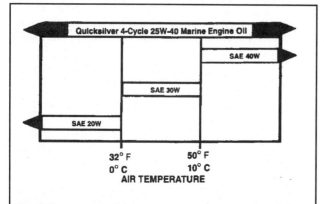

Fig. 86 The ambient temperature dictates the required viscosity of engine oil

5. Insert the dipstick fully into the tube and remove it again. Hold the dipstick horizontal and read the level on the dipstick. The level should always be at the upper limit. If the oil level is below the upper limit, sufficient oil should be added to restore the proper level of oil in the crankcase. Most dipsticks are marked with ADD and FULL or OP RANGE gradations.

6. See "Engine Oil Recommendations" for the proper viscosity and type of oil.

7. Oil is added through the filler port cap in the top of the valve cover. Add oil slowly and check the level frequently to prevent overfilling the engine.

※※ WARNING

Do not overfill the engine. If the engine is overfilled, the crankshaft will whip the engine oil into a foam causing loss of lubrication and severe engine damage.

OIL & FILTER CHANGE  MODERATE

◆ See Figures 87, 88, 89, 90 and 91

A few precautions can make the messy job of oil and filter maintenance much easier. By placing oil absorbent pads, available at industrial supply stores, into the area below the engine, you can prevent oil spillage from reaching the bilge.

It is a good idea to warm the engine oil first so it will flow better; and the contaminates in the bottom of the pan are suspended in the oil. This is accomplished by starting the engine and allowing it to reach normal operating temperature.

Changing engine oil is sometimes complicated by the location of the drain plug. Most boats equipped with inboard engines use an evacuation pump to remove the used engine oil through the dipstick tube. If you don't have a permanently mounted oil suction pump in your engine compartment, you may want to consider installing one. This pump sucks waste oil out through either the dipstick tube or a connection on the oil drain plug and is available from your Mercruiser dealer or from any number of aftermarket sources. They come in a variety of configurations: motorized, hand-pumped or attached to an ordinary household drill.

The maintenance interval for oil and filter change is every 100 hours of engine operation or annually, whichever comes first. This interval should be strictly kept; in fact we recommend cutting the interval in half!

■ **Since you will be hanging into the engine compartment, gather all the tools and spare parts necessary for the job. Don't forget plenty of rags to clean up any spills, and most importantly, remember a container or plastic milk jug to drain the oil into.**

1. Start the engine and run it until it reaches normal operating temperature. Turn the engine off and remove the dipstick.

2. Connect the evacuation pump hose to the dipstick tube or insert it into the tube. Position the other hose, or the outlet at the bottom of the pump in your container. Keep in mind that the fast flowing oil, which will spill out of the pump hose will flow with enough force that it could miss the container and end up all over the deck or in the bilge. Position the container accordingly and be ready to move it if necessary. Some models are equipped with a quick drain oil hose attachment; be sure to pull the tether through the bilge drain before removing the drain plug from the hose.

※※ CAUTION

Use caution around the hot oil; when at operating temperature, it is hot enough to cause a severe burn.

3. Allow the oil to drain until nothing but a few drops come out of the pump. It should be noted that depending on the angle of the engine, some oil may be left in the crankcase. This is normal and should not cause the engine harm.

4. Remove evacuation pump and reinsert the dipstick..

5. Position a drain pan, or a cut-down milk jug under the oil filter. Some filters are mounted horizontally and some vertically. In either case, there is usually oil left in the filter. When the filter is removed, oil will flow out of the engine and the filter. If you are not prepared, you will have a mess on your

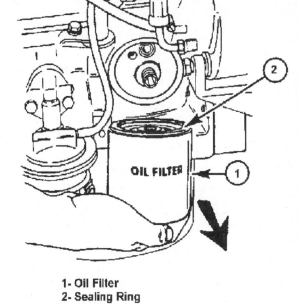

1- Oil Filter
2- Sealing Ring

Fig. 87 Removing the oil filter

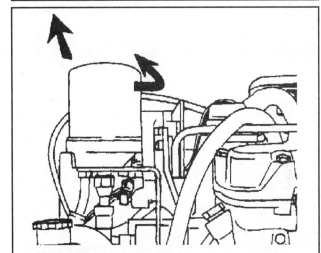

Fig. 88 Some filters are upside down—make sure you have a container and plenty of rags available

Fig. 89 Before installing the new filter, wipe the gasket mating surface clean

hands. It may also be necessary to use a funnel or fashion some type of drain shield to guide the oil into the drain pan. This can be a simple as a recycled oil bottle with the bottom cut off or an elaborate creation made of tin. In any case, its purpose is to prevent oil from spilling all over the engine and bilge.

6. To remove the filter, you will probably need an oil filter strap wrench (available at almost any place that sells marine or automotive parts and fluids). Heat from the engine tends to tighten even a properly installed filter and makes it difficult to remove. Place the wrench on the filter as close to the engine as possible, while still leaving room to work. This will put the wrench at the strongest part of the filter (near the threaded end) and prevent crushing the filter. Loosen the filter with the wrench using a counterclockwise turning motion.

■ **Some models may utilize a remote oil filter.**

7. Once loosened, wrap a rag around the filter and unscrew it from the boss on the engine. Make sure that the drain pan and shield are positioned properly before you start unscrewing the filter. Should some of the hot oil happen to get on your hands and burn you, dump the filter into the drain pan.

8. Wipe the base of the mounting boss with a clean, dry lint-free rag. If the filter is installed vertically, you may want to fill the filter about half way with oil prior to installation. This will prevent oil starvation when you fire the engine up again. Pre-filling the oil filter is usually not possible with horizontally mounted filters. Smear a little bit of fresh oil on the filter gasket to help it seat properly on the engine.

9. Install the filter and tighten it approximately a quarter-turn after it contacts the mounting boss (always follow the filter manufacturer's instructions). This usually equals "hand tight." Using a wrench to tighten the filter is not required.

10. If any oil has gotten into the bilge, remove it using an oil absorbent pad. These pads are specially formulated to only absorb oil and will not soak up any water in the bilge. Perform a visual inspection and make sure all connections are tight.

11. Carefully remove the drain pan from under the oil filter and transfer the oil into a suitable container for recycling.

12. Refill the engine with the proper quantity and quality of oil immediately (see the Capacities chart). You may laugh at the severity of this warning, but if you wait to refill the engine and someone unknowingly tries to start it, severe and costly engine damage will result.

13. Refill the engine crankcase slowly through the filler cap on the valve cover. Use a funnel as necessary to prevent spilling oil. Check the level often. You may notice that it usually takes less than the amount of oil listed in the Capacities chart to refill the crankcase. This is only until the engine is started and the oil filter is filled with oil.

14. To make sure the proper level is obtained, run the engine to normal operating temperature, turn the engine **OFF**, allow the oil to drain back into the oil pan and recheck the level after about 5 minutes. Top off the oil to the correct mark on the dipstick.

Fig. 90 Dip a finger into a fresh bottle of oil, and lubricate the gasket of the new filter

Fig. 91 Tighten the new filter 1/2 to 3/4 of a turn after the gasket contacts the mating surface. Do not use a wrench to tighten the filter!

Power Steering

FLUID LEVEL

◆ **See Figures 92 and 93**

■ **Power steering fluid level should always be checked with the engine hot if at all possible.**

1. Start the engine and run it until it reaches normal operating temperature.

2. Turn the engine OFF and place the drive unit so that it is straight back.

3. Locate the dipstick cap on top of the power steering fluid reservoir. Unscrew the cap and remove the dipstick. Wipe down the dipstick with a clean rag and reinsert it into the reservoir making sure that the cap is screwed all the way down.

4. Remove the dipstick again and check that the fluid level is up to the FULL HOT mark (FULL COLD if the level is being checked cold). If the level is between the FULL mark and the indent on the stick, it's OK. Below the indent or hash mark and you must add fluid. Use either Quicksilver Power Steering Fluid or Dexron® III ATF fluid. Do Not Overfill!

5. It is always a good idea to bleed the system after adding fluid.

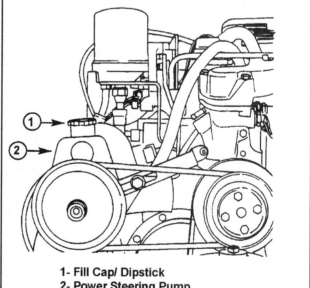

1- Fill Cap/ Dipstick
2- Power Steering Pump

Fig. 92 Typical power steering pump

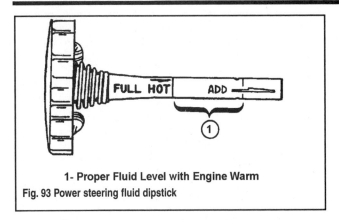

1- Proper Fluid Level with Engine Warm

Fig. 93 Power steering fluid dipstick

BLEEDING THE SYSTEM

■ **It is important that all air be removed from the system after filling. If air is left in the system, the fluid in the pump may foam during operation causing discharge or spongy steering.**

1. With the engine stopped and the drive unit positioned straight back, check that the fluid level is at the FULL COLD mark on the dipstick.
2. Turn the steering wheel from lock to lock several times in succession and then check the fluid level again. Add fluid if necessary.
3. Install the cap/dipstick onto the reservoir, start the engine and run it at a fast idle (1000–1500 rpm) until it reaches normal operating temperature. At the same time you should be turning the wheel from lock to lock again. Turn off the engine and remove the reservoir cap allowing any foam in the pump to escape.
4. Reinsert the dipstick and check the fluid again, making sure that it is at the FULL HOT mark this time. Add fluid as necessary.
5. If the fluid is still foamy, repeat the procedure several times until the foam is gone and the level remains constant.

Cooling System

All engines covered in this manual utilize one of two cooling systems, or variations thereof. Simply, there is a Seawater Cooling system and a Closed Seawater systems are just that...they utilize the water that the boat is operating in to cool and lubricate the drive and engine. Water is drawn in through the stern drive unit and circulated to, and through, the engine and its components.

Closed cooling systems are actually two systems working together; a seawater system and a closed system consisting of antifreeze. The two systems work together in a variety of ways.

Complete information on individual system operation, draining and component repair procedures is detailed in Section 6. In this section we will only deal with checking fluid levels and flushing the systems.

LEVEL CHECK

Closed Cooling System

◆ **See Figure 94**

■ **Always allow the engine to cool down before checking the coolant level. If possible, check cold. Opening a pressure cap with the engine hot can cause a violent discharge resulting in severe injury. If the engine is anything but cold always remove the cap a quarter turn at a time in order to allow any residual pressure to escape slowly.**

1. Remove the cap at the top of the heat exchanger. Coolant should be visible at the bottom of the filler neck, or within 1 in. (25mm) of the bottom of the neck.
2. Reinstall the cap on the heat exchanger making sure that seats on its stops on the filler neck.

3. Start the engine and run it until it reaches normal operating temperature. Turn off the engine and check the coolant level is to the FULL line on the side of the coolant recovery tank.
4. Fill the system with a 50/50 mixture of distilled water and ethylene glycol antifreeze.

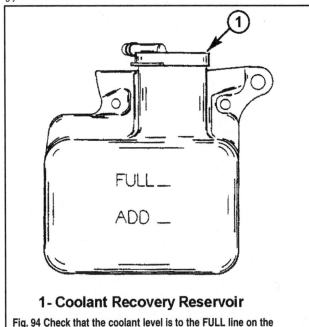

1- Coolant Recovery Reservoir

Fig. 94 Check that the coolant level is to the FULL line on the coolant recovery tank when the engine is hot.

FLUSHING THE SYSTEM

■ **If your boat is operated in saltwater, heavy mineral fresh water or severely polluted water, flush the cooling system regularly—after each usage if at all possible! Always flush the system before draining and/or winter storage.**

Boat Out Of Water

◆ **See Figures 95 and 96**

✳✳ CAUTION

NEVER run the engine when the boat is out of the water without water being supplied to the stern drive.

1. Connect a flushing attachment over the water intake openings in the stern drive gear housing. These devices are available at your local MerCruiser dealer or through a variety of aftermarket suppliers.
2. Connect a garden hose between the flushing attachment and a water spigot.
3. If the boat is equipped with a seawater pick-up pump, disconnect the inlet hose and attach an additional hose to the inlet neck—you'll need an adaptor for this.
4. Open the spigot(s) slowly, no more than half way, and allow the drive and cooling system to fill completely. You'll know the system is full when water begins to flow out of the drive unit, or through the propeller on closed systems.

✳✳ CAUTION

Never allow the water to reach the flushing device at anything more than low pressure.

5. With the drive in Neutral, start the engine and let it idle for 10 minutes or until the water being discharged is clear and then turn the engine off.

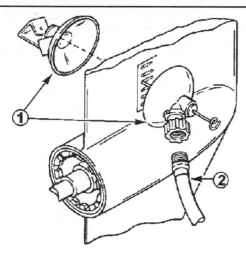

1- Flushing Attachment
2- Water Hose

Fig. 95 Attach a flushing device to the drive unit when flushing the cooling system

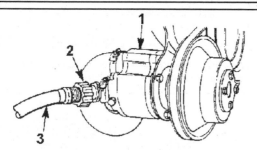

1- Seawater Pickup Pump
2- Adaptor
3- Tap Water Hose

Fig. 96 Disconnect the inlet hose and attach a hose to the inlet neck on models with a seawater pick-up pump

Never run the engine above 1500 rpm with the flushing device attached and always keep an eye on the temperature gauge incase the engine begins to overheat.

6. Turn off the water from the spigot(s), disconnect the hose(s) and remove the flushing attachment.

Boat In Water

EXC. 1998-2001 454 MAG MPI HORIZON

◆ See Figures 95 and 96

☀☀ CAUTION

If your boat has a seacock (a water inlet valve), it must remain closed during this procedure in order to prevent water from flowing back into the boat. If your boat does not have a seacock, locate the water inlet hose at the seawater pick-up pump, disconnect it and plug it. We highly recommend that you leave a note to yourself in the vicinity of the ignition key reminding of the fact that this procedure has been done so that you, or someone else doesn't start the engine after flushing without reopening the seacock or reconnecting the inlet hose. Seem silly? Do it anyway!

1. Raise the stern drive unit to the full UP position.
2. Connect a flushing attachment over the water intake openings in the stern drive gear housing. These devices are available at your local

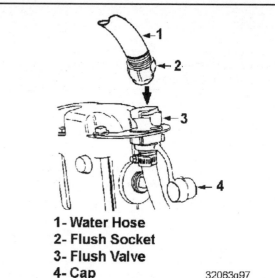

1- Water Hose
2- Flush Socket
3- Flush Valve
4- Cap

32063g97

Fig. 97 These engines come equipped with a flush socket and valve

MerCruiser dealer or through a variety of aftermarket suppliers.
3. Connect a garden hose between the flushing attachment and a water spigot.
4. Lower the stern drive unit to the full IN/DOWN position.
5. If the boat is equipped with a seawater pick-up pump, disconnect the inlet hose, plug it and attach an additional hose to the inlet neck—you'll need an adaptor for this.
6. Open the spigot(s) slowly, no more than half way, and allow the drive and cooling system to fill completely. You'll know the system is full when water begins to flow out of the drive unit, or through the propeller on closed systems.

☀☀ CAUTION

Never allow the water to reach the flushing device at anything more than low pressure.

7. With the drive in Neutral, start the engine and let it idle for 10 minutes or until the water being discharged is clear and then turn the engine off. Never run the engine above 1500 rpm with the flushing device attached and always keep an eye on the temperature gauge incase the engine begins to overheat.
8. Turn off the water from the spigot(s).
9. Raise the drive unit to the UP position again, disconnect the hose and remove the flushing attachment.
10. Lower the unit and make sure you open the seacock and reconnect the inlet hose (don't forget to unplug it first!).

1998-2001 454 MAG MPI HORIZON (INBOARD)

◆ See Figure 97

■ This procedure should not be used on boats that are equipped with waterlift exhaust collectors or mufflers.

1. Do not, under any circumstance, start the engine while performing this procedure!
2. Locate the flush valve connected to a black hose on the exhaust elbow. There should be flush socket attachment with a blue cap connected to it.
3. Remove the socket from the cap, connect a garden hose and then plug into the flush valve.
4. Turn the hose spigot to its highest pressure and allow the engine to flush for 5 minutes.
5. Turn off the water and remove the hose/socket from the valve.
6. Unscrew the socket from the hose and insert it back into the cap.

Transmission Fluid

FLUID LEVEL

The transmission fluid level should be checked on a weekly basis. Always fill using hydraulic oil (Mobil 424) or Dexron® III ATF fluid. DO NOT MIX THE TWO.

◆ **See Figures 98, 99 and 100**

Warm

1. Start the engine and run it until it reaches normal operating temperature. The transmission fluid must be at least 190 degrees F, as the incorrect temperature can affect oil level greatly.
2. Move the shifter to Neutral and turn off the engine.
3. Locate the dipstick on the transmission and remove it. Wipe the dipstick with a clean, lint-free rag and reinsert it into the transmission.
4. Remove the dipstick and check that the level is up to the FULL mark. If not, carefully add the appropriate fluid through the dipstick hole until the correct level is reached. On Hurth down angle transmissions, locate the transmission oil filter on the other side, turn it counterclockwise while pulling up and remove the filter (later models may have a set screw). Add fluid here. After adding fluid, insert the filter and turn it clockwise while pushing down until it seats fully.

Cold

We figure that if you can check the fluid level without having to start the engine every time, you'll probably be more likely to check it regularly so here's a way you can do just that.

1. Follow the procedure for checking the oil with the engine hot and make sure that the level is correct.
2. Allow the boat to sit overnight and then remove the dipstick with the engine/transmission cold.
3. Take note of the oil level, wipe of the dipstick with a clean, lint-free rag and scribe a mark on the stick where the oil level had been.
4. Although this does not obviate the need to check the fluid level when hot on a regular basis, it allows you a good, quick idea that your level is OK in between normal checks.

Stern Drive Unit

FLUID LEVEL

Check the fluid level after the first 20 hours of operation and then at least every 50 hours of operation or once a season, whichever comes first on

Alphas. It's never a bad idea to check your unit more frequently, particularly if you use your boat in severe service conditions. On Bravos and Blackhawks check the fluid level weekly. MerCruiser recommends using only Quicksilver Premium Blend Gear Lube for all units except the early Alpha One, which should use Quicksilver Super-Duty Lower Unit Lubricant and all Blackhawks with Mag engines, which should use only Quicksilver High Performance Gear Lube. Never substitute regular automotive grease.

Units Without An Oil Reservoir

ALPHA ONE AND BLACKHAWK

◆ **See Figures 101 and 102**

> ❋❋ **CAUTION**
>
> **Never remove the oil vent plug when the drive unit is hot. Fluid levels should always be checked with the unit cold.**

1. Position the drive unit in the full IN/DOWN position so that the anti-ventilation plate is level.
2. Locate the oil vent plug on the side of the drive shaft housing (port on Alpha; starboard on Blackhawk). Unscrew and then remove the plug and its washer.
3. Check that the oil level is up to the bottom edge of the vent hole. If the level is satisfactory, check the condition of the washer and then install it and the vent plug. It's a good idea to replace the washer on a regular basis regardless of its visible condition.
4. If the level is low, DO NOT add oil through the vent hole. Reinstall the vent plug and washer to create an air lock in the drive unit case.
5. Locate the oil fill/drain plug on the bottom of the unit and remove it.
6. Quickly install a lube pump into the hole, remove the vent plug again and then pump in lubricant through the fill/drain plug until an air-free stream of lubricant comes out the vent hole (no bubbles).

> ❋❋ **WARNING**
>
> **The unit should never require more than 2 oz. of lubricant. If more is required, you've got an oil leak and the unit should not be operated until it is found and fixed.**

7. With the lube pump still attached, reinstall the oil vent plug and then quickly remove the pump and install the fill/drain plug. Tighten the plugs on the Blackhawk to 30–50 inch lbs. (3–6 Nm).
8. Clean any excess oil from the housing. Recheck the oil level a final time.

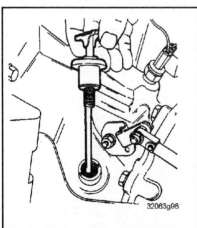

Fig. 98 Check the transmission fluid with the dipstick (Borg Warner and Velvet)

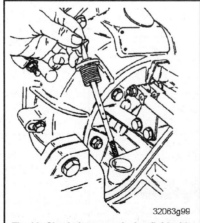

Fig. 99 Check the transmission fluid with the dipstick (Hurth/ZF)

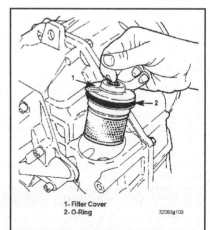

1- Filter Cover
2- O-Ring

Fig. 100 Always add fluid through the oil filter hole on Hurth/ZF transmissions

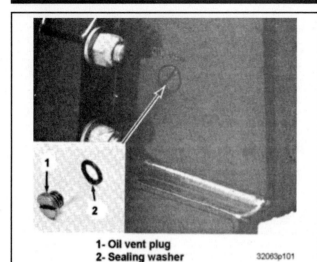

1- Oil vent plug
2- Sealing washer

32063p101

Fig. 101 Remove the vent plug to check the fluid level—early Alpha One

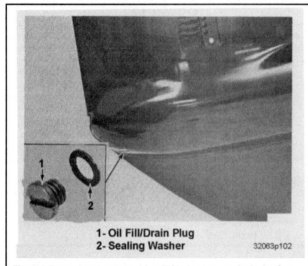

1- Oil Fill/Drain Plug
2- Sealing Washer

32063p102

Fig. 102 Removing the stern drive fill/drain plug—early Alpha One

ALPHA ONE GENERATION II AND 1992-98 BRAVO ONE, TWO & THREE

◆ **See Figures 103 and 104**

1. Position the drive unit in the full IN/DOWN position so that the anti-ventilation plate is level.

2. Remove the oil level dipstick found on top of the drive shaft housing and check that it is up to the line on the stick. If the level is satisfactory, check the condition of the washer and then install it and the dipstick. It's a good idea to replace the washer on a regular basis regardless of its visible condition.

3. If the level is low, DO NOT add oil through the dipstick hole. Reinstall the dipstick and washer to create an air lock in the drive unit case.

4. Locate the oil fill/drain plug on the bottom of the unit and remove it.

5. Quickly install a lube pump into the hole, remove the vent plug on the upper housing (Port on Alpha; Starboard on Bravo) and then pump in lubricant through the fill/drain plug until an air-free stream of lubricant comes out the vent hole (no bubbles).

✳✳ CAUTION

The unit should never require more than 2 oz. of lubricant. If more is required, you've got an oil leak and the unit should not be operated until it is found and fixed.

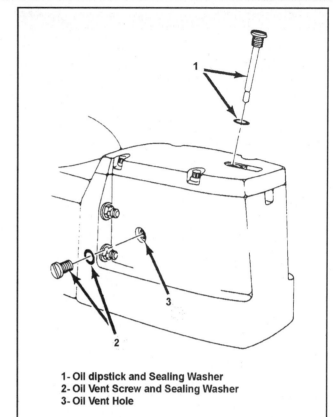

1- Oil dipstick and Sealing Washer
2- Oil Vent Screw and Sealing Washer
3- Oil Vent Hole

Fig. 103 Dipstick and vent plug locations on the Alpha One Generation II

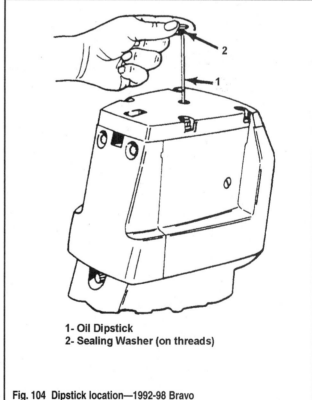

1- Oil Dipstick
2- Sealing Washer (on threads)

Fig. 104 Dipstick location—1992-98 Bravo

6. With the lube pump still attached, reinstall the oil vent plug and then quickly remove the pump and install the fill/drain plug. Both plugs should be tightened to 17 inch lbs. (2 Nm) on the Alpha; 30–50 inch lbs. (3–6 Nm) on the Bravos.

7. Clean any excess oil from the housing. Recheck the oil level on the dipstick again and then recheck it a final time after the next use.

Units With An Oil Reservoir

◆ See Figures 105 and 106

ALPHA ONE

✳✳ CAUTION

Fluid levels should always be checked with the stern drive unit cold.

■ If the oil in the reservoir is milky brown in color or you are continually adding fluid, there is a good chance there is a leak in the unit. The stern drive unit should not be operated until the leak is found and corrected.

1. Check that the fluid level in the reservoir is up to the FULL mark.
2. If below the FULL mark, unscrew the reservoir cap and slowly add fluid until the level reaches the FULL mark. Do not overfill.

✳✳ CAUTION

The unit should never require more than 2 oz. of lubricant. If more is required, you've got an oil leak and the unit should not be operated until it is found and fixed.

ALPHA ONE GENERATION II, BLACKHAWK AND BRAVO ONE, TWO & THREE

◆ See Figures 105 and 107

✳✳ CAUTION

Fluid levels should always be checked with the stern drive unit cold.

■ If the oil in the reservoir is milky brown in color or you are continually adding fluid, there is a good chance there is a leak in the unit. The stern drive unit should not be operated until the leak is found and corrected.

1. Check that the fluid level in the reservoir is up to the FILL TO LINE mark on the reservoir decal. Some models may also have a dipstick located in the top of the drive housing – DO NOT REMOVE DIPSTICK. DO NOT CHECK OIL LEVEL WITH THIS DIPSTICK.
2. If below the mark, unscrew the reservoir cap and slowly add fluid until the level reaches the FILL TO LINE mark. Do not overfill. Always coat the cap seal with a little oil before reinstalling it; do not over-tighten the cap—one quarter turn past initial seating is fine.

✳✳ CAUTION

The unit should never require more than 2 oz. of lubricant. If more is required, you've got an oil leak and the unit should not be operated until it is found and fixed.

Fig. 105 Most reservoirs have a sticker like this

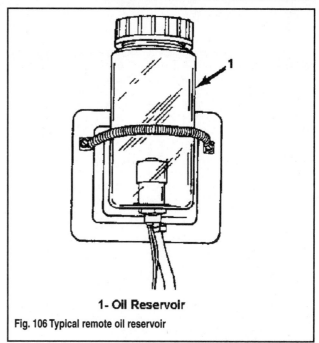

1- Oil Reservoir

Fig. 106 Typical remote oil reservoir

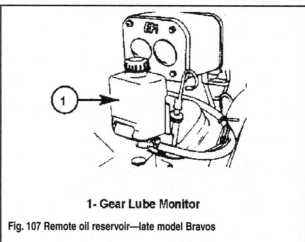

1- Gear Lube Monitor

Fig. 107 Remote oil reservoir—late model Bravos

CHECKING FOR WATER

Units Without An Oil Reservoir

It is a good idea to check the unit periodically for water contamination. Water in the drive oil usually indicates a bad seal somewhere on the unit and should be corrected immediately.

ALL MODELS

1. With the engine off and the stern drive unit cold, trim the drive to the full UP position.
2. Remove the oil fill/drain plug and take a small sample of the lubricant—a teaspoon worth is more than enough.
3. If the oil sample looks milky brown then there is likely a leaking seal in the drive unit and it should be located and replaced before operating the boat again.
4. Most units covered in the manual have magnetic fill/drain plugs, so be sure to check that the end of the plug is free of any metal filings. If you find any metallic particles on the plug it can be an indication of greater problems within the drive unit.

Units With An Oil Reservoir

ALL MODELS

> ✳✳ **CAUTION**
>
> Fluid levels should always be checked with the stern drive unit cold.

Take a look at the oil reservoir, if the oil in the reservoir is milky brown in color there is a good chance there is a leak in the unit. The stern drive unit should not be operated until the leak is found and corrected.

If a Bravo or Blackhawk unit has sat overnight, the condition of the oil should be checked via the fill/drain plug as detailed under Units Without An Oil Reservoir. If your unit does not have a plug at the bottom of the housing, it can be found behind the prop in the bearing carrier. The prop may have to be removed and the drive unit should be in the full DOWN position.

DRAIN AND REFILL   ② ◁ MODERATE

The oil in the stern drive should be changed at least every 100 hours of operation or once a season, whichever comes first. Its never a bad idea to change the fluid in your unit more frequently, particularly if you use your boat in severe service conditions. MerCruiser recommends using only Quicksilver Premium Blend Gear Lube for all units except the early Alpha One, which should use Quicksilver Super-Duty Lower Unit Lubricant, and the Blackhawk with Mag engines, which uses only Quicksilver High Performance Gear Lube. Never substitute regular automotive grease.

Units Without An Oil Reservoir

ALPHA, BLACKHAWK AND 1992-98 BRAVO ONE, TWO & THREE

◆ See Figures 108, 109 and 110

> ✳✳ **CAUTION**
>
> Stern drive oil should always be changed with the unit cold.

■ Take a look at the oil as it is being drained, if the oil is milky brown in color there is a good chance there is a leak in the unit. The stern drive unit should not be operated until the leak is found and corrected.

1. Trim the stern drive to the full OUT position for models with the fill/drain plug in the bottom of the lower housing. On models with the plug found in the propeller bearing housing, trim the stern drive to the full IN/DOWN position.

2. Remove the upper oil vent plug from the side of the drive shaft housing (Port on Alpha; Starboard on Bravo/Blackhawk).

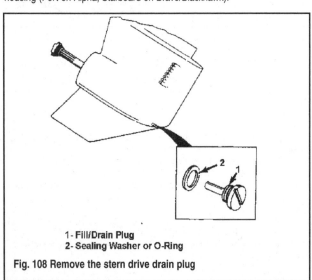

1- Fill/Drain Plug
2- Sealing Washer or O-Ring

Fig. 108 Remove the stern drive drain plug

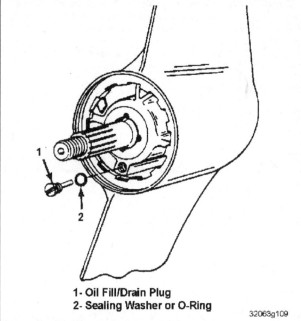

1- Oil Fill/Drain Plug
2- Sealing Washer or O-Ring

32063g109

Fig. 109 On some models, the drain plug is behind the propeller, in the bearing housing

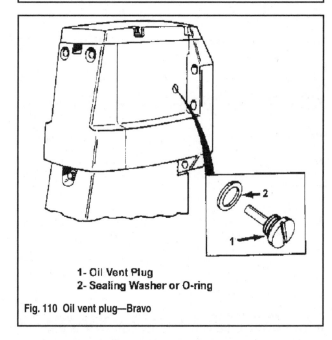

1- Oil Vent Plug
2- Sealing Washer or O-ring

Fig. 110 Oil vent plug—Bravo

3. Position an oil drain plan or an old plastic milk jug under the drain hole on the bottom of the drive unit and then remove the fill/drain plug, keeping a slight inward pressure on it until it is completely unthreaded.

4. When the old oil has completely drained, trim the drive back to the IN/DOWN position and install a suitable lubricant pump into the drain hole. On models with the prop drain hole, move the drive to the full OUT position for a minute to drain any remaining oil in the housing and move it back to the DOWN position.

5. Pump the proper lubricant into the drive through the fill/drain hole until it comes out of the vent hole with no bubbles.

6. With the lube pump still attached, reinstall the oil vent plug, tighten it to 17 inch lbs. (2 Nm) on Alphas, or 30–50 inch lbs. (3–6 Nm) on Bravos/Blackhawks and then quickly remove the pump and install the fill/drain plug and washer; tighten it to the same specs as the vent plug.

7. Clean any excess oil on the housing. Check the oil level a final time and then recheck it again after the first use..

Units With An Oil Reservoir

ALPHA ONE

◆ See Figure 111

✳✳ CAUTION

Stern drive oil should always be changed with the unit cold.

■ Take a look at the oil as it is being drained, if the oil is milky brown in color there is a good chance there is a leak in the unit. The stern drive unit should not be operated until the leak is found and corrected.

1. Trim the stern drive to the full OUT position.
2. Position an oil drain plan or an old plastic milk jug under the drain hole on the bottom of the drive unit and then remove the fill/drain plug, keeping a slight inward pressure on it until it is completely unthreaded.
3. Unbolt the oil reservoir from its mounting bracket, unscrew the cap and pour out all oil into a suitable container.
4. Remove the oil line and its adapter at the upper oil vent plug on the Port side of the drive shaft housing.
5. When the old oil has completely drained, trim the drive back to the IN/DOWN position and install a suitable lubricant pump into the drain hole
6. Pump the proper lubricant into the drive through the fill/drain hole until it comes out of the vent hole with no bubbles.
7. With the lube pump still attached, plug the oil vent hole and then quickly remove the pump and install the fill/drain plug and washer.
8. Unplug the oil vent hole and reinstall the oil line adapter making sure to tighten it securely.
9. Reinstall the oil reservoir and fill it with lubricant to the FULL line. Screw the cap on tightly.
10. Loosen the oil line adapter fitting at the drive housing and allow oil to escape until it is free of air and bubbles. Retighten the fitting securely.
11. Clean any excess oil from the housing. Check the oil level in the reservoir a final time and then recheck it again after the first use..

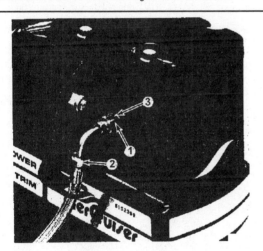

1- Adaptor
2- Hose
3- Oil Vent Hole

Fig. 111 Removing the oil line and adapter

ALPHA ONE GENERATION II, BLACKHAWK AND 1992-98 BRAVO ONE, TWO & THREE

◆ See Figures 103 and 104

✳✳ CAUTION

Stern drive oil should always be changed with the unit cold.

■ Take a look at the oil as it is being drained, if the oil is milky brown in color there is a good chance there is a leak in the unit. The stern drive unit should not be operated until the leak is found and corrected.

1. Trim the stern drive to the full OUT position for models with the fill/drain plug in the bottom of the lower housing. On models with the plug found in the propeller bearing housing, trim the stern drive to the full IN/DOWN position.
2. Unbolt the oil reservoir from its mounting bracket, unscrew the cap and pour out all oil into a suitable container. Clean the reservoir and install it back into the mounting bracket.
3. Position an oil drain plan or an old plastic milk jug under the drain hole on the bottom of the drive unit (or under the prop housing) and then remove the fill/drain plug, keeping a slight inward pressure on it until it is completely unthreaded.
4. Remove the oil vent plug on the side of the drive shaft housing (Port on Alpha; Starboard on Bravo/Blackhawk).
5. When the old oil has completely drained, trim the drive back to the IN/DOWN position and install a suitable lubricant pump into the drain hole. On models with the prop drain hole, move the drive to the full OUT position for a minute to drain any remaining oil in the housing and move it back to the DOWN position.
6. Pump the proper lubricant into the drive through the fill/drain hole until it comes out of the vent hole with no bubbles.
7. With the lube pump still attached, plug the oil vent hole and then quickly remove the pump and install the fill/drain plug and washer, tightening to 17 inch lbs. (2 Nm) on Alphas, or 30–50 inch lbs. (3–6 Nm) on Bravos/Blackhawks.
8. Unplug the oil vent hole.
9. Fill the oil reservoir with lubricant until oil begins to seep out the vent hole and then install the vent plug, tightening to the same specification as the drain plug. Fill the reservoir to the FILL TO LINE mark. Screw the cap on tightly.
10. Clean any excess oil from the housing. Check the oil level in the reservoir a final time and then recheck it again after the first use..

1999-2001 BRAVO ONE, TWO & THREE

◆ See Figures 108, 109 and 110

✳✳ CAUTION

Stern drive oil should always be changed with the unit cold.

■ Take a look at the oil as it is being drained, if the oil is milky brown in color there is a good chance there is a leak in the unit. The stern drive unit should not be operated until the leak is found and corrected.

1. Trim the stern drive to the full OUT position for Bravo Two and Three models with the fill/drain plug in the bottom of the lower housing. On Bravo One models with the plug found in the propeller bearing housing, trim the stern drive to the full IN/DOWN position.
2. Unbolt the gear lube monitor from its mounting bracket, unscrew the cap and pour out all oil into a suitable container. Clean the reservoir and install it back into the mounting bracket. Do not refill it yet!
3. Position an oil drain plan or an old plastic milk jug under the drain hole on the bottom of the drive unit (or under the prop housing) and then remove the fill/drain plug, keeping a slight inward pressure on it until it is completely unthreaded.
4. Remove the oil vent plug on the starboard side of the drive shaft housing.
5. When the old oil has completely drained, trim the drive back to the IN/DOWN position and install a suitable lubricant pump into the drain hole. On models with the prop drain hole, move the drive to the full OUT position for a minute to drain any remaining oil in the housing and move it back to the DOWN position.
6. Pump the proper lubricant into the drive through the fill/drain hole until it reaches the bottom of the vent hole. Install the vent plug and tighten it to 40 inch lbs. (4.5 Nm).
7. Continue pumping oil into the drive until there is about 1 inch in the gear lube monitor and then quickly remove the pump and install the fill/drain plug and washer, tightening to 40 inch lbs. (4.5 Nm).
8. Fill the gear lube monitor to the FILL TO LINE mark. Screw the cap on tightly, but do not overtighten.
9. Clean any excess oil from the housing. Check the oil level in the reservoir a final time and then recheck it again after the first use.

Power Trim Pump

FLUID LEVEL

Alpha One

Two types of pumps were available on the units, an Oildyne or a Prestolite. Fluid levels in both pumps should be checked every 100 hours of operation or once a season, whichever comes first. Both pumps use SAE 10W-30 or 10W-40 motor oil.

OILDYNE

◆ See Figure 112

1. Position the stern drive in the full IN/DOWN position.
2. Remove the fill/vent screw form the top of the housing and wipe the dipstick clean with a lint-free rag. Reinsert it into the housing without screwing it in.
3. Wait a few minutes, remove it again and check that the fluid level is between the ADD and FULL marks on the dipstick. If necessary, add oil through the filler hole, a little at a time, until the dipstick read the correct level. DO NOT overfill it!
4. Purge and air that has gotten into the system by raising and lowering the drive unit 2 times and then recheck the fluid level.
5. Insert the dipstick and thread it in until it seats completely. Now back it out one complete turn.

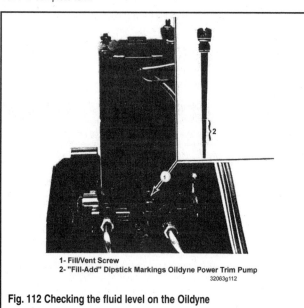

1- Fill/Vent Screw
2- "Fill-Add" Dipstick Markings Oildyne Power Trim Pump
32063g112

Fig. 112 Checking the fluid level on the Oildyne

PRESTOLITE

◆ See Figure 113

1. Position the stern drive in the full IN/DOWN position.
2. Loosen and remove the fill screw from the corner of the pump housing.
3. Check that the oil is up to the bottom of the threads in the hole.If necessary, add SAE 10W-30 or 10W-40 slowly through the hole until the level is correct.
4. With the fill screw still out, move the drive through its full range 6–10 times in order to purge any air that may have entered the system. Lower the drive fully and recheck the fluid level again; repeat the process if necessary.
5. Install the fill screw and tighten it securely. Locate the vent screw (next to the fill screw), screw it down until it seats fully and then back it out 2 full turns so the system is properly vented.

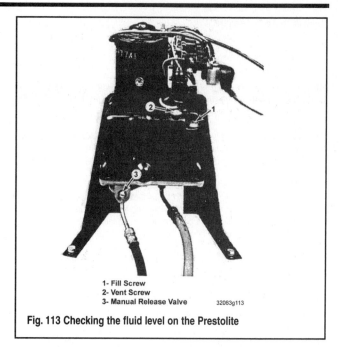

1- Fill Screw
2- Vent Screw
3- Manual Release Valve   32063g113

Fig. 113 Checking the fluid level on the Prestolite

Alpha One Generation II and Bravo One, Two & Three

◆ See Figure 114

Fluid levels should be checked weekly. Use Quicksilver Power Trim and Steering fluid. SAE 10W-30 or 10W-40 motor oil may be used also.
1. Move the stern drive to the full IN/DOWN position.
2. Check that the oil level is between the MAX and MIN marks on the side of the reservoir. Add oil through the filler hole, no higher than the bottom of the filler neck.
3. Move the drive unit through its full range 6–10 times in order to purge any air that may be in the system. Lower the drive a final time and recheck the oil level; repeat the process if necessary.
4. Install the fill cap. Some models utilize a vent screw located at the bottom of the pump body, just above the reservoir; if equipped, tighten the screw fully and then back it out 2 complete turns so that the system is properly vented. Models without a vent screw use a vented fill cap, make sure that the filler neck seal is removed so that the cap vents properly.

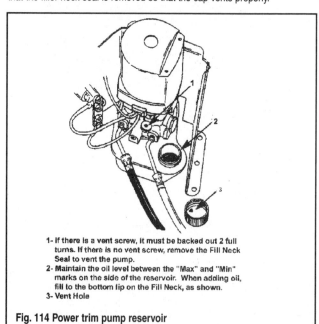

1- If there is a vent screw, it must be backed out 2 full turns. If there is no vent screw, remove the Fill Neck Seal to vent the pump.
2- Maintain the oil level between the "Max" and "Min" marks on the side of the reservoir. When adding oil, fill to the bottom lip on the Fill Neck, as shown.
3- Vent Hole

Fig. 114 Power trim pump reservoir

LUBRICATION POINTS

Throttle Cable

◆ See Figures 115, 116, 117, 118, 119, 120, 121 and 122

Lubricate the cable pivot points every 100 hours or 6 months, whichever comes first when the vessel is being operated in fresh water. In salt water, lubricate every 50 hours or 2 months whichever comes first. Always use SAE 20 or 30 weight engine oil, or Quicksilver 2-4-C Marine Lubricant.

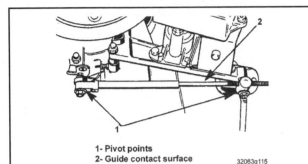

1- Pivot points
2- Guide contact surface

32063a115

Fig. 115 Lubricate the throttle cable pivot points here 1992-97 V8 engines w/2bbl

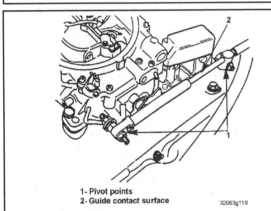

1- Pivot points
2- Guide contact surface

32063g116

Fig. 116 Lubricate the throttle cable pivot points here V8 engines w/4bbl

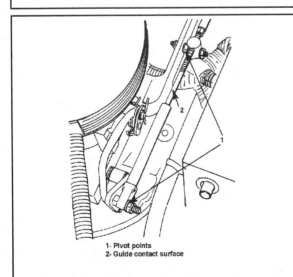

1- Pivot points
2- Guide contact surface

Fig. 117 Lubricate the throttle cable pivot points here 1992-97 V8 engines w/TBI

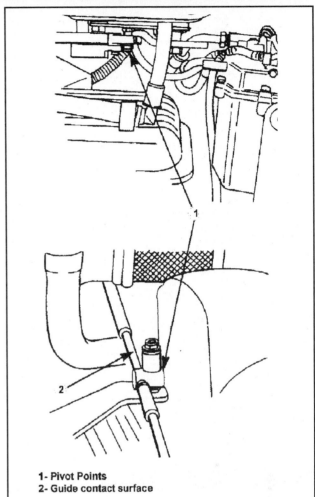

1- Pivot Points
2- Guide contact surface

Fig. 118 Lubricate the throttle cable pivot points here 1992-97 V8 engines w/MPI

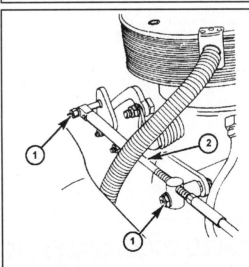

1- Pivot Points
2- Guide contact surface

Fig. 119 Lubricate the throttle cable pivot points here 1998-2001 5.0L, 5.7L and 6.2L engines w/EFI

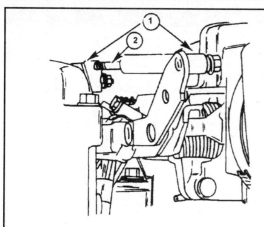

1- Pivot Points
2- Guide contact surface

Fig. 120 Lubricate the throttle cable pivot points here 1998-2001 7.4L engines w/MPI

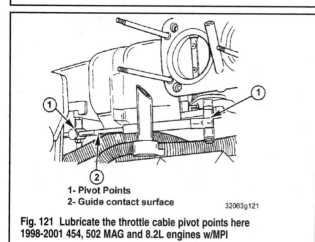

1- Pivot Points
2- Guide contact surface

32063g121

Fig. 121 Lubricate the throttle cable pivot points here 1998-2001 454, 502 MAG and 8.2L engines w/MPI

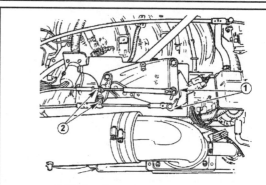

Fig. 122 Lubricate the throttle cable pivot points here 8.1L engines

Shift Cable and Transmission Linkage Pivot Points

◆ **See Figures 123, 124, 125, 126 and 127**

Lubricate the cable pivot points every 100 hours or 6 months, whichever comes first when the vessel is being operated in fresh water. In salt water, lubricate every 50 hours or 2 months whichever comes first. Always use SAE 20 or 30 weight engine oil, or Quicksilver 2-4-C Marine Lubricant.

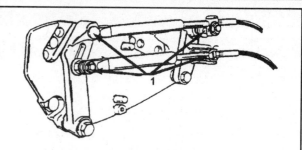

1- Lubricate with SAE 20 or 30 engine oil

Fig. 123 Lubricate the shift cable pivot points here—Alpha

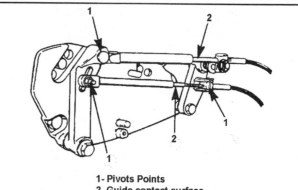

1- Pivots Points
2- Guide contact surface

Fig. 124 Lubricate the shift cable pivot points here—Bravo

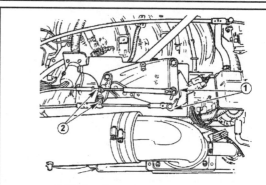

1- Pivot points
2- Guide contact surface

Fig. 125 Lubricate the shift cable pivot points here—8.1L engines

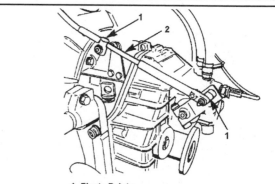

1- Pivots Points
2- Guide contact surface

Fig. 126 Lubricate the shift cable pivot points here Hurth transmissions

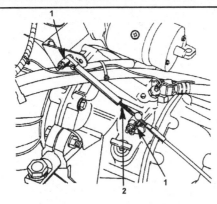

1- Pivot Points
2- Guide contact surface

Fig. 127 Lubricate the shift cable pivot points here Borg Warner inboards

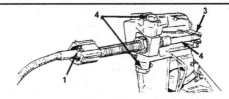

1- Steering cable grease fitting - 2-4-C Marine Lubricant
2- Steering - Cable end and exposed portion - Special Lubricant 101
3- Pivot Points - SAE 20 or 30 engine oil
4- Pivot Bolts - Special Lubricant 101

Fig. 128 Lubricate the steering system at these locations Alpha w/manual steering

Steering System

◆ See Figures 128, 129, 130 and 131

■ Always check the steering system and its components for loose, missing or damaged components and fittings before performing your lubrication procedures.

On Alpha models, lubricate the various steering system components every 50 hours or 2 months, whichever comes first when the vessel is being operated in fresh water. In salt water, lubricate every 25 hours or 30 days whichever comes first.

On Blackhawk and 1990-98 Bravo models, lubricate the system components every 100 hours or 6 months, whichever comes first when the vessel is being operated in fresh water. In salt water, lubricate every 50 hours or 2 months whichever comes first.

On 1999-2001 Bravo models, lubricate the system components every 100 hours or once a season, whichever comes first.

The transom end of the steering cable must be fully retracted into the cable housing before applying Quicksilver 2-4-C Marine Lube to the fitting.

The steering cable end should be lubricated with Quicksilver Special Lube 101, as should the pivot bolt.

Lubricate all pivot points on the system with SAE 20 or 30 motor oil.

The control valve grease fitting takes Quicksilver 2-4-C Marine Lube.

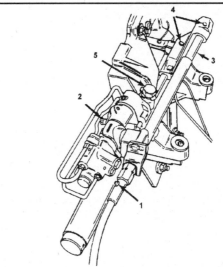

1- Steering cable grease fitting - 2-4-C Marine Lubricant
2- Control valve grease fitting - 2-4-C Marine Lubricant
3- Steering cable end - Special Lubricant 101
4- Pivot Points - SAE 20 or 30 engine oil
5- Pivot Bolts - Special Lubricant 101 (top & bottom)

Fig. 129 Lubricate the steering system at these locations Alpha w/power steering

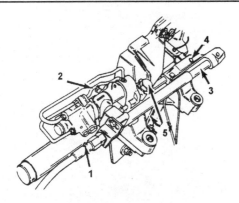

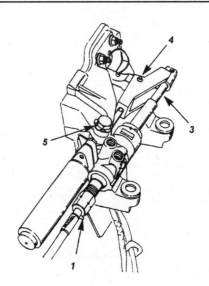

1- Steering cable grease fitting - 2-4-C Marine Lubricant (See warning)
2- Control valve grease fitting - 2-4-C Marine Lubricant
3- Steering cable end - Special Lubricant 101
4- Pivot point - Sae 20 or 30 engine oil
5- Pivot bolts - Special Lubricant 101

Fig. 130 Lubricate the steering system at these locations—1992-98 Bravo

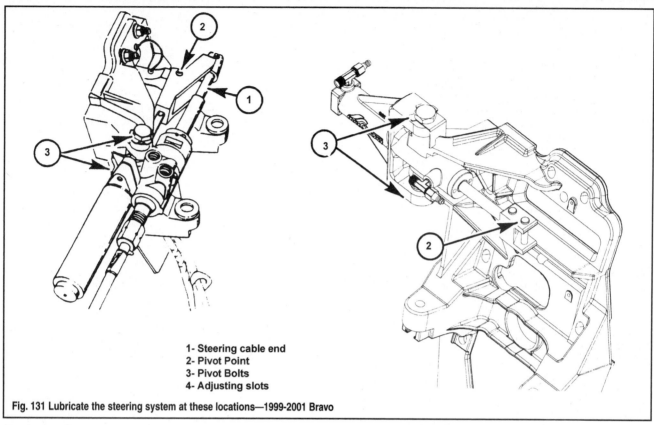

1- Steering cable end
2- Pivot Point
3- Pivot Bolts
4- Adjusting slots

Fig. 131 Lubricate the steering system at these locations—1999-2001 Bravo

Tie Bar Pivot Points

◆ See Figure 132

On Alpha units, lubricate all pivot points at least every 50 hours or 6 months, whichever comes first.. On Blackhawk and Bravo units, lubricate all pivot points at least every 100 hours or once a season, whichever comes first.

Use only SAE 20 or 30 weight engine oil.

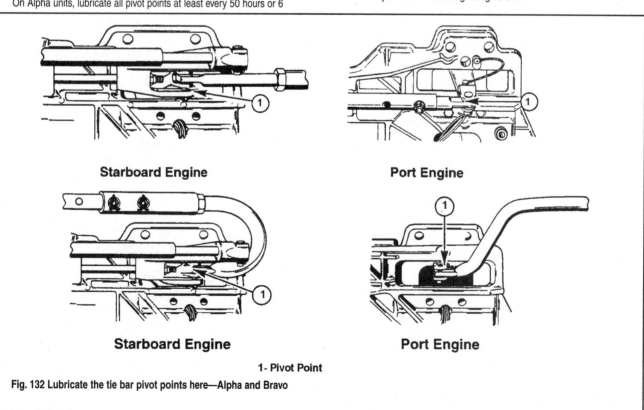

Starboard Engine

Port Engine

Starboard Engine

Port Engine

1- Pivot Point

Fig. 132 Lubricate the tie bar pivot points here—Alpha and Bravo

Transom and Gimbal Assembly, Hinge Pins and Gimbal Bearing

◆ **See Figures 133, 134 and 135**

On Alpha, Blackhawk and 1990–98 Bravo units, lubricate every 100 hours or 6 months, whichever comes first when the vessel is being operated in fresh water. In salt water, lubricate every 50 hours or 2 months whichever comes first. Always use Quicksilver 2-4-C Marine Lubricant.

On 1999–2001 Bravo Standard units, lubricate every 100 hours or once a season, whichever comes first. On Horizon models, lubricate every 200 hours or every third season, whichever comes first. Always use Quicksilver 2-4-C Marine Lubricant.

Alpha One and early Bravo models utilize a grease fitting on the back of the transom assembly. Generation II and later Bravo units are equipped with self-lubing bushings and no longer utilize this fitting.

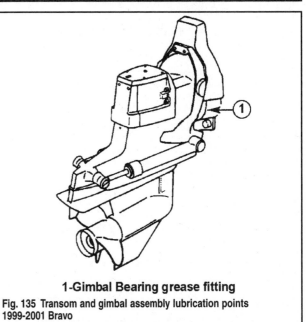

1-Gimbal Bearing grease fitting

Fig. 135 Transom and gimbal assembly lubrication points 1999-2001 Bravo

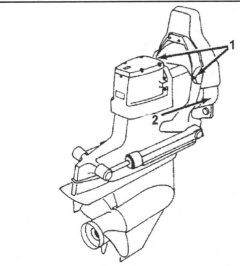

1- Hinge Pins - (One on each side)
2- Gimball Bearing - grease with
2-4-C Marine Lubricant

Fig. 133 Transom and gimbal assembly lubrication points—Alpha

Engine Coupler/U-Joint Splines

◆ **See Figures 136, 137, 138, 139, 140, 141 and 142**

Lubricate every 100 hours or once a season whichever comes first. On Horizon models, lubricate every 200 hours or every third season, whichever comes first. Always use Quicksilver Engine Coupler Spline grease.

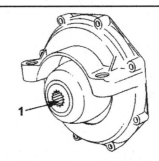

1- Quicksilver engine coupler spline grease

Fig. 136 Lubricate the coupler splines here Alpha w/o a sealed coupler

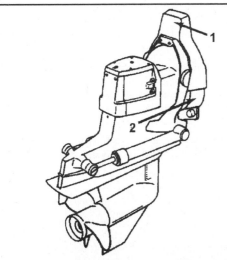

1- Swivel shaft grease fitting - 2-4-C marine lubricant (on earlier models)
2- Gimball Bearing grease fitting - 2-4-C marine lubricant

Fig. 134 Transom and gimbal assembly lubrication points 1992-98 Bravo

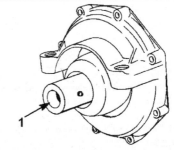

1- Quicksilver engine coupler spline grease

Fig. 137 Lubricate the coupler splines here with the grease fitting Alpha with a sealed coupler

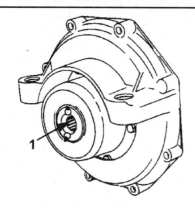

1- Quicksilver engine coupler spline grease

Fig. 138 Lubricate the coupler splines here with the grease fitting Bravo

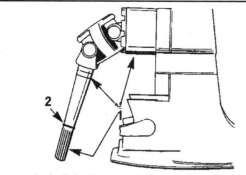

1- Quicksilver engine coupler spline grease
2- O-ring on modems with sealed steel hub coupler

32063g139

Fig. 139 Lubricate the U-joint shaft/splines here—Alpha

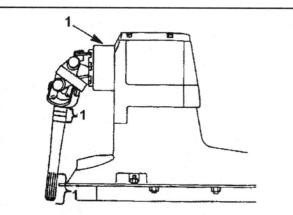

1- Quicksilver engine coupler spline grease

Fig. 140 Lubricate the U-joint shaft/splines here—Bravo

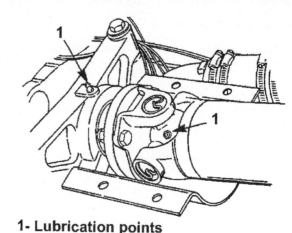

1- Lubrication points

Fig. 141 Lubricate the transom-end U-joints here inboards w/extended driveshafts

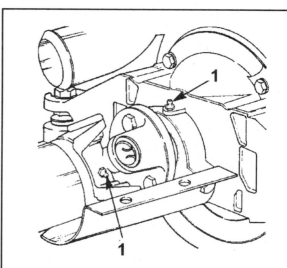

1-Lubrication points

Fig. 142 Lubricate the engine-end U-joints here inboards w/extended driveshafts

Propeller Shaft

◆ See Figures 143, 144, 145 and 146

On Alpha, Blackhawk and 1990-98 Bravo units, lubricate every 100 hours or 6 months, whichever comes first when the vessel is being operated in fresh water. In salt water, lubricate every 50 hours or 2 months whichever comes first. Always use Quicksilver Special Lube 101, Quicksilver 2-4-C Marine Lubricant or Quicksilver Perfect Seal. All three are listed in order of their relative effectiveness.

On 1999-2001 Bravo units, lubricate every 4 months when the vessel is being operated in fresh water. In salt water, lubricate every 2 months. Always use Quicksilver Special Lube 101 or Quicksilver 2-4-C Marine Lubricant. The two are listed in order of their relative effectiveness.

Remove the propeller and apply lubricant to the shaft splines.

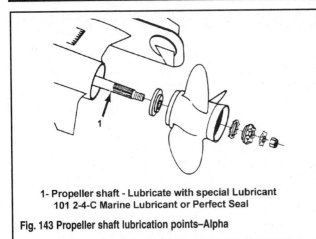

1- Propeller shaft - Lubricate with special Lubricant
101 2-4-C Marine Lubricant or Perfect Seal

Fig. 143 Propeller shaft lubrication points–Alpha

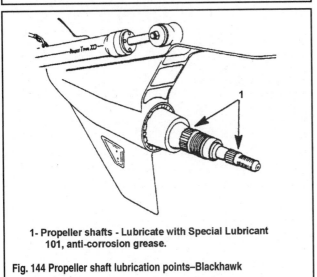

1- Propeller shafts - Lubricate with Special Lubricant
101, anti-corrosion grease.

Fig. 144 Propeller shaft lubrication points–Blackhawk

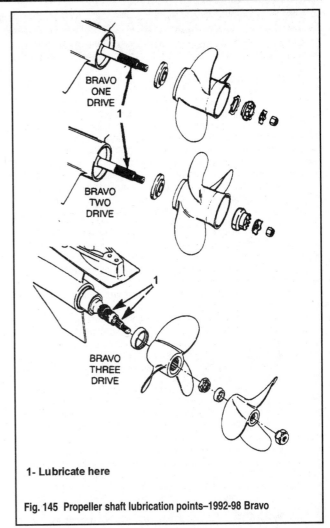

1- Lubricate here

Fig. 145 Propeller shaft lubrication points–1992-98 Bravo

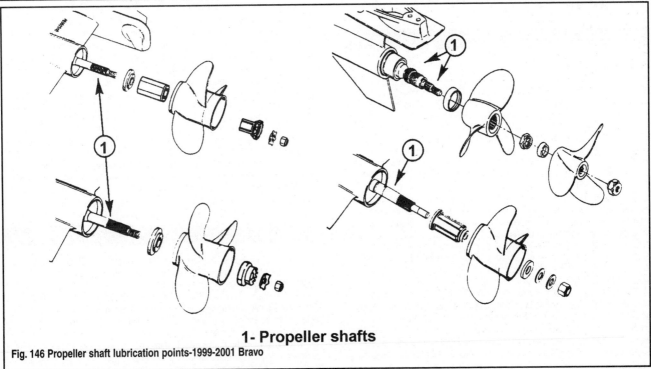

1- Propeller shafts

Fig. 146 Propeller shaft lubrication points-1999-2001 Bravo

BOAT MAINTENANCE

Inside The Boat

Probably the biggest surprise for boat owners is the extent to which mold and mildew develop in a boats interior. Preventing this growth is a two-fold process. First, the boat's interior should be thoroughly cleaned. Second, ventilation should be provided to allow adequate air circulation.

Properly cleaning the boat includes removing as much as possible from it. The less there is on a boat, the less there is to attract mildew. Clothing, foul weather gear, shoes, books, charts, paper goods, leather, bedding, curtains, food stuffs, first aid supplies, odds and ends, etc., should be taken home. They're all fertile soil for mildew, as is dirt, grease, soap scum, etc.

The next step is to vacuum, scrub and polish every surface, especially galley and head surfaces. Use a polish that leaves a protective coating on which it's hard for mold to get a foothold. Mildew-preventive sprays are available for carpets and furniture. The cleaner and more highly polished the insides of lockers, cabinets, refrigerators, icemakers, drawers and shower stalls are when you leave them, the less likely it is you'll find mold. Marine stores sell cleaners and polishes specifically for boats and for the marine environment. Many people find that regular household products are satisfactory. It's the elbow grease that counts.

If dirt and grease are the growing medium, stagnant air is the fertilizer for mold and mildew (mildew is a form of mold). The more freely air can circulate in the interior of your boat the less conducive conditions will be for the growth of mold. Openings or vents at each end near the top can be fashioned to let air in. Flaps or stovepipe elbows can be used to keep rain out. Hatches and windows can then be left partially open. Visiting the boat on dry, sunny days and opening it up as much as possible is a great way to let fresh air in. Do this whenever possible.

The best way to protect navigation and communications equipment from deteriorating is to remove the units from the boat. Wrap them in towels, not plastic, and take them home. Television sets, VCRs and stereos should also be removed to keep them safe and to keep cold and dampness from affecting them adversely. Before storing electronics equipment, clean off the terminals and the connecting plugs and spray them with a moisture-displacing lubricant that specifically says it's intended for use on electronics. Antennas should also be stored in a safe, dry place to protect them from both the elements and accidental damage. All terminal blocks, junction blocks, fuse holders and the back of electrical panels should be cleaned and sprayed with an appropriate protectant. Remove any corrosion that has developed. Anything that could be easily stolen, such as anchors, flare guns, binoculars, etc., is best taken home.

The Boats Exterior

◆ See Figure 147

Fiberglass reinforced plastic hulls are tough, durable, and highly resistant to impact. However, like any other material they can be damaged. One of the advantages of this type of construction is the relative ease with which it may be repaired. Because of its break characteristics, and the simple techniques used in restoration, these hulls have gained popularity throughout the world. From the most congested urban marina, to isolated lakes in wilderness areas, to the severe cold of northern seas, and in sunny tropic remote rivers of primitive islands or continents, fiberglass boats can be found performing their daily task with a minimum of maintenance.

A fiberglass hull has almost no internal stresses. Therefore, when the hull is broken or stove-in, it retains its true form. It will not dent to take an out-of-shape set. When the hull sustains a severe blow, the impact will either be absorbed by deflection of the laminated panel or the blow will result in a definite, localized break. In addition to hull damage, bulkheads, stringers, and other stiffening structures attached to the hull may also be affected and therefore, should be checked. Repairs are usually confined to the general area of the rupture.

The best way to care for a fiberglass hull is to first wash it thoroughly. Immediately after hauling the boat, while the bottom is still wet, is best, if possible. Remove any growth that has developed on the bottom. Use a pressure cleaner or a stiff brush to remove barnacles, grass, and slime. Pay particular attention to the waterline area. A scraper of some sort may be needed to attack tenacious barnacles. Pot scrubbers work well. Attend to any blisters; don't wait!

Remove cushions and all weather curtains and enclosures and take them home. Make sure they are clean before storing them, and don't store them tightly rolled. After washing the topsides, remove any stains that have developed with one of the fiberglass stain removers sold in marine stores. For stubborn stains, wet-sanding with 600-grit paper may be necessary. Remove oxidation and stains from metal parts. Apply a coat of wax to everything.

Fig. 147 The best way to care for a fiberglass hull is to wash it thoroughly to remove any growth that has developed on the bottom

BELOW WATERLINE

◆ See Figures 148, 149, 150 and 151

A foul bottom can seriously affect boat performance. This is one reason why racers, large and small, both powerboat and sail, are constantly giving attention to the condition of the hull below the waterline.

In areas where marine growth is prevalent, a coating of vinyl, anti-fouling bottom paint should be applied. If growth has developed on the bottom, it can be removed with a solution of Muriatic acid applied with a brush or swab and then rinsed with clear water. Always use rubber gloves when working with Muriatic acid and take extra care to keep it away from your face and hands. The fumes are toxic. Therefore, work in a well-ventilated area, or if outside, keep your face on the windward side of the work.

Barnacles have a nasty habit of making their home on the bottom of boats that have not been treated with anti-fouling paint. Actually they will not harm the fiberglass hull, but can develop into a major nuisance.

If barnacles or other crustaceans have attached themselves to the hull, extra work will be required to bring the bottom back to a satisfactory condition. First, if practical, put the boat into a body of fresh water and allow it to remain for a few days. A large percentage of the growth can be removed in this manner. If this remedy is not possible, wash the bottom thoroughly with a high-pressure fresh water source and use a scraper. Small particles of hard shell may still hold fast. These can be removed with sandpaper.

Anodes

The idea behind anodes is simple: When dissimilar metals are dunked in water and a small current is leaked between or amongst them, the less-noble metal (galvanically speaking) is sacrificed.

The zinc alloy that the anodes are made of is designed to be less noble than the aluminum alloy your drive unit is made from. If there's any electrolysis and there almost always is, the inexpensive zinc anodes are consumed in lieu of the expensive drive.

These zincs need a little attention in order to do their job. Make sure they're there, solidly attached to a clean mounting site and not covered with any kind of paint or wax.

Fig. 148 In areas where marine growth is prevalent, a coating of vinyl, anti-fouling bottom paint should be applied

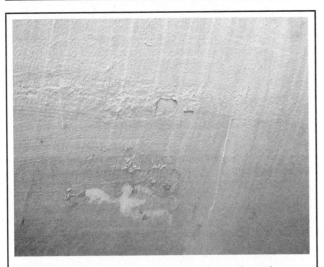

Fig. 149 This anti-foul paint has seen better days and must be replaced

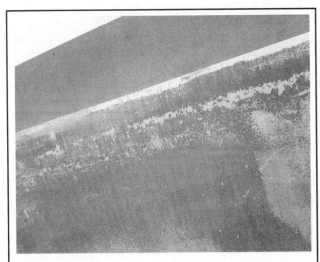

Fig. 150 This hull is in even worse condition and should be sand blasted and repainted

Periodically inspect them to make sure they haven't eroded too much. At a certain point in the erosion process, the mounting holes start to enlarge, which is when the zinc might fall off. Obviously, once that happens your drive no longer has any protection.

Please refer to the section on corrosion found later in this manual for a complete discussion of anodes and Mercruiser's MerCathode system.

SERVICING

◆ **See Figures 151, 152, 153, 154 and 155**

Depending on what kind of drive your boat has, you might have any number of zincs. Regardless of the number, there are some fundamental rules to follow that will give your boat's sacrificial anodes the ability to do the best job protecting your boat's underwater hardware that they can.

The first thing to remember is that zincs are electrical components and like all electrical components, they require good clean connections. So after you've undone the mounting hardware and removed last year's zincs, you want to get the zinc mounting sites clean and shiny.

Get a piece of coarse emery cloth or some 80-grit sandpaper. Thoroughly rough up the areas where the zincs attach (there's often a bit of corrosion residue in these spots). Make sure to remove every trace of corrosion.

Zincs are attached with stainless steel machine screws that thread into the mounting for the zincs. Over the course of a season, this mounting hardware is inclined to loosen. Mount the zincs and tighten the mounting hardware securely. Tap the zincs with a hammer hitting the mounting screws squarely. This process tightens the zincs and allows the mounting hardware to become a bit loose in the process. Now, do the final tightening. This will insure your zincs stay put for the entire season.

Fig. 151 What a trim tab should look like when it's in good condition

Fig. 152 Although many stern drives use the trim tab as an anode . .

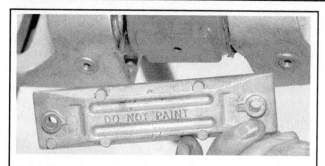

Fig. 153 . . . other types of anodes are also used throughout the drive, like this one mounted on a stern bracket . .

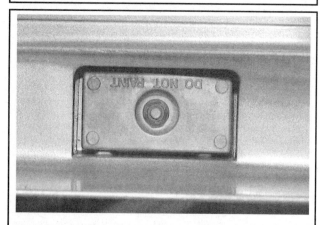

Fig. 154 . . . and this one mounted on a lower unit

Fig. 155 Most anodes are easily removed by looseing and removing their attaching fasteners

Fig. 156 Such extensive erosion of a trim tab compared with a new tab suggests an electrolysis problem or complete disregard for periodic maintenance

INSPECTION

MODERATE ②

◆ See Figure 156

If you use your boat in salt water, and your zincs never wear, inspect them carefully. Paint or wax on zincs prevents them from working properly. They must be left bare. If the zincs are installed properly and not painted or waxed, inspect around them for signs of corrosion. If corrosion is found, strip it off immediately and repaint with a rust inhibiting paint. If in doubt, replace the zincs.

On the other hand, if your zinc seems to erode in no time at all, this may be a symptom of the zinc itself. Each manufacturer uses a specific blend of metals in their zincs. If you are using zincs with the wrong blend of metals, they may erode more quickly or leave you with diminished protection.

Batteries

Difficulty in starting accounts for almost half of the service required on boats each year. A survey by a major engine parts company indicated that roughly one third of all boat owners experienced a "won't start" condition in a given year. When an engine won't start, most people blame the battery when, in fact, it may be that the battery has run down in a futile attempt to start an engine with other problems.

Maintaining your battery in peak condition may be thought of as either tune-up or maintenance material. Most wise boaters will consider it to be both. A complete check up of the electrical system in your boat at the beginning of the boating season is a wise move. Continued regular maintenance of the battery will ensure trouble free starting on the water.

Complete battery service procedures are included in this section. The following are a list of basic electrical system service checks that should be performed.

- Check the battery for solid cable connections
- Check the battery and cables for signs of corrosion damage
- Check the battery case for damage or electrolyte leakage
- Check the electrolyte level in each cell
- Check to be sure the battery is fastened securely in position
- Check the battery's state of charge and charge as necessary
- Check battery voltage while cranking the starter. Voltage should remain above 9.5 volts
- Clean the battery, terminals and cables
- Coat the battery terminals with dielectric grease or terminal protector
- Check the tension on the alternator belt.

Batteries that are not maintained on a regular basis can fall victim to parasitic loads (small current drains which are constantly drawing current from the battery, like clocks, small lights, etc.). Normal parasitic loads may drain a battery on boat that is in storage and not used frequently. Boats that have additional accessories with increased parasitic load may discharge a battery sooner. Storing a boat with the negative battery cable disconnected or battery switch turned **OFF** will minimize discharge due to parasitic loads.

CLEANING

EASY ①

Keep the battery clean; as a film of dirt can help discharge a battery that is not used for long periods. A solution of baking soda and water mixed into a paste may be used for cleaning, but be careful to flush this off with clear water.

■ Do not let any of the solution into the filler holes on non-sealed batteries. Baking soda neutralizes battery acid and will de-activate a battery cell.

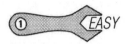

CHECKING SPECIFIC GRAVITY

The electrolyte fluid (sulfuric acid solution) contained in the battery cells will tell you many things about the condition of the battery. Because the cell plates must be kept submerged below the fluid level in order to operate, maintaining the fluid level is extremely important. In addition, because the specific gravity of the acid is an indication of electrical charge, testing the fluid can be an aid in determining if the battery must be replaced. A battery in a boat with a properly operating charging system should require little maintenance, but careful, periodic inspection should reveal problems before they leave you stranded.

✳✳ CAUTION

Battery electrolyte contains sulfuric acid. If you should splash any on your skin or in your eyes, flush the affected area with plenty of clear water. If it lands in your eyes, get medical help immediately.

As stated earlier, the specific gravity of a battery's electrolyte level can be used as an indication of battery charge. At least once a year, check the specific gravity of the battery. It should be between 1.20 and 1.26 on the gravity scale. Most parts stores carry a variety of inexpensive battery testing hydrometers. These can be used on any non-sealed battery to test the specific gravity in each cell.

Conventional Battery

◆ See Figure 157

A hydrometer is required to check the specific gravity on all batteries that are not maintenance-free. The hydrometer has a squeeze bulb at one end and a nozzle at the other. Battery electrolyte is sucked into the hydrometer until the float or pointer is lifted from its seat. The specific gravity is then read by noting the position of the float/pointer. If gravity is low in one or more cells, the battery should be slowly charged and checked again to see if the gravity has come up. Generally, if after charging, the specific gravity of any two cells varies more than 50 points (0.50), the battery should be replaced, as it can no longer produce sufficient voltage to guarantee proper operation.

Check the battery electrolyte level at least once a month, or more often in hot weather or during periods of extended operation. Electrolyte level can be checked either through the case on translucent batteries or by removing the cell caps on opaque-case types. The electrolyte level in each cell should be kept filled to the split ring inside each cell, or the line marked on the outside of the case.

■ **Never use mineral water or water obtained from a well. The iron content in these types of water is too high and will shorten the life of or damage the battery.**

If the level is low, add only distilled water through the opening until the level is correct. Each cell is separate from the others, so each must be checked and filled individually. Distilled water should be used, because the chemicals and minerals found in most drinking water are harmful to the battery and could significantly shorten its life.

If water is added in freezing weather, the battery should be warmed to allow the water to mix with the electrolyte. Otherwise, the battery could freeze.

Maintenance-Free Batteries

Although some maintenance-free batteries have removable cell caps for access to the electrolyte, the electrolyte condition and level is usually checked using the built-in hydrometer "eye." The exact type of eye varies between battery manufacturers, but most apply a sticker to the battery itself explaining the possible readings. When in doubt, refer to the battery manufacturer's instructions to interpret battery condition using the built-in hydrometer.

The readings from built-in hydrometers may vary, however a green eye usually indicates a properly charged battery with sufficient fluid level. A dark eye is normally an indicator of a battery with sufficient fluid, but one that

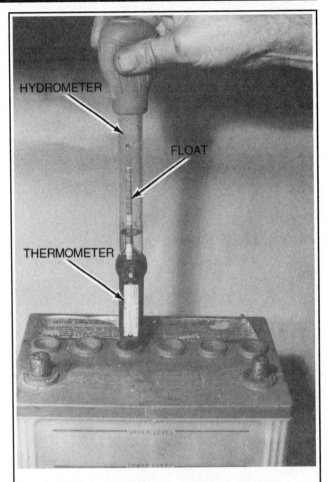

Fig. 157 The best way to determine the condition of a battery is to test the electrolyte with a battery hydrometer

may be low in charge. In addition, a light or yellow eye is usually an indication that electrolyte supply has dropped below the necessary level for battery (and hydrometer) operation. In this last case, sealed batteries with an insufficient electrolyte level must usually be discarded.

BATTERY TERMINALS

◆ See Figures 158 and 159

At least once a season, the battery terminals and cable clamps should be cleaned. Loosen the clamps and remove the cables, negative cable first. On batteries with top mounted posts, the use of a puller specially made for this purpose is recommended. These are inexpensive and available in most parts stores.

Clean the cable clamps and the battery terminal with a wire brush, until all corrosion, grease, etc., is removed and the metal is shiny. It is especially important to clean the inside of the clamp thoroughly (a wire brush is useful here), since a small deposit of foreign material or oxidation there will prevent a sound electrical connection and inhibit either starting or charging. It is also a good idea to apply some dielectric grease to the terminal, as this will aid in the prevention of corrosion.

After the clamps and terminals are clean, reinstall the cables, negative cable last; do not hammer the clamps onto battery posts. Tighten the clamps securely, but do not distort them. To retard corrosion, give the clamps and terminals a thin external coating of clear polyurethane paint after installation.

Check the cables at the same time that the terminals are cleaned. If the insulation is cracked or broken, or if its end is frayed, the cable should be replaced with a new one of the same length and gauge.

Fig. 158 A battery post cleaner is used to clean the battery posts...

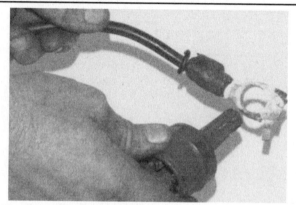

Fig. 159 ...and the battery terminals

BATTERY & CHARGING SAFETY PRECAUTIONS

Always follow these safety precautions when charging or handling a battery.

1. Wear eye protection when working around batteries. Batteries contain corrosive acid and produce explosive gas a byproduct of their operation. Acid on the skin should be neutralized with a solution of baking soda and water made into a paste. In case acid contacts the eyes, flush with clear water and seek medical attention immediately.

2. Avoid flame or sparks that could ignite the hydrogen gas produced by the battery and cause an explosion. Connection and disconnection of cables to battery terminals is one of the most common causes of sparks.

3. Always turn a battery charger **OFF**, before connecting or disconnecting the leads. When connecting the leads, connect the positive lead first, then the negative lead, to avoid sparks.

4. When lifting a battery, use a battery carrier or lift at opposite corners of the base.

5. Ensure there is good ventilation in a room where the battery is being charged.

6. Do not attempt to charge or load-test a maintenance-free battery when the charge indicator dot is indicating insufficient electrolyte.

7. Disconnect the negative battery cable if the battery is to remain in the boat during the charging process.

8. Be sure the ignition switch is **OFF** before connecting or turning the charger **ON**. Sudden power surges can destroy electronic components.

9. Use proper adapters to connect the charger leads to batteries with non-conventional terminals.

BATTERY CHARGERS

Before using any battery charger, consult the manufacturer's instructions for its use. Battery chargers are electrical devices that change Alternating Current (AC) to a lower voltage of Direct Current (DC) that can be used to charge a marine battery. There are two types of battery chargers—manual and automatic.

A manual battery charger must be physically disconnected when the battery has come to a full charge. If not, the battery can be overcharged, and possibly fail. Excess charging current at the end of the charging cycle will heat the electrolyte, resulting in loss of water and active material, substantially reducing battery life.

■ **As a rule, on manual chargers, when the ammeter on the charger registers half the rated amperage of the charger, the battery is fully charged. This can vary, and it is recommended to use a hydrometer to accurately measure state of charge.**

Automatic battery chargers have an important advantage—they can be left connected (for instance, overnight) without the possibility of overcharging the battery. Automatic chargers are equipped with a sensing device to allow the battery charge to taper off to near zero as the battery becomes fully charged. When charging a low or completely discharged battery, the meter will read close to full rated output. If only partially discharged, the initial reading may be less than full rated output, as the charger responds to the condition of the battery. As the battery continues to charge, the sensing device monitors the state of charge and reduces the charging rate. As the rate of charge tapers to zero amps, the charger will continue to supply a few milliamps of current—just enough to maintain a charged condition.

REPLACING BATTERY CABLES

Battery cables don't go bad very often, but like anything else, they can wear out. If the cables on your boat are cracked, frayed or broken, they should be replaced.

When working on any electrical component, it is always a good idea to disconnect the negative (–) battery cable. This will prevent potential damage to many sensitive electrical components

Always replace the battery cables with one of the same length, or you will increase resistance and possibly cause hard starting. Smear the battery posts with a light film of dielectric grease, or a battery terminal protectant spray once you've installed the new cables. If you replace the cables one at a time, you won't mix them up.

■ **Any time you disconnect the battery cables, it is recommended that you disconnect the negative (–) battery cable first. This will prevent you from accidentally grounding the positive (+) terminal when disconnecting it, thereby preventing damage to the electrical system.**

Before you disconnect the cable(s), first turn the ignition to the **OFF** position. This will prevent a draw on the battery which could cause arcing. When the battery cable(s) are reconnected (negative cable last), be sure to check all electrical accessories are all working correctly.

STORAGE

If the boat is to be laid up for the winter or for more than a few weeks, special attention must be given to the battery to prevent complete discharge or possible damage to the terminals and wiring. Before putting the boat in storage, disconnect and remove the batteries. Clean them thoroughly of any dirt or corrosion and then charge them to full specific gravity reading. After they are fully charged, store them in a clean cool dry place where they will not be damaged or knocked over, preferably on a couple blocks of wood. Storing the battery up off the deck, will permit air to circulate freely around and under the battery and will help to prevent condensation.

Never store the battery with anything on top of it or cover the battery in such a manner as to prevent air from circulating around the filler caps. All batteries, both new and old, will discharge during periods of storage, more so if they are hot than if they remain cool. Therefore, the electrolyte level and the specific gravity should be checked at regular intervals. A drop in the specific gravity reading is cause to charge them back to a full reading.

In cold climates, care should be exercised in selecting the battery storage area. A fully-charged battery will freeze at about 60 degrees below zero. A discharged battery, almost dead, will have ice forming at about 19 degrees above zero.

WINTER STORAGE

Taking extra time to store the boat properly at the end of each season will increase the chances of satisfactory service at the next season. Remember, storage is the greatest enemy of a marine engine. In a perfect world the unit should be run on a monthly basis. The steering and shifting mechanism should also be worked through complete cycles several times each month. But who lives in a perfect world!

For most of us, if a small amount of time is spent in winterizing our beloved boats, the reward will be satisfactory performance, increased longevity and greatly reduced maintenance expenses.

Winter Storage Checklist

Proper winterizing involves adequate protection of the unit from physical damage, rust, corrosion, and dirt. The following steps provide guide to winterizing your marine engine at the end of a season.

1. Always keep a note or a checklist of just what maintenance needs to be done prior to starting the engine in the spring.
2. Replace the oil and oil filter. Normal combustion produces corrosive acids that are absorbed by the oil. Leaving dirty oil in the engine for an extended time allows these acids to attack and damage bearing surfaces.
3. Change the engine oil and filter. Used oil contains harmful acids and contaminants that should not be allowed to go to work on the engine all winter long. The drive/transmission oil should also be changed.
4. Replace the fuel filter elements—draining any water from the filter bowls.
5. Keeping the fuel tank full over the winter will help cut down on condensation. Using fuel additives, like Quicksilver Gasoline stabilizer, that control bacterial growth or prevent gelling in cold climates are also popular with many boaters.
6. Bleed the fuel system of air and replace the water separating fuel filter.
   a. Flush the seawater circuit to remove corrosive salts.
   b. Drain the seawater system, taking special care to empty all the low spots.
   c. On closed cooling systems, replace the coolant with a clean, fresh, 50/50 mixture of water and antifreeze. Always mix the antifreeze and water mixture prior to pouring it into the engine.
7. Inspect all hoses for signs of softening, cracking or bulging, especially those routinely exposed to high heat. Check hose clamps for tightness and corrosion.
8. Close the fuel shut-off valve, if equipped, to stop all flow of fuel from the tanks to the engine.
9. Remove the flame arrestor and start the engine. Run the engine at fast idle and squirt about 8 oz. of Quicksilver Storage Seal into the carburetor. Just as the engine is starting to sputter, squirt the remaining 2 oz. of Seal into the carb and allow the engine to die. You can also use SAE 20 motor oil.
10. It's also a good idea to remove the spark plugs and pour about an ounce of Storage Seal into each cylinder. You can also use SAE 20 motor oil. Crank the engine with the starter for a few seconds to make sure that the oil coats the cylinder bores completely.
11. On EFI engines, remove the water separating fuel filter and pour in about 2 oz. of 2 cycle outboard oil. Reinstall the filter, start the engine and run it until it dies. Install a new filter.
12. Perform all the lubrication procedures detailed in this section.
13. Loosen the adjusting bolts or tensioners on all drive or serpentine belts do the tension is relieved.
14. Fully charge the batteries. Disconnect all leads. Unattended, a battery naturally discharges over a period of several weeks. The electrolyte on a discharged battery can freeze at 20 degrees F (-7 degrees C), so keep the batteries fully charged or, better still, remove them to a warmer storage area. Small automatic trickle chargers work well.
15. Treat battery and cable terminals with petroleum jelly, silicone grease, or a heavy-duty corrosion inhibitor.
16. Protect external surfaces with a heavy-duty corrosion inhibitor.
17. Grease all greaseable points on the drivetrain.
18. Lightly coat the alternator and starter with a light lubricant to disperse the water. Loosen the alternator belt tension.
19. Ensure that all drain holes on the stern drive are open by inserting a piece of wire into them.
20. Make sure the stern drive is in the full IN/DOWN position.
21. Cover the engine with a waterproof sheet in case there are any leaks from above. Some boaters will take the time to seal all openings to the engine (air inlet, breathers, exhaust), however this traps moisture and may do more harm than good.
22. Place a checklist in a handy spot to remind you of just what maintenance needs to be done prior to starting the engine in the spring.
23. If you visit the boat during the winter, additional protection can be achieved using the starter to turn the engine over and circulate oil to the bearings and cylinder walls.
24. If the boat is to be stored out of the water, always remove the drain plug(s) if equipped. We suggest you attach the plug to the ignition key as added insurance that you remember to reinstall it in the spring–sound silly? Why take the chance!

■ **Special inhibiting oils are available that provide greater protection. Use these to replace the standard engine oil; run the engine briefly to coat all surfaces, and then drain the oil. Protection remains good, provided the engine is not turned.**

25. If the engine cannot be fully winterized, replace oil, coolant, and all filters and run the engine up to operating temperatures monthly.

SPRING COMMISSIONING

Satisfactory performance and maximum enjoyment can be realized if a little time is spent in commissioning your boat in the spring. Assuming you have followed the steps we recommended to winterize your vessel (in addition to any the manufacturer specifies) and the unit has been properly stored, a minimum amount of work should be required to prepare it for use.

After performing the spring commissioning and testing the boat on the water, it is a good idea to perform a full tune-up. Remember, you are relying on your engine to get you where you want to go. Treat it good now and it will treat you good later.

Spring Commissioning Checklist

The following steps outline a logical sequence of tasks to be performed before starting your engine for the first time in a new season.

1. Pick up the checklist you made to remind yourself of just what maintenance needs to be done prior to starting the engine in the spring. You did remember to write yourself a checklist....right?
2. INSTALL THE DRAIN PLUG IF REMOVED!
3. Remove the cover placed over the engine last winter. Unseal any engine openings (air inlet, breathers, exhaust) previously sealed.
4. Replace all zincs.
5. If you took our advice, you removed the battery for the winter. While it was in storage, you should have kept it fully charged. It should be ready to go. So install the battery and connect the battery cables. Treat battery and cable terminals with petroleum jelly, silicone grease, or a heavy-duty corrosion inhibitor. Capacity test the batteries.
6. Tighten the alternator and other belts.
7. As you did in the winter, inspect all hoses for signs of softening, cracking or bulging, especially those routinely exposed to high heat. Check hose clamps for tightness and corrosion.
8. Ensure that exhaust manifolds are tight.
9. On closed system engines, check the condition of the coolant mixture with a coolant tester and adjust the mixture by adding antifreeze or water.
10. Bleed the fuel system of air.
11. Start the engine and allow it to reach operating temperature.
12. Once running, check the oil pressure, the raw water discharge and the engine for oil and water leaks.
13. Drain water from the filter bowls. If an excessive amount of water is noted in the fuel system, you may want to consider a fuel conditioner or replacing the old fuel with fresh fuel.
14. After testing the boat on the water, it is a good idea to change the engine oil and filter. The drive/transmission oil should also be changed. This eliminates the adverse effects of moisture in the oil.

FIRING ORDERS

◆ See Figures 161, 162, 163, 164 and 165

■ To avoid confusion, ALWAYS label the spark plug wires with a piece of tape before disconnecting them from the plug or distributor cap.

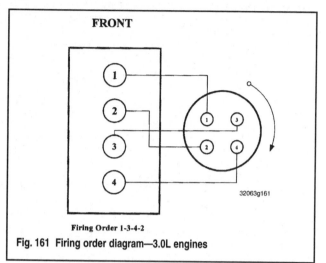

Firing Order 1-3-4-2

Fig. 161 Firing order diagram—3.0L engines

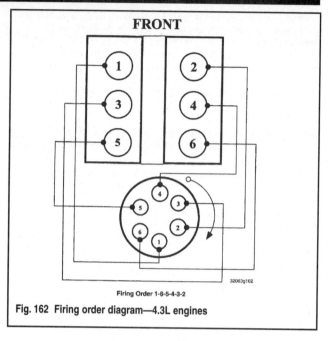

Firing Order 1-6-5-4-3-2

Fig. 162 Firing order diagram—4.3L engines

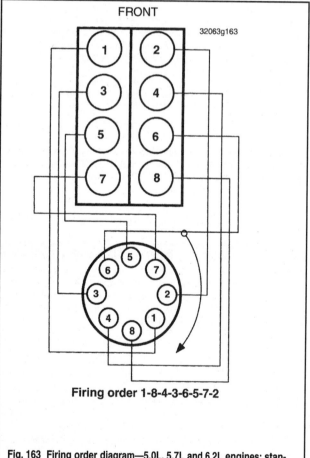

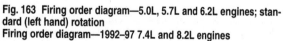

Firing order 1-8-4-3-6-5-7-2

**Fig. 163 Firing order diagram—5.0L, 5.7L and 6.2L engines; standard (left hand) rotation
Firing order diagram—1992–97 7.4L and 8.2L engines**

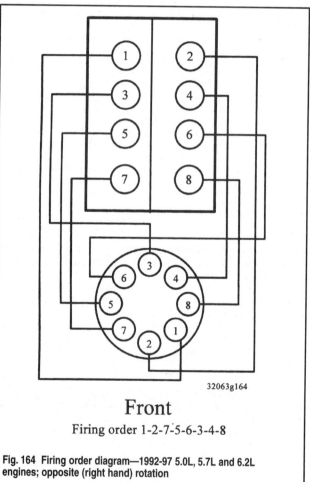

Front
Firing order 1-2-7-5-6-3-4-8

Fig. 164 Firing order diagram—1992-97 5.0L, 5.7L and 6.2L engines; opposite (right hand) rotation

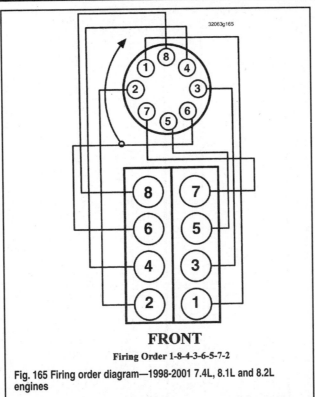

FRONT

Firing Order 1-8-4-3-6-5-7-2

Fig. 165 Firing order diagram—1998-2001 7.4L, 8.1L and 8.2L engines

SPECIFICATIONS

CAPACITIES

Year	Model	Engine Crankcase (With Filter) Qts. (L) ①	Cooling System Closed Qts. (L)	Seawater Qts. (L) ②	Alpha	Bravo 1	Bravo 2	Bravo 3	Blackhawk	Inboard Trans. Qts. (L)
							Stern Drive Qts. (L) ③			
1992	3.0L/LX	4.0 (3.8)	9.0 (8.5)	9.0 (8.5)	2.0 (1.9) ⑦	-	-	-	-	-
	185R/MR, 205MR, 4.3L	4.5 (4.28)	20 (19)	14 (13)	4.5 (4.28)	-	-	-	-	-
	5.0L/LX	5.5 (5.2)	20 (19)	15 (14.1)	2.0 (1.9)	-	-	-	-	-
	5.7L, 350 Mag	5.5 (5.2)	20 (19)	15 (14.1)	2.0 (1.9)	-	3.25 (3.0)	-	-	-
	5.7L Blue Water	4.0 (3.8)	20 (19)	15 (14.1)	-	-	-	-	-	⑤
	350 Magnum	5.0 (4.7)	20 (19)	15 (14.1)	2.0 (1.9)	-	-	-	-	-
	7.4L, 454 Mag	7.0 (6.6)	28 (26.5)	20 (19)	-	2.75 (2.6)	3.25 (3.0)	-	-	⑧
	8.2L, 502 Mag	8.0 (7.5)	28 (26.5)	20 (19)	-	-	-	-	-	⑧
1993	3.0L/LX	4.0 (3.8)	9.0 (8.5)	9.0 (8.5)	2.0 (1.9)	-	-	-	-	-
	185R/MR, 205MR, 4.3L	4.5 (4.28)	20 (19)	14 (13)	4.5 (4.28)	-	-	-	-	-
	5.0L/LX	5.5 (5.2)	20 (19)	15 (14.1)	2.0 (1.9)	-	3.25 (3.0)	-	-	-
	5.7L , 350 Mag	5.5 (5.2)	20 (19)	15 (14.1)	2.0 (1.9)	-	-	-	-	⑤
	5.7L Blue Water	4.0 (3.8)	20 (19)	15 (14.1)	-	-	-	-	-	-
	350 Magnum	5.0 (4.7)	20 (19)	15 (14.1)	2.0 (1.9)	-	-	-	-	-
	7.4L, 454 Mag	7.0 (6.6)	28 (26.5)	20 (19)	-	2.75 (2.6)	3.25 (3.0)	3.0 (2.8)	-	⑧
	8.2L, 502 Mag	8.0 (7.5)	28 (26.5)	20 (19)	-	2.75 (2.6)	-	-	-	⑧
1994	3.0L/LX	4.0 (3.8)	9.0 (8.5)	9.0 (8.5)	2.0 (1.9)	-	-	-	-	-
	4.3L/LX	4.5 (4.3)	20 (19)	15 (14.1)	2.0 (1.9)	-	-	-	-	-
	5.0L/LX, 350 Mag	5.5 (5.2)	20 (19)	15 (14.1)	2.0 (1.9)	-	-	-	-	-
	5.7L, 350 Mag	5.5 (5.2)	20 (19)	15 (14.1)	2.0 (1.9)	-	3.25 (3.0)	3.0 (2.8)	-	⑤
	5.7L Blue Water	4.0 (3.8)	20 (19)	15 (14.1)	-	-	-	-	-	-
	350 Magnum	5.0 (4.7)	20 (19)	15 (14.1)	2.0 (1.9)	-	-	-	-	-
	7.4L , 454 Mag	7.0 (6.6)	28 (26.5)	20 (19)	-	2.75 (2.6)	3.25 (3.0)	3.0 (2.8)	-	⑧
	8.2L, 502 Mag	8.0 (7.5)	28 (26.5)	20 (19)	-	2.75 (2.6)	-	-	④	⑧
1995	3.0LX	4.0 (3.8)	9.0 (8.5)	9.0 (8.5)	2.0 (1.9)	-	-	-	-	-
	4.3L/LX	4.5 (4.3)	20 (19)	15 (14.1)	2.0 (1.9)	-	-	-	-	-
	5.0L/LX	5.5 (5.2)	20 (19)	15 (14.1)	2.0 (1.9)	-	3.25 (3.0)	-	-	-
	5.7L/EFI, 350 Mag	5.5 (5.2)	20 (19)	15 (14.1)	2.0 (1.9)	2.75 (2.6)	3.25 (3.0)	3.0 (2.8)	-	-

CAPACITIES (cont'd)

Year	Model	Engine Crankcase (With Filter) Qts. (L) [1]	Cooling System Closed Qts. (L)	Cooling System Seawater Qts. (L) [2]	Stern Drive Qts. (L) [3]					Inboard Trans. Qts. (L)
					Alpha	Bravo 1	Bravo 2	Bravo 3	Blackhawk	
1995 (cont'd)	5.7L Blue Water	4.0 (3.8)	20 (19)	15 (14.1)	-	-	-	-	-	[5]
	350 Magnum EFI/MP	5.0 (4.7)	20 (19)	15 (14.1)	-	2.75 (2.6)	-	3.0 (2.8)	-	-
	7.4L/EFI, 454 Mag	7.0 (6.6)	28 (26.5)	20 (19)	-	2.75 (2.6)	3.25 (3.0)	3.0 (2.8)	-	[8]
	8.2L, 502 Mag	8.0 (7.5)	28 (26.5)	20 (19)	-	2.75 (2.6)	3.25 (3.0)	3.0 (2.8)	-	[8]
1996	3.0LX	4.0 (3.8)	9.0 (8.5)	9.0 (8.5)	2.0 (1.9)	-	-	-	-	-
	4.3L/LX	4.5 (4.3)	20 (19)	15 (14.1)	2.0 (1.9)	-	3.25 (3.0)	3.0 (2.8)	-	-
	5.7L/LX/EFI, 350 Mag	5.5 (5.2)	20 (19)	15 (14.1)	2.0 (1.9)	2.75 (2.6)	3.25 (3.0)	3.0 (2.8)	-	-
	5.7L	4.0 (3.8)	20 (19)	15 (14.1)	-	-	-	-	-	[5]
	350 Magnum	5.0 (4.7)	20 (19)	15 (14.1)	-	2.75 (2.6)	-	3.0 (2.8)	-	-
	7.4L/LX/MPI, 454 Mag	7.0 (6.6)	28 (26.5)	20 (19)	-	2.75 (2.6)	3.25 (3.0)	3.0 (2.8)	[4]	[8]
	8.2L, 502 Mag	8.0 (7.5)	28 (26.5)	20 (19)	-	2.75 (2.6)	3.25 (3.0)	3.0 (2.8)	[4]	[8]
1997	3.0LX	4.0 (3.8)	9.0 (8.5)	9.0 (8.5)	2.0 (1.9)	-	-	-	-	-
	4.3LX/LXH Gen+, 262	4.5 (4.3)	20 (19)	15 (14.1)	2.0 (1.9)	-	3.25 (3.0)	3.0 (2.8)	-	-
	5.7L/LX/EFI, 350 Mag	5.5 (5.2)	20 (19)	15 (14.1)	2.0 (1.9)	2.75 (2.6)	3.25 (3.0)	3.0 (2.8)	-	-
	5.7L	4.0 (3.8)	20 (19)	15 (14.1)	-	-	-	-	-	[5]
	350 Magnum	5.0 (4.7)	20 (19)	15 (14.1)	-	2.75 (2.6)	-	3.0 (2.8)	-	-
	7.4L/LX/MPI/EFI, 454	7.0 (6.6)	28 (26.5)	20 (19)	-	2.75 (2.6)	3.25 (3.0)	3.0 (2.8)	[4]	[8]
	8.2L, 502 Mag	8.0 (7.5)	28 (26.5)	20 (19)	-	2.75 (2.6)	3.25 (3.0)	3.0 (2.8)	[4]	[8]
1998	3.0L Alpha	4.0 (3.8)	9.0 (8.5)	9.0 (8.5)	2.0 (1.9)	-	-	-	-	-
	4.3L/LH/EFI	4.5 (4.3)	20 (19)	15 (14.1)	2.0 (1.9)	2.75 (2.6)	3.25 (3.0)	3.0 (2.8)	-	-
	5.0L/EFI	5.5 (5.2)	20 (19)	15 (14.1)	2.0 (1.9)	2.75 (2.6)	3.25 (3.0)	3.0 (2.8)	-	-
	5.7L/EFI, 350 Mag	5.5 (5.2)	20 (19)	15 (14.1)	2.0 (1.9)	2.75 (2.6)	3.25 (3.0)	3.0 (2.8)	-	[6]
	6.2 MPI	5.5 (5.2)	20 (19)	15 (14.1)	-	2.75 (2.6)	3.25 (3.0)	3.0 (2.8)	-	[7]
	7.4L, 454, 8.2L, 502	7.0 (6.6)	18 (17)	20 (19)	-	2.75 (2.6)	3.25 (3.0)	3.0 (2.8)	2.5 (2.4)	[9]
	8.1S, 496 Mag	9.0 (8.5)	19 (18)	20 (19)	-	2.75 (2.6)	3.25 (3.0)	3.0 (2.8)	2.5 (2.4)	[10]
1999	3.0L Alpha	4.0 (3.8)	9.0 (8.5)	9.0 (8.5)	2.0 (1.9)	-	-	-	-	-
	4.3L/LH/EFI	4.5 (4.3)	20 (19)	15 (14.1)	2.0 (1.9)	2.75 (2.6)	3.25 (3.0)	3.0 (2.8)	-	-
	5.0L/EFI	5.5 (5.2)	20 (19)	15 (14.1)	2.0 (1.9)	2.75 (2.6)	3.25 (3.0)	3.0 (2.8)	-	-

32063c02

CAPACITIES (cont'd)

Year	Model	Engine Crankcase (With Filter) Qts. (L) ①	Cooling System Closed Qts. (L)	Cooling System Seawater Qts. (L) ②	Stern Drive Qts. (L) ③ Alpha	Bravo 1	Bravo 2	Bravo 3	Blackhawk	Inboard Trans. Qts. (L)
1999 (cont'd)	5.7L/EFI, 350 Mag	5.5 (5.2)	20 (19)	15 (14.1)	2.0 (1.9)	2.75 (2.6)	3.25 (3.0)	3.0 (2.8)	-	⑥
	6.2 MPI	5.5 (5.2)	20 (19)	15 (14.1)	-	2.75 (2.6)	3.25 (3.0)	3.0 (2.8)	-	⑦
	7.4L, 454, 8.2L, 502	7.0 (6.6)	18 (17)	20 (19)	-	2.75 (2.6)	3.25 (3.0)	3.0 (2.8)	-	⑨
	8.1S, 496 Mag	9.0 (8.5)	19 (18)	20 (19)	-	2.75 (2.6)	3.25 (3.0)	3.0 (2.8)	-	⑩
2000	3.0L Alpha	4.0 (3.8)	9.0 (8.5)	9.0 (8.5)	-	-	-	-	-	-
	4.3L/LH/EFI	4.5 (4.3)	20 (19)	15 (14.1)	2.0 (1.9)	-	3.25 (3.0)	3.0 (2.8)	-	-
	5.0L/EFI	5.5 (5.2)	20 (19)	15 (14.1)	2.0 (1.9)	2.75 (2.6)	3.25 (3.0)	3.0 (2.8)	-	-
	5.7L/EFI, 350 Mag	5.5 (5.2)	20 (19)	15 (14.1)	2.0 (1.9)	2.75 (2.6)	3.25 (3.0)	3.0 (2.8)	-	⑥
	6.2 MPI	5.5 (5.2)	20 (19)	15 (14.1)	-	2.75 (2.6)	3.25 (3.0)	3.0 (2.8)	-	⑦
	7.4L, 454, 8.2L, 502	7.0 (6.6)	18 (17)	20 (19)	-	2.75 (2.6)	3.25 (3.0)	3.0 (2.8)	-	⑨
	8.1S, 496 Mag	9.0 (8.5)	19 (18)	20 (19)	-	2.75 (2.6)	3.25 (3.0)	3.0 (2.8)	-	⑩
2001	3.0L	4.0 (3.8)	9.0 (8.5)	9.0 (8.5)	-	-	-	-	-	-
	4.3L/L/EFI	4.5 (4.3)	20 (19)	15 (14.1)	2.0 (1.9)	-	104 (3076)	96 (2839)	-	-
	5.0L/EFI	5.5 (5.2)	20 (19)	15 (14.1)	2.0 (1.9)	2.75 (2.6)	104 (3076)	96 (2839)	-	-
	5.7L/EFI, 350 Mag	5.5 (5.2)	20 (19)	15 (14.1)	2.0 (1.9)	2.75 (2.6)	104 (3076)	96 (2839)	-	⑥
	6.2 MPI	5.5 (5.2)	20 (19)	15 (14.1)	-	2.75 (2.6)	104 (3076)	96 (2839)	-	⑦
	7.4L, 454, 8.2L, 502	7.0 (6.6)	18 (17)	20 (19)	-	2.75 (2.6)	104 (3076)	96 (2839)	-	⑨
	8.1S, 496 Mag	9.0 (8.5)	19 (18)	20 (19)	-	2.75 (2.6)	104 (3076)	96 (2839)	-	⑩

32063c03

CAPACITIES (cont'd)

Year	Model	Engine Crankcase (With Filter) Qts. (L) [1]	Cooling System Closed Qts. (L)	Cooling System Seawater Qts. (L) [2]	Stern Drive Qts. (L) [3]					Inboard Trans. Qts. (L)
					Alpha	Bravo 1	Bravo 2	Bravo 3	Blackhawk	

[1] All capacities are approximate. Always use the dipstick to determine exact quantity of oil required

[2] Seawater cooling system capacities are for winterization only

[3] All capacities are approximate. On models without a reservoir, fill to the bottom of the vent hole. On models with a reservoir, fill to the 'FULL line.

[4] Round Monitor: 3.25 (3.0)
Square monitor: 2.25 (1.8)
W/O monitor: .75 (0.5)

[5] All capacities are approximate. Always use the dipstick to determine exact quantity of oil required
BW (1.0:1 :2 (1.9); 1.5:1 - 3.0:1: 3.0 (2.9); V-drive: 4.5 (4.8)
Hurth: 4.5 (4.8)

[6] All capacities are approximate. Always use the dipstick to determine exact quantity of oil required
Velvet 71C: 1.5 (1.33); 72, V: 3.0 (2.75)
Walter: 0.75 (0.5)
Hurth 630V: 4.25 (4.0); 630A: 3.25 (3.0); 800A: 5.75 (5.5)

[7] All capacities are approximate. Always use the dipstick to determine exact quantity of oil required
Velvet 71C, 72C:2.25 (2.13)
Walter: 0.75 (0.5)
Hurth 630V: 4.25 (4.0); 630A: 3.25 (3.0); 800A: 5.75 (5.5)

[8] All capacities are approximate. Always use the dipstick to determine exact quantity of oil required
BW: 2.0 (1.9)
Hurth 630A: 4.5 (4.0); 800A: 4.25 (4.2); 630V: 5.0 (4.7)

[9] All capacities are approximate. Always use the dipstick to determine exact quantity of oil required
Velvet 72C: 1.5 (1.3); 500A/V: 3.0 (2.75)
Hurth: 4.25 (4.0)
Walter: 0.75 (0.5)

[10] All capacities are approximate. Always use the dipstick to determine exact quantity of oil required
Velvet 5000A: 2.75 (2.6); 5000V: 3.5 (3.3)
ZF/Hurth 630A: 3.25 (3.0); 630V: 4.25 (4.0); 800A: 5.75 (5.5)

32063c04

TUNE-UP SPECIFICATIONS

Year	Model	Spark Plug Champion	NGK	Gap In. (mm)	Idle Timing (rpm)	Idle Speed Rpm (In Gear)	Idle Mixture (Turns Out)	Oil Pressure @ 2000 rpm psi	Fuel Pressure @ 1000 rpm psi	Compression Pressure psi
1992	3.0L	RV8C	BR6FS	.035 (0.9)	①	650-700	1 1/4	30-60	5-6.5	140
	3.0LX	RS9YC	BPR6EFS	.035 (0.9)	①	650-700	1 1/4	30-60	5-6.5	140
	185MR/4.3L	RV8C	-	.035 (0.9)	8 BTDC	650-700	1 1/4	30-55	3-7	180
	205MR/4.3L	RV8C	-	.035 (0.9)	8 BTDC	650-700	2-3	30-55	3-7	180
	5.0L/LX	RV15Y	BR6FS	.035 (0.9)	8 BTDC	650-700	1 1/4	30-60	3-7	150
	5.7L Alpha One/Bravo	RV15Y	BR6FS	.035 (0.9)	8 BTDC	650-700	1 1/4	30-60	3-7	150
	5.7L Blue Water	RV15Y	BR6FS	.035 (0.9)	8 BTDC	650-700	1 1/4	30-60	3-7	150
	350 Magnum	RV15Y	BR6FS	.035 (0.9)	8 BTDC	650-700	1 1/4	30-60	3-7	150
	7.4L	RV15YC4	BR6FS	.035 (0.9)	8 BTDC	650-700	1 1/4	30-70	3-7	150
	7.4L Blue Water	RV15YC4	BR6FS	.035 (0.9)	8 BTDC	650-700	1 1/4	30-70	3-7	150
	454 Magnum	RV15YC4	BR6FS	.035 (0.9)	8 BTDC	650-700	1 1/4	30-70	3-7	150
	8.2L Blue Water	RV15YC4	BR6FS	.035 (0.9)	8 BTDC	650-700	1 1/4	30-70	3-7	150
	502 Magnum	RV15YC4	BR6FS	.035 (0.9)	8 BTDC	650-700	1 1/4	30-70	3-7	150
1993	3.0L	RV8C	BR6FS	.035 (0.9)	①	650-700	1 1/4	30-60	5-6.5	140
	3.0LX	RS9YC	BPR6EFS	.035 (0.9)	①	650-700	1 1/4	30-60	5-6.5	140
	185MR/4.3L	RV8C	-	.035 (0.9)	8 BTDC	650-700	1 1/4	30-55	3-7	180
	205MR/4.3L	RV8C	-	.035 (0.9)	8 BTDC	650-700	2-3	30-55	3-7	180
	5.0L/LX	RV15Y	BR6FS	.035 (0.9)	8 BTDC	650-700	1 1/4	30-60	3-7	150
	5.7L Alpha One/Bravo	RV15Y	BR6FS	.035 (0.9)	8 BTDC	650-700	1 1/4	30-60	3-7	150
	5.7L Blue Water	RV15Y	BR6FS	.035 (0.9)	8 BTDC	650-700	1 1/4	30-60	3-7	150
	350 Magnum	RV15Y	BR6FS	.035 (0.9)	8 BTDC	650-700	1 1/4	30-60	3-7	150
	7.4L	RV15YC4	BR6FS	.035 (0.9)	8 BTDC	650-700	1 1/4	30-70	3-7	150
	7.4L Blue Water	RV15YC4	BR6FS	.035 (0.9)	8 BTDC	650-700	1 1/4	30-70	3-7	150
	454 Magnum	RV15YC4	BR6FS	.035 (0.9)	8 BTDC	650-700	1 1/4	30-70	3-7	150
	8.2L Blue Water	RV15YC4	BR6FS	.035 (0.9)	8 BTDC	650-700	1 1/4	30-70	3-7	150
	502 Magnum	RV15YC4	BR6FS	.035 (0.9)	8 BTDC	650-700	1 1/4	30-70	3-7	150
1994	3.0L	RV8C	BR6FS	.035 (0.9)	①	650-700	1 1/4	30-60	5-6.5	140
	3.0LX	RS9YC	BPR6EFS	.035 (0.9)	①	650-700	1 1/4	30-60	5-6.5	140
	4.3L	RV8C	BR6FS	.035 (0.9)	8 BTDC	650-700	1 1/4	30-55	3-7	180

32063c05

TUNE-UP SPECIFICATIONS (cont'd)

Year	Model	Spark Plug Champion	Spark Plug NGK	Gap In. (mm)	Idle Timing (rpm)	Idle Speed Rpm (In Gear)	Idle Mixture (Turns Out)	Oil Pressure @ 2000 rpm psi	Fuel Pressure @ 1000 rpm psi	Compression Pressure psi
1994 (cont'd)	4.3LX Alpha One	RV8C	BR6FS	.035 (0.9)	8 BTDC	650-700	2	30-55	3-7	180
	5.0L/LX Alpha One	RV15Y	BR6FS	.035 (0.9)	8 BTDC	650-700	1 1/4	30-60	3-7	150
	5.7L Alpha One/Bravo	RV15Y	BR6FS	.035 (0.9)	8 BTDC	650-700	1 1/4	30-60	3-7	150
	5.7L Blue Water	RV15Y	BR6FS	.035 (0.9)	8 BTDC	650-700	1 1/4	30-60	3-7	150
	350 Magnum	RV15Y	BR6FS	.035 (0.9)	8 BTDC	650-700	1 1/4	30-60	3-7	150
	7.4L	RV15YC4	BR6FS	.035 (0.9)	8 BTDC	650-700	1 1/4	30-70	3-7	150
	7.4L Blue Water	RV15YC4	BR6FS	.035 (0.9)	8 BTDC	650-700	1 1/4	30-70	3-7	150
	454 Magnum	RV15YC4	BR6FS	.035 (0.9)	8 BTDC	650-700	1 1/4	30-70	3-7	150
	454 Magnum EFI	RV15YC4	BR6FS	.035 (0.9)	8 BTDC	550-650	NA	30-70	37	150
	8.2L Blue Water	RV15YC4	BR6FS	.035 (0.9)	8 BTDC	650-700	1 1/4	30-70	3-7	150
	502 Magnum EFI	RV15YC4	BR6FS	.035 (0.9)	8 BTDC	650-750	NA	30-70	37	150
1995	3.0LX	RS9YC	BPR6EFS	.035 (0.9)	①	650-700	1 1/4	30-60	5-6.5	140
	4.3L Alpha One	RV8C	BR6FS	.035 (0.9)	8 BTDC	650-700	1 1/4	30-55	3-7	180
	4.3LX Alpha One	RV8C	BR6FS	.035 (0.9)	8 BTDC	650-700	2	30-55	3-7	180
	5.0L/LX Alpha/Bravo	RV15Y	BR6FS	.035 (0.9)	8 BTDC	650-700	1 1/4	30-60	3-7	150
	5.7L Alpha/Bravo	RV15Y	BR6FS	.035 (0.9)	8 BTDC	650-700	1 1/4	30-60	3-7	150
	5.7L EFI Alpha/Bravo	RV15Y	BR6FS	.035 (0.9)	8 BTDC	650-700	NA	30-60	3-7	150
	5.7L Blue Water	RV15Y	BR6FS	.035 (0.9)	8 BTDC	650-700	1 1/4	30-60	3-7	150
	350 Magnum Alpha	RV15Y	BR6FS	.035 (0.9)	8 BTDC	650-700	1 1/4	30-60	3-7	150
	7.4L	RV15YC4	BR6FS	.035 (0.9)	8 BTDC	650-700	1 1/4	30-70	3-7	150
	7.4L Blue Water	RV15YC4	BR6FS	.035 (0.9)	8 BTDC	650-700	1 1/4	30-70	3-7	150
	454 Magnum EFI	RV15YC4	BR6FS	.035 (0.9)	8 BTDC	550-650	NA	30-70	37	150
	8.2L Blue Water	RV15YC4	BR6FS	.035 (0.9)	8 BTDC	650-700	1 1/4	30-70	3-7	150
	502 Magnum EFI/MP	RV15YC4	BR6FS	.035 (0.9)	8 BTDC	650-750	NA	30-70	37	150
1996	3.0LX	RS9YC	BPR6EFS	.035 (0.9)	①	650-700	1 1/4	30-60	5-6.5	140
	4.3L Alpha One	RV8C	BR6FS	.035 (0.9)	8 BTDC	650-700	1 1/4	30-55	3-7	180
	4.3LX Alpha One	RV8C	BR6FS	.035 (0.9)	8 BTDC	650-700	2	30-55	3-7	180
	5.7L/LX	RV15Y	BR6FS	.035 (0.9)	8 BTDC	650-700	1 1/4	30-60	3-7	150
	5.7LX EFI	RV15Y	BR6FS	.035 (0.9)	8 BTDC	650-700	NA	30-60	3-7	150
	350 Magnum	RV15Y	BR6FS	.035 (0.9)	8 BTDC	650-700	1 1/4	30-60	3-7	150
	350 Magnum MPI Gen+	RV15Y	BR6FS	.035 (0.9)	8 BTDC	650-700	NA	30-60	3-7	150

32063c06

TUNE-UP SPECIFICATIONS (cont'd)

Year	Model	Spark Plug Champion	NGK	Gap In. (mm)	Idle Timing (rpm)	Idle Speed Rpm (In Gear)	Idle Mixture (Turns Out)	Oil Pressure @ 2000 rpm psi	Fuel Pressure @ 1000 rpm psi	Compression Pressure psi
1996 (cont'd)	7.4L	RV15YC4	BR6FS	.035 (0.9)	8 BTDC	650-700	1 1/4	30-70	3-7	150
	7.4L Blue Water	RV15YC4	BR6FS	.035 (0.9)	8 BTDC	650-700	1 1/4	30-70	3-7	150
	7.4L/LX MPI	RV15YC4	BR6FS	.035 (0.9)	8 BTDC	550-650	NA	30-70	37	150
	454 Magnum	RV15YC4	BR6FS	.035 (0.9)	8 BTDC	650-700	1 1/4	30-70	37	150
	454 Magnum MPI	RV15YC4	BR6FS	.035 (0.9)	8 BTDC	550-650	NA	30-70	37	150
	8.2L	RV15YC4	BR6FS	.035 (0.9)	8 BTDC	650-700	1 1/4	30-70	3-7	150
	502 Magnum MPI	RV15YC4	BR6FS	.035 (0.9)	8 BTDC	650-750	NA	30-70	37	150
1997	3.0LX	RS9YC	BPR6EFS	.035 (0.9)	①	650-700	1 1/4	30-60	5-6.5	140
	4.3LX Gen+	RV8C	BR6FS	.035 (0.9)	8 BTDC	650-700	1 1/4	30-55	3-7	180
	4.3LXH Gen +	RV8C	BR6FS	.035 (0.9)	8 BTDC	650-700	2	30-55	3-7	180
	262 Magnum EFI Gen+	RS12YC	BPR6EFS	.045 (1.1)	8 BTDC	600	NA	30 min.	30	100 min.
	5.7L/LX	RV15Y	BR6FS	.035 (0.9)	8 BTDC	650-700	1 1/4	30-60	3-7	150
	5.7L inboard	RV15Y	BR6FS	.035 (0.9)	8 BTDC	650-700	1 1/4	30-60	3-7	150
	5.7L/LX EFI/EFI Gen+	RV15Y	BR6FS	.035 (0.9)	8 BTDC	650-700	NA	30-60	30	150
	350 Magnum	RV15Y	BR6FS	.035 (0.9)	8 BTDC	650-700	1 1/4	30-60	3-7	150
	350 Magnum EFI Gen+	RV15Y	BR6FS	.035 (0.9)	8 BTDC	650-700	NA	30-60	30	150
	350 Magnum MPI Gen+	RV15Y	BR6FS	.035 (0.9)	8 BTDC	650-700	NA	30-60	30	150
	7.4L	RV15YC4	BR6FS	.035 (0.9)	8 BTDC	650-700	1 1/4	30-70	3-7	150
	7.4L Blue Water	RV15YC4	BR6FS	.035 (0.9)	8 BTDC	650-700	1 1/4	30-70	3-7	150
	7.4L/LX MPI/EFI	RV15YC4	BR6FS	.035 (0.9)	8 BTDC	550-650	NA	30-70	37	150
	454 Magnum	RV15YC4	BR6FS	.035 (0.9)	8 BTDC	650-700	1 1/4	30-70	3-7	150
	454 Magnum MPI	RV15YC4	BR6FS	.035 (0.9)	8 BTDC	550-650	NA	30-70	37	150
	8.2L MPI	RV15YC4	BR6FS	.035 (0.9)	8 BTDC	650-700	NA	30-70	37	150
	502 Magnum MPI	RV15YC4	BR6FS	.035 (0.9)	8 BTDC	650-750	NA	30-70	37	150
1998	3.0L Alpha	RS12YC	BPR6EFS	.035 (0.9)	②	700 ④	1 1/4	30	6-8	100 min.
	4.3L/LH	RS12YC	BPR6EFS	.045 (1.1)	10 BTDC	650 ④	1 1/4	30 min.	3-7	100 min.
	4.3L EFI	RS12YC	BPR6EFS	.045 (1.1)	8 BTDC	600 ④	NA	30 min.	30	100 min.
	5.0L	RS12YC	BPR6EFS	.045 (1.1)	10 BTDC	650 ④	1 1/4	30 min.	3-7	100 min.
	5.0L EFI	RS12YC	BPR6EFS	.045 (1.1)	8 BTDC	600 ④	NA	30 min.	30	100 min.
	5.7L	RS12YC	BPR6EFS	.045 (1.1)	10 BTDC	650 ④	1 1/4	30 min.	3-7	100 min.

32063c07

TUNE-UP SPECIFICATIONS (cont'd)

Year	Model	Spark Plug Champion	Spark Plug NGK	Gap In. (mm)	Idle Timing (rpm)	Idle Speed Rpm (In Gear)	Idle Mixture (Turns Out)	Oil Pressure @ 2000 rpm psi	Fuel Pressure @ 1000 rpm psi	Compression Pressure psi
1998 (cont'd)	5.7L inboard	RSY9C	BPR6EFS	.045 (1.1)	10 BTDC	650 ④	1 1/4	30 min.	3-7	100 min.
	5.7L EFI	RS12YC	BPR6EFS	.045 (1.1)	8 BTDC	600 ④	NA	30 min.	30	100 min.
	350 Mag MPI	RS12YC	BPR6EFS	.045 (1.1)	8 BTDC	600 ④	NA	30 min.	30	100 min.
	350 Mag MPI inboard	RSY9C	BPR6EFS	.045 (1.1)	8 BTDC	600 ④	NA	30 min.	30	100 min.
	6.2 MPI	RS12YC	BPR6EFS	.045 (1.1)	8 BTDC	600 ④	NA	30 min.	30	100 min.
	7.4L MPI	RV12YC	BPR6EFS	.045 (1.1)	8 BTDC	600	NA	30 min.	43	100 min.
	454 Mag MPI	RV15YC4	BR6FS	.040 (1.0)	8 BTDC	600	NA	30 min.	43	100 min.
	454 Mag MPI Horizon	-	BR6FZX	.035 (0.9)	8 BTDC	600	NA	30 min.	43	100 min.
	8.1S HO/Horizon	③	③	.060 (1.52)	NA	650	NA	60	6	100 min.
	496 Mag/Mag HO	③	③	.060 (1.52)	NA	650	NA	60	6	100 min.
	8.2L MPI	RV15YC4	BR6FS	.040 (1.0)	8 BTDC	600	NA	30 min.	43	100 min.
	502 Mag MPI	RV15YC4	BR6FS	.040 (1.0)	8 BTDC	600	NA	30 min.	43	100 min.
1999	3.0L Alpha	RS12YC	BPR6EFS	.035 (0.9)	②	700 ④	1 1/4	30	6-8	100 min.
	4.3L/LH	RS12YC	BPR6EFS	.045 (1.1)	10 BTDC	650 ④	1 1/4	30 min.	3-7	100 min.
	4.3L EFI	RS12YC	BPR6EFS	.045 (1.1)	8 BTDC	600 ④	NA	30 min.	30	100 min.
	5.0L	RS12YC	BPR6EFS	.045 (1.1)	10 BTDC	650 ④	1 1/4	30 min.	3-7	100 min.
	5.0L EFI	RS12YC	BPR6EFS	.045 (1.1)	8 BTDC	600 ④	NA	30 min.	30	100 min.
	5.7L	RS12YC	BPR6EFS	.045 (1.1)	10 BTDC	650 ④	1 1/4	30 min.	3-7	100 min.
	5.7L inboard	RSY9C	BPR6EFS	.045 (1.1)	10 BTDC	650 ④	1 1/4	30 min.	3-7	100 min.
	5.7L EFI	RS12YC	BPR6EFS	.045 (1.1)	8 BTDC	600 ④	NA	30 min.	30	100 min.
	350 Mag MPI	RS12YC	BPR6EFS	.045 (1.1)	8 BTDC	600 ④	NA	30 min.	30	100 min.
	350 Mag MPI inboard	RSY9C	BPR6EFS	.045 (1.1)	8 BTDC	600 ④	NA	30 min.	30	100 min.
	6.2 MPI	RS12YC	BPR6EFS	.045 (1.1)	8 BTDC	600 ④	NA	30 min.	30	100 min.
	7.4L MPI	RV12YC	BPR6EFS	.045 (1.1)	8 BTDC	600	NA	30 min.	43	100 min.
	454 Mag MPI	RV15YC4	BR6FS	.040 (1.0)	8 BTDC	600	NA	30 min.	43	100 min.
	454 Mag MPI Horizon	-	BR6FZX	.035 (0.9)	8 BTDC	600	NA	30 min.	43	100 min.
	8.1S HO/Horizon	③	③	.060 (1.52)	NA	650	NA	60	6	100 min.

32063c08

TUNE-UP SPECIFICATIONS (cont'd)

Year	Model	Spark Plug Champion	Spark Plug NGK	Gap In. (mm)	Idle Timing (rpm)	Idle Speed Rpm (In Gear)	Idle Mixture (Turns Out)	Oil Pressure @ 2000 rpm psi	Fuel Pressure @ 1000 rpm psi	Compression Pressure psi
1999 (cont'd)	496 Mag/Mag HO	③	③	.060 (1.52)	NA	650	NA	60	6	100 min.
	8.2L MPI	RV15YC4	BR6FS	.040 (1.0)	8 BTDC	600	NA	30 min.	43	100 min.
	502 Mag MPI	RV15YC4	BR6FS	.040 (1.0)	8 BTDC	600	NA	30 min.	43	100 min.
2000	3.0L Alpha	RS12YC	BPR6EFS	.035 (0.9)	②	700 ④	1 1/4	30	6-8	100 min.
	4.3L/LH	RS12YC	BPR6EFS	.045 (1.1)	10 BTDC	650 ④	1 1/4	30 min.	3-7	100 min.
	4.3L EFI	RS12YC	BPR6EFS	.045 (1.1)	8 BTDC	600 ④	NA	30 min.	30	100 min.
	5.0L	RS12YC	BPR6EFS	.045 (1.1)	10 BTDC	650 ④	1 1/4	30 min.	3-7	100 min.
	5.0L EFI	RS12YC	BPR6EFS	.045 (1.1)	8 BTDC	600 ④	NA	30 min.	30	100 min.
	5.7L	RS12YC	BPR6EFS	.045 (1.1)	10 BTDC	650 ④	1 1/4	30 min.	3-7	100 min.
	5.7L inboard	RSY9C	BPR6EFS	.045 (1.1)	10 BTDC	650 ④	1 1/4	30 min.	3-7	100 min.
	5.7L EFI	RS12YC	BPR6EFS	.045 (1.1)	8 BTDC	600 ④	NA	30 min.	30	100 min.
	350 Mag MPI	RS12YC	BPR6EFS	.045 (1.1)	8 BTDC	600 ④	NA	30 min.	30	100 min.
	350 Mag MPI inboard	RSY9C	BPR6EFS	.045 (1.1)	8 BTDC	600 ④	NA	30 min.	30	100 min.
	6.2 MPI	RS12YC	BPR6EFS	.045 (1.1)	8 BTDC	600 ④	NA	30 min.	30	100 min.
	7.4L MPI	RV12YC	BPR6EFS	.045 (1.1)	8 BTDC	600	NA	30 min.	43	100 min.
	454 Mag MPI	RV15YC4	BR6FS	.040 (1.0)	8 BTDC	600	NA	30 min.	43	100 min.
	454 Mag MPI Horizon	-	BR6FZX	.035 (0.9)	8 BTDC	600	NA	30 min.	43	100 min.
	8.1S HO/Horizon	③	③	.060 (1.52)	NA	650	NA	60	6	100 min.
	496 Mag/Mag HO	③	③	.060 (1.52)	NA	650	NA	60	6	100 min.
	8.2L MPI	RV15YC4	BR6FS	.040 (1.0)	8 BTDC	600	NA	30 min.	43	100 min.
	502 Mag MPI	RV15YC4	BR6FS	.040 (1.0)	8 BTDC	600	NA	30 min.	43	100 min.
2001	3.0L Alpha	RS12YC	BPR6EFS	.035 (0.9)	②	700 ④	1 1/4	30	6-8	100 min.
	4.3L/LH	RS12YC	BPR6EFS	.045 (1.1)	10 BTDC	650 ④	1 1/4	30 min.	3-7	100 min.
	4.3L EFI	RS12YC	BPR6EFS	.045 (1.1)	8 BTDC	600 ④	NA	30 min.	30	100 min.
	5.0L	RS12YC	BPR6EFS	.045 (1.1)	10 BTDC	650 ④	1 1/4	30 min.	3-7	100 min.
	5.0L EFI	RS12YC	BPR6EFS	.045 (1.1)	8 BTDC	600 ④	NA	30 min.	30	100 min.
	5.7L	RS12YC	BPR6EFS	.045 (1.1)	10 BTDC	650 ④	1 1/4	30 min.	3-7	100 min.
	5.7L inboard	RSY9C	BPR6EFS	.045 (1.1)	10 BTDC	650 ④	1 1/4	30 min.	3-7	100 min.

32063c09

TUNE-UP SPECIFICATIONS (cont'd)

Year	Model	Spark Plug Champion	Spark Plug NGK	Gap In. (mm)	Idle Timing (rpm)	Idle Speed Rpm (In Gear)	Idle Mixture (Turns Out)	Oil Pressure @ 2000 rpm psi	Fuel Pressure @ 1000 rpm psi	Compression Pressure psi
2001 (cont'd)	5.7L EFI	RS12YC	BPR6EFS	.045 (1.1)	8 BTDC	600 ④	NA	30 min.	30	100 min.
	350 Mag MPI	RS12YC	BPR6EFS	.045 (1.1)	8 BTDC	600 ④	NA	30 min.	30	100 min.
	350 Mag MPI inboard	RSY9C	BPR6EFS	.045 (1.1)	8 BTDC	600 ④	NA	30 min.	30	100 min.
	6.2 MPI	RS12YC	BPR6EFS	.045 (1.1)	8 BTDC	600 ④	NA	30 min.	30	100 min.
	7.4L MPI	RV12YC	BPR6EFS	.045 (1.1)	8 BTDC	600	NA	30 min.	43	100 min.
	454 Mag MPI	RV15YC4	BR6FS	.040 (1.0)	8 BTDC	600	NA	30 min.	43	100 min.
	454 Mag MPI Horizon	-	BR6FZX	.035 (0.9)	8 BTDC	600	NA	30 min.	43	100 min.
	8.1S HO/Horizon	③	③	.060 (1.52)	NA	650	NA	60	6	100 min.
	496 Mag/Mag HO	③	③	.060 (1.52)	NA	650	NA	60	6	100 min.
	8.2L MPI	RV15YC4	BR6FS	.040 (1.0)	8 BTDC	600	NA	30 min.	43	100 min.
	502 Mag MPI	RV15YC4	BR6FS	.040 (1.0)	8 BTDC	600	NA	30 min.	43	100 min.

ATDC After Top Dead Center
BTDC Before Top Dead Center
① 8 BTDC except models w/EST - 1 BTDC
② Up to serial number 0L096999: 1 BTDC
   Serial number 0L0097000 thru 0L034099: 1 ATDC
   Serial number 0L0341000 and above: 2 ATDC
③ Denso TJ14R-P15
④ See procedure

32063c10

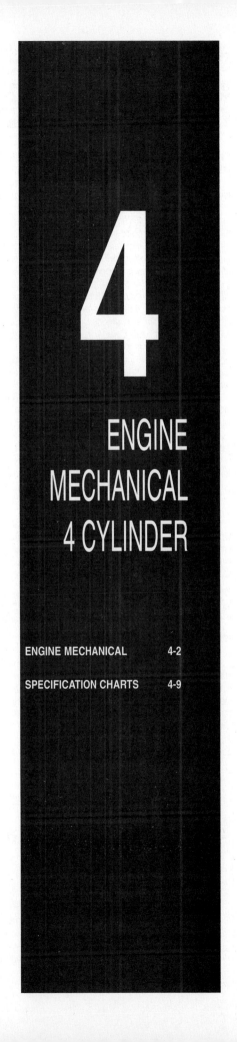

4

ENGINE
MECHANICAL
4 CYLINDER

ENGINE MECHANICAL

General Information

The MerCruiser 3.0L, 181 cubic inch displacement engine is manufactured by GMC and has been a favorite combination when mated to the Mercruiser stern drive. This engine is used in two models known as the 3.0L and the 3.0LX. The basic differences between these models are in the carburetion and ignition systems; otherwise they are identical.

This in-line four-cylinder powerplant uses a full pressure lubrication system with a disposable flow thru oil filter cartridge. Oil pressure is furnished by a gear-type oil pump, driven by the distributor, which is driven by a helical gear on the camshaft. A regulator on the oil pump controls the amount of oil pressure output. The oil pump scavenges oil from the bottom of the oil pan and feeds it through the oil filter and then to the main oil gallery in the block. Drilled passages in the block and crankshaft distribute oil to the camshaft and crankshaft to lubricate the rod, main and camshaft bearings. The main oil gallery also feeds oil to the valve lifters, which pump oil up through the hollow pushrods to the rocker arms to lubricate the valve train in the cylinder head. Cylinder numbering and firing order is identified in the illustrations at the end of Section Three.

All 3.0L engines are left hand (counterclockwise) rotation when viewed from the stern of the boat. This does not necessarily indicate that your prop rotation is the same—always check them both!

NEVER, NEVER attempt to use standard automotive parts when replacing anything on your engine. Due to the uniqueness of the environment in which they are operated in, and the levels at which they are operated at, marine engines require different versions of the same part; even if they look the same. Stock and most aftermarket automotive parts will not hold up for prolonged periods of time under such conditions. Automotive parts may appear identical to marine parts, but be assured, Mercury marine parts are specially manufactured to meet Mercury marine specifications. Most marine items are super heavy-duty units or are made from special metal alloy to combat against a corrosive salt water atmosphere.

Mercury marine electrical and ignition parts are extremely critical. In the United States, all electrical and ignition parts manufactured for marine application must conform to stringent U.S. Coast Guard requirements for spark or flame suppression. A spark from a non-marine cranking motor solenoid could ignite an explosive atmosphere of gasoline vapors in an enclosed engine compartment.

Engine Identification

◆ See Figure 1

All engines can be identified by a code number stamped into the lower starboard side of the engine block, aft of the distributor mounting hole—the last two letters of the code designate the engine. 1992-97 3.0L engines are identified with RM or RJ, while 3.0LX engines will use an RN or RK. 1998-2001 carry RX on engines below serial number 0L096999 (w/12 3/4 in. flywheel), and RP on engines with serial number 0L097000 and above (w/14 in. flywheel). This code is stamped on all original equipment engines and all partial replacement engines (but NOT on replacement cylinder blocks).

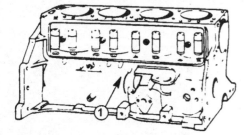

1- 4 cylinder-in-line engine code (next to distributor)

32064ag01

Fig. 1 The engine identification code can be found here. The last two letters are the engine designation

In the event that the engine serial number plates or stickers are missing from the engine, this is a good way to ensure that you know the exact engine in your vessel when ordering parts, etc.

Engine

OEM ③ DIFFICULT

REMOVAL & INSTALLATION

◆ See Figures 2, 3, 4, 5, 6, 7 and 8

1. Remove the stern drive unit as detailed in Section Eight.
2. Open or remove the engine hatch cover.
3. Disconnect the battery cables (negative first) at the battery and then disconnect them from the engine block and starter.
4. Disconnect the fuel inlet line at the fuel pump and quickly plug it. Make sure you have rags handy as there will be some spillage.
5. Pull the cotter pin from the clevis, remove the lock nut and disconnect the shift assist assembly.
6. Disconnect the shift cables and position them out of the way.
7. Disconnect the throttle cable at the throttle lever on the unit and position it out of the way.
8. Loosen the hose clamp on the engine wiring harness connector and unplug the instrumentation wiring harness. Label it and move it aside.
9. Carefully loosen and remove the two hydraulic lines at the power steering control valve. Plug the hose ends and the control valve fittings. Tag the lines and fittings and move them aside.
10. Loosen the hose clamps and disconnect the exhaust elbow bellows from the upper pipe. Remove it.
11. Tag and disconnect the engine ground wire at the stud.
12. Tag and disconnect the two trim position wires at the sender unit.
13. Losen the hose clamp and remove the water inlet line at the water tube.
14. On models equipped with the MerCathode system, tag and disconnect the wires at the controller.
15. Tag and disconnect any remaining lines, wires or hoses at the engine.
16. Attach a suitable engine hoist to the lifting eyes and take up any line slack until it is just taught.

■ **DO NOT use the front lifting eye attached to the thermostat housing.**

17. Locate the front engine mount(s) and remove the two 3/8 inch lag bolts.
18. Locate the rear engine mount and remove the two mounting bolts.
19. Slowly and carefully, lift out the engine. Try not to hit the power steering control valve on later models.
To install:

■ **An engine alignment tool (#91-57797A3) is necessary to reinstall the engine. Even if the mounts have not been removed from the engine it is still a good idea to re-align the unit.**

20. Apply Quicksilver Engine Coupler grease to the splines of the coupler.
21. Position a plastic stop nut into the slot at the bottom of the inner transom plate support and then lay a fiber washer over each of the thruholes in the transom plate.
22. Install a double wound lockwasher into the fiber washer and then lower the engine into the compartment so that it just rests on the washers and the transom plate. Make sure that the holes line up and DO NOT release the hoist tension yet.
23. Slide a large steel washer and a metal spacer onto the mounting bolt and then run them through the mount and transom plate. Tighten the bolts to 35-40 ft. lbs (47-54 Nm).
24. Slide the alignment tool into the center of the gimbal housing assembly bearing and then into the engine coupler splines—you may have to pivot the bearing slightly. Use the hoist to raise (or lower) the front of the engine until the tool slides completely into the coupler (with no binding).

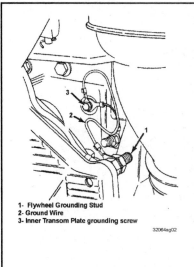

1- Flywheel Grounding Stud
2- Ground Wire
3- Inner Transom Plate grounding screw

32064ag02

Fig. 2 Tag and disconnect the engine ground wire

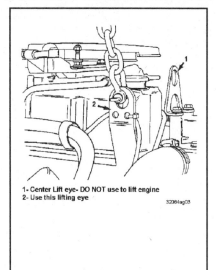

1- Center Lift eye- DO NOT use to lift engine
2- Use this lifting eye

32064ag03

Fig. 3 Attach an engine hoist at the lifting eyes

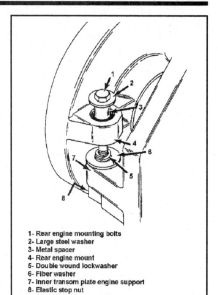

1- Rear engine mounting bolts
2- Large steel washer
3- Metal spacer
4- Rear engine mount
5- Double wound lockwasher
6- Fiber washer
7- Inner transom plate engine support
8- Elastic stop nut

Fig. 4 Details of the rear engine mount

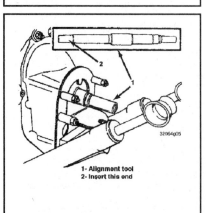

1- Alignment tool
2- Insert this end

32064g05

Fig. 5 Slide and alignment tool through the gimbal bearing...

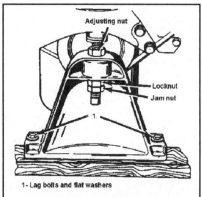

Adjusting nut

Locknut
Jam nut

1- Lag bolts and flat washers

Fig. 6 ...and then adjust the front mount so that the tool slides freely between the gimbal housing and the coupler

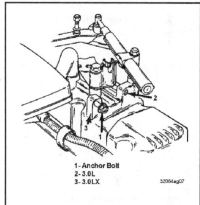

1- Anchor Bolt
2- 3.0L
3- 3.0LX

32064ag07

Fig. 7 There are two anchor studs for the throttle cable (1992-97)

25. Loosen the jam and lock nuts on the front mount(s) and then turn the adjusting nut (top of the mount) so that the mount base sits true on the stringer. Check that the alignment tool still slides freely between the engine and gimbal housing. Install the mount base lag bolts and tighten them securely. Tighten the adjusting nut and the jam nut.

26. Recheck that the alignment tool still slides freely and then remove it. Remove the engine hoist.

27. Reconnect the exhaust bellows and tighten the clamps securely.

28. Reconnect the water inlet hose and tighten the hose clamp.

29. Carefully, and quickly, connect the power steering lines. Tighten the large fitting to 20-25 ft. lbs. (27-34 Nm) and the small fitting to 96-108 inch lbs.

(11-12 Nm). Don't forget to bleed the system when you are finished with the installation.

30. Reconnect the trim position sender leads, the engine ground wire, the battery cables and all other wires, lines of hoses that were disconnected during removal.

31. Unplug the fuel line and pump fitting and reconnect them.

32. Connect the MerCathode lines to the controller and coat them with liquid neoprene. If you forgot to tag the wires as we suggested, refer to the illustration for proper hook-up.

33. Install and adjust the throttle cable. Make sure that the barrel positions the cable so that it is routed BELOW the anchor stud or else it will interfere with the shift cables. There are two anchor studs, one for the 3.0L and one for the 3.0LX. Please refer to the Fuel Systems section later in this manual for further details.

34. Install and adjust the shift cables. Tighten the nuts until they bottom out on the washer and then back them out 1/2 turn. Please refer to the Stern Drive section later in this manual for further details.

35. Check and refill all fluids and go have fun!

Cylinder Head Cover

MODERATE

REMOVAL & INSTALLATION

1. Open or remove the engine compartment hatch. Disconnect the negative battery cable.

2. Loosen the clamp and remove the crankcase ventilation hose at the

---

Mercathode
A + - R

① ② ③ ④

1- ORANGE wire- from electrode on transom assy
2- RED/PURPLE wire- connect other end to Pos (+) battery
3- BLACK wire- from engine harness
4- BROWN Wire- from electrode on transom assy

32064ag08

Fig. 8 Proper MerCathode controller hook-up

cover. Carefully move it out of the way.

3. Tag and disconnect the shift cut-out switch leads at the terminal block.

4. If your engine has a spark plug wire retainer attached to the cover, unclip the wires or remove the retainer.

5. Remove the fuel line at the pump and plug it, and the fitting, to avoid spills and system contamination. Have some rags nearby, as there will be some seepage either way. Move the line out of the way of the cover.

6. Loosen the cover mounting bolts and lift off the cylinder head cover. Take note of any harness or hose retainers and clips that might be attached to certain of the mounting bolts; you need to make sure they go back in the same place.

To install:

7. Clean the cylinder head and cover mounting surfaces of any residual gasket material with a scraper or putty knife.

8. Position a new gasket on the cylinder head and then position the cover (don't forget the J-clips!). Tighten the mounting bolts to 40 inch lbs. (4.5 Nm). Make sure any retainers or clips that were removed are back in their original positions.

9. Reconnect the fuel line to the pump. Check for leaks now, and after you restart the engine.

10. Connect the crankcase ventilation hose and the cut-out switch leads. Check that there were no ther wires or hoses you may have repositioned in order to gain access to the cover.

11. Connect the battery cables.

Rocker Arms and Push Rods

REMOVAL & INSTALLATION

② MODERATE

◆ **See Figure 9**

1. Open or remove the engine hatch cover and disconnect the negative battery cable. Remove the cylinder head cover as detailed previously.

2. Bring the piston in the No. 1 cylinder to TDC. If servicing only one arm, bring the piston in that cylinder to TDC. The No. 1 cylinder on 3.0L engines is the first cylinder at the front of the engine.

3. Loosen and remove the rocker arm nuts and lift out the balls. Lift the arm itself off of the mounting stud and pull out the pushrod. It is very important to keep each cylinders component parts together as an assembly. We suggest drilling a set of holes in a 2x4 and positioning the pieces in the holes.

To install:

4. Clean and inspect the rocker assemblies.

5. Coat all bearing surfaces of the rocker assembly with engine oil.

6. Slide the push rods into their holes. Make sure that each rod seats in its socket on the lifter.

7. Position the rocker arm over the stud so that the cupped side rides on the push rod. Slide the ball over the stud, install the nut and tighten it securely.

8. Adjust the valves and install the cylinder head cover.

9. Connect the battery cable and check the idle speed.

VALVE ADJUSTMENT

② MODERATE

◆ **See Figure 10**

These engines utilize hydraulic valve lifters, although there is no need for periodic valve adjustment, it is necessary to perform a preliminary adjustment after any work on the valvetrain/rocker assembly. All adjustment should be undertaken while the lifter is on the base circle of the camshaft lobe for that particular cylinder. This means the opposite side of the pointy part of each lobe.

1. Rotate the crankshaft, or bump the engine with the starter until the No. 1 cylinder is at TDC. Note that the notch or mark on the damper pulley will be lined up with the **0** mark on the timing scale. Be careful here though, this could mean that either the No. 1 or the No. 4 piston is at TDC. Place your hand on the No. 1 cylinder's valve and check that it does not move as

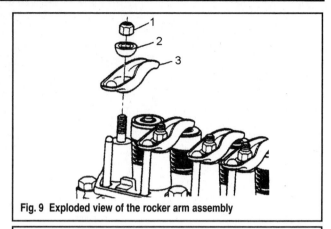

Fig. 9 Exploded view of the rocker arm assembly

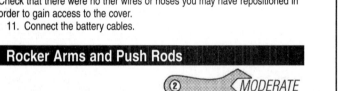

Fig. 10 Wiggle the push rod slowly while tightening the rocker adjusting nut

the mark on the pulley is approaching the **0** mark on the tab. If it does not move, you're ready to proceed; if it does move, you are on the No. 4 cylinder and need to rotate the engine an additional full turn. This is important so make sure you've gotten it right!

2. Now that the No. 1 cylinder is at TDC, you can adjust the following valves:
- No. 1 cylinder: intake and exhaust
- No. 2 cylinder: intake
- No. 3 cylinder: exhaust
- No. 4 cylinder: intake

3. Loosen the adjusting nut on the rocker until you can feel lash (play in the push rod) and then tighten the nut until the lash has been removed. Carefully jiggle the push rod while tightening the nut until it won't move any-more—this is zero lash. Tighten the nut an additional full turn to set the lifter and then you're done; 1998-2001 engines require that the nut be tightened and additional 3/4 turn from zero lash. Perform this procedure on each of the valves listed above.

4. Slowly rotate the engine an additional full turn and this will bring the No. 4 piston to TDC. The pulley notch/mark should once again be in line with the **0** on the timing tab. You can now adjust the remaining valves:
- No. 2 cylinder: exhaust
- No. 3 cylinder: intake
- No. 4 cylinder: exhaust

Combination Manifold

These engines incorporate the intake and exhaust manifolds into one unit called a combination manifold. It is serviced as a unit.

REMOVAL & INSTALLATION

② MODERATE

3.0L Engines Only

1. Open or remove the engine compartment hatch. Disconnect the negative battery cable.

2. Drain all water from the engine, manifold and exhaust elbow as detailed in the Cooling System section of this manual.

3. Remove the flame arrestor and set it aside.

4. Disconnect the throttle cable from the carburetor, remove the anchor bolt and position the cable out of the way. Mark which anchor stud the cable was attached to for installation.

5. Disconnect the fuel line at the carburetor and fuel pump. Plug the pump inlet fitting.

6. Remove the crankcase ventilation hose. Tag and disconnect any vacuum or electrical lines and then remove the carburetor.

7. Disconnect the shift cables.

8. Disconnect the water inlet line.

9. Tag and disconnect any wires or hoses that may interfere with manifold removal.

10. Loosen the clamps and remove the exhaust pipe at the elbow bellows.

11. Loosen and remove the manifold mounting bolts from the center outward and remove the manifold from the cylinder head—you may have to provide a little friendly persuasion.

■ **You may find that removing the alternator will allow for easier access to some of the manifold mounting nuts.**

12. Remove the exhaust elbow

To install:

13. Carefully clean all residual gasket material from the head, manifold and elbow mating surfaces with a scraper or putty knife. Inspect all gasket surfaces for scratches, cuts or other imperfections.

14. Position a new gasket on the cylinder head and install the manifold; making sure that everything is aligned properly. Tighten all nuts until they are just tight and then torque them to 23 ft. lbs. (33 Nm), starting in the center and working your way out to the ends of the manifold. The outer bolts on 1998-2001 engines should only be tightened to 17 ft. lbs. (25 Nm).

15. Position a new gasket on the manifold, making sure that the indents line up and then install the elbow. Tighten the mounting bolts to 25 ft. lbs. (34 Nm). Connect the exhaust pipe/bellows and tighten the clamps securely.

16. Connect the water inlet line and the shift cables.

17. Install the carburetor and reconnect the throttle cable, choke wire and any vacuum lines. Remember which anchor stud the throttle was attached to.

18. Connect the fuel line to the carb and the fuel pump (remember to unplug the pump fitting) and then install the flame arrestor.

19. Make sure that any miscellaneous lines or hoses that you may have moved or disconnected during removal are reconnected and routed properly.

20. Fill the system with water, connect the battery cable and start the engine. When the engine reaches normal operating temperature, turn it off and re-torque the manifold bolts.

Oil Pan

REMOVAL & INSTALLATION

◆ **See Figure 11**

■ **More times than not this procedure will require the removal of the engine. Your boat and its unique engine installation will determine this, but the procedure is almost always easier with the engine removed from the boat.**

1. Remove the engine as previously detailed in this section.

2. If you haven't already drained the engine oil, do it now. Make sure you have a container and lots of rags available.

3. Remove the starter motor as detailed in the Electrical section of this manual.

4. Loosen and remove the oil pan retaining bolts and nuts, starting with the center bolts and working out toward the pan ends. Lightly tap the pan with a rubber mallet to break the seal and then lift it off the cylinder block. If your engine stand will allow for rotating the engine, you'll find that this will be easier with the pan facing up.

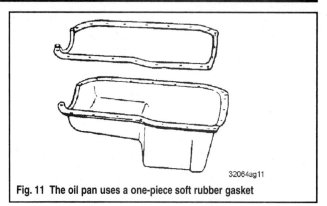

Fig. 11 The oil pan uses a one-piece soft rubber gasket

To install:

5. Clean the pan mating surfaces of any residual gasket material with a scraper or putty knife. Make sure that no old gasket material has been pressed into the retaining bolt holes in the pan, block or front cover. Clean the pan itself thoroughly with solvent.

6. Position a new pan gasket onto the pan being very careful to line up all the holes—do not use RTV sealant with this gasket.

7. Move the pan and gasket onto the block. It is very important that you ensure all the holes line up correctly; sometimes a few bolts inserted through the pan and gasket will help the gasket stay in place.

8. Install all bolts and nuts finger tight and then tighten them to 80 inch lbs. (9 Nm), except the pan-to-front cover bolts which should be tightened to 45 inch lbs. (5 Nm). On 1998-2001 models, tighten the 1/4-20 bolts to 80 inch lbs. (9 Nm) and the 5/16-18 bolts to 165 inch lbs. (19 Nm). Remember to start with the center bolts and work outward toward the ends of the pan.

9. Install the starter motor and then install the engine (if removed). Run the engine up to normal operating temperature, shut it off and check the pan for any leaks.

Oil Pump

The two-piece oil pump utilizes two pump gears and a pressure regulator valve. A baffled pick-up tube is press-fit into the body of the pump.

REMOVAL & INSTALLATION

◆ **See Figure 12**

1. Remove the oil pan as previously detailed. Remember that you probably need to remove the engine for this procedure.

2. Loosen and remove the pump pick-up tube support bracket bolt. The tube is pressed into the pump housing and should not be removed unless replacement is necessary.

3. Loosen and remove the two pump mounting bolts and lift off the pump assembly.

4. Check that the pump and block mating surfaces are clean and then position the pump over the block so that the pump drive shafts are aligned with the distributor tang. Make sure that the flange covers the alignment bushing. Do not use a gasket or RTV sealant.

5. Tighten the pump mounting bolts to 120 inch lbs. (14 Nm). Position the pick-up tube bracket and tighten the bolt to 60 inch lbs. (7 Nm).

6. Install the oil pan and engine.

Torsional Damper

REMOVAL & INSTALLATION

◆ **See Figures 13 and 14**

1. If the engine is in the boat, install an engine hoist and tighten the chain so that the engine's weight is removed from the front engine mount.

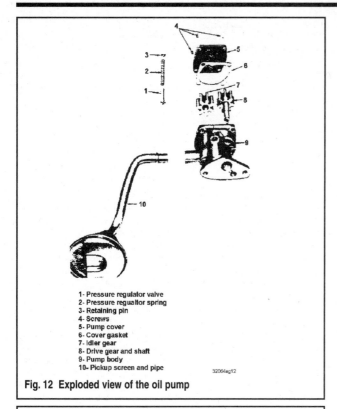

1- Pressure regulator valve
2- Pressure regualtor spring
3- Retaining pin
4- Screws
5- Pump cover
6- Cover gasket
7- Idler gear
8- Drive gear and shaft
9- Pump body
10- Pickup screen and pipe

32064ag12

Fig. 12　Exploded view of the oil pump

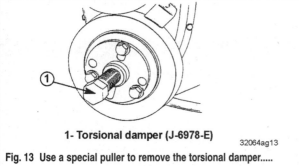

1- Torsional damper (J-6978-E)

32064ag13

Fig. 13　Use a special puller to remove the torsional damper.....

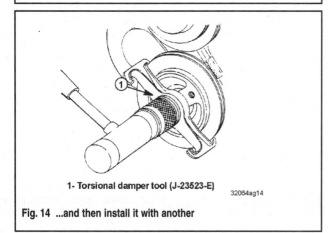

1- Torsional damper tool (J-23523-E)

32064ag14

Fig. 14　...and then install it with another

Do not use the lifting eye attached to the thermostat housing. Remove the front engine mount.

2. Remove the drive or serpentine belt as detailed in Section Three.

3. Remove the damper retaining bolt and install special tool J-6978-E onto the damper. Tighten the tool press bolt and remove the damper. MerCruiser suggests that you do not use a conventional gear puller for this procedure.

To install:

4. Coat the front cover oil seal lip with clean engine oil and then install the damper with a proper installation tool. Be sure that you tread the tool into the crankshaft at least 1/2 inch to protest the threads. In a pinch you can use a block of wood and a plastic mallet, but be careful that the pulley does not shift on its mountings while you're hammering.

5. Install the retaining bolt and tighten it to 50 ft. lbs. (68 Nm) on 1992-97 engines; 70 ft. lbs. (95 Nm) on 1998-2001 engines. Tighten the drive pulley to 35 ft. lbs (48 Nm).

6. Install the drive/serpentine belt and make sure that it is adjusted properly.

7. Install the front mount and unhook the engine hoist.

Front Cover, Oil Seal, Camshaft and Gear

OEM

REMOVAL & INSTALLATION　③ DIFFICULT

◆ **See Figures 15, 16, 17, 18 and 19**

■ **This procedure may require engine removal, depending upon your particular boat. If necessary, remove the engine as detailed previously in this section.**

1. Open the drain valves and drain the coolant from the block and exhaust manifold. Loosen the alternator and power steering brackets (or the idler pulley) to provide slack, and then remove the drive or serpentine belts. Remove the water circulation pump.

2. Secure the engine with a hoist and remove the front engine mount.

3. Remove the cylinder head cover and gasket as detailed previously in this section.

4. Loosen the valve rocker arm nuts until the pivot rocker arm clears the pushrod.

5. Mark the position of the distributor rotor to the distributor housing and remove the distributor.

6. Remove the ignition coil and side cover gasket. Take time to set up a system to keep the push rods and valve lifters in order, to ensure each will be installed back into the exact location from which it was removed. Withdraw each push rod and valve lifter in order.

7. Remove the bolts securing the alternator pulley, power steering pulley, and the harmonic balancer to the torsional damper.

8. Remove the torsional damper as detailed previously in this section.

9. Remove the two screws through the oil pan into the front cover. Loosen the remaining cover bolts and then pull the cover forward just a bit.

10. Use a sharp knife and cut the oil pan front seal flush with the cylinder block at both sides of the cover, as shown in the accompanying illustration.

11. Remove the front cover and attached portion of the oil pan front seal. Remove the front cover gasket. Clean all gasket material from the cover and block mating surfaces.

12. Pry the oil seal from the timing gear cover with a drift or prybar.

13. Rotate the camshaft gear until the holes in the gear are aligned with the thrust plate screws. Remove the two screws. Carefully withdraw the camshaft gear and camshaft by pulling the gear straight forward and the shaft out of the block.

14. Check the gear and thrust plate end-play. This clearance should be 0.003-0.008 in. (0.08-0.2032mm) max. If the decision is made to replace the camshaft gear, or the thrust plate, the gear must be pressed from the shaft. Gear removal from the camshaft requires the use of camshaft gear removal tool P/N J-791. Place the end of the removal tool onto the table of an arbor press, and then press the shaft free of the gear.

15. The thrust plate must be positioned to prevent the woodruff key in the shaft from damaging the shaft when the shaft is pressed out of the gear. Also, be sure to support the end of the gear or the gear will be seriously damaged.

16. To assemble the camshaft parts, first firmly support the camshaft at the back of the front journal in an arbor press. Next, place the gear spacer ring and the thrust plate over the end of the shaft, and install the Woodruff key in the shaft keyway. Install the camshaft gear and press it onto the shaft until it bottoms against the gear spacer ring.

17. Insert an alignment tool (J-23042) through the oil seal hole. Coat a new gasket with Perfect Seal and install the front cover. Tighten the bolts to 30 ft. lbs. (3.5 Nm).

18. Installation of the remaining components is in the reverse order of removal.

Cylinder Head

REMOVAL & INSTALLATION

③ ◁DIFFICULT

◆ See Figures 19 and 20

1. Drain the water from the cylinder block and manifold.

2. Remove the fuel line support brackets. Disconnect the fuel line at the carburetor and fuel pump, plug the fitting holes and remove the line.

3. Remove the combination manifold as previously detailed in this section; you can leave the carburetor/throttle body attached if you like.

4. Disconnect the coolant hoses at the thermostat housing and move them out of the way. Have some rags available, as there will still be some coolant/water in the hoses.

5. Tag and disconnect the temperature sending unit lead at the thermostat housing, loosen the mounting bolts and then remove the housing and thermostat.

6. Tag and disconnect all wires at the ignition coil and then remove the mounting bracket bolt and lift off the coil.

7. Tag and disconnect the spark plug wires at the plugs; move them out of the way. Although not necessary, it's a good idea to remove the plugs themselves also.

8. Remove the circuit breaker bracket and then unbolt and remove the engine lifting bracket.

9. Remove the cylinder head cover and rocker assemblies as detailed previously in this section.

10. Loosen the cylinder head bolts, from the center bolts and working out to the ends of the head and then carefully lift the head off the block. You may need to persuade it with a rubber mallet—be careful! Set the head down carefully; do not sit it on cement.

To install:

11. Carefully, and thoroughly, remove all residual head gasket material from the cylinder head and block mating surfaces with a scraper or putty knife. Check that the mating surfaces are free of any nicks or cracks. Make sure there is no dirt or old gasket material in any of the bolt holes. Refer to the Engine Rebuilding section found later in this manual for complete details on inspection and refurbishing procedures.

12. Apply Perfect Seal to both sides of a new gasket and position the gasket over the cylinder block dowel pins. DO NOT use automotive-type steel gaskets.

13. Position the cylinder head over the dowels in the block. Coat the threads of the head bolts with Perfect Seal and install them finger tight. It never hurts to use new bolts, although its not necessary. Tighten the bolts, a little at a time, in the sequence illustrated, until the proper tightening torque is achieved.

14. Install rocker assemblies and the cylinder head cover.

15. Install the circuit breaker and engine lifting brackets. Install the spark plugs if they were removed and then connect the plug wires.

16. Install the coil and reconnect all the electrical leads.

17. Install the thermostat housing, the coolant hoses and the temperature sending lead.

18. Install the manifold and connect the fuel line. Don't forget to remove the fitting plugs.

19. Add coolant/water, connect the battery and check the oil. Start the engine and run it for a while to ensure that everything is operating properly. Keep an eye on the temperature gauge.

20. Re-tighten the cylinder head bolts after 20 hours of operation.

Fig. 15 Slide a sharp knife between the front cover and the cylinder block to cut the oil pan gasket

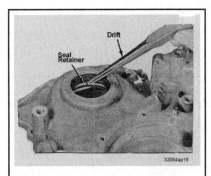

Fig. 16 Remove the front oil seal with a drift

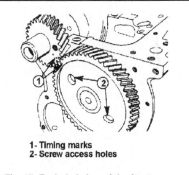

1- Timing marks
2- Screw access holes

Fig. 17 Exploded view of the front cover

1- Timing marks
2- Screw access holes

Fig. 18 Line up the timing gear marks and the access holes should rest over the thrust plate screws (engine upside down)

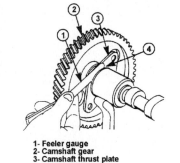

1- Feeler gauge
2- Camshaft gear
3- Camshaft thrust plate
4- Take measurement here

Fig. 19 Checking the camshaft endplay

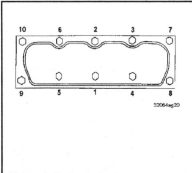

Fig. 20 Cylinder head tightening sequence

Rear Main Seal

It is not necessary to remove the engine or rear main bearing cap when removing the one-piece oil seal on these engines although you may find it easier to do just that.

REMOVAL & INSTALLATION

◆ See Figures 21 and 22

1. Remove the flywheel housing and cover as detailed later in this section.
2. Remove the engine coupler and flywheel from the engine as detailed later in this section.
3. Insert a small prybar into one of the three slots in the edge of the seal retainer and slowly pry the seal out of the retainer.
4. Thoroughly clean the retainer surface.
5. Apply Perfect Seal to the inner mating surface of the seal retainer. Spread a small amount of grease around the outside edge of a new seal and position it over its slot in the retainer.
6. Position a seal driver (J-26817-A) over the seal and strike it with a mallet until the seal is fully seated in its bore.
7. Install the flywheel and engine coupler. Install the cover and flywheel housing.

Flywheel

REMOVAL & INSTALLATION

1. Remove the engine from the boat as detailed previously in this section.
2. Loosen the retaining bolts and lift off the flywheel housing (bell housing in automotive parlance).
3. Loosen the three engine coupler mounting bolts and then remove the coupler. Don't lose the rubber bumpers
4. Loosen the six flywheel mounting bolts gradually and as you would the lug nuts on your car or truck—that is, in a diagonal star pattern.
5. Thoroughly clean the flywheel mating surface and check it for any nicks, cracks or gouges. Check for any broken teeth.
6. Install the flywheel over the dowel on the crankshaft and tighten the mounting bolts to 65 ft. lbs. (88 Nm) on 1992-97 engines; 75 ft. lbs. (100 Nm) on 1998-2001 engines. Once again use the star pattern while tightening the bolts.
7. Insert the rubber bumpers into the coupler. Position the engine coupler and tighten the mounting bolts to 40 ft. lbs. (54 Nm) on 1990-97 engines; 35 ft. lbs. (48 Nm) on later engines. If you are re-using the old bolts, make sure you coat them with Loctite® 271. Lubricate the coupler splines with Quicksilver 2-4-C Marine Lubricant.
8. Install the flywheel housing and tighten the bolts to 30 ft. lbs. (41 Nm).
9. Install the engine.

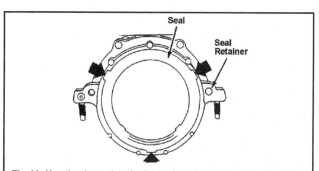

Fig. 22 Use the three slots in the seal retainer when removing the rear main seal

Fig. 21 Remove the cylinder head carefully

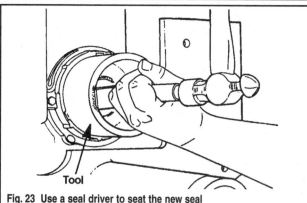

Fig. 23 Use a seal driver to seat the new seal

TORQUE SPECIFICATIONS

Component		ft. lbs.	Nm
Camshaft sprocket bolts		80 ①	9
Connecting rod cap nuts		45	61
Cylinder head bolts		90 ②	122
Distributor clamp bolt		20	27
Engine coupler-to-flywheel		35	47
Flywheel			
	Housing-to-block	21	28
	To crankshaft bolts	65	88
Front cover bolts		30 ①	3.4
Front mount-to-block		21	28
Main bearing cap bolts		65	88
Manifold-to-cylinder head (1992-97)		23 ③	31
Manifold-to-cylinder head (1998-2001)			
	Center	23 ③	31
	Outer	17	25
Oil Pan			
	Crankcase bolts	80 ①	9
	Front cover bolts	45 ①	5
	Studs to oil seal or crankcase	15 ①	1.7
Oil pump			
	Cover	72 ①	8
	Block	10	14
	Pick-up	60 ①	7
Push rod cover bolts		40 ①	4.5
Rocker arm cover bolts		40 ①	4.5
Rocker arm nuts		④	
Rear crank oil seal retainer nuts			
	1992-97	135 ①	15
	1998-2001	14	17
Spark plugs		22	30
Starter motor		37	50
Timing gear cover		72 ①	8
Torsional damper bolt (1992-97)		50	68
Torsional damper bolt (1998-2001)		70	95
Water pump		15	20

① Specification is inch lbs.

② Retighten after 20 hours of operation

③ Retighten after the initial engine start

④ Adjust rockers to zero lash and tighten bolt
and additional full turn (1992-97); or an
additional 3/4 turn (1998-2001)

32064ac01

ENGINE SPECIFICATIONS

Component			Standard (in.) ④	Metric (mm) ④
Cylinder Bore				
Diameter			3.9995-4.0025	101.588-101.790
Out-of-round				
	Production		0.0005 Max	0.012 Max
	Service Limit		0.002 Max	0.05 Max
Taper				
	Production			
		Thrust side	0.0005 Max	0.012 Max
		Relief side	0.0005 Max	0.012 Max
	Service Limit		0.001 Max	0.02 Max
Piston				
Clearance				
	Production		0.0025-0.0035	0.064-0.088
	Service Limit		0.0035	0.08
Compression Rings Groove clearance				
	Production			
		Top	0.0012-0.0029	0.030-0.073
		2nd	0.0012-0.0029	0.030-0.073
	Service Limit		0.003 Max	0.09 Max
Gap				
	Production			
		Top	0.010-0.020	0.254-0.508
		2nd	0.017-0.025 ⑤	0.432-0.635 ⑤
	Service Limit		0.035 Max	0.88 Max
Oil Ring				
	Groove clearance			
		Production	0.001-0.006	0.026-0.152
		Service Limit	0.007 Max	0.17 Max
	Gap			
		Production	0.010-0.030	0.254-0.762
		Service Limit	0.040 Max	1.01 Max
Piston pin				
	Diameter		0.9270-0.9271	23.546-23.550
	Clearance			
		Production	0.0003-0.0006	0.008-0.016
		Service Limit	0.001 Max	0.02 Max
	Interference fit		0.0008-0.0019	0.020-0.050
Crankshaft				
Main Journal				
	Diameter		2.2979-2.2994	58.367-58.404
	Taper			
		Production	0.0002 Max	0.005 Max
		Service Limit	0.001 Max	0.02 Max

32064c02

ENGINE SPECIFICATIONS
(cont'd)

Component			Standard (in.) ④	Metric (mm) ④
Crankshaft (cont'd)	Out-of-round			
		Production	0.0002 Max	0.005 Max
		Service Limit	0.001 Max	0.02 Max
Main Bearing Clearance				
	Production			
		#1 thru 4	0.001-0.0024	0.025-0.060
		#5	0.0016-0.0035	0.041-0.088
	Service Limit			
		#1 thru 4	0.001-0.0025	0.03-0.06
		#5	0.002-0.0035	0.05-0.08
Crankshaft end play			0.002-0.006	0.05-0.15
Crankpin				
	Diameter		2.0980-2.0995	53.289-53.327
	Taper			
		Production	0.0003 Max	0.007 Max
		Service Limit	0.001 Max	0.02 Max
	Out-of-round			
		Production	0.0002 Max	0.051 Max
		Service Limit	0.001 Max	0.02 Max
Connecting rod				
	Bearing clearance			
		Production	0.0017-0.0027	0.044-0.686
		Service Limit	0.003 Max	0.07 Max
	Side clearance		0.006-0.017	0.153-0.4318
Camshaft				
Lobe Lift				
	Intake		0.2529 Max	6.425 Max
	Exhaust		0.2529 Max	6.425 Max
Journal diameter			1.8677-1.8697	47.440-47.490
End play			0.003-0.008	0.08-0.2032
Cylinder Head				
Gasket surface flatness			①	①
Valve system				
Lifter			Hydraulic	Hydraulic
Rocker arm ratio			1.75:1	
Tappet gap colapse			②	②
Face Angle				
	Intake		45 deg.	
	Exhaust		45 deg.	
Seat Angle			46 deg.	
Seat runout			0.002 Max	0.05 Max

32064c04

ENGINE SPECIFICATIONS
(cont'd)

Component			Standard (in.) ④	Metric (mm) ④
Valve System (cont'd)				
Seat width				
	Intake		0.0625	1.6
	Exhaust		0.07	1.8
Stem clearance				
	Production			
		Intake	0.0010-0.0027	0.025-0.069
		Exhaust	0.0007-0.0027	0.018-0.0686
	Service Limit			
		Intake	0.003 Max	0.09 Max
		Exhaust	0.004 Max	0.11 Max
Spring w/internal damper removed	Free length		0.063	52
	Pressure			
		Valve open	208-222 ft. lbs. @ 1.22 in.	282-300 Nm @ 32mm
		Valve close	100-110 ft. lbs.. @ 1.610 in.	136-149 Nm @ 41mm
	Installed height ③			
		Intake	1.66	42
		Exhaust	1.66	42
Internal damper			None	None

32064c04

① 0.003 in. (0.07mm) across any six inches; or 0.007 in. (0.15mm) overall

② One half to one full turn tighter from zero lash

③ +/- 0.031 in. (0.8mm)

④ Unless otherwise noted

⑤ 1998-2001:0.010-0.020 in. (0.254-0.762mm)

5

ENGINE MECHANICAL V6 & V8

ENGINE MECHANICAL

General Information

NEVER, NEVER attempt to use standard automotive parts when replacing anything on your engine. Due to the uniqueness of the environment in which they are operated in, and the levels at which they are operated at, marine engines require different versions of the same part; even if they look the same. Stock and most aftermarket automotive parts will not hold up for prolonged periods of time under such conditions. Automotive parts may appear identical to marine parts, but be assured, Mercury marine parts are specially manufactured to meet Mercury marine specifications. Most marine items are super heavy-duty units or are made from special metal alloy to combat against a corrosive salt water atmosphere.

Mercury marine electrical and ignition parts are extremely critical. In the United States, all electrical and ignition parts manufactured for marine application must conform to stringent U.S. Coast Guard requirements for spark or flame suppression. A spark from a non-marine cranking motor solenoid could ignite an explosive atmosphere of gasoline vapors in an enclosed engine compartment.

4.3L V6 ENGINES

The MerCruiser 4.3L, 262 cubic inch displacement V6 engine is manufactured by GMC. This engine is used in numerous models known as the 4.3L, 4.3LX, 4.3LH and in 1997, the 262 Magnum.

The 4.3L model is equipped with a 2-barrel carburetor, while the 4.3LX and LH models are equipped with a 4-barrel carburetor. Throttle body fuel injection was introduced on the 262 Magnum in 1997 and continues on the 4.3L EFI models through current year.

The lubrication system and component locations on this engine is virtually identical to the larger V8 engines, except for having only three cylinders in each bank.

A balance shaft is mounted above the camshaft on 1993 and later models and extends the entire length of the block and is supported on each end by a bearing. The balance shaft is driven by gears on the end of the camshaft and equalizes the dynamic forces known as harmonic vibrations minimizing engine vibration during routine operation of a V6 engine. Cylinder numbering and firing order is identified in the illustrations at the end of Section Three.

All 4.3L engines are left hand (counterclockwise) rotation when viewed from the stern of the boat. This does not necessarily indicate that your prop rotation is the same- always check them both!

V8 Engines

The MerCruiser 5.0L, 305 cubic inch, 5.7L, 350 cubic inch and 6.2L, 377 cubic inch displacement (small block) V8 engines are manufactured by GMC. These engines are used in the following configurations:
- 5.0L
- 5.0LX
- 5.0L EFI
- 5.7L
- 5.7L EFI
- 5.7LX
- 5.7LX EFI
- 350 MAGNUM
- 350 MAGNUM EFI or MPI
- 6.2L MPI
- 6.2L Black Scorpion

The MerCruiser 7.4L, 454 cubic inch, 8.1L 496 cubic inch and 8.2L, 502 cubic inch displacement (big block) V8 engines are also manufactured by GMC. These engines are used in the following configurations:
- 7.4L
- 7.4L EFI
- 7.4L MPI
- 7.4LX EFI
- 7.4LX MPI
- 8.1S/S HO
- 8.2L
- 8.2L MPI
- 454 Magnum/Magnum EFI/MP
- 496 Mag/Mag HO
- 502 Magnum/Magnum EFI/MP

The lubrication system is a force fed type where oil is supplied under full pressure to the crankshaft, main and connecting rod bearings, camshaft bearings and the valve lifters. Oil flow from the valve lifters is metered and pumped by the lifter through the hollow core pushrods to lubricate the rocker

arms and valve train. All other components are lubricated by gravity and splash methods.

The oil pump is mounted on the rear main bearing cap and is driven by an extension shaft from the distributor-driven by the camshaft. Oil is drawn into the pump through the oil pick-up tube and screen. Should the screen become clogged, a relief valve in the screen will open and allow oil to be drawn into the pump.

Once oil reaches the pump, the pump forces oil through the lubrication system. A spring-loaded relief valve in the oil pump limits the maximum pump output pressure.

The pressurized oil flows out the pump through a full-flow disposable oil filter cartridge. On engines equipped with an oil cooler, the oil flows through the filter, out to the oil cooler via hoses and then returns to the block. Should the oil filter and/or cooler become clogged, a by-pass valve will open allowing the pressurized oil to by-pass the filter and cooler.

Some of the oil is then routed to the No. 5 crankshaft main bearing, the remainder of the oil pressure is routed to the main oil gallery. The main oil gallery is located above the camshaft and runs the full length of the block. Oil from the main gallery is routed through individual passages to the camshaft bearings, No's 1, 2, 3, and 4, crankshaft main bearings and the lifter galley's on each side of the block.

Holes in the camshaft bearings and crankshaft main bearings align with the holes in the block for oil flow. Grooves in the bearings allow oil to flow between the bearing and the component.

Oil in the lifter galleries is forced into each hydraulic lifter through a hole in the side of the lifter. Oil flowing through the lifter must pass through a metering valve in each of the lifters. The metered volume of oil then flows up through the hollow push-rods to the valve rockers. A small hole in the rocker arm allows oil to lubricate the valve train bearing surfaces. All excess oil drains back to the oil pan through oil return holes in the cylinder head.

A baffle plate or "splash pan" mounted below the main bearing caps prevents excess oil being thrown off the crankshaft from aerating the oil in the oil pan.

The distributor shaft and gear is lubricated by oil in the starboard lifter gallery. The timing chain and gears are lubricated with oil flowing out the front of the No. 1 main bearing journal. The mechanical fuel pump and pushrod is lubricated with oil thrown off the camshaft eccentric.

All V8 engines are left hand (counterclockwise) rotation when viewed from the stern of the boat except NR and MR inboards which are right hand (clockwise). This does not necessarily indicate that your prop rotation is the same- always check them both!

Cylinder numbering and firing order is identified in the illustrations at the end of Section Three.

Engine Identification

ALL ENGINES EXC. 8.1L V8

◆ See Figure 1

All engines can be identified by a code number stamped into the starboard side of the front of the engine block, just below the cylinder head-the last two letters of the code designate the engine:
- CD: 1992-97 5.7L Bravo, 5.7L EFI, 350 Magnum (stern drive & inboard)
- FH: 1992-97 8.2L inboard
- FJ: 1992-97 502 Magnum/EFI
- HH: 1998 and later 8.2L inboard
- HJ: 502 Mag MPI
- JK: 1992-93 4.3L engines w/o a balance shaft
- LG: 1993-97 4.3L
- LH: 1993-97 4.3LX
- LJ: 1998-2001 4.3L
- LK: 1998-2001 4.3LH and 4.3L EFI
- MB: 1992-97 5.7L Alpha and 5.7 Competition Ski inboard w/LH rotation
- MD: 1992-97 5.7L Bravo, 5.7L EFI, 350 Magnum (stern drive & inboard)
- MH: 1998-2001 5.0L EFI, 5.7L/5.7L EFI (stern drive & ski), 350 Mag MPI (stern drive & ski,

Black Scorpion
- MHA: 6.2L MPI (stern drive) and 6.2 Black Scorpion
- MK: 1998-2001 5.7L/350 Mag MPI (inboard)
- MKA: 6.2L MPI (inboard)
- MR: 1992-97 5.7L Competition Ski inboard w/RH rotation

- NN: 1992-97 5.7L inboard w/LH rotation
- NR: 1992-97 5.7L inboard w/RH rotation
- UA: 1998 and later 454 Mag MPI (stern drive)
- UB: 1992-97 7.4L Bravo III
- UF: 1998 and later 454 Mag MPI (inboard)
- XA: 1992-97 454 Magnum/EFI (stern drive and inboard)
- XW: 7.4L/MPI (stern drive)
- XX: 1992-97 7.4L Bravo III
- XY: 7.4L inboard
- ZA: 1998-2001 5.0L
- ZC: 1992-97 5.0L
- ZD: 1992-97 5.0LX

This code is stamped on all original equipment engines and all partial replacement engines (but NOT on replacement cylinder blocks).

In the event that the engine serial number plates or stickers are missing from the engine, this is a good way to ensure that you know the exact engine in your vessel when ordering parts, etc.

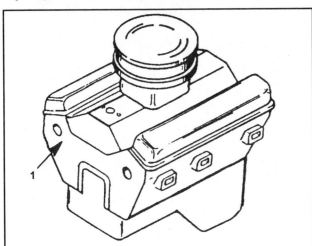

1- Location of GM engine code(front starboard side, near cylinder head mating surface)

32065g01

Fig. 1 The engine identification code can be found here on all engines except the 8.1L V8. The last two letters are the engine designation

8.1L V8 ENGINES

◆ **See Figure 2**

The engine code '8.1L' can be found stamped on the port side of the cylinder block

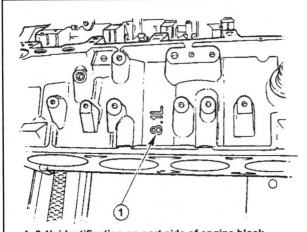

1- 8.1L identification on port side of engine block

32065g02

Fig. 2 The engine identification code can be found here on the 8.1L V8

OEM ③ DIFFICULT

REMOVAL & INSTALLATION

V6 Engines
V8 Engines (Stern Drive W/O Extension Driveshaft)

◆ **See Figures 3, 5, 6, 7, 8, 9, 10, 11, 12 and 13**

1. Remove the stern drive unit as detailed in Section Eight.
2. Open or remove the engine hatch cover.
3. Disconnect the battery cables (negative first) at the battery and then disconnect them from the engine block and starter.
4. Disconnect the fuel inlet line at the fuel pump and quickly plug it. Make sure you have rags handy as there will be some spillage.
5. Disconnect the throttle cable at the throttle lever on the carburetor/throttle body unit and position it out of the way.
6. Loosen the hose clamp on the engine wiring harness connector and unplug the instrument wiring harness. Label it and move it aside.
7. Tag and disconnect the two trim position wires at the sender unit.
8. Tag and disconnect the engine harness wires at the shift cut-out switch harness. When you pull back the sleeves, they should be Black and White/Green, although sometimes the W/G will be Gray.
9. On models equipped with the MerCathode system, tag and disconnect the wires at the controller.
10. Loosen the hose clamp and remove the water inlet line at the gimbal housing.
11. Loosen the hose clamps and disconnect the exhaust elbow bellows from the upper pipe. Remove them
12. Remove the shift cables at the shift plate and move them out of the way. Don't lose the hardware.
13. Carefully loosen and remove the two hydraulic lines at the power steering control valve on the transom. Plug the hose ends and the control valve fittings. Tag the lines and fittings and move them aside.
14. Tag and disconnect the engine ground wire at the stud.
15. Tag and disconnect any remaining lines, wires or hoses at the engine.
16. Attach a suitable engine hoist to the lifting eyes and take up any line slack until it is just taught.

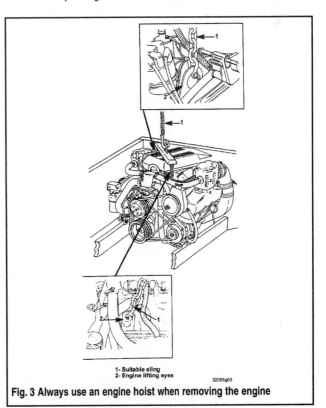

1- Suitable sling
2- Engine lifting eyes

32065g03

Fig. 3 Always use an engine hoist when removing the engine

Fig. 5 Removing the V8 engine

■ DO NOT use the front lifting eye attached to the thermostat housing at this point.

17. Locate the front engine mount(s) and remove the two lag bolts.
18. Locate the rear engine mount and remove the two mounting bolts.
19. Slowly and carefully, lift out the engine. Try not to hit the power steering control valve.

To install:

■ An engine alignment tool (#91-57797A3) is necessary to reinstall the engine. Even if the mounts have not been removed from the engine it is still a good idea to re-align the unit.

20. Check that the fiber washers on top of the transom plate mounting holes are still in position. If look worn or damaged in any way, replace them. Remember they are glued in place.
21. Install a double wound lockwasher into the fiber washer and transom plate holes.
22. Pop the rear engine mount locknuts into their slots on the bottom of the transom plate mounting holes.
23. Apply some spline grease to the engine coupler splines and then lubricate the exhaust bellows with soapy water- this will allow it to slip over the pipe much easier.
24. Lower the engine into the compartment so that it just rests on the washers and the transom plate. Make sure that the holes line up and DO NOT release the hoist tension yet. Make sure at this point that the exhaust

Fig. 6 Always install the engine on a stand after removal

pipes are in alignment.

25. Slide a large steel washer and a metal spacer onto the mounting bolt and then run them through the mount and transom plate. Tighten the bolts to 35-40 ft. lbs (47-54 Nm).
26. Lower the front of the engine until the front mount just rests on the stringers. Unhook the hoist and connect the chain to the center lifting eye on the thermostat housing.
27. Slide the alignment tool into the center of the gimbal housing assembly bearing and then into the engine coupler splines-you may have to pivot the bearing slightly. Use the hoist to raise (or lower) the front of the engine until the tool slides completely into the coupler (with no binding). DO NOT force the tool through the bearing under any circumstances. Slowly raise (or lower) the front of the engine until the tool slides freely
28. Loosen the jam and lock nuts on the front mount(s) and then turn the adjusting nut (top of the mount) so that the mount base sits true on the stringer. Check that the alignment tool still slides freely between the engine and gimbal housing. Install the mount base lag bolts and tighten them securely. Tighten the adjusting nut and the jam nut.
29. Recheck that the alignment tool still slides freely and then remove it. Remove the engine hoist.

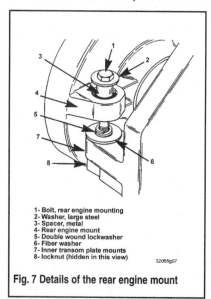

1- Bolt, rear engine mounting
2- Washer, large steel
3- Spacer, metal
4- Rear engine mount
5- Double wound lockwasher
6- Fiber washer
7- Inner transom plate mounts
8- locknut (hidden in this view)

32065g07

Fig. 7 Details of the rear engine mount

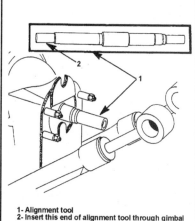

1- Alignment tool
2- Insert this end of alignment tool through gimbal housing assembly

32065g08

Fig. 8 Slide an alignment tool through the gimbal bearing...

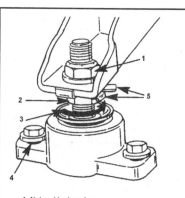

1- Nut and lockwasher
2- Adjustment nut
3- Turn adjustment nut in this direction (counterclockwise) to raise front of engine
4- Slotted hole toward front of engine
5- tab washer

32065g09

Fig. 9 ...and then adjust the front mount so that the tool slides freely between the gimbal housing and the coupler

30. Reconnect the exhaust bellows and tighten the clamps securely. Refer to the illustration for positioning.

31. Reconnect the water inlet hose and tighten the hose clamp.

32. Carefully, and quickly, connect the power steering lines. Tighten the large fitting to 20-25 ft. lbs. (27-34 Nm) and the small fitting to 96-108 inch lbs. (11-12 Nm). On later models, tighten all fittings to 23 ft. lbs. (31 Nm). Don't forget to bleed the system when you are finished with the installation.

33. Reconnect the trim position sender leads, the engine ground wire, the battery cables and all other wires, lines of hoses that were disconnected during removal.

34. Unplug the fuel line and pump fitting and reconnect them.

35. Connect the MerCathode lines to the controller and coat them with liquid neoprene. If you forgot to tag the wires as we suggested, refer to the illustration for proper hook-up.

36. Install and adjust the throttle cable. Make sure that the barrel positions the cable so that there are no kinks or sharp bends in the cable. Make sure that the cable does not come in contact with any other moving parts. Please refer to the Fuel Systems section later in this manual for further details.

37. Install and adjust the shift cables. Please refer to the Stern Drive section later in this manual for further details.

38. Check and refill all fluids and go have fun!

V8 Engines (Stern Drive W/Driveshaft Extension)

◆ See Figures 3, 10, 11, 12, 13, 14, 15 and 16

1. Open or remove the engine hatch cover.

2. Disconnect the battery cables (negative first) at the battery and then disconnect them from the engine block and starter.

3. Disconnect the fuel inlet line at the fuel pump and quickly plug it. Make sure you have rags handy as there will be some spillage.

4. Disconnect the throttle cable at the throttle lever on the unit and position it out of the way.

5. Loosen the hose clamp on the engine wiring harness connector and unplug the instrument wiring harness. Label it and move it aside.

6. Tag and disconnect the two trim position wires at the sender unit.

7. Tag and disconnect the engine harness wires at the shift cut-out switch harness. When you pull back the sleeves, they should be Black and White/Green, although sometimes the W/G will be Gray.

8. On models equipped with the MerCathode system, tag and disconnect the wires at the controller.

Fig. 11 Reconnecting the instrument harness

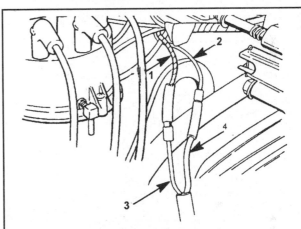

1- Brown/White (from engine harness)
2- Black (from engine harness)
3- Black (from transom assembly)
4- Black (from transom assembly)

32065g12

Fig. 12 Make sure the trim position sender leads are connected properly

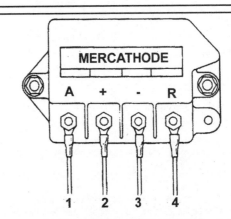

1- Orange wire- from electrode on transom assy
2- Red/Purple wire- connect (other end) to positive (+) battery terminal
3- Black wire- from engine harness
4- Brown wire- from electrode on transom assy

Fig. 13 Proper MerCathode controller hook-up

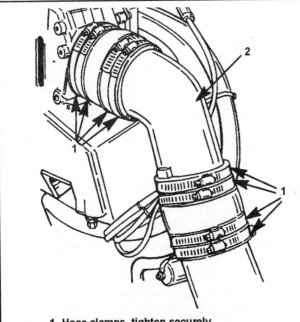

1- Hose clamps- tighten securely
2- Exhaust tube- Long tube, port side
          - Short tube, starboard side

Fig. 10 Proper hose clamp location when installing the exhaust bellows (exc. 8.1L V8)

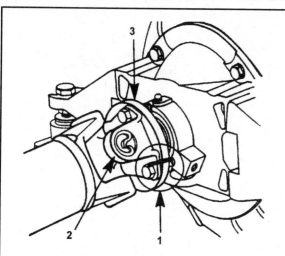

1- Matching marks on flange and drive shaft connection
2- Extension drive shaft U-Joint yoke
3- flange, output

Fig. 14 Matchmark the driveshaft and output shaft flanges before removal

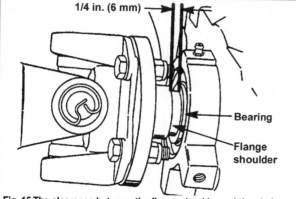

1/4 in. (6 mm)

Bearing

Flange shoulder

Fig. 15 The clearance between the flange shoulder and the shaft housing bearing should be checked here.

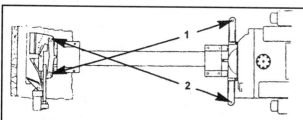

Fig. 16 Measure here and adjust the engine until they are equal

9. Loosen the hose clamp and remove the water inlet line at the engine.
10. Loosen the hose clamps and disconnect the exhaust elbow bellows from the upper pipe. Remove them.
11. Remove the shift cables at the shift plate and move them out of the way. Don't lose the hardware.
12. Carefully loosen and remove the two hydraulic lines at the power steering control valve on the transom. Plug the hose ends and the control valve fittings. Tag the lines and fittings and move them aside.
13. Tag and disconnect the engine ground wire at the stud.
14. Tag and disconnect any remaining lines, wires or hoses at the engine.
15. Loosen the seven driveshaft shield mounting bolts and lift out the upper and lower shields at the engine end of the shaft.
16. Paint matchmarks across the driveshaft U-joint yoke and engine output shaft flanges where they mate and then disconnect the driveshaft from the output shaft.

17. Attach a suitable engine hoist to the lifting eyes and take up any line slack until it is just taught.

■ **DO NOT use the front lifting eye attached to the thermostat housing at this point.**
18. Locate the front engine mount(s) and remove the two lag bolts.
19. Locate the rear engine mount and remove the two mounting bolts.

■ **Do not tamper with the engine mount adjustment bolts or you will need to check the engine/shaft alignment on installation.**

20. Slowly and carefully, lift out the engine. Try not to hit the power steering control valve.
To install:
21. Lower the engine into the compartment so that it just rests on the washers and the transom plate or stringers. Make sure that the holes line up and DO NOT release the hoist tension yet. Make sure at this point that the exhaust pipes are in alignment.
22. Lubricate the U-joints with 2-4-C Marine lubricant.
23. Reconnect the driveshaft to the engine output shaft and tighten the bolts to 50 ft. lbs. (68 Nm). Be certain that the alignment marks applied earlier are in complete alignment.
24. Release the tension on the engine hoist and then slide the engine slowly until you can obtain 1/4 (6mm) clearance between the flange shoulder and the extension shaft housing bearing. Measure the diagonal distance between the two points shown in the illustration; move the front or rear sides of the engine slightly until the two measurements are equal. Recheck the driveshaft clearance. If OK, move on. If not, repeat this procedure again until the clearance is correct and the measurements are equal.
25. Tighten the engine mount bolts securely.
26. Coat the driveshaft shield bolts with Loctite and install the lower shield. Tighten the three bolts to 30 ft. lbs. (41 Nm). Install the top shield and tighten the four bolts to 30 ft. lbs. (41 Nm).
27. Reconnect the exhaust bellows and tighten the clamps securely. Refer to the illustration for positioning.
28. Reconnect the water inlet hose and tighten the hose clamp securely.
29. Carefully, and quickly, connect the power steering lines. Tighten the large fitting to 20-25 ft. lbs. (27-34 Nm) and the small fitting to 96-108 inch lbs. (11-12 Nm). On later models, tighten all fittings to 23 ft. lbs. (31 Nm). Don't forget to bleed the system when you are finished with the installation.
30. Reconnect the trim position sender leads, the engine ground wire, the battery cables and all other wires, lines of hoses that were disconnected during removal.
31. Unplug the fuel line and pump fitting and reconnect them.
32. Connect the MerCathode lines to the controller and coat them with liquid neoprene. If you forgot to tag the wires as we suggested, refer to the illustration for proper hook-up.
33. Install and adjust the throttle cable. Make sure that the barrel positions the cable so that there are no kinks or sharp bends in the cable. Make sure that the cable does not come in contact with any other moving parts. Please refer to the Fuel Systems section later in this manual for further details.
34. Install and adjust the shift cables. Please refer to the Stern Drive section later in this manual for further details.
35. Check and refill all fluids and go have fun!

V8 Engines (Inboard)

◆ **See Figures 3, 13, 17, 18, 19, 20, 21, 22, 23 and 24**

1. Open or remove the engine hatch cover.
2. Disconnect the battery cables (negative first) at the battery and then disconnect them from the engine block and starter.
3. Disconnect the fuel inlet line at the fuel pump and quickly plug it. Make sure you have rags handy as there will be some spillage.
4. Disconnect the throttle cable at the throttle lever on the unit and position it out of the way.
5. Loosen the hose clamp on the engine wiring harness connector and unplug the instrument wiring harness. Label it and move it aside.
6. Loosen the hose clamps and disconnect the exhaust elbow bellows from the upper pipe. Remove them
7. Remove the shift cables at the shift plate and move them out of the way. Don't lose the hardware.
8. Disconnect and remove the seawater inlet hose at the engine.
9. Tag and disconnect the engine ground wire at the stud.
10. Tag and disconnect any remaining lines, wires or hoses at the engine.

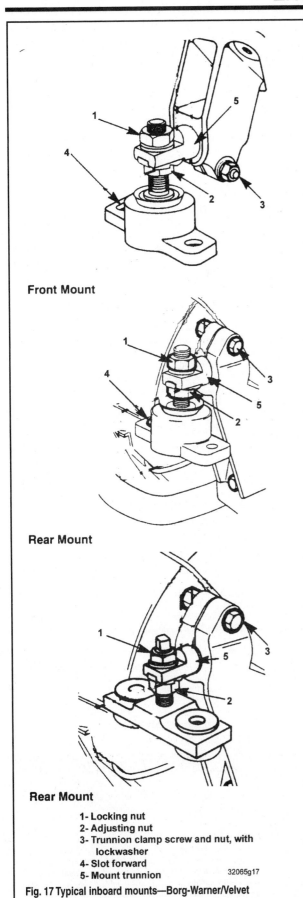

Front Mount

Rear Mount

Rear Mount

1- Locking nut
2- Adjusting nut
3- Trunnion clamp screw and nut, with lockwasher
4- Slot forward
5- Mount trunnion

32065g17

Fig. 17 Typical inboard mounts—Borg-Warner/Velvet

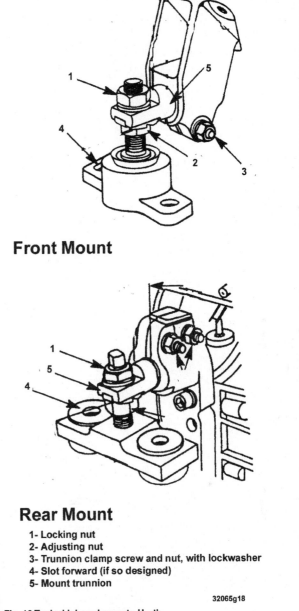

Front Mount

Rear Mount

1- Locking nut
2- Adjusting nut
3- Trunnion clamp screw and nut, with lockwasher
4- Slot forward (if so designed)
5- Mount trunnion

32065g18

Fig. 18 Typical inboard mounts-Hurth

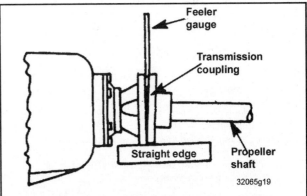

Fig. 19 Transmission and mating surfaces must have a uniform gap when measured in four positions—Borg-Warner/Velvet

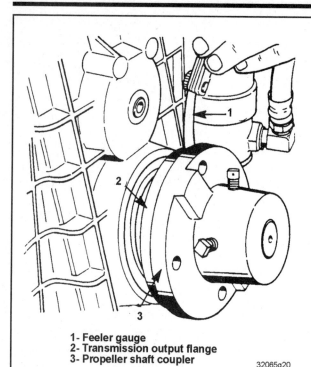

1- Feeler gauge
2- Transmission output flange
3- Propeller shaft coupler

32065g20

Fig. 20 Transmission and mating surfaces must have a uniform gap when measured in four positions—Hurth

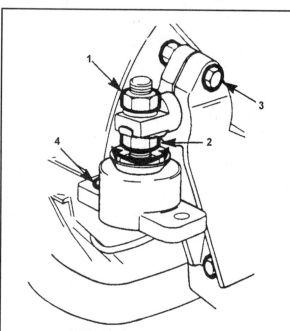

1- Locking nut
2- Adjusting nut
3- Clamping screws and nuts, with lockwashers (two each on some mounts)
4- Slot forward (if so designed) - NOT slotted on this style rear mount)

32065g21

Fig. 21 Typical inboard mount with adjustment nut and slot Borg-Warner/Velvet

11. Disconnect the propeller shaft coupler at the transmission output flange.

12. Attach a suitable engine hoist to the lifting eyes and take up any line slack until it is just taught.

■ **DO NOT use the front lifting eye attached to the thermostat housing at this point.**

13. Locate the front engine mount(s) and remove the two lag bolts.

14. Locate the rear engine mount and remove the mounting bolts.

15. Slowly and carefully, lift out the engine. Try not to hit the power steering control valve.

To install:

16. Loosen the stop nut and move the adjustment nuts on all mounts so that they are in the center of the adjustment range.

17. All mounts (except on the 350 Magnum Tournament Ski) have one slotted bolt hole and this should be facing forward.

18. Lower the engine into the compartment so that it just rests on the washers and the transom plate or stringers. Make sure that the holes line up and DO NOT release the hoist tension yet. Make sure that there is at least 1/4 in. (6mm) of up or down adjustment at all mounts.

19. Position the engine so that the transmission output flange and the propeller shaft coupler are in alignment and firmly mated together. Adjust the engine bed height to true the alignment-do not use the mount adjustment screws yet.

20. Make sure that the mounts and their bolt holes are still lined up and then install the mounting bolts. Tighten them securely.

21. Center the propeller shaft in the log. Press it down as far as it will go and then pull it up as far as it will go; position the shaft at the apparent center point of its vertical movement. Now do the same thing horizontally.

22. Make certain that the shaft coupler and the engine output flange align exactly. Most engines have a shoulder on the shaft coupler that should engage the recess on the output flange with no resistance.

23. Check that the gap between the two mating surfaces is 0.003 in. (0.07mm) when checked at 90 degree positions on the couplers (12, 3, 6 and 9 o'clock). If the gap is other than suggested, move the adjustment nuts on the engine mounts until you achieve the proper clearance. Always turn the front and rear mount adjusters equally to maintain level on your engine.

24. Once you are comfortable that everything is in alignment, tighten the mount bracket bolts to 50 ft. lbs. (68 Nm). Bend the tab on the washer over onto the flat of the mounting bolt head.

25. Make sure that the mount trunnion shaft is not exposed more than 3/4 in. (20mm) from the mount bracket.

26. Reconnect the propeller shaft to the transmission output shaft and tighten the bolts to 50 ft. lbs. (68 Nm). Some shafts may have setscrews and the shaft should be dimpled at their locations. Safety wire the set screws after securely tightening them.

27. Reconnect the water inlet hose and tighten the hose clamp securely.

28. Unplug the fuel line and pump fitting and reconnect them.

29. Connect the MerCathode lines to the controller and coat them with liquid neoprene. If you forgot to tag the wires as we suggested, refer to the illustration for proper hook-up.

30. Install and adjust the throttle cable. Make sure that the barrel positions the cable so that there are no kinks or sharp bends in the cable. Make sure that the cable does not come in contact with any other moving parts. Please refer to the Fuel Systems section later in this manual for further details.

31. Install and adjust the shift cables. Please refer to the Transmission section later in this manual for further details.

32. Check and refill all fluids and go have fun!

Cylinder Head Cover

REMOVAL & INSTALLATION

◆ **See Figures 25 and 26**

■ In order to perform this procedure efficiently, we recommend removing the exhaust manifold in order to have sufficient working room to remove the cylinder head cover. Although not completely necessary, it's worth the extra effort to avoid the aggravation of working around the manifolds. Please refer to the manifold procedure later in this section.

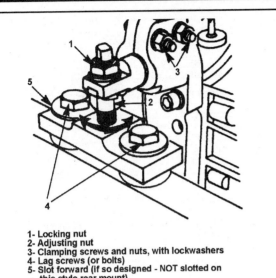

1- Locking nut
2- Adjusting nut
3- Clamping screws and nuts, with lockwashers
4- Lag screws (or bolts)
5- Slot forward (if so designed - NOT slotted on
   this style rear mount)

Fig. 22 Typical inboard mount with adjustment nut and slot-Hurth

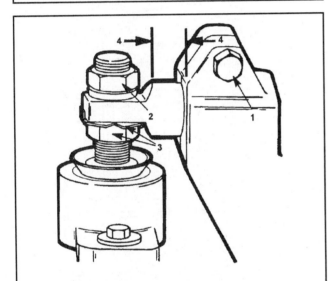

1- Tighten clamping screw and nut on all four mount
   brackets to 50 lb ft (68 Nm)
2- Tighten locking nut on all four mounts securely
3- Bend one of the tabs on tab washer down onto flat
   of adjusting nut
4- Maximum extension of large diameter of trunnion -
   3/4 in. (20 mm)

Fig. 23 The adjusting trunnion should not extend from the mount-ing bracket more than 3/4 in. —Borg-Warner/Velvet

1. Open or remove the engine compartment hatch. Disconnect the negative battery cable.

2. Loosen the clamp and remove the crankcase ventilation hose at the cover. Carefully move it out of the way.

3. Tag and disconnect any lines, leads or hoses that might be in the way of removal. In some instances you may be able to simply secure them out of the way without disconnecting them-you be the judge.

4. If your engine has a spark plug wire retainer attached to the cover, unclip the wires or remove the retainer and then tag and disconnect the plug wires at the spark plugs.

5. On the 8.1L V8, remove the coolant reservoir and the remote oil lines.

6. Loosen the cover mounting bolts and lift off the cylinder head cover. Take note of any harness or hose retainers and clips that might be attached to certain of the mounting bolts; you need to make sure they go back in the same place.

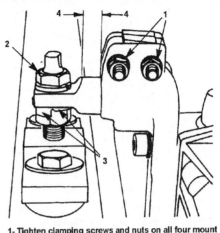

1- Tighten clamping screws and nuts on all four mount
   brackets to 50 lb ft (68nm)
2- Tighten locking nut on all four mounts securely
3- Bend one of the tabs on tab washer down onto flat
   of adjusting nut
4- Maximum extension of large diameter of trunnion
   3/4 in. (20 mm)

32065g24

Fig. 24 The adjusting trunnion should not extend from the mounting bracket more than 3/4 in.—Hurth

To install:
7. Clean the cylinder head and cover mounting surfaces of any residual gasket material with a scraper or putty knife.

8. Position a new gasket on the cylinder head and then position the cover (don't forget the J-clips!). Tighten the mounting bolts to:
- 45 inch lbs. (5 Nm) on non-Gen II V6 engines
- 50 inch lbs. (5.5 Nm) on 1993-97 Gen II V6 engines
- 106 inch lbs. (12 Nm) on 1998 and later V6 engines and the 8.1L V8
- 90 inch lbs. (10 Nm) on 1992-97 V8 engines
- 106 inch lbs. (12 Nm) on 1998-2001 5.0L/5.7L/6.2L V8 engines
- 71 inch lbs. (8 Nm) on 1998-2001 7.4L/8.2L V8 engines

Make sure any retainers or clips that were removed are back in their original positions.

9. Connect the crankcase ventilation hose and any other lines or hoses that may have been disconnected. Check that there were no other wires or hoses you may have repositioned in order to gain access to the cover.

10. Install the coolant tank and remote oil lines on the 8.1L.

11. Install the exhaust manifold.

12. Connect the battery cables.

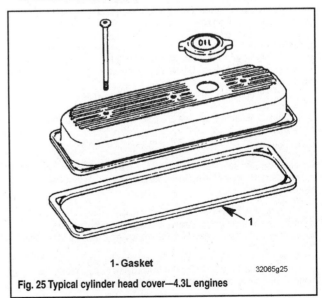

1- Gasket

32065g25

Fig. 25 Typical cylinder head cover—4.3L engines

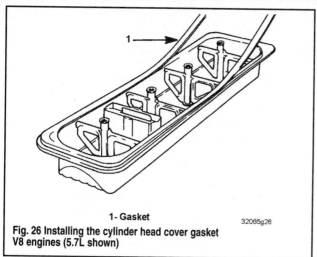

1- Gasket   32065g26

Fig. 26 Installing the cylinder head cover gasket V8 engines (5.7L shown)

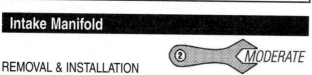

Intake Manifold

REMOVAL & INSTALLATION

MODERATE

All Engines Exc. 8.1L V8

◆ See Figures 27, 28, 29, 30, 31, 32, 33, 34 and 35

1. Open or remove the engine compartment hatch. Disconnect the negative battery cable.

2. Loosen the clamp and remove the crankcase ventilation hose at the cylinder head covers. Carefully move them out of the way.

3. Drain all water from the cylinder block and manifolds.

4. Tag and disconnect all water hoses at the manifold and thermostat housing. Have some rags handy, since there will still be some water in them. Carefully move them out of the way. Don't miss the circulating pump bypass hose on late model big blocks.

5. Remove the flame arrestor and then disconnect the throttle cable at the carburetor/throttle body. Disconnect the fuel line also and plug the line end and the carb/throttle body/fuel rail fitting. If you have a non-flexible line, disconnect it at the fuel pump also. Move both the cable and fuel line out of the way.

6. Tag and disconnect any lines, leads or hoses that might be in the way of removal. In some instances you may be able to simply secure them out of the way without disconnecting them-you be the judge.

7. If your engine has a spark plug wire retainer attached to the cylinder head cover, unclip the wires or remove the retainer and then tag and disconnect the plug wires at the spark plugs.

8. Remove the distributor cap with the leads still connected. Mark the position of the distributor rotor to the distributor body, loosen the clamp bolt and remove the distributor. Please refer to the Electrical section for further details on distributor removal. DO NOT turn the engine over once the distributor has been removed.

9. Tag and disconnect any leads at the ignition module and move them out of the way.

10. Tag and disconnect the wire at the oil sending unit and then remove the unit itself.

11. Loosen and remove the manifold mounting bolts in the reverse order of the tightening sequence and then remove the manifold. There is a good likelihood you will need to pry the manifold off the block; be very careful that you don't scratch or mar the mating surfaces on the block, manifold or heads.

To install:

12. Carefully remove all remaining gasket material from the manifold mating surfaces with a scraper or putty knife. Be careful that you don't accidentally drop any old gasket into the crankcase or intake ports on the cylinder head.

13. Inspect the manifold and all mating surfaces for any cracks or nicks.

14. Apply Perfect Seal to both sides of each new gasket and position them on the cylinder heads. Note the following precautions:

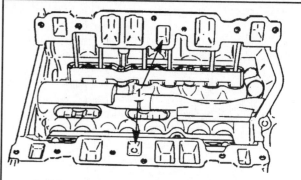

1- Exhaust crossover port opening in gasket   32065g27

Fig. 27 Certain engines utilize an extra opening in the gasket for the exhaust cross-over port—V6

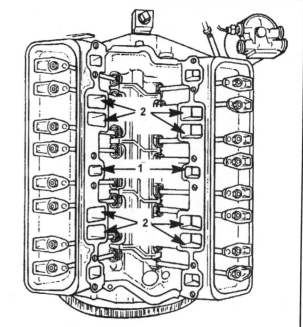

1- Exhaust crossover port opening in gasket
2- Intake valve ports

32065g28

Fig. 28 Certain engines utilize an extra opening in the gasket for the exhaust cross-over port—1992-97 V8

15. Manifold gaskets are identical for both sides; always make sure that they are installed with the marked side facing UP.

    a. On V6 or 1992-97 V8 engines, if you have an engine with a 2 bbl carburetor make sure that you remove the metal insert from the gasket going on the starboard side cylinder head before you coat with sealer and lay into position. This will allow for heat pipe clearance on the intake manifold.

    b. When ordering your gaskets, make sure you get the right ones for your engine. Engines that utilize an automatic choke require an additional opening in the gasket for the exhaust cross-over port in the intake manifold. Without this hole the choke will not operate correctly and could cause rough operation.

16. Apply a 3/4 in. (5mm) bead of RTV sealant to the forward and aft edges of the cylinder block mating surface. Make sure that you run the bead at least a 1/2 in. up onto the gaskets.

■ **There is an oil sending unit hole on the edge of many rear cylinder block mating surfaces. DO NOT get RTV sealer into this hole!**

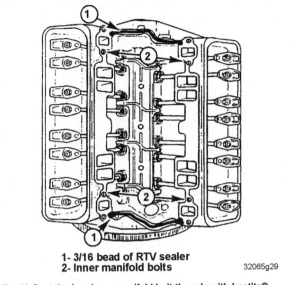

1- 3/16 bead of RTV sealer
2- Inner manifold bolts

32065g29

**Fig. 29 Coat the four inner manifold bolt threads with Loctite®
1998-2001 V8**

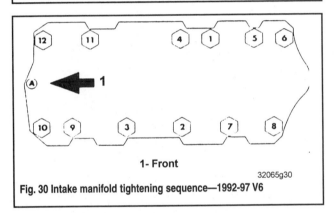

1- Front

32065g30

Fig. 30 Intake manifold tightening sequence—1992-97 V6

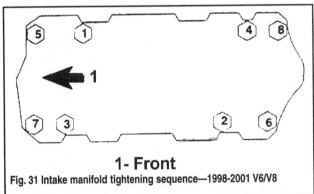

1- Front

Fig. 31 Intake manifold tightening sequence—1998-2001 V6/V8

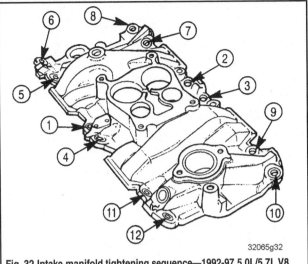

32065g32

Fig. 32 Intake manifold tightening sequence—1992-97 5.0L/5.7L V8

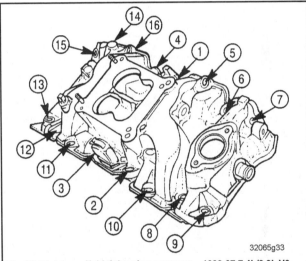

32065g33

Fig. 33 Intake manifold tightening sequence—1992-97 7.4L/8.2L V8

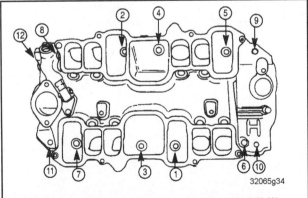

32065g34

Fig. 34 Intake manifold tightening sequence—1998-2001 7.4L V8

17. Install the manifold into place so that all the bolt holes line up, insert the bolts and tighten them to:
- 30 ft. lbs. (41 Nm) on 1992-93 V6 models w/o a balance shaft
- 35 ft. lbs. (48 Nm) on 1993-97 V6 models with a balance shaft and V8 engines (exc. 454/502 Mag EFI)
- 25-30 ft. lbs (34-41 Nm) on 1992-97 454/502 Mag EFI V8s
- 11 ft. lbs. (15 Nm) for 1998-2001 V6 engines and the 6.2L V8
- 18 ft. lbs. (24 Nm) on 1998-2001 5.0L/5.7L V8s
- 35 ft. lbs. (48 Nm) on 1998-2001 454/502/8.2L Mag MPI V8 engine
- 30 ft. lbs (40 Nm) on 1998-2001 7.4L MPI V8 engines

Check the Torque Specifications chart at the end of this section for any further details on tightening needs. Tighten all bolts in the order shown in the illustration. On 1998-2001 V8 engines, make sure you coat the threads of the four inner bolts with Loctite.

18. Install the oil sending unit and connect the wire.

19. Connect the ignition module leads and install the distributor as detailed in the Electrical section. Put the cap back on and reconnect the plug wires to the spark plugs.

20. Install all water hoses and tighten their clamps securely.

21. Install the fuel line at the carburetor/throttle body and fuel pump. Make sure you remove any plugs you may have inserted on removal.

22. Install the throttle cable and adjust it as detailed in the Fuel System section. Install the flame arrestor.

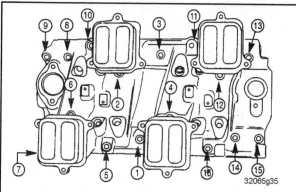

Fig. 35 Intake manifold tightening sequence—1998-2001 454/502 Mag/8.2L V8

23. Connect all other lines, leads and hoses that may have been removed to facilitate manifold removal.

24. Connect the battery cable and start the engine. Check the ignition timing and idle speed. Check all hoses and seals for leaks.

8.1L V8 Engines

◆ See Figures 36, 37 and 38

1. Open or remove the engine compartment hatch. Disconnect the negative battery cable.

2. Drain all water from the cylinder block and manifolds.

3. Remove the gear lube monitor.

4. Remove the flame arrestor and then disconnect the throttle cable at the throttle body. Disconnect the fuel line also and plug the line end and the throttle body. If you have a non-flexible line, disconnect it at the fuel pump also. Move both the cable and fuel line out of the way.

5. Tag and disconnect any lines, leads or hoses that might be in the way of removal. In some instances you may be able to simply secure them out of the way without disconnecting them-you be the judge.

6. If your engine has a spark plug wire retainer attached to the cylinder head cover, unclip the wires or remove the retainer and then tag and disconnect the plug wires at the spark plugs.

7. Remove the shift plate assembly.

8. Remove the ECM bracket.

9. Loosen and remove the manifold mounting bolts in the reverse order of the tightening sequence and then remove the manifold. There is a good likelihood you will need to pry the manifold off the block; be very careful that you don't scratch or mar the mating surfaces on the block, manifold or heads.

To install:

10. Carefully remove all remaining gasket material from the manifold mating surfaces with a scraper or putty knife. Be careful that you don't accidentally drop any old gasket into the crankcase or intake ports on the cylinder head.

11. Inspect the manifold and all mating surfaces for any cracks or nicks.

12. Manifold gaskets are identical for both sides; always make sure that they are installed with the marked side facing UP.

13. Apply a 3/16 in. (5mm) bead of RTV sealant to the forward and aft edges of the cylinder block mating surface and then position the two neoprene gaskets.

14. Apply a small dab of RTV sealer to the ends of the neoprene gaskets and then set the two manifold gaskets into place making sure that the bolt holes align.

15. Install the manifold into place so that all the bolt holes line up, insert the bolts and tighten them to 106 inch lbs. (12 Nm). Tighten all bolts in the order shown in the illustration.

16. Reconnect the plug wires to the spark plugs.

17. Install all water hoses and tighten their clamps securely.

18. Install the fuel line at the throttle body. Make sure you remove any plugs you may have inserted on removal.

19. Install the throttle cable and adjust it as detailed in the Fuel System section. Install the flame arrestor.

20. Connect all other lines, leads and hoses that may have been removed to facilitate manifold removal.

21. Connect the battery cable and start the engine. Check the ignition timing and idle speed. Check all hoses and seals for leaks.

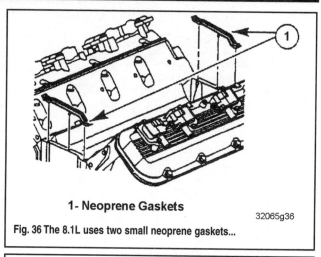

1- Neoprene Gaskets

Fig. 36 The 8.1L uses two small neoprene gaskets...

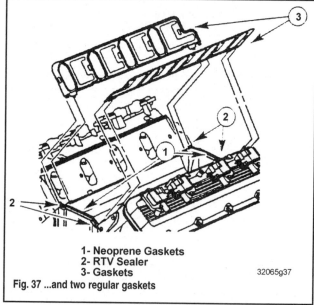

1- Neoprene Gaskets
2- RTV Sealer
3- Gaskets
Fig. 37 ...and two regular gaskets

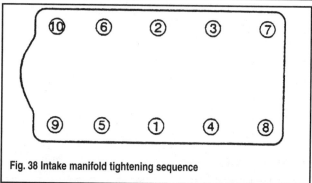

Fig. 38 Intake manifold tightening sequence

Rocker Arms and Push Rods

REMOVAL & INSTALLATION

MODERATE

1992-93 V6 Engines (W/O Balance Shaft)
5.0L, 5.7L And 6.2LV8 Engines

1. Open or remove the engine hatch cover and disconnect the negative battery cable. Remove the cylinder head cover as detailed previously.

2. Bring the piston in the No. 1 cylinder to TDC. If servicing only one arm, bring the piston in that cylinder to TDC. The No. 1 cylinder on is the first cylinder at the port side of the engine.

3. Loosen and remove the rocker arm nuts and lift out the balls. Lift the arm itself off of the mounting stud and pull out the pushrod. It is very important to keep each cylinder's component parts together as an assembly. We suggest drilling a set of holes in a 2 x 4 and positioning the pieces in the holes.

To install:

4. Clean and inspect the rocker assemblies.

5. Coat all bearing surfaces of the rocker assembly with engine oil.

6. Slide the push rods into their holes. Make sure that each rod seats in its socket on the lifter.

7. Position the rocker arm over the stud so that the cupped side rides on the push rod. Slide the ball over the stud, install the nut and tighten it until zero lash is present.

8. Adjust the valves as detailed.

9. Install the cylinder head cover, connect the battery cable and check the idle speed.

1993-00 V6 Engines (w/Balance Shaft), 7.4L, 8.1L & 8.2L V8 Engines

1. Open or remove the engine hatch cover and disconnect the negative battery cable. Remove the cylinder head cover as detailed previously.

2. Bring the piston in the No. 1 cylinder to TDC. If servicing only one arm, bring the piston in that cylinder to TDC. The No. 1 cylinder on is the first cylinder at the port side of the engine.

3. Loosen and remove the rocker arm nuts and lift out the balls. Lift the arm itself off of the mounting stud and pull out the pushrod. It is very important to keep each cylinder's component parts together as an assembly. We suggest drilling a set of holes in a 2 X 4 and positioning the pieces in the holes.

To install:

4. Clean and inspect the rocker assemblies.

5. Coat all bearing surfaces of the rocker assembly with engine oil.

6. Slide the push rods into their holes. Make sure that each rod seats in its socket on the lifter.

7. Position the rocker arm over the stud so that the cupped side rides on the push rod. Slide the ball over the stud, install the nut and tighten it to:

- 20 ft. lbs. (27 Nm) on 1993-97 V6 engines
- 22 ft. lbs. (30 Nm) on 1998 and later V6 engines
- 40 ft. lbs. (54 Nm) on 7.4L and 1998-00 454/502/8.2L V8 engines
- 45 ft. lbs. (61 Nm) on 1992-97 454/502 Mag/EFI V8 engines
- 19 ft. lbs. (25 Nm) on 8.1L V8

On the 8.1L engine, make sure that each cylinder is at TDC and the valves are closed.

8. No additional adjustment of the valves is necessary as the lash is set automatically when the rocker is tightened to specifications.

9. Install the cylinder head cover, connect the battery cable and check the idle speed.

VALVE ADJUSTMENT

4.3L Engines (w/o Balance Shaft), 5.0L, 5.7L & 6.2L V8 Engines

These engines utilize hydraulic valve lifters, although there is no need for periodic valve adjustment, it is necessary to perform a preliminary adjustment after any work on the valve train/rocker assembly. All adjustment should be undertaken while the lifter is on the base circle of the camshaft lobe for that particular cylinder. This means the opposite side of the pointy part of each lobe.

1. Rotate the crankshaft, or bump the engine with the starter until the No. 1 cylinder is at TDC. Note that the notch or mark on the damper pulley will be lined up with the **0** mark on the timing scale. Be careful here though, this could mean that either the No. 1 or the No. 4 (No. 6 on V8) piston is at TDC. Place your hand on the No. 1 cylinder's valve and check that it does not move as the mark on the pulley is approaching the **0** mark on the tab. If it does not move, you're ready to proceed; if it does move, you are on the No. 4 (No.6, V8) cylinder and need to rotate the engine an additional full turn. This is important so make sure you've gotten it right!

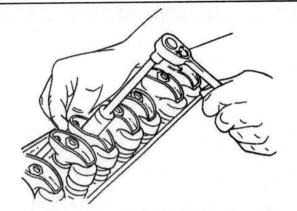

Fig. 39 Wiggle the push rod slowly while tightening the rocker arm adjusting nut

2. Now that the No. 1 cylinder is at TDC, you can adjust the following valves on V6 engines:
- No. 1 cylinder: intake and exhaust
- No. 2 cylinder: intake
- No. 3 cylinder: intake
- No. 5 cylinder: exhaust
- No. 6 cylinder: exhaust

Or these valves on V8 engines (LH rotation):
- No. 1 cylinder: intake and exhaust
- No. 2 cylinder: intake
- No. 3 cylinder: exhaust
- No. 4 cylinder: exhaust
- No. 5 cylinder: intake
- No. 7 cylinder: intake
- No. 8 cylinder: exhaust

Or these valves on V8 engines (RH rotation:
- No. 1 cylinder: intake and exhaust
- No. 2 cylinder: exhaust
- No. 3 cylinder: intake
- No. 4 cylinder: intake
- No. 5 cylinder: exhaust
- No. 7 cylinder: exhaust
- No. 8 cylinder: exhaust

3. Loosen the adjusting nut on the rocker until you can feel lash (play in the push rod) and then tighten the nut until the lash has been removed. Carefully jiggle the push rod while tightening the nut until it won't move anymore - this is zero lash. Tighten the nut an additional full turn to set the lifter and then you're done. Perform this procedure on each of the valves listed above.

4. Slowly rotate the engine an additional full turn and this will bring the No. 4 (No. 6, V8) piston to TDC. The pulley notch/mark should once again be in line with the **0** on the timing tab. You can now adjust the remaining valves on V6 engines:
- No. 2 cylinder: exhaust
- No. 3 cylinder: exhaust
- No. 4 cylinder: intake and exhaust
- No. 5 cylinder: intake
- No. 6 cylinder: intake

Or these valves on V8 engines (LH rotation):
- No. 2 cylinder: exhaust
- No. 3 cylinder: intake
- No. 4 cylinder: intake
- No. 5 cylinder: exhaust
- No. 6 cylinder: intake and exhaust
- No. 7 cylinder: exhaust
- No. 8 cylinder: intake

Or these valves on V8 engines (RH rotation):
- No. 2 cylinder: intake
- No. 3 cylinder: exhaust
- No. 4 cylinder: exhaust
- No. 5 cylinder: intake
- No. 6 cylinder: intake and exhaust
- No. 7 cylinder: intake
- No. 8 cylinder: exhaust

Exhaust Manifold

REMOVAL & INSTALLATION

② MODERATE

All Engines Exc. 8.1L V8

◆ **See Figures 40, 41, 42 and 43**

1. Open or remove the engine compartment hatch. Disconnect the negative battery cable.
2. Drain all water and/or coolant from the engine, manifold and exhaust elbow as detailed in the Cooling System section of this manual.
3. Disconnect all water/coolant lines at the manifold and move them out of the way.
4. Disconnect the exhaust pipe bellows at the elbow and move it out of the way. Take note of where all the hose clamps were situated.
If removing the starboard manifold:
5. Disconnect the shift cables on stern drive models and move them out of the way.
6. Tag and disconnect the electrical lead at the shift cut-out switch on V6 and V8 Alpha models and then remove the shift plate assembly on the elbow.
7. Remove the water separating fuel filter bracket bolt (if equipped) and position it out of the way.
8. Remove the remote oil filter and bracket from the port manifold on V8 engines.
9. Loosen the manifold retaining bolts from the center outward and then pry off the manifold/elbow assembly.
10. If necessary, loosen the four retaining bolts and remove the exhaust elbow from the manifold. If equipped with a riser, remove this as well
To install:
11. Carefully clean all residual gasket material from the head, manifold and elbow mating surfaces with a scraper or putty knife. Inspect all gasket surfaces for scratches, cuts or other imperfections.
12. Position a new gasket on the cylinder head and install the manifold; making sure that everything is aligned properly. Tighten all nuts until they are just tight and then torque them to:
  • 20 ft. lbs. (27 Nm) on early V6 models w/o a balance shaft
  • 33 ft. lbs. (45 Nm) on the 6.2L V8 engines
  • 25 ft. lbs. (34 Nm) on all other engines
Starting in the center, and working your way out to the ends of the manifold.

■ **On 1998-2001 5.0L/5.7L/6.2L V8 engines it is unlikely that the replacement gaskets will look like the original ones. When installing these gaskets apply a 1/8 in. (3mm) bead of Loctite 510 to both sides of the gasket as shown in the illustration and then install the components immediately. Do not run the engine for a few hours after installation until the sealant sets up.**

13. Position a new gasket on the manifold, making sure that the indents line up (on seawater cooled engines) and then install the elbow. Tighten the mounting bolts to 25 ft. lbs. (34 Nm), 20 ft. lbs. (27 Nm) on early V6 models w/o a balance shaft. Connect the exhaust pipe/bellows and tighten the clamps securely. Models with a riser may also have spacers between the top of the riser and the elbow.

■ **Late model 7.4L/8.2L V8 engines may have an open (four slots) gasket rather than a restrictor gasket (two slots/two holes). It is OK to replace restrictor gaskets with open gaskets as long as the same type is used on both manifolds.**

14. If you removed the manifold or elbow plugs for some reason, make sure that the threads are coated with Perfect Seal before screwing them back in.
15. Install the oil filter and bracket if removed.
16. Install the fuel filter and bracket if removed.
17. Install the shift plate onto the starboard manifold (if removed), connect the lead to the cut-out switch and then install the shift cables. Adjust them as detailed later in this manual.
18. Connect the water/coolant hoses and tighten the clamps securely.
19. Make sure that any miscellaneous lines or hoses that you may have moved or disconnected during removal are reconnected and routed properly.

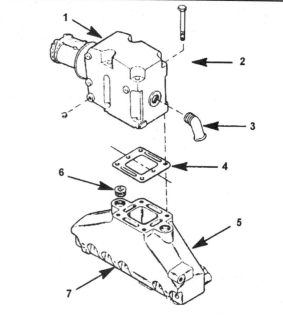

1- Exhaust elbow assembly - starboard side shown
2- Bolts
3- Fitting (45 degree) or fitting (straight)
4- Gasket - Seawater cooled (align as shown)
5- Exhaust manifold assembly
6- Plug
7- Bolts

32065g40

Fig. 40 Exhaust manifold and elbow—seawater cooled

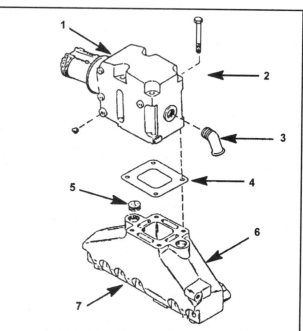

1- Exhaust elbow assembly - starboard side show
2- bolts
3- Fitting (45 degree) or fitting (straight)
4- Manifold separator gasket
5- Exhaust manifold assembly
6- Plug
7- Bolt

32065g41

Fig. 41 Exhaust manifold and elbow—closed cooling system

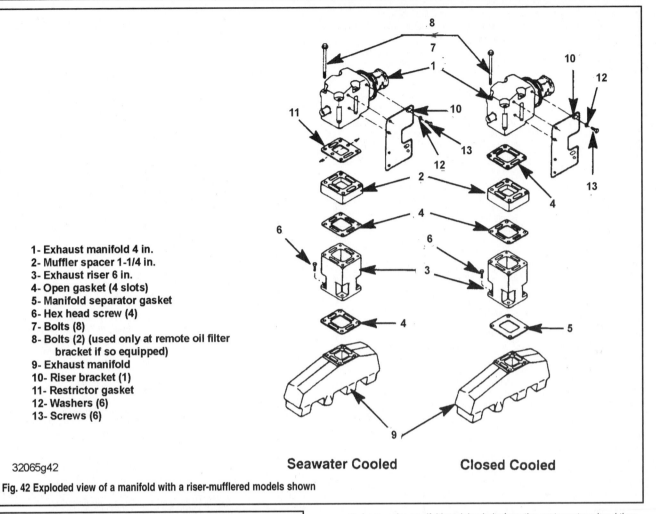

1- Exhaust manifold 4 in.
2- Muffler spacer 1-1/4 in.
3- Exhaust riser 6 in.
4- Open gasket (4 slots)
5- Manifold separator gasket
6- Hex head screw (4)
7- Bolts (8)
8- Bolts (2) (used only at remote oil filter
    bracket if so equipped)
9- Exhaust manifold
10- Riser bracket (1)
11- Restrictor gasket
12- Washers (6)
13- Screws (6)

32065g42

Seawater Cooled   **Closed Cooled**

Fig. 42 Exploded view of a manifold with a riser-mufflered models shown

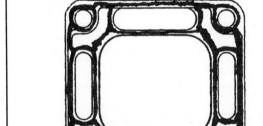

32065g43

Fig. 43 Coat the gasket on 1998 and later—5.0L/5.7L/6.2L V8 engines—with Loctite as shown

20. Fill the system with water or coolant, connect the battery cable and start the engine. When the engine reaches normal operating temperature, turn it off and re-torque the manifold bolts.

8.1L V8 Engines

1. Open or remove the engine compartment hatch. Disconnect the negative battery cable.
2. Drain all water and/or coolant from the engine, manifold and exhaust elbow as detailed in the Cooling System section of this manual.
3. Disconnect all water/coolant lines at the manifold and move them out of the way.
4. Disconnect the exhaust elbow. Take note of where all the hose clamps were situated.

5. Loosen the manifold retaining bolts from the center outward and then pry off the manifold.

To install:

6. Carefully clean all residual gasket material from the head, manifold and elbow mating surfaces with a scraper or putty knife. Inspect all gasket surfaces for scratches, cuts or other imperfections.
7. Position a new gasket on the cylinder head and install the manifold; making sure that everything is aligned properly. Coat the threads of the nuts with Loctite 271 and then tighten them until they are just tight and then torque them to 26 ft. lbs. (35 Nm) starting in the center and working your way out to the ends of the manifold.
8. Install a new elbow gasket and then install the elbow. Coat the threads with Loctite 242 and then tighten the stud nuts to 12 ft. lbs. (16 Nm)
9. Connect the water/coolant hoses and tighten the clamps securely.
10. Make sure that any miscellaneous lines or hoses that you may have moved or disconnected during removal are reconnected and routed properly.
11. Fill the system with water or coolant, connect the battery cable and start the engine. When the engine reaches normal operating temperature, turn it off and re-torque the manifold bolts.

Oil Pan

REMOVAL & INSTALLATION ③ ◁DIFFICULT

All Engines Exc. 8.1L V8

◆ See Figures 44, 45 and 46

■ More times than not, this procedure will require the removal of the engine. Your boat and its unique engine installation will determine this, but the procedure is almost always easier with the engine removed from the boat.

1. Remove the engine as previously detailed in this section.
2. If you haven't already drained the engine oil, do it now. Make sure you have a container and lots of rags available.
3. Certain models may require removing the outlet hose at the seawater pump to provide easier access (particularly if you are attempting this without removing the engine).
4. Remove the oil dipstick and then remove the dipstick tube(s). On inboard engines, make sure you note the dipstick tube banjo fitting position and which tube goes where.
5. Loosen and remove the oil pan retaining bolts and nuts, starting with the center bolts and working out toward the pan ends. Lightly tap the pan with a rubber mallet to break the seal and then lift it off the cylinder block. If your engine stand will allow for rotating the engine, you'll find that this will be easier with the pan facing up.

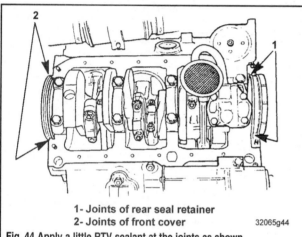

1- Joints of rear seal retainer
2- Joints of front cover

32065g44

Fig. 44 Apply a little RTV sealant at the joints as shown

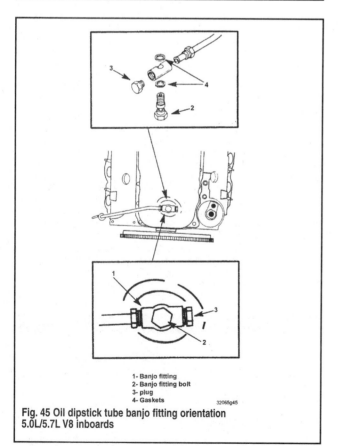

1- Banjo fitting
2- Banjo fitting bolt
3- plug
4- Gaskets

32065g45

Fig. 45 Oil dipstick tube banjo fitting orientation
5.0L/5.7L V8 inboards

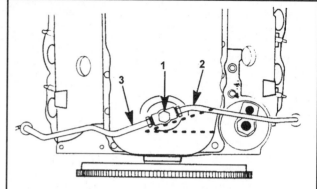

1- Factory positioned fitting for tubes (Do not move)
2- Port tube
3- Starboard tube

32065g46

Fig. 46 Refer to this for fitting the dipstick tubes
7.4L/8.2L V8 inboards

To install:
6. Clean the pan mating surfaces of any residual gasket material with a scraper or putty knife. Make sure that no old gasket material has been pressed into the retaining bolt holes in the pan, block or front cover. Clean the pan itself thoroughly with solvent.
7. Apply a small dab of RTV sealer to the joints on either side of the rear oil seal retainer and front cover, position a new pan gasket onto the pan being very careful to line up all the holes-do not use RTV sealant with this gasket other than where noted.
8. Move the pan and gasket onto the block; don't dawdle here because the RTV sealant applied in the previous step sets up very quickly. It is very important that you ensure all the holes line up correctly; sometimes a few bolts inserted through the pan and gasket will help the gasket stay in place.
9. Install all bolts and nuts finger tight and then tighten the 1/4-20 bolts to 80 inch lbs. (9 Nm) and the 5/16-18 nuts to 165 inch lbs. (19 Nm). On 1998 and later V8 engines, tighten the nuts to 18 ft. lbs. (25 Nm) and the bolts (if equipped) to 106 inch lbs. (12 Nm). Remember to start with the center bolts and work outward toward the ends of the pan.
10. Install the dipstick tube(s) and dipstick and then install the engine (if removed). Run the engine up to normal operating temperature, shut it off and check the pan for any leaks.
11. If you have a 5.0L/5.7L inboard, install the dipstick tube, but leave the banjo fitting at the bottom of the pan loose. Install the retaining bracket on the manifold and tighten it securely. Insert the plug into the banjo and then tighten the bolt to 180 inch lbs. (20 Nm).

8.1L V8 Engines

■ **More times than not, this procedure will require the removal of the engine. Your boat and its unique engine installation will determine this, but the procedure is almost always easier with the engine removed from the boat.**

1. Remove the engine as previously detailed in this section.
2. If you haven't already drained the engine oil, do it now. Make sure you have a container and lots of rags available.
3. Remove the oil dipstick and then remove the dipstick tube. Make sure you note the dipstick tube banjo fitting position on the bottom of the pan.
4. Remove the cool fuel cell on the port side of the engine.
5. Loosen and remove the oil pan retaining bolts and nuts, starting with the center bolts and working out toward the pan ends. Lightly tap the pan with a rubber mallet to break the seal and then lift it off the cylinder block. If your engine stand will allow for rotating the engine, you'll find that this will be easier with the pan facing up.

To install:
6. Clean the pan mating surfaces of any residual gasket material with a scraper or putty knife. Make sure that no old gasket material has been pressed into the retaining bolt holes in the pan, block or front cover. Clean the pan itself thoroughly with solvent.

7. Apply a small dab of RTV sealer to the joints on either side of the rear oil seal retainer and front cover, position a new pan gasket onto the pan being very careful to line up all the holes-do not use RTV sealant with this gasket other than where noted.

8. Move the pan and gasket onto the block; don't dawdle here because the RTV sealant applied in the previous step sets up very quickly. It is very important that you ensure all the holes line up correctly; sometimes a few bolts inserted through the pan and gasket will help the gasket stay in place.

9. Install all bolts finger tight and then tighten them to 19 ft. lbs. (25 Nm). Remember to start with the center bolts and work outward toward the ends of the pan.

10. Install the dipstick tube and dipstick and then install the engine (if removed). Run the engine up to normal operating temperature, shut it off and check the pan for any leaks.

Oil Pump

The two-piece oil pump utilizes two pump gears and a pressure regulator valve. A baffled pick-up tube is press-fit into the body of the pump.

REMOVAL & INSTALLATION

◆ See Figures 47, 48 and 49

1. Remove the oil pan as previously detailed. Remember that you probably need to remove the engine for this procedure.

2. V8 engines utilize an oil baffle plate; remove the 5 retaining nuts and lift out the baffle (this should not be necessary on the 8.1L V8).

3. Remove the spacers under the baffle on the 6.2L engine.

4. Loosen and remove the pump mounting bolts and lift off the pump assembly.

5. Check that the pump and block mating surfaces are clean and then position the pump over the block so that the pump extension shaft is aligned with the distributor drive shaft. Do not use a gasket or RTV sealant.

6. Tighten the pump mounting bolts to 65 ft. lbs. (88 Nm); 70 ft. lbs. (95 Nm) on 1992-97 7.4L and 8.2L V8 engines, and 56 ft. lbs. (75 Nm) on the 8.1L.

7. Install the oil baffle on V8 engines and tighten the nuts to 25 ft. lbs. (34 Nm). Use Loctite 271 for the spacers on the 6.2L engine.

8. Install the oil pan and engine.

Crankshaft Pulley/Torsional Damper

REMOVAL & INSTALLATION

◆ See Figure 50

1. On certain engine installations you may have to remove the engine mount. Install an engine hoist and tighten the chain so that the engine's weight is removed from the front engine mount. Do not use the lifting eye attached to the thermostat housing. Remove the front engine mount.

2. Remove the drive or serpentine belt(s) as detailed in Section Three.

3. Remove the bolts and pull off the drive pulley and/or the water pump pulley.

4. Remove the damper retaining bolt and install special tool J-23523-E onto the damper. Tighten the tool press bolt and remove the damper; don't lose the crankshaft key. MerCruiser suggests that you do not use a conventional gear puller for this procedure.

To install:

5. Inspect the crank key and then install it into the shaft.

6. Coat the front cover oil seal lip with clean engine oil and then install the damper with a proper installation tool. Be sure that you thread the tool into the crankshaft at least 1/2 inch to protest the threads. In a pinch you can use a block of wood and a plastic mallet, but be careful that the pulley does not shift on its mountings while you're hammering.

7. Install the plate, thrust bearing and nut on to the rod. Tighten the nut until it bottoms out and then remove the tool.

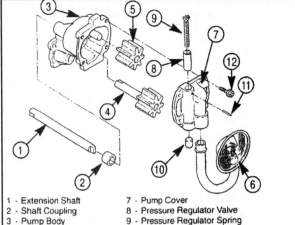

1 - Extension Shaft	7 - Pump Cover
2 - Shaft Coupling	8 - Pressure Regulator Valve
3 - Pump Body	9 - Pressure Regulator Spring
4 - Drive Gear and Shaft	10- Plug
5 - Idler Gear	11- Retaining Pin
6 - Pickup Screen and Pipe	12- Screws

32065g47

Fig. 47 Exploded view of the oil pump (big block similar)

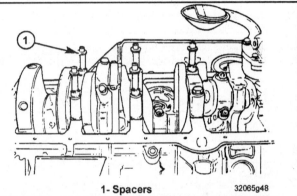

1- Spacers   32065g48
Fig. 48 There are spacers under the baffle plate on the 6.2L engine

Fig. 49 Oil pump and pick-up screen properly installed on a typical V8

8. Install the retaining bolt and tighten it to:
- 70 ft. lbs. (95 Nm) on 1992-97 V6 engines
- 74 ft. lbs. (100 Nm) on 1998 and later V6 engines
- 60 ft. lbs. (81 Nm) on 5.0L/5.7L/6.2L V8 engines
- 90 ft. lbs. (122 Nm) on 1992-97 7.4L/8.2L V8 engines
- 110 ft. lbs. (149 Nm) on 1998-2001 7.4L/8.2L V8 engines
- 188 ft. lbs. (225 Nm) on the 8.1L V8

Squirt a little RTV sealant into the crankshaft keyway to guard against oil seepage. Tighten the drive pulley to 35 ft. lbs (48 Nm), 30 ft. lbs. (81 Nm) on V6 engines with no balance shaft.

9. Install the drive/serpentine belt and make sure that it is adjusted properly.

10. Install the front mount and unhook the engine hoist if necessary.

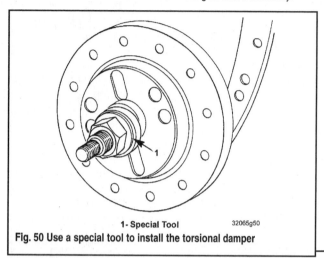

1- Special Tool   32065g50

Fig. 50 Use a special tool to install the torsional damper

Front Cover and Oil Seal

REMOVAL & INSTALLATION   ③ DIFFICULT

◆ **See Figures 51 and 52**

■ **This procedure may require engine removal, depending upon your particular boat. If necessary, remove the engine as detailed previously in this section.**

1. Open the drain valves and drain the coolant from the block and exhaust manifold. Loosen the alternator and power steering brackets (or the idler pulley) to provide slack, and then remove the drive or serpentine belts. Remove the water circulation pump.

2. Remove the heat exchanger and crossover on the 8.1L V8.

3. Remove the torsional damper as detailed previously in this section.

4. Tag and disconnect the cam position sensor lead on the 8.1L and then remove the sensor from the cover.

5. Remove the oil pan as detailed previously in this section.

6. Loosen the mounting bolts and remove the front cover. If the oil seal needs replacement, press it out from the back (inner) side of the cover with a punch. Remove the front cover gasket.

To Install:

7. Clean all gasket material from the cover and block mating surfaces with a scraper of putty knife. Be careful not to knock any pieces of gasket in the timing assembly.

8. If you removed the oil seal, install a new one with the lip toward the inside of the cover. Position a support under the seal and cover and then press the seal into the cover with the proper tool. Check the inside of the seal before installation; if there are helical grooves on the inner seal surface it can only be used on left hand rotation engines, if the inner surface is smooth it may be used on any engine.

9. Coat both sides of a new gasket with Perfect Seal and then position the gasket onto the engine. Install the cover so that all the bolt holes line up; there are dowel pins on the cylinder block that will help alignment. Tighten the bolts to:
   - 80 inch lbs. (9 Nm) on V6 models w/o a balance shaft.
   - 100 inch lbs. (11 Nm) on V6 engines with a balance shaft
   - 100 inch lbs. (11 Nm) on 1992-97 5.0L/5.7L/6.2L V8s
   - 106 inch lbs. (12 Nm) on 1998 and later 5.0L/5.7L/6.2L V8s and the 8.1L V8
   - 120 inch lbs. (14 Nm) on 1992-97 7.4L/8.2L V8s
   - 89 inch lbs. (10 Nm) on 1998 and later 7.4L/8.2L V8s

10. Install the oil pan and torsional damper.

11. Install the camshaft position sensor and reconnect the electrical lead (8.1L).

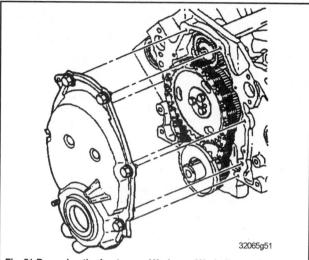

32065g51

Fig. 51 Removing the front cover-V6 shown, V8 similar

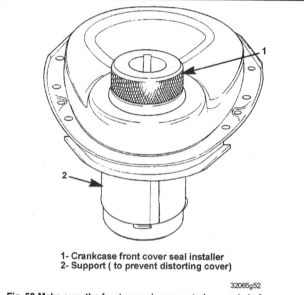

1- Crankcase front cover seal installer
2- Support (to prevent distorting cover)

32065g52

Fig. 52 Make sure the front cover is supported securely before pressing in the oil seal

12. Install the heat exchanger and crossover (8.1L)

13. Install the crankshaft pulley and pull the belts back on. Check their tension adjustment.

14. Install the water circulation pump and connect the hose.

15. Install the engine if removed. Add oil and water/coolant, start the engine and check for any leaks.

Timing Chain and Sprockets/Gears

OEM ③ DIFFICULT

REMOVAL & INSTALLATION

◆ **See Figure 53 and 54**

■ **Certain inboard engines (RH rotation) utilize timing gears rather than then more common timing chain and sprocket arrangement.**

1. Remove the crankshaft pulley and torsional damper as previously detailed in this section.

2. Remove the oil pan as previously detailed in this section.

3. Remove the front cover as previously detailed in this section.

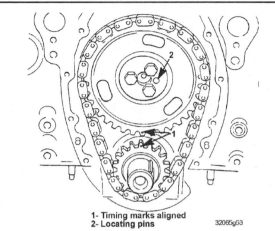

1- Timing marks aligned
2- Locating pins   32065g53

Fig. 53 The two marks on each of the timing sprockets must be in alignment before removing or installing the timing chain

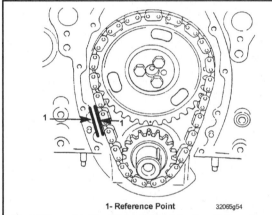

1- Reference Point   32065g54

Fig. 54 When checking timing chain deflection, find a reference point on the cylinder block and use it for each measurement

4. Look carefully at the camshaft and crankshaft sprockets/gears-you should notice a small indent on the front edge of one of the teeth on each sprocket/gear. Bump the engine over until these two marks are in alignment as shown in the illustration, a remote starter will work or you can screw the damper bolt back into the crankshaft and turn it.

5. Dab a little paint across one of the chain links and the camshaft sprocket. Loosen the camshaft sprocket retaining bolts (three), grasp the sprocket on each side with the chain still attached and wiggle it off the shaft. It should come off readily, but if not, tap the bottom edge lightly with a rubber mallet.

6. Mount a gear puller (J-5825-A) over the crankshaft pulley and pull it off the shaft.

To Install:

7. Clean the chain and sprockets/gears in solvent and let them air dry. Check the chain for wear and damage, making sure there are no loose or cracked links. Check the sprockets or gears for cracked or worn teeth.

8. Install the crankshaft sprocket/gear onto the shaft with an installation tool.

9. On engines with a timing chain, install the timing chain onto the camshaft sprocket so that the paint marks made during removal match up. If they do, and you've moved the engine, the timing marks on the two sprockets should also. Hold the sprocket/chain in both hands so the chain is hanging down, engage the chain around the crankshaft sprocket and then slide the cam sprocket/chain onto the camshaft. Do not force it! Tighten the three mounting bolts to:

- 18 ft. lbs. (24 Nm) on V6 engines with a balance shaft
- 20 ft. lbs. (27 Nm) on V6 engines w/o a balance shaft.
- 18 ft. lbs. (24 Nm) on 5.0L/5.7L/6.2L V8 engines

- 25 ft. lbs. (34 Nm) on 1992-97 7.4L/8.2L V8 engines
- 22 ft. lbs. (30 Nm) on 1998-2001 7.4L/8.1L/8.2L V8 engines

10. On engines with timing gears, position the camshaft gear so that the timing mark is aligned with the one on the crank gear, align the dowel on the end of the camshaft with the hole in the gear and press it onto the camshaft. Do not force it. Tighten the bolts to 18 ft. lbs. (24 Nm). Check the gear backlash and runout.

11. On engines with timing chains, rotate the camshaft slightly so that it creates tension on one side of the timing chain (either side if OK). Find a reference point on the same side of the cylinder block as the side that the timing chain is tight on and then measure from this point to the outer edge of the chain.

12. Rotate the camshaft in the opposite direction until the other side of the chain is tight. Press the inner side of the chain outward until it stops and then measure from your reference point on the cylinder block (obviously, do this from the same side of the chain as you did in the previous step) to the outer edge of the chain. This is timing chain deflection and it should be no more than 3/4 inch (19mm). If it is, replace the chain.

13. Install the front cover, the oil pan and the torsional damper.

■ **Remember that when the timing marks were aligned properly, the No. 4 (or No. 6, V8) cylinder was at TDC so if you are reinstalling the distributor the rotor should be at the No. 4 (No. 6) post on the cap, NOT at the No. 1 post.**

Balance Shaft

REMOVAL & INSTALLATION   ③ *DIFFICULT*

1993-2001 4.3L V6 Engines

◆ **See Figures 55, 56 and 57**

1. Remove the intake manifold as previously detailed in this section.

2. Remove the front cover and timing chain as previously detailed in this section.

3. Rotate the balance shaft until the marks on its driven gear and the camshaft drive gear are aligned.

4. Fashion a small wedge out of an old piece of wood and jam it in between the teeth of the balance shaft driven gear and the camshaft drive gear to hold the shafts from turning. Using a Torx socket (you are probably going to have to buy this one), remove the driven gear bolt.

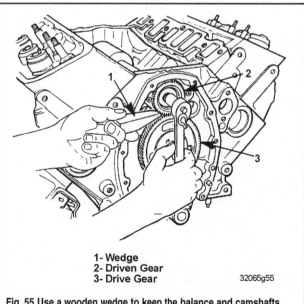

1- Wedge
2- Driven Gear
3- Drive Gear   32065g55

Fig. 55 Use a wooden wedge to keep the balance and camshafts from rotating while loosening the retaining bolt

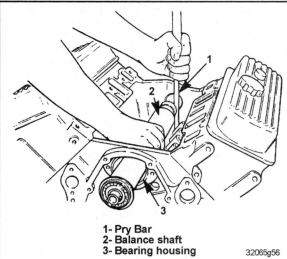

1- Pry Bar
2- Balance shaft
3- Bearing housing

32065g56

Fig. 56 Carefully lever the balance shaft forward and out of the cylinder block

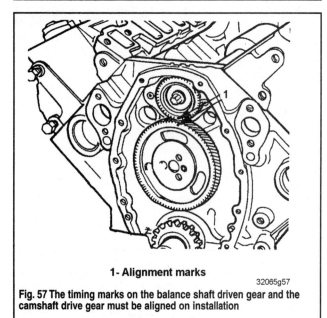

1- Alignment marks

32065g57

Fig. 57 The timing marks on the balance shaft driven gear and the camshaft drive gear must be aligned on installation

5. Pull the retaining stud from the camshaft drive gear and remove the gear itself.

6. Remove the two thrust plate mounting bolts from the balance shaft and lift off the plate. These are Torx fasteners also.

7. Insert a suitable prybar between the rear of the shaft and the cylinder block and carefully shimmy the shaft forward and out of the bearing housing in the rear of the block. Slide the entire shaft out through the front of the block. Be careful not to knock any debris into the inside of the cylinder block.

To install:

8. Clean the balance shaft in solvent and let it dry completely. Inspect the rear bearing for wear or damage. Inspect the front bearing for excessive side play on the shaft, wear or damage. Check the bearing bore in the front of the cylinder block for scoring.

9. Lubricate the front and rear bearing with motor oil and then carefully slide the shaft through the front bore until it feeds into the rear bearing. Tap the front edge of the shaft with a plastic mallet until the retaining ring on the front bearing seats against the cylinder block.

10. Install the thrust plate and tighten the two Torx fasteners to 120 inch lbs. (14 Nm).

11. Position the driven gear onto the balance shaft with the bolt finger tight. Make sure that the timing mark is at the bottom of the gear.

12. Install the camshaft drive gear so that the mark aligns with the one on the balance shaft driven gear and tighten the stud to 120 inch lbs. (16 Nm).

13. Remove the balance shaft bolt again and coat the threads with Loctite. Screw it in and tighten it to 15 ft. lbs. (20 Nm); and then turn the bolt an additional 35 degrees.

14. Install the timing chain and front cover.

15. Install the intake manifold.

Camshaft

③ ◁DIFFICULT

REMOVAL & INSTALLATION

◆ **See Figure 58**

1. Remove the cylinder head covers and rocker assemblies as previously detailed in this section.

2. Remove the lifter restrictor mounting bolts and lift out the restrictors (if equipped). Matchmark the lifters to their bores and then lift out the valve lifters with a small magnet and store them in a rack with labels so they can be reinstalled in their original locations.

3. Remove the front cover and timing chain/sprockets (or gears) as previously detailed in this section.

4. Remove the fuel pump and push rod on small block V8 engines.

5. Remove the inner camshaft drive gear on V6 engines with a balance shaft as detailed in the Balance Shaft section.

6. Remove the thrust plate/retainer on V6 and 5.7L/6.2L/8.1L V8 engines.

7. Thread two (three on the 8.1L) 5/16-18 x 5 inch bolts into the camshaft bolt holes and carefully pull the camshaft out of the cylinder block.

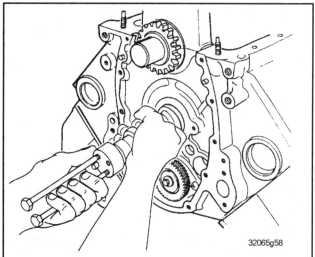

32065g58

Fig. 58 Make sure you don't cant the camshaft during removal or installation. Note the two 5 in. bolts screwed into the end of the shaft (V6 shown)

You may have to wiggle it back and forth a bit, so be sure you don't lean it up or down or else you could damage the bearings.

To install:

8. Inspect the camshaft as detailed in the Engine Rebuilding section.

9. Reinstall the long bolts and coat the journals with motor oil; coat the camshaft lobes with GM Cam and Lifter Pre-Lube or the equivalent. Carefully insert the shaft into the cylinder block and slide it all the way in.

10. Remove the installation bolts and install the thrust plate/retainer (if equipped) and tighten the bolts to 106 inch lbs. (12 Nm).

11. Install the fuel pump and push rod on V8 engines.

12. Install the drive gear, timing chain and front cover.

13. Drop the lifters back into their original bores so that they are aligned with the matchmarks and then install the restrictors. Tighten the mounting bolts to 12 ft. lbs. (16 Nm).

14. Install the rocker assemblies and the cylinder heads covers.

Cylinder Head

REMOVAL & INSTALLATION

③ *DIFFICULT*

All Engines Exc. 8.1L V8

◆ **See Figures 59, 60 and 61**

1. Drain the water from the cylinder block and manifold.
2. Remove the fuel line support brackets. Disconnect the fuel line at the carburetor and fuel pump, plug the fitting holes and remove the line.
3. Remove the intake and exhaust manifolds as previously detailed in this section; you can leave the carburetor/throttle body attached to the intake manifold if you like.
4. Tag and disconnect the spark plug wires at the plugs; move them out of the way. Although not necessary, it's a good idea to remove the plugs themselves also.
5. Remove the cylinder head cover and rocker assemblies as detailed previously in this section.
6. Remove or relocate any components or connections that may interfere with the removal of an individual cylinder head.
7. Loosen the cylinder head bolts, from the center bolts and working out to the ends of the head and then carefully lift the head off the block. You may need to persuade it with a rubber mallet-be careful! Set the head down carefully; do not sit it on cement.
To install:
8. Carefully, and thoroughly, remove all residual head gasket material from the cylinder head and block mating surfaces with a scraper or putty knife. Check that the mating surfaces are free of any nicks or cracks. Make sure there is no dirt or old gasket material in any of the bolt holes. Refer to the Engine Rebuilding section found later in this manual for complete details on inspection and refurbishing procedures.
9. Apply a THIN coating of Perfect Seal to both sides of a new ribbed stainless steel gasket and position the gasket over the cylinder block dowel pins. If your engine uses a graphite composition gasket or is a 7.4L/8.2L, do not use any sealer. DO NOT use automotive-type steel gaskets.
10. Position the cylinder head over the dowels in the block. Coat the threads of the head bolts with Perfect Seal and install them finger tight. It never hurts to use new bolts, although it's not necessary. Tighten the bolts, a little at a time, in the sequence illustrated, until the proper tightening torque is achieved. On 1992-97 engines the first step should be 20 ft. lbs. (27 Nm), the second step should be 40 ft. lbs. (54 Nm), or 50 ft. lbs. (68 Nm) on the 7.4L/8.2L and the last step should be to spec-please refer to the Torque Specifications chart at the end of this section.
11. Install rocker assemblies and the cylinder head cover. Don't forget the baffle plate and restrictors on the 5.7L/350 Magnum engines.
12. Install the spark plugs if they were removed and then connect the plug wires.
13. Install the manifolds and connect the fuel line. Don't forget to remove the fitting plugs.
14. Install or connect any other components removed to facilitate getting the head off.
15. Add coolant/water, connect the battery and check the oil. Start the engine and run it for a while to ensure that everything is operating properly. Keep an eye on the temperature gauge.
16. It never hurts to re-tighten the cylinder head bolts again after 20 hours of operation.

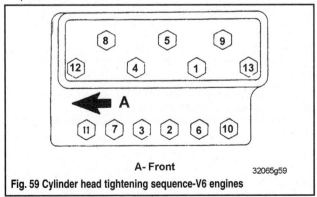

A- Front   32065g59

Fig. 59 Cylinder head tightening sequence-V6 engines

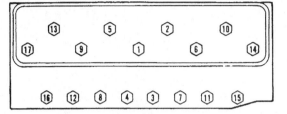

Fig. 60 Cylinder head tightening sequence-5.0L/5.7L/6.2L V8 engines

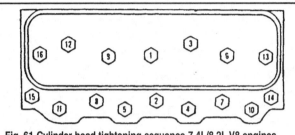

Fig. 61 Cylinder head tightening sequence-7.4L/8.2L V8 engines

8.1L V8 Engines

◆ **See Figures 62 and 63**

1. Drain the water from the cylinder block and manifold.
2. Remove the intake and exhaust manifolds as previously detailed in this section; you can leave the carburetor/throttle body attached to the intake manifold if you like.
3. Tag and disconnect the spark plug wires at the plugs; move them out of the way. Although not necessary, it's a good idea to remove the plugs themselves also.
4. Remove the heat exchanger and coolant crossover.
5. Remove the front and rear engine lifting hooks. Just behind the front hook are hoses for the air actuated drain system, remove them.
6. Remove the alternator bracket.
7. Remove the cylinder head cover and rocker assemblies as detailed previously in this section.
8. Remove or relocate any components or connections that may interfere with the removal of an individual cylinder head.
9. Loosen the cylinder head bolts, from the center bolts and working out to the ends of the head and then carefully lift the head off the block. You may need to persuade it with a rubber mallet-be careful! Set the head down carefully; do not sit it on cement.
To install:
10. Carefully, and thoroughly, remove all residual head gasket material from the cylinder head and block mating surfaces with a scraper or putty knife. Check that the mating surfaces are free of any nicks or cracks. Make sure there is no dirt or old gasket material in any of the bolt holes. Refer to the Engine Rebuilding section found later in this manual for complete details on inspection and refurbishing procedures.
11. Position the cylinder head over the dowels in the block. Coat the threads of the head bolts with Perfect Seal and install them finger tight. It never hurts to use new bolts, although it's not necessary. Tighten the bolts, a little at a time, in the sequence illustrated, until the proper tightening torque is achieved as detailed in the Torque Specifications chart at the end of this section.
12. Install rocker assemblies and the cylinder head cover.
13. Install the spark plugs if they were removed and then connect the plug wires.
14. Install the manifolds and connect the fuel line. Don't forget to remove the fitting plugs.
15. Install or connect any other components removed to facilitate getting the head off.
16. Add coolant/water, connect the battery and check the oil. Start the engine and run it for a while to ensure that everything is operating properly. Keep an eye on the temperature gauge.
17. It never hurts to re-tighten the cylinder head bolts again after 20 hours of operation.

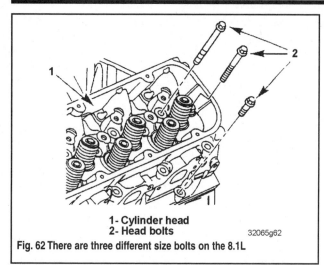

1- Cylinder head
2- Head bolts

32065g62

Fig. 62 There are three different size bolts on the 8.1L

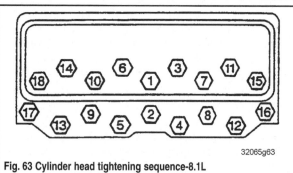

32065g63

Fig. 63 Cylinder head tightening sequence-8.1L

Rear Main Seal

REMOVAL & INSTALLATION

4.3L V6 Engines Without Balance Shaft

◆ See Figures 64 and 65

These engines utilize a two-piece rear main seal. The seal can be removed without removing the crankshaft. You will need to remove the engine for this procedure.

1. Remove the engine as detailed previously in this section.
2. Remove the oil pan and pump as detailed previously in this section.
3. Loosen the retaining bolts and remove the rear main bearing cap. Carefully insert a small prybar and remove the lower half of the seal. Do not damage the seal seating surface.
4. Using a hammer and a small drift, tap on end of the upper seal until it starts to protrude form the other side of the race. Grab the protruding end with pliers and pull out the remaining seal half.
5. Check that you have the correct new seal. Seals with a hatched inner surface can only be used on left hand rotation engines, smooth seals can be used on any engine. Coat the lip and bead with motor oil. Keep oil away from the seal parting surfaces.
6. Your seal kit should come with an installation tool, if not, take a 0.004 in. feeler gauge and cut each side back about a half inch so that you're left with an 11/64 inch point. Bend the tool into the gap between the crankshaft and the seal seating surface. This will be your "shoe horn".
7. Position the upper half of the seal (lip facing the engine) between the crank and the tool so that the seal's bead is in contact with the tool tip. Roll the seal around the crankshaft using the tool as a guide until each end is flush with the cylinder block. Remove the tool.

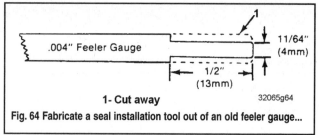

1- Cut away

32065g64

Fig. 64 Fabricate a seal installation tool out of an old feeler gauge...

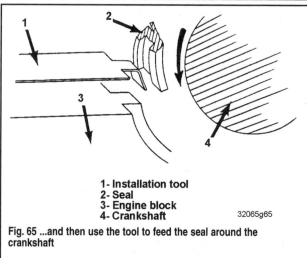

1- Installation tool
2- Seal
3- Engine block
4- Crankshaft

32065g65

Fig. 65 ...and then use the tool to feed the seal around the crankshaft

8. Insert the lower seal half into the main bearing cap with the lip facing the cap. Start it so that one end is slightly below the edge of the cap and then use the tool to shimmy the seal all the way in until both edges are flush with the edge of the cap. Remove the tool.
9. Make sure that the cap/block mating surfaces and the seal ends are free of any oil and then apply a small amount of Perfect Seal to the block just behind where the upper seal ends are.
10. Install the bearing cap and tighten the bolts to 11 ft. lbs. (15 Nm). Tap the crankshaft rearward and then forward to line up the bearing thrust surfaces and then tighten the bolts to 70 ft. lbs. (95 Nm).
11. Install the oil pump and pan.
12. Install the engine.

V6 Engines With Balance Shaft
V8 Engines

◆ See Figures 66 and 67

It is not necessary to remove the engine, oil pan or rear main bearing cap when removing the one-piece oil seal on these engines although you may find it easier to do just that.

1. Remove the flywheel housing and cover as detailed later in this section.
2. Remove the engine coupler and flywheel from the engine as detailed later in this section.
3. Insert a small prybar into one of the three slots in the edge of the seal retainer and slowly pry the seal out of the retainer. Not all 7.4L/8.1L/8.2L engines will necessarily have slots, for that matter they will not have a seal retainer at all; if not, be careful not to nick the block or bearing cap when inserting the prybar.
4. Thoroughly clean the retainer surface.
5. Apply Perfect Seal to the inner mating surface of the seal retainer. Spread a small amount of grease around the outside edge of a new seal and position it over its slot in the retainer.
6. Position a seal driver (J-26817-A) over the seal and strike it with a mallet until the seal is fully seated in its bore.
7. Install the flywheel and engine coupler. Install the cover and flywheel housing.

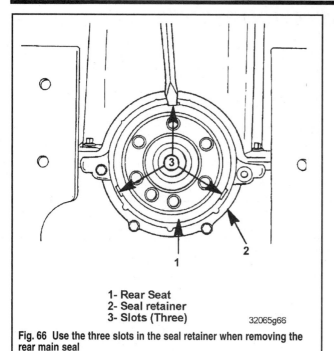

1- Rear Seat
2- Seal retainer
3- Slots (Three)

32065g66

Fig. 66 Use the three slots in the seal retainer when removing the rear main seal

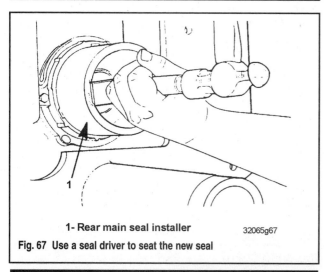

1- Rear main seal installer

32065g67

Fig. 67 Use a seal driver to seat the new seal

Rear Main Seal Retainer

REMOVAL & INSTALLATION

 DIFFICULT

V6 Engines With Balance Shaft
5.0L, 5.7L And 6.2L V8 Engines

◆ See Figure 68

1. Remove the oil pan as previously detailed in this section.
2. Loosen and then remove the six retainer fasteners and lift out the seal retainer.
3. Clean all mating surfaces of old gasket material with a scraper or putty knife.
4. Coat the inner lip of the oil seal with motor oil and then install the retainer with a new gasket. Tighten the nuts and bolts to 133 inch lbs. (15 Nm) on all engines except the 1998 and later V8s which should be tightened to 106 inch lbs. (12 Nm).
5. Install the oil pan.

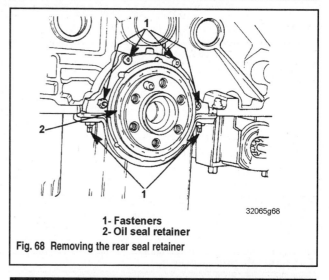

1- Fasteners
2- Oil seal retainer

32065g68

Fig. 68 Removing the rear seal retainer

Flywheel

REMOVAL & INSTALLATION

 DIFFICULT

V6 Engines Without Balance Shaft

1. Remove the engine from the boat as detailed previously in this section.
2. Loosen the retaining bolts and lift off the flywheel housing (bell housing in automotive parlance).
3. Loosen the flywheel/coupler mounting bolts gradually and as you would the lug nuts on your car or truck-that is, in a diagonal star pattern.
4. Thoroughly clean the flywheel mating surface and check it for any nicks, cracks or gouges. Check for any broken teeth.
5. Install the flywheel/coupler over the dowel on the crankshaft and tighten the mounting bolts to 40 ft. lbs. (54 Nm). Once again use the star pattern while tightening the bolts.
6. Install the flywheel housing and tighten the bolts to 30 ft. lbs. (41 Nm).
7. Install the engine.

V6 Engines With Balance Shaft
V8 Engines

1. Remove the engine from the boat as detailed previously in this section.
2. Remove the transmission on inboard models.
3. Loosen the retaining bolts and lift off the flywheel housing (bell housing in automotive parlance).
4. Loosen the engine coupler (stern drive) or drive plate (inboard) mounting bolts and then remove the coupler. Don't lose the rubber bumpers (Alpha).
5. Loosen the flywheel mounting bolts gradually and as you would the lug nuts on your car or truck-that is, in a diagonal star pattern.
6. Thoroughly clean the flywheel mating surface and check it for any nicks, cracks or gouges. Check for any broken teeth.
7. Install the flywheel over the dowel on the crankshaft and tighten the mounting bolts to 75 ft. lbs. (100 Nm); 70 ft. lbs. (95 Nm) on 1992-97 7.4L/8.2L engines, 65 ft. lbs. (88 Nm) on later big blocks. Once again use the star pattern while tightening the bolts.
8. Insert the rubber bumpers into the coupler (Alpha only). Position the engine coupler/drive plate and tighten the mounting bolts to 35 ft. lbs. (48 Nm). If you are re-using the old bolts, make sure you coat them with Loctite® 271. Lubricate the coupler splines with Quicksilver 2-4-C Marine Lubricant.
9. Install the flywheel housing and tighten the bolts to 30 ft. lbs. (41 Nm). Tighten the cover bolts to 80 inch lbs. (9 Nm).
10. Install the transmission if removed.
11. Install the engine.

TORQUE SPECIFICATIONS - 4.3L V6 ENGINES

Component		ft. lbs.	Nm
Balance Shaft			
	Drive Gear Stud	10	16
	Driven Gear Bolt	15 ③	20 ③
	Thrust Plate	10	16
Camshaft			
	Thrust Plate	106 ①	12
	Sprocket bolts	20	27
Connecting rod cap nuts			
	1992-93 (w/o balance shaft)	45	61
	1993-2001 (w/balance shaft)	20 ④	27 ④
Crankshaft pulley bolt		40	54
Cylinder head bolts			
	1992-97	65	88
	1998-2001	⑤	⑤
Distributor clamp bolt			
	1992-93 (w/o balance shaft)	20	27
	1993-2001 (w/balance shaft)	25	34
Engine coupler-to-flywheel			
	1992-93 (w/o balance shaft)	40	54
	1993-2001 (w/balance shaft)	35	48
Engine mounts			
	Front	30	41
	Rear	40	54
Flywheel			
	Housing cover	80 ①	9
	Housing-to-block	30	41
	To crankshaft bolts	75	100
Front cover bolts			
	1992-93 (w/o balance shaft)	80 ①	9
	1993-2001 (w/balance shaft)	124 ①	14
Lifter restrictor plate		12	16
Main bearing cap bolts			
	1992-93 (w/o balance shaft)	70	95
	1993-2001 (w/balance shaft)	75	100
Manifold	Exhaust	20	27
	Intake		
	1992-93 (w/o balance shaft)	40	54
	1993-97 (w/balance shaft)	35	48
	1998-2001	11	15
Oil Pan			
	Screws	80 ①	9
	Nuts	165 ①	19
	Drain plug	15	20

TORQUE SPECIFICATIONS - 4.3L V6 ENGINES (cont'd.)

Component		ft. lbs.	Nm
Oil pump			
	Cover	80 ①	9
	Block	65	88
Rocker arm cover bolts			
	1992-93 (w/o balance shaft)	45	5
	1993-97 (w/balance shaft)	50 ①	5.5
	1998-2001	106 ①	12
Rocker arm nuts			
	1993-97 (w/balance shaft)	20	27
	1998-2001	20	27
Rear crank oil seal retainer nuts		135 ①	15
Seawater pump			
	Brace	30	41
	Bracket	30	41
Spark plugs		15	20
Starter motor		50	68
Timing gear cover		124 ①	14
Torsional damper bolt			
	1992-93 (w/o balance shaft)	60	81
	1993-2001 (w/balance shaft)	40	54
Water circulating pump		30	41
Water temperature sender		20	27

① Specification is inch lbs.

② Retighten after 20 hours of operation

③ Then tighten an additional 35 degrees

④ Then tighten an additional 60 degrees (1998-2001, 70 degrees)

⑤ Step One: 22 ft. lbs. (30 Nm)
   Step Two: Short bolt - additional 55 degrees; medium bolt - additional 65
      degrees; long bolt - additional 75 degrees

TORQUE SPECIFICATIONS - 5.0L, 5.7L AND 6.2L V8 ENGINES

Component			ft. lbs.	Nm
Camshaft				
		Thrust Plate	106 ①	12
		Sprocket bolts	18	24
Connecting rod cap nuts				
	1992-97		45	61
	1998-2001			
		5.0L/5.7L	20 ③	27 ③
		6.2L	36	49
Crankshaft pulley bolt			35	48
Cylinder head bolts				
	1992-97		65	88
	1998-2001		⑤	⑤
Cylinder head cover				
	1992-97		90 ①	10
	1998-2001		106 ①	12
Distributor clamp bolt				
	1992-97		25	34
	1998-2001		18	25
Drive plate-to-flywheel			35	48
Engine coupler-to-flywheel			35	48
Engine mounts				
	Front		30	41
	Rear			
		Stern drive	40	54
		Inboard	50	68
Flywheel				
		Housing cover	80 ①	9
		Housing-to-block	30	41
		To crankshaft bolts	75	100
Front cover bolts				
	1992-97		100 ①	11
	1998-2001		106 ①	12
Lifter restrictor plate				
	1992-97		12	16
	1998-2001		18	25
Main bearing cap bolts				
	1992-97		80	109
	1998-2001		77 ②	105 ②
Manifold				
	Exhaust			
		5.0L/5.7L	25	34
		6.2L	33	45

TORQUE SPECIFICATIONS - 5.0L, 5.7L AND 6.2L V8 ENGINES
(cont'd)

Component			ft. lbs.	Nm
Manifold (cont'd)				
	Intake			
		1992-97	35	48
		1998-2001 5.0L/5.7L	18	24
		1998-2001 6.2L	133 ①	15
Oil Pan				
	Screws		80 ①	9
	Nuts			
		1992-97	165 ①	19
		1998-2001	106 ①	12
	Drain plug		15	20
Oil pump				
	Baffle		25	34
	Cover		80 ①	9
	Block		65	88
Rear crank oil seal retainer nuts				
	1992-97		133 ①	15
	1998-2001		106 ①	12
Seawater pump				
	Brace		30	41
	Bracket		30	41
Spark plugs				
	1992-97		15	20
	1998-2001		15 ④	20 ④
Starter motor			50	68
Torsional damper bolt			60	81
Water circulating pump				
	1992-97		30	41
	1998-2001		33	45
Water temperature sender			20	27

① Specification is inch lbs.

② With angular torque: 15 ft. lbs. (20 Nm) on 1st pass with a final pass of an additional 73 degrees

③ Then tighten an additional 55 degrees

④ New heads: 22 ft. lbs. (30 Nm)

⑤ Step One: 22 ft. lbs. (30 Nm)
   Step Two: Short bolt - additional 55 degrees; medium bolt - additional 65 degrees; long bolt - additional 75 degrees

TORQUE SPECIFICATIONS - 7.4L, 8.1L AND 8.2L V8 ENGINES

Component				ft. lbs.	Nm
Camshaft					
	Retainer bolts				
		8.1L		106 ①	12
	Sprocket bolts				
		1992-97		25mmm	34
		1998-2001		22	30
Connecting rod cap					
	1992-97				
		3/8 in.		50	68
		7/16 in.		73	99
	1998-2001				
		7.4L/8.2L			
			3/8 in.	47	64
			7/16 in.	73	99
		8.1L		22 ⑥	30 ⑥
Crankshaft pulley bolt				35	48
Cylinder head bolts					
	1992-97				
		7.4L		85 ②	115 ②
		454/502 Mag		90 ②	124 ②
	1998-2001				
		7.4L/8.2L			
			Short	85 ③	115 ③
			Long	92 ③	125 ③
		8.1L		⑦	⑦
Cylinder head cover					
	1992-97			90 ①	10
	1998-2001				
		7.4L/8.2L		71 ①	8
		8.1L		106 ①	12
Distributor clamp bolt					
	1992-97			20	27
	1998-2001			24	33
Drive plate-to-flywheel				35	48
Engine coupler-to-flywheel				35	48
Engine mounts					
	Front			30	41
	Rear				
		Stern drive			
			7.4L/8.2L	40	54
			8.1L	38	51
		Inboard		50	68
Flywheel					
	Cover			80 ①	9
	Housing			30	41

TORQUE SPECIFICATIONS - 7.4L, 8.1L AND 8.2L V8 ENGINES
(cont'd)

Component				ft. lbs.	Nm
Flywheel (cont'd)	Crank bolts				
		1992-97		70	95
		1998-2001			88
			7.4L/8.2L	65	88
			8.1L	74	100
Front cover bolts					
		1992-97		120 ①	14
		1998-2001			
			7.4L/8.2L	89 ①	10
			8.1L	106 ①	12
Lifter restrictor plate					
			8.1L	19	25
Main bearing cap bolts					
		1992-97		110	149
		1998-2001			
			7.4L	102	138
			454/502/8.2L	110	149
			8.1L	22 ⑧	30 ⑧
Manifold					
	Exhaust				
		7.4L/8.2L		25	34
		8.1L		26	35
	Intake				
		1992-97		35 ④	48
		1998-2001			
		7.4L			
			Lower	124 ①	14
			Upper	30	40
		454/502/8.2L		35	48
		8.1L		106 ①	12
Oil Pan					
	Bolts				
		1992-97		165 ①	20
		1998-2001		18	25
	Drain plug				
		1992-97		15	20
		1998-2001			
			7.4L/8.2L	20	27
			8.1L	21	28
Oil pump					
	Baffle			25	34
	Cover				
		1992-97		80 ①	9

32065c74

TORQUE SPECIFICATIONS - 7.4L, 8.1L AND 8.2L V8 ENGINES
(cont'd)

Component			ft. lbs.	Nm
Oil pump (cont'd)		1998-2001	106 ①	12
	Block			
		1992-97	70	95
		1998-2001		
		7.4L/8.2L	65	90
		8.1L	56	75
Rocker arm nuts				
		1998-2001		
		7.4L/8.2L	40	54
		8.1L	19	25
Seawater pump				
	Brace		30	41
	Bracket			
		1992-97	30	41
		1998-2001	45	61
Spark plugs				
	1992-97		15	20
	1998-2001			
		7.4L/8.2L	15 ⑤	20 ⑤
		8.1L	22	30
Starter motor			50	68
Thermostat housing				
	7.4L/8.2L		30	41
	8.1L		12	16
Torsional damper bolt				
		1992-97	90	122
		1998-2001		
		7.4L/8.2L	110	149
		8.1L	189	255
Water circulating pump				
	1992-97		35	48
	1998-2001			
		7.4L/8.2L	30	41
		8.1L	37	50

① Specification is inch lbs.

② Figure is final torque. 1st step: 20 ft. lbs. (27 Nm), 2nd step: 50 ft. lbs (68 Nm)

③ Figure is final torque. 1st step: 30 ft. lbs. (40 Nm), 2nd step: 59 ft. lbs (80 Nm)

④ 502 Magnum EFI engines: 25-30 ft. lbs. (34-41 Nm)

⑤ New cylinder head: 22 ft. lbs. (30 Nm)

⑥ Figure is for 1st pass. Tighten an additional 90 degrees

⑦ 1st pass: 22 ft. lbs. (30 Nm); 2nd pass: tighten an additional 120 degrees; Final pass:
long & medium bolts (#'s 1,2,3,6,7,8,9,10,11,14,15,16,17,18) - tighten an additional 60 degrees,
short bolts (#'s 4,5,12,13) - tighten an additional 30 degrees

⑧ Figure is for 1st pass. Tighten an additional 90 degrees

ENGINE SPECIFICATIONS - 4.3L V6 ENGINES

Component		Standard (in.) ①	Metric (mm) ①
Balance Shaft			
Front bearing journal			
	1993-97	1.1812-1.1815	30.002-30.010
	1998-2001	2.1648-2.1654	55.985-55.001
Rear bearing journal			
	1993-97	1.4209-1.4215	36.091-36.106
	1998-2001	1.4994-1.5000	38.084-38.100
Rear bearing I.D.		1.5014-1.503	37.525-37.575
Rear bearing O.D.		1.875-1.876	46.876-46.900
Camshaft			
End play			
	1992-93 w/o balance shaft	0.003-0.008	0.08-0.2032
	1993-97 w/balance shaft	0.004-0.012	0.11-0.30
	1998-2001	0.001-0.009	0.0254-0.2286
Journal diameter			
	1992-93 w/o balance shaft	1.8677-1.8697	47.440-47.490
	1993-2001 w/balance shaft	1.8682-1.8692	47.452-47.478
Journal out-of-round		0.001 Max	0.025 Max
Lobe Lift (+/- 0.002)			
	Intake		
	1992-93 w/o balance shaft	0.2529	6.425
	1993-97 w/balance shaft	0.269	6.892
	1998-2001	0.286-0.290	7.26-7.36
	Exhaust		6.934
	1992-93 w/o balance shaft	0.2529	6.425
	1993-97 w/balance shaft	0.273	6.934
	1998-2001	0.292-0.296	7.42-7.52
Runout		0.002 Max	0.051 Max
Connecting rod			
Bearing clearance			
	Production		
	1992-93 w/o balance shaft	0.0017-0.0027	0.044-0.0686
	1993-2001	0.0013-0.0035	0.0330-0.0889
	Service Limit	0.003 Max	0.07 Max
Crankpin Journal			
	Diameter		
	1992-93 w/o balance shaft	2.0980-2.0995	53.289-53.327
	1993-2001	2.2487-2.2497	57.1170-57.1423
Taper			
	Production		
	1992-93 w/o balance shaft	0.0003 Max	0.008
	1993-97 w/balance shaft	0.0005 Max	0.0127 Max
	1998-2001	0.0003 Max	0.008
	Service Limit	0.001 Max	0.02 Max

ENGINE SPECIFICATIONS - 4.3L V6 ENGINES (cont'd)

Component			Standard (in.) ①	Metric (mm) ①
Connecting Rod				
Out-of-round				
	Production			
		1992-93 w/o bal.	0.0002 Max	0.0007 Max
		1993-97 w/bal. shaft	0.0005 Max	0.0127 Max
		1998-2001	0.0002 Max	0.0007 Max
	Service Limit		0.001 Max	0.02 Max
Side clearance				
		1992-93 w/o bal. shaft	0.006-0.017	0.152-0.44
		1993-97 w/ba. shaft	0.006-0.014	0.152-0.356
		1998-2001	0.006-0.017	0.152-0.44
Crankshaft				
Crankshaft end play				
		1992-97	0.002-0.006	0.05-0.15
		1998-2001	0.002-0.008	0.05-0.20
Main Bearing Clearance				
Production				
	1992-93 w/o bal.shaft			
		#1, 2, 3, 4	0.001-0.0024	0.025-0.060
		#5	0.0016-0.0035	0.041-0.088
	1993-2001			
		#1	0.0008-0.0020	0.0203-0.0508
		#2, 3, 4	0.0011-0.0023	0.0279-0.0584
Service Limit				
	1992-93 w/o bal. shaft			
		#1, 2, 3, 4	0.001-0.0025	0.03-0.06
		#5	0.002-0.0035	0.05-0.08
	1993-97 w/bal. shaft			
		#1	0.0010-0.0015	0.03-0.06
		#2, 3	0.0010-0.0025	0.03-0.06
		#4	0.0025-0.0035	0.07-0.08
	1998-2001		0.001-0.002	0.025-0.06
Main Journal				
Diameter				
	1992-93 w/o bal.		2.2979-2.2994	58.367-58.404
	1993-97 w/bal.			
		#1	2.4484-2.4493	62.1894-62.2122
		#2, 3	2.4481-2.4490	62.1817-62.2046
		#4	2.4479-2.4488	62.1767-62.1995
	1998-2001			
		#1	2.4488-2.4495	62.199-62.217
		#2, 3	2.4485-2.4494	62.191-62.215
		#4	2.4479-2.4489	62.179-62.203

ENGINE SPECIFICATIONS - 4.3L V6 ENGINES (cont'd)

Component			Standard (in.) ①	Metric (mm) ①
Crankshaft (cont'd)				
Taper				
	Production			
		1992-97	0.0002 Max	0.005 Max
		1998-2001	0.0003 Max	0.007 Max
	Service Limit		0.001 Max	0.02 Max
Out-of-round				
		Production	0.0002 Max	0.005 Max
		Service Limit	0.001 Max	0.02 Max
Cylinder Bore				
Diameter				
		1992-93 w/o balance shaft	3.9995-4.0025	101.588-101.790
		1993-2001 w/balance shaft	4.0007-4.0017	101.618-101-643
Out-of-round				
	Production			
		1992-93 w/o balance shaft	0.0005 Max	0.0127 Max
		1993-97 w/balance shaft	0.001 Max	0.025 Max
		1998-2001	0.0005 Max	0.0127 Max
	Service Limit		0.002 Max	0.05 Max
Taper				
Production				
	Thrust side			
	Relief side			
		1992-93 w/o balance shaft	0.0005 Max	0.012 Max
		1993-2001 w/balance shaft	0.001 Max	0.025 Max
Service Limit			0.001 Over Prod.	0.025 Over Prod.
Cylinder Head				
Gasket surface flatness				
		1992-93 w/o balance shaft	0.007 Overall Max	0.15 Overall Max
		1993-97 w/balance shaft	0.004 Overall Max	0.10 Overall Max
		1998-2001	0.010 Overall Max	0.254 Overall Max
Flywheel				
	Runout		0.008 Max	0.203 Max
Piston				
Clearance				
	Production			
		1992-93 w/o balance shaft	0.0025-0.0035	0.064-0.088
		1993-97 w/balance shaft	0.0007-0.0017	0.0178-0.0431
		1998-2001	0.0007-0.002	0.017-0.05
	Service Limit			
		1992-93 w/o balance shaft	0.0035 Max	0.08 Max
		1993-97 w/balance shaft	0.0027 Max	0.07 Max
		1998-2001	0.002 Max	0.60 Max

ENGINE SPECIFICATIONS - 4.3L V6 ENGINES (cont'd)

Component			Standard (in.) ①	Metric (mm) ①
Piston (cont'd)				
Compression Rings				
Groove clearance				
Production				
	Top:			
		1992-93 w/o bal. shaft	0.0012-0.0029	0.030-0.073
		1993-97 w/bal. shaft	0.0012-0.0032	0.0305-0.0813
		1998-2001	0.02-0.06	0.508-1.524
	2nd:			
		1992-93 w/o bal. shaft	0.0012-0.0029	0.030-0.073
		1993-97 w/bal. shaft	0.0012-0.0032	0.0305-0.0813
		1998-2001	0.04-0.08	1.016-2.032
Service Limit				
		1992-93 w/o bal. shaft	0.003 Max	0.09 Max
		1993-97 w/bal. shaft	High limit prod. +0.001	High limit prod. +0.025
		1998-2001	0.004 Max	0.10 Max
Gap				
Production				
	Top			
		1992-97	0.010-0.020	0.254-0.508
		1998-2001	0.010-0.016	0.25-0.40
	2nd:			
		1992-93 w/o bal. shaft	0.017-0.025	0.432-0.635
		1993-97 w/bal. shaft	0.010-0.025	0.254-0.635
		1998-2001	0.018-0.026	0.46-0.66
Service Limit				
		1992-93 w/o bal. shaft	0.035 Max	0.88 Max
		1993-97 w/bal. shaft	High limit prod. +0.010	High limit prod. +0.254
		1998-2001	0.06-0.035 Max	1.52-0.88 Max
Oil Ring				
Groove clearance				
	Production			
		1992-93 w/o bal. shaft	0.001-0.006	0.026-0.152
		1993-2001 w/bal. shaft	0.002-0.007	0.051-0.177
	Service Limit			High limit prod. +0.025
		1992-93 w/o bal. shaft	0.007 Max	0.17 Max
		1993-97 w/bal. shaft	High limit prod. +0.001	High limit prod. +0.025
		1998-2001	0.02-0.09 Max	0.50-2.03 Max
Gap				
	Production			
		1992-93 w/o bal. shaft	0.010-0.030	0.254-0.762
		1993-97 w/bal. shaft	0.015-0.055	0.381-1.397
		1998-2001	0.015-0.050	0.381-1.27

ENGINE SPECIFICATIONS - 4.3L V6 ENGINES (cont'd)

Component			Standard (in.) ①	Metric (mm) ①
Piston (cont'd)				
	Service Limit			
		1992-93 w/o bal. shaft	0.040 Max	1.01 Max
		1993-97 w/bal. shaft	High limit prod. +0.010	High limit prod. +0.254
		1998-2001	0.009-0.065	0.23-1.65
Piston pin				
	Diameter			
		1992-93 w/o bal. shaft	0.9270-0.9271	23.546-23.550
		1993-97 w/bal. shaft	0.9270-0.9273	23.546-23.553
		1998-2001	0.9267-0.9271	23.545-23.548
Clearance				
	Production			
		1992-93 w/o bal. shaft	0.0003-0.0006	0.008-0.016
		1993-2001 w/bal. shaft	0.0002-0.0007	0.0051-0.0177
	Service Limit		0.001 Max	0.025 Max
Interference fit				
		1992-93 w/o bal. shaft	0.0008-0.0019	0.020-0.050
		1993-2001 w/bal. shaft	0.0008-0.0016	0.0203-0.04060
Valve system				
Damper				
	1992-93		None	
	1993-2001			
		Free length	1.86	47
		Number of coils	4	
Face Angle				
		Intake	45 deg.	
		Exhaust	45 deg.	
Lash				
		1992-93 w/o bal. shaft	Tighten 1 turn from 0 lash	
		1993-2001 w/bal. shaft	Fixed	
Lift				
	Intake			
		1993-97 w/bal. shaft	0.394	10
		1998-2001	0.414	10.52
	Exhaust			
		1993-97 w/bal. shaft	0.404	10.26
		1998-2001	0.428	10.87
Lifter			Roller Hydraulic	
Rocker arm ratio				
		1992-93 w/o bal. shaft	1.75:1	
		1993-2001 w/bal. shaft	1.50:1	

ENGINE SPECIFICATIONS - 4.3L V6 ENGINES (cont'd)

Component			Standard (in.) ①	Metric (mm) ①
Valve system				
Seat				
Angle			46 deg.	
Runout			0.002 Max	0.051 Max
Width				
	Intake			
		1992-93 w/o bal. shaft	0.0625	1.6
		1993-97 w/bal. shaft	0.0625	1.6
		1998-2001	0.035-0.060	0.89-1.52
	Exhaust			
		1992-93 w/o bal. shaft	0.063-0.078	1.6-2.0
		1993-97 w/bal. shaft	0.07	1.8
		1998-2001	0.062-0.093	1.58-2.38
Spring w/damper removed Free length			2.03	51.6
Pressure				
	Valve open			
		1992-93 w/o bal. shaft	208-222 ft. lbs @ 1.22 in.	282-300 Nm @ 31mm
		1993-97 w/bal. shaft	194-206 lbs. @ 1.25 in.	863-916 N @ 31.75mm
		1998-2001	187-203 lbs. @ 1.27 in.	832-903 N @ 32.25MM
	Valve closed			
		1992-93 w/o bal. shaft	100-110 ft. lbs. @ 1.61 in.	136-149 Nm @ 41mm
		1993-2001 w/bal. shaft	76-84 lbs. @ 1.70 in.	338-374 N @ 43.16mm
Installed height				
	Intake			
		1992-93 w/o bal. shaft	1.66	42
		1993-97 w/bal. shaft	1.7	43
		1998-2001	1.78	45.2
	Exhaust			
		1992-93 w/o bal. shaft	1.66	42
		1993-97 w/bal. shaft	1.7	43
		1998-2001	1.69-1.71	42.9-43.43
Stem clearance Production				
	Intake		0.0010-0.0027	0.0254-0.0686
	Exhaust			
		1992-93 w/o bal. shaft	0.0007-0.0027	0.018-0.0686
		1993-2001 w/bal. shaft	0.0010-0.0027	0.0254-0.0686
Service Limit				
	Intake			
		1992-97	0.003 Max	0.09 Max
		1998-2001	0.001 Max	0.025 Max

ENGINE SPECIFICATIONS - 4.3L V6 ENGINES (cont'd)

Component			Standard (in.) ①	Metric (mm) ①
Valve system (cont'd)				
	Exhaust			
		1992-97	0.004 Max	0.11 Max
		1998-2001	0.002 Max	0.51 Max
Stem diameter				
	Intake		0.341	8.66
	Exhaust			
		1992-97	0.372	9.45
		1998-2001	0.341	8.66
Valve diameter				
	Intake		1.94	49.28
	Exhaust		1.5	38.1

① Unless otherwise noted

ENGINE SPECIFICATIONS - 5.0L, 5.7L AND 6.2L V8 ENGINES

Component			Standard (in.) ①	Metric (mm) ①
Camshaft				
Journal diameter				
		1992-97	1.8682-1.8692	47.452-47.478
		1998-2001	1.8677-1.8661	47.440-47.490
Journal out-of-round			0.001 Max	0.025 Max
Lobe Lift (+/- 0.002)				
	Intake			
		1992-97	0.263 ②	6.68 ②
	Exhaust			
		1992-97	0.269 ③	6.833 ③
Runout				
		1992-97	0.002 Max	0.051 Max
		1998-2001	0.0026 Max	0.06 Max
Connecting rod				
Bearing clearance				
	Production		0.0013-0.0035	0.0330-0.0889
	Service Limit			
		1992-97	0.003 Max	0.07 Max
		1998-2001	0.001-0.003	0.025-0.076
Crankpin Journal				
	Diameter			
		1992-97	2.0988-2.0998	53.3095-53.3349
		1998-2001	2.0977-2.0997	53.284-53.334
Taper				
	Production			
		1992-97	0.0005 Max	0.0127 Max
		1998-2001	0.0003 Max	0.007 Max
	Service Limit		0.001 Max	0.02 Max
Out-of-round				
	Production			
		1992-97	0.0005 Max	0.0127 Max
		1998-2001	0.0003 Max	0.007 Max
	Service Limit		0.001 Max	0.02 Max
Side clearance				
		Production	0.0013-0.0035	0.0330-0.0889
		Service Limit	0.001 Max	0.02 Max
Crankshaft				
	End play			
		1992-97	0.002-0.006	0.05-0.15
		1998-2001	0.002-0.008	0.05-0.20

ENGINE SPECIFICATIONS - 5.0L, 5.7L AND 6.2L V8 ENGINES (cont'd)

Component			Standard (in.) ①	Metric (mm) ①
Main Bearing Clearance Production				
	1992-97			
		#1	0.0008-0.0020	0.0203-0.0508
		#2, 3, 4	0.0011-0.0023	0.0279-0.0584
		# 5	0.0017-0.0032	0.0432-0.0813
	1998-2001			
		#1	0.0007-0.0022	0.018-0.053
		#2, 3, 4	0.0009-0.0024	0.022-0.061
		# 5	0.0010-0.0027	0.025-0.069
Service Limit				
	1992-97			
		#1	0.0010-0.0015	0.03-0.06
		#2, 3, 4	0.0010-0.0025	0.03-0.06
		# 5	0.0025-0.0035	0.07-0.08
	1998-2001			
		#1	0.0010-0.0020	0.025-0.051
		#2, 3, 4	0.0010-0.0025	0.025-0.064
		# 5	0.0015-0.0029	0.038-0.076
Main Journal Diameter				
	1992-97			
		#1	2.4484-2.4493	62.1894-62.2122
		#2, 3, 4	2.4481-2.4490	62.1817-62.2046
		#5	2.4479-2.4488	62.1767-62.1995
	1998-2001			
		#1	2.4483-2.4492	62.189-62.212
		#2, 3, 4	2.4480-2.4490	62.181-62.207
		#5	2.4479-2.4490	62.177-62.207
Taper				
	Production		0.0002 Max	0.005 Max
	Service Limit		0.001 Max	0.02 Max
Out-of-round				
	Production		0.0002 Max	0.005 Max
	Service Limit		0.001 Max	0.02 Max
Cylinder Bore Diameter				
	1992-97			
		5.0L	3.7350-3.7385	94.8690-94.9579
		5.7L/350 Mag	3.9995-4.0025	101.5873-101.6635
	1998-2001			
		5.0L	3.7350-3.7384	94.881-94.958

ENGINE SPECIFICATIONS - 5.0L, 5.7L AND 6.2L V8 ENGINES (cont'd)

Component			Standard (in.) ①	Metric (mm) ①
Cyl. Bore (cont'd)	5.7L/350/6.2L		4.00-4.001	101.618-101.643
Out-of-round				
		Production	0.001 Max	0.025 Max
		Service Limit	0.002 Max	0.05 Max
Taper				
	Production			
		Thrust side	0.0005 Max	0.012 Max
		Relief side	0.001 Max	0.025 Max
	Service Limit		0.001 Over prod.	0.025 Over prod.
Cylinder Head				
Surface flatness				
	1992-97		0.007 Overall Max	0.15 Overall Max
	1998-2001			
		Ex. Man. deck	0.0019	0.05
		Eng. block deck	0.0039	0.1
		In. man. deck	0.0039	0.1
Flywheel				
		Runout	0.008 Max	0.203 Max
Piston				
Clearance				
	Production			
		1992-97	0.0007-0.0017	0.0178-0.0431
		1998-2001	0.0007-0.002	0.017-0.05
	Service Limit			
		1992-97	0.0027 Max	0.07 Max
		1998-2001	0.0007-0.0026 Max	0.018-0.068 Max
Compression Rings				
Groove clearance				
Production				
	Top:			
		1992-97	0.0012-0.0032	0.0305-0.0813
		1998-2001	0.0012-0.0027	0.030-0.070
	2nd:			
		1992-97	0.0012-0.0032	0.0305-0.0813
		1998-2001	0.0015-0.003	0.040-0.080
Service Limit				
	1992-97		High limit prod. +0.001	High limit prod. +0.025
	1998-2001			
		Top:	0.0012-0.0035	0.030-0.090
		2nd:	0.0015-0.004	0.040-0.100
Gap				
	Production		0.010-0.020	0.254-0.508

ENGINE SPECIFICATIONS - 5.0L, 5.7L AND 6.2L V8 ENGINES (cont'd)

Component			Standard (in.) ①	Metric (mm) ①
Piston (cont'd)	Service Limit			
	1992-97		High limit prod. +0.010	High limit prod. +0.254
	1998-2001			
		Top:	0.010-0.025 ⑤	0.25-0.65 ⑤
		2nd:	0.018-0.035 ⑤	0.46-0.90 ⑤
Oil Ring Groove clearance Production			0.002-0.007	0.051-0.177
Service Limit				
	1992-97		High limit prod. +0.001	High limit prod. +0.025
	1998-2001			
		5.0L	0.002-0.008	0.051-0.22
		5.7L/350/6.2L	0.002-0.0076	0.051-0.195
Gap Production				
	1992-97		0.015-0.055	0.381-1.397
	1998-2001		0.009-0.029	0.25-0.76
Service Limit				
	1992-97		High limit prod. +0.010	High limit prod. +0.254
	1998-2001			
		5.0L	0.009-0.035	0.25-0.89
		5.7L/350 Mag	0.009-0.030 ⑥	0.25-0.785 ⑥
Piston pin	Diameter			
	1992-97		0.9269-0.9271	23.5458-23.5534
	1998-2001		0.9269-0.9270	23.545-23.548
Clearance	Production			
	1992-97		0.0002-0.0007	0.0051-0.0177
	1998-2001		0.0005-0.0009	0.013-0.023
	Service Limit			
	1992-97		0.001 Max	0.025 Max
	1998-2001		0.0005-0.00098 Max	0.013-0.025 Max
Interference fit			0.0008-0.0016	0.0203-0.04060
Valve system Damper	Free length			
	1992-97		1.86	47
	1998-2001		2.019	51.3
	# of coils		4	
Face Angle				
	Intake		45 deg.	
	Exhaust		45 deg.	

ENGINE SPECIFICATIONS - 5.0L, 5.7L AND 6.2L V8 ENGINES (cont'd)

Component		Standard (in.) ①	Metric (mm) ①
Valve (cont'd)			
Lash		Tighten 1 turn from 0 lash	
Lift			
Intake			
1992-97		0.394	10
1998-2001			
	5.0L/5.7L/350	0.2744-0.2783	6.97-7.07
	6.2L	0.3114-0.3153	7.91-8.01
Exhaust			
1992-97	1992-97	0.404	10.26
1998-2001			
	5.0L/5.7L/350	0.2834-0.2874	7.20-7.30
	6.2L	0.3160-0.3200	8.03-8.13
Lifter			
	1992-97	Flat Hydraulic ④	
	1998-2001	Roller Hydraulic	
Rocker arm ratio		1.50:1	
Seat			
Angle		46 deg.	
Runout		0.002 Max	0.051 Max
Width			
Intake			
1992-97		0.031-0.063	0.0794-1.587
1998-2001			
	5.0L/5.7L/350	0.035-0.070	0.0401-0.649
	6.2L	0.040-0.065	1.02-1.65
Exhaust			
1992-97		0.063-0.094	1.587-2.381
1998-2001	5.0L/5.7L/350	0.065-0.098	0.059-0.100
	6.2L	0.059-0.100	1.50-2.56
Spring w/damper removed			
Free length			
	1992-97	2.03	51.6
	1998-2001	2.019	51.3
Pressure			
Valve open			
	1992-97	194-206 lbs. @ 1.25 in.	863-916 N @ 31.75mm
	1998-2001	187-203 lbs. @ 1.27 in.	832-903 N @ 32.25mm
	Valve closed	76-84 lbs. @ 1.70 in.	338-374 N @ 43.16mm
Installed height			
Intake			
	1992-97	1.7	43
	1998-2001	1.68-1.70	42.92-43.43
Exhaust			
	1992-97	1.7	43

ENGINE SPECIFICATIONS - 5.0L, 5.7L AND 6.2L V8 ENGINES (cont'd)

Component			Standard (in.) ①	Metric (mm) ①
Valve (cont'd)				
Installed height	1998-2001		1.68-1.70	42.92-43.43
Stem clearance				
Production				
	Intake		0.0010-0.0027	0.0254-0.0686
	Exhaust		0.0010-0.0027	0.0254-0.0686
Service Limit				
	Intake			
	1992-97		0.004 Max	0.094 Max
	1998-2001		0.0010-0.0037	0.025-0.094
	Exhaust			
	1992-97		0.005 Max	0.11 Max
	1998-2001			
		5.0L	0.0010-0.0037	0.025-0.094
		5.7L/350/6.2L	0.0010-0.0076	0.025-0.194
Stem diameter				
	Intake		0.341	8.66
	Exhaust			
	1992-97		0.372	9.45
	1998-2001		0.341	8.66
Valve diameter				
	Intake			
	5.0L		1.84	46.74
	5.7L/350/6.2L		1.94	49.28
	Exhaust		1.5	38.1

① Unless otherwise noted
② 5.7L engines with roller lifters and steel camshaft: 0.287 in. (6.680)
③ 5.7L engines with roller lifters and steel camshaft: 0.300 in. (6.833)
④ Engines with steel camshafts use roller lifters
⑤ 6.2L: Top - 0.012-0.027; 2nd - 0.020-0.037
⑥ 6.2L:0.0011-0.032 in. (0.28-0.81mm)

ENGINE SPECIFICATIONS - 7.4L, 8.1L AND 8.2L V8 ENGINES

Component			Standard (in.) ①	Metric (mm) ①
Camshaft				
Journal diameter				
	7.4L/8.2L		1.9482-1.9492	49.485-49.509
	8.1L		1.9477-1.9479	49.472-49.522
Out-of-round			0.001 Max	0.025 Max
Lobe Lift				
	Intake			
	1992-97			
		7.4L	0.271 ②	6.88 ②
		454/502/8.2L	0.300 ②	7.62 ②
	1998-2001			
		7.4L	0.282 ②	7.163 ②
		8.1S/496 Mag	0.282	7.163
		8.1S/496 HO	0.3	7.62
		454/502/8.2L	0.301 ②	7.65 ②
	Exhaust			
	1992-97			
		7.4L	0.282 ②	7.163 ②
		454/502/8.2L	0.300 ②	7.62 ②
	1998-2001			
		7.4L	0.284 ②	7.21 ②
		8.1S/496 Mag	0.284	7.22
		8.1S/496 HO	0.3	7.62
		454/502/8.2L	0.301 ②	7.65 ②
Runout			0.002 Max	0.051 Max
Connecting rod				
Clearance				
	Production		0.0011-0.0029	0.0279-0.0735
	Service Limit			
	1992-97		0.003 Max	0.07 Max
	1998-2001			
		7.4L	0.001 Max	0.0254 Max
		454/502/8.2L	0.003 Max	0.07 Max
Journal				
	Diameter		2.1990-2.1996	55.855-55.870
Taper				
	Production			
		7.4L/8.2L	0.0005 Max	0.0127 Max
		8.1L	0.0004	0.0102
	Service Limit		0.001 Max	0.02 Max

ENGINE SPECIFICATIONS - 7.4L, 8.1L AND 8.2L V8 ENGINES (cont'd)

Component			Standard (in.) ①	Metric (mm) ①
Connecting rod (cont'd)				
Out-of-round				
	Production			
		7.4L/8.2L	0.0005 Max	0.0127 Max
		8.1L	0.0004	0.0102
	Service Limit		0.001 Max	0.02 Max
Side clearance				
		7.4L/8.2L	0.013-0.023	0.35-0.58
		8.1L	0.015-0.027	0.384-0.686
Crankshaft				
	End play			
		1992-97	0.006-0.010	0.15-0.25
		1998-2001		
		7.4L/8.1L	0.005-0.011	0.127-0.279
		454/502/8.2L	0.006-0.010	0.15-0.25
Main Bearing Clearance Production				
		#1, 2, 3, 4		
		7.4L/8.2L	0.0017-0.0030	0.0431-0.0762
		8.1L	0.0011-0.0024	0.028-0.061
		# 5	0.0025-0.0038	0.0635-0.0965
Service Limit				
	1992-97			
		#1		
		7.4L	0.001-0.003	0.03-0.07
		454/502/8.2L	0.0010-0.0015	0.025-0.038
		#2, 3, 4		
		7.4L	0.001-0.003	0.03-0.07
		454/502/8.2L	0.0010-0.0025	0.025-0.064
		# 5	0.0025-0.0040	0.064-0.102
	1998-2001			
		#1		
		7.4L	0.0010-0.0030	0.025-0.076
		8.2L	0.0010-0.0015	0.025-0.038
		#2, 3, 4		
		7.4L	0.0010-0.0030	0.025-0.076
		8.2L	0.0010-0.0025	0.025-0.064
		# 5	0.0025-0.0040	0.064-0.102
Main Journal Diameter			2.7482-2.7489	69.804-69.822

ENGINE SPECIFICATIONS - 7.4L, 8.1L AND 8.2L V8 ENGINES (cont'd)

Component			Standard (in.) ①	Metric (mm) ①
Crankshaft (cont'd)				
Taper				
	Production			
	454/502/8.2L		0.0002 Max	0.005 Max
	7.4L/8.1L			
		1992-97	0.0002 Max	0.005 Max
		1998-2001	0.0004 Max	0.010 Max
	Service Limit		0.001 Max	0.02 Max
Out-of-round				
	Production			
	454/502/8.2L		0.0002 Max	0.005 Max
	7.4L8.1L			
		1992-97	0.0002 Max	0.005 Max
		1998-2001	0.0004 Max	0.010 Max
	Service Limit		0.001 Max	0.02 Max
Cylinder Bore				
Diameter				
	7.4L/8.1L		4.2500-4.2507	107.95-107.97
	454 Mag		4.2451-4.2525	107.83-108.01
	502 Mag/8.2L		4.4655-4.4662	113.42-113.44
Out-of-round				
	Production		0.001 Max	0.025 Max
	Service Limit		0.002 Max	0.05 Max
Taper				
	Production			
		Thrust side	0.0005 Max	0.012 Max
		Relief side	0.001 Max	0.025 Max
	Service Limit		0.001 Over prod.	0.025 Over prod.
Cylinder Head				
Surface flatness				
1992-97			0.007 Overall Max	0.15 Overall Max
1998-2001				
	7.4L/8.1L		0.004 Max	0.102 Overall Max
	454/502/8.2L		0.007 Overall Max	0.15 Overall Max
Flywheel				
	Runout		0.008 Max	0.203 Max
Piston				
Clearance				
	Production			
	7.4L		0.0030-0.0042 ③	0.076-0.107 ③
	454 Mag		0.0025-0.0037	0.064-0.094
	502 Mag/8.2L		0.0040-0.0057	0.102-0.145

ENGINE SPECIFICATIONS - 7.4L, 8.1L AND 8.2L V8 ENGINES (cont'd)

Component			Standard (in.) ①	Metric (mm) ①
Piston (cont'd)				
Service Limit				
	1992-97			
		7.4L	0.005 Max	0.12 Max
		454 Mag	0.0075 Max	0.15 Max
		502 Mag/8.2L	0.0065 Max	0.16 Max
	1998-2001			
		7.4L	0.0018 Max	0.046 Max
		454 Mag	0.0075 Max	0.15 Max
		502 Mag/8.2L	0.0065 Max	0.16 Max
Compression rings				
Clearance				
Production				
	Top:			
		7.4L/8.1L	0.0012-0.0029	0.031-0.074
		454/502/8.2L	0.0017-0.0032	0.044-0.081
	2nd:			
		7.4L/8.1L	0.0012-0.0029	0.031-0.074
		454/502/8.2L	0.0017-0.0032	0.044-0.081
Service Limit			High limit +0.001	High limit +0.025
Gap				
Production				
	Top:			
		7.4L	0.010-0.018	0.25-0.46
		8.1L	0.010-0.016	0.25-0.41
		8.2L	0.011-0.021	0.28-0.53
	2nd:			
		7.4L	0.016-0.024	0.41-0.61
		8.1L	0.018-0.026	0.45-0.65
		8.2L	0.016-0.026	0.41-0.66
Service Limit			High limit +0.010	High limit +0.254
Oil Ring				
Clearance				
		Production	0.0050-0.0065	0.127-0.165
		Service Limit		
		7.4L	High limit +0.001	High limit +0.025
		8.2L	High limit +0.005	High limit +0.127
Gap				
		Production		
		7.4L/8.2L	0.010-0.030	0.254-0.762
		8.1L	0.0098-0.0099	0.249-0.759
		Service Limit		
		7.4L	High limit +0.001	High limit +0.025
		8.2L	High limit +0.005	High limit +0.127

ENGINE SPECIFICATIONS - 7.4L, 8.1L AND 8.2L V8 ENGINES (cont'd)

Component				Standard (in.) ①	Metric (mm) ①
Piston pin					
	Diameter				
			7.4L	0.9895-0.9897	25.132-25.137
			8.1L	1.04-1.0401	26.416-26.419
			454/502/8.2L	0.9895-0.9898	25.132-25.140
	Clearance				
		Production			
			7.4L	0.0002-0.0007	0.0050-0.0177
			8.1L	0.0004-0.0007	0.010-0.017
			454/502/8.2L	0.00025-0.00035	0.0064-0.0088
		Service Limit		0.001 Max	0.025 Max
Interference fit				0.0008-0.0016	0.0203-0.04060
	7.4L				
			1992-97	0.0008-0.0016	0.0203-0.04060
			1998-2001	0.0021-0.0031	0.0533-0.0787
	8.1L			0.0002-0.0007	0.049-0.020
	454/502/8.2L			0.0008-0.0016	0.0203-0.04060
Valve system					
Damper					
		Free length			
			454/502/8.2L	1.86	47
		# of coils		4	
Face Angle					
		Intake		45 deg.	
		Exhaust		45 deg.	
Lash				Fixed	
Lift					
	Intake				
		1992-97			
			7.4L	0.461	11.709
			454/502/8.2L	0.51	12.954
	Exhaust				
		1992-97			
			7.4L	0.479	12.167
			454/502/8.2L	0.51	12.954
Lifter				Hydraulic	
Rocker arm ratio				1.7:1	
Seat					
	Angle			46 deg.	
	Runout			0.002 Max	0.051 Max
	Width				
		Intake		0.031-0.063	0.8-1.6
		Exhaust		0.063-0.094	1.6-2.3

ENGINE SPECIFICATIONS - 7.4L, 8.1L AND 8.2L V8 ENGINES (cont'd)

Component			Standard (in.) ①	Metric (mm) ①
Valves (cont'd)				
Spring w/damper removed Free length			1.86	47.2
Pressure				
	Valve open			
		7.4L		
		1992-97	195-215 lbs.@1.40 in.	867-956 N@35.5mm
		1998-2001	238-262 lbs.@1.35 in.	1059-1165 N@34.2mm
		8.1L	189-207 lbs.@1.34 in.	840-920 N@33.98mm
		454/502/8.2L	316 lbs.@1.34 in.	1406 N@35.1mm
	Valve closed			
		7.4L		
		1992-97	74-86 lbs.@1.80 in.	329-382 N@45.7mm
		1998-2001	71-79 lbs.@1.84 in.	316-351 N@46.69mm
		8.1L	58-64 lbs.@1.84 in.	267-293 N@46.68mm
		454/502/8.2L	110 lbs.@1.88 in.	489 N@47.8mm
Installed height				
		7.4L		
		1992-97	1.875	47.6
		1998-2001	1.838	46.7
		8.1L	1.838-1.869	46.7-47.5
		454/502/8.2L	1.88	47.7
Stem clearance Production				
	Intake			
		7.4L		
		1992-97	0.0010-0.0027	0.0254-0.0686
		1998-2001	0.0012-0.0029	0.0304-0.0736
		8.1L	0.0010-0.0029	0.0254-0.0736
		454/502/8.2L	0.0010-0.0027	0.0254-0.0686
	Exhaust		0.0012-0.0029	0.0304-0.0736
		7.4L/8.1L		
		1992-97	0.0012-0.0029	0.0304-0.0736
		1998-2001	0.0012-0.0031	0.0304-0.0787
		454/502/9.2L	0.0012-0.0029	0.0304-0.0736
Service Limit				
	Intake			
		7.4L		
		1992-97	0.001	0.02
		1998-2001	0.004	0.094
		454	0.003	0.076
		502/8.2L	0.004	0.094

ENGINE SPECIFICATIONS - 7.4L, 8.1L AND 8.2L V8 ENGINES (cont'd)

Component				Standard (in.) ①	Metric (mm) ①
Valves (cont'd)					
	Exhaust				
		7.4L			
			1992-97	0.002	0.05
			1998-2001	0.005	0.125
		454		0.004	0.102
		502/8.2L		0.005	0.125
Stem diameter					
	Intake			0.372	9.45
	Exhaust			0.372	9.45
Valve diameter					
	Intake				
		1992-97			
			7.4L	1.84	46.74
			8.2L	1.94	49.28
	Exhaust				
		1992-97		1.5	38.1

① Unless otherwise noted

② +/- 0.002 in. (0.051mm)

③ 1998-2001: 0.0018-0.0030 in. (0.0457-0.0762mm)

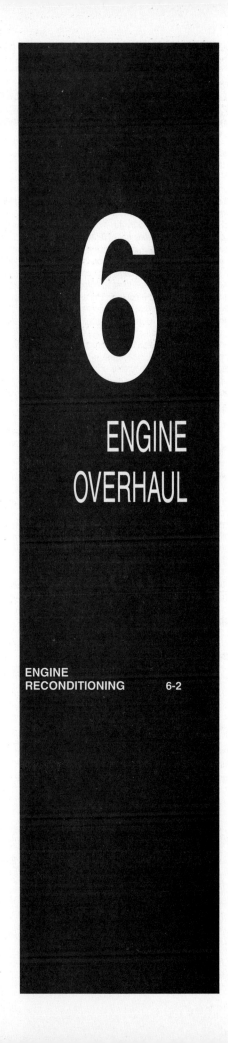

6

ENGINE
OVERHAUL

ENGINE
RECONDITIONING   6-2

ENGINE RECONDITIONING

Determining Engine Condition

Anything that generates heat and/or friction will eventually burn or wear out (for example, a light bulb generates heat, therefore its life span is limited). With this in mind, a running engine generates tremendous amounts of both; friction is encountered by the moving and rotating parts inside the engine and heat is created by friction and combustion of the fuel. However, the engine has systems designed to help reduce the effects of heat and friction and provide added longevity. The oiling system reduces the amount of friction encountered by the moving parts inside the engine, while the cooling system reduces heat created by friction and combustion. If either system is not maintained, a break-down will be inevitable. Therefore, you can see how regular maintenance can affect the service life of your engine. If you do not drain, flush and refill your cooling system at the proper intervals, deposits will begin to accumulate, thereby reducing the amount of heat it can extract from the coolant. The same applies to your oil and filter; if it is not changed often enough it becomes laden with contaminates and is unable to properly lubricate the engine. This increases friction and wear.

There are a number of methods for evaluating the condition of your engine. A compression test can reveal the condition of your pistons, piston rings, cylinder bores, head gasket(s), valves and valve seats. An oil pressure test can warn you of possible engine bearing, or oil pump failures. Excessive oil consumption, evidence of oil in the engine air intake area and/or bluish smoke from the exhaust may indicate worn piston rings, worn valve guides and/or valve seals.

COMPRESSION TEST

◆ See Figure 1

A noticeable lack of engine power, excessive oil consumption and/or poor fuel mileage measured over an extended period are all indicators of internal engine wear. Worn piston rings, scored or worn cylinder bores, blown head gaskets, sticking or burnt valves, and worn valve seats are all possible culprits. A check of each cylinder's compression will help locate the problem.

■ **A screw-in type compression gauge is more accurate than the type you simply hold against the spark plug hole.**
Although it takes slightly longer to use, it's worth the effort to obtain a more accurate reading.

1. Make sure that the proper amount and viscosity of engine oil is in the crankcase, then ensure the battery is fully charged.
2. Warm-up the engine to normal operating temperature, then shut the engine **OFF**.
3. Disable the ignition system.
4. Label and disconnect all of the spark plug wires from the plugs.
5. Thoroughly clean the cylinder head area around the spark plug ports, then remove the spark plugs.
6. Set the throttle plate to the fully open (wide-open throttle) position. You can block the throttle linkage open for this, or you can have an assistant operate the throttle lever in the boat.

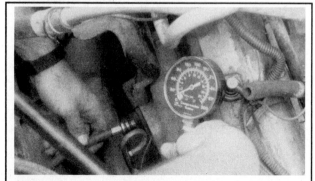

Fig. 1 A screw-in type compression gauge is more accurate and easier to use without an assistant

7. Install a screw-in type compression gauge into the No. 1 spark plug hole until the fitting is snug.

✳✳ WARNING

Be careful not to crossthread the spark plug hole.

8. According to the tool manufacturer's instructions, connect a remote starting switch to the starting circuit.
9. With the ignition switch in the **OFF** position, use the remote starting switch to crank the engine through at least five compression strokes (approximately 5 seconds of cranking) and record the highest reading on the gauge.
10 Repeat the test on each cylinder, cranking the engine approximately the same number of compression strokes and/or time as the first.
11. Compare the highest readings from each cylinder to that of the others. The indicated compression pressures are considered within specifications if the lowest reading cylinder is within 75 percent of the pressure recorded for the highest reading cylinder. For example, if your highest reading cylinder pressure was 150 psi (1034 kPa), then 75 percent of that would be 113 psi (779 kPa). So the lowest reading cylinder should be no less than 113 psi (779 kPa).
12. If a cylinder exhibits an unusually low compression reading, pour a tablespoon of clean engine oil into the cylinder through the spark plug hole and repeat the compression test. If the compression rises after adding oil, it means that the cylinder's piston rings and/or cylinder bore are damaged or worn. If the pressure remains low, the valves may not be seating properly (a valve job is needed), or the head gasket may be blown near that cylinder. If compression in any two adjacent cylinders is low, and if the addition of oil doesn't help raise compression, there is leakage past the head gasket. Oil and coolant in the combustion chamber, combined with blue or constant white smoke from the exhaust, are symptoms of this problem. However, don't be alarmed by the normal white smoke emitted from the exhaust during engine warm-up or from cold weather operation. There may be evidence of water droplets on the engine dipstick and/or oil droplets in the cooling system if a head gasket is blown.

OIL PRESSURE TEST

Check for proper oil pressure at the sending unit passage with an externally mounted mechanical oil pressure gauge (as opposed to relying on a factory installed dash-mounted gauge). A tachometer may also be needed, as some specifications may require running the engine at a specific rpm.

1. With the engine cold, locate and remove the oil pressure sending unit.
2. Following the manufacturer's instructions, connect a mechanical oil pressure gauge and, if necessary, a tachometer to the engine.
3. Start the engine and allow it to idle.
4. Check the oil pressure reading when cold and record the number. You may need to run the engine at a specified rpm, so check the specifications.
5. Run the engine until normal operating temperature is reached.
6. Check the oil pressure reading again with the engine hot and record the number. Turn the engine **OFF**.
7. Compare your hot oil pressure reading to that given in the chart. If the reading is low, check the cold pressure reading against the chart. If the cold pressure is well above the specification, and the hot reading was lower than the specification, you may have the wrong viscosity oil in the engine. Change the oil, making sure to use the proper grade and quantity, then repeat the test.

Low oil pressure readings could be attributed to internal component wear, pump related problems, a low oil level, or oil viscosity that is too low. High oil pressure readings could be caused by an overfilled crankcase, too high of an oil viscosity or a faulty pressure relief valve.

Buy or Rebuild

Now that you have determined that your engine is worn out, you must make some decisions. The question of whether or not an engine is worth rebuilding is largely a subjective matter and one of personal worth. Is the engine a popular one, or is it an obsolete model? Are parts available? Will it get acceptable gas mileage once it is rebuilt? Is the vessel it's being put into worth keeping? Would it be less expensive to buy a new engine, have your engine rebuilt by a pro, rebuild it yourself or buy a used engine? Or would it be simpler and less expensive to buy another boat? If you have considered all these matters and more, and have still decided to rebuild the engine, then it is time to decide how you will rebuild it.

■ **The editors at Seloc feel that most engine machining should be performed by a professional machine shop. Don't think of it as wasting money, rather, as an assurance that the job has been done right the first time. There are many expensive and specialized tools required to perform such tasks as boring and honing an engine block or having a valve job done on a cylinder head. Even inspecting the parts requires expensive micrometers and gauges to properly measure wear and clearances. Also, a machine shop can deliver to you clean, and ready to assemble parts, saving you time and aggravation. Your maximum savings will come from performing the removal, disassembly, assembly and installation of the engine and purchasing or renting only the tools required to perform the above tasks. Depending on the particular circumstances, you may save 40 to 60 percent of the cost doing these yourself.**

A complete rebuild or overhaul of an engine involves replacing all of the moving parts (pistons, rods, crankshaft, camshaft, etc.) with new ones and machining the non-moving wearing surfaces of the block and heads. Unfortunately, this may not be cost effective. For instance, your crankshaft may have been damaged or worn, but it can be machined undersize for a minimal fee.

So, as you can see, you can replace everything inside the engine, but, it is wiser to replace only those parts which are really needed, and, if possible, repair the more expensive ones. Later we will break the engine down into its two main components: the cylinder head and the engine block. We will discuss each component, and the recommended parts to replace during a rebuild on each.

Engine Overhaul Tips

Most engine overhaul procedures are fairly standard. In addition to specific parts replacement procedures and specifications for your individual engine, this is also a guide to acceptable rebuilding procedures. Examples of standard rebuilding practice are given and should be used along with specific details concerning your particular engine.

Competent and accurate machine shop services will ensure maximum performance, reliability and engine life. In most instances it is more profitable for the do-it-yourself mechanic to remove, clean and inspect the component, buy the necessary parts and deliver these to a shop for actual machine work.

Much of the assembly work (crankshaft, bearings, piston rods, and other components) is well within the scope of the do-it-yourself mechanic's tools and abilities. You will have to decide for yourself the depth of involvement you desire in an engine repair or rebuild.

TOOLS

The tools required for an engine overhaul or parts replacement will depend on the depth of your involvement. With a few exceptions, they will be the tools found in a mechanic's tool kit. More in-depth work will require some or all of the following:
- A dial indicator (reading in thousandths) mounted on a universal base
- Micrometers and telescope gauges
- Jaw and screw-type pullers
- Scraper
- Valve spring compressor
- Ring groove cleaner
- Piston ring expander and compressor
- Ridge reamer
- Cylinder hone or glaze breaker
- Plastigage®
- Engine stand

The use of most of these tools is illustrated in the procedures. Many can be rented for a one-time use from a local parts jobber or tool supply house specializing in marine or automotive work.

Occasionally, the use of special tools is called for. See the information on Special Tools and the Safety Notice in the front of this manual before substituting another tool.

OVERHAUL TIPS

Aluminum has become extremely popular for use in engines, due to its low weight. Observe the following precautions when handling aluminum parts:
- Never hot tank aluminum parts (the caustic hot tank solution will eat the aluminum.
- Remove all aluminum parts (identification tag, etc.) from engine parts prior to the tanking.
- Always coat threads lightly with engine oil or anti-seize compounds before installation, to prevent seizure.
- Never overtighten bolts or spark plugs especially in aluminum threads.

When assembling the engine, any parts that will be exposed to frictional contact must be prelubed to provide lubrication at initial start-up. Any product specifically formulated for this purpose can be used, but engine oil is not recommended as a prelube in most cases.

When semi-permanent (locked, but removable) installation of bolts or nuts is desired, threads should be cleaned and coated with Loctite® or another similar, commercial non-hardening marine sealant.

CLEANING

◆ **See Figures 2, 3, 4, and 5**

Before the engine and its components are inspected, they must be thoroughly cleaned. You will need to remove any engine varnish, oil sludge and/or carbon deposits from all of the components to insure an accurate inspection. A crack in the engine block or cylinder head can easily become overlooked if hidden by a layer of sludge or carbon.

Most of the cleaning process can be carried out with common hand tools and readily available solvents or solutions. Carbon deposits can be chipped away using a hammer and a hard wooden chisel. Old gasket material and varnish or sludge can usually be removed using a scraper and/or cleaning solvent. Extremely stubborn deposits may require the use of a power drill with a wire brush. If using a wire brush, use extreme care around any critical machined surfaces (such as the gasket surfaces, bearing saddles, cylinder bores, etc.). USE OF A WIRE BRUSH IS NOT RECOMMENDED ON ANY ALUMINUM COMPONENTS. Always follow any safety recommendations given by the manufacturer of the tool and/or solvent. You should always wear eye protection during any cleaning process involving scraping, chipping or spraying of solvents.

An alternative to the mess and hassle of cleaning the parts yourself is to drop them off at a local marina or machine shop (or even an automotive garage). They will, more than likely, have the necessary equipment to properly clean all of the parts for a nominal fee.

✳✳ CAUTION

Always wear eye protection during any cleaning process involving scraping, chipping or spraying of solvents.

Remove any oil galley plugs, freeze plugs and/or pressed-in bearings and carefully wash and degrease all of the engine components including the fasteners and bolts. Small parts such as the valves, springs, etc., should be placed in a metal basket and allowed to soak. Use pipe cleaner type brushes, and clean all passageways in the components. Use a ring expander and remove the rings from the pistons. Clean the piston ring grooves with a special tool or a piece of broken ring. Scrape the carbon off of the top of the piston. You should never use a wire brush on the pistons. After preparing all of the piston assemblies in this manner, wash and degrease them again.

✳✳ WARNING

**Use extreme care when cleaning around the cylinder head valve seats. A mistake or slip may cost you a new seat.
When cleaning the cylinder head, remove carbon from the**

Fig. 2 Use a gasket scraper to remove the old gasket material from the mating surfaces

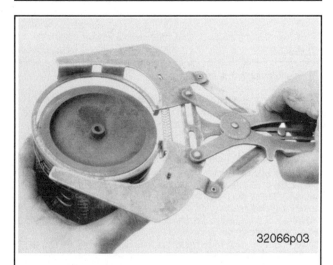

32066p03

Fig. 3 Use a ring expander tool to remove the piston rings

32066p04

Fig. 4 Clean the piston ring grooves using a ring groove cleaner tool, or . . .

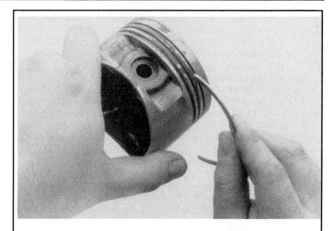

Fig. 5. . . use a piece of an old ring to clean the grooves. Be careful, the ring can be quite sharp

combustion chamber with the valves installed. This will avoid damaging the valve seats.

REPAIRING
DAMAGED THREADS

 MODERATE

◆ **See Figures 6, 7, 8, 9 and 10**

Several methods of repairing damaged threads are available. Heli-Coil® (shown here), Keenserts® and Microdot® are among the most widely used. All involve basically the same principle—drilling out stripped threads, tapping the hole and installing a prewound insert—making welding, plugging and oversize fasteners unnecessary.

Two types of thread repair inserts are usually supplied: a standard type for most inch coarse, inch fine, metric course and metric fine thread sizes and a spark lug type to fit most spark plug port sizes. Consult the individual tool manufacturer's catalog to determine exact applications. Typical thread repair kits will contain a selection of prewound threaded inserts, a tap (corresponding to the outside diameter threads of the insert) and an installation tool. Spark plug inserts usually differ because they require a tap equipped with pilot threads and a combined reamer/tap section. Most manufacturers also supply blister-packed thread repair inserts separately in addition to a master kit containing a variety of taps and inserts plus installation tools.

Before attempting to repair a threaded hole, remove any snapped, broken or damaged bolts or studs. Penetrating oil can be used to free frozen

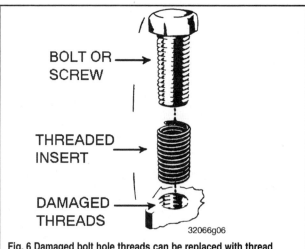

BOLT OR SCREW →

THREADED INSERT →

DAMAGED THREADS →

32066g06

Fig. 6 Damaged bolt hole threads can be replaced with thread repair inserts

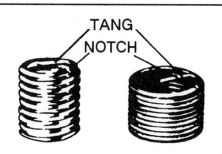

Fig. 7 Standard thread repair insert (left), and spark plug thread insert

threads. The offending item can usually be removed with locking pliers or using a screw/stud extractor. After the hole is clear, the thread can be repaired, as shown in the series of accompanying illustrations and in the kit manufacturer's instructions.

Engine Preparation

To properly rebuild an engine, you must first remove it from the vessel, then disassemble and diagnose it. Ideally you should place your engine on an engine stand. This affords you the best access to the engine components. Follow the manufacturer's directions for using the stand with your particular engine. Remove the flywheel or coupler before installing the engine to the stand.

Now that you have the engine on a stand, and assuming that you have drained the oil and coolant from the engine, it's time to strip it of all but the necessary components. Before you start disassembling the engine, you may want to take a moment to draw some pictures, or fabricate some labels or containers to mark the locations of various components and the bolts and/or studs which fasten them. Modern day engines use a lot of little brackets and clips which hold wiring harnesses and such, and these holders are often mounted on studs and/or bolts that can be easily mixed up. The manufacturer spent a lot of time and money designing your engine/boat, and they wouldn't have wasted any of it by haphazardly placing brackets, clips or fasteners on the boat. If it's present when you disassemble it, put it back when you assemble, you will regret not remembering that little bracket which holds a wire harness out of the path of a rotating part.

You should begin by unbolting any accessories still attached to the engine, such as the water pump, power steering pump, alternator, etc. Then, unfasten any manifolds (intake or exhaust) which were not removed during the engine removal procedure. Finally, remove any covers remaining on the engine such as the rocker arm, front or timing cover and oil pan. Some front covers may require the balancer and/or crank pulley to be removed beforehand. The idea is to reduce the engine to the bare necessities (cylinder head(s), valve train, engine block, crankshaft, pistons and connecting rods),

plus any other 'in block' components such as oil pumps, balance shafts and auxiliary shafts.

Finally, remove the cylinder head(s) from the engine block and carefully place on a bench. Disassembly instructions for each component follow later.

Cylinder Head

There are two basic types of cylinder heads used on today's engines: the Overhead Valve (OHV) and the Overhead Camshaft (OHC). The latter can also be broken down into two subgroups: the Single Overhead Camshaft (SOHC) and the Dual Overhead Camshaft (DOHC). Generally, if there is only a single camshaft on a head, it is just referred to as an OHC head. Also, an engine with an OHV cylinder head is also known as a pushrod engine.

Most cylinder heads these days are made of an aluminum alloy due to its light weight, durability and heat transfer qualities. However, cast iron was the material of choice in the past, and is still used on many engines today. Whether made from aluminum or iron, all cylinder heads have valves and seats. Most use two valves per cylinder, while the more hi-tech engines will utilize a multi-valve configuration using 3, 4 and even 5 valves per cylinder. When the valve contacts the seat, it does so on precision machined surfaces, which seals the combustion chamber. All cylinder heads have a valve guide for each valve. The guide centers the valve to the seat and allows it to move up and down within it. The clearance between the valve and guide can be critical. Too much clearance and the engine may consume oil, lose vacuum and/or damage the seat. Too little, and the valve can stick in the guide causing the engine to run poorly if at all, and possibly causing severe damage. The last component all cylinder heads have are valve springs. The spring holds the valve against its seat. It also returns the valve to this position when the valve has been opened by the valve train or camshaft. The spring is fastened to the valve by a retainer and valve locks (sometimes called keepers). Aluminum heads will also have a valve spring shim to keep the spring from wearing away the aluminum.

An ideal method of rebuilding the cylinder head would involve replacing all of the valves, guides, seats, springs, etc. with new ones. However, depending on how the engine was maintained, often this is not necessary. A major cause of valve, guide and seat wear is an improperly tuned engine. An engine that is running too rich, will often wash the lubricating oil out of the guide with gasoline, causing it to wear rapidly. Conversely, an engine which is running too lean will place higher combustion temperatures on the valves and seats allowing them to wear or even burn. Springs fall victim to the operating habits of the individual. A driver who often runs the engine rpm to the redline will wear out or break the springs faster then one that stays well below it. Unfortunately, 'hours of operation' takes it toll on all of the parts. Generally, the valves, guides, springs and seats in a cylinder head can be machined and re-used, saving you money. However, if a valve is burnt, it may be wise to replace all of the valves, since they were all operating in the same environment. The same goes for any other component on the cylinder head. Think of it as an insurance policy against future problems related to that component.

Unfortunately, the only way to find out which components need replacing, is to disassemble and carefully check each piece. After the cylinder head(s) are disassembled, thoroughly clean all of the components.

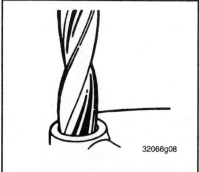

32066g08

Fig. 8 Drill out the damaged threads with the specified size bit. Be sure to drill completely through the hole or to the bottom of a blind hole

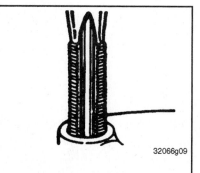

32066g09

Fig. 9 Using the kit, tap the hole in order to receive the thread insert. Keep the tap well oiled and back it out frequently to clean out the chips

Fig. 10 Screw the insert onto the installer tool until the tang engages the slot. Thread the insert into the hole until it is 1/4-1/2 turn below the top surface, then remove the tool and break off the tang using a punch

DISASSEMBLY

◆ **See Figures 11, 12, 13, 14, 15, 16, 17, 18, 19, 20, 21 and 22**

Before disassembling the cylinder head, you may want to fabricate some containers to hold the various parts, as some of them can be quite small (such as keepers) and easily lost. Also keeping yourself and the components organized will aid in assembly and reduce confusion. Where possible, try to maintain a components original location; this is especially important if there is not going to be any machine work performed on the components.

1. If you haven't already removed the rocker arms and/or shafts, do so now.
2. Position the head so that the springs are easily accessed.
3. Use a valve spring compressor tool, and relieve spring tension from the retainer.

■ **Due to engine varnish, the retainer may stick to the valve locks. A gentle tap with a hammer may help to break it loose.**

4. Remove the valve locks from the valve tip and/or retainer. A small magnet may help in removing the locks.
5. Lift the valve spring(s), tool and all, off of the valve stem.
6. Remove the valve seal from the stem and guide. If the seal is difficult to remove with the valve in place, try removing the valve first, then the seal.
7. Position the head to allow access for withdrawing the valve.

■ **Cylinder heads that have seen a lot of miles and/or abuse may have mushroomed the valve lock grove and/or tip, causing difficulty in removal of the valve. If this has happened, use a metal file to carefully remove the high spots around the lock grooves and/or tip. Only file it enough to allow removal.**

8. Remove the valve from the cylinder head.
9. If equipped, remove the valve spring shim. A small magnetic tool or screwdriver will aid in removal.
10. Repeat Steps 3 though 9 until all of the valves have been removed.

INSPECTION

Now that all of the cylinder head components are clean, it's time to inspect them for wear and/or damage. To accurately inspect them, you will need some specialized tools:

• 0-1 in. micrometer for the valves

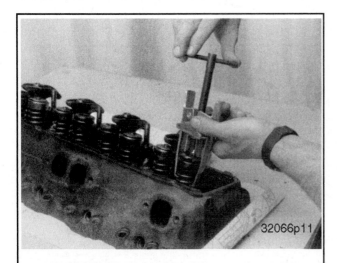

Fig. 11 When removing a valve spring, use a compressor tool to relieve the tension from the retainer...

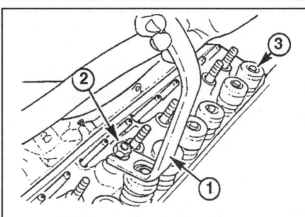

1- Valve spring compressor
2- Rocker arm nut
3- Valve locks

32066g12

Fig. 12 ...you may also find a compressor that looks like this

Fig. 13 A small magnet will help in removal of the valve locks

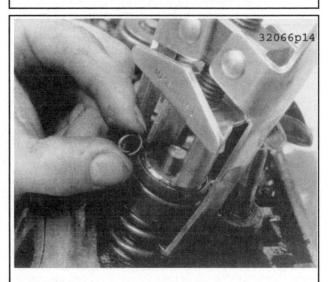

32066p14

Fig. 14 Be careful not to lose the small valve locks (keepers)

Fig. 15 Remove the valve seal from the valve stem—O-ring type seal shown

Fig. 16 Removing an umbrella/positive type seal

Fig. 17 Invert the cylinder head and withdraw the valve from the valve guide bore

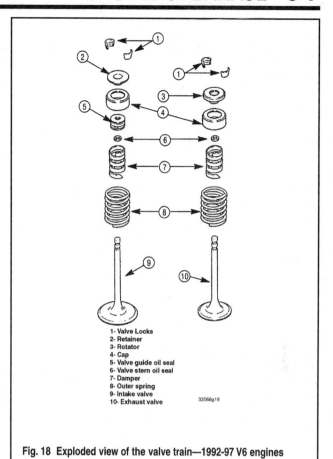

1- Valve Locks
2- Retainer
3- Rotator
4- Cap
5- Valve guide oil seal
6- Valve stern oil seal
7- Damper
8- Outer spring
9- Intake valve
10- Exhaust valve

32066g18

Fig. 18 Exploded view of the valve train—1992-97 V6 engines

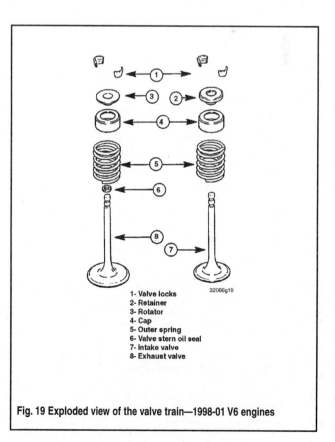

1- Valve locks
2- Retainer
3- Rotator
4- Cap
5- Outer spring
6- Valve stern oil seal
7- Intake valve
8- Exhaust valve

32066g19

Fig. 19 Exploded view of the valve train—1998-01 V6 engines

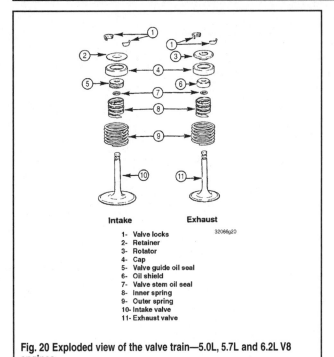

Intake   Exhaust

1- Valve locks
2- Retainer
3- Rotator
4- Cap
5- Valve guide oil seal
6- Oil shield
7- Valve stem oil seal
8- Inner spring
9- Outer spring
10- Intake valve
11- Exhaust valve

Fig. 20 Exploded view of the valve train—5.0L, 5.7L and 6.2L V8 engines

1- Valve Locks
2- Retainer
3- Oil Shield seal
4- Oil shield
5- Outer spring
6- Damper shield
7- Roatator
8- Intake valve
9- Exhaust valve

Fig. 20 Exploded view of the valve train—7.4L and 8.1L V8 engines (1998-01 engines do not have damper shields)

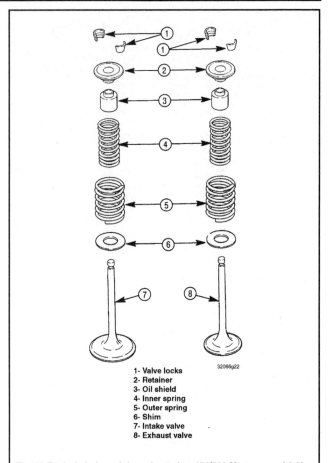

1- Valve locks
2- Retainer
3- Oil shield
4- Inner spring
5- Outer spring
6- Shim
7- Intake valve
8- Exhaust valve

Fig. 22 Exploded view of the valve train—454/502 Magnum and 8.2L V8 engines

- A dial indicator or inside diameter gauge for the valve guides
- A spring pressure test gauge

If you do not have access to the proper tools, you may want to bring the components to a shop that does.

Valves

◆ See Figures 23 and 24

The first thing to inspect are the valve heads. Look closely at the head, margin and face for any cracks, excessive wear or burning. The margin is the best place to look for burning. It should have a squared edge with an even width all around the diameter. When a valve burns, the margin will look melted and the edges rounded. Also inspect the valve head for any signs of tulipping. This will show as a lifting of the edges or dishing in the center of the head and will usually not occur to all of the valves. All of the heads should look the same, any that seem dished more than others are probably bad. Next, inspect the valve lock grooves and valve tips. Check for any burrs around the lock grooves, especially if you had to file them to remove the valve. Valve tips should appear flat, although slight rounding with high mileage engines is normal. Slightly worn valve tips will need to be machined flat. Last, measure the valve stem diameter with the micrometer.

Measure the area that rides within the guide, especially towards the tip where most of the wear occurs. Take several measurements along its length and compare them to each other. Wear should be even along the length with little to no taper. If no minimum diameter is given in the specifications, then the stem should not read more than 0.001 in. (0.025mm) below the unworn area of the valve stem. Any valves that fail these inspections should be replaced.

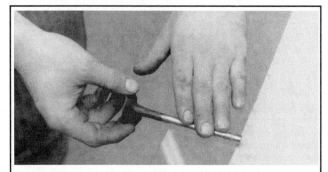

Fig. 23 Valve stems may be rolled on a flat surface to check for bends

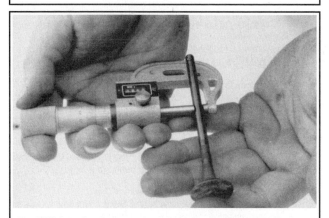

Fig. 24 Use a micrometer to check the valve stem diameter

Springs, Retainers and Valve Locks

◆ See Figures 25 and 26

The first thing to check is the most obvious, broken springs. Next check the free length and squareness of each spring. If applicable, insure to distinguish between intake and exhaust springs. Use a ruler and/or carpenter's square to measure the length. A carpenter's square should be used to check the springs for squareness. If a spring pressure test gauge is available, check each springs rating and compare to the specifications chart. Check the readings against the specifications given. Any springs that fail these inspections should be replaced.

The spring retainers rarely need replacing, however they should still be checked as a precaution. Inspect the spring mating surface and the valve lock retention area for any signs of excessive wear. Also check for any signs of cracking. Replace any retainers that are questionable.

Valve locks should be inspected for excessive wear on the outside contact area as well as on the inner notched surface. Any locks which appear worn or broken and its respective valve should be replaced.

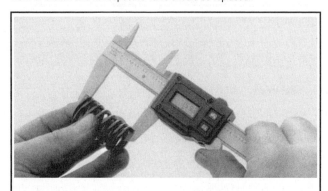

Fig. 25 Use a caliper to check the valve spring free-length

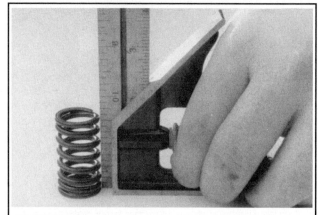

Fig. 26 Check the valve spring for squareness on a flat surface; a carpenter's square can be used

Cylinder Head

There are several things to check on the cylinder head: valve guides, seats, cylinder head surface flatness, cracks and physical damage.

VALVE GUIDES

◆ See Figure 27

Now that you know the valves are good, you can use them to check the guides, although a new valve, if available, is preferred. Before you measure anything, look at the guides carefully and inspect them for any cracks, chips or breakage. Also if the guide is a removable style (as in most aluminum heads), check them for any looseness or evidence of movement. All of the guides should appear to be at the same height from the spring seat. If any seem lower (or higher) from another, the guide has moved. Mount a dial indicator onto the spring side of the cylinder head. Lightly oil the valve stem and insert it into the cylinder head. Position the dial indicator against the valve stem near the tip and zero the gauge. Grasp the valve stem and wiggle towards and away from the dial indicator and observe the readings. Mount the dial indicator 90 degrees from the initial point and zero the gauge and again take a reading. Compare the two readings for an out of round condition. Check the readings against the specifications given. An Inside Diameter (I.D.) gauge designed for valve guides will give you an accurate valve guide bore measurement. If the I.D. gauge is used, compare the readings with the specifications given. Any guides that fail these inspections should be replaced or machined.

VALVE SEATS

A visual inspection of the valve seats should show a slightly worn and pitted surface where the valve face contacts the seat. Inspect the seat carefully for severe pitting or cracks. Also, a seat that is badly worn will be recessed into the cylinder head. A severely worn or recessed seat may need to be replaced. All cracked seats must be replaced. A seat concentricity gauge, if available, should be used to check the seat run-out. If run-out exceeds specifications the seat must be machined (if no specification is given use 0.002 in. or 0.051mm).

CYLINDER HEAD SURFACE FLATNESS

◆ See Figures 28 and 29

After you have cleaned the gasket surface of the cylinder head of any old gasket material, check the head for flatness.

Place a straightedge across the gasket surface. Using feeler gauges, determine the clearance at the center of the straightedge and across the cylinder head at several points. Check along the centerline and diagonally on the head surface. If the warpage exceeds 0.003 in. (0.076mm) within a 6.0 in. (15.2cm) span, or 0.006 in. (0.152mm) over the total length of the head, the cylinder head must be resurfaced. After resurfacing the heads of a V-type engine, the intake manifold flange surface should be checked, and if necessary, milled proportionally to allow for the change in its mounting position.

Fig. 27 A dial gauge may be used to check valve stem-to-guide clearance; read the gauge while moving the valve stem

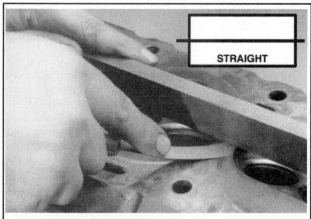

Fig. 28 Check the head for flatness across the center of the head surface using a straightedge and feeler gauge

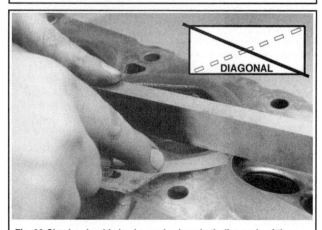

Fig. 29 Checks should also be made along both diagonals of the head surface

CRACKS AND PHYSICAL DAMAGE

Generally, cracks are limited to the combustion chamber, however, it is not uncommon for the head to crack in a spark plug hole, port, outside of the head or in the valve spring/rocker arm area. The first area to inspect is always the hottest: the exhaust seat/port area.

A visual inspection should be performed, but just because you don't see a crack does not mean it is not there. Some more reliable methods for inspecting for cracks include Magnaflux®, a magnetic process or Zyglo®, a dye penetrant. Magnaflux® is used only on ferrous metal (cast iron) heads. Zyglo® uses a spray on fluorescent mixture along with a black light to reveal the cracks. It is strongly recommended to have your cylinder head checked professionally for cracks, especially if the engine was known to have

overheated and/or leaked or consumed coolant. Contact a local shop for availability and pricing of these services.

Physical damage is usually very evident. For example, a broken mounting ear from dropping the head or a bent or broken stud and/or bolt. All of these defects should be fixed or, if unrepairable, the head should be replaced.

REFINISHING & REPAIRING

Many of the procedures given for refinishing and repairing the cylinder head components must be performed by a machine shop. Certain steps, if the inspected part is not worn, can be performed yourself inexpensively. However, you spent a lot of time and effort so far, why risk trying to save a couple bucks if you might have to do it all over again?

Valves

Any valves that were not replaced should be refaced and the tips ground flat. Unless you have access to a valve grinding machine, this should be done by a machine shop. If the valves are in extremely good condition, as well as the valve seats and guides, they may be lapped in without performing machine work.

It is a recommended practice to lap the valves even after machine work has been performed and/or new valves have been purchased. This insures a positive seal between the valve and seat.

LAPPING THE VALVES

■ Before lapping the valves to the seats, read the rest of the cylinder head procedure to insure that any related parts are in acceptable enough condition to continue.

■ Before any valve seat machining and/or lapping can be performed, the guides must be within factory recommended specifications.

1. Invert the cylinder head.
2. Lightly lubricate the valve stems and insert them into the cylinder head in their numbered order.
3. Raise the valve from the seat and apply a small amount of fine lapping compound to the seat.
4. Moisten the suction head of a hand-lapping tool and attach it to the head of the valve.
5. Rotate the tool between the palms of both hands, changing the position of the valve on the valve seat and lifting the tool often to prevent grooving.
6. Lap the valve until a smooth, polished circle is evident on the valve and seat.
7. Remove the tool and the valve. Wipe away all traces of the grinding compound and store the valve to maintain its lapped location.

✳✳ WARNING

Do not get the valves out of order after they have been lapped. They must be put back with the same valve seat with which they were lapped.

Springs, Retainers and Valve Locks

There is no repair or refinishing possible with the springs, retainers and valve locks. If they are found to be worn or defective, they must be replaced with new (or known good) parts.

Cylinder Head

Most refinishing procedures dealing with the cylinder head must be performed by a machine shop. Read the procedures below and review your inspection data to determine whether or not machining is necessary.

VALVE GUIDE

■ If any machining or replacements are made to the valve guides, the seats must be machined.

Unless the valve guides need machining or replacing, the only service to perform is to thoroughly clean them of any dirt or oil residue.

There are only two types of valve guides used on automobile engines: the replaceable-type (all aluminum heads) and the cast-in integral-type (most cast iron heads). There are four recommended methods for repairing worn guides.

- Knurling
- Inserts
- Reaming oversize
- Replacing

Knurling is a process in which metal is displaced and raised, thereby reducing clearance, giving a true center, and providing oil control. It is the least expensive way of repairing the valve guides. However, it is not necessarily the best, and in some cases, a knurled valve guide will not stand up for more than a short time. It requires a special knurlizer and precision reaming tools to obtain proper clearances. It would not be cost effective to purchase these tools, unless you plan on rebuilding several of the same cylinder head.

Installing a guide insert involves machining the guide to accept a bronze insert. One style is the coil-type which is installed into a threaded guide. Another is the thin-walled insert where the guide is reamed oversize to accept a split-sleeve insert. After the insert is installed, a special tool is then run through the guide to expand the insert, locking it to the guide. The insert is then reamed to the standard size for proper valve clearance.

Reaming for oversize valves restores normal clearances and provides a true valve seat. Most cast-in type guides can be reamed to accept an valve with an oversize stem. The cost factor for this can become quite high as you will need to purchase the reamer and new, oversize stem valves for all guides which were reamed. Oversizes are generally 0.003 to 0.030 in. (0.076 to 0.762mm), with 0.015 in. (0.381mm) being the most common.

To replace cast-in type valve guides, they must be drilled out, then reamed to accept replacement guides. This must be done on a fixture which will allow centering and leveling off of the original valve seat or guide, otherwise a serious guide-to-seat misalignment may occur making it impossible to properly machine the seat.

Replaceable-type guides are pressed into the cylinder head. A hammer and a stepped drift or punch may be used to install and remove the guides. Before removing the guides, measure the protrusion on the spring side of the head and record it for installation. Use the stepped drift to hammer out the old guide from the combustion chamber side of the head. When installing, determine whether or not the guide also seals a water jacket in the head, and if it does, use the recommended sealing agent. If there is no water jacket, grease the valve guide and its bore. Use the stepped drift, and hammer the new guide into the cylinder head from the spring side of the cylinder head. A stack of washers the same thickness as the measured protrusion may help the installation process.

VALVE SEATS

■ Before any valve seat machining can be performed, the guides must be within factory recommended specifications.

■ If any machining or replacements were made to the valve guides, the seats must be machined.

If the seats are in good condition, the valves can be lapped to the seats, and the cylinder head assembled. See the valves procedures for instructions on lapping.

If the valve seats are worn, cracked or damaged, they must be serviced by a machine shop. The valve seat must be perfectly centered to the valve guide, which requires very accurate machining.

CYLINDER HEAD SURFACE

If the cylinder head is warped, it must be machined flat. If the warpage is extremely severe, the head may need to be replaced. In some instances, it may be possible to straighten a warped head enough to allow machining. In either case, contact a professional machine shop for service.

CRACKS AND PHYSICAL DAMAGE

Certain cracks can be repaired in both cast iron and aluminum heads.

For cast iron, a tapered threaded insert is installed along the length of the crack. Aluminum can also use the tapered inserts, however welding is the preferred method. Some physical damage can be repaired through brazing or welding. Contact a machine shop to get expert advice for your particular dilemma.

ASSEMBLY

The first step for any assembly job is to have a clean area in which to work. Next, thoroughly clean all of the parts and components that are to be assembled. Finally, place all of the components onto a suitable work space and, if necessary, arrange the parts to their respective positions.

1. Lightly lubricate the valve stems and insert all of the valves into the cylinder head. If possible, maintain their original locations.
2. If equipped, install any valve spring shims which were removed.
3. If equipped, install the new valve seals, keeping the following in mind:
- If the valve seal presses over the guide, lightly lubricate the outer guide surfaces.
- If the seal is an O-ring type, it is installed just after compressing the spring but before the valve locks.
4. Place the valve spring and retainer over the stem.
5. Position the spring compressor tool and compress the spring.
6. Assemble the valve locks to the stem.
7. Relieve the spring pressure slowly and insure that neither valve lock becomes dislodged by the retainer.
8. Remove the spring compressor tool.
9. Repeat Steps 2 through 8 until all of the springs have been installed.

Engine Block

GENERAL INFORMATION

A thorough overhaul or rebuild of an engine block would include replacing the pistons, rings, bearings, timing belt/chain assembly and oil pump. For OHV engines also include a new camshaft and lifters. The block would then have the cylinders bored and honed oversize (or if using removable cylinder sleeves, new sleeves installed) and the crankshaft would be cut undersize to provide new wearing surfaces and perfect clearances. However, your particular engine may not have everything worn out. What if only the piston rings have worn out and the clearances on everything else are still within factory specifications? Well, you could just replace the rings and put it back together, but this would be a very rare example. Chances are, if one component in your engine is worn, other components are sure to follow, and soon. At the very least, you should always replace the rings, bearings and oil pump. This is what is commonly called a "freshen up".

Cylinder Ridge Removal

Because the top piston ring does not travel to the very top of the cylinder, a ridge is built up between the end of the travel and the top of the cylinder bore.

Pushing the piston and connecting rod assembly past the ridge can be difficult, and damage to the piston ring lands could occur. If the ridge is not removed before installing a new piston or not removed at all, piston ring breakage and piston damage may occur.

■ It is always recommended that you remove any cylinder ridges before removing the piston and connecting rod assemblies. If you know that new pistons are going to be installed and the engine block will be bored oversize, you may be able to forego this step. However, some ridges may actually prevent the assemblies from being removed, necessitating its removal.

There are several different types of ridge reamers on the market, none of which are inexpensive. Unless a great deal of engine rebuilding is anticipated, borrow or rent a reamer.

1. Turn the crankshaft until the piston is at the bottom of its travel.
2. Cover the head of the piston with a rag.
3. Follow the tool manufacturers instructions and cut away the ridge, exercising extreme care to avoid cutting too deeply.
4. Remove the ridge reamer, the rag and as many of the cuttings as possible. Continue until all of the cylinder ridges have been removed.

DISASSEMBLY

◆ **See Figures 30 and 31**

The engine disassembly instructions following assume that you have the engine mounted on an engine stand. If not, it is easiest to disassemble the engine on a bench or the floor with it resting on the bell housing or transmission mounting surface. You must be able to access the connecting rod fasteners and turn the crankshaft during disassembly. Also, all engine covers (timing, front, side, oil pan, whatever) should have already been removed. Engines which are seized or locked up may not be able to be completely disassembled, and a core (salvage yard) engine should be purchased.

If not done during the cylinder head removal, remove the pushrods and lifters, keeping them in order for assembly. Remove the timing gears and/or timing chain assembly, then remove the oil pump drive assembly and withdraw the camshaft from the engine block. Remove the oil pick-up and pump assembly. If equipped, remove any balance or auxiliary shafts. If necessary, remove the cylinder ridge from the top of the bore. See the cylinder ridge removal procedure.

Rotate the engine over so that the crankshaft is exposed. Use a number punch or scribe and mark each connecting rod with its respective cylinder number. The cylinder closest to the front of the engine is always number 1. However, depending on the engine placement, the front of the engine could either be the flywheel or damper/pulley end. Generall,y the front of the engine faces the bow. Use a number punch or scribe and also mark the main bearing caps from front to rear with the front most cap being number 1 (if there are five caps, mark them 1 through 5, front to rear).

✳✳ WARNING

Take special care when pushing the connecting rod up from the crankshaft because the sharp threads of the rod bolts/studs will score the crankshaft journal. Insure that special plastic caps are installed over them, or cut two pieces of rubber hose to do the same.

Again, rotate the engine, this time to position the number one cylinder bore (head surface) up. Turn the crankshaft until the number one piston is at the bottom of its travel, this should allow the maximum access to its connecting rod. Remove the number one connecting rods fasteners and cap and place two lengths of rubber hose over the rod bolts/studs to protect the crankshaft from damage. Using a sturdy wooden dowel and a hammer, push the connecting rod up about 1 in. (25mm) from the crankshaft and remove the upper bearing insert. Continue pushing or tapping the connecting rod up until the piston rings are out of the cylinder bore. Remove the piston and rod by hand, put the upper half of the bearing insert back into the rod, install the cap with its bearing insert installed, and hand-tighten the cap fasteners. If the parts are kept in order in this manner, they will not get lost and you will be able to tell which bearings came form what cylinder if any problems are discovered and diagnosis is necessary. Remove all the other piston assemblies in the same manner. On V-style engines, remove all of the pistons from one bank, then reposition the engine with the other cylinder bank head surface up, and remove that banks piston assemblies.

The only remaining component in the engine block should now be the crankshaft. Loosen the main bearing caps evenly until the fasteners can be turned by hand, then remove them and the caps. Remove the crankshaft from the engine block. Thoroughly clean all of the components.

INSPECTION

Now that the engine block and all of its components are clean, it's time to inspect them for wear and/or damage. To accurately inspect them, you will need some specialized tools:

- Two or three separate micrometers to measure the pistons and crankshaft journals
- A dial indicator
- Telescoping gauges for the cylinder bores
- A rod alignment fixture to check for bent connecting rods

If you do not have access to the proper tools, you may want to bring the components to a shop that does.

Fig. 30 Place rubber hose over the connecting rod studs to protect the crankshaft and cylinder bores from damage

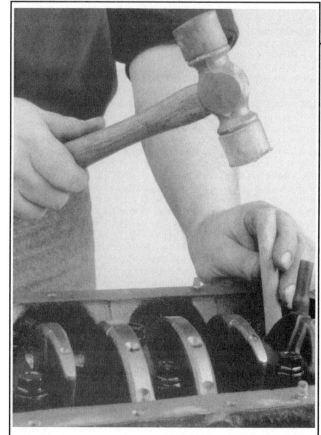

Fig. 31 Carefully tap the piston out of the bore using a wooden dowel

Generally, you shouldn't expect cracks in the engine block or its components unless it was known to leak, consume or mix engine fluids, it was severely overheated, or there was evidence of bad bearings and/or crankshaft damage. A visual inspection should be performed on all of the components, but just because you don't see a crack does not mean it is not there. Some more reliable methods for inspecting for cracks include Magnaflux®, a magnetic process or Zyglo®, a dye penetrant. Magnaflux® is used only on ferrous metal (cast iron). Zyglo® uses a spray on fluorescent mixture along with a black light to reveal the cracks. It is strongly recommended to have your engine block checked professionally for cracks, especially if the engine was known to have overheated and/or leaked or consumed coolant. Contact a local shop for availability and pricing of these services.

Engine Block

ENGINE BLOCK BEARING ALIGNMENT

Remove the main bearing caps and, if still installed, the main bearing inserts. Inspect all of the main bearing saddles and caps for damage, burrs or high spots. If damage is found, and it is caused from a spun main bearing, the block will need to be align-bored or, if severe enough, replacement. Any burrs or high spots should be carefully removed with a metal file.

Place a straightedge on the bearing saddles, in the engine block, along the centerline of the crankshaft. If any clearance exists between the straightedge and the saddles, the block must be align-bored.

Align-boring consists of machining the main bearing saddles and caps by means of a flycutter that runs through the bearing saddles.

DECK FLATNESS

The top of the engine block where the cylinder head mounts is called the deck. Insure that the deck surface is clean of dirt, carbon deposits and old gasket material. Place a straightedge across the surface of the deck along its centerline and, using feeler gauges, check the clearance along several points. Repeat the checking procedure with the straightedge placed along both diagonals of the deck surface. If the reading exceeds 0.003 in. (0.076mm) within a 6.0 in. (15.2cm) span, or 0.006 in. (0.152mm) over the total length of the deck, it must be machined. Always check the Specification chart for your specific engine.

CYLINDER BORES

◆ **See Figure 32**

The cylinder bores house the pistons and are slightly larger than the pistons themselves. A common piston-to-bore clearance is 0.0015-0.0025 in. (0.0381mm-0.0635mm). Refer to the Specification chart for your specific engine. Inspect and measure the cylinder bores. The bore should be checked for out-of-roundness, taper and size. The results of this inspection will determine whether the cylinder can be used in its existing size and condition, or a rebore to the next oversize is required (or in the case of removable sleeves, have replacements installed).

The amount of cylinder wall wear is always greater at the top of the cylinder than at the bottom. This wear is known as taper. Any cylinder that has a taper of 0.0012 in. (0.305mm) or more, must be rebored. Measurements are taken at a number of positions in each cylinder: at the top, middle and bottom and at two points at each position; that is, at a point 90 degrees from the crankshaft centerline, as well as a point parallel to the crankshaft centerline. The measurements are made with either a special dial indicator or a telescopic gauge and micrometer. If the necessary precision tools to check the bore are not available, take the block to a machine shop and have them mike it. Also if you don't have the tools to check the cylinder bores, chances are you will not have the necessary devices to check the pistons, connecting rods and crankshaft. Take these components with you and save yourself an extra trip.

For our procedures, we will use a telescopic gauge and a micrometer. You will need one of each, with a measuring range which covers your cylinder bore size.

1. Position the telescopic gauge in the cylinder bore, loosen the gauges lock and allow it to expand.

■ **Your first two readings will be at the top of the cylinder bore, then proceed to the middle and finally the bottom, making a total of six measurements.**

2. Hold the gauge square in the bore, 90 degrees from the crankshaft centerline, and gently tighten the lock. Tilt the gauge back to remove it from the bore.

3. Measure the gauge with the micrometer and record the reading.

4. Again, hold the gauge square in the bore, this time parallel to the crankshaft centerline, and gently tighten the lock. Again, you will tilt the gauge back to remove it from the bore.

5. Measure the gauge with the micrometer and record this reading. The difference between these two readings is the out-of-round measurement of the cylinder.

6. Repeat Steps 1 through 5, each time going to the next lower position,

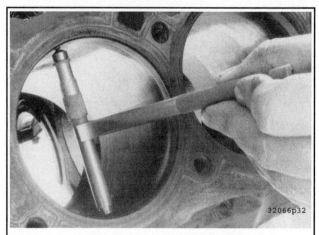

Fig. 32 Use a telescoping gauge to measure the cylinder bore diameter—take several readings within the same bore

until you reach the bottom of the cylinder. Then go to the next cylinder, and continue until all of the cylinders have been measured.

The difference between these measurements will tell you all about the wear in your cylinders. The measurements which were taken 90 degrees from the crankshaft centerline will always reflect the most wear. That is because at this position is where the engine power presses the piston against the cylinder bore the hardest. This is known as thrust wear. Take your top, 90 degree measurement and compare it to your bottom, 90 degree measurement. The difference between them is the taper. When you measure your pistons, you will compare these readings to your piston sizes and determine piston-to-wall clearance.

Crankshaft

Inspect the crankshaft for visible signs of wear or damage. All of the journals should be perfectly round and smooth. Slight scores are normal for a used crankshaft, but you should hardly feel them with your fingernail. When measuring the crankshaft with a micrometer, you will take readings at the front and rear of each journal, then turn the micrometer 90 degrees and take two more readings, front and rear. The difference between the front-to-rear readings is the journal taper and the first-to-90 degree reading is the out-of-round measurement. Generally, there should be no taper or out-of-roundness found, however, up to 0.0005 in. (0.0127mm) for either can be overlooked. Also, the readings should fall within the factory specifications for journal diameters.

If the crankshaft journals fall within specifications, it is recommended that it be polished before being returned to service. Polishing the crankshaft insures that any minor burrs or high spots are smoothed, thereby reducing the chance of scoring the new bearings.

Pistons and Connecting Rods

PISTONS

◆ **See Figure 33**

The piston should be visually inspected for any signs of cracking or burning (caused by hot spots or detonation), and scuffing or excessive wear on the skirts. The wrist pin attaches the piston to the connecting rod. The piston should move freely on the wrist pin, both sliding and pivoting. Grasp the connecting rod securely, or mount it in a vise, and try to rock the piston back and forth along the centerline of the wrist pin. There should not be any excessive play evident between the piston and the pin. If there are C-clips retaining the pin in the piston then you have wrist pin bushings in the rods. There should not be any excessive play between the wrist pin and the rod bushing. Normal clearance for the wrist pin is approx. 0.001-0.002 in. (0.025mm-0.051mm). Please refer to the Specification chart for your specific engine.

Use a micrometer and measure the diameter of the piston, perpendicular to the wrist pin, on the skirt. Compare the reading to its original cylinder measurement obtained earlier. The difference between the two readings is

Fig. 33 Measure the piston's outer diameter, perpendicular to the wrist pin, with a micrometer

the piston-to-wall clearance. If the clearance is within specifications, the piston may be used as is. If the piston is out of specification, but the bore is not, you will need a new piston. If both are out of specification, you will need the cylinder rebored and oversize pistons installed. Generally if two or more pistons/bores are out of specification, it is best to rebore the entire block and purchase a complete set of oversize pistons.

CONNECTING ROD

You should have the connecting rod checked for straightness at a machine shop. If the connecting rod is bent, it will unevenly wear the bearing and piston, as well as place greater stress on these components. Any bent or twisted connecting rods must be replaced. If the rods are straight and the wrist pin clearance is within specifications, then only the bearing end of the rod need be checked. Place the connecting rod into a vice, with the bearing inserts in place, install the cap to the rod and torque the fasteners to specifications. Use a telescoping gauge and carefully measure the inside diameter of the bearings. Compare this reading to the rods original crankshaft journal diameter measurement. The difference is the oil clearance. If the oil clearance is not within specifications, install new bearings in the rod and take another measurement. If the clearance is still out of specifications, and the crankshaft is not, the rod will need to be reconditioned by a machine shop.

■ **You can also use Plastigage® to check the bearing clearances. The assembling procedure has complete instructions on its use.**
Camshaft

Inspect the camshaft and lifters/followers as described earlier.

Bearings

All of the engine bearings should be visually inspected for wear and/or damage. The bearing should look evenly worn all around with no deep scores or pits. If the bearing is severely worn, scored, pitted or heat blued, then the bearing, and the components that use it, should be brought to a machine shop for inspection. Full-circle bearings (used on most camshafts, auxiliary shafts, balance shafts, etc.) require specialized tools for removal and installation, and should be brought to a machine shop for service.

Oil Pump

■ **The oil pump is responsible for providing constant lubrication to the whole engine and so it is recommended that a new oil pump be installed when rebuilding the engine.**

Completely disassemble the oil pump and thoroughly clean all of the components. Inspect the oil pump gears and housing for wear and/or damage. Insure that the pressure relief valve operates properly and there is no binding or sticking due to varnish or debris. If all of the parts are in proper working condition, lubricate the gears and relief valve, and assemble the pump.

REFINISHING

◆ **See Figure 34**

Almost all engine block refinishing must be performed by a machine shop. If the cylinders are not to be rebored, then the cylinder glaze can be removed with a ball hone. When removing cylinder glaze with a ball hone, use a light or penetrating type oil to lubricate the hone. Do not allow the hone to run dry as this may cause excessive scoring of the cylinder bores and wear on the hone. If new pistons are required, they will need to be installed to the connecting rods. This should be performed by a machine shop as the pistons must be installed in the correct relationship to the rod or engine damage can occur.

Pistons and Connecting Rods

◆ **See Figure 35**

Only pistons with the wrist pin retained by C-clips are serviceable by the home-mechanic. Press fit pistons require special presses and/or heaters to remove/install the connecting rod and should only be performed by a machine shop.

All pistons will have a mark indicating the direction to the front of the engine and the must be installed into the engine in that manner. Usually it is a notch or arrow on the top of the piston, or it may be the letter F cast or stamped into the piston.

C-CLIP TYPE PISTONS

1. Note the location of the forward mark on the piston and mark the connecting rod in relation.
2. Remove the C-clips from the piston and withdraw the wrist pin.

■ **Varnish build-up or C-clip groove burrs may increase the difficulty of removing the wrist pin. If necessary, use a punch or drift to carefully tap the wrist pin out.**

3. Insure that the wrist pin bushing in the connecting rod is usable, and lubricate it with assembly lube.
4. Remove the wrist pin from the new piston and lubricate the pin bores on the piston.
5. Align the forward marks on the piston and the connecting rod and install the wrist pin.
6. The new C-clips will have a flat and a rounded side to them. Install both C-clips with the flat side facing out.
7. Repeat all of the steps for each piston being replaced.

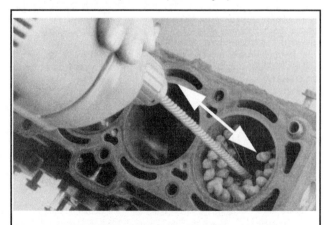

Fig. 34 Use a ball type cylinder hone to remove any glaze and provide a new surface for seating the piston rings

Fig. 35 Most pistons are marked to indicate positioning in the engine (usually a mark means the side facing the front)

ASSEMBLY

Before you begin assembling the engine, first give yourself a clean, dirt free work area. Next, clean every engine component again. The key to a good assembly is cleanliness.

Mount the engine block into the engine stand and wash it one last time using water and detergent (dishwashing detergent works well). While washing it, scrub the cylinder bores with a soft bristle brush and thoroughly clean all of the oil passages. Completely dry the engine and spray the entire assembly down with an anti-rust solution such as WD-40® or similar product. Take a clean lint-free rag and wipe up any excess anti-rust solution from the bores, bearing saddles, etc. Repeat the final cleaning process on the crankshaft. Replace any freeze or oil galley plugs which were removed during disassembly.

Crankshaft

◆ See Figures 36, 37, 38, 39 and 40

1. Remove the main bearing inserts from the block and bearing caps.
2. If the crankshaft main bearing journals have been refinished to a definite undersize, install the correct undersize bearing. Be sure that the bearing inserts and bearing bores are clean. Foreign material under inserts will distort bearing and cause failure.
3. Place the upper main bearing inserts in bores with tang in slot.

■ The oil holes in the bearing inserts must be aligned with the oil holes in the cylinder block.

4. Install the lower main bearing inserts in bearing caps.
5. Clean the mating surfaces of block and rear main bearing cap.
6. Carefully lower the crankshaft into place. Be careful not to damage bearing surfaces.
7. Check the clearance of each main bearing by using the following procedure:
   a. Place a piece of Plastigage® or its equivalent, on bearing surface across full width of bearing cap and about 1/4 in. off center.
   b. Install cap and tighten bolts to specifications. Do not turn crankshaft while Plastigage® is in place.
   c. Remove the cap. Using the supplied Plastigage® scale, check width of Plastigage® at widest point to get maximum clearance. Difference between readings is taper of journal.
   d. If clearance exceeds specified limits, try a 0.001 in. or 0.002 in. undersize bearing in combination with the standard bearing. Bearing clearance must be within specified limits. If standard and 0.002 in. undersize bearing does not bring clearance within desired limits, refinish crankshaft journal, then install undersize bearings.
8. Install the rear main seal.

9. After the bearings have been fitted, apply a light coat of engine oil to the journals and bearings. Install the rear main bearing cap. Install all bearing caps except the thrust bearing cap. Be sure that main bearing caps are installed in original locations. Tighten the bearing cap bolts to specifications.
10. Install the thrust bearing cap with bolts finger-tight.
11. Pry the crankshaft forward against the thrust surface of upper half of bearing.
12. Hold the crankshaft forward and pry the thrust bearing cap to the rear. This aligns the thrust surfaces of both halves of the bearing.
13. Retain the forward pressure on the crankshaft. Tighten the cap bolts to specifications.
14. Measure the crankshaft end-play as follows:
    a. Mount a dial gauge to the engine block and position the tip of the gauge to read from the crankshaft end.
    b. Carefully pry the crankshaft toward the rear of the engine and hold it there while you zero the gauge.
    c. Carefully pry the crankshaft toward the front of the engine and read the gauge.
    d. Confirm that the reading is within specifications. If not, install a new thrust bearing and repeat the procedure. If the reading is still out of specifications with a new bearing, have a machine shop inspect the thrust surfaces of the crankshaft, and if possible, repair it.
15. Rotate the crankshaft so as to position the first rod journal to the bottom of its stroke.

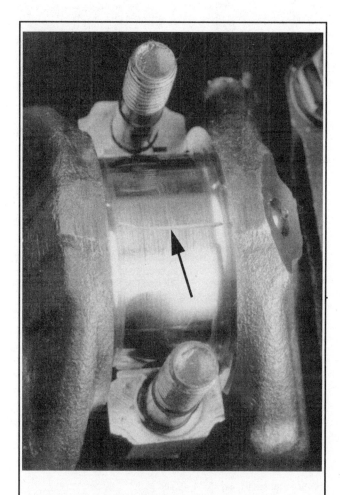

Fig. 36 Apply a strip of gauging material to the bearing journal, then install and torque the cap

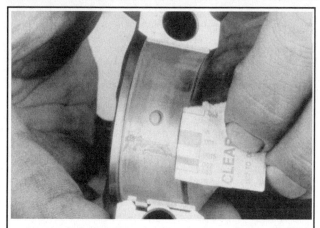

Fig. 37 After the cap is removed again, use the scale supplied with the gauging material to check the clearance

Fig.38 A dial gauge may be used to check crankshaft end-play...

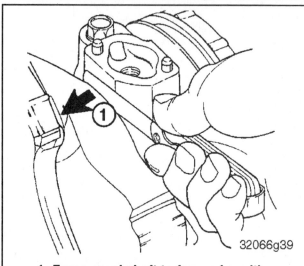

32066g39

1- Force crankshaft to forward position

Fig. 39....or you can use a feeler gauge

Pistons and Connecting Rods

◆ **See Figures 41, 42, 43, 44, 45 and 46**

1. Before installing the piston/connecting rod assembly, oil the pistons, piston rings and the cylinder walls with light engine oil. Install connecting rod bolt protectors or rubber hose onto the connecting rod bolts/studs. Also perform the following:

    a. Select the proper ring set for the size cylinder bore.

    b. Position the ring in the bore in which it is going to be used.

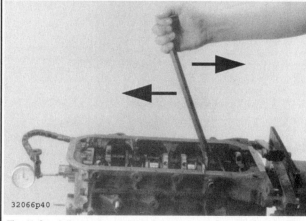

Fig. 40 Carefully pry the crankshaft back and forth while reading the dial gauge for end-play

    c. Push the ring down into the bore area where normal ring wear is not encountered.

    d. Use the head of the piston to position the ring in the bore so that the ring is square with the cylinder wall. Use caution to avoid damage to the ring or cylinder bore.

    e. Measure the gap between the ends of the ring with a feeler gauge. Ring gap in a worn cylinder is normally greater than specification. If the ring gap is greater than the specified limits, try an oversize ring set.

    f. Check the ring side clearance of the compression rings with a feeler gauge inserted between the ring and its lower land according to specification. The gauge should slide freely around the entire ring circumference without binding. Any wear that occurs will form a step at the inner portion of the lower land. If the lower lands have high steps, the piston should be replaced.

2. Unless new pistons are installed, be sure to install the pistons in the cylinders from which they were removed. The numbers on the connecting rod and bearing cap must be on the same side when installed in the cylinder bore. If a connecting rod is ever transposed from one engine or cylinder to another, new bearings should be fitted and the connecting rod should be numbered to correspond with the new cylinder number. The notch on the piston head goes toward the front of the engine.

3. Install all of the rod bearing inserts into the rods and caps.

4. Install the rings to the pistons. Install the oil control ring first, then the second compression ring and finally the top compression ring. Use a piston ring expander tool to aid in installation and to help reduce the chance of breakage.

5. Make sure the ring gaps are properly spaced around the circumference of the piston. Fit a piston ring compressor around the piston and slide the piston and connecting rod assembly down into the cylinder bore, pushing it in with the wooden hammer handle. Push the piston down until it is only slightly below the top of the cylinder bore. Guide the connecting rod onto the crankshaft bearing journal carefully, to avoid damaging the crankshaft.

6. Check the bearing clearance of all the rod bearings, fitting them to the crankshaft bearing journals. Follow the procedure in the crankshaft installation above.

7. After the bearings have been fitted, apply a light coating of assembly oil to the journals and bearings.

8. Turn the crankshaft until the appropriate bearing journal is at the bottom of its stroke, then push the piston assembly all the way down until the connecting rod bearing seats on the crankshaft journal. Be careful not to allow the bearing cap screws to strike the crankshaft bearing journals and damage them.

9. After the piston and connecting rod assemblies have been installed, check the connecting rod side clearance on each crankshaft journal.

10. Prime and install the oil pump and the oil pump intake tube.

11. Install the auxiliary/balance shaft/assembly if equipped.

CAMSHAFT, LIFTERS AND TIMING ASSEMBLY

1. Install the camshaft.
2. Install the lifters/followers into their bores.
3. Install the timing gears/chain assembly.

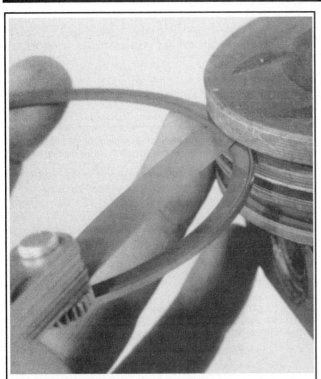

Fig. 41 Checking the piston ring-to-ring groove side clearance using the ring and a feeler gauge

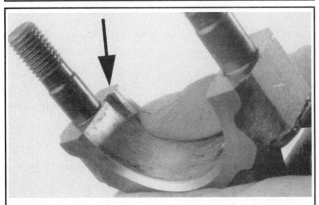

Fig. 42 The notch on the side of the bearing cap matches the tang on the bearing insert

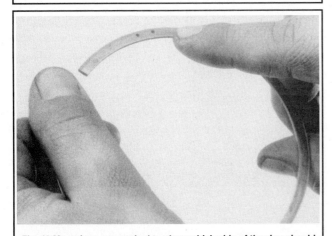

Fig. 43 Most rings are marked to show which side of the ring should face up when installed to the piston

Fig. 44 Install the piston and rod assembly into the block using a ring compressor and the handle of a hammer

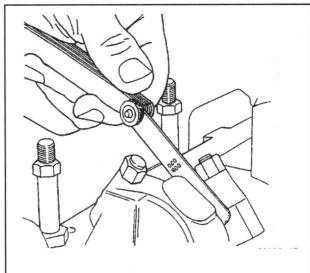

Fig. 45 Checking the connecting rod side clearance

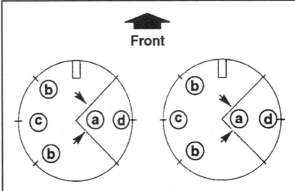

Front

1-3-5-7 Cylinders   **2-4-6-8 Cylinders**

1- Oil ring spacer gap (tang in hole or slot within arc)
2- Oil ring rail gaps
3- 2nd compression ring gap
4- Top compression ring gap

32066g46

Fig. 46 Piston ring gap location

CYLINDER HEAD(S)

1. Install the cylinder head(s) using new gaskets.
2. Assemble the rest of the valve train (pushrods and rocker arms and/or shafts).

Engine Start-up and Break-in

STARTING THE ENGINE

Now that the engine is installed and every wire and hose is properly connected, go back and double check that all coolant and vacuum hoses are connected. Check that your oil drain plug is installed and properly tightened. If not already done, install a new oil filter onto the engine. Fill the crankcase with the proper amount and grade of engine oil. Fill the cooling system with a 50/50 mixture of coolant/water on models with a closed system.

1. Connect the battery.
2. Start the engine. Keep your eye on your oil pressure indicator; if it does not indicate oil pressure within 10 seconds of starting, turn the engine off.

✳✳ WARNING

Damage to the engine can result if it is allowed to run with no oil pressure. Check the engine oil level to make sure that it is full. Check for any leaks and if found, repair the leaks before continuing. If there is still no indication of oil pressure, you may need to prime the system.

3. Confirm that there are no fluid leaks (oil or other).
4. Allow the engine to reach normal operating temperature.
5. At this point you can perform any necessary checks or adjustments, such as checking the ignition timing.
6. Install any remaining components that were removed.

BREAKING IT IN

Make the first hours on the new engine, easy ones. Vary the speed but do not accelerate hard. Most importantly, do not lug the engine, and avoid sustained high speeds until at least 20 hours. Check the engine oil and coolant levels frequently. Expect the engine to use a little oil until the rings seat. Change the oil and filter at 20 hours and then follow the normal maintenance intervals from there out.

KEEP IT MAINTAINED

Now that you have just gone through all of that hard work, keep yourself from doing it all over again by thoroughly maintaining it. Not that you may not have maintained it before, heck you could have had a couple of thousand hours on it before doing this. However, you may have bought the vehicle used, and the previous owner did not keep up on maintenance. Which is why you just went through all of that hard work. See?

7

FUEL SYSTEMS
CARBURETORS

FUEL AND COMBUSTION

Fuel

Fuel recommendations have become more complex as the chemistry of modern gasoline changes. The major driving force behind the changes in gasoline chemistry is the search for additives to replace lead as an octane booster and lubricant. These new additives are governed by the types of emissions they produce in the combustion process. Also, the replacement additives do not always provide the same level of combustion stability, making a fuel's octane rating less meaningful.

In the 1960's and 1970's, leaded fuel was common. The lead served two functions. First, it served as an octane booster (combustion stabilizer) and second, in 4-stroke engines, it served as a valve seat lubricant. For 2-stroke engines, the primary benefit of lead was to serve as a combustion stabilizer. Lead served very well for this purpose, even in high heat applications.

Today, all lead has been removed from the refining process. This means that the benefit of lead as an octane booster has been eliminated. Several substitute octane boosters have been introduced in the place of lead. While many are adequate in an engine, most do not perform nearly as well as lead did, even though the octane rating of the fuel is the same.

OCTANE RATING

A fuel's octane rating is a measurement of how stable the fuel is when heat is introduced. Octane rating is a major consideration when deciding whether a fuel is suitable for a particular application. For example, in an engine, we want the fuel to ignite when the spark plug fires and not before, even under high pressure and temperatures. Once the fuel is ignited, it must burn slowly and smoothly, even though heat and pressure are building up while the burn occurs. The unburned fuel should be ignited by the traveling flame front, not by some other source of ignition, such as carbon deposits or the heat from the expanding gasses. A fuel's octane rating is known as a measurement of the fuel's anti-knock properties (ability to burn without exploding).

Usually a fuel with a higher octane rating can be subjected to a more severe combustion environment before spontaneous or abnormal combustion occurs. To understand how two gasoline samples can be different, even though they have the same octane rating, we need to know how octane rating is determined.

The American Society of Testing and Materials (ASTM) has developed a universal method of determining the octane rating of a fuel sample. The octane rating you see on the pump at a fuel dock is known as the pump octane number. Look at the small print on the pump. The rating has a formula. The rating is determined by the R+M/2 method. This number is the average of the research octane reading and the motor octane rating.

• The Research Octane Rating is a measure of a fuel's anti-knock properties under a light load or part throttle conditions. During this test, combustion heat is easily dissipated.

• The Motor Octane Rating is a measure of a fuel's anti-knock properties under a heavy load or full throttle conditions, when heat buildup is at maximum.

VAPOR PRESSURE

Fuel vapor pressure is a measure of how easily a fuel sample evaporates. Many additives used in gasoline contain aromatics. Aromatics are light hydrocarbons distilled off the top of a crude oil sample. They are effective at increasing the research octane of a fuel sample but can cause vapor lock (bubbles in the fuel line) on a very hot day. If you have an inconsistent running engine and you suspect vapor lock, use a piece of clear fuel line to look for bubbles, indicating that the fuel is vaporizing.

One negative side effect of aromatics is that they create additional combustion products such as carbon and varnish. If your engine requires high octane fuel to prevent detonation, de-carbon the engine more frequently with an internal engine cleaner to prevent ring sticking due to excessive varnish buildup.

ALCOHOL-BLENDED FUELS

When the Environmental Protection Agency mandated a phase-out of the leaded fuels in January of 1986, fuel suppliers needed an additive to improve the octane rating of their fuels. Although there are multiple methods currently employed, the addition of alcohol to gasoline seems to be favored because of its favorable results and low cost. Two types of alcohol are used in fuel today as octane boosters, methanol (wood alcohol) or ethanol (grain alcohol).

When used as a fuel additive, alcohol tends to raise the research octane of the fuel. There are, however, some special considerations due to the effects of alcohol in fuel.

• Since alcohol contains oxygen, it replaces gasoline without oxygen content and tends to cause the air/fuel mixture to become leaner.

• On older engines, the leaching affect of alcohol may, in time, cause fuel lines and plastic components to become brittle to the point of cracking. Unless replaced, these cracked lines could leak fuel, increasing the potential for hazardous situations.

• When alcohol blended fuels become contaminated with water, the water combines with the alcohol then settles to the bottom of the tank. This leaves the gasoline on a top layer.

■ **Modern fuel lines and plastic fuel system components have been specially formulated to resist alcohol leaching effects.**

HIGH ALTITUDE OPERATION

At elevated altitudes there is less oxygen in the atmosphere than at sea level. Less oxygen means lower combustion efficiency and less power output. As a general rule, power output is reduced three percent for every thousand feet above sea level.

On carbureted engines, re-jetting for high altitude does not restore lost power, it simply corrects the air-fuel ratio for the reduced air density and makes the most of the remaining available power. The most important thing to remember when re-jetting for high altitude is to reverse the jetting when returning to sea level. If the jetting is left lean when you return to sea level conditions, the correct air/fuel ratio will not be achieved and possible engine damage may occur.

RECOMMENDATIONS

According to the fuel recommendations that come with your engine, there are only a few engines in the product line that requires more than 87 octane. Most MerCruiser engines need only 87 octane or less. An 89 octane rating generally means middle grade unleaded. Premium unleaded is more stable under severe conditions but also produces more combustion products. Therefore, when using premium unleaded, more frequent de-carboning is necessary.

Combustion

In a high heat environment like an modern engine, the fuel must be very stable to avoid detonation. If any parameters affecting combustion change suddenly (the engine runs lean for example), uncontrolled heat buildup will occurs very rapidly.

The combustion process is affected by several interrelated factors. This means that when one factor is changed, the other factors also must be changed to maintain the same controlled burn and level of combustion stability.

• Compression—determines the level of heat buildup in the cylinder when the air-fuel mixture is compressed. As compression increases, so does the potential for heat buildup

• Ignition Timing—determines when the gasses will start to expand in relation to the motion of the piston. If the ignition timing is too advanced, gasses will be ignited and begin to expand too soon, such as they would during pre-ignition. The motion of the piston opposes the expansion of the gasses, resulting in extremely high combustion chamber pressures and heat. If the ignition timing is retarded, the gases are ignited later in relation to piston position. This means that the piston has already traveled back down the bore toward the bottom of the cylinder, resulting in less usable power.

• Fuel Mixture—determines how efficient the burn will be. A rich mixture burns slower than a lean one. If the mixture is too lean, it can't become explosive. The slower the burn, the cooler the combustion chamber, because pressure buildup is gradual.

• Fuel Quality (Octane Rating)—determines how much heat is necessary to ignite the mixture. Once the burn is in progress, heat is on the rise.

The unburned poor quality fuel is ignited all at once by the rising heat instead of burning gradually as a flame front of the burn passing by. This action results in detonation (pinging).

There are two types of abnormal combustion—pre-ignition and detonation.

- Pre-ignition—occurs when the air-fuel mixture is ignited by some incandescent source other than the correctly timed spark from the spark plug.
- Detonation—occurs when excessive heat and or pressure ignites the air/fuel mixture rather than the spark plug. The burn becomes explosive.

CARBURETED FUEL SYSTEM

Troubleshooting

In general, anything that can cause abnormal heat buildup can be enough to push an engine over the edge to abnormal combustion, if any of the four basic factors previously discussed are already near the danger point, for example, excessive carbon buildup raises the compression and retains heat as glowing embers.

Troubleshooting fuel systems requires the same techniques used in other areas. A thorough, systematic approach to troubleshooting will pay big rewards. Build your troubleshooting checklist, with the most likely offenders at the top. Use your experience to adjust your list for local conditions. Everyone has been tempted to jump into the carburetor on a vague hunch. Pause a moment and review the facts when this urge occurs.

In order to accurately troubleshoot a carburetor or fuel system problem, you must first verify that the problem is fuel related. Many symptoms can have several different possible causes. Be sure to eliminate mechanical and electrical systems as the potential fault. Carburetion is the number one cause of most engine problems but there are other possibilities.

One of the toughest tasks with a fuel system is the actual troubleshooting. Several tools are at your disposal for making this process very simple. A timing light works well for observing carburetor spray patterns. Look for the proper amount of fuel and for proper atomization in the two fuel outlet areas (main nozzle and bypass holes). The strobe effect of the lights helps you see in detail the fuel being drawn through the throat of the carburetor. On multiple carburetor engines, always attach the timing light to the cylinder you are observing so the strobe doesn't change the appearance of the patterns. If you need to compare two cylinders, change the timing light hookup each time you observe a different cylinder.

Pressure testing fuel pump output can determine whether the fuel spray is adequate and if the fuel pump diaphragms are functioning correctly. A pressure gauge placed between the fuel pump(s) and the carburetor(s) will test the entire fuel delivery system. Normally a fuel system problem will show up at high speed where the fuel demand is the greatest. A common symptom of a fuel pump output problem is surging at wide open throttle but normal operation at slower speeds. To check the fuel pump output, install the pressure gauge and accelerate the engine to wide-open throttle. Observe the pressure gauge needle. It should always swing up the value indicated in the specification charts and remain steady. This reading would indicate a system that is functioning properly.

If the needle gradually swings down toward zero, fuel demand is greater than the fuel system can supply. This reading isolates the problem to the fuel delivery system (fuel tank or line). To confirm this, an auxiliary tank should be installed and the engine re-tested. Be aware that a bad anti-siphon valve on a built-in tank can create enough restriction to cause a lean condition and serious engine damage.

If the needle movement becomes erratic, suspect a ruptured diaphragm in the fuel pump.

To check for air entering the fuel system, install a clear fuel hose between the fuel screen and fuel pump. If air is in the line, check all fittings back to the boat's fuel tank.

Spark plug tip appearance is a good indication of combustion efficiency. The tip should be a light tan. A White insulator or small beads on the insulator indicate too much heat. A dark or oil fouled insulator indicates incomplete combustion. To properly read spark plug tip appearance, run the engine at the RPM you are testing for about 15 seconds and then immediately turn the engine OFF without changing the throttle position.

Reading spark plug tip appearance is also the proper way to test jet verifications in high altitude.

COMMON PROBLEMS

Fuel Delivery

Many times fuel system troubles are caused by a plugged fuel filter, a defective fuel pump or by a leak in the line from the fuel tank to the fuel pump. A defective choke may also cause problems. Would you believe, an majority of starting troubles which are traced to the fuel system are the result of an empty fuel tank or aged sour fuel.

Sour Fuel

Under average conditions (temperate climates), fuel will begin to break down in about four months. A gummy substance forms in the bottom of the fuel tank and in other areas. The filter screen between the tank and the carburetor and small passages in the carburetor will become clogged. The gasoline will begin to give off an odor similar to rotten eggs. Such a condition can cause the owner much frustration, time in cleaning components and the expense of replacement or overhaul parts for the carburetor.

Even with the high price of fuel, removing gasoline that has been standing unused over a long period of time is still the easiest and least expensive preventative maintenance possible. In most cases, this old gas can be used without harmful effects in an automobile using regular gasoline.

A gasoline preservative will keep the fuel fresh for up to twelve months. These products are available in most areas under various trade names.

Choke Problems

When the engine is hot, the fuel system can cause starting problems. After a hot engine is shut down, the temperature inside the fuel bowl may rise to 200 degrees F and cause the fuel to actually boil. All carburetors are vented to allow this pressure to escape to the atmosphere. However, some of the fuel may percolate over the high-speed nozzle.

If the choke should stick in the open position, the engine will be hard to start. If the choke should stick in the closed position, the engine will flood, making it very difficult to start.

In order for this raw fuel to vaporize enough to burn, considerable air must be added to lean out the mixture. Therefore, the only remedy is to remove the spark plugs, ground the leads, crank the engine through about ten revolutions, clean the plugs, reinstall the plugs and start the engine.

If the needle valve and seat assembly is leaking, an excessive amount of fuel may enter the reed housing in the following manner. After the engine is shut down, the pressure left in the fuel line will force fuel past the leaking needle valve. This extra fuel will raise the level in the fuel bowl and cause fuel to overflow into the reed housing.

A continuous overflow of fuel into the reed housing may be due to a sticking inlet needle or to a defective float, which would cause an extra high level of fuel in the bowl and overflow into the reed housing.

Rough Engine Idle

If an engine does not idle smoothly, the most reasonable approach to the problem is to perform a tune-up to eliminate such areas as:
- Defective points
- Faulty spark plugs
- Timing out of adjustment

Other problems that can prevent an engine from running smoothly include:
- An air leak in the intake manifold
- Uneven compression between the cylinders

Of course any problem in the carburetor affecting the air/fuel mixture will also prevent the engine from operating smoothly at idle speed. These problems usually include:
- Too high a fuel level in the bowl
- A heavy float
- Leaking needle valve and seat
- Defective automatic choke
- Improper adjustments for idle mixture or idle speed

Excessive Fuel Consumption

Excessive fuel consumption can be the result of any one of four conditions or a combination of all.
- Inefficient engine operation.
- Faulty condition of the hull, including excessive marine growth.
- Poor boating habits of the operator.
- Leaking or out-of-tune carburetor.

Application Chart

Year	Model	Engine L/Cu. In.	Engine Type	Carburetor Type	Fuel Delivery
1992	3.0L/LX	3.0/181	L4	MerCarb	2 bbl
	185MR/4.3L	4.3/262	V6	MerCarb	2 bbl
	205MR/4.3L	4.3/262	V6	Weber WFB	4 bbl
	5.0L	5.0/305	V8	MerCarb	2 bbl
	5.0LX	5.0/305	V8	Rochester 4MV	4 bbl
	5.7L	5.7/350	V8	Rochester 4MV	4 bbl
	350 Magnum	5.7/350	V8	Weber WFB	4 bbl
	7.4L	7.4/454	V8	Weber WFB	4 bbl
	454 Magnum	7.4/454	V8	Weber WFB	4 bbl
	8.2L	8.2/502	V8	Weber WFB	4 bbl
	502 Magnum	8.2/502	V8	Weber WFB	4 bbl
1993	3.0L/LX	3.0/181	L4	MerCarb	2 bbl
	185MR/4.3L	4.3/262	V6	MerCarb	2 bbl
	205MR/4.3L	4.3/262	V6	Weber WFB	4 bbl
	5.0L	5.0/305	V8	MerCarb	2 bbl
	5.0LX	5.0/305	V8	Rochester 4MV	4 bbl
	5.7L	5.7/350	V8	Rochester 4MV	4 bbl
	350 Magnum	5.7/350	V8	Weber WFB	4 bbl
	7.4L	7.4/454	V8	Weber WFB	4 bbl
	454 Magnum	7.4/454	V8	Weber WFB	4 bbl
	8.2L	8.2/502	V8	Weber WFB	4 bbl
	502 Magnum	8.2/502	V8	Weber WFB	4 bbl
1994	3.0L/LX	3.0/181	L4	MerCarb	2 bbl
	4.3L	4.3/262	V6	MerCarb	2 bbl
	4.3LX	4.3/262	V6	Weber WFB	4 bbl
	5.0L	5.0/305	V8	MerCarb	2 bbl
	5.0LX	5.0/305	V8	Rochester 4MV	4 bbl
	5.7L	5.7/350	V8	Weber WFB	4 bbl
	350 Magnum	5.7/350	V8	Rochester 4MV	4 bbl
	7.4L	7.4/454	V8	Weber WFB	4 bbl
	454 Magnum	7.4/454	V8	Weber WFB	4 bbl
	8.2L	8.2/502	V8	Weber WFB	4 bbl
1995	3.0LX	3.0/181	L4	MerCarb	2 bbl
	4.3L	4.3/262	V6	MerCarb	2 bbl
	4.3LX	4.3/262	V6	Weber WFB	4 bbl
	5.0L	5.0/305	V8	MerCarb	2 bbl
	5.0LX	5.0/305	V8	Weber WFB	4 bbl
	5.7L	5.7/350	V8	Rochester 4MV	4 bbl

Application Chart

Year	Model	Engine L/Cu. In.	Engine Type	Carburetor Type	Fuel Delivery
1995	350 Magnum	5.7/350	V8	Weber WFB	4 bbl
	7.4L	7.4/454	V8	Weber WFB	4 bbl
	8.2L	8.2/502	V8	Weber WFB	4 bbl
1996	3.0LX	3.0/181	L4	MerCarb	2 bbl
	4.3L	4.3/262	V6	MerCarb	2 bbl
	4.3LX	4.3/262	V6	Weber WFB	4 bbl
	5.7L	5.7/350	V8	MerCarb	2 bbl
				Rochester 4MV	4 bbl
	350 Magnum	5.7/350	V8	Weber WFB	4 bbl
	7.4L	7.4/454	V8	Weber WFB	4 bbl
	454 Magnum	7.4/454	V8	Weber WFB	4 bbl
	8.2L	8.2/502	V8	Weber WFB	4 bbl
1997	3.0LX	3.0/181	L4	MerCarb	2 bbl
	4.3LX Gen+	4.3/262	V6	MerCarb	2 bbl
	4.3LXH Gen +	4.3/262	V6	Weber WFB	4 bbl
	5.7L	5.7/350	V8	MerCarb	2 bbl
				Rochester 4MV	4 bbl
	350 Magnum	5.7/350	V8	Weber WFB	4 bbl
	7.4L	7.4/454	V8	Weber WFB	4 bbl
	454 Magnum	7.4/454	V8	Weber WFB	4 bbl
1998	3.0L Alpha	3.0/181	L4	MerCarb	2 bbl
	4.3L	4.3/262	V6	MerCarb	2 bbl
	4.3LH	4.3/262	V6	Weber WFB	4 bbl
	5.0L	5.0/305	V8	MerCarb	2 bbl
	5.7L	5.7/350	V8	MerCarb	2 bbl
1999	3.0L Alpha	3.0/181	L4	MerCarb	2 bbl
	4.3L	4.3/262	V6	MerCarb	2 bbl
	4.3LH	4.3/262	V6	Weber WFB	4 bbl
	5.0L	5.0/305	V8	MerCarb	2 bbl
	5.7L	5.7/350	V8	MerCarb	2 bbl
2000	3.0L Alpha	3.0/181	L4	MerCarb	2 bbl
	4.3L	4.3/262	V6	MerCarb	2 bbl
	4.3LH	4.3/262	V6	Weber WFB	4 bbl
	5.0L	5.0/305	V8	MerCarb	2 bbl
	5.7L	5.7/350	V8	MerCarb	2 bbl
2001	3.0L	3.0/181	L4	MerCarb	2 bbl
	4.3L	4.3/262	V6	MerCarb	2 bbl
	4.3LH	4.3/262	V6	Weber WFB	4 bbl
	5.0L	5.0/305	V8	MerCarb	2 bbl
	5.7L	5.7/350	V8	MerCarb	2 bbl

32067c1

If the fuel consumption suddenly increases over what could be considered normal, then the cause can probably be attributed to the engine or boat and not the operator.

Marine growth on the hull can have a very marked effect on boat performance. This is why sailboats always try to have a haul-out as close to race time as possible.

While you are checking the bottom, take note of the propeller condition. A bent blade or other damage will definitely cause poor boat performance.

If the hull and propeller are in good shape, then check the fuel system for possible leaks. Check the line between the fuel pump and the carburetor while the engine is running and the line between the fuel tank and the pump when the engine is not running. A leak between the tank and the pump many times will not appear when the engine is operating, because the suction created by the pump drawing fuel will not allow the fuel to leak. Once the engine is turned off and the suction no longer exists, fuel may begin to leak.

If a minor tune-up has been performed and the spark plugs, points (if equipped) and timing are properly adjusted, then the problem most likely is in the carburetor and an overhaul is in order.

Check the needle valve and seat for leaking. Use extra care when making any adjustments affecting the fuel consumption, such as the float level or automatic choke.

Engine Surge

If the engine operates as if the load on the boat is being constantly increased and decreased, even though an attempt is being made to hold a constant engine speed, the problem can most likely be attributed to the fuel pump or a restriction in the fuel line between the tank and the carburetor.

COMBUSTION RELATED PISTON FAILURES

When an engine has a piston failure due to abnormal combustion, fixing the mechanical portion of the engine is the easiest part of the equation. The hard part is determining what caused the problem in order to prevent a repeat failure. Think back to the four basic areas that affect combustion to find the cause of the failure.

Since you probably removed the cylinder head, inspect the failed piston and look for excessive deposit buildup that could raise compression or retain heat in the combustion chamber. Statically check the wide open throttle timing. Be sure that the timing is not over advanced. It is a good idea to seal these adjustments with paint to detect tampering.

Look for a fuel restriction that could cause the engine to run lean. Don't forget to check the fuel pump, fuel tank and lines, especially if a built in tank is used. Be sure to check the anti-siphon valve on built in tanks. If everything else looks good, the final possibility is poor quality fuel.

Operation

BASIC FUNCTIONS

◆ See Figure 1

Traditional carburetor theory often involves a number of laws and principles. The diagram illustrates several carburetor basics. If you blow air across a straw inserted into a container of liquid, a pressure drop is created in the straw column. As the liquid in the column is expelled, an atomized mixture (air and fuel droplets) is created. In a carburetor this is mostly air and a little fuel.

The actual ratio of air to fuel differs with engine conditions but is usually from 15 parts air to one part fuel at optimum cruise to as little as 7 parts air to one part fuel at full choke.

Using our example, what if the top of the container is covered and sealed around the straw, what will happen? No flow. This is typical of a clogged carburetor bowl vent. If the base of the straw is clogged or restricted what will happen? No flow or low flow. This represents a clogged main jet. If the liquid in the glass is lowered and you blow through the straw with the same force what will happen? Not as much fuel will flow. A lean condition occurs. If the fuel level is raised and you blow again at the same velocity what happens? The result is a richer mixture.

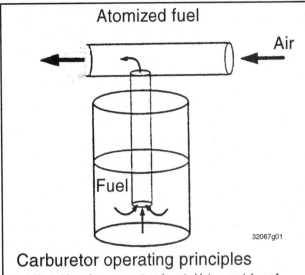

Carburetor operating principles

Fig. 1 If you blow air across a straw inserted into a container of liquid, a pressure drop is created in the straw column. As the liquid in the column is expelled, an atomized mixture (air and fuel droplets) is created

FUEL & AIR METERING

The carburetor is merely a metering device for mixing fuel and air in the proper proportions for efficient engine operation.

Float Systems

◆ See Figure 2

A small chamber in the carburetor serves as a fuel reservoir. A float valve admits fuel into the reservoir to replace the fuel consumed by the engine. If the carburetor has more than one reservoir, the fuel level in each reservoir (chamber) is controlled by identical float systems.

Fuel level in each chamber is extremely critical and must be maintained accurately. Accuracy is obtained through proper adjustment of the floats. This adjustment will provide a balanced metering of fuel to each cylinder at all speeds.

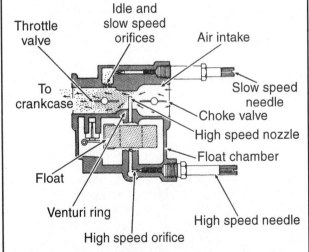

Fig. 2 Fuel flow through a venturi, showing principal and related parts controlling intake and outflow

Following the fuel through its course, from the fuel tank to the combustion chamber of the cylinder, will provide an appreciation of exactly what is taking place. In order to start the engine, the fuel must be moved from the tank to the carburetor by a fuel pump installed in the fuel line.

After the engine starts, the fuel passes through the pump to the carburetor. All systems have some type of filter installed somewhere in the line between the tank and the carburetor. Most engine also have a filter as an integral part of the carburetor.

At the carburetor, the fuel passes through the inlet passage to the needle and seat and then into the float chamber (reservoir). A float in the chamber rides up and down on the surface of the fuel. After fuel enters the chamber and the level rises to a predetermined point, a tang on the float closes the inlet needle and the flow entering the chamber is cut off. When fuel leaves the chamber as the engine operates, the fuel level drops and the float tang allows the inlet needle to move off its seat and fuel once again enters the chamber. In this manner, a constant reservoir of fuel is maintained in the chamber to satisfy the demands of the engine at all speeds.

A fuel chamber vent hole is located near the top of the carburetor body to permit atmospheric pressure to act against the fuel in each chamber. This pressure assures an adequate fuel supply to the various operating systems.

Air/Fuel Mixture

◆ See Figure 3

A suction effect is created each time the piston moves upward in the cylinder. This suction draws air through the throat of the carburetor. A restriction in the throat, called a venturi, controls air velocity and has the effect of reducing air pressure at this point.

The difference in air pressures at the throat and in the fuel chamber, causes the fuel to be pushed out of metering jets extending down into the fuel chamber. When the fuel leaves the jets, it mixes with the air passing through the venturi. This fuel/air mixture should then be in the proper proportion for burning in the cylinders for maximum engine performance.

In order to obtain the proper air/fuel mixture for all engine speeds, some models have high and low speed jets. These jets have adjustable needle valves that are used to compensate for changing atmospheric conditions. In almost all cases, the high-speed circuit has fixed high-speed jets and is not adjustable.

A throttle valve controls the flow of air/fuel mixture drawn into the combustion chambers. A cold engine requires a richer fuel mixture to start and during the brief period it is warming to normal operating temperature. A choke valve is placed ahead of the metering jets and venturi. As this valve begins to close, the volume of air intake is reduced, thus enriching the mixture entering the cylinders. When this choke valve is fully closed, a very rich fuel mixture is drawn into the cylinders.

The throat of the carburetor is usually referred to as the barrel. Carburetors with single, double or four barrels have individual metering jets, needle valves, throttle and choke plates for each barrel. Single and two barrel carburetors are fed by a single float and chamber.

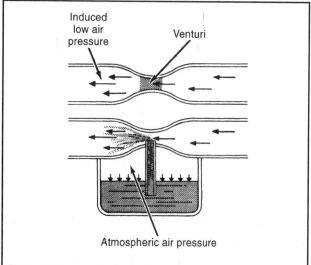

Fig. 3 Air flow principle of a modern carburetor

Induced low air pressure

Venturi

Atmospheric air pressure

CARBURETOR CIRCUITS

The following section illustrates the circuit functions and locations of a typical marine carburetor.

Starting Circuit

◆ See Figure 4

The choke plate is closed, creating a partial vacuum in the venturi. As the piston rises, negative pressure in the crankcase draws the rich air-fuel mixture from the float bowl into the venturi and on into the engine.

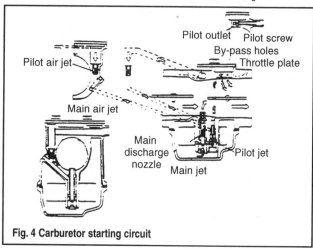

Pilot outlet　Pilot screw
By-pass holes
Pilot air jet　　　　　　Throttle plate
Main air jet
Main discharge nozzle
Main jet　　Pilot jet

Fig. 4 Carburetor starting circuit

Low Speed Circuit

◆ See Figure 5

Zero-one-eighth throttle, when the pressure in the crankcase is lowered, the air-fuel mixture is discharged into the venturi through the pilot outlet because the throttle plate is closed. No other outlets are exposed to low venturi pressure. The fuel is metered by the pilot jet. The air is metered by the pilot air jet. The combined air-fuel mixture is regulated by the pilot air screw.

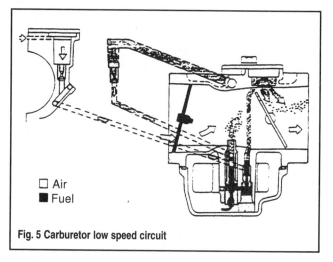

☐ Air
■ Fuel

Fig. 5 Carburetor low speed circuit

Mid-Range Circuit

◆ See Figure 6

One-eighth-three-eighths throttle, as the throttle plate continues to open, the air-fuel mixture is discharged into the venturi through the bypass holes. As the throttle plate uncovers more bypass holes, increased fuel flow results because of the low pressure in the venturi. Depending on the model, there could be two, three or four bypass holes.

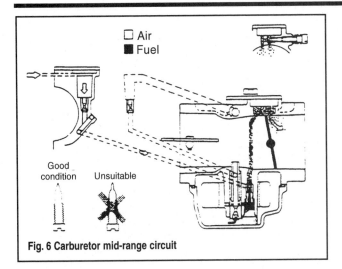

Fig. 6 Carburetor mid-range circuit

High Speed Circuit

◆ **See Figure 7**

Three-eighths-wide-open throttle, as the throttle plate moves toward wide open, we have maximum air flow and very low pressure. The fuel is metered through the main jet and is drawn into the main discharge nozzle. Air is metered by the main air jet and enters the discharge nozzle, where it combines with fuel. The mixture atomizes, enters the venturi and is drawn into the engine.

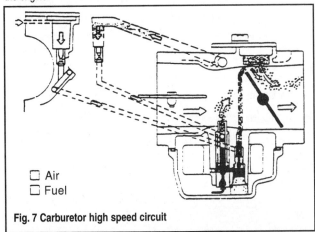

Fig. 7 Carburetor high speed circuit

MerCarb 2bbl Carburetor

DESCRIPTION

◆ **See Figure 8**

The MerCarb carburetor used on 3.0L, 4.3L, 5.0L and 5.7L engines is a two barrel carburetor with a separate fuel feed for each venturi. It has one idle adjustment screw and is equipped with an electric choke. A removable venturi cluster—attached to the float bowl assembly—has the calibrated main well tubes and pump jets built into it. The venturi cluster is serviced as a unit.

A part number and a date code are embossed on the carburetor, as shown in the accompanying illustration.

The upper number is the Part Number.

The lower number is the date code and is interpreted as follows:

First digit is the year of manufacturer. In the illustration "2" equals 1992; "6" would designate 1996.

The second digit is the month with "3" designating March, "9" designating September, etc. " X" designates October, "Y"designates November and, you guessed it, "Z" is for December.

The third and fourth digits indicate the day of the month, with the first nine days preceded by a "0". Thus "01" would be the first day; "08" the eighth day, etc.

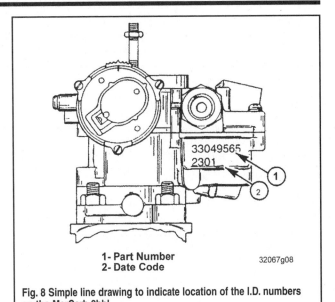

1- Part Number
2- Date Code

Fig. 8 Simple line drawing to indicate location of the I.D. numbers on the MerCarb 2bbl

Complete carburetor adjustment specifications can be found at the end of this section.

✱✱ WARNING

Always disconnect the battery cables before attempting to work on the fuel system. Never smoke or allow open flame near the engine— this sounds like an obvious precaution, but you'd be surprised at how many people forget!

REMOVAL & INSTALLATION

■ No matter how much they look alike, marine carburetors are completely different from automotive carburetors. Never substitute an automotive carburetor for the one on your engine! Venting procedures are not the same and an automotive carburetor could allow dangerous vapors to escape into the bilge. Don't even think about it.

1. Open the engine hatch or remove the covers and then disconnect the battery cables. Turn the fuel petcock OFF and/or shut down the fuel supply at the tank.
2. Remove the flame arrestor after disconnecting the vent hose. It's always a good idea to plug the throttle bores with a clean, lint-free cloth.
3. Disconnect the throttle cable at the carburetor and carefully move it out of the way.
4. Using two open-end wrenches, hold the fuel inlet nut at the carburetor securely and loosen the fuel line nut. Disconnect the two and carefully move the line out of the way. Plug both the inlet and line open ends to prevent fuel seepage.
5. Disconnect the fuel pump sight tube at the unit (3.0L engines only).
6. Tag and disconnect the electric choke lead.
7. Loosen and remove the carburetor mounting nuts/washers and lift the unit off the manifold. Plug the opening with a clean lint-free cloth.

To Install:
8. Clean the mating surfaces thoroughly of all residual gasket material, position a new gasket and then install the carburetor. Tighten the nuts to 20 ft. lbs. (27 Nm) on the 3.0L or 132 inch lbs. (15 Nm) on the other engines. Hopefully you remembered to remove the rag!
9. Reconnect the fuel line to the inlet nut after removing the plugs and tighten it to 18 ft. lbs. (24 Nm). Don't forget to use two wrenches.
10. Connect the plastic sight tube line (on the 3.0L) and then connect the choke lead.
11. Install and adjust the throttle cable as detailed later in this section.
12. Install the flame arrestor and reconnect the vent line.
13. Connect the battery cables. Start the engine and ensure there are no fuel leaks; shut down the engine immediately if there are. Check and adjust the idle speed and mixture.

Accelerator Pump Lever

◆ See Figure 9

The accelerator lever is equipped with 3 separate holes, allowing you to change the amount of fuel being delivered to the engine via the accelerator pump. The hole closest to the pump lever's shaft will deliver the full amount of fuel. The center hole will provide approximately 0.5cc less fuel with each stroke, while the hole furthest from the shaft will provide a flow of approximately 1.0cc less per stroke.

Idle Speed & Mixture

Please refer to Section 3 for adjustment procedures.

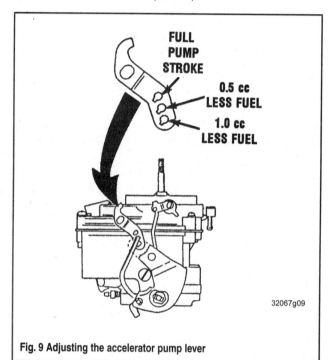

Fig. 9 Adjusting the accelerator pump lever

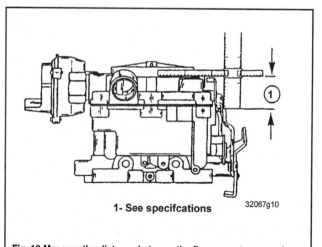

Fig. 10 Measure the distance between the flame arrestor mounting surface and the top of the pump rod

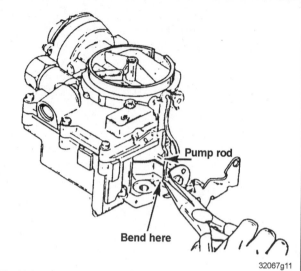

Fig. 11 Bend the pump rod until the measurement is within specifications

Pump Rod

◆ See Figures 10 and 11

1. Back out the idle speed screw until it is no longer in contact with the idle cam.
2. Ensure that the throttle valves are closed completely and then measure the distance between the top of the flame arrestor mounting surface on the carburetor body and the top of the pump rod. Grasp the rod with a pair of needle nose pliers and bend it (carefully) until the distance is equal to 1 5/32 in. (29mm).

Throttle Cable

◆ See Figures 12, 13 and 14

1. With the remote control lever in Neutral (idle), install the throttle cable end guide onto the lever and then push the barrel slightly toward the throttle

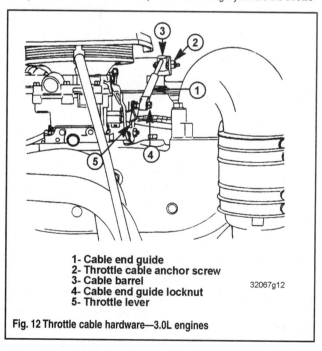

1- Cable end guide
2- Throttle cable anchor screw
3- Cable barrel
4- Cable end guide locknut
5- Throttle lever

Fig. 12 Throttle cable hardware—3.0L engines

lever end so as to preload the cable slightly and remove any slack. Adjust the barrel so that it aligns with the anchor stud.

2. Secure the throttle cable with its hardware and then tighten the locknut on the cable end guide until it just touches the cable, and then back it out one full turn on the 3.0L and 1/2 turn on 4.3L, 5.0L and 5.7L engines.

3. Tighten the throttle cable anchor screw securely—do not overtighten.

4. Move the remote control lever to full throttle position. The throttle valves should be open all the way and the throttle shaft lever should be in contact with the body of the carburetor.

5. Move the lever back to the Neutral position; the throttle lever should touch the idle speed adjustment screw.

Choke Setting

◆ See Figure 15

Loosen the three choke housing cover screws and rotate the cover 2 index marks clockwise past the larger center index mark. Tighten the housing cover screws securely. On 1992-97 V8 engines, the scribed mark on the housing cover should be aligned with the long case mark on the housing.

Choke Unloader

◆ See Figures 16 and 17

1. Remove the flame arrestor.
2. Hold the throttle valves so they are open fully.
3. Press down gently on the choke plate and then slide a 0.080 in. (0.2mm) drill bit between the upper edge of the choke plate and the air horn assembly—the bit should just fit through. If not, bend the tang on the throttle lever until it does.

Float Level And Drop

◆ See Figures 18, 19, 20, 21 and 22

1. Following the disassembly procedure detailed later, remove the air horn from the carburetor.
2. Turn the air horn assembly upside down and check that the float pivots freely on the pin. Raise the float and let it fall—do not force it!
3. Using a standard carburetor gauge, measure the distance between the air horn gasket and the toe of the float it should be within the specification given at the end of this section—3/8 in. (10mm) for solid needle, and 9/16 in. (14mm) for spring loaded needles. Carefully bend the float arm with needle nose pliers to achieve the correct measurement. Make sure the float stays in alignment.
4. Its usually a good idea to check the float drop now. Turn the air horn right side up and allow the float to hang down freely.
5. Using the same gauge, measure the distance between the gasket and toe once again. If not 1-3/32 in. (27mm), bend the float tang with needle nose pliers to achieve the correct measurement.
6. Recheck both measurements one more time.
7. Reinstall the air horn and carburetor.

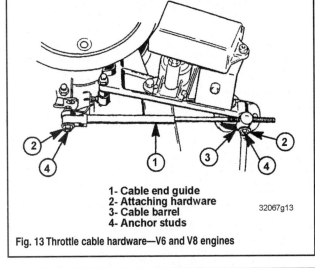

1- Cable end guide
2- Attaching hardware
3- Cable barrel
4- Anchor studs

32067g13

Fig. 13 Throttle cable hardware—V6 and V8 engines

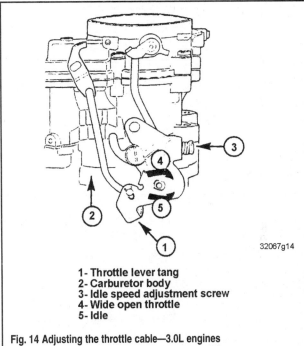

1- Throttle lever tang
2- Carburetor body
3- Idle speed adjustment screw
4- Wide open throttle
5- Idle

32067g14

Fig. 14 Adjusting the throttle cable—3.0L engines

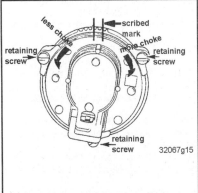

less choke
scribed mark
more choke
retaining screw
retaining screw
retaining screw

32067g15

Fig. 15 The choke setting should be 2 marks to the right of the center index mark—3.0L, 4.3L and 1998-01 V8 engines

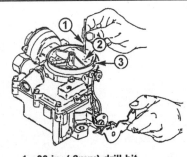

1- .08 in. (.2mm) drill bit
2- Choke pin
3- Air horn

32067g16

Fig. 16 Insert a drill bit between the choke plate and the air horn to check the choke unloader

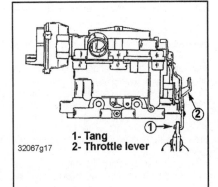

32067g17

1- Tang
2- Throttle lever

Fig. 17 Bend the throttle lever tang to adjust the unloader if the bit does not fit properly

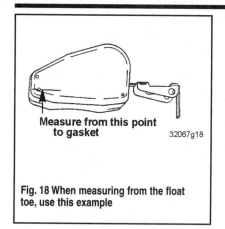

Measure from this point to gasket

32067g18

Fig. 18 When measuring from the float toe, use this example

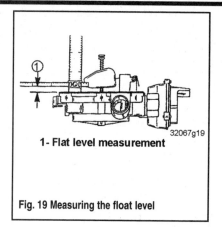

1- Flat level measurement

32067g19

Fig. 19 Measuring the float level

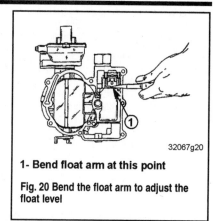

1- Bend float arm at this point

32067g20

Fig. 20 Bend the float arm to adjust the float level

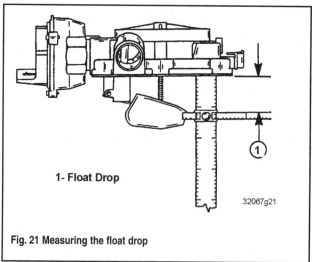

1- Float Drop

32067g21

Fig. 21 Measuring the float drop

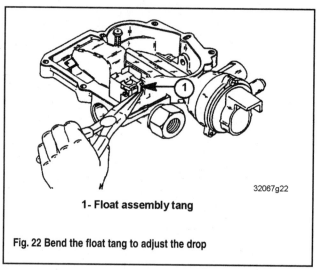

1- Float assembly tang

32067g22

Fig. 22 Bend the float tang to adjust the drop

DISASSEMBLY

◆ **See Figures 23, 24, 25, 26, 27, 28, 29, 30, 31, 32, 33, 34 and 35**

■ **Always be certain that your carburetor rebuild kit is for marine applications.**

1. Remove the carburetor from the manifold and mount it securely on a workbench.
2. Remove the choke cover, the lever and the housing.
3. Remove the fuel inlet nut and then pull out the filter.
4. Remove the retaining clip at the bottom of the of the pump rod. Pivot the rod until it's retaining ear align with the slot on the pump shaft and lever and then pull off the rod.
5. Remove the idle cam screw and then remove the choke rod in the same manner as the accelerator pump rod above.
6. Remove the seven air horn screws and carefully lift the horn off the float bowl. Position the air horn upside down on the bench, remove the float hinge pin and lift out the float.
7. Remove the air horn gasket and baffle. Lift out the needle assembly from the seat. Insert a screwdriver into the seat slots and screw out the seat.
8. Loosen the accelerator pump set screw, slide the pump shaft and lever assembly out of the air horn and remove the entire accelerator pump. Slide the retaining clip and washer from the pump shaft and disconnect the pump from the lever.
9. Lift the accelerator pump return spring out of the pump well.
10. Remove the power valve assembly and then pull out the main metering jets.
11. Remove the three mounting screws for the venturi cluster and lift the cluster and gasket straight out of the housing. Be very careful not to damage the brass tubes attached to the bottom of the cluster; they are permenantly pressed into the cluster and are not replaceable.

13. Reach into the housing with needle nose pliers and pop out the check ball spring retainer. Turn the assembly over and allow the check ball and spring to fall out.

CLEANING & INSPECTION

Never use a wire brush or drill to clean jet passages or tubes in the carburetor.

Never allow the carburetor to soak in a cleaner bath for more than two hours. In fact, we recommend using spray cleaner.

Never immerse the float assembly, needle, accelerator pump plunger or fuel filter in cleaner. Wipe them carefully with a clean cloth.

Otherwise clean all allowable parts with carb cleaner and then dry with compressed air if at all possible.

Blow out and through all passages to ensure there is no foreign material clogging them.

Check the float needle and seat, if either is worn or damaged, replace them as a matched set.

Check the float assembly and hinge pin for wear or damage, replace as necessary.

Check the pump plunger, return spring, piston spring, idle mixture needle and all levers and linkages for wear or damage. Replace as necessary.

Check the throttle valve shaft for excessive looseness in the throttle body. Check that the valve and shaft do not bind through their range of operation and that the valve opens and closes fully. Replace the assembly if it fails any of these tests.

Check the choke valve lever and shaft for excessive looseness in the air horn. Check that the valve, lever and shaft do not bind through their range of operation and that the valve opens and closes fully. Replace the air horn assembly if it fails any of these tests.

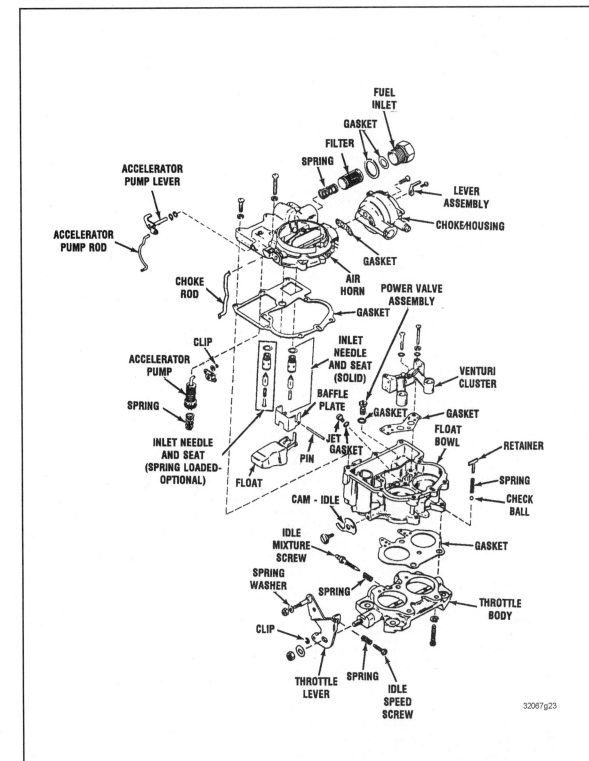

Fig. 23 Exploded view of the MerCarb 2bbl carburetor—3.0L and 1992-97 V6 and V8 engines

32067g23

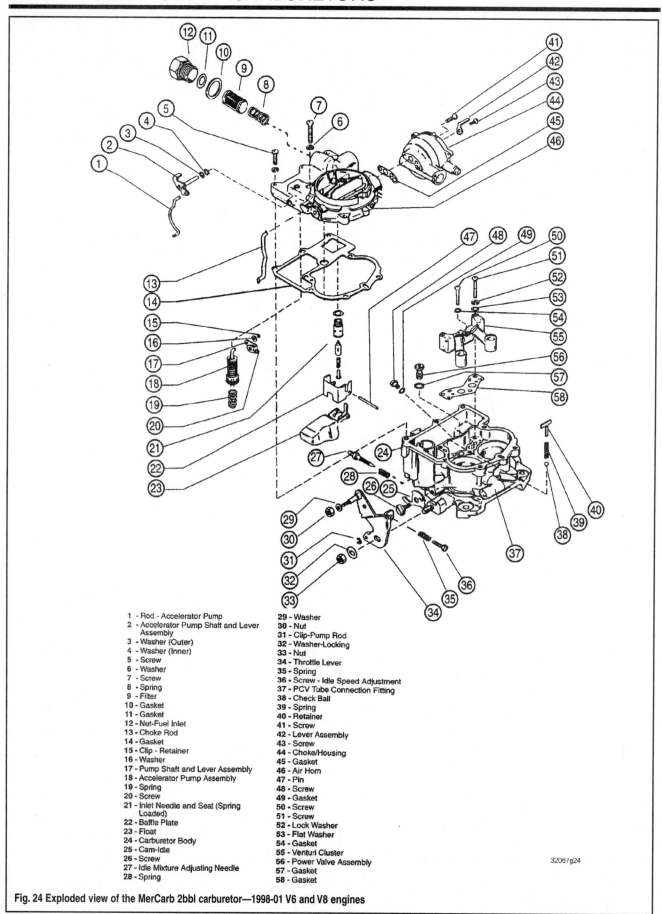

1 - Rod - Accelerator Pump
2 - Accelerator Pump Shaft and Lever Assembly
3 - Washer (Outer)
4 - Washer (Inner)
5 - Screw
6 - Washer
7 - Screw
8 - Spring
9 - Filter
10 - Gasket
11 - Gasket
12 - Nut-Fuel Inlet
13 - Choke Rod
14 - Gasket
15 - Clip - Retainer
16 - Washer
17 - Pump Shaft and Lever Assembly
18 - Accelerator Pump Assembly
19 - Spring
20 - Screw
21 - Inlet Needle and Seat (Spring Loaded)
22 - Baffle Plate
23 - Float
24 - Carburetor Body
25 - Cam-Idle
26 - Screw
27 - Idle Mixture Adjusting Needle
28 - Spring

29 - Washer
30 - Nut
31 - Clip-Pump Rod
32 - Washer-Locking
33 - Nut
34 - Throttle Lever
35 - Spring
36 - Screw - Idle Speed Adjustment
37 - PCV Tube Connection Fitting
38 - Check Ball
39 - Spring
40 - Retainer
41 - Screw
42 - Lever Assembly
43 - Screw
44 - Choke/Housing
45 - Gasket
46 - Air Horn
47 - Pin
48 - Screw
49 - Gasket
50 - Screw
51 - Screw
52 - Lock Washer
53 - Flat Washer
54 - Gasket
55 - Venturi Cluster
56 - Power Valve Assembly
57 - Gasket
58 - Gasket

32067g24

Fig. 24 Exploded view of the MerCarb 2bbl carburetor—1998-01 V6 and V8 engines

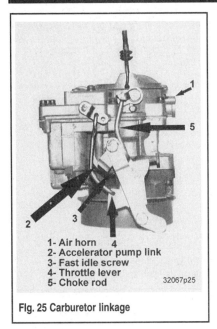

1- Air horn
2- Accelerator pump link
3- Fast idle screw
4- Throttle lever
5- Choke rod

32067p25

Fig. 25 Carburetor linkage

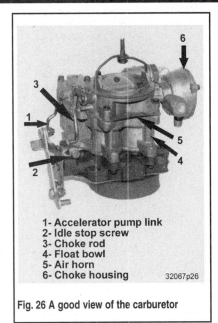

1- Accelerator pump link
2- Idle stop screw
3- Choke rod
4- Float bowl
5- Air horn
6- Choke housing

32067p26

Fig. 26 A good view of the carburetor

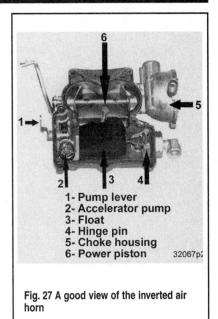

1- Pump lever
2- Accelerator pump
3- Float
4- Hinge pin
5- Choke housing
6- Power piston

32067p2

Fig. 27 A good view of the inverted air horn

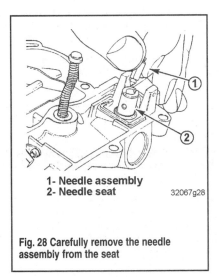

1- Needle assembly
2- Needle seat

32067g28

Fig. 28 Carefully remove the needle assembly from the seat

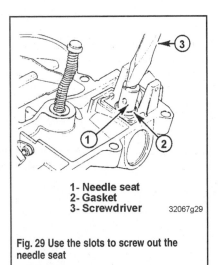

1- Needle seat
2- Gasket
3- Screwdriver

32067g29

Fig. 29 Use the slots to screw out the needle seat

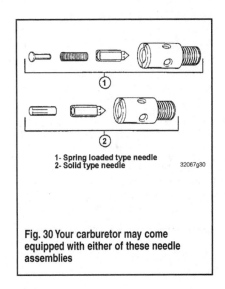

1- Spring loaded type needle
2- Solid type needle

32067g30

Fig. 30 Your carburetor may come equipped with either of these needle assemblies

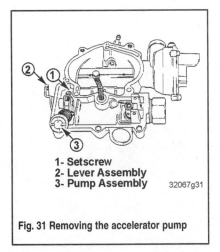

1- Setscrew
2- Lever Assembly
3- Pump Assembly

32067g31

Fig. 31 Removing the accelerator pump

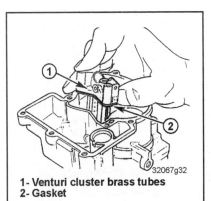
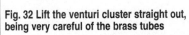

32067g32

1- Venturi cluster brass tubes
2- Gasket

Fig. 32 Lift the venturi cluster straight out, being very careful of the brass tubes

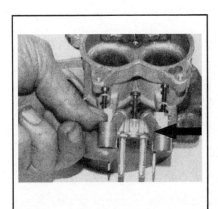

Fig. 33 A close up view of the venturi cluster

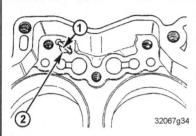

1- Spring retainer
2- Spring and check ball (not shown)

Fig. 34 Remove the retainer before removing the accelerator pump check ball and spring...

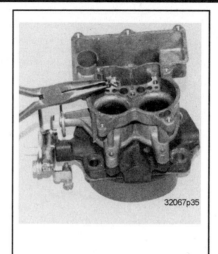

Fig. 35 ...with needle nose pliers

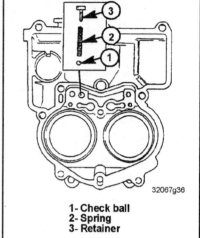

1- Check ball
2- Spring
3- Retainer

Fig. 36 Install the accelerator pump check ball and spring as shown

ASSEMBLY

◆ See Figures 36, 37, 38 and 39

1. Install the check ball and spring into the passage. Clip the retainer into its slots.

2. Slide a new gasket over the venturi tubes and carefully slide the venturi into the carburetor. Using a flat washer and new fiber washer on the center screw and a flat and lock washer on the two outer screws, install the mounting screws and tighten them securely.

3. Install the main metering jets and gaskets.

4. Install the power valve with a new gasket and tighten it securely.

5. Position the accelerator pump spring into the pump well. If the pump was disconnected from the lever, re-attach it with the retaining clip and then insert the assembly into the air horn (don't forget the washer). Align the pump lever hole with the shaft assembly and then slide it into the lever until its shoulder is resting on the lever. Tighten the set-screw.

6. Drop the needle seat and gasket into position and tighten it securely. Position the needle into the seat and then install the baffle and gasket.

7. Install the float assembly and slide in the hinge pin, making sure the float pivots smoothly through its range. Check the float level and drop.

8. Carefully drop the air horn assembly into the float bowl, making sure that accelerator pump slides correctly into the fuel well.

9. Insert the seven mounting screws and tighten them securely, a little at a time.

10. Slide the end of the choke rod into the choke lever assembly.

11. Pop the idle cam onto the choke rod and attach it to the float bowl assembly using the screw so that the cam moves freely without binding.

12. Pop the ear end of the accelerator pump rod into the pump shaft and then insert the other end of the rod into the hole in the throttle lever. Slide the retaining clip over the end.

13. Position the choke housing onto the air horn and tighten the two screws securely. Install the choke lever and tighten the screw.

14. Install the choke cover so that the hook on the choke coil is engaged with the choke lever. Rotate the cover until the index marks align and then rotate the cover clockwise two marks. Tighten the three screws securely.

15. Install the carburetor to the manifold and tighten the nuts to 20 ft. lbs. (27 Nm).

Weber WFB 4bbl Carburetor

DESCRIPTION

◆ See Figure 40

The Weber 4bbl carburetor used on some 4.3L engines and some 1992-97 V8 engines, has its serial number embossed on the lower mounting flange of the carburetor.

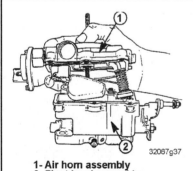

1- Air horn assembly
2- Float bowl assembly

Fig. 37 When installing the air horn assembly always make sure the accelerator pump is positioned in the fuel well correctly

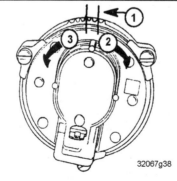

1- Accelerator pump rod
2- Throttle lever
3- Retainer Clip

Fig. 38 Make sure the index marks line up, as mentioned, before tightening the choke housing cover

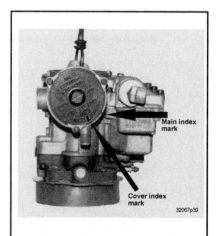

Main index mark

Cover index mark

Fig. 39 A good view of the choke housing

The Weber WFB is unique in that the main body and flange are cast as one piece. There are two separate float circuits.

Certain 4.3L models may come equipped with a special emissions carburetor, designated the Bodensee Emissions version through 1997, and then the SAV1 Emissions version from 1998 on. Additionally, early V8 models used something similar called DESA. In a nutshell the differences are a Positive Crankcase Ventilation (PCV) system, a Ported Vacuum Switch (PVS) circuit and a higher temperature thermostat. Also, the idle mixture screws are sealed. These units are readily identified by the sealed mixture screws and the two vacuum fittings just above them on the carburetor body.

Mechanical procedures following are for both versions of the Weber carburetor.

✳✳ WARNING

Always disconnect the battery cables before attempting to work on the fuel system. Never smoke or allow open flame near the engine—this sounds like an obvious precaution, but you'd be surprised at how many people forget!

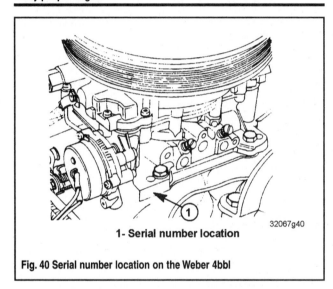

1- Serial number location

32067g40

Fig. 40 Serial number location on the Weber 4bbl

REMOVAL & INSTALLATION

■ No matter how much they look alike, marine carburetors are completely different from automotive carburetors. Never substitute an automotive carburetor for the one on your engine! Venting procedures are not the same and an automotive carburetor could allow dangerous vapors to escape into the bilge. Don't even think about it.

4.3L Engines

1. Open the engine hatch or remove the covers and then disconnect the battery cables. Turn the fuel petcock OFF and/or shut down the fuel supply at the tank.
2. Remove the flame arrestor after disconnecting the vent hose. It's always a good idea to plug the throttle bores with a clean, lint-free cloth.
3. Disconnect the throttle cable hardware from the throttle bracket and lever anchor studs. Remove the cable and carefully move it out of the way.
4. Using two open-end wrenches, hold the fuel inlet nut at the carburetor securely and loosen the fuel line nut. Disconnect the two and carefully move the line out of the way. Plug both the inlet and line open ends to prevent fuel seepage.
5. Tag and disconnect the electric choke lead.
6. Tag and disconnect the two vacuum lines at the carburetor body on Emissions versions.
7. Loosen and remove the carburetor mounting nuts/washers and lift the unit off the manifold. Plug the opening with a clean lint-free cloth.

To Install:

8. Clean the mating surfaces thoroughly of all residual gasket material, position a new gasket and then install the carburetor. Certain models may use an adaptor or wedge plate; install this first. Tighten the nuts to 132 inch lbs. (15 Nm). Hopefully you remembered to remove the rag!
9. Reconnect the fuel line to the inlet line after removing the plugs and tighten it to 18 ft. lbs. (24 Nm). Don't forget to use two wrenches.
10. Connect the two vacuum leads.
11. Connect the choke lead.
12. Install and adjust the throttle cable as detailed later in this section.
13. Install the flame arrestor and reconnect the vent line.
14. Connect the battery cables. Start the engine and ensure there are no fuel leaks; shut down the engine immediately if there are. Check and adjust the idle speed and mixture.

5.0L and 5.7L Engines

1. Open the engine hatch or remove the covers and then disconnect the battery cables. Turn the fuel petcock OFF and/or shut down the fuel supply at the tank.
2. Remove the flame arrestor after disconnecting the vent hose. It's always a good idea to plug the throttle bores with a clean, lint-free cloth.
3. Disconnect the throttle cable hardware from the throttle bracket and lever anchor studs. Remove the cable and carefully move it out of the way.
4. Remove the retaining clip and disconnect the choke linkage rod.
5. Disconnect and plug the fuel pump sight tube.
6. Using two open-end wrenches, hold the fuel inlet nut at the carburetor securely and loosen the fuel line nut. Disconnect the two and carefully move the line out of the way. Plug both the inlet and line open ends to prevent fuel seepage.
7. Loosen and remove the carburetor mounting nuts/washers and lift the unit off the manifold. Plug the opening with a clean lint-free cloth. Remove the anchor/wedge plate from the manifold if so equipped.

To Install:

8. Clean the mating surfaces thoroughly of all residual gasket material, position a new gasket and then install the carburetor. Certain models may use an adaptor or wedge plate; install this first. Tighten the nuts to 132 inch lbs. (15 Nm). Hopefully you remembered to remove the rag!
9. Reconnect the fuel line to the inlet line after removing the plugs and tighten it to 18 ft. lbs. (24 Nm). Don't forget to use two wrenches.
10. Connect the sight tube.
11. Connect the choke linkage rod and slide in the retaining clip.
12. Install and adjust the throttle cable as detailed later in this section.
13. Install the flame arrestor and reconnect the vent line.
14. Connect the battery cables. Start the engine and ensure there are no fuel leaks; shut down the engine immediately if there are. Check and adjust the idle speed and mixture.

ADJUSTMENT

Accelerator Pump Lever

◆ See Figure 41

The accelerator lever is equipped with 3 separate holes, allowing you to change the amount of fuel being delivered to the engine via the accelerator pump. The hole closest to the pump lever's pivot point will deliver the full amount of fuel. The center hole will provide approximately 0.5cc less fuel with each stroke, while the hole furthest from the shaft will provide a flow of approximately 1.0cc less per stroke.

Accelerator Pump

◆ See Figures 42 and 43

1. Back out the idle speed screw until it is no longer in contact with the throttle lever.
2. Ensure that the throttle valves are closed fully and then measure the distance between the top of the carburetor and the bottom of the S-link on the plunger. Grasp the pump linkage with a pair of needle nose pliers and bend it (carefully) until the distance is equal to 7/16 (11mm).

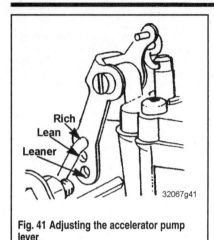

Fig. 41 Adjusting the accelerator pump lever

Fig. 42 Measure the distance between the top of the carburetor and the bottom of the S-link

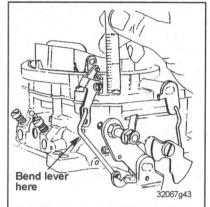

Fig. 43 Bend the pump linkage until the measurement is within specifications

Choke Pull-Off

◆ See Figures 44 and 45

1. Remove the flame arrestor.
2. Press in on the vacuum diaphragm and then, still holding it, press on the choke plate lightly to see how far it will close. Slide a 1/8 in. (3.5mm) drill bit between the upper edge of the choke plate and the air horn assembly—the bit should just fit through. If not, bend the pull-off linkage carefully until it does. 1992-97 5.0L and 5.7L V8 engines will require a 3/16 in. (4.5mm) drill bit; while 7.4L and 8.2L V8 engines will require a 15/64 in. (6mm) drill bit.

Electric Choke

◆ See Figure 46

Loosen the three choke housing screws and rotate the housing until the center mark on the housing and the mark on the carburetor body are in alignment.

Float Level And Drop

◆ See Figures 18, 47, 48, 49 and 50

1. Following the disassembly procedure detailed later, remove the bowl cover from the carburetor.
2. Turn the cover assembly upside down and check that the float pivots freely on the pin. Raise the float and let it fall—do not force it!
3. Using a standard carburetor gauge, measure the distance between the bottom of the bowl cover gasket (remember its upside down, so this will be the top) and the toe of the float it should be within the specification given

at the end of this section—1-9/32 in. (33mm). Carefully bend the float arm with needle nose pliers to achieve the correct measurement. Make sure the float stays in alignment.

4. Its usually a good idea to check the float drop now. Turn the bowl cover assembly right side up and allow the float to hang down freely.
5. Using the same gauge, measure the distance between the bottom of the bowl cover and toe once again. If not 2 in. (51mm), bend the float tab with needle nose pliers to achieve the correct measurement.
6. Recheck both measurements one more time.
7. Reinstall the bowl cover and arrestor.

Idle Speed & Mixture

Please refer to Section 3 for adjustment procedures.

Ported Vacuum Switch
EMISSIONS VERSIONS ONLY

◆ See Figures 53 and 54

1. With the engine cold, start it and disconnect the vacuum line at the carburetor as shown. Cover the hose end with your finger and confirm that you feel no vacuum.
2. Connect the hose and allow the engine to run until it reaches normal operating temperature. Disconnect the hose again, cover the end with your finger and check that there is now vacuum. If so, the PVS is operating properly; if not, check that all hoses are properly connected and not plugged or cracked.
3. Reconnect the hose and make sure the engine is still running and at normal operating temperature. Disconnect the two PVS lines at the carburetor.

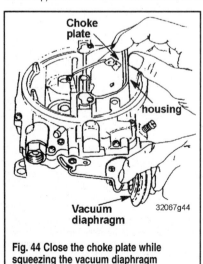

Fig. 44 Close the choke plate while squeezing the vacuum diaphragm

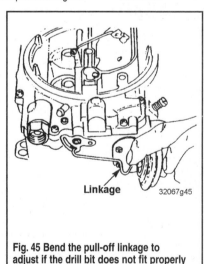

Fig. 45 Bend the pull-off linkage to adjust if the drill bit does not fit properly

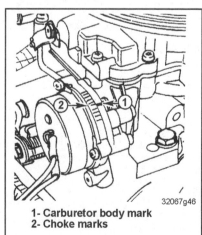

1- Carburetor body mark
2- Choke marks

Fig. 46 The marks on the choke housing and carburetor body should be aligned

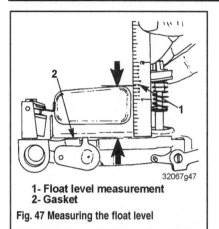

1- Float level measurement
2- Gasket

Fig. 47 Measuring the float level

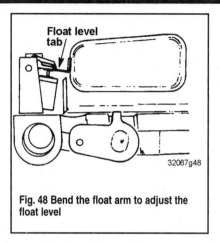

Fig. 48 Bend the float arm to adjust the float level

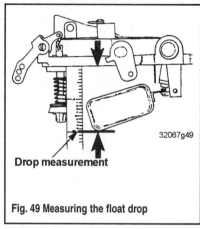

Fig. 49 Measuring the float drop

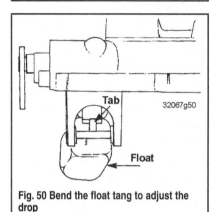

Fig. 50 Bend the float tang to adjust the drop

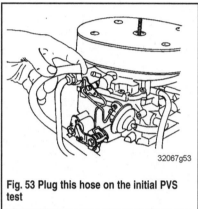

Fig. 53 Plug this hose on the initial PVS test

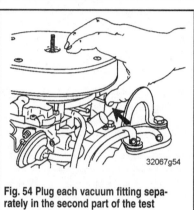

Fig. 54 Plug each vacuum fitting separately in the second part of the test

4. Connect a tachometer as per the manufacturer's instructions. The engine should be running at no more than 775 rpm. Plug one of the two vacuum line fittings at the carburetor and check that the engine speed increases. Now block the other fitting and look for the same results.

5. If vacuum is not found on either fitting, you may have a plugged port. Clean all ports and repeat the test.

Throttle Cable

◆ **See Figures 55, 56, 57 and 58**

1. With the remote control lever in Neutral (idle), install the throttle cable end guide onto the lever and then push the barrel slightly toward the throttle lever end so as to preload the cable slightly and remove any slack. Adjust the barrel so that it aligns with the anchor stud.

2. Secure the throttle cable with its hardware and then tighten the locknut on the cable end guide until it just touches the cable, and then back it out 1 complete turn on 1992-97 models, or 1/2 turn on 1998-01 models.

3. Tighten the throttle cable anchor screw securely—do not overtighten.

4. Move the remote control lever to the full throttle position. The throttle valves should be open all the way and the throttle shaft lever should be in contact with the body of the carburetor.

5. Move the lever back to the Neutral position; the throttle lever should touch the idle speed adjustment screw.

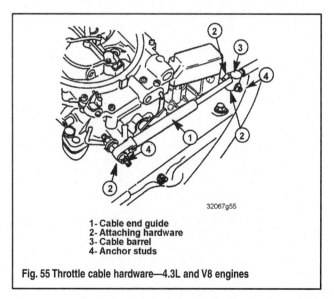

1- Cable end guide
2- Attaching hardware
3- Cable barrel
4- Anchor studs

Fig. 55 Throttle cable hardware—4.3L and V8 engines

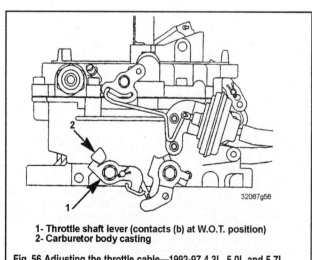

1- Throttle shaft lever (contacts (b) at W.O.T. position)
2- Carburetor body casting

Fig. 56 Adjusting the throttle cable—1992-97 4.3L, 5.0L and 5.7L engines

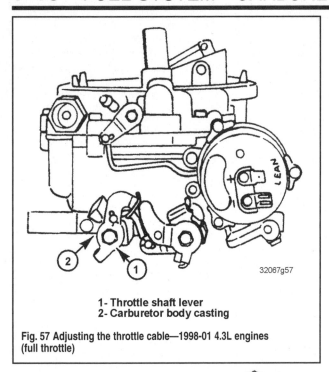

1- Throttle shaft lever
2- Carburetor body casting

Fig. 57 Adjusting the throttle cable—1998-01 4.3L engines (full throttle)

1- Throttle lever
2- Idle speed adjustment screw

Fig. 58 Adjusting the throttle cable—1998-01 4.3L engines (idle)

DISASSEMBLY

② ◁ MODERATE

◆ **See Figures 59, 60, 61, 62, 63, 64, 65, 66, 67, 68, 69, 70, 71, 72, 73,74, 75, 76, 77 and 78**

■ **Always be certain that your carburetor rebuild kit is for marine applications.**

1. Remove the carburetor from the manifold and mount it securely on a workbench.

2. Pull out the retaining clip at the end of the accelerator pump linkage and then disconnect the linkage.

3. Pull out the retaining clip at the end of the choke plate linkage and then disconnect the linkage.

4. If you intend to service the choke diaphragm, remove the vacuum line from the diaphragm, otherwise leave it alone. Likewise, remove the two diaphragm bracket mounting screws (#25 Torx) and remove it and the linkage.

5. Loosen the metering rod cover screws (#15 Torx) and lift off the covers. Certain models may be equipped with air deflectors, which will come off with the covers. Carefully lift out the metering rods and their springs. Rods are not interchangeable so make sure you identify and store the rods correctly for reinstallation. Also, the springs are color-coded and not interchangeable.

6. Remove the nine bowl cover mounting screws (#25 Torx) and lift off the assembly, disconnecting the choke linkage at the same time. Certain models may only have eight mounting screws. Always make sure to pull the top half straight up so as not to damage the accelerator pump or floats.

7. Raise the float slightly, grasp the float pin with needle nose pliers and pull it out. Lift out the float. Do not mix up the two floats—use a marker to identify right and left.

8. Lift out the two inlet needles from under the float attachments. Loosen the inlet seat and then lift the seat, gasket and filter from the housing as a unit. Make sure to keep the needles and seats together and marked as to which recess they came out of.

9. Pull up the gasket that should still be attached to the bottom of the bowl cover. Unscrew the accelerator pump lever and then lift out the pump assembly itself. Reach into the pump housing in the lower half of the carburetor and remove the pump spring.

10. Look into the carburetor and take note of which venturi clusters have 'distribution tags' attached and also of the ID number stamped on each one.

Loosen the mounting screws (#25 Torx) and lift out each set of clusters. Make sure you lift them straight up and remove the gasket.

11. Repeat the above procedure for the secondary venturi clusters and then lift out the secondary air valve and weight assembly from underneath the clusters.

12. Loosen the two pump jet housing screws and pull out the housing and its gasket. At the bottom of the housing recess you should see a check ball and weight, or a check ball and spring—whichever it is, remove it(them).

13. Grasp the float bowl baffle plates with needle nose pliers and slide them up and out.

14. Remove the primary and secondary jets (#25 Torx). Make sure you note their location in the carb body and their sizes PRIOR to removal!

15. Turn each mixture screw in until it seats itself and not the number of turns (for installation); now you can remove the screws but be careful to note which one came from which hole.

CLEANING & INSPECTION

② ◁ MODERATE

Never use a wire brush or drill to clean jet passages or tubes in the carburetor.

Never allow the carburetor to soak in a cleaner bath for more than two hours. In fact, we recommend using spray cleaner.

Never immerse the float assembly, needle, accelerator pump plunger or fuel filter in cleaner. Wipe them carefully with a clean cloth.

Otherwise clean all allowable parts with carb cleaner and then dry with compressed air if at all possible.

Blow out and through all passages to ensure there is no foreign material clogging them.

Check the float needle and seat, if either is worn or damaged, replace them as a matched set.

Check the float assembly and hinge pin for wear or damage, replace as necessary.

Check the pump plunger, return spring, piston spring, idle mixture needle and all levers and linkages for wear or damage. Replace as necessary.

Check the throttle valve shaft for excessive looseness in the throttle body. Check that the valve and shaft do not bind through their range of operation and that the valve opens and closes fully. Replace the assembly if it fails any of these tests.

Check the choke valve lever and shaft for excessive looseness in the air horn. Check that the valve, lever and shaft do not bind through their range of operation and that the valve opens and closes fully. Replace the air horn assembly if it fails any of these tests.

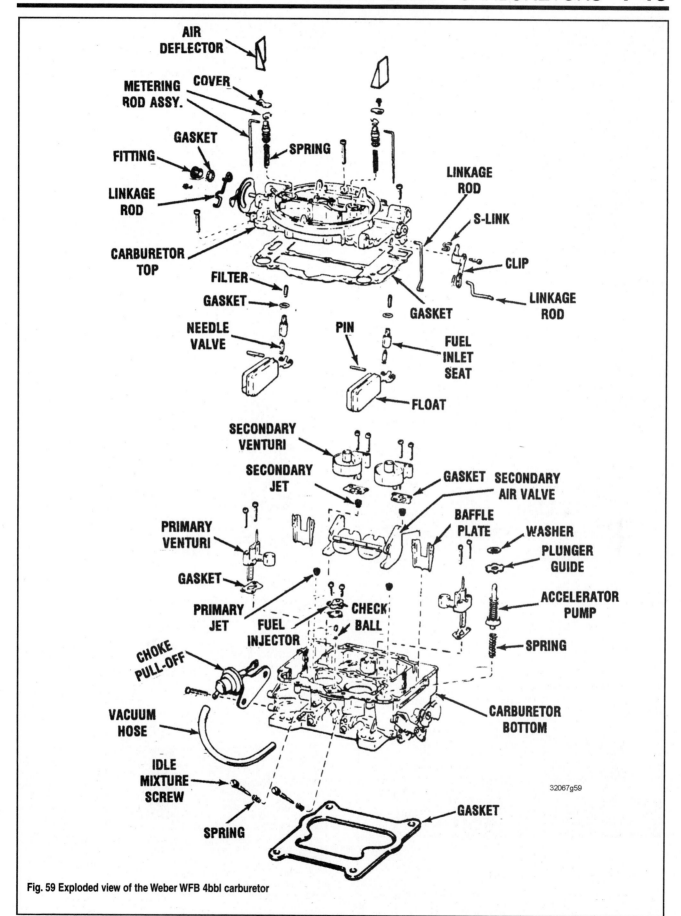

Fig. 59 Exploded view of the Weber WFB 4bbl carburetor

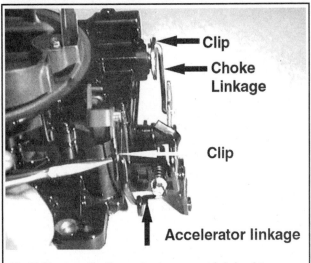

Fig. 60 Disconnecting the accelerator pump and choke plate linkages

Fig. 61 Unscrew the fuel metering rod cover...

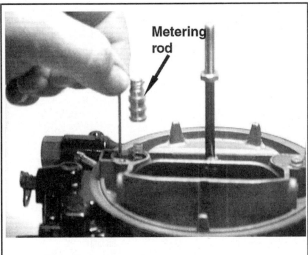

Fig. 62 Lift out the metering rod...

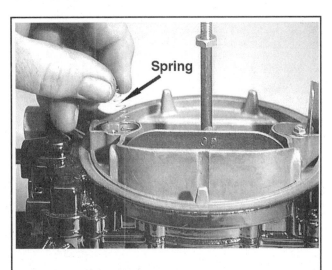

Fig. 63...and then the spring

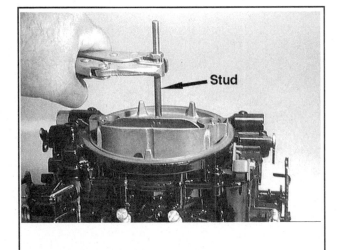

Fig. 64 Remove the flame arrestor stud

Fig. 65 Accelerator pump linkage

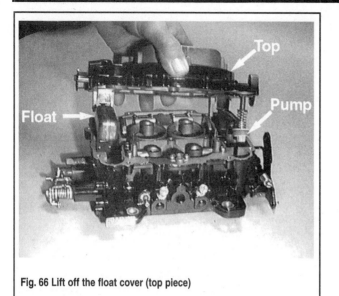

Fig. 66 Lift off the float cover (top piece)

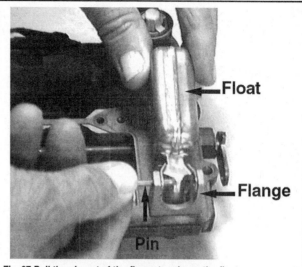

Fig. 67 Pull the pin out of the flange to release the float

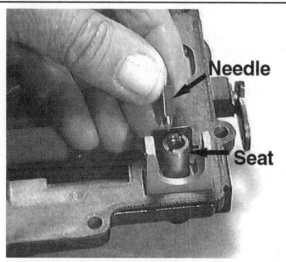

Fig. 68 Lift the needle valve out of the seat...

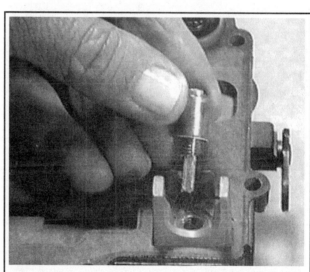

Fig. 69 ...and then unscrew and remove the seat, gasket and filter as an assembly

Fig. 70 Remove the accelerator pump from the top half and the spring from the bottom

Fig. 71 Always check the 'distribution tabs' before removing the venturi clusters

Fig. 72 Removing the secondary venturi cluster

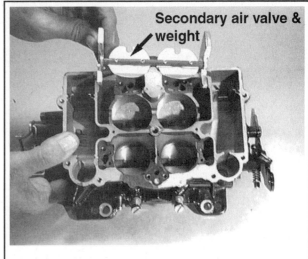

Fig. 73 Lift out the secondary air valve and weight

Fig. 74 Remove the pump jet housing...

Fig. 75 ...and then lift out the check ball

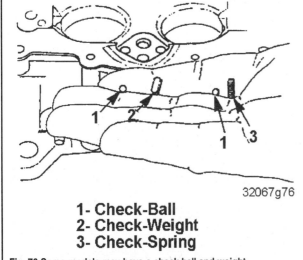

32067g76

1- Check-Ball
2- Check-Weight
3- Check-Spring

Fig. 76 Some models may have a check ball and weight, while others may have a ball and spring

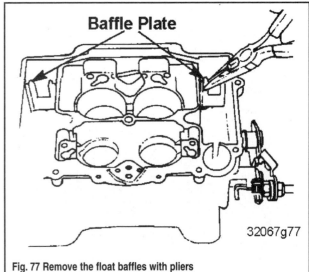

32067g77

Fig. 77 Remove the float baffles with pliers

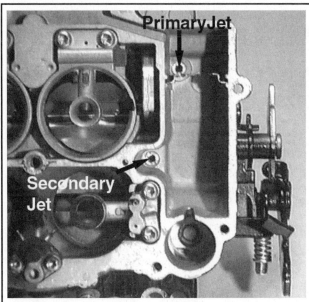

Fig. 78 Remove the primary and secondary jets

Fig. 79 Install the idle mixture screw and spring

ASSEMBLY   *MODERATE*

◆ See Figures 79, 80 and 81

1. Install the idle mixture screws and tighten them until they just seat themselves. Now back them out to the number of turns you noted in the disassembly section.

2. Install the primary and secondary jets. Make sure you put them back into the same positions that you noted during removal.

3. Slide the float bowl baffle plates into position.

4. Install the check ball into the recess at the bottom of the pump jet housing. Slide the check weight of spring in on top of the ball. Never mix and match—if a spring came out, don't pop in a weight, or vice versa. Position a pump housing gasket and drop in the housing. Tighten the two screws securely.

5. Position the secondary air valve and weight assembly into the recess and then install the secondary venturi clusters with a new gasket. Tighten the two mounting screws on each cluster securely.

6. Repeat the procedure above and install the primaries.

7. Slip the accelerator pump spring into the housing on the lower carburetor body and then install the pump itself into the bowl cover. Remember to get the washer and guide into place first. Slip the end of the S-link on the accelerator pump lever into the hole on the end of the pump plunger and then install and tighten the lever pivot screw. Before you move on, check that the lever moves the pump plunger freely with no binding..

8. Position a new gasket onto the mating surface of the bowl cover. Press a new inlet filter into the bottom of the seat, install the gasket and press the assembly into the recess. Install the correct needles into their seats.

9. Position the floats (on the sides if reusing the old ones) and slide in the hinge pins. Carefully turn over the bowl cover and install it onto the lower half of the carburetor. Check that the gasket is still in the correct position and then tighten the nine mounting screws securely.

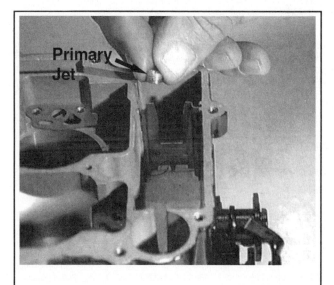

Fig. 80 Installing the primary jet

10. Install the metering rod springs into their holes on top of the carburetor—remember to check the specifications for the correct color. Drop the rods themselves into their original holes on top of the springs. Push down on them lightly to make sure the plunger is working properly and then install the covers and air deflectors (if equipped).

11. Reconnect the free end of the choke pull-off linkage and then install the diaphragm assembly. Reconnect the vacuum line to the rear of the diaphragm. Connect the choke plate linkage and press in the retaining clip.

12. Install the accelerator pump linkage end into the hole it came out of. If you've forgotten, refer to the correct procedure in Adjustments.

13. Install the carburetor.

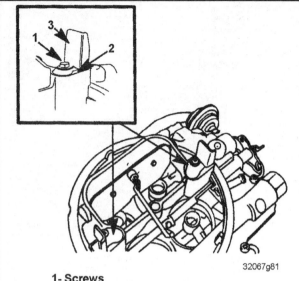

1- Screws
2- Metering rod cover (s)
3- Air deflectors (if so equippped)

32067g81

Fig. 81 When installing the metering rod covers, make sure not to forget the air deflectors if equipped.

Rochester 4MV 4bbl Carburetor

DESCRIPTION

The Rochester 4MV Quadrajet carburetor has two stages. The primary (fuel inlet) side has small 1-3/8 in. bores with a triple venturi setup equipped with plain-tube nozzles. The carburetor operates much the same as other carburetors using the venturi principle. The triple venturi, plus the small primary bores, makes for a more stable and finer fuel control during idle and partial throttle operation. When the throttle is partially open, the fuel metering is accomplished with tapered metering rods, positioned by a vacuum-responsive piston and operating in specially designed jets.

The secondary side has two large, 2-1/4 in., bores. These large bores, when added to the primary side bores, provide enough air capacity to meet most engine requirements. The air valve is used in the secondary side for metering control and backs-up the primary bores to meet air and fuel demands of the engine.

The secondary air valve operates the tapered metering rods. These rods move in orifice plates and thus control fuel flow from the secondary nozzles in direct relation to the air flowing through the secondary bores.

The float bowl is designed to avoid problems of fuel spillage during sharp turns of the boat which could result in engine cutout and delayed fuel flow. The bowl reservoir is smaller than most four-barrel carburetors to reduce fuel evaporation during hot engine shut down.

The float system has one pontoon float and fuel valve which makes servicing much easier than on some other model carburetors. A fuel filter is located in the float bowl ahead of the float needle valve. This filter is easily removed for cleaning or replacement.

The throttle body is made of aluminum as part of a weight-reduction program and also to improve heat transfer away from the fuel bowl and prevent the fuel from "percolating" during hot engine shut down. A heat insulator gasket is used between the throttle body and bowl to help prevent "percolating".

✳✳ WARNING

Always disconnect the battery cables before attempting to work on the fuel system. Never smoke or allow open flame near the engine—this sounds like an obvious precaution, but you'd be surprised at how many people forget!

REMOVAL & INSTALLATION

■ No matter how much they look alike, marine carburetors are completely different from automotive carburetors. Never substitute an automotive carburetor for the one on your engine! Venting procedures are not the same and an automotive carburetor could allow dangerous vapors to escape into the bilge. Don't even think about it.

1. Open the engine hatch or remove the covers and then disconnect the battery cables. Turn the fuel petcock OFF and/or shut down the fuel supply at the tank.
2. Remove the flame arrestor after disconnecting the vent hose. It's always a good idea to plug the throttle bores with a clean, lint-free cloth.
3. Disconnect the throttle cable hardware from the throttle bracket and lever anchor studs. Remove the cable and carefully move it out of the way.
4. Remove the retaining clip and disconnect the choke spring rod.
5. Disconnect and plug the fuel pump sight tube.
6. Using two open end wrenches, hold the fuel inlet nut at the carburetor securely and loosen the fuel line nut. Disconnect the two and carefully move the line out of the way. Plug both the inlet and line open ends to prevent fuel seepage.
7. Loosen and remove the carburetor mounting nuts/washers and lift the unit off the manifold. Plug the opening with a clean lint-free cloth. Remove the cable bracket.

To Install:

8. Clean the mating surfaces thoroughly of all residual gasket material, position a new gasket and then install the carburetor. Certain models may use an adaptor or wedge plate, install this first. Tighten the nuts to 132 inch lbs. (15 Nm). Hopefully you remembered to remove the rag!
9. Reconnect the fuel line to the inlet line after removing the plugs and tighten it to 18 ft. lbs. (24 Nm). Don't forget to use two wrenches.
10. Connect the sight tube.
11. Connect the choke spring rod and slide in the retaining clip.
12. Install and adjust the throttle cable as detailed later in this section.
13. Install the flame arrestor and reconnect the vent line.
14. Connect the battery cables. Start the engine and ensure there are no fuel leaks; shut down the engine immediately if there are. Check and adjust the idle speed and mixture.

ADJUSTMENT

Accelerator Pump

◆ See Figures 82 and 83

1. Back out the idle speed screw until it is no longer in contact with the throttle lever.
2. Move the pump rod to the inner hole on the lever if its not already there.
3. Ensure that the throttle valves are closed fully and then measure the distance between the top of the carburetor (flame arrestor mounting surface) and the top of the pump plunger stem. Grasp the tip of the pump lever with a pair of needle nose pliers and bend it (carefully) until the distance is equal to 23/64 in. (9.1mm) while supporting the lever.

Air Valve Dashpot

◆ See Figure 84

Push in on the vacuum break stem until the diaphragm is seated and the air valve is completely closed. Bend the rod at the bend until the gap between the rod end and the lever is 0.025 in. (0.64mm).

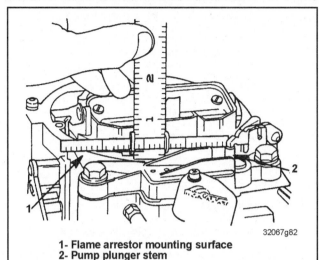

1- Flame arrestor mounting surface
2- Pump plunger stem

Fig. 82 Measure the distance between the top of the carburetor and the top of the plunger

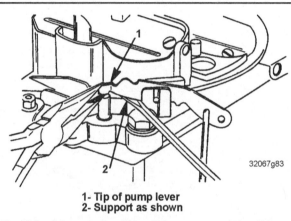

1- Tip of pump lever
2- Support as shown

Fig. 83 Bend the pump lever tip until the measurement is within specifications

Air Valve Spring Windup

◆ See Figures 85 and 86

This procedure will require the use of a gram scale.
1. Rotate the gram scale until the air valve just begins to open. The scale should read 70<en dash>90 grams on the 5.0LX, or 75-85 grams on the 5.7L Comp Ski.
2. If adjustment is required, hold the tension spring screw with a screwdriver and loosen the Allen head screw. Turn the tension screw clockwise to increase the tension and counterclockwise to decrease it.
3. Still holding the tension screw with the screwdriver, tighten the Allen screw securely.

Choke Coil Rod

◆ See Figures 87, 88 and 87

1. Remove the flame arrestor.
2. Gently push the choke coil rod down into the housing.
3. The top of the choke coil rod must be even with the bottom of the hole in the vacuum break choke lever.
4. Bend the rod mid way down at the kink to adjust it.

Float Level

◆ See Figure 89

1. Remove the flame arrestor.
2. Start the engine and let it idle while filling the float bowl. Insert a float gauge into the hole next to the flame arrestor stud.
3. Tap the gauge down lightly and then take a reading at the top of the carburetor casing. If not 15/64 in. (6mm) you will need to adjust the float as detailed in the Assembly procedures following.

Idle Speed & Mixture

Please refer to Section 3 for adjustment procedures.

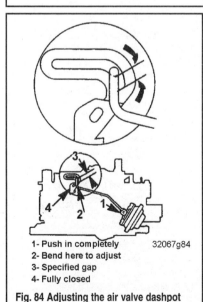

1- Push in completely
2- Bend here to adjust
3- Specified gap
4- Fully closed

Fig. 84 Adjusting the air valve dashpot

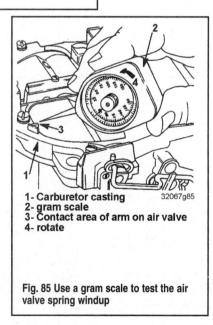

1- Carburetor casting
2- gram scale
3- Contact area of arm on air valve
4- rotate

Fig. 85 Use a gram scale to test the air valve spring windup

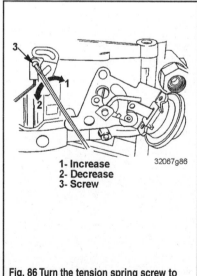

1- Increase
2- Decrease
3- Screw

Fig. 86 Turn the tension spring screw to adjust the spring tension

Ported Vacuum Switch

EMISSIONS VERSIONS ONLY

1. With the engine cold, start it and disconnect the vacuum line at the carburetor as shown. Cover the hose end with your finger and confirm that you feel no vacuum.

2. Connect the hose and allow the engine to run until it reaches normal operating temperature. Disconnect the hose again, cover the end with your finger and check that there is now vacuum. If so, the PVS is operating properly; if not, check that all hoses are properly connected and not plugged or cracked.

3. Reconnect the hose and make sure the engine is still running and at normal operating temperature. Disconnect the two PVS lines at the carburetor.

4. Connect a tachometer as per the manufacturer's instructions. The engine should be running at no more than 775 rpm. Plug one of the two vacuum line fittings at the carburetor and check that the engine speed increases. Now block the other fitting and look for the same results.

5. If vacuum is not found on either fitting, you may have a plugged port. Clean all ports and repeat the test.

Throttle Cable

◆ See Figure 90

1. With the remote control lever in Neutral (idle), install the throttle cable end guide onto the lever and then push the barrel slightly toward the throttle lever end so as to preload the cable slightly and remove any slack. Adjust the barrel so that it aligns with the anchor stud.

2. Secure the throttle cable with its hardware and then tighten the lock-nut on the cable end guide until it just touches the cable, and then back it out 1 complete turn

3. Tighten the throttle cable anchor screw securely—do not over-tighten.

4. Move the remote control lever to the full throttle position. The throttle valves should be open all the way and the throttle shaft lever should be in contact with the body of the carburetor.

5. Move the lever back to the Neutral position; the throttle lever should touch the idle speed adjustment screw.

Vacuum Break

◆ See Figures 91 and 92

1. Press in on the vacuum break control diaphragm until it is fully seated.

2. Rotate the vacuum break choke lever counterclockwise until the left tang comes in contact with the vacuum break rod. The choke rod must be at the bottom of the slot in the choke lever.

3. Bend the tang on the choke lever until you can achieve an 5/64 inch gap in the choke valve when checked with a drill bit.

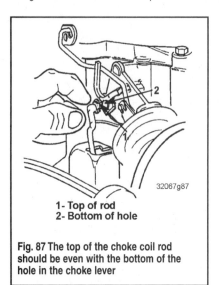

1- Top of rod
2- Bottom of hole

Fig. 87 The top of the choke coil rod should be even with the bottom of the hole in the choke lever

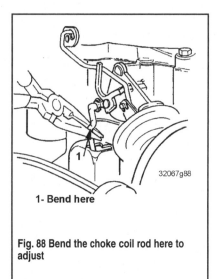

1- Bend here

Fig. 88 Bend the choke coil rod here to adjust

Read here

Fig. 89 Read the float gauge where it intersects the top of the carburetor casing

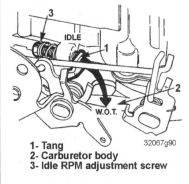

IDLE

W.O.T.

1- Tang
2- Carburetor body
3- Idle RPM adjustment screw

Fig. 90 Adjusting the throttle cable— 1992-97 5.0L and 5.7L engines

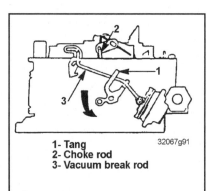

1- Tang
2- Choke rod
3- Vacuum break rod

Fig. 91 Rotate the lever until the tang comes in contact with the vacuum break rod

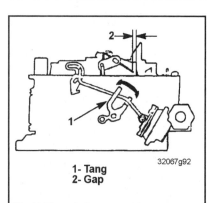

1- Tang
2- Gap

Fig. 92 Measure the gap and then bend the tang to adjust

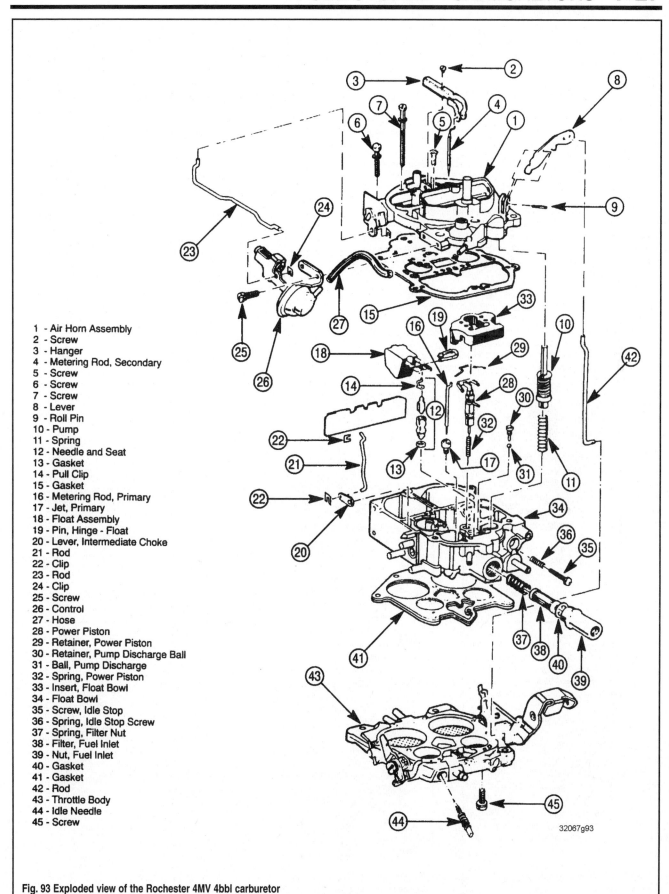

1 - Air Horn Assembly
2 - Screw
3 - Hanger
4 - Metering Rod, Secondary
5 - Screw
6 - Screw
7 - Screw
8 - Lever
9 - Roll Pin
10 - Pump
11 - Spring
12 - Needle and Seat
13 - Gasket
14 - Pull Clip
15 - Gasket
16 - Metering Rod, Primary
17 - Jet, Primary
18 - Float Assembly
19 - Pin, Hinge - Float
20 - Lever, Intermediate Choke
21 - Rod
22 - Clip
23 - Rod
24 - Clip
25 - Screw
26 - Control
27 - Hose
28 - Power Piston
29 - Retainer, Power Piston
30 - Retainer, Pump Discharge Ball
31 - Ball, Pump Discharge
32 - Spring, Power Piston
33 - Insert, Float Bowl
34 - Float Bowl
35 - Screw, Idle Stop
36 - Spring, Idle Stop Screw
37 - Spring, Filter Nut
38 - Filter, Fuel Inlet
39 - Nut, Fuel Inlet
40 - Gasket
41 - Gasket
42 - Rod
43 - Throttle Body
44 - Idle Needle
45 - Screw

32067g93

Fig. 93 Exploded view of the Rochester 4MV 4bbl carburetor

DISASSEMBLY

② ◁MODERATE

◆ **See Figures 93, 94, 95, 96, 97, 98 and 99**

■ **Always be certain that your carburetor rebuild kit is for marine applications.**

1. Remove the carburetor.

2. Remove the retaining clip from the vacuum-break link at the vacuum-break diaphragm. Disconnect the vacuum-break link from the vacuum-break assembly. Gently push apart the retaining ears of the bracket to release the vacuum-break canister.

3. Place the carburetor on the workbench in the upright position. If servicing an older carburetor, remove the idle vent valve attaching screw, and then remove the idle vent valve assembly. Remove the clip from the upper end of the choke shutter rod, disconnect the choke rod from the upper choke shaft lever, and then remove the choke rod from the bowl.

4. Drive back the roll pin at the upper end of the accelerator pump lever, and then disconnect the pump rod and lever.

5. Hold the air valve wide open and then remove the secondary metering rods by tilting and sliding the rods from the holes in the hanger.

6. Remove the nine air horn-to-bowl attaching screws. Lift straight up on the air horn and remove it, taking care not to bend the accelerator pump and air bleed tubes sticking out from the air horn.

7. Remove the dashpot piston from the air-valve link by rotating the bend through the hole, and then remove the dashpot from the air horn by rotating the bend through the air horn. Further disassembly of the air horn is not necessary. Do not remove the air valves, air valve shaft, and secondary metering rod hangers because they are calibrated. Do not attempt to remove the high-speed air bleeds and accelerating well tubes because they are pressed into position.

8. Remove the accelerator pump piston from the pump well. Release the air horn gasket from the dowels on the secondary side of the bowl, and then pry the gasket from around the power piston and primary metering rods. Remove the pump return spring from the pump well.

9. Pull off the insert block, remove the plastic filler from over the float valve. Use a pair of needle-nosed pliers and pull straight up on the metering rod hanger directly over the power piston and remove the power piston and the primary metering rods. Disconnect the tension spring from the top of each rod and then rotate the rod and remove the metering rods from the power piston. Pull up just a bit on the float assembly hinge pin until the pin can be removed by sliding it toward the pump well.

10. Insert a screwdriver into the slots on the fuel inlet seat and remove the seat and gasket.

11. Remove both primary metering rod jets. Remove the pump discharge check ball retainer and the check ball.

12. Remove the baffle plates from the secondary side of the bowl.

13. Disconnect the vacuum hose from the tube connection on the bowl and from the vacuum break assembly. Remove the retaining screw, and then lift the assembly from the float bowl.

14. Remove the fuel inlet filter retaining nut, gasket, filter, and spring.

15. Remove the throttle body by taking out the two throttle body-to-bowl attaching screws, and then lift off the insulator gasket.

16. Remove the idle mixture screws and springs. Take care not to damage the secondary throttle valves.

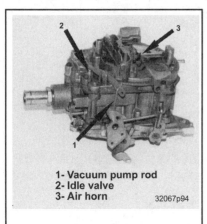

1- Vacuum pump rod
2- Idle valve
3- Air horn
32067p94

Fig. 93 Remove the vacuum pump rod

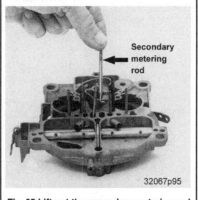

Secondary metering rod
32067p95

Fig. 95 Lift out the secondary metering rod

Accelerator pump piston
32067p96

Fig. 96 Lift the accelerator pump piston and spring out of the recess

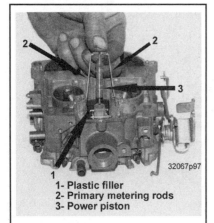

1- Plastic filler
2- Primary metering rods
3- Power piston
32067p97

Fig. 97 Lift out the metering rods and power piston

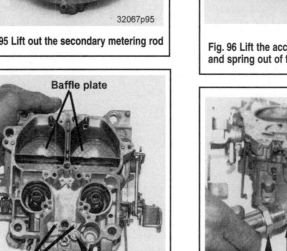

Baffle plate
Primary metering jets
Check ball retainer
32067p98

Fig. 98 Remove the primary jets and then remove the check ball retainer and ball

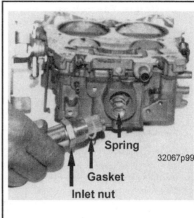

Spring
Gasket
Inlet nut
32067p99

Fig. 99 Remove the fuel inlet nut and associated parts

CLEANING & INSPECTION

Never use a wire brush or drill to clean jet passages or tubes in the carburetor.

Never allow the carburetor to soak in a cleaner bath for more than two hours. In fact, we recommend using spray cleaner.

Never immerse the float assembly, needle, accelerator pump plunger or fuel filter in cleaner. Wipe them carefully with a clean cloth.

Otherwise clean all allowable parts with carb cleaner and then dry with compressed air if at all possible.

Blow out and through all passages to ensure there is no foreign material clogging them.

Check the float needle and seat, if either is worn or damaged, replace them as a matched set.

Check the float assembly and hinge pin for wear or damage, replace as necessary.

Check the pump plunger, return spring, piston spring, idle mixture needle and all levers and linkages for wear or damage. Replace as necessary.

Check the throttle valve shaft for excessive looseness in the throttle body. Check that the valve and shaft do not bind through their range of operation and that the valve opens and closes fully. Replace the assembly if it fails any of these tests.

Check the choke valve lever and shaft for excessive looseness in the air horn. Check that the valve, lever and shaft do not bind through their range of operation and that the valve opens and closes fully. Replace the air horn assembly if it fails any of these tests.

ASSEMBLY

◆ **See Figures 100, 101, 102, 103, 104, 105, 106, 107, 108, 109, 110, 111, 112 and 113**

1. Turn the idle mixture adjusting screws in until they are barely seated, and then back them out two to three turns as a rough adjustment at this time. Never turn the adjusting screws down tight into their seats or they will be damaged.

2. Place a new insulator gasket on the bowl with the holes in the gasket indexed over the two dowels. Install, and then tighten the throttle body-to-bowl screws evenly.

3. Install the fuel inlet filter spring, filter, new gasket, and inlet nut. Tighten the nut.

4. Install the baffle plates in the secondary side of the bowl with the notches facing up.

5. Grasp the choke rod so that the intermediate choke lever is at the bottom. Suspend the assembly into the float bowl so that the hole with the flat side aligns with the hole in the side of the float bowl. Now insert the

choke shaft on the vacuum break through the hole until it engages the flat-sided hole in the choke lever. Install the screw and tighten it securely.

6. Install the primary main metering jets. Install a new float needle seat and gasket. Install the pump discharge check ball and retainer in the passage next to the pump well.

7. Install the pull clip on the needle with the open end toward the front of the bowl. Install the float by sliding the float lever under the pull clip from the front to the back. Hold the float assembly by the toe and with the float lever in the pull clip, install the retaining pin from the pump well side. Take care not to distort the pull clip.

8. Measure the distance from the top of the float bowl gasket surface, with the gasket removed, to the top of the float at the toe end. Check to be sure the retaining pin is held firmly in place and the tang of the float is seated on the float needle when making the measurement. Check your measurement with the Specifications chart at the end of this section (15/64 in.). Carefully bend the float up or down until the correct measurement is reached.

9. Install the power piston spring in the power piston well. Install the primary metering jets, if they were removed during disassembly. Be sure the tension spring is connected to the top of each metering rod. Install the power-piston assembly in the well with the metering jets; the retainer should be flush with the casing. A sleeve around the piston holds the piston in place during assembly.

10. Install the plastic filler over the float needle. Press it down firmly until it is seated.

Place the accelerator pump return spring in the pump well. Install the air horn gasket around the primary metering rods and piston. Install the gasket on the secondary side of the bowl with the holes in the gasket indexed over the two dowels. Install the accelerating pump plunger in the pump well.

11. Install the secondary metering rods. Hold the air valve wide open and check to be sure the rods are positioned with their upper ends through the hanger holes and pointing toward each other. Hanger size may vary from the specifications chart due to inconsistencies in carburetor castings—the hole location may be different so make sure they are the same size as those originally installed.

12. Slowly position the air horn assembly on the bowl and carefully insert the secondary metering rods, the high-speed air bleeds, and the accelerating well tubes through the holes of the air horn gasket. Never force the air horn assembly onto the float bowl. Such action may distort the secondary metering plates. If the air horn assembly moves slightly sideways it will center the metering rods in the metering plates. Install the attaching screw.

13. Connect the pump rod to the inner hole of the pump lever and secure it with a spring clip. Position the lever into the casting mount on the carburetor and press the roll pin through the lever with a screwdriver.

14. Connect the bottom end of the choke shutter rod into the intermediate choke lever and secure it with a clip. Install the upper end of the rod into the choke blade lever and then press on the retaining clip.

15. Install the actuating rod into the air valve lever and then swivel the other end into the vacuum break arm and press on the retaining clip. Reconnect the vacuum line between the break and the carburetor.

16. Install the carburetor.

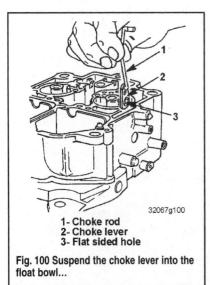

1- Choke rod
2- Choke lever
3- Flat sided hole

Fig. 100 Suspend the choke lever into the float bowl...

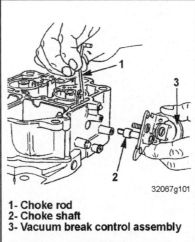

1- Choke rod
2- Choke shaft
3- Vacuum break control assembly

Fig. 101 ...and then slide the vacuum break in until it engages the lever

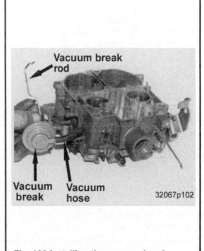

Fig. 102 Installing the vacuum break

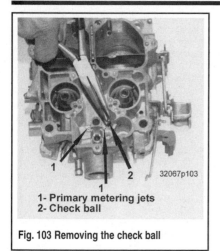

1- Primary metering jets
2- Check ball

Fig. 103 Removing the check ball

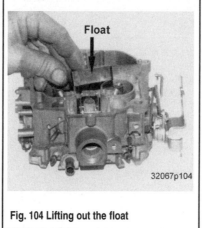

Float

Fig. 104 Lifting out the float

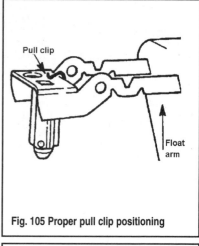

Pull clip

Float arm

Fig. 105 Proper pull clip positioning

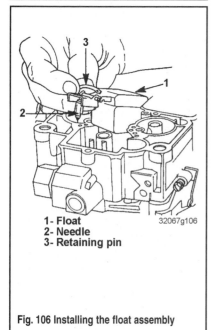

1- Float
2- Needle
3- Retaining pin

Fig. 106 Installing the float assembly

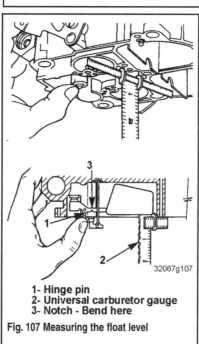

1- Hinge pin
2- Universal carburetor gauge
3- Notch - Bend here

Fig. 107 Measuring the float level

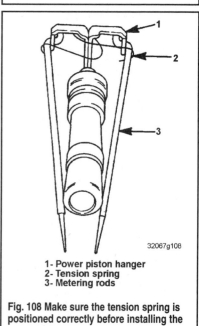

1- Power piston hanger
2- Tension spring
3- Metering rods

Fig. 108 Make sure the tension spring is positioned correctly before installing the power piston

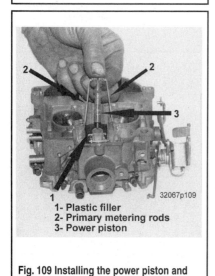

1- Plastic filler
2- Primary metering rods
3- Power piston

Fig. 109 Installing the power piston and metering rods

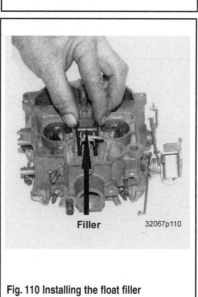

Filler

Fig. 110 Installing the float filler

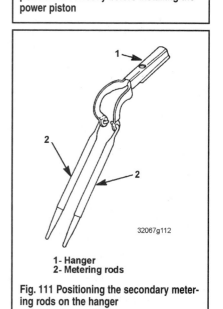

1- Hanger
2- Metering rods

Fig. 111 Positioning the secondary metering rods on the hanger

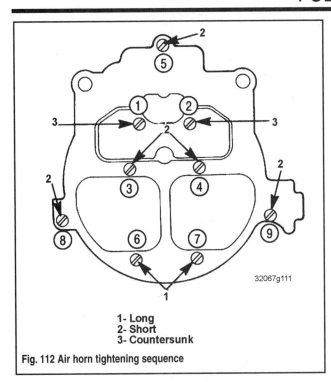

1- Long
2- Short
3- Countersunk

Fig. 112 Air horn tightening sequence

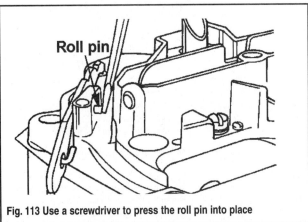

Fig. 113 Use a screwdriver to press the roll pin into place

Mechanical Fuel Pump

DESCRIPTION

◆ See Figures 114 and 115

The engines covered in this manual utilize a mechanical-type fuel pump driven off the camshaft except for the big blocks (7.4L/8.2L) which have the fuel pump driven by an eccentric cam on the seawater pump shaft.

The fuel filter on the 3.0L pump is an integral part of the pump as indicated in the accompanying exploded diagram. All other engines utilize a small filter at the fuel tank, and, in almost all instances, a separate water separating fuel filter in-line.

This fuel pump is equipped with a yellow sight tube. Any fuel in the sight tube indicates the diaphragm has been ruptured. The fuel pump cannot be repaired with any type of "kit". Due to the hazardous condition indicated by fuel in the sight tube, the pump must be replaced with a complete pump assembly, at once.

The fuel pump sucks gasoline from the fuel tank and delivers it to the carburetor in sufficient quantities, under pressure, to satisfy engine demands under all operating conditions.

The pump is operated by a two-part rocker arm. The outer part rides on an eccentric on the camshaft and is held in constant contact with the camshaft by a strong return spring. The inner part is connected to the fuel pump diaphragm by a short connecting rod. As the camshaft rotates, the rocker arm moves up and down. As the outer part of the rocker arm moves downward, the inner part moves upward, pulling the fuel diaphragm upward. This upward movement compresses the diaphragm spring and creates a vacuum in the fuel chamber below the diaphragm. The vacuum causes the outlet valve to close and permits fuel from the gas tank to enter the chamber by way of the fuel filter and the inlet valve.

Now, as the eccentric on the camshaft allows the outer part of the rocker arm to move upward, the inner part moves downward, releasing its hold on the connecting rod. The compressed diaphragm spring then exerts pressure on the diaphragm, which closes the inlet valve and forces fuel out through the outlet valve to the carburetor.

Because the fuel pump diaphragm(s) is(are) moved downward only by the diaphragm spring, the pump delivers fuel to the carburetor only when the pressure in the outlet line is less than the pressure exerted by the diaphragm spring. This lower pressure condition exists when the carburetor float needle valve is unseated and the fuel passages from the pump into the carburetor float chamber are open.

When the needle valve is closed and held in place by the pressure of the fuel on the float, the pump builds up pressure in the fuel chamber until it overcomes the pressure of the diaphragm spring. This pressure almost stops movement of the diaphragm until more fuel is needed in the carburetor float bowl.

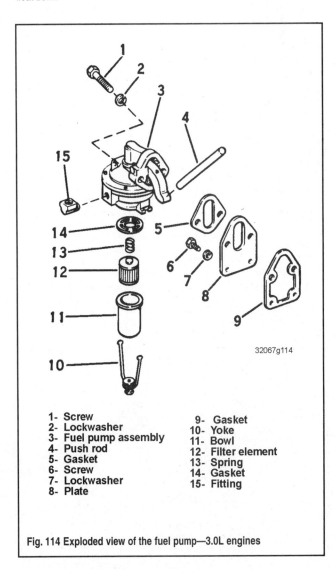

1- Screw
2- Lockwasher
3- Fuel pump assembly
4- Push rod
5- Gasket
6- Screw
7- Lockwasher
8- Plate
9- Gasket
10- Yoke
11- Bowl
12- Filter element
13- Spring
14- Gasket
15- Fitting

Fig. 114 Exploded view of the fuel pump—3.0L engines

REMOVAL & INSTALLATION

■ **Always have a fire extinguisher handy when working on any part of the fuel system. Remember, a very small amount of fuel vapor in the bilge, has the tremendous potential explosive power.**

1. Open or remove the engine hatch/covers. Disconnect the battery cables.

2. Position a container under the pump and remove the fuel inlet and outlet lines. It is important to use two open-end wrenches; one to hold the fitting nut and the other to loosen the line nut. Plug the line with a golf tee or something similar to prevent additional fuel spillage. Take care not to spill fuel on a hot engine, because such fuel may ignite.

3. Disconnect the sight tube.

4. Loosen the two pump mounting screws and carefully pull the pump out. Scrape any old gasket material from the pump and block mating surfaces. The pump pushrod may fall out of position when removing the pump; to prevent this, slip a small screwdriver in behind the pump and support the pushrod. Check the pump pushrod for damage or wear.

5. Coat BOTH sides of the new gasket with Perfect Seal and position it on the pump. Swab the end of the pushrod with grease and insert the pump into the cylinder block so that the rod is riding on the camshaft eccentric. Coat the mounting bolts with Perfect Seal and then tighten them to 20 ft. lbs. (27 Nm) on the 3.0L; 25 ft. lbs (34 Nm) on the 1992-97 V8.

6. Coat the threads of the fuel line pump fittings with #592 Loctite Pipe Sealant, thread them into the pump base until they are finger tight, and then tighten an additional 1-3/4 to 2-1/4 turns with a wrench. Do not over-tighten and do not use Teflon tape.

7. Install the fuel lines and tighten securely while holding the pump fittings with another wrench.

8. Install the sight tube, remove the container and connect the battery cables. Start the engine and check for fuel leaks.

PRESSURE TEST

◆ **See Figure 116**

1. Open or remove the engine hatch/covers. Disconnect the battery cables.

2. Position a container under the fuel line connection at the carburetor and remove the fuel inlet line. It is important to use two open-end wrenches; one to hold the fitting nut and the other to loosen the line nut. Plug the line with a golf tee or something similar to prevent additional fuel spillage.

3. Thread a 'tee' or a fuel pressure connector (#91-18078) into the fitting nut on the carburetor and then reconnect the fuel line to the other end.

4. Connect a fuel pressure test gauge to the remaining fitting on the connector.

5. Connect the battery cables, start the engine and let it idle. The fuel pressure should be 5-1/4 to 6-1/4 psi on the 3.0L; 3 to 7 psi on the 1992-97 V8.

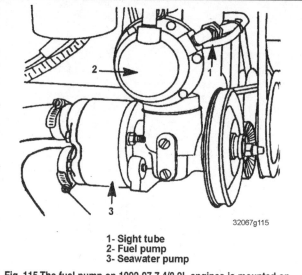

1- Sight tube
2- Fuel pump
3- Seawater pump

Fig. 115 The fuel pump on 1992-97 7.4/8.2L engines is mounted on the seawater pump

6. Slowly run the engine up to 1,800 rpm and check that the pressure remains the same.

7. If pressure varies, replace the pump.

8. Turn off the engine, remove the pressure gauge and fuel line and then remove the connector. Reinstall the fuel line to the fitting and tighten to 18 ft. lbs. (24 Nm).

FUEL LINE TEST

The fuel line, from the tank to the fuel pump, can be quickly tested by disconnecting the existing fuel line at the fuel pump and connecting a six-gallon portable tank and fuel line. This simple substitution eliminates the fuel tank and fuel lines in the boat. Now, start the engine and check the performance.

If the problem has been corrected, the fuel system between the fuel pump inlet and the fuel tank is at fault. This area includes the fuel line, the fuel pickup in the tank, the fuel filter, anti-siphon valve, the fuel tank vent, and excessive foreign matter in the fuel tank, and loose fuel fittings sucking air into the system. Improper size fuel fittings can also restrict fuel flow.

Possible cause of fuel problems may be deterioration of the inside lining of the fuel line which may cause some of the lining to develop a blockage similar to the action of a check valve. Therefore, if the fuel line appears the least bit questionable, replace the entire line.

Another possible restriction in the fuel line may be caused by some heavy object lying on the line—a tackle box, etc.

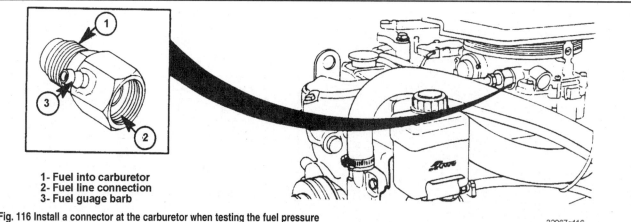

1- Fuel into carburetor
2- Fuel line connection
3- Fuel guage barb

Fig. 116 Install a connector at the carburetor when testing the fuel pressure

Electric Fuel Pump

DESCRIPTION

All 4.3L engines and 1998-01 5.0L/5.7L engines utilize an electric fuel pump with their carbureted fuel systems.

The electric fuel pump system includes the fuel tank/s, a water separator fuel filter, the electric fuel pump and a carburetor.

When the ignition switch on the control panel is turned to the ON position, the fuel pump is energized. Operation of the pump draws fuel from the fuel tank through the water separator fuel filter and is then pushed on through the fuel line to the carburetor. Efficient operation of the carburetor supplies sufficient fuel to the engine under all loads and rpm speeds.

Electric fuel pumps are not repairable.

REMOVAL & INSTALLATION

◆ See Figure 117

■ Always have a fire extinguisher handy when working on any part of the fuel system. Remember, a very small amount of fuel vapor in the bilge, has the tremendous potential explosive power.

1. Open or remove the engine hatch/covers. Disconnect the battery cables.

2. Position a container under the pump and remove the fuel inlet (large) and outlet (small) lines. It is important to use two open-end wrenches; one to hold the fuel pump fitting nut and the other to loosen the line nut. Plug the line with a golf tee or something similar to prevent additional fuel spillage. Take care not to spill fuel on a hot engine, because such fuel may ignite.

3. Unplug the electrical harness at the pump connector and move it out of the way.

4. Grasp the pump in the center of its body and pull it out of its holding bracket. Remove the two inlet and outlet fitting with a wrench on the fitting nut and one on the pump flat.

5. Install the small rubber grommet onto the outlet side of the pump. Slide a new O-ring over the fitting and screw it into the pump. Hold the pump on its flat and then tighten the fitting to 84 inch lbs. (9.5 Nm).

6. Install the large rubber grommet onto the inlet side of the pump. Slide a new O-ring over the fitting and screw it into the pump. Hold the pump on its flat and then tighten the fitting to 96 inch lbs. (11 Nm).

7. Press the pump into the bracket making sure the large grommet is at the bottom. Connect the wiring harness.

8. Coat the threads of the lower fuel line pump fittings with #592 Loctite Pipe Sealant, thread them into the pump base until they are finger tight, and then tighten an additional 1-3/4 to 2-1/4 turns with a wrench. Do not over-tighten and do not use Teflon tape. Repeat this procedure for the upper line.

9. Remove the container and connect the battery cables. Start the engine and check for fuel leaks.

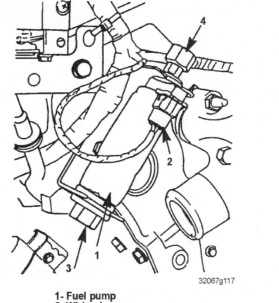

32067g117

1- Fuel pump
2- Wiring harness
3- Inlet coupling (large)
4- Outlet coupling (small)

Fig. 117 The electric fuel pump is usually located on the starboard side of the engine, near the front

PRESSURE TEST

1. Open or remove the engine hatch/covers. Disconnect the battery cables.

2. Position a container under the fuel line connection at the carburetor (or the outlet side of the fuel pump) and remove the fuel inlet line. It is important to use two open-end wrenches; one to hold the fitting nut and the other to loosen the line nut. Plug the line with a golf tee or something similar to prevent additional fuel spillage.

3. Thread a 'tee' or a fuel pressure connector (#91-18078) into the fitting nut on the carburetor and then reconnect the fuel line to the other end.

4. Connect a fuel pressure test gauge to the remaining fitting on the connector.

5. Connect the battery cables, start the engine and let it idle. The fuel pressure should be 3-7 psi.

6. If pressure varies, replace the pump.

7. Turn off the engine, remove the pressure gauge and fuel line and then remove the connector. Reinstall the fuel line to the fitting and tighten to 18 ft. lbs. (24 Nm).

Carburetor Specifications
Measurements are in.(mm) unless otherwise noted

Carburetor	Type	Float Level	Float Drop	Pump Rod	Choke Setting	Choke Unloader	Idle Mixture Screw	Float Weight	Primary Jet	Power Valve	Venturi I.D.
MerCarb	2bbl (35mm)										
	4.3L	①	1 3/32(27)	1 5/32(29)	Marks aligned	0.080 (0.2)	1 1/4 turns	9g Max	1.6 ②	0.75 ②	421
	5.0L	①	1 3/32(27)	1 5/32(29)	Marks aligned	0.080 (0.2)	1 1/4 turns	9g Max	1.85 ②	1.0 ②	⑤
	2bbl (36mm)	①	1 3/32(27)	1 5/32(29)	2 marks lean	0.080 (0.2)	1 1/4 turns	9g Max	1.45 ②	2 x 0.60 ②	470
	2bbl (43mm) ③	①	1 3/32(27)	1 5/32(29)	2 marks lean	0.080 (0.2)	Factory	9g Max	1.55 ②	4 x 0.65 ②	171
	2bbl (43mm) ④										
	4.3L	9/16(14)	1 3/32(27)	Middle hole	2 marks lean	0.080 (0.2)	1 1/4 turns	9g Max	1.55 ②	0.74 ②	472
	5.0L	11/32(9)	15/16(24)	Middle hole	2 marks lean	0.080 (0.2)	1 1/2 turns	9g Max	1.65 ②	0.90 ②	476
	5.7L	11/32(9)	15/16(24)	Middle hole	2 marks lean	0.080 (0.2)	1 1/2 turns	9g Max	1.65 ②	0.90 ②	475

① Solid needle: 3/8(10); spring loaded needle:9/16(14)
② Measurement is in mm
③ 1992-97

④ 1998-01
⑤ Carb. #1389-9670:407; #3310-806082:463

Carburetor Specifications
All measurements in.(mm) unless other wise noted

Carburetor	Type		Float Level	Pump Rod	Air Valve Dashpot	Vacuum Break	Accel. Pump	Primary Jet	Air Valve Spring Wind-Up	Primary Metering Rods	Secondary Metering Rods
Rochester 4MV	4bbl										
		5.0L	15/64(6)	Inner	0.025(0.64)	0.080(2)	23/64(9)	0.069	70-90 gr	0.042	CL
		5.7L	15/64(6)	Inner	0.025(0.64)	0.080(2)	23/64(9)	0.069	75-85 gr	0.042	CH

32067c3

Carburetor Specifications
All measurements in.(mm) unless other wise noted

Carburetor	Type		Float Level	Float Drop	Pump Rod	Choke Pull-Off	Accel. Pump	Primary Jet	Secondary Jet	Metering Rod Number	Metering Rod Spring
Weber WFB											
	4bbl (1992-97)										
		4.3L	1 9/32(33)	2.0(51)	#3 from end	3/16(4.5)	7/16(11)	0.089	0.089	16-6857	Yellow
		5.7L/350	1 9/32(33)	2.0(51)	#3 from end	3/16(4.5)	7/16(11)	①	②	③	④
		7.4L/8.2L	1 9/32(33)	2.0(51)	#3 from end	15/64(6)	7/16(11)	⑤	⑥	⑦	Pink
	4bbl (1998-01)		1 9/32(33)	2.0(51)	#3 from end	1.28(32.5)	7/16(11)	0.092	0.089	16-686647	Green
Bodensee	4bbl (1992-97)		1 9/32(33)	2.0(51)	#3 from end	1/8(3.5)	7/16(11)	0.086	0.08	16-656457	Natural
SAV1	4bbl (1998-01)		1 9/32(33)	2.0(51)	#3 from end	NA	7/16(11)	0.087	0.086	16-656457	Natural

① #3310-805484 (9661): 0.104
#3310-806761 (9665): 0.101
#3310-806761 (9665S): 0.101
#3310-816343A1 (9770): 0.101
#3310-806970 (9770SA): 0.101
#3310-806545 (9663): 0.089

② #3310-805484 (9661): 0.101
#3310-806761 (9665): 0.095
#3310-806761 (9665S): 0.095
#3310-816343A1 (9770): 0.09
#3310-806970 (9770SA): 0.092
#3310-806545 (9663): 0.092

③ #3310-805484 (9661): 16-7147
#3310-806761 (9665): 16-7052
#3310-806761 (9665S): 16-7052
#3310-816343A1 (9770): 16-6852
#3310-806970 (9770SA): 16-6850
#3310-806545 (9663): 16-656357

④ #3310-805484 (9661): Natural
#3310-806761 (9665): Pink
#3310-806761 (9665S): Pink
#3310-816343A1 (9770): Natural
#3310-806970 (9770SA): Natural
#3310-806545 (9663): Natural

⑤ #3310-818659 (9772): 0.107
#3310-806969 (9780S): 0.104
#3310-805569 (9777): 0.107
#3310-816917 (9773):0.107
#3310-806755 (9779S): 0.106
#3310-805341 (9776): 0.110
#3310-806791 (9776S): 0.110
#3310-817693 (9774): 0.107

⑥ #3310-818659 (9772): 0.098
#3310-806969 (9780S): 0.098
#3310-805569 (9777): 0.098
#3310-816917 (9773):0.107

#3310-806755 (9779S): 0.092
#3310-805341 (9776): 0.101
#3310-806791 (9776S): 0.101
#3310-817693 (9774): 0.098

⑦ #3310-818659 (9772): 16-6542
#3310-806969 (9780S): 16-757347
#3310-805569 (9777): 16-6542
#3310-816917 (9773): 16-6542
#3310-806755 (9779S): 16-757347
#3310-805341 (9776): 16-7147
#3310-806791 (9776S): 16-7147
#3310-817693 (9774): 16-6542

32067c4

8

FUEL SYSTEM FUEL INJECTION

FUEL AND COMBUSTION

Fuel

Fuel recommendations have become more complex as the chemistry of modern gasoline changes. The major driving force behind the changes in gasoline chemistry is the search for additives to replace lead as an octane booster and lubricant. These new additives are governed by the types of emissions they produce in the combustion process. Also, the replacement additives do not always provide the same level of combustion stability, making a fuel's octane rating less meaningful.

In the 1960's and 1970's, leaded fuel was common. The lead served two functions. First, it served as an octane booster (combustion stabilizer) and second, in 4-stroke engines, it served as a valve seat lubricant. For 2-stroke engines, the primary benefit of lead was to serve as a combustion stabilizer. Lead served very well for this purpose, even in high heat applications.

Today, all lead has been removed from the refining process. This means that the benefit of lead as an octane booster has been eliminated. Several substitute octane boosters have been introduced in the place of lead. While many are adequate in an engine, most do not perform nearly as well as lead did, even though the octane rating of the fuel is the same.

OCTANE RATING

A fuel's octane rating is a measurement of how stable the fuel is when heat is introduced. Octane rating is a major consideration when deciding whether a fuel is suitable for a particular application. For example, in an engine, we want the fuel to ignite when the spark plug fires and not before, even under high pressure and temperatures. Once the fuel is ignited, it must burn slowly and smoothly, even though heat and pressure are building up while the burn occurs. The unburned fuel should be ignited by the traveling flame front, not by some other source of ignition, such as carbon deposits or the heat from the expanding gasses. A fuel's octane rating is known as a measurement of the fuel's anti-knock properties (ability to burn without exploding).

Usually a fuel with a higher octane rating can be subjected to a more severe combustion environment before spontaneous or abnormal combustion occurs. To understand how two gasoline samples can be different, even though they have the same octane rating, we need to know how octane rating is determined.

The American Society of Testing and Materials (ASTM) has developed a universal method of determining the octane rating of a fuel sample. The octane rating you see on the pump at a fuel dock is known as the pump octane number. Look at the small print on the pump. The rating has a formula. The rating is determined by the R+M/2 method. This number is the average of the research octane reading and the motor octane rating.

• The Research Octane Rating is a measure of a fuel's anti-knock properties under a light load or part throttle conditions. During this test, combustion heat is easily dissipated.

• The Motor Octane Rating is a measure of a fuel's anti-knock properties under a heavy load or full throttle conditions, when heat buildup is at maximum.

VAPOR PRESSURE

Fuel vapor pressure is a measure of how easily a fuel sample evaporates. Many additives used in gasoline contain aromatics. Aromatics are light hydrocarbons distilled off the top of a crude oil sample. They are effective at increasing the research octane of a fuel sample but can cause vapor lock (bubbles in the fuel line) on a very hot day. If you have an inconsistent running engine and you suspect vapor lock, use a piece of clear fuel line to look for bubbles, indicating that the fuel is vaporizing.

One negative side effect of aromatics is that they create additional combustion products such as carbon and varnish. If your engine requires high-octane fuel to prevent detonation, de-carbon the engine more frequently with an internal engine cleaner to prevent ring sticking due to excessive varnish buildup.

ALCOHOL-BLENDED FUELS

When the Environmental Protection Agency mandated a phase-out of the leaded fuels in January of 1986, fuel suppliers needed an additive to improve the octane rating of their fuels. Although there are multiple methods currently employed, the addition of alcohol to gasoline seems to be favored because of its favorable results and low cost. Two types of alcohol are used in fuel today as octane boosters, methanol (wood alcohol) or ethanol (grain alcohol).

When used as a fuel additive, alcohol tends to raise the research octane of the fuel. There are, however, some special considerations due to the effects of alcohol in fuel.

• Since alcohol contains oxygen, it replaces gasoline without oxygen content and tends to cause the air/fuel mixture to become leaner.

• On older engines, the leaching affect of alcohol may, in time, cause fuel lines and plastic components to become brittle to the point of cracking. Unless replaced, these cracked lines could leak fuel, increasing the potential for hazardous situations.

• When alcohol blended fuels become contaminated with water, the water combines with the alcohol then settles to the bottom of the tank. This leaves the gasoline on a top layer.

■ **Modern fuel lines and plastic fuel system components have been specially formulated to resist alcohol leaching effects.**

RECOMMENDATIONS

According to the fuel recommendations that come with your engine, there are only a few engines in the product line that requires more than 87 octane. Most MerCruiser engines need only 87 octane or less. An 89 octane rating generally means middle grade unleaded. Premium unleaded is more stable under severe conditions but also produces more combustion products. Therefore, when using premium unleaded, more frequent de-carboning is necessary.

Combustion

In a high heat environment like an modern engine, the fuel must be very stable to avoid detonation. If any parameters affecting combustion change suddenly (the engine runs lean for example), uncontrolled heat buildup will occurs very rapidly.

The combustion process is affected by several interrelated factors. This means that when one factor is changed, the other factors also must be changed to maintain the same controlled burn and level of combustion stability.

• Compression—determines the level of heat buildup in the cylinder when the air-fuel mixture is compressed. As compression increases, so does the potential for heat buildup

• Ignition Timing—determines when the gasses will start to expand in relation to the motion of the piston. If the ignition timing is too advanced, gasses will be ignited and begin to expand too soon, such as they would during pre-ignition. The motion of the piston opposes the expansion of the gasses, resulting in extremely high combustion chamber pressures and heat. If the ignition timing is retarded, the gases are ignited later in relation to piston position. This means that the piston has already traveled back down the bore toward the bottom of the cylinder, resulting in less usable power.

• Fuel Mixture—determines how efficient the burn will be. A rich mixture burns slower than a lean one. If the mixture is too lean, it can't become explosive. The slower the burn, the cooler the combustion chamber, because pressure buildup is gradual.

• Fuel Quality (Octane Rating)—determines how much heat is necessary to ignite the mixture. Once the burn is in progress, heat is on the rise. The unburned poor quality fuel is ignited all at once by the rising heat instead of burning gradually as a flame front of the burn passing by. This action results in detonation (pinging).

There are two types of abnormal combustion—pre-ignition and detonation.

• Pre-ignition—occurs when the air-fuel mixture is ignited by some incandescent source other than the correctly timed spark from the spark plug.

• Detonation—occurs when excessive heat and or pressure ignites the air/fuel mixture rather than the spark plug. The burn becomes explosive.

In general, anything that can cause abnormal heat buildup can be enough to push an engine over the edge to abnormal combustion, if any of the four basic factors previously discussed are already near the danger point, for example, excessive carbon buildup raises the compression and retains heat as glowing embers.

ELECTRONIC FUEL INJECTION

Description And Operation

◆ **See Figures 1, 2, 3, 4 and 5**

The purpose of this section is to describe—in layman's terms whenever possible—the Throttle Body Fuel Injection (TBI) and Multi-Port Fuel Injection (MP/MPI) systems installed on many engines covered in this manual.

Visual inspections, and simple tests that may be performed using only basic shop test equipment. Again, we emphasize: specialized test equipment, hours of training and considerable experience is required to perform detailed service on a fuel injection system which is beyond the scope of this manual.

The first fuel injection system was introduced in Europe over 60 years ago—in 1932 on diesel truck engines. In the beginning, the system and individual components were quite expensive. Over the years, state-of-the-art microprocessors (commonly referred to as "computer chips"), have lowered the cost of electronically controlled fuel injection systems. Today, the price of EFI is getting close to the cost of modern carbureted systems.

An electronic fuel injection system is quite different from a carburetor system—even though they appear similar (particularly TBI). The fuel injection system has a more efficient delivery of fuel to the cylinders than can be obtained with standard carburetor operation.

The EFI system provides a means of fuel distribution by precisely controlling the air/fuel mixture and under all operating conditions for, as near as possible, complete combustion.

This is accomplished by using and Electronic Control Module (ECM), a small 'on-board' microcomputer that receives electrical inputs from various sensors about engine operating conditions. The ECM uses these inputs to modify fuel delivery to achieve, as near as possible, an ideal air/fuel ratio of 14.7:1.

The ECM program automatically signals the fuel injectors in the throttle body assembly to provide the correct quantity of fuel for a wide range of operating conditions. Several sensors are used to determine existing operating conditions and the ECM then signals the injectors to provide the precise amount of fuel.

The ECM has a "learning" capability. That is, if the battery is disconnected for any reason, the learning process has to begin all over again.

On TBI systems, the TBI assembly is located on the intake manifold, much like a carburetor, where air and fuel are distributed through a bore in the throttle body. Air for combustion is controlled by a throttle valve connected to the throttle linkage. Fuel is supplied by two injectors mounted in the TBI assembly, their metering tips are located directly above the throttle valve.

On MP/MPI systems, there is a throttle body located on the intake manifold, much like a carburetor except that only air is distributed and metered through the bores on the throttle body Air for combustion is controlled by a throttle valve connected to the throttle linkage. Fuel is supplied by individual injectors mounted in the intake manifold and attached to a fuel rail assembly.

Each injector is "pulsed" or "timed" to open or close by an electronic signal from the ECM. While constantly receiving input from various sensors, the ECM performs high speed calculations of engine fuel requirements and then "pulses the injectors open or closed.

MODES OF OPERATION

The ECM receives signals from several sensors, and then responds by delivering the proper amount of fuel to the cylinders under one of several conditions. These conditions are labeled "modes" and are controlled by the ECM, as described in the following short sections.

Starting Mode

When the ignition switch is rotated to the crank position, the ECM energizes the fuel pump relay and the pump builds up pressure. The ECM then checks the Engine Coolant Temperature (ECT) sensor and the Throttle Position (TP) sensor. From these incoming signals, the ECM determines the correct air/fuel ratio for starting the engine. The ECM controls the amount of fuel delivered in the starting mode by changing the length of time the injectors are turned on and off. This is accomplished by "pulsing" the injectors briefly.

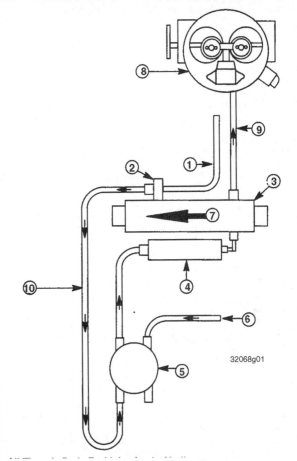

1- **Diaphragm rupture line to flame arrestor**
2- **Fuel pressure regulator**
3- **Fuel cooler**
4- **Electric fuel pump**
5- **Water separating fuel filter**
6- **Fuel from tank**
7- **Direction of water flow**
8- **Throttle body unit**
9- **Fuel line to throttle body**
10- **Excess fuel return to water separating fuel filter**

32068g01

Fig. 1 TBI system diagram—models with Cool Fuel system

Clear Flood Mode

A flooded engine can be cleared by opening the throttle between 50% of its travel. Once this is done, the ECM shuts down the fuel injectors and no fuel is delivered to the cylinders. The ECM will hold this injector rate as long as the throttle remains 50–75% open and engine speed is below 300 rpm. If the throttle position changes to greater than 75% or slightly less than 50%, the ECM will return to the starting mode.

Run Mode

When the engine is first started and operating above 300 rpm, the system operates in the Run Mode. The ECM will calculate the desired air/fuel ratio based on rpm and input from the MAP, IAT and ECT sensors. Higher engine load, from the MAP, and colder engine temperature, from the ECT requires more fuel, or a richer air/fuel ratio.

Acceleration Mode

If the ECM receives rapid change signals from the TP sensor, the ECM will provide extra fuel by increasing the injector pulse width.

Fuel Cut-off Mode

When the ignition switch is in the OFF position, no fuel is delivered to the cylinders by the injectors. Therefore, "dieseling" is prevented. If the ECM does not receive a distributor reference pulse—the engine is not operating— no fuel pulses are delivered to the injectors.

The fuel cutoff mode is also enabled at high engine rpm. This feature is an over-speed protection for the engine. Now, when cutoff is in effect due to high rpm, injector pulses will resume after engine rpm drops below the maximum OEM rpm specification.

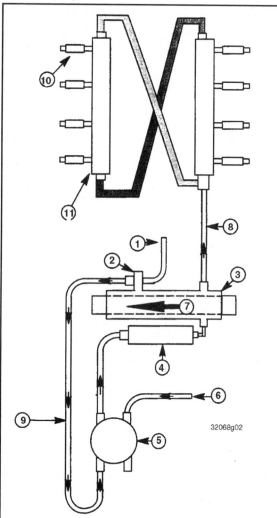

32068g02

1- Vacuum line to intake manifold base
2- Fuel pressure regulator
3- Fuel cooler
4- Electric fuel pump
5- Water separating fuel filter
6- Fuel from tank
7- Direction of water flow
8- Fuel line to fuel rail
9- Excess fuel return to water separating fuel filter
10- Fuel injectors
11- Fuel rail

Fig. 2 MP/MPI system diagram—5.7L/6.2L models with Cool Fuel system (MEFI-1/2)

Deceleration Mode

The Idle Air Control (IAC) valve provides additional air to the system when the throttle is rapidly released, causing the engine to move to idle, preventing it from dying.

Rev-Limit Mode

A fuel cut-off function is enabled when the engine hits a specified high rev limit. Fuel delivery to the injectors is cut off at a certain high rpm and then resumed when the engine falls below that rpm again.

Load Anticipation Mode

MEFI-3 INBOARDS ONLY

This function helps inboard engine during shifting. An electrical signal from the neutral safety switch on the transmission notifies the ECM if the switch is open or closed—in Neutral the switch is closed (grounded), and when the shift lever is moved, putting the transmission into gear the switch opens. As the boat is shifted into gear the signal will cause the ECM to add bypass air mixture with the IAC valve, while removing the extra air mixture when the boat is shifted back into Neutral.

Moving Desired RPM Mode

MEFI-3 INBOARDS ONLY

This mode will increase the desired rpm at idle to a specified point according to the throttle position. When the TP sensor is showing a closed throttle setting, the ECM will utilize the IAC valve and the Ignition Control (IC) to maintain the specified rpm. This will smooth out the transition from idle to full throttle and also help maintain constant engine speeds between 600 and 1200 rpm.

RPM Reduction Mode

1994–97 V8

The rpm reduction mode operates as an engine protection feature. The reduction mode permits normal fuel injection up to OEM specifications of approximately 2800 rpm. The ECM can recognize a change of state in a discrete input identifying an abnormal condition affecting engine operation such as low stern drive gear lube quantity; low engine oil pressure; and high coolant temperature.

SUBSYSTEMS

Fuel Metering System

As the name suggests, the fuel metering system "meters" the correct amount of fuel delivered to the cylinders through the intake manifold under all engine operating conditions. Fuel is delivered by the throttle body (TBI) or the individual injector (MP/MPI) and is controlled by the ECM.

Fuel Supply

Naturally, the fuel supply will begin at the boat's fuel tank. From the fuel tank, the fuel is drawn through a water separating fuel filter and then moved by the fuel pump to the Vapor Separator Tank (VST) or the Cool Fuel (CF) System. All models utilize an electric fuel pump found in the VST or CF, while certain early models also have a mechanical pump mounted on the engine. An additional electric fuel boost pump is used on 8.1L engine. A pressure regulator is standard on all models, while certain early V6 engines engine may have two.

Water Separating Fuel Filter

The water separator is a "typical" unit designed to prevent moisture from continuing on through the fuel lines and eventually through the injectors into the cylinders.

Mechanical Fuel Pump

Small Block Engine (1995–97 5.0L, 5.7L)—the mechanical fuel pump is driven by the engine camshaft eccentric. An internal diaphragm draws fuel in on one side and discharges it out the opposite side to the vapor separator tank.

Big Block Engine (1994–97 7.4L, 8.2L)—the fuel pump is driven by the "sea water" pump mounted on the forward starboard side of the engine.

Electric Fuel Pump

This fuel pump is located internally in the vapor separator tank or the cool fuel system casing. When the ignition switch is moved to the ON position, the ECM will energize the fuel pump relay to ON, but only for a couple seconds. The fuel pump almost instantly pressurizes the fuel system. As soon as the ignition switch is moved to the START position, the ECM energizes the fuel pump relay again and the fuel pump begins to operate.

Now, if the ECM fails to receive ignition reference pulses—indicating the engine is either cranking or actually operating—the ECM de-energizes the fuel pump relay and the fuel pump will stop.

An inoperative fuel pump relay can cause an "Engine Cranks, But Fails To Operate" condition.

The pump is capable of providing more fuel than is required by the injectors at WOT (wide open throttle). A pressure regulator is an integral part of the system maintaining fuel to the injectors at a predetermined regulated pressure. Excess fuel not required by the injectors is returned to the fuel separator tank by a separate fuel line.

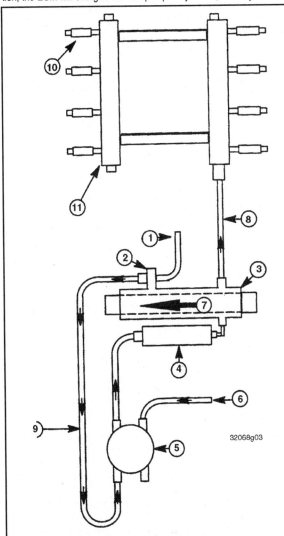

1- **Vacuum line to intake manifold base**
2- **Fuel pressure regulator**
3- **Fuel cooler**
4- **Electric fuel pump**
5- **Water separating fuel filter**
6- **Fuel from tank**
7- **Direction of water flow**
8- **Fuel line to fuel rail**
9- **Excess fuel return to water separating fuel filter**
10- **Fuel injectors**
11- **Fuel rail**

Fig. 3 MP/MPI system diagram—5.7L/6.2L models with Cool Fuel system (MEFI-3)

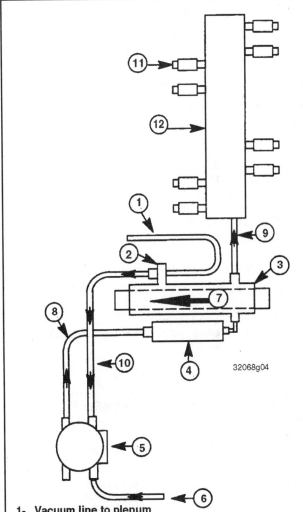

1- **Vacuum line to plenum**
2- **Fuel pressure regulator**
3- **Fuel cooler**
4- **Electric fuel pump**
5- **Water separating fuel filter**
6- **Fuel line from tank**
7- **Direction of water flow**
8- **Fuel line to fuel pump**
9- **Fuel line to fuel rail**
10- **Excess fuel return line to water separating fuel filter**
11- **Fuel injectors**
12- **Fuel rail**

Fig. 4 MP/MPI system diagram—7.4L/8.2L models with Cool Fuel system

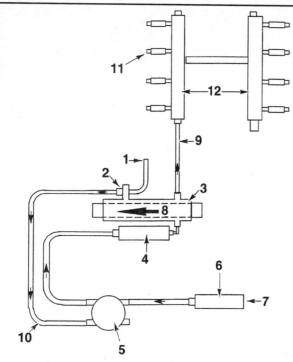

1 - **Vacuum line to intake manifold base**
2 - **Fuel pressure regulator**
3 - **Fuel Cooler**
4 - **Electric fuel pump**
5 - **Water separating fuel filter**
6 - **Fuel boost pump**
7 - **Fuel from tank**
8 - **Direction of water flow**
9 - **Fuel line to fuel rail**
10- **Excess fuel return to water separating fuel filter**
11- **Fuel injectors**
12- **Fuel rail**

Fig. 5 MP/MPI system diagram—8.1L models with Cool Fuel system

Vapor Separator Tank

1994–97 V8 ENGINES

The vapor separator tank contains the internal electric fuel pump that pressurizes fuel to the throttle body. The tank also collects all excess fuel returned from the injectors through the pressure regulator.

A float valve on the fuel inlet line will keep the fuel level in the tank at a pre-determined height at all times. As the fuel is consumed by the engine and the level begins to drop, the float valve opens and allows fresh fuel to enter the tank. This float valve operates identically to a float in a carburetor.

Cool Fuel System

This system is used on all V6 and 1998–01 V8 engines in place of the vapor separator tank system. It consists of a fuel cooler, pressure regulator and a fuel pump inside a box located on the lower port side of the engine. Early models using this system also had an additional regulator mounted in the fuel rail that was used to dampen fuel system pulsation. Additionally, certain 7.4L engines (L29) also had a second regulator mounted on the fuel rail; this one has no function at all in the system, the return port is plugged and there is a hose connected to the vacuum port for draining fuel into the engine if the diaphragm fails. 8.1L engines also utilize a fuel boost pump between the fuel tank and the water separating filter.

Throttle Body

MODELS WITH TBI

The throttle body consists of the following assemblies:
• Fuel Meter Cover—also houses the pressure regulator on models with vapor separator tank or the fuel damper on models with cool fuel..
• Fuel Meter Body—two fuel injectors.
• Throttle Body.
• Throttle Valves—two throttle valves controlling air flow into the cylinders.
• IAC (Idle Air Control) valve.
• TP (Throttle Position) sensor.

From the above list of components, it can easily be appreciated why the throttle body is considered one of the most critical items in the injection system. If the proper amount of air and fuel are not injected into the cylinders, the engine will not operate efficiently or may even fail to start.

MODELS WITH MP/MPI

The throttle body assembly is attached to the intake manifold or plenum and controls air flow to the engine, thus controlling engine output. Throttle valves within the assembly are controlled by the throttle cables. At idle, the valves are almost completely closed, while they open wider in proportion to the amount of throttle applied by the operator.

Fuel Rail

MODELS WITH MP/MPI ONLY

The fuel rail positions the injectors in the intake manifold, distributes fuel evenly to each injector and integrates the pressure regulator into the entire metering system.

Fuel Pressure Regulator

◆ See Figures 6 and 7

The pressure regulator is a diaphragm operated relief valve with fuel pump pressure on one side, and regulator spring pressure on the other side. The purpose of the regulator is to maintain a constant pressure differential across the injectors at all times under all conditions.

When the ignition is ON, and the engine is not operating, fuel pressure is about 30 psi (207 kPa). A low/hi pressure condition would result in poor engine performance.

The regulator is located in the throttle body assembly on TBI models with a vapor separator tank, while on MP/MPI models with a VST it is mounted on the fuel rail. On models with cool fuel, TBI or MP/MPI, it can be found on the cool fuel assembly itself.

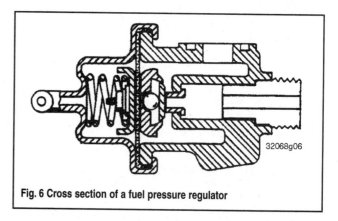

Fig. 6 Cross section of a fuel pressure regulator

Fuel Damper

◆ See Figure 8

This unit acts as an equalization device to reduce and stabilize the fuel pressure spikes created by the fuel injectors.

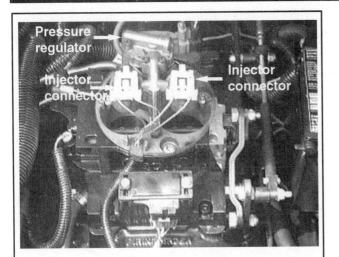

Fig. 7 TBI engines with VST used a pressure regulator in the fuel meter cover, models with Cool Fuel used a damper in the same spot

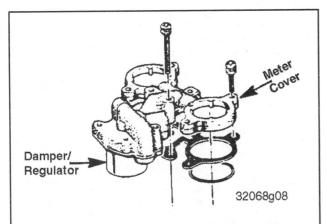

Fig. 8 The fuel damper is found in the throttle body on V6 and 1998–01 V8 TBI models. This is also where the pressure regulator is on the early V8 TBI engines with VST systems

Fuel Injectors

◆ See Figures 9 and 10

TBI ENGINES

Each injector is a solenoid-operated device, controlled by the ECM, that meters pressurized fuel into the intake manifold. When the injector's solenoid is energized by the ECM, a pintle valve opens allowing fuel to flow past the valve and through the injector tip. The tip has holes that control the fuel flow, generating a conical spray pattern of atomized fuel at the injector tip. Fuel is directed at the intake valve, becoming further atomized and then vaporized prior to entering the combustion chamber.

MP/MPI ENGINES

Each injector is a solenoid-operated device, controlled by the ECM, that meters pressurized fuel into an individual engine cylinder. When the injector's solenoid is grounded by the ECM, a ball valve opens allowing fuel to flow past the valve and through a flow director plate. The director plate has six holes that control the fuel flow, generating a conical spray pattern of atomized fuel at the injector tip.

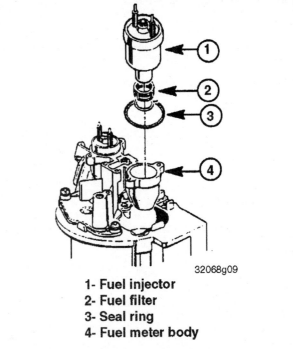

1- Fuel injector
2- Fuel filter
3- Seal ring
4- Fuel meter body

Fig. 9 On TBI engines, there are two fuel injectors in the throttle body

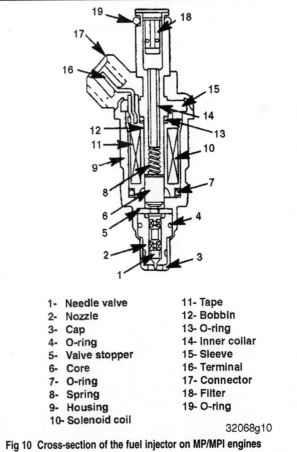

1- Needle valve	11- Tape
2- Nozzle	12- Bobbin
3- Cap	13- O-ring
4- O-ring	14- Inner collar
5- Valve stopper	15- Sleeve
6- Core	16- Terminal
7- O-ring	17- Connector
8- Spring	18- Filter
9- Housing	19- O-ring
10- Solenoid coil	

Fig 10 Cross-section of the fuel injector on MP/MPI engines

Intake Air Temperature Sensor (IAT)

This sensor is a thermistor mounted on the underside of the intake plenum. A thermistor is a resistor capable of changing value based on temperature. Low coolant temperature produces a high resistance and a high temperature causes low resistance.

Idle Air Control Valve (IAC)

◆ **See Figure 11**

The IAC valve assembly is mounted in the throttle body and controls engine idle speed. At the same time, the assembly prevents stalls due to changes in engine load.

Operation of the IAC can best be described as a device to control bypass air around the throttle valves. This feature is accomplished by moving a conical valve know as a "pintle" inward towards a seat—to decrease air flow or outward, away from the seat—to increase air flow. In this manner, a controlled amount of air moves around the throttle valve.

If rpm is too low, more air is bypassed around the throttle valve to increase rpm. If rpm is too high, less air is bypassed around the throttle valve to decrease rpm.

The ECM moves the IAC valve in small increments. These increments can only be measured using expensive special scan tool test equipment plugged into a DLC (Data Link Connector).

During engine idle speed, the proper position of the IAC valve is calculated by the ECM, and is based on coolant temperature and engine rpm. If the rpm drops below specification and the throttle valve is closed, the ECM senses a near stall condition and calculates a new valve position to prevent the stall.

Understand—engine idle speed is a function of total air flow into the cylinders based on IAC valve "pintle" position to maintain the desired idle speed for all engine operating conditions and loads.

The minimum idle air rate is set at the factory with a stop screw. This setting allows sufficient air flow by the throttle valves to cause the IAC valve "pintle" to be positioned a calibrated number of steps (counts) from the seat during "controlled" idle operation.

Throttle Position Sensor (TP)

The TP sensor is a potentiometer connected to the throttle shaft on the throttle body. One end of the sensor is connected to 5-volts from the ECM and the other end is connected to ECM ground.

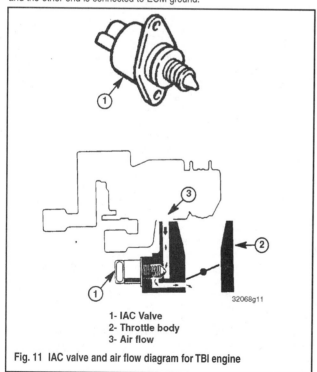

1- IAC Valve
2- Throttle body
3- Air flow

Fig. 11 IAC valve and air flow diagram for TBI engine

A third wire is connected directly to the ECM to measure the voltage from the TP sensor. The voltage output of the TP sensor changes when the throttle valve angle is changed.

When the throttle position is closed, the voltage output of the TP sensor is low—about 1/2-volt. When the throttle valve is opened, the output increases and at WOT the output voltage should be close to 5 volts.

Therefore, the ECM can determine fuel delivery requirements based on throttle valve angle—operator demand.

Engine Coolant Temperature Sensor (ECT)

◆ **See Figure 12**

This sensor is a thermistor immersed in the engine coolant passageway. A thermistor is a resistor capable of changing value based on temperature. Low coolant temperature produces a high resistance and a high temperature causes low resistance.

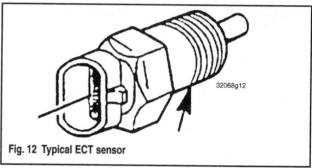

Fig. 12 Typical ECT sensor

Manifold Absolute Pressure Sensor (MAP)

◆ **See Figure 13**

The MAP sensor is a pressure transducer capable of measuring the changes in the intake manifold pressure. Pressure changes are the result of engine load and speed changes. MAP is the opposite of what is measured with a vacuum gauge.

When manifold pressure is high, vacuum is low— requires more fuel and of course the opposite is true. A low pressure—higher vacuum requires less fuel. The MAP sensor is also used to measure barometric pressure under certain conditions. This feature permits the ECM to automatically adjust for changes in operating altitude.

The ECM uses the MAP sensor to control fuel delivery and ignition timing.

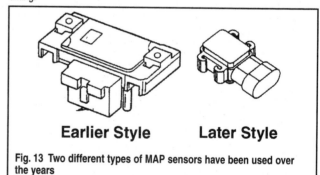

Earlier Style　　**Later Style**

Fig. 13 Two different types of MAP sensors have been used over the years

Knock Sensor (KS)

◆ **See Figure 14**

On a V6 or 1998–01 small block V8 engines with TBI, the KS is mounted on the engine block drain "Y" fitting located on the lower starboard side of the engine block. On a MP/MPI V8 engines, the KS sensor is mounted on the block, next to the starter motor. V6 models and 1998–01 V8 models with MP/MPI using the MEFI-3 ECM do not have a KS on the block, instead it is incorporated into the ECM.

The ECM uses this signal in calculating ignition timing.

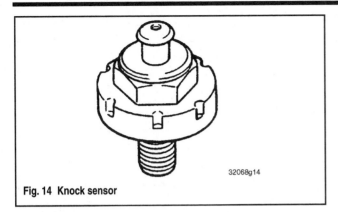

Fig. 14 Knock sensor

Knock Sensor Module

◆ See Figure 15

The KS module contains solid-state circuitry monitoring the KS AC voltage signal. If no spark knock is present an 8–10 volt signal is sent to the ECM. If a spark knock is present, the module will remove the 8–10 volt signal to the ECM.

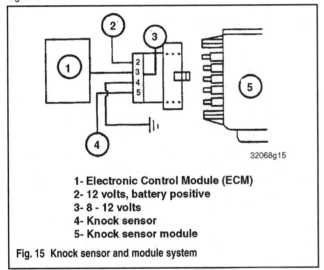

1- Electronic Control Module (ECM)
2- 12 volts, battery positive
3- 8 - 12 volts
4- Knock sensor
5- Knock sensor module

Fig. 15 Knock sensor and module system

Distributor Reference Signal (DIST REF)

The Hi Reference signal is supplied to the ECM by way of the "REF" line from the High Energy Ignition (HEI). The Hi Reference signal is used by the ECM to determine engine speed. This pulse type signal input creates the timing signal for pulsing of the fuel injections as well as the Ignition Control (IC) functions.

Discrete Switch Inputs

Discrete switch inputs are utilized by the EFI system to identify abnormal conditions that could affect engine and stern drive operation. Normally these switches are at rest in an open position. When one of the discrete switches changes state from open to closed, the ECM will detect a change in the voltage value and responds by placing the engine into the power reduction mode.

The power reduction mode is an engine protection feature that allows the operator limited engine power up to 2800 rpm. Should the operator attempt to surpass this rpm limit, the ECM will reduce fuel and spark timing until the engine speed drops to approximately 1200 rpm. The power reduction mode will allow the operator maneuvering power but eliminates the possibility of high rpm engine damage, until the discrete switch fault condition is corrected.

Discrete switches are used to detect the critical engine and stern drive operations—namely Engine Oil Pressure and Stern Drive Fluid Level. Engine Coolant Temperature (ECT) sensor is part of the engine protection mode, but the sensor is not a discrete switch input to the ECM.

Engine Control Module

◆ See Figure 16

The ECM is the control center of the fuel injection system, monitoring input information from all sensors and then turning this information into system commands that affect engine performance.

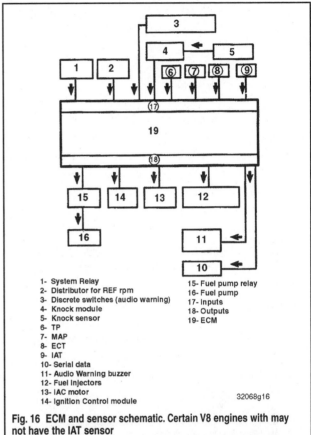

1- System Relay
2- Distributor for REF rpm
3- Discrete switches (audio warning)
4- Knock module
5- Knock sensor
6- TP
7- MAP
8- ECT
9- IAT
10- Serial data
11- Audio Warning buzzer
12- Fuel Injectors
13- IAC motor
14- Ignition Control module
15- Fuel pump relay
16- Fuel pump
17- Inputs
18- Outputs
19- ECM

Fig. 16 ECM and sensor schematic. Certain V8 engines with may not have the IAT sensor

Relieving Fuel Pressure

✳✳ CAUTION

To reduce the risk of fire and personal injury, it is necessary to relieve the fuel system pressure before servicing any fuel system component. If this procedure is not performed, fuel may be sprayed out of a connection under extreme pressure. Always keep a dry chemical fire extinguisher near the work area when serving the fuel system.

Disconnect the electrical connection at the electric fuel pump and move the harness out of the way. Turn the ignition key and crank the engine for ten seconds to relieve any residual pressure in the system. If the engine starts, don't worry, allow it to run until it dies out.

Cool Fuel System

REMOVAL & INSTALLATION   ② ◀ *MODERATE*

◆ See Figures 17, 18 and 19

1. Remove the engine compartment hatch and disconnect the battery cables.
2. If your boat has a fuel tank shut-off valve, close it and remove the inlet line at the water separating fuel filter. If not, remove the line and quickly plug it. Make sure you've got a container and plenty of rags available.

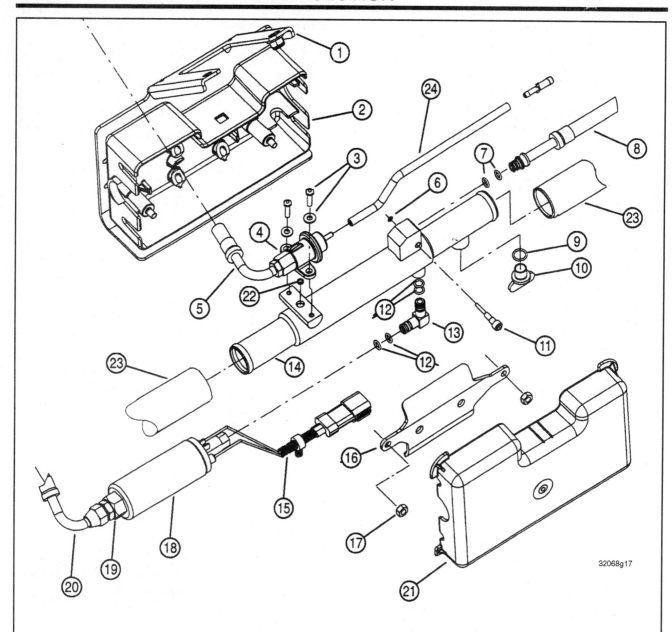

32068g17

1 - Bracket
2 - Cover Base
3 - Screw and Washer (2)
4 - Fuel Pressure Regulator
5 - Return Fuel Line
6 - Retaining Ring
7 - O-Rings (2)
8 - Fuel Line To Throttle Body
9 - Gasket
10 - Drain Plug
11 - Stepped Screw
12 - O-Rings (4)
13 - Elbow

14 - Fuel Cooler
15 - Fuel Pump Wiring Harness
16 - Retainer Bracket
17 - Nut (2)
18 - Electric Fuel Pump
19 - Inlet Fitting
20 - Fuel Line Inlet
21 - Cover
22 - Filter
23 - Seawater Hoses (Hose Clamps Not Shown)
24 - Vacuum Hose

Fig. 17 Exploded view of the Cool Fuel System

3. Drain the seawater system as detailed in Section 9.

4. Loosen the hose clamps and disconnect the seawater hoses at each end of the cool fuel assembly.

5. Disconnect and plug the fuel lines at the water separating filter and the throttle body/fuel rail. On 8.1L engines, disconnect and plug the fuel inlet line at the boost pump.

6. Remove the cover from the cool fuel assembly.

7. Tag and disconnect the electrical lead at the fuel pump.

8. Tag and disconnect the vacuum hose at the pressure regulator.

9. Remove the two upper engine mount bracket nuts that attach the assembly bracket to the engine and lift out the entire assembly.

10. Remove the two nuts from the cool fuel retaining bracket and lift the bracket and cooler assembly out of the cover base.

11. Remove the throttle body/fuel rail fuel line from the fuel cooler and then remove the return fuel line at the pressure regulator. The cooler-to-throttle body/fuel rail line utilizes a stepped screw to retain the line that is held in position by a retainer ring—do not remove the ring or the screw.

12. Disconnect the fuel pump elbow fitting (with pump still attached) from the bottom of the fuel cooler. Remove the pump/elbow and then remove the elbow from the pump.

13. Remove the two mounting screws and detach the pressure regulator from the fuel cooler tube (don't lose the small filter in the tube recess under the regulator!).

14. Remove the seawater drain plug (with seal) from the cooler tube.

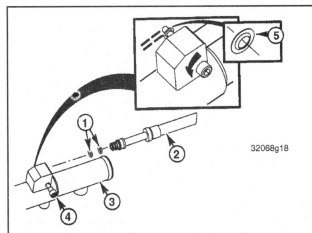

1- Fuel line O-Rings
2- Fuel line, cooler-to-throttle body
3- Fuel cooler
4- Stepped screw - loosened to accept fuel line
5- Retainer ring

Fig. 18 When removing the line from the throttle body/fuel rail, be careful not to remove the stepped screw or the retaining ring

To install:

15. Install the small filter into the recess under the pressure regulator on the cooler tube so that the conical side is facing into the hole. Position the regulator on the tube and tighten the two bolts and washers to 53 inch lbs. (5.8 Nm). Hold the fitting on the regulator with a wrench and connect the fuel outlet line to the regulator, tightening it securely.

16. Install two O-rings onto the cool fuel tube end of the throttle body/fuel rail fuel line. Loosen the stepped screw (be careful not to remove it), coat the O-rings with dish soap and insert the end of the fuel line into the cooler orifice. Hand-tighten the screw and the torque to 81 inch lbs. (9 Nm). Do not over-tighten this screw!

17. Position two O-rings over the threads on each side of the fuel pump elbow fitting, lubricate the rings with dish soap and then spin the elbow onto the fuel pump. Install the entire assembly onto the fuel cooler.

18. Position the fuel cooler assembly into the cover base. Position the retainer bracket over the assembly and take note of where it contacts the cooler and fuel pump, remove the bracket and apply a thin coating of thermal grease to the bracket contact area. Reinstall the bracket over the assembly, coat the mounting bolts with Loctite and tighten the nuts to 50 inch lbs. (5.6 Nm).

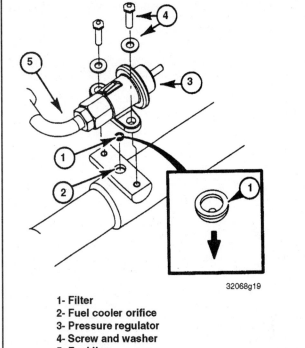

1- Filter
2- Fuel cooler orifice
3- Pressure regulator
4- Screw and washer
5- Fuel line

Fig. 19 Removing the pressure regulator. Don't lose the little filter underneath it!

19. Install the drain plug and tighten it securely.

20. Hold the filter nut with a wrench and then install the fuel lines to the water separating fuel filter.

21. Making sure that the fuel lines are routed properly, position the cool fuel assembly on the engine mount studs and then tighten the mounting nuts to 30 ft. lbs. (41 Nm).

22. Route the cooler-to-throttle body/fuel rail fuel line around to the back of the engine and then connect it to the throttle body or fuel rail. Remember to use a second wrench on the filter nut and to remove the plug you inserted at removal. Please refer to the Fuel Rail procedures later in this section for complete details on this connection.

23. Reconnect the vacuum line at the pressure regulator. Snap in the fuel pump harness connector at the pump.

24. Connect the seawater hoses to the cooler and tighten the clamps securely.

25. Reconnect the seawater inlet line and open the seacock.

26. Remove the plug and reconnect the fuel tank supply line. Open the fuel shut-off valve.

27. Connect the battery cables. Start the engine and check for any water or fuel leaks. Install the cool fuel system cover.

Fuel Boost Pump (Electric)

REMOVAL & INSTALLATION
2   MODERATE

8.1L Engines Only

1. Open or remove the engine hatch/cover and relieve the fuel system pressure. Disconnect the battery cable.

2. Loosen the line nut and disconnect the fuel tank inlet line at the pump. Plug the line.

3. Tag and disconnect the electrical harness.

4. Disconnect the fuel line between the pump and the water separating filter base.

5. Press the pump toward the engine and pop it out of the bracket. Remove the fuel line.

To install:

6. Position the pump so that the grommets are lined up with the bracket and then press it into the bracket arms.

7. Attach the fuel lines and plug in the electrical harness.

8. Connect the battery cables, start the engine and check for fuel leaks.

Fuel Meter Cover

REMOVAL & INSTALLATION

2 *MODERATE*

TBI Engines Only

◆ See Figures 20

■ Do not remove the four mounting screws for the pressure regulator/damper installed in the fuel meter cover.

1. Open or remove the engine hatch/cover and relieve the fuel system pressure. Disconnect the battery cable.

2. Remove the flame arrestor and then squeeze the plastic tabs on the injector electrical connectors and pull straight up until they come apart. Position the leads out of the way and make sure you tag them for correct reassembly.

3. Loosen and remove the meter cover mounting screws and carefully lift the cover straight up and off of the throttle body assembly. Watch for the damper seal and the outlet passage and cover gaskets—they may be hanging from the bottom of the cover.

4. Never immerse the cover in solvents or cleaner as damage to the damper diaphragm could occur. Inspect the mating surfaces of the throttle body assembly and the damper and cover for damage—use a magnifying glass if necessary! If any damage or scoring is found, replace the assembly.

5. Position a new damper seal in the throttle body cavity and then position the outlet passage and cover gaskets on the assembly.

6. Drop the meter cover into position and then slip in the mounting screws. Make sure they are coated with locking compound and that the short screws go into the holes next to the injectors. Tighten all screws to 28 inch lbs. (3 Nm).

7. Pop the injector lead connectors back on and connect the battery cables. Turn the ignition key to ON so that the fuel pump is energized and check for any fuel leaks. Do not actually start the engine until confirming that there are no leaks at the throttle body or fuel lines. Install the flame arrestor.

Fuel Injectors

REMOVAL & INSTALLATION

2 *MODERATE*

TBI Engines

◆ See Figures 21, 22 and 23

1. Open or remove the engine hatch/cover and relieve the fuel system pressure. Disconnect the battery cable.

2. Remove the flame arrestor and then squeeze the plastic tabs on the injector electrical connectors and pull straight up until they come apart. Position the leads out of the way and make sure you tag them for correct reassembly.

3. Loosen and remove the meter cover mounting screws and carefully lift the cover straight up and off of the throttle body assembly. Watch for the damper/regulator seal and the outlet passage and cover gaskets—they may be hanging from the bottom of the cover.

4. Make sure that the old meter cover gasket is still in place. Lay a small metal rod across the throttle body and then insert a small pry bar under the injector lip and carefully pry the injector out and up using the metal rod as a fulcrum.

To install:

5. Inspect the injectors for damage or clogging. MerCruiser recommends replacing damaged injectors. Always replace injectors with new ones that are the identical part number (as shown on the injector itself). All injectors are uniquely calibrated for flow rates so check the part number on the one being replaced and replace it with the same.

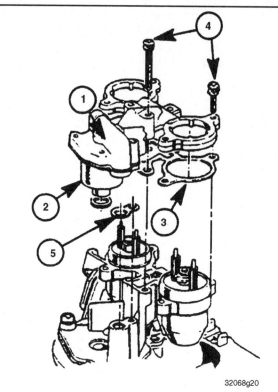

32068g20

1- Fuel meter cover
2- Fuel damper
3- Gaskets (regulator passages)
4- Screws
5- Fuel meter outlet gasket

Fig. 20 When removing the fuel meter cover, keep track of the gaskets and seals. On early models, the damper is a pressure regulator

6. Slide the filter screen over the bottom of the injector and then install a new lower O-ring after soaking it in soapy water.

7. Soak a new upper O-ring in soapy water and install it into the fuel meter body.

8. Align the raised lug on the injector body with the notch in the meter cavity and install the injector. The injector terminals should now be parallel with the throttle shaft.

9. Install the fuel meter cover as detailed above.

MP/MPI Engines

Please refer to the Fuel Rail and Injectors procedure detailed later in this section.

Fuel Pressure Damper

REMOVAL & INSTALLATION

2 *MODERATE*

1998–01 454/502 Mag/8.2L MPI V8 Engines

1. Remove the fuel rail as detailed later in this section.

2. Remove the two damper mounting screws and nuts and lift off the damper. Make sure to remove the passage plug and the filter.

3. Press a new filter into the rail. Coat new O-rings in soapy water and install them on the damper.

4. Position the damper on the rail, coat the threads of the mounting screws with Loctite and then tighten the nuts to 88–124 inch lbs. (10–14 Nm).

Install the fuel rail.

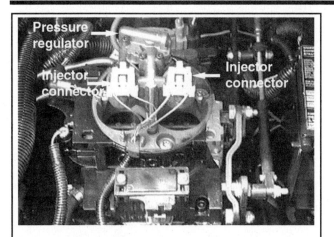

Fig. 21 Remove the connectors...

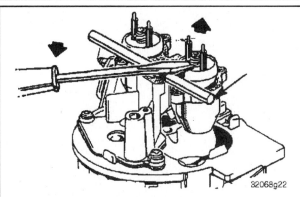

Fig. 22 ...and then carefully pry the injector out of the fuel meter body

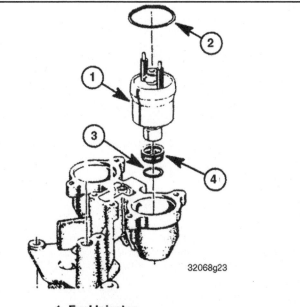

1- Fuel Injector
2- Upper O-Ring
3- Lower O-Ring
4- Fuel filter

Fig. 23 Installing the injector into the meter body

Fuel Pressure Regulator

REMOVAL & INSTALLATION

V6 and 1998–01 V8 Engines

Please refer to the Cool Fuel System procedures in this section for pressure regulator removal. The regulator is incorporated into the assembly and removal is detailed in the overall procedure.

1994–97 V8 Engines With MP/MPI

◆ **See Figure 24**

1. Remove the fuel rail as detailed in this section.
2. Remove the two mounting bolts and pull the regulator out of the rail.
3. Use a small awl and carefully pry the regulator filter out of the fuel rail on 1994–97 big blocks.
4. Inspect the regulator mounting surface for damage or irregularities. Clean the filter and blow it dry with compressed air.
5. Carefully press the filter into the fuel rail. Make sure it is completely dry.
6. Soak two new O-rings in soapy water and then position them into the regulator-to-rail bore and over the fuel line nipple.
7. Slide the regulator into position on the rail, coat the threads of the mounting bolts with Loctite and slide them through the holes in the rail. Tighten the nuts to 88–124 inch lbs. (10–14 Nm).
8. Install the fuel rail. Turn the ignition switch to ON without starting the engine, wait 2 seconds and then turn it OFF. Repeat this 4 times with 10 seconds in between each cycle so as to prime the fuel system and check for any leaks.

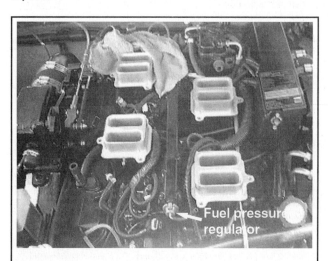

Fig. 24 Early MPI engines had the regulator located at the end of the fuel rail

Fuel Pump Relay

REMOVAL & INSTALLATION

◆ **See Figure 25**

Locate the relay in the engine electrical box. Tag and disconnect the electrical lead. Remove the relay from the bracket. Do not soak the relay in any solvent of cleaner.

Fig. 25 The fuel pump relay is located in the engine electrical box, near the ECM

Fuel Pump—Electric

REMOVAL & INSTALLATION

V6 and 1998–01 V8 Engines

■ **8.1L engines utilize an additional electric pump between the tank and filter, this pump is detailed in the Fuel Boost Pump procedure earlier in this section.**

Please refer to the Cool Fuel System procedures later in this section for fuel pump removal. The pump is incorporated into the assembly and removal is detailed in the overall procedure.

1994–97 V8 Engines

Please refer to the Vapor Separator Tank procedures later in this section for fuel pump removal. The pump is incorporated into the assembly and removal is detailed in the overall procedure.

Fuel Pump—Manual

REMOVAL & INSTALLATION

1994-97 V8 Engines

Sometimes referred to as a fuel lift pump, this is a mechanical pump that helps the fuel get from the tank to the VST. Removal and installation procedures for these pumps are identical to the standard mechanical fuel pump procedures detailed in the Carburetor section of this manual.

Fuel Rail And Injectors

REMOVAL & INSTALLATION

MP/MPI Engines Only

1995–97 350 MAGNUM

◆ **See Figure 26**

1. Open or remove the engine hatch/cover and relieve the fuel system pressure. Disconnect the battery cable.
2. Remove the flame arrestor and throttle body as detailed in this section.

3. Tag and disconnect the fuel lines from the port and starboard side rails. Make sure you have plenty of rags and a container available for the inevitable spillage. Plug the lines.
4. Tag and disconnect the electrical lead at each injector. Move them out of the way.
5. Remove the four mounting screws on each side and lift out the rail and injectors. Remove the injectors form the rail.
 To install:
6. Inspect the injector seating recess in the fuel rail for pitting, nicks, burrs or any other irregularities. Replace the rail assembly if necessary.
7. Soak new injector O-rings in soapy water for a minute and then press them into their respective recesses on the intake manifold.
8. Position each rail over the injectors and then carefully press the rail down until all the injector are seated properly. Install the retaining bolts and tighten them to 105 inch lbs. (12 Nm).
9. Reconnect the fuel lines at the rails and tighten the fitting nuts to 25 ft. lbs. (31 Nm).
10. Reconnect the electrical leads at each injector.
11. Install the throttle body and arrestor. Connect the battery cables, turn the ignition ON and check for leaks.

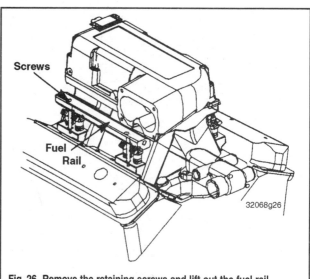

Fig. 26 Remove the retaining screws and lift out the fuel rail

1998–01 350 MAG/6.2L (BLACK SCORPION)

◆ **See Figures 27 and 28**

1. Open or remove the engine hatch/cover and relieve the fuel system pressure. Disconnect the battery cable.
2. Relieve the fuel system pressure.
3. Remove the flame arrestor and throttle body as detailed in this section.
4. Remove any lines or leads connected to the intake plenum. Remove the six retaining bolts and two nuts (under the front side) and then lift off the intake plenum.
5. Remove the distributor cap and position it out of the way with all wires still attached.
6. Carefully loosen and remove the fuel lines at the aft end of each rail. There should be a rail-to-rail fuel line on each side and a cool fuel-to-rail lines, additionally, on the starboard side. Make sure you have a suitable container and plenty of rags available for the inevitable spills. Plug the line endings.
7. Loosen the connector fittings and remove the two rail-to-rail fuel lines at the front of the engine.
8. Tag and disconnect the electrical lead at each harness. Move them out of the way.
9. Remove the rail mounting screws. Grasp a rail at each end and then slowly rock it back and forth while pulling up on it until it (and the injectors) come free of the manifold. Don't be surprised if some injectors stay in the manifold.
10. Carefully pop the injectors out of their fittings in the rail or manifold. If you are not replacing them, please be sure to keep them in order.

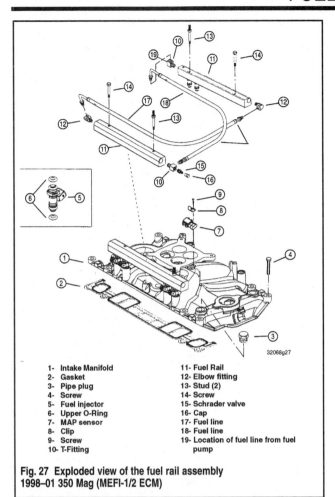

1- Intake Manifold
2- Gasket
3- Pipe plug
4- Screw
5- Fuel injector
6- Upper O-Ring
7- MAP sensor
8- Clip
9- Screw
10- T-Fitting

11- Fuel Rail
12- Elbow fitting
13- Stud (2)
14- Screw
15- Schrader valve
16- Cap
17- Fuel line
18- Fuel line
19- Location of fuel line from fuel pump

Fig. 27 Exploded view of the fuel rail assembly 1998–01 350 Mag (MEFI-1/2 ECM)

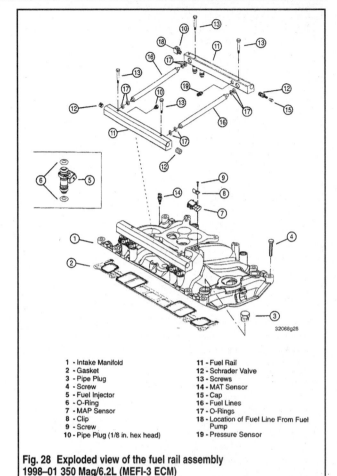

1 - Intake Manifold
2 - Gasket
3 - Pipe Plug
4 - Screw
5 - Fuel Injector
6 - O-Ring
7 - MAP Sensor
8 - Clip
9 - Screw
10 - Pipe Plug (1/8 in. hex head)

11 - Fuel Rail
12 - Schrader Valve
13 - Screws
14 - MAT Sensor
15 - Cap
16 - Fuel Lines
17 - O-Rings
18 - Location of Fuel Line From Fuel Pump
19 - Pressure Sensor

Fig. 28 Exploded view of the fuel rail assembly 1998–01 350 Mag/6.2L (MEFI-3 ECM)

To install:

11. Inspect the injectors for damage. Clean the rails with solvent and dry with compressed air if possible; if not, make sure they are completely dry prior to installation. If you are replacing the injectors, make sure that the new ones have identical part numbers.

12. Slide the upper and lower seals (preferably new ones) over each end of the injector. Soak new O-rings in soapy water and position them on the injectors. Position the injectors and them pop them into their respective seats in the manifold—or you can do the same except with the fuel rail.

13. Position each rail (with or without the injectors) and rock it back and forth while pushing down on it until the injectors are all seated properly in the manifold and the rail. Repeat this on the other side.

14. Tighten the mounting screws to 105 inch lbs. (12 Nm).

15. Check that the rail-to-rail lines are still routed correctly and them hand tighten them to the connector fittings. Install the aft cool fuel-to-rail line and then tighten all fittings to 18 ft. lbs. (24 Nm).

16. Replace the distributor cap.

17. Position a new plenum gasket on the manifold so that the word **FRONT** is facing up and then install the plenum. Tighten the screws and nuts in a diagonal pattern to 150 inch lbs. (17 Nm).

18. Install the throttle body and arrestor. Connect the battery cables and check for fuel leaks.

1994–97 7.4L/8.2L V8 ENGINES
1998–01 454/502 Mag/8.2L MPI V8 ENGINES

◆ See Figure 29

1. Open or remove the engine hatch/cover and relieve the fuel system pressure. Disconnect the battery cable.

2. Remove the flame arrestor, throttle body and intake plenum as detailed in this section.

3. Tag and disconnect the electrical harness at each injector, moving all of them out of the way.

4. Loosen the screw and remove the fuel line retainer from the rail.

5. Making sure you have a container and plenty of rags available, remove the inlet and outlet fuel lines from the rail. Plug the lines and keep track of the O-rings. There are some 1997 models that may only have one fuel line.

6. Remove the rail mounting screws and lift out the fuel rail. Half the injectors are going to come out with the rail while the other half will stay in the manifold. Face the engines and lift out the rail, slowly and carefully, so that the front right injector pops from the rail and stays in the manifold. Next, pop the front left injector from the manifold so that it stays in the rail. Work your way to the rear of the engine in this way until the rail is free. The left bank of injectors should be in the rail and the right side injectors should be in the manifold.

7. Remove all injectors from the rail and the intake manifold.

To install:

8. Inspect the injector seating recess in the fuel rail for pitting, nicks, burrs or any other irregularities. Replace the rail assembly if necessary.

9. Soak new injector O-rings in soapy water for a minute and then press the four left bank injector into the rail. Next, do the same with the right bank injectors except pop them into the intake manifold—remember, left and right in this instance is as you face the front of the engine.

10. Now grasp the assembly and position it so you can pop the first injector (at the rear of the engine) on the left side into the manifold. Next would be the first injector on the right side and it should pop into the fuel rail. Repeat this procedure down the line from the rear to the front until everything is in place.

11. Install the rail mounting screws and tighten them to 105 inch lbs. (12 Nm).

12. Soak new O-rings for the fuel lines in soapy water and position them on the ends of the lines. Press the lines into the fuel rail and install the retainer, tightening the bolt to 81 inch lbs. (9 Nm).

13. Reconnect the electrical harness at each injector and install the plenum.

14. Install the throttle body and flame arrestor. Connect the battery cables, turn the ignition ON and check for leaks.

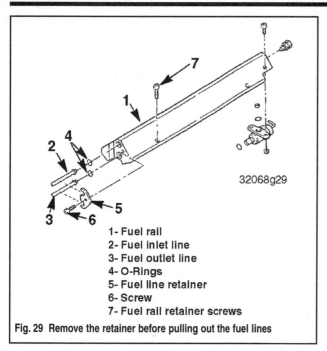

1- Fuel rail
2- Fuel inlet line
3- Fuel outlet line
4- O-Rings
5- Fuel line retainer
6- Screw
7- Fuel rail retainer screws

32068g29

Fig. 29 Remove the retainer before pulling out the fuel lines

1998–01 7.4L/8.1L V8 ENGINES

◆ **See Figures 30 and 31**

1. Open or remove the engine hatch/cover and relieve the fuel system pressure. Disconnect the battery cable.
2. Remove the flame arrestor, throttle body and intake plenum as detailed in this section.
3. Disconnect the fuel line at the fuel rail.
4. Tag and disconnect the electrical lead at the injector harness. Do the same at each injector.
5. Remove the screws, nut, MAP sensor retaining clip and stud from the rail. Lift out the rail/injector assembly from the manifold.
6. Pop off the injector retaining clip and remove each injector from the fuel rail.

To install:

7. Inspect the injector seating recess in the fuel rail for pitting, nicks, burrs or any other irregularities. Replace the rail assembly if necessary.
8. Soak new injector O-rings in soapy water for a minute and then press the injectors into the rail.
9. Pop the retaining clip into place over each injector and then connect the electrical lead.
10. Lubricate the remaining O-rings, position them in the manifold and then install the rail/injector assembly. Install the screws, nut and stud. Tighten securely.
11. Install the plenum, throttle body and arrestor.

OEM ② MODERATE

INJECTOR BALANCE TEST

All injectors should have the same amount of pressure drop after being turned on and off for a controlled amount of time. Any injector with a drop of 1.5 psi or more, above or below the average drop across all injectors should be replaced. This procedure will require the use of an injector balance tester, available through your Mercury dealer or selected aftermarket suppliers. Be sure the tester has a pulse width of at least 200–400 milliseconds; particularly if you are testing a 7.4L or 8.2L engine. At the very least, the tester must have a pulse width that is capable of dropping the fuel rail pressure to half of the normal operating pressure.

1. The engine should be cool, or at least have been shut down for 10 minutes prior to testing.
2. Relieve the pressure in the fuel system and then remove the intake plenum.
3. Turn OFF the ignition and connect a fuel pressure gauge to the tap at the rear of the fuel rail.

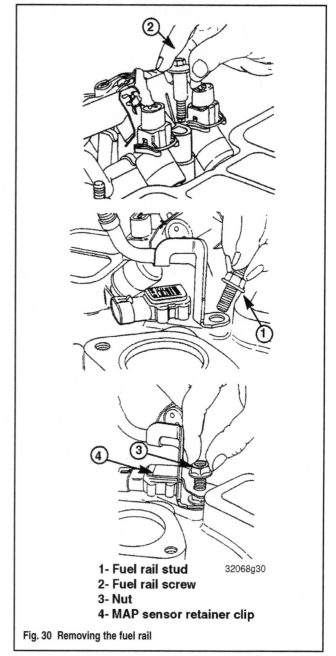

1- Fuel rail stud   32068g30
2- Fuel rail screw
3- Nut
4- MAP sensor retainer clip

Fig. 30 Removing the fuel rail

32068g31

1- Fuel rail
2- Retainer clip
3- Fuel injector

Fig. 31 Installing the injector

4. Tag and disconnect the electrical lead at each injector and move them out of the way.

5. Connect the injector tester (or scan tool) to an injector with the adaptor harness that should have come with the tool.

6. Make sure that the ignition has been OFF for at least ten seconds and then turn it ON. The fuel pump should run for about 2 seconds.

7. Connect a length of tubing to the vent valve on the pressure gauge and put the other end into a suitable container. Open the valve to bleed off any air in the system and repeat until you are confident that there is no longer any air in the gauge.

8. Turn the ignition OFF for 10 seconds and then turn it ON. Repeat this procedure several times to get the fuel pressure up to its maximum. Record this reading. Energize the tester and record the pressure at its lowest point—you may notice a slight increase in pressure after the low point, don't worry about this as its normal.

9. Subtract the second reading from the first to get your pressure drop figure. Repeat the procedure for each injector and compare the readings. Good injectors will have almost identical readings. Retest any injector that deviates from the average more than 1.5 psi (up or down). If it still fails on the retest, replace the injector.

10. Disconnect the tester and the pressure gauge.

11. Connect all the leads at the injectors and reinstall the plenum.

Throttle Body

REMOVAL & INSTALLATION

 ②◀ *MODERATE*

TBI Engines

◆ **See Figure 32**

1. Open or remove the engine hatch/cover and relieve the fuel system pressure. Disconnect the battery cable.

2. Remove the flame arrestor and then squeeze the plastic tabs on the injector electrical connectors and pull straight up until they come apart. Position the leads out of the way and make sure you tag them for correct reassembly.

3. Disconnect the throttle cable and carefully move it out of the way.

4. Unplug the electrical connectors at the TP sensor and the IAC valve. Tag each lead and move them out of the way.

5. Hold a container under the connections and then remove the fuel inlet and outlet lines at the throttle body. Have some rags available to wipe up any spills. Plug the fuel lines to prevent any further spillage.

6. Remove the mounting screws and lift the throttle body off of the adapter plate. Cover the manifold opening with a rag.

To install:

7. Remove the TP sensor, IAC valve, fuel injectors and the meter cover and then thoroughly clean the throttle body assembly in an immersion type cleaner. Dry with compressed air if available and make certain all passages are clean.

8. Inspect all mating and casting surfaces for damage or cracks and scoring.

9. Install the sensor, valve, injectors and meter cover as detailed elsewhere in this section.

10. Position a new gasket on the adapter plate, install the throttle body and tighten the mounting screws to 30 ft. lbs. (40 Nm).

11. Connect and adjust the throttle linkage.

12. Connect the fuel lines and tighten the flange nuts to 23 ft. lbs. (31 Nm).

13. Connect the TP sensor and IAC valve electrical leads.

14. Pop the injector lead connectors back on and connect the battery cables. Turn the ignition key to ON so that the fuel pump is energized and check for any fuel leaks. Do not actually start the engine until confirming that there are no leaks at the throttle body or fuel lines. Install the flame arrestor.

MP/MPI Engines

1994–97 ENGINES
1998–01 454/502 Mag/8.2L ENGINES OEM

◆ **See Figures 33 and 34**

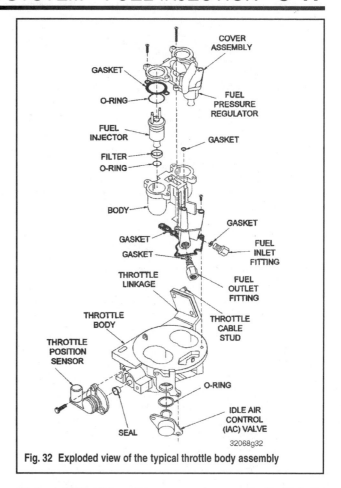

Fig. 32 Exploded view of the typical throttle body assembly

Labels: COVER ASSEMBLY, GASKET, O-RING, FUEL PRESSURE REGULATOR, FUEL INJECTOR, GASKET, FILTER, O-RING, BODY, GASKET, GASKET, GASKET, FUEL INLET FITTING, THROTTLE LINKAGE, FUEL OUTLET FITTING, THROTTLE BODY, THROTTLE CABLE STUD, THROTTLE POSITION SENSOR, O-RING, SEAL, IDLE AIR CONTROL (IAC) VALVE, 32068g32

■ **A special Stud Driver will be necessary to remove the throttle body.**

1. Open or remove the engine hatch/cover and relieve the fuel system pressure. Disconnect the battery cable.

2. Remove four mounting nuts and then remove the flame arrestor.

3. Disconnect the throttle linkage.

4. Using a stud driver tool, remove the 4 throttle body mounting studs. Turn the assembly over on its side and disconnect the TP sensor and IAC valve electrical leads. Disconnect the sight tube. Be sure you tag them all to avoid confusion on installation. Lift out the throttle body and set it down in a holding fixture to avoid damage to the valves. Stuff a clean, lint-free rag into the plenum opening.

5. Remove the TP sensor and IAC valve with their O-rings from the assembly if necessary.

To install:

6. Carefully clean the throttle bore and valves. Do not use anything containing methyl ethyl ketone! If you didn't remove the TP sensor and IAC valve, make sure that you get no solvent on them during cleaning. CAREFULLY scrape any gasket material off the mating surfaces. Make sure all passages are free of dirt and completely dry before installation.

7. Install the sensor and valve. Connect the sight tube. Position the throttle body near the plenum, reconnect the electrical leads and then position the assembly on the plenum studs. Don't forget to use a new gasket.

8. Coat the threads of the mounting studs with Loctite 242 (don't use anything stronger) and install them with a driver. Tighten each one to 165 inch lbs. (18 Nm).

9. Connect the throttle cable. Secure the cable barrel with a locknut and tighten it securely. Secure the end guide to the lever stud and then tighten the locknut until it just touches the end guide. If your boat utilizes Quicksilver Zero Effort Controls, make sure that the cable mounting stud is in the forward-most hole on the throttle lever.

10. Install the flame arrestor, coat the studs with Loctite 242 and then tighten the mounting nuts to 50 inch lbs. (6 Nm).

11. Reconnect the battery cables.

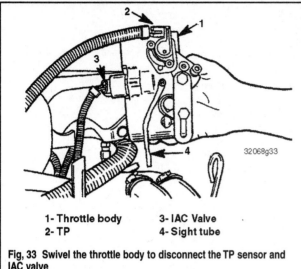

1- Throttle body   3- IAC Valve
2- TP   4- Sight tube

Fig, 33 Swivel the throttle body to disconnect the TP sensor and IAC valve

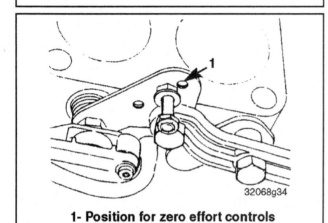

1- Position for zero effort controls

Fig. 34 If your boat uses Zero Effort controls, male sure the throttle cable stud is in the forward hole

1998–01 5.7L/6.2L/7.4L/8.1L ENGINES

◆ **See Figures 35 and 36**

1. Open or remove the engine hatch/cover and relieve the fuel system pressure. Disconnect the battery cable.
2. Remove mounting nut(s) and then remove the flame arrestor. Certain models will have a bracket holding the arrestor on.
3. Disconnect the throttle linkage.
4. Tag and disconnect the wiring harness for the IAC valve and TP sensor. On the 8.1L, disconnect the IAC valve hose.
5. Remove the 3 throttle body mounting bolts (later model Black Scorpions may have a bolt and nuts). Lift out the throttle body and set it down in a holding fixture to avoid damage to the valves. Stuff a clean, lint-free rag into the plenum opening.
6. Remove the TP sensor and IAC valve with their O-rings from the assembly if necessary.
To install:
7. Carefully clean the throttle bore and valves. Do not use anything containing methyl ethyl ketone! If you didn't remove the TP sensor and IAC valve, make sure that you get no solvent on them during cleaning. CAREFULLY scrape any gasket material off the mating surfaces. Make sure all passages are free of dirt and completely dry before installation.
8. Install the sensor and valve. Position the throttle body and new gasket on the adaptor (350 Mag/6.2 MPI) or plenum and tighten the bolts to 75 inch lbs. (8.5 Nm) except on the 8.1L which must be tightened to 89 inch lbs. (10 Nm). Later model Black Scorpions will have two gaskets and also a plate—the plate goes against the plenum.

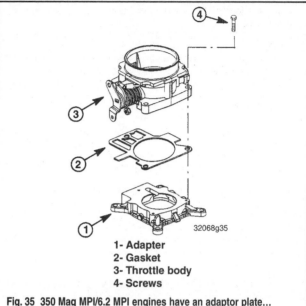

1- Adapter
2- Gasket
3- Throttle body
4- Screws

Fig. 35 350 Mag MPI/6.2 MPI engines have an adaptor plate...

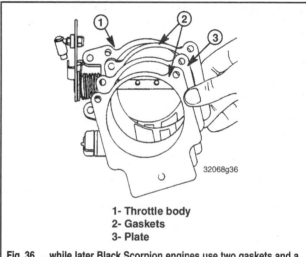

1- Throttle body
2- Gaskets
3- Plate

Fig. 36 ...while later Black Scorpion engines use two gaskets and a plate

9. Connect the IAC and TP leads. Connect the hose on the 8.1L.
10. Connect the throttle cable.
11. Install the flame arrestor.
12. Reconnect the battery cables.

THROTTLE CABLE ADJUSTMENT   ②  MODERATE

◆ **See Figures 37 and 38**

1. Install the cable end guide onto the throttle lever and then pre-load the shift cable by pushing the cable barrel end slightly toward the end of the throttle lever. Adjust the barrel so that it aligns with the hole in the anchor plate.
2. Secure the throttle cable with its hardware and tighten the nut securely. Tighten the locknut until it contacts the other nut and then back it off 1/2 turn on V6 and 1998–01 V8 engines with TBI, 1 turn on 1994–97 V8 engines with TBI. Do not back it off at all on 1994–97 V8 engines with MP/MPI.
3. Place the remote control lever in the wide open throttle position and check that the throttle plates are open all the way. If so, move the lever back to the idle position.

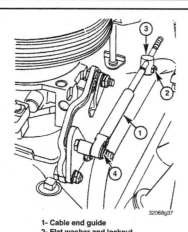

1- Cable end guide
2- Flat washer and locknut
3- Cable barrel
4- Flat washer and locknut

Fig. 37 Adjusting the throttle cable on V6 and 1998–01 V8 stern drive engines with TBI

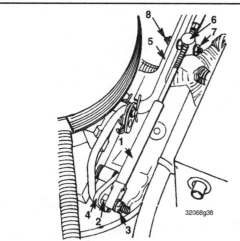

1- Cable end guide
2- Throttle lever stud
3- Locknut and flat washer - tighten until nut
   bottoms out, then back off 1/2 turn
4- Throttle lever
5- Throttle bracket
6- Cable barrel
7- Flat washer and locknut
8- Throttle cable anchor stud

Fig. 38 Adjusting the throttle cable on 1994–97 V8 engines and 1998–01 inboard V8 engines—TBI

Throttle Body Adapter Plate

REMOVAL & INSTALLATION

② ◀MODERATE

TBI Engines Only

◆ See Figure 39

1. Open or remove the engine hatch/cover and relieve the fuel system pressure. Disconnect the battery cable.

2. Remove the flame arrestor and then squeeze the plastic tabs on the injector electrical connectors and pull straight up until they come apart. Position the leads out of the way and make sure you tag them for correct reassembly.

3. Remove the throttle body as detailed in the preceding procedure. Lift the gasket off the adapter if it hasn't already come up with the throttle body.

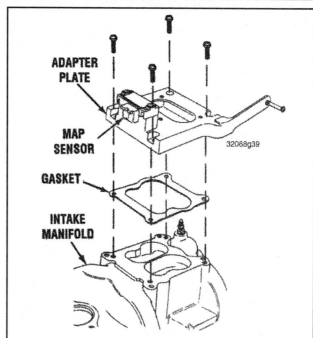

Fig. 39 Removing the throttle body adapter plate. Not all models will have the MAP sensor on the plate

4. Remove the four mounting screws and lift off the adapter plate. Plug the intake manifold with a clean rag.

5. Install the adapter plate with a new gasket and tighten the mounting screw to 15 ft. lbs (19 Nm) on the V6 and 1998–01 V8, 30 ft. lbs. (40 Nm) on the 1994–97 V8.

6. Install the throttle body.

Vapor Separator Tank (VST)

REMOVAL & INSTALLATION

② ◀MODERATE

1994–97 V8 Engines Only

◆ See Figure 40 and 41

TANK ASSEMBLY

◆ See Figures 42 and 43

■ 454 Magnums with serial numbers higher than OF130438 and 502 Magnums with numbers higher than OF128962 should have fuel lines that look similar to figure A. If your lines look like figure B, check with an authorized Mercury representative and ask about Technical Service Bulletin 92-26.

1. Open or remove the engine hatch/cover and relieve the fuel system pressure. Disconnect the battery cable.

2. Disconnect the electrical connector at the fuel pump.

3. Tag and disconnect the fuel lines at the VST. Make sure you have rags and a container handy; plug the lines to prevent additional spillage.

4. Loosen and remove the mounting bolt and lift out the VST. Be careful not to lose the bushing and grommet when removing the bolt.

To install:

5. Coat the threads of the mounting bolt with Loctite and then install the tank, Make sure that the bushings and grommet are in position and tighten the bolt to 105 inch lbs. (12 Nm).

6. Remove the fuel line plugs and then reconnect them. Tighten the fitting nuts to 23 lbs. (31 Nm).

7. Connect the pump electrical lead and the battery cables.

8. Without starting the engine, turn the ignition ON for 2 seconds and then OFF for 2 seconds. Do this four time to prime the fuel system and then start the engine and check for leaks.

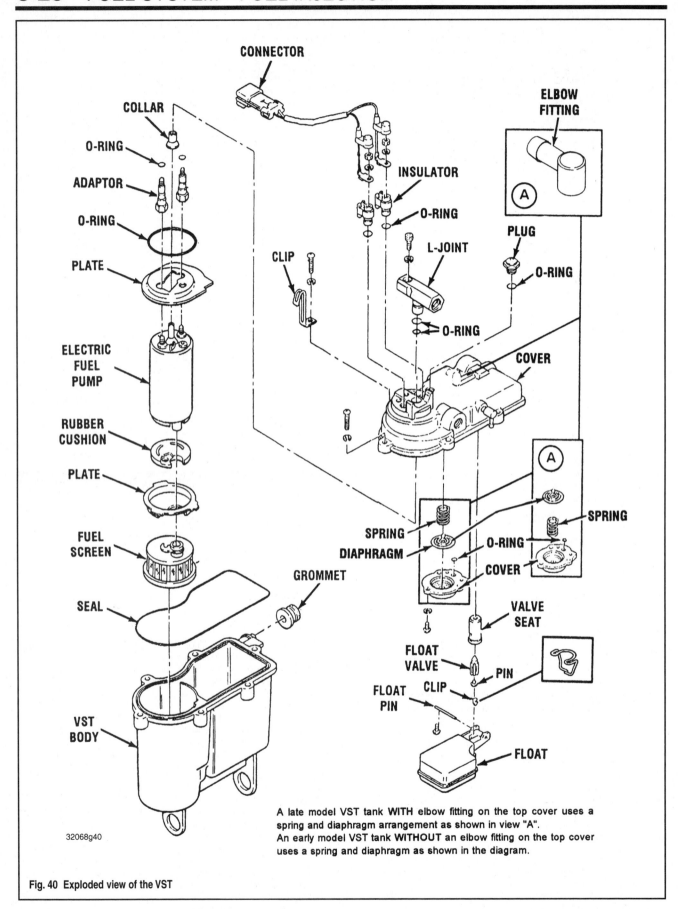

CONNECTOR

COLLAR

O-RING

ADAPTOR

O-RING

PLATE

ELECTRIC
FUEL
PUMP

RUBBER
CUSHION

PLATE

FUEL
SCREEN

SEAL

VST
BODY

ELBOW
FITTING

A

INSULATOR

O-RING

CLIP

L-JOINT

PLUG

O-RING

O-RING

COVER

A

SPRING

DIAPHRAGM

O-RING

SPRING

COVER

GROMMET

VALVE
SEAT

FLOAT
VALVE

PIN

FLOAT
PIN

CLIP

FLOAT

32068g40

A late model VST tank **WITH** elbow fitting on the top cover uses a
spring and diaphragm arrangement as shown in view "A".
An early model VST tank **WITHOUT** an elbow fitting on the top cover
uses a spring and diaphragm as shown in the diagram.

Fig. 40 Exploded view of the VST

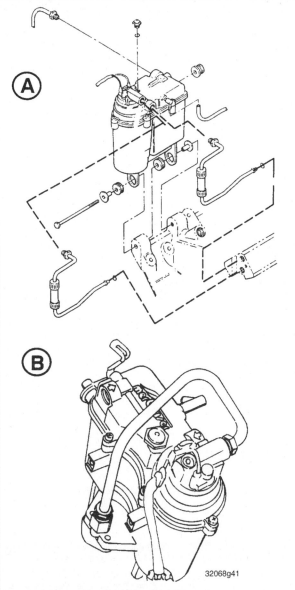

Fig. 41 Early 454/502 Magnums had a replacement fuel line set up

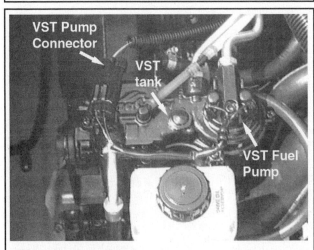

Fig. 42 A better view of the VST

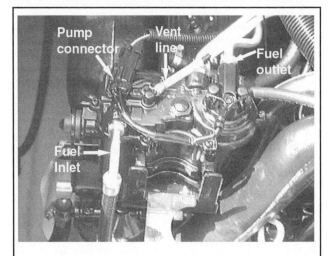

Fig. 43 Most fuel leaks will occur after assembly if care is not taken with these items

FUEL PUMP

◆ See Figures 44 and 45

1. Open the engine compartment hatch or covers and disconnect the battery cables. Have plenty of rags and a container available to take care of any spills.

2. Unplug the pump harness connector at the top of the vapor separator tank.

3. Relieve the pressure in the fuel system.

4. Tag and disconnect the fuel lines and remove the electrical line from the retaining clip. Don't forget to plug the lines!

5. Loosen and remove the retaining screw from the L-Joint and lift out the joint.

6. Carefully pry each side of the electrical connector cover up and over the retaining tabs, remove the connector retaining nuts and disconnect the electrical connectors (2) from the top of the tank.

7. Remove the tank cover screws and remove the cover and retaining clip.

8. Carefully remove the pump from the cover. Remove the O-ring, pump plate, 2 adapters and the collar from the top of the pump.

9. Disconnect the filter from the bottom of the pump and then remove the retaining plate and rubber cushion.

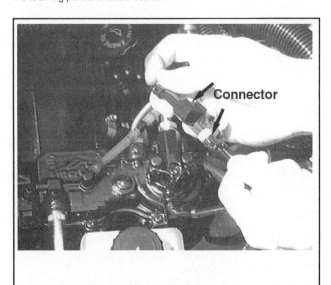

Fig. 44 Disconnect the fuel pump electrical lead

To install:

10. If replacing the pump, make sure that the replacement is the same part number as the outgoing pump.

11. Position the rubber cushion on the pump and then install the pump plate so that the cutout in the plate aligns with the inlet on the pump.

12. Install the filter.

13. Install the 2 adaptors, collar, plate and O-ring onto the top of the pump, making sure that the relief valve on the pump fits through the hole in the plate.

14. Slide the pump assembly into the cover.

15. Coat the threads of the cover mounting with Loctite and install the cover assembly. Make sure you tighten the cover bolts to 6 inch lbs. (8 Nm) as detailed in sequence illustrated.

16. Connect the pump electrical connectors, snap the connector covers into place and pop the electrical line into the clip.

17. Install the L-joint into the cover and tighten the mounting screw securely.

18. Connect the fuel lines and tighten the fitting nuts to 23 ft. lbs. (31 Nm).

19. Reconnect the harness connector and the battery cables. Without starting the engine, cycle the ignition switch ON and OFF four times to prime the fuel system. Start the engine and check for any fuel leaks.

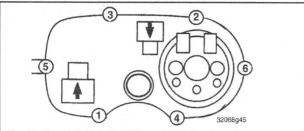

Fig. 45 Always tighten the VST cover in this order

FLOAT, NEEDLE AND SEAT

1. Follow Steps 1–7 of the Fuel Pump procedure.

2. Remove the float arm pin retaining clip and lift off the float and needle assembly.

3. Disassemble the float and needle assembly and then pull the seat from the cover.

To install:

4. Clean the components with carburetor cleaner. Inspect components for wear or damage and replace as necessary.

5. Press the valve seat into the VST cover so that it's fully seated.

6. Assemble the components and install the assembly, securing with the retainer clip.

7. Coat the threads of the cover mounting with Loctite and install the cover assembly. Make sure you tighten the cover bolts to 6 inch lbs. (8 Nm) as detailed in sequence illustrated.

8. Connect the pump electrical connectors, snap the connector covers into place and pop the electrical line into the clip.

9. Install the L-joint into the cover and tighten the mounting screw securely.

10. Connect the fuel lines and tighten the fitting nuts to 23 ft. lbs. (31 Nm).

11. Reconnect the harness connector and the battery cables. Without starting the engine, cycle the ignition switch ON and OFF four times to prime the fuel system. Start the engine and check for any fuel leaks.

DIAPHRAGM ASSEMBLY

1. Remove the VST cover and float assembly as detailed above.

2. Loosen and remove the diaphragm cover and pull off the cover, O-ring and diaphragm. Don't lose the spring under the diaphragm.

■ If your VST cover has an elbow on it, you'll find that the spring is between the cover and the diaphragm, please refer to the exploded view.

3. Clean everything with carburetor cleaner and inspect for wear or damage, replacing as necessary.

4. Install the components and tighten the cover retaining screws securely.

5. Install the float assembly and VTC cover.

Electronic Control Module (ECM)

REMOVAL & INSTALLATION ② *MODERATE*

◆ See Figures 46, 47 and 48

■ Static electricity can severely damage the ECM! Take great care when removing the module that you never touch the connector pins on either side of the casing.

1. Locate the ECM in the electrical box on the side of the engine and carefully disconnect the two harness leads (3 on the 8.1L) going into each side of the module.

2. Loosen the mounting screws and lift out the ECM. Clean only with a clean, lint-free dry cloth. Check all connector pins for straightness and corrosion.

3. Install the ECM and tighten the mounting screws securely.

4. Reconnect the electrical lead at each end; J1 is the front connector and J2 is the rear.

Engine Coolant Temperature Sensor (ECT)

REMOVAL & INSTALLATION ② *MODERATE*

◆ See Figures 49, 50 and 51

1. Disconnect the battery cables.

2. Tag and disconnect the electrical connector at the sensor and move it out of the way.

3. Unscrew the ECT from the thermostat housing or crossover tube and remove it. Clean the sensor with a dry cloth; make sure to remove any excess sealant from the threads.

4. Screw the sensor into the housing until it is hand tight and then turn it an additional 2-1/2 turns.

5. Connect the electrical lead. Connect the battery cables.

Idle Air Control Valve (IAC)

REMOVAL & INSTALLATION ② *MODERATE*

◆ See Figures 52, 53 and 54

Except 8.1L Engines

1. Disconnect the battery cables.

2. Remove the flame arrestor and the throttle body.

3. Tag and disconnect the electrical connector at the valve and move it out of the way.

4. Unscrew the mounting screws and remove the valve. Remove the O-ring from the valve and throw it away. Clean the valve mating surfaces, pintle valve seat and air passage with carburetor cleaner—do not push or pull on the valve pintle. Don't be concerned if you notice shiny spots on the pintle or seat.

■ If replacing the IAC valve, make sure that the new one has the same pintle shape and diameter.

5. Install a new O-ring onto the valve.

6. Install the valve and tighten the screws to 20 inch lbs. (2 Nm).

7. Install the throttle body.

8. Connect the electrical lead. Connect the battery cables. Turn the ignition key to the ON position for 10 seconds, turn the key to the OFF position for 10 seconds. Start the engine and check that the idle is OK.

Fig. 46 The ECM and the relays are all located in the electrical box

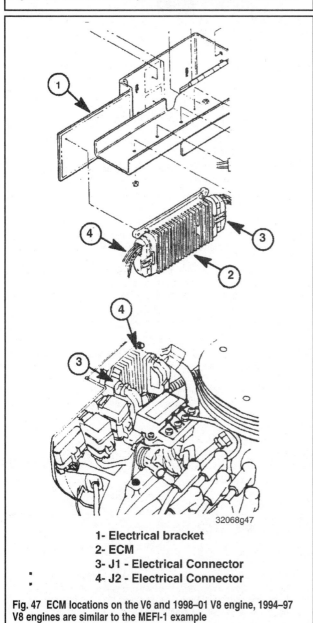

1- Electrical bracket
2- ECM
3- J1 - Electrical Connector
4- J2 - Electrical Connector

Fig. 47 ECM locations on the V6 and 1998–01 V8 engine, 1994–97 V8 engines are similar to the MEFI-1 example

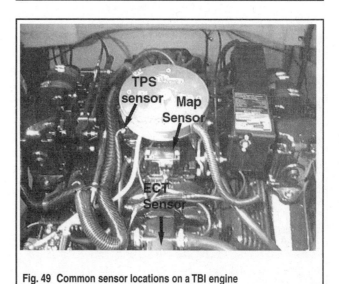

32068g48

1- Connector A
2- Connector B
3- Connector C
4- ECM

Fig. 48 8.1L engines will have three connectors

Fig. 49 Common sensor locations on a TBI engine

Fig. 50 Common sensor locations on an MP/MPI engine

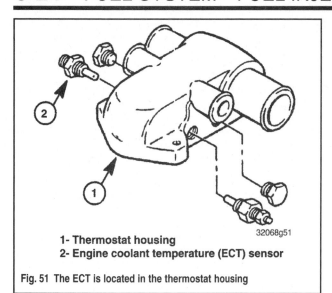

1- Thermostat housing
2- Engine coolant temperature (ECT) sensor

32068g51

Fig. 51 The ECT is located in the thermostat housing

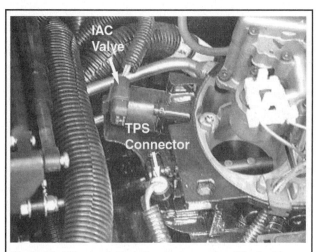

Fig. 52 The IAC valve is located on the lower edge of the throttle body (TBI)

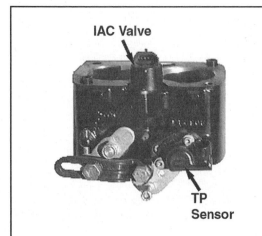

Fig. 53 IAC valve location on MP/MPI engines

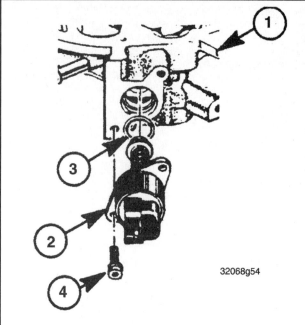

1- Throttle body
2- Idle air control (IAC) valve
3- O-Ring
4- Screws

32068g54

Fig. 54 Installing the IAC valve. Always use a new O-ring (models w/TBI shown; MPI similar)

8.1L Engines

1. Remove the engine hatch or cover and disconnect the battery cables.
2. Loosen the hose clamp and slide the IAC hose from the fitting on the throttle body.
3. Locate the valve at the back of the engine and disconnect the electrical lead. Unscrew the 2 hex bolts and lift out the valve.
4. Remove the gasket from the valve and throw it away. Clean the valve mating surfaces, pintle valve seat and air passage with carburetor cleaner— do not push or pull on the valve pintle. Don't be concerned if you notice shiny spots on the pintle or seat.

■ **If replacing the IAC valve, make sure that the new one has the same pintle shape and diameter.**

5. Position a new gasket onto the valve.
6. Install the valve and tighten the screws to 20 inch lbs. (2 Nm). Connect the electrical lead.
7. Reconnect the vacuum hose, connect the battery cables, start the engine and ensure proper idle.

Intake Air Temperature Sensor (IAT)

REMOVAL & INSTALLATION  MODERATE

TBI/MP/MPI w/MEFI-3 ECM Only

◆ **See Figure 55**

1. Remove the intake plenum as detailed earlier in this section.
2. Tag and disconnect the electrical lead and unscrew the sensor.
3. Clean the sensor with a dry cloth and inspect for signs of damage.
4. Install the sensor into the plenum and tighten it securely (no more than 2-1/2 turns).
5. Install the plenum and connect the electrical lead.

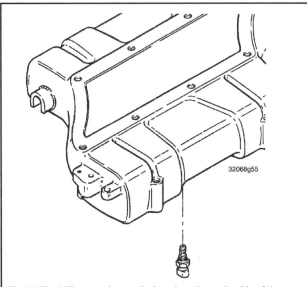

Fig. 55 The IAT sensor is usually found on the underside of the intake plenum

Knock Sensor (KS)

REMOVAL & INSTALLATION

◆ See Figures 50, 56 and 57

■ Many 1998–01 7.4L MPI engines will have two sensors, one on each side of the engine.

1. Disconnect the battery cables.
2. Tag and disconnect the electrical connector at the sensor and move it out of the way.
3. Unscrew the sensor from the Y-connector (exc. 1994-97 V8 and 1998-01 350 Mag MPI/6.2/7.4L/8.1L/8.2L MPI, which don't use a "Y") and remove the sensor. Clean the sensor with a dry cloth; especially the threads. If replacing the sensor, make sure the new one is the identical part number.
4. Install the sensor on the upper side of the Y-fitting on engines so equipped and tighten it to 12–16 ft. lbs. (16–22 Nm). Do not use any thread sealer or locking solution. Proper torque is imperative to ensure precise performance
5. Connect the electrical lead.

Manifold Absolute Pressure Sensor (MAP)

REMOVAL & INSTALLATION

◆ See Figures 49, 50, 58 and 59

■ On 1998–01 7.4L MPI engines the sensor is attached to the manifold itself. Remove the fuel rail as detailed previously and then pull out the MAP sensor.

1. Disconnect the battery cables.
2. Tag and disconnect the electrical connector at the sensor and move it out of the way.
3. Unscrew the mounting screw(s) and remove the sensor. Clean the sensor with a dry cloth; make sure that the sensor seal is in good condition, if not replace it.
4. Install the sensor seal and then install the sensor and tighten the screw to 44–62 inch lbs. (5–7 Nm).
5. Connect the electrical lead and the battery cables.

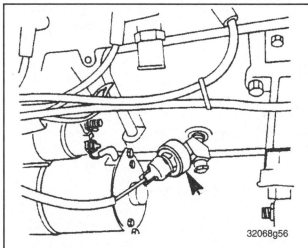

Fig. 56 The KS sensor is located on the engine block, just ahead of the starter motor on most engines. Some models use a "Y" fitting...

Fig. 57 ...while others do not

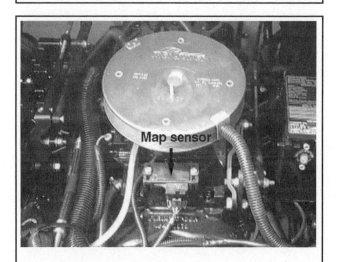

Fig. 58 The MAP sensor is located on the lower edge of the throttle body adapter plate on TBI engines...

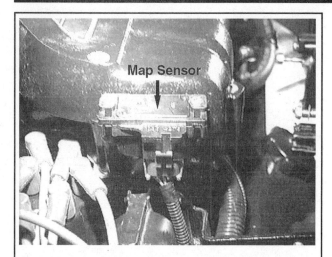

Fig. 59 ...while on MP/MPI engines it is attached to the intake plenum

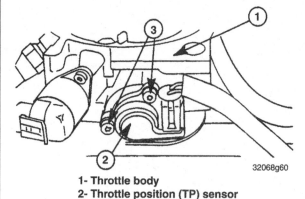

32068g60

1- Throttle body
2- Throttle position (TP) sensor
3- Screws

Fig. 60 The TP sensor is located on the lower edge of the throttle body (TBI shown)

Throttle Position Sensor (TPI)

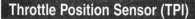  MODERATE

REMOVAL & INSTALLATION

◆ See Figures 49, 50, 52, 53 and 60

1. Disconnect the battery cables.
2. Remove the flame arrestor. Although not absolutely necessary, you may find it beneficial to remove the throttle body for better access to the sensor.
3. Tag and disconnect the electrical connector at the sensor and move it out of the way.

4. Unscrew the mounting screws and remove the sensor. Clean the sensor with a dry cloth; make sure that the sensor is in good condition (no wear, cracks or damage), if not replace it.

■ If replacing the sensor with a new one, make sure to use the two new screws that came with the package.

5. Install the sensor and tighten the screws to 20 inch lbs. (2 Nm). Coat the threads with Loctite 242 and don't forget the washers!
6. Connect the electrical lead. Start the engine and check the sensor output voltage. It should be approximately 0.7V at idle and 4.5V at wide open throttle.

SYSTEM DIAGNOSIS

◆ See Figure 61

Prior to performing any diagnostics on the EFI system, a diagnostic circuit check must first be completed. This is an organized approach to identifying a problem created by a malfunction in the electronic engine control management system and can only be performed with the correct scan tool—a Rinda Scan Tool or a Quicksilver Digital Diagnostic Tool. After hooking the tool up properly and finding that the on-board diagnostic system if functioning correctly and there are no codes displayed, the accompanying table should be used as a reference for what a normally functioning system should display. Use only the parameters listed in the chart.

If codes are displayed, move on to the next section.

PRECAUTIONS

◆ See Figures 62 and 63

Good shop practice, as well as good judgment requires all the following practices be observed.

1. The negative battery cable must be disconnected before any ECM system component is removed.
2. Always check to be sure the battery cables are securely connected, before starting the engine.
3. Never separate the battery from the on-board electrical system while the engine is operating.
4. Never separate the battery feed wire from the charging system while the engine is operating.
5. Disconnect the battery from the boat electrical system before starting to charge the battery.
6. Check to be sure all cable harnesses are securely connected and the battery terminals are clean and the cables securely connected.

Scan Position		Units	Data Value
BARO		Volts	3-5
Battery		Volts	12-14.5
Coolant Temperature		Deg. F/C	150-170 (66-77)
Engine Overtemp		OK/Overheat	OK
Fuel Consumption		GPH	1 to 2
Idle Air Control Valve			
	Follower	Counts (Steps)	0
	Minimum	Counts (Steps)	0-40
	Normal	Counts (Steps)	0-40
Injector			
	On Time Cranking	Msec	2.5-3.5
	Pulse Width	Msec	2-3
Knock Retard		Deg.	0
Lanyard Stop Mode		OFF/ON	OFF
Manifold Air Temperature		Deg. F/C	Varies w/ambient
MAP		Volts	1-3
Memory Calibration Check		Sum of Check	Varies
Oil Pressure			
	I/O	OK/LOW	OK
	Transmission	OK/LOW	OK
Rpm			
	Desired	Rpm	600-700
	Normal	Rpm	600-650
Spark Advance		Deg.	10-30
Throttle			
	Angle	0-100%	0-1
	Position	Volts	0.4-0.8

32068c61

Fig. 61 Normal specifications for a scan tool

7. Never connect or disconnect the wiring harness at the ECM when the ignition switch is in the on position.

8. Before any electric arc welding is attempted, disconnect the battery leads and the ECM connector/s.

9. Never direct a steam-cleaning nozzle at ECM components. Such action will cause damage or corrosion the component terminals.

10. Do not use any test equipment not specified in the diagnostic charts. Such equipment may give an incorrect reading and/or may actually damage good components.

11. A digital voltmeter with a rating of 10 megohms input impedance must be used when taking any voltage measurements.

12. A "low-amphere rated" test light must be used when a test light is specified for a test. Never use a high-amphere rated test light. Power the setup with the boat battery. If the ammeter indicates less than 3/10 amp current flow, the test light is safe to use.

A final word here. If a test light with 100 mA or less is used, a faint glow may show, when the test actually states "no light".

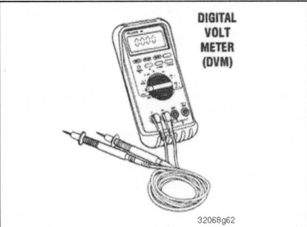

Fig. 62 A quality multi-meter (DVOM) is necessary when performing diagnostic tests…

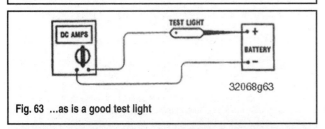

Fig. 63 …as is a good test light

General Diagnostic Tests

■ **Please refer to wiring schematics at the end of this section for correct pin and circuit locations.**

ON-BOARD DIAGNOSTIC SYSTEM CHECK

1. With the ignition switch in the ON position and the engine OFF, connect the tester to the diagnostic link connector (DLC) in the electrical box and switch it to the Normal mode. The malfunction indicator lamp (MIL) should come on, if not please refer to the following procedure for "No MIL or DLC Data".

2. If the tester flashes a Code 12, check circuit 451 for a short to ground. Otherwise, switch the tester to the Service mode and confirm that it flashes Code 12. If it doesn't, check the "MIL on Steady…" procedure following.

3. Once you've gotten your Code 12, switch the tester back to the Normal mode and start the engine. If the engine continues to run, move to

the next step; if it won't run, move to the "Engine Cranks But Won't Run" procedure.

4. Turn off the engine but leave the ignition switch in the ON position. Switch the tester back to Service and check for any stored codes. Refer to the DTC charts at the end of this section.

NO MIL OR NO DLC DATA

■ **Please refer to wiring schematics at the end of this section for correct pin and circuit locations.**

When a tester is connected to the DLC it should receive voltage via circuit 440 (terminal **F**) and be grounded through circuit 419 (**E**). There should always be a steady MIL when the ignition is ON (engine stopped). Obviously, always check for bad connections or frayed wires. It's also a good idea to check that the indicator bulb is not burned out.

TBI And 1998–01 MP/MPI Engines

1. Remove the CodeMate tester or scan tool and turn the ignition switch to ON (engine OFF). Connect a test light to ground and touch the probe to terminal **F** of the DLC. If the light goes on move to Step 2; if it does not come on, circuit 440 is open or has a short to ground.

2. Now connect the test light to the battery positive and touch the probe to terminal **E** of the DLC. If the light goes on, your tester is bad; if it does not come on, go to Step 3.

3. With the ignition OFF, disconnect the front and rear (J1 and J2) connectors at the ECM. Connect a multi-meter (DVOM) and measure the resistance between ECM circuit 419 and terminal **E** on the DLC. If less than an ohm, go to Step 4; if more than an ohm, circuit 419 is open.

4. If you have a scan tool, go to Step 5; if you don't have access to a scan tool, replace your tester.

5. Check the ECM/DLC fuse. If it's OK, go to Step 6; if its blown, you have a short to ground somewhere in the battery feed circuit.

6. With the ignition switch OFF and the ECM still disconnected, connect a test light to ground and touch the other probe to ECM circuit 440. If the light comes on, go to Step 7; if it doesn't, circuit 440 is open or shorted to ground.

7. Turn the ignition switch back ON and now touch the probe to ECM circuit 439. If the light comes on and your initial problem was no DLC data (while using the scan tool), circuit 461 is open or shorted to ground, or you've got a bad ground on the ECM; if the light doesn't come on, go to Step 8.

8. Check the ignition/injection fuses(s). If OK, perform the "EFI System/Ignition Relay Check" test following; if the fuse is blown, there is a short to ground in circuit 439.

1994–97 MP/MPI Engines

1. Remove the CodeMate tester or scan tool and turn the ignition switch to ON (engine OFF). Connect a test light to ground and touch the probe to terminal **F** of the DLC. If the light goes on, move to Step 2; if it does not come on, circuit 440 is open or has a short to ground.

2. Now connect the test light to the battery positive and touch the probe to terminal **E** of the DLC. If the light goes on, your tester is bad; if it does not come on, go to Step 3.

3. With the ignition OFF, disconnect the rear (J2) connector at the ECM. Connect a multi-meter (DVOM) and measure the resistance between ECM circuit 419 (J2-31) and terminal **E** on the DLC. If resistance is close to an ohm, check the system circuits for bad connections or frayed wires, otherwise replace the ECM; if not close to 1 ohm, check for a short in circuit 419.

MIL ON STEADY—WILL NOT FLASH DTC 12

■ **Please refer to wiring schematics at the end of this section for correct pin and circuit locations.**

When the CodeMate tester is installed, it receives voltage through circuit 440 (terminal **E**) and is grounded through circuit 419 (**F**). When the ignition is ON and the engine OFF there should always be a steady MIL. Obviously, always check for bad connections or frayed wires. It's also a good idea to check that the indicator bulb is not burned out.

TBI And 1998–01 MP/MPI Engines

1. With the ignition switch ON and the engine OFF, switch the CodeMate tester to Normal mode. If the MLC flashes DTC 12, circuit 451 has a short to ground; if no DTC 12, go to Step 2.

2. With the ignition switch still in the ON position, switch the tester to the Service mode. If the MIL flashes DTC 12, look for bad connections in the circuit; if DTC 12 does not flash, go to Step 3.

3. Turn the ignition OFF and disconnect the two connectors at the ECM. Turn the ignition back ON. If the MIL comes ON, circuit 419 has a short to ground; if the light does not come on, go to Step 4.

4. Turn the ignition OFF and connect a jumper wire between terminals **A** and **B** at the DLC. Connect a test light between circuit 451 and battery positive. If the light goes ON, you'll need to verify that the tester is working properly by hooking it up to a known good system—we know its unlikely that you have another engine available, so all we can suggest is running down to your local dealer and asking them to confirm its OK. If the light does not go on, circuits 450 and/or 451 are open.

1994–97 MP/MPI Engines

1. With the ignition switch still in the ON position, switch the tester to the Service mode. If the MIL flashes DTC 12, look for bad connections in the 461 circuit, otherwise replace the ECM; if DTC 12 does not flash, go to Step 2.

2. Turn the ignition OFF and disconnect the rear (J2) connector at the ECM. Turn the ignition back ON. If the MIL comes ON, circuit 419 has a short to ground; if the light does not come on, go to Step 3.

3. Turn the ignition OFF and connect a jumper wire between terminals **A** and **B** at the DLC. Connect a test light between circuit 451 (J1-7) and battery positive. If the light goes ON, replace the ECM. If the light does not go on, circuits 450 and/or 451 are open.

ENGINE CRANKS BUT WILL NOT RUN

■ **Please refer to wiring schematics at the end of this section for correct pin and circuit locations.**

The distributor ignition system and the fuel injector circuit are both supplied voltage via the EFI system relay. From the relay, circuit 902 delivers voltage to the injector/ECM fuse and to the coil. The following test assumes a properly functioning battery and adequate fuel supply. Is also a good idea to check all connections and wires before hand.

TBI And 1998–01 MP/MPI Engines

1. Ensure that the ignition has been in the OFF position for at least 10 seconds and then switch it ON. If the fuel pump runs for 2 seconds, go to Step 2; if not, move on to the "Fuel System Electrical Test" procedure following.

2. Check for secondary ignition spark. If adequate, go to Step 3; if not, move on to the "Ignition System Check" procedure following.

3. Install a fuel pressure gauge as per the manufacturer's instructions or as detailed in elsewhere in this manual. Turn the ignition OFF for 10 seconds and then turn the ignition ON and let the pump run for about 2 seconds. Check the fuel pressure while the pump is running. If above 25 psi, go to Step 4; if below 25 psi, move on to the "Fuel System Diagnosis" procedure.

4. Turn the ignition OFF and disconnect the two ECM connectors. Connect a multi-meter (DVOM) and measure the resistance between ECM circuits 467 and 468. If more than an ohm, go to Step 5; if less than an ohm, circuits 467 or 468 have a short to ground.

5. Check the resistance across each injector. If higher than an ohm, go to Step 6; if less than 1 ohm, look for a short to ground at the injector, if none replace the injector.

6. Reconnect the injectors, turn the ignition OFF and disconnect the ECM. Connect a test light to ground and touch the probes to circuits 467 and 468. If the light comes on, go to Step 7; if it doesn't, circuits 467 or 468 have an open connection.

7. Disconnect the injectors and turn the ignition ON. Connect a test light to ground and touch the probe to circuits 467 and 468 on the ECM side of the injector harness. Test the harness on each side of the engine, if the light goes on you have a short.

1994–97 MP/MPI Engines

Before performing this test, check that the lanyard stop switch is not activated and that the TP and ECT sensors are working properly.

1. Move the lanyard stop switch to the RUN position. If OK, go to Step 2; if not, go to Step 6.

2. Check for spark at the plug wires while cranking the engine. If you have spark, go to Step 3; if no spark, refer to the Ignition System Checks detailed in Section 10.

3. Disconnect all injectors. Connect a test light and touch the probe to the injector harness connector (one from each bank); crank the engine. If the light blinks, go to Step 4; if it doesn't blink, go to the Lanyard Stop Circuit check later in this section.

4. Connect a multi-meter (DVOM) and check the resistance across each injector. If resistance is 12 ohms (+/- 0.4), go to Step 5; if not 12 ohms, replace any injector not to specification.

5. Turn the ignition OFF and connect a fuel pressure gauge. Check the pressure while cranking the engine. If at least 36 psi, check for bad plugs, water or contamination in the fuel system or a bad MAP sensor; if pressure is lower, go to the Fuel System Diagnosis test later in this section.

6. Check for spark at the plug wires while cranking the engine. If you have spark, go to Step 7; if no spark, check that there is power to the ignition system, if there is voltage, refer to the Ignition System Checks detailed in Section 10.

7. With the ignition OFF, disconnect the 4-way connector at the distributor. Connect a scan tool if available, if not, connect a tachometer. Connect a test light to battery positive, turn the ignition ON and touch the probe to the harness connector terminal **C**. If the scan tool or tachometer indicate any rpm, check for a bad connection, if none, replace the ignition module; if no rpm is indicated, circuit 430 may be open or shorted to ground, if not, replace the ECM.

FUEL SYSTEM DIAGNOSIS

■ **Please refer to wiring schematics at the end of this section for correct pin and circuit locations.**

TBI And 1998–01 MP/MPI Engines

1. Install a fuel pressure gauge. Turn the ignition switch OFF and then ON. The fuel pump should run for about 2 seconds. If the fuel pressure is less than 25 psi, go to Step 2; if higher than 25 psi, go to Step 4.

2. Try and start the engine. If it starts, go to Step 3; if it won't start, go to Step 5.

3. With the engine still idling, connect a vacuum gauge to the pressure regulator and apply 10 in. of vacuum. If the pressure drops by about 5 psi, the fuel supply is probably restricted, check the pump and lines; if the pressure doesn't drop to the above spec, replace the fuel pressure regulator.

4. Was the any fuel pressure at all? If there was, go to Step 5; if no pressure, move to the "Fuel System Electrical Test" procedure.

5. Did the system register pressure and then drop off to no pressure? If it does, go to Step 6; if it doesn't, check for a restriction in the fuel lines.

6. Turn the ignition OFF, disconnect the fuel line between the pump and the throttle body or fuel rail, and plug it. Turn the ignition On and check to see if pressure holds. If it does, replace the fuel pressure regulator; if it doesn't hold, check the pump, fuel lines and inlet filter for leaks. Check that the battery isn't low. If everything is OK, replace the fuel pump.

1994–97 MP/MPI Engines

1. Install a fuel pressure gauge. Turn the ignition switch OFF for 10 seconds and then ON. The fuel pump should run for about 2 seconds. Check the pressure while the pump is running and then after it stops; it should hold steady after the pump stops. If the fuel pressure is 34–38 psi, go to Step 2; if not within specification, go to Step 4.

2. Start the engine and allow it to idle. If the fuel pressure is lower than the reading in Step 1 by 3–10 psi, the system is OK; if not, go to Step 3.

3. With the engine still idling, connect a vacuum gauge to the pressure regulator and apply 10 in. of vacuum to the pressure regulator. If the pressure drops by 3–10 psi, you have a vacuum leak at the regulator; if it doesn't drop to the above spec, replace the fuel pressure regulator.

4. Was the any fuel pressure at all? If there was pressure while the pump was energized, but it didn't hold, go to Step 5. If pressure was below

34 psi, go to Step 8. If above 38 psi, go to Step 10. If there was no pressure at all, go to Step 7.

5. Turn the ignition OFF for 10 seconds and then turn it back ON. Block the pressure fuel line with a special shut-off valve. If the pressure holds, check for fuel system leaks or a bad VST; if it doesn't hold, go to Step 6.

6. Turn the ignition OFF for 10 seconds and then turn it back ON. Block the return fuel line with a special shut-off valve. If the pressure holds, replace the pressure regulator; if it doesn't hold, check for a bad or leaking injector.

7. Turn the ignition OFF and apply 12 volts to the fuel pump connector (gray wire, terminal **A**). If the pump runs, check for a clogged fuel line, otherwise replace the pump; if it doesn't run, check for an open in circuit 120 or a bad pump ground, otherwise replace the pump.

8. Check for clogged or restricted fuel lines, If they are OK, go to Step 9.

9. Turn the ignition OFF and block the return line with a shut off valve. Turn the ignition ON. If the pressure goes above 38 psi, replace the pressure regulator; if there is pressure but its less then 34 psi, replace the fuel pump.

10. Disconnect the fuel return line. Connect a flexible plastic hose to the end of the line and insert the other end in a small container of gasoline. Turn the ignition ON and check the pressure after 2 seconds. If it's above 38 psi, check for restriction between the regulator and the point where the return line was disconnected, if none, replace the regulator. If the pressure is 34–38 psi, check for a restriction in the line between the disconnect and the VST.

FUEL SYSTEM ELECTRICAL TEST

■ **Please refer to wiring schematics at the end of this section for correct pin and circuit locations.**

The fuel system receives voltage via a system relay on circuit 902. It is protected by an inline 15 amp fuse that then sends the voltage on to the pump relay via circuit 339 (terminal 30). The ECM turns the pump relay on via circuit 465; it will remain on as long as the engine is cranking or running, and the ECM continues to receive a reference pulse. If no pulse is received, the ECM will de-energize the pump within 2 seconds after the ignition switch is turned ON or the engine is stopped.

Always check for bad connections and frayed wires. The following test also assumes an uncontaminated fuel supply.

1. Turn the ignition OFF, remove the fuel pump relay and then turn the ignition back ON. Connect a test light to ground and touch the probe to the pump relay harness terminal 30. If the light goes on, move to Step 2; if it doesn't, go to Step 5.

2. Turn the ignition OFF again and connect a fused jumper wire between terminals 30 and 87 on the pump relay connector. Turn the ignition ON. If the fuel pump runs, go to Step 3; if it doesn't energize, check that circuits 120 or 150 may be open. If they're OK, replace the fuel pump.

3. Turn the ignition off and disconnect the jumper wire. Connect a test light to battery positive and touch the probe to terminal 86 on the pump relay connector. If the light comes on, go to Step 4; if not, circuit 450 is open.

4. Now touch the probe to terminal 85 and turn the ignition ON. If the light goes on for 2 seconds and then goes out, check the fuel system for a blocked filter, vapor lock, plugged or disconnected lines or hoses, and make sure there is fuel in the tank; if it doesn't, you've either got a bad connection at the ECM J2 terminal 9 or circuit 465 is open.

5. Check the fuse between the system relay and pump relay. If it's OK, circuits 339 or 902 are open; if blown, look for a short to ground in circuit 339 or 120. Also check for contaminated fuel. If everything looks good, replace the fuel pump. And also the bad fuse!

EFI SYSTEM/IGNITION RELAY CHECK

■ **Please refer to wiring schematics at the end of this section for correct pin and circuit locations.**

The battery supplies voltage to terminal 30 of the system relay. When the ignition switch is turned on, battery voltage is also supplied, through the

ignition, to terminal 86 on the system relay. The pull-in coil is then energized which creates a magnetic field and then closes the system relay contacts. Voltage and current are then supplies to the ignition control module, injectors, ECM and pump relay via circuit 902.

Always check for bad connections and frayed wires. The following test also assumes an uncontaminated fuel supply.

1. Turn the ignition OFF and remove the EFI system relay connector. Turn the ignition ON, connect a test light to ground and touch the probe to terminals **30** and **86** on the relay harness connector. If the light goes on at both terminals, move to the next step; if it doesn't, check for an open or short to ground on the circuit where it didn't light (circuit 2 or 3).

3. Now connect the test light to battery positive and touch the probe to terminal **85**. If it lights, check the relay connector for a poor contact otherwise replace the relay; if it doesn't light, circuit 150 is open to ground.

IGNITION SYSTEM CHECK

Please refer to Section 10 for complete testing on the ignition system.

IDLE AIR CONTROL VALVE FUNCTIONAL TEST

■ **Please refer to wiring schematics at the end of this section for correct pin and circuit locations.**

The ECM controls idle speed to a pre-set, desired rpm based on input from sensors and the actual engine rpm. Four separate circuits are used to move the IAC valve; said movement varying the amount of air allowed to bypass the throttle plates. Idle speed then is controlled, via the ECM, by the position of the IAC valve.

1. With the engine at normal operating temperature, turn it off, connect a tachometer and then restart it and allow the idle to stabilize. Record the idle speed and turn off the engine for 10 seconds. Disconnect the IAC harness connector, restart the engine and check the rpm. If the second recorded idle speed is higher than the first by 200 rpm or more, go to Step 2; if not, go to Step 3.

2. Reconnect the IAC harness. If the idle speed returns to within 75 rpm of the originally recorded speed in Step 1 within 30 seconds, the IAC circuit is functioning properly; if it does not, go to the next step.

3. Turn off the ignition for 10 seconds. Disconnect the IAC again and restart the engine. Connect a test light to ground and then touch the probe to each of the four IAC harness terminals. If the light blinks on each terminal, check for a bad IAC connection somewhere in the circuit, otherwise replace the valve; if it doesn't blink, look for an open or shorted circuit on the terminal(s) that didn't blink. Also, check for any bad connections at the ECM.

LANYARD STOP CIRCUIT CHECK

This circuit is a safety feature that ensures that the engine will stop in the event that the operator is removed from a safe control position during normal vessel operation. This is an open switch that is physically connected to the operator by a tether—if the tether is pulled out, the switch closes, signaling the ECM to shut down the engine. The following test assumes that the engine will crank, but not start.

1. Is the switch in its normal position? If it is, go to Step 2; if not, move it to the correct position and attempt to start the engine again.

2. Turn the ignition OFF and disconnect the harness connector. Connect a multi-meter (DVOM) and measure the resistance between circuit 942 (pin A) and ground (pin B) on the connector. Now measure it on the harness side. Resistance should be less than 5 ohms on the connector side and infinite on the harness side. If it is, refer to the Ignition System checks in Section 10; if not within specifications, go to Step 3.

3. If resistance was lower on the ECM side of the connector, go to Step 4; if lower on the switch side of the connector, check for a bad switch. If OK, look for a short to ground in circuit 942.

4. Disconnect the front (J1) connector at the ECM. Measure resistance between pin J1-21 and ground. If infinite, replace the ECM; if not, check for a short to ground in circuit 942.

Diagnostic Trouble Codes (DTC)

◆ See Figures 64, 65 and 66

Operational problems during everyday engine operation are recognized by the ECM and a diagnostic trouble code is created to identify the particular problem. The ECM will retain this code, or a combination thereof, in its memory until such a time as you access it, read it and then clear it. If the reason for the code has not been repaired prior to clearing the code it will reset itself again shortly after the engine begins to operate.

Remember that DTCs do not necessarily indicate the specific problem or component, merely the area from which the problem is originating. Code charts for your particular engine can be found at the end of this section; while fault and function tests for each code are also found following the section on reading and clearing codes. You will find that more times than not, the source of the problem is a bad connection or frayed wire—particularly where intermittent codes are seen. Follow the steps in the tests carefully and you will find that diagnostics is not always as difficult as you may have thought.

Wiring schematics for all individual components or systems can be found with the fault/function test procedures. Complete system schematics and ECM symptoms charts can be found at the end of this section, while ECM connector pin locations are detailed here; remember there are two ECM harness connectors, J1 is the front connector and J2 is at the rear.

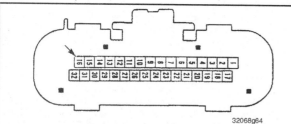

32068g64

Fig. 64 ECM connector pin locations—please remember that not all pins are used on all engines so make sure you refer to the pin location charts at the end of this section

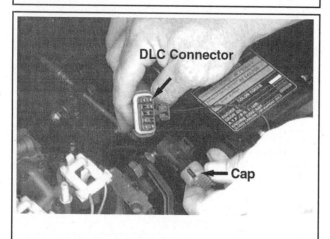

Fig. 65 Data Link Connector (DLC)...

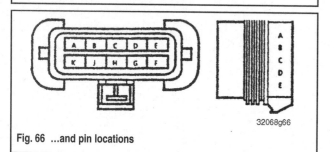

32068g66

Fig. 66 ...and pin locations

READING CODES

All trouble codes stored in the ECM are displayed by means of a series of flashes and pauses; the number of flashes represents the number of the code, with the first and second digits being separated by a short pause. Each code will be flashed three times with a long pause separating the repeat of the code each time. For instance, DTC 12 would be represented as "flash, pause, flash-flash, long pause; and then it would repeat this cycle two more times. Count the number of flashes you observe and determine your DTC. Diagnostic trouble codes can be pulled with either a Scan tool or a CodeMate tester. When using a scan tool, simply follow the manufacturer's instructions. If you are using a code tool:

1. With the ignition switch in the ON position and the engine OFF, connect the tester to the diagnostic link connector (DLC) in the electrical box and switch it to the Normal mode. The malfunction indicator lamp (MIL) should come on and flash a Code 12 (3 times), indicating the diagnostic system is operating properly. If Code 12 is not displayed, please refer to the following procedure for "No MIL or DLC Data".

2. Shortly after displaying Code 12, the system will flash any stored codes from the system; three times each and in ascending order if more than one code has been stored. If Code 12 continues being displayed then you have no stored codes (or an intermittent one).

CLEARING CODES

Once again, if using a scan tool, follow the tool manufacturer's instructions. If using a CodeMate tester:

1. Ensure that the battery is fully charged.
2. Connect the tester to the DLC in the electrical box and switch the ignition key to the ON position.
3. Select the Service mode on the tester and then move the throttle through its full range, from idle to full throttle and then back to idle.
4. Switch out of Service mode on the tester, start the engine and let it run for fifteen seconds.
5. Turn the ignition switch to OFF for five seconds, switch the tester back to Service mode and then turn the ignition switch ON again.
6. There should no longer be any coded present. If there are, check the battery again and then perform the procedure one more time. If codes are still apparent at the end of the second go-around (we're assuming that you fixed the code's problem before attempting to clear it!), refer to the appropriate troubleshooting or diagnostic charts. If the battery is not at full charge, you should hear the audio warning buzzer come on after engine start-up.

CODE 14

Engine Coolant Temperature Sensor (ECT)

◆ See Figures 67 and 68

The ECT sensor utilizes a thermistor to control the signal voltage being sent to the ECM. The ECM then applies specified voltage back through the 410 circuit to the sensor. When the coolant is cold, the sensor resistance is high. As the coolant warms, resistance lessens and voltage drops.

TBI AND 1998–01 V6 AND 5.7L/6.2L V8 MP/MPI ENGINES W/SCAN TOOL

1. Hook up the scan tool as per the manufacturer's instructions.
2. Turn the ignition ON. If the tool displays a coolant temperature higher than 266° F (130° C), turn OFF the ignition and disconnect the ECT harness. Turn the ignition back ON. If the tool displays a temperature less than –22° F (-30° C), look for a bad electrical connection or replace the sensor. If the temperature shown is not less than the above, the 410 circuit lead has a short and it must be found and repaired.
3. If the temperature shown in Step 2 is not higher than 266° F, turn off the ignition and disconnect the ECT harness. Connect terminals A and B with a jumper wire and then turn the ignition back ON. If the tool displays a temperature higher than –22° F (-30° C), look for a bad electrical connection or replace the sensor. If the temperature shown is less than the above, the 410 circuit lead or the sensor has a short and it must be found and repaired.

TBI AND 1998–01 V6 AND 5.7L/6.2L V8 MP/MPI ENGINES W/O SCAN TOOL

1. With the ignition OFF, disconnect the ECT sensor harness connector. Turn the ignition ON (but don't start the engine) and then connect a multi-meter (DVOM) across the **A** and **B** sensor harness terminals.

2. If the voltage is above 4 volts, look for a bad electrical connection or replace the sensor.

3. If the voltage is below 4 volts, connect the positive lead on the multi-meter to terminal B and the negative lead to a good ground.

4. If voltage is above 4 volts, you've got a bad ground somewhere on the 814 circuit (terminal A) or a bad connection at the ECM.

5. If the voltage is not above 4 volts, remove the multi-meter and turn the ignition ON. Connect a test light to the positive battery terminal and then touch terminal B. If the light does not go ON, check for an open connection in the 410 circuit or a bad connection at the ECM.

6. If the light goes ON, disconnect the J-1 connector at the ECM on MEFI-1/MEFI-2 engines or the J-2 connector on MEFI-3 engines. If the light goes ON, you've got a bad ground in the 410 circuit. If it does not go ON, you've either got a bad ground in the 410 circuit or it's shorted to sensor ground.

7.4L/8.1L/8.2L MP/MPI ENGINES W/SCAN TOOL

1. Hook up the scan tool as per the manufacturer's instructions.

2. Turn the ignition ON. If the tool displays a coolant temperature higher than 266º F (130º C), go to Step 3; if less than
–22º F (-30º C), go to Step 4.

3. Turn off the ignition and disconnect the ECT harness. Turn the ignition ON. If the tool displays a temperature higher than –22º F
(-30º C), the 410 circuit has a short to ground, if not, replace the ECM. If the temperature shown is less than the above, replace the sensor.

4. Turn off the ignition and disconnect the ECT harness. Connect terminals A and **B** with a jumper wire and then turn the ignition back ON. If the tool displays a temperature higher than 266º F (130º C), look for a bad electrical connection or replace the sensor. If the temperature shown is less than the above, the 410 circuit is open or the sensor has a ground open and it must be found and repaired. Otherwise, replace the ECM.

7.4L/8.1L/8.2L MP/MPI ENGINES W/O SCAN TOOL

1. Turn the ignition OFF and disconnect the ECT sensor harness. Connect a multi-meter (DVOM) and turn the ignition ON. If voltage across the **A** and **B** terminals is greater than 4 volts, check for bad connections or frayed wires. If less than 4 volts, go to the next step.

2. Connect the positive meter lead to the harness terminal **B** (circuit 814) and the negative lead to a good engine ground. If voltage is over 4 volts, check circuit 814 for a bad ground or a bad connection at the ECM; if OK, replace the ECM. If less than 4 volts, go to the next step.

3. Remove the multi-meter and connect a test light to the battery positive. Touch the probe to harness terminal **B** (circuit 410). If the light goes on, move to the next step. If it doesn't go on, check for an open on circuit 410 or a bad connection at the ECM. Otherwise replace the ECM.

4. With the test light still connected, disconnect the front (J1) harness at the ECM. If the light goes on, circuit 410 is shorted to ground. If it doesn't go on, circuit 410 is shorted to the sensor ground or the ECM will require replacement.

CODE 15

Engine Coolant Temperature Sensor (ECT)

◆ **See Figures 67 and 68**

The ECT sensor utilizes a thermistor to control the signal voltage being sent to the ECM. The ECM then applies specified voltage back through the 410 circuit to the sensor. When the coolant is cold, the sensor resistance is high. As the coolant warms, resistance lessens and voltage drops.

TBI AND 1998–01 MP/MPI ENGINES W/SCAN TOOL

1. Hook up the scan tool as per the manufacturer's instructions.

2. Turn the ignition ON. If the tool displays a coolant temperature higher than 266º F (130º C), turn OFF the ignition and disconnect the ECT harness. Turn the ignition back ON. If the tool displays a temperature less than –22º F (-30º C), look for a bad electrical connection or replace the sensor. If the temperature shown is not less than the above, the 410 circuit lead has a short and it must be found and repaired.

3. If the temperature shown in Step 2 is not higher than 266º F, turn off the ignition and disconnect the ECT harness. Turn the ignition ON and connect a multi-meter (DVOM) across terminals **A** and **B**. If the reading is above 4 volts, look for a bad electrical connection or replace the sensor. If the reading is below 4 volts, the 410 circuit lead has a short and it must be found and repaired.

TBI AND 1998–01 MP/MPI ENGINES W/O SCAN TOOL

1. With the ignition OFF, disconnect the ECT sensor harness connector. Turn the ignition ON (but don't start the engine) and then connect a multi-meter (DVOM) across the **A** and **B** terminals at the sensor.

2. If the voltage is above 4 volts, look for a bad electrical connection or replace the sensor.

3. If the reading is below 4 volts, the 410 circuit lead has a short and it must be found and repaired.

CODE 21

Throttle Position Sensor (TP)

◆ **See Figure 69**

The TP sensor provides voltage signal to the ECM that varies with the opening and closing of the throttles. Signal voltage will vary from approximately 0.5 volts at idle to slightly more than 4 volts at wide open throttle.

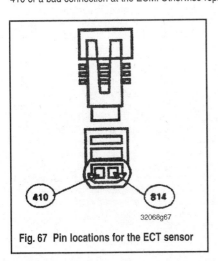

Fig. 67 Pin locations for the ECT sensor

°F	°C	Ohms
210	100	185
160	70	450
100	38	1800
70	20	3400
40	4	7500
20	-7	13500
0	-18	25000
-40	-40	100700

Temp.-to-Resistance Values

32068g68

Fig. 68 ECT sensor values chart

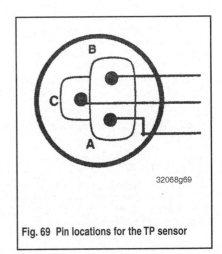

32068g69

Fig. 69 Pin locations for the TP sensor

TBI AND 1998–01 V6 AND 5.7L/6.2L V8 MP/MPI ENGINES W/SCAN TOOL

1. Connect a scan tool as per the manufacturer's instructions.
2. With the throttle closed and the ignition switch ON, check the reading on the tool. If over 4 volts go to Step 4; if under 0.3 volts, go to Step 3.
3. Turn the ignition OFF, disconnect the TP sensor electrical lead and run a jumper wire between harness terminals **A** and **C**. Turn the ignition ON. If the tool reads higher than 4 volts, replace the TP sensor; if it reads less than 4 volts, go to Step 5.
4. Turn the ignition OFF, disconnect the TP sensor connector and then turn the ignition back to ON. If the tool reads over 4 volts, 417 circuit is shorted to voltage; if the tool reads less than 4 volts, go to Step 5.
5. Turn the ignition switch OFF, disconnect the TP sensor harness connector and then turn the ignition back to ON. Connect a multi-meter (DVOM) between harness terminal **A** and **B**. If the reading is over 4 volts, go to Step 6; if the reading is less than 4 volts, go to Step 9.
6. Connect the multi-meter between harness terminals **A** and **C**. If the reading is higher than 4 volts, go to Step 7; if the reading is less than 4 volts, go to Step 8.
7. Turn the ignition switch OFF. Connect a test light to the positive battery terminal. Touch the harness terminal **C**. If the light goes ON, go to Step 10; if the light does not come on, replace the TP sensor.
8. Connect the multi-meter between harness terminal **C** and an engine ground. If the reading is higher than 4 volts, check 417 circuit for a short to voltage; if the reading is less than 4 volts, 417 circuit is open or you have a bad connection at the ECM.
9. Connect the multi-meter between harness terminal **A** and an engine ground. If the reading is higher than 4 volts, 813 circuit is open or you have a bad connection at the ECM; if the reading is less than 4 volts, 416 circuit is open or shorted to ground, or you have a bad connection at the ECM.
10. Disconnect the ECM and touch the test light to harness terminal **C**. If the light comes ON, 417 circuit is shorted to ground; if the light doesn't go ON, 417 circuit is open or you have a bad connection at the ECM.

1994–97 7.4L/8.2L AND 1998–01 454/502 Mag/8.2L MP/MPI ENGINES W/SCAN TOOL

1. Connect a scan tool as per the manufacturer's instructions.
2. With the throttle closed and the ignition switch ON, check the reading on the tool is over 4 volts or less than 0.36 volts. If so, go to Step 3 if the voltage was less than 0.36, or Step 4 if it was above 4 volts; if not, check the system circuit for bad connections or frayed wires.
3. Turn the ignition OFF, disconnect the TP sensor electrical lead and run a jumper wire between harness terminals **A** and **C**. Turn the ignition ON. If the tool reads higher than 4 volts, replace the TP sensor; if it reads less than 4 volts, go to Step 7.
4. Turn the ignition OFF, disconnect the TP sensor connector and then turn the ignition back to ON. If the tool reads over 4 volts, 417 circuit is shorted to voltage or you have a bad connection at the ECM; if both are OK, replace the ECM. if the tool reads less than 4 volts, go to Step 6.
5. Turn the ignition switch OFF, disconnect the TP sensor harness connector and then turn the ignition back to ON. Connect a multi-meter (DVOM) between harness terminal **A** and **B**. If the reading is over 4 volts, go to Step 6; if the reading is less than 4 volts, go to Step 9.
6. Connect the multi-meter between harness connector terminals **A** and **C**. If the reading is higher than 4 volts, replace the sensor. If the reading is less than 4 volts, check circuit 813 for an open sensor ground or check for bad ECM connections; if both are OK, replace the ECM.
7. Connect the multi-meter between harness terminal **A** and a good ground. Turn the ignition ON. If the reading is higher than 4 volts, 416 circuit is open or you have a bad connection at the ECM; if both are OK, replace the ECM. If the reading is less than 4 volts, 417 circuit is open or shorted to ground, or you have a bad connection at the ECM; if both are OK, replace the ECM.

1998–01 7.4L/8.1L MP/MPI ENGINES W/SCAN TOOL

1. Connect a scan tool as per the manufacturer's instructions.
2. With the throttle closed and the ignition switch OFF, check the reading on the tool is over 4 volts. If so, go to Step 3; if not, check the system circuit for bad connections or frayed wires.

3. Turn the ignition OFF, disconnect the TP sensor connector and then turn the ignition back to ON. If the tool reads less than 0.3 volts, go to Step 4; if over 0.3 volts, 417 circuit is shorted to voltage or you have a bad connection at the ECM; if both are OK, replace the ECM.
4. Connect the multi-meter between harness connector terminals **A** and **C**. If the reading is less than 4 volts, replace the sensor. If the reading is higher than 4 volts, check circuit 813 for an open sensor ground or check for bad ECM connections; if both are OK, replace the ECM.

TBI AND 1998–01 V6 AND 5.7L/6.2L V8 MP/MPI ENGINES W/O SCAN TOOL

1. With the ignition switch OFF, disconnect the TP sensor harness connector and then turn the ignition ON. Connect a multi-meter (DVOM) between the **A** and **B** harness terminals.
2. If the reading is over 4 volts, connect the multi-meter between harness terminals **A** and **C**.
3. If the reading is over 4 volts, turn the ignition OFF and connect a test light to the positive battery cable. Touch the probe to the **C** harness terminal. If the light doesn't come ON, replace the TP sensor. If the light comes on, disconnect the ECM and touch the harness terminal **C** with the probe. If the light comes ON, 417 circuit is shorted to ground; if the light doesn't come ON, 417 is an open circuit or you have a bad connection at the ECM.
4. If the reading in Step 3 is under 4 volts, connect the multi-meter between terminal **C** and engine ground. If the reading is over 4 volts, 417 circuit is shorted to voltage; if the reading is less than 4 volts, 417 is an open circuit or you have a bad connection at the ECM.
5. If the reading in Step 2 is less than 4 volts, connect a multi-meter between harness connector **A** and a good engine ground. If the reading is over 4 volts, 813 circuit is open or you have a bad connection at the ECM; if less than 4 volts, 416 circuit is open or shorted to ground, or you have a bad connection at the ECM.

1994–97 7.4L/8.2L AND 1998–01 454/502 Mag/8.2L MP/MPI ENGINES W/O SCAN TOOL

1. With the ignition switch OFF, disconnect the TP sensor harness connector and then turn the ignition ON. Connect a multi-meter (DVOM) between the **A** and **B** harness terminals. If the reading is over 4 volts, go to Step 3; if under 4 volts, go to Step 2.
2. Connect the meter between harness terminal **A** and an engine ground. If it shows over 4 volts, circuit 813 is open or you have bad connection at the ECM; if both are OK, replace the ECM. If under 4 volts, circuit 416 is either shorted to ground or open, or there is a bad connection at the ECM; if both are OK, replace the ECM.
3. Connect the meter between harness terminals **A** and **C**. If the reading is over 4 volts, go to Step 4; if under 4 volts, go to Step 6.
4. Turn the ignition OFF and connect a test light to the battery positive. Touch the probe to harness terminal **C**. If the light goes on, move to Step 5; if not, replace the TP sensor.
5. With the test light still connected to the battery, disconnect both ECM connectors and touch the probe to **C** again. If it lights, check for a short to ground in circuit 417, if it doesn't, replace the ECM.
6. Connect a multi-meter (DVOM) between harness terminal **C** and an engine ground. If voltage is over 4 volts, circuit 417 is shorted to voltage. If not over 4 volts, check for an open at circuit 417 or bad connections at the ECM. If both are OK, replace the ECM.

1998–01 7.4L/8.1L MP/MPI ENGINES W/O SCAN TOOL

1. With the ignition OFF, connect a code tester and switch it to the Normal mode. Disconnect the TP sensor harness, start the engine and allow it to idle for 2 minutes or until the tool indicates a stored tool. Turn the ignition ON, switch the tool to the Service mode and check the code. If DTC 22 comes up, check the system for bad connections or frayed wires, if everything is OK replace the sensor. If no DTC 22, go to the next step.
2. With the sensor harness disconnected and the ignition ON. Connect a multi-meter (DVOM) and check voltage across harness terminals **A** and **B**. If over 4 volts, check circuit 417 for a short to voltage, if nothing is found, look for a bad connection at the ECM, otherwise replace the ECM. If not over 4 volts, check circuit 813 for an open. If none is found, check the ECM connections and then replace the ECM.

CODE 22

Throttle Position Sensor (TP)

◆ **See Figure 70**

The TP sensor provides voltage signal to the ECM that varies with the opening and closing of the throttles. Signal voltage will vary from approximately 0.5 volts at idle to slightly more than 4 volts at wide open throttle.

TBI AND 1998–01 MP/MPI ENGINES (EXC. 7.4L) W/SCAN TOOL

1. Connect a scan tool as per the manufacturer's instructions.
2. With the throttle closed and the ignition switch ON, check the reading on the tool. If less than 0.3 volts, go to Step 3.
3. With the ignition OFF, disconnect the TP sensor harness connector and connect a jumper wire between terminal **A** and **B**. Turn the ignition ON. If the reading is greater than 4 volts, check for bad connections or replace the TP sensor; if less than 4 volts, go to Step 4.
4. With the ignition OFF, connect a multi-meter between terminal **A** and a good ground. If the tool indicates sensor voltage greater than the specified value, 417 circuit has is open or shorted to ground, or you have a bad connection at the ECM; if less than the value, 416 circuit has is open or shorted to ground, or you have a bad connection at the ECM.

1998–01 7.4L/8.1L MPI W/SCAN TOOL

1. Connect a scan tool as per the manufacturer's instructions.
2. With the throttle closed and the ignition switch OFF, check the reading on the tool. If less than 0.36 volts, go to Step 3. If higher then 0.36 volts, check each system for bad connections or frayed wires.
3. With the ignition OFF, disconnect the TP sensor harness connector and connect a jumper wire between harness terminals **A** and **B**. Turn the ignition ON. If the reading is greater than 4 volts, check for bad connections or replace the TP sensor; if less than 4 volts, go to Step 4.
4. With the ignition OFF, connect a multi-meter (DVOM) between terminal **A** and a good ground. If the tool indicates sensor voltage greater than 4 volts, 417 circuit has is open or shorted to ground, or you have a bad connection at the ECM; if everything is OK, replace the ECM. If less than 4 volts, 416 circuit has is open or shorted to ground, or you have a bad connection at the ECM. Also check circuit 416E to the MAP sensor for a short to ground. If everything is OK, replace the ECM

TBI AND 1998–01 MP/MPI ENGINES W/O SCAN TOOL

1. Connect a CodeMate Tester and switch it to the Normal mode.
2. Turn the ignition switch OFF and disconnect the TP sensor harness connector.
3. Connect a jumper wire between terminals **A** and **C**. Start the engine and allow it to idle for two minutes or until the tester indicates a stored code.
4. Turn the engine OFF, but leave the ignition switch in the ON position. Switch the tester to the Service mode and note the trouble code. If DTC 21, check all connections or replace the TP sensor; if no DTC 21, 417 circuit is open or shorted to ground; if everything is OK, you have a bad connection at the ECM. If you have a 7.4L MPI engine and got no DTC 21, go to the next step.
5. Disconnect the jumper wire and connect a multi-meter (DVOM). Measure the voltage across harness terminals **A** and **B**. If the reading is over 4 volts, 417 circuit is open or shorted to ground; if everything is OK, you have a bad connection at the ECM. If the reading is under 4 volts, check

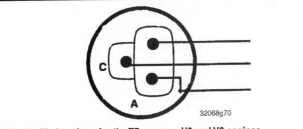

Fig. 70 Pin locations for the TP sensor—V6 and V8 engines (TBI with MEFI-2/3)

circuit 416 for an open or short to ground. Also check circuit 416E to the MAP sensor for a short to ground. If no problems are found, replace the ECM.

CODE 23

Intake Air Temperature Sensor (IAT)

◆ **See Figure 71**

The IAT sensor utilizes a thermistor to control a voltage signal to the ECM. Resistance is high when intake air is cold so the ECM will see a high voltage. As the air warms, resistance lessens and voltage drops. At normal operating temperature, voltage should be 1.5–2.0 volts.

TBI AND 1998—01 V6 AND 5.7L/6.2L/7.4L/8.1L MP/MPI W/MEFI-3 ECM

1. If you are using a scan tool, go to Step 2; if not, go to Step 4.
2. Connect a scan tool as per the manufacturer's instructions. Turn the ignition switch to ON. If the tool displays a temperature less than –22⁰ F (-30⁰ C), go to Step 3; if the temperature is higher than specified, check all system circuits for bad connections or frayed wires.
3. Turn the ignition OFF and disconnect the harness at the IAT sensor. Connect a jumper wire between harness terminals **A** and **C**. Turn the ignition ON. If the tool shows a temperature above 266⁰ F (130⁰ C), check all system circuits for bad connections or frayed wires, if OK replace the sensor; if the temperature is below 266 degrees, circuits 472 or 813 are open.
4. Turn the ignition OFF and disconnect the harness at the sensor. Turn the ignition ON and connect a multi-meter (DVOM) across the harness terminals. If voltage is higher than 4 volts, check all system circuits for bad connections or frayed wire, otherwise replace the sensor; if lower than 4 volts, go to Step 5.
5. Connect the positive meter lead to harness terminal **B** and the negative lead to a good ground. If the voltage is over 4 volts, circuit 813 is open ground or you have a bad connection at the ECM; if under 4 volts, circuit 472 is open or you have a bad connection at the ECM.

1994–97 7.4L/8.2L AND 1998–01 454/502 Mag/8.2L MP/MPI W/SCAN TOOL

1. Connect the tool and turn the ignition switch ON. The tool should display an IAT temperature of less than –22⁰ F (-30⁰ C) or higher than 266⁰ F (130⁰ C). If it does, go to the next step; if not, check all system circuit connections and look for frayed wires.
2. If the temperature was less than –22⁰ F, go to Step 3; if higher than 266⁰ F, go to Step 5.
3. Turn the ignition OFF and disconnect the IAT sensor. Connect a jumper wire between the two harness terminals and turn the ignition ON. If the tool shows a temperature higher then 266⁰ F, check all connections or replace the sensor; if it's not higher, go to Step 4.
4. Turn the ignition OFF and connect the jumper wire between circuit 472 and an engine ground. If the tool shows a temperature above 266 ⁰ F, you have an open sensor ground circuit; if OK, replace the ECM. If below 266⁰ F, circuit 472 is open or you need to replace the ECM.
5. Turn the ignition OFF and disconnect the IAT sensor. Turn the ignition back ON and check the scan tool. If it shows less than –22⁰ F, replace the sensor. If not, check if circuit 472 is shorted to ground, otherwise replace the ECM.

1994–97 7.4L/8.2L AND 454/502 Mag/8.2L MP/MPI W/O SCAN TOOL

1. Disconnect the IAT sensor and turn the ignition ON. Connect a multi-meter (DVOM) and check voltage across the sensor harness terminals. If over 4 volts, go to Step 2; if under 4 volts, go to Step 3.
2. Now check the resistance across the terminals. If less than 25,000 ohms, check the signal circuit for shorting to voltage; otherwise, check all circuits for bad connections or frayed wires. If over 25,000 ohms, replace the sensor.
3. Check the voltage between the harness connector signal circuit (427) and ground. If above 4 volts, check for a bad sensor ground circuit or bad connections in the circuit; if OK, replace the ECM. If under 4 volts, look for an open signal circuit (427) or bad connections in the circuit; if OK, replace the ECM.

32068g70

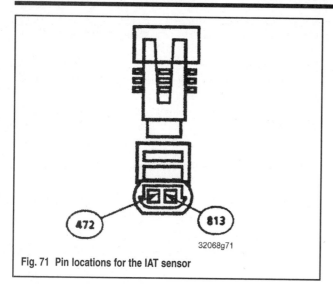

Fig. 71 Pin locations for the IAT sensor

CODE 25

Intake Air Temperature Sensor (IAT)

◆ **See Figure 71**

The IAT sensor utilizes a thermistor to control a voltage signal to the ECM. Resistance is high when intake air is cold so the ECM will see a high voltage. As the air warms, resistance lessens and voltage drops. At normal operating temperature, voltage should be 1.5–2.0 volts.

TBI AND 1998–01 MP/MPI ENGINES W/MEFI-3 ECM

1. If you are using a scan tool, go to Step 2; if not, go to Step 4.
2. Connect a scan tool as per the manufacturer's instructions. Turn the ignition switch to ON. If the tool displays a temperature greater than 266° F (130° C), go to Step 3; if the temperature is less than specified, check all system circuits for bad connections or frayed wires.
3. Turn the ignition OFF and disconnect the harness at the IAT sensor. Turn the ignition ON. If the tool shows a temperature below -22° F (-30° C), check all system circuits for bad connections or frayed wires, if OK replace the sensor; if the temperature is above -22ºF, circuit 472 is open and will require repair, if not open you have a bad connection at the ECM.
4. Turn the ignition OFF and disconnect the harness at the sensor. Turn the ignition ON and connect a multi-meter (DVOM) across the harness terminals. If voltage is higher than 4 volts, check all system circuits for bad connections or frayed wire, otherwise replace the sensor; if lower than 4 volts, circuit 472 is open or you have a bad connection at the ECM

CODE 33

Manifold Absolute Pressure Sensor (MAP)

◆ **See Figures 72 and 73**

The MAP sensor responds to changes in pressure in the manifold (vacuum). This information is sent to the ECM as signal voltage and will vary from 1.0–2.0 volts at idle to about 4.0–5.0 volts at full throttle. The ECM compensates for a failing Map sensor by defaulting to a MAP value program that varies with rpm. Circuit 416 provides a reference value of about 5 volts to the MAP sensor, while 814 is the ground. Circuit 432 carries the signal to the ECM.

TBI AND 1998–01 V6 AND 5.7L/6.2L/7.4L/8.1L V8 MP/MPI W/SCAN TOOL

1. Connect a scan tool as per the manufacturer's instructions.
2. With the ignition switch OFF, install a vacuum gauge to the manifold. Start the engine and increase the idle to 1000 rpm with the engine in Neutral. If the reading is steady and above 14 in. Hg., go to Step 3; if the reading fluctuates or is below 14 in. Hg., look for and repair a vacuum leak.

3. With the engine idling, check the voltage reading on the scan tool. If sensor voltage is greater than 4 volts, go to Step 4; if less than 4 volts, check all sensor circuits for bad connections or scraped wires.
4. With the ignition switch OFF, disconnect the sensor wiring harness. Turn the ignition ON (engine OFF). If the tool reads less than 1 volt, go to Step 5; if the tool reads more than 1 volt, you have a short to voltage in 432 circuit or a bad connection at the ECM.
5. With the ignition OFF, connect a multi-meter (DVOM) between the harness terminal **A** and **C**. Turn the ignition switch to the ON position. If sensor voltage is higher than 4 volts, check for a leaky or damaged vacuum fitting on the sensor, otherwise replace the sensor; if less than 4 volts, 814 circuit is open or you have a bad connection at the ECM.
6. With the ignition OFF install a vacuum gauge to the manifold, start the engine and increase the idle to 1000 rpm. If the gauge reads steady at 14 in. Hg. or higher, go to Step 7; if it fluctuates or is lower, you have vacuum leak somewhere.
7. Connect a CodeMate Tester and switch it to the Normal mode. Turn the ignition switch OFF and disconnect the MAP sensor harness connector. Start the engine and allow it to idle for two minutes or until the tester indicates a stored code. Turn the engine OFF, but leave the ignition switch in the ON position. Switch the tester to the Service mode and note the trouble code. If code 34, check for a leaking or damaged vacuum fitting on the sensor or replace the MAP sensor; if no code 34, go to Step 8.
8. With the Map sensor harness disconnected, turn the ignition switch ON. Connect a multi-meter (DVOM) between the harness terminals **A** and **C**. If the reading is higher than 4 volts, you have a short to voltage in 432 circuit or a bad connection at the ECM; if lower than 4 volts, 814 circuit is open or you have a bad ECM connection.

1994–97 7.4L/8.2L AND 1998–01 454/502 Mag/8.2L MP/MPI W/SCAN TOOL

1. Connect a scan tool as per the manufacturer's instructions. Start the engine and allow it to idle. The tool should show sensor voltage of less than 1 volt or more than 4 volts. If it does and its less then 1 volt, go to Step 2; if more than 4 volts, go to Step 4. Otherwise, check all system circuits for bad connections or frayed wires.
2. Turn the engine OFF and disconnect the sensor electrical lead. Connect a jumper wire between harness terminals **B** and **C**. Start the engine. If the tool displays voltage above 4 volts, check circuit 432 for a short to ground. If OK, replace the sensor. If less than 4 volts, go to Step 3.
3. Turn the ignition OFF, disconnect the scan tool and connect a multi-meter (DVOM) between harness terminals **A** and **C**. Turn the ignition ON. If the meter shows more than 4 volts, check circuit 432 for an open or short to ground, also check for bad connections at the ECM. If both are OK, replace the ECM. If the meter is less than 4 volts, check circuit 416 for an open ECM connection. If OK, replace the ECM.
4. Turn the ignition OFF and disconnect the electrical lead at the sensor. Start the engine. If the tool displays less than 1 volt, go to Step 5. If more than 1 volt, check circuit 432 for a short to voltage or a bad ECM connection. If both are OK, replace the ECM.
5. Turn the ignition OFF, disconnect the scan tool and connect a multi-meter between harness terminals **A** and **C**. If voltage is above 4 volts, replace the sensor. If under 4 volts, circuit 814 is open.

TBI AND 1998–01 V6 AND 5.7L/6.2L/7.4L/8.1L V8 MP/MPI W/O SCAN TOOL

1. With the ignition OFF install a vacuum gauge to the manifold, start the engine and increase the idle to 1000 rpm. If the gauge reads steady at 14 in. Hg. or higher, go to Step 2; if it fluctuates or is lower, you have vacuum leak somewhere.
2. Connect a CodeMate Tester and switch it to the Normal mode. Turn the ignition switch OFF and disconnect the MAP sensor harness connector. Start the engine and allow it to idle for two minutes or until the tester indicates a stored code. Turn the engine OFF, but leave the ignition switch in the ON position. Switch the tester to the Service mode and note the trouble code. If DTC 34, check for a leaking or damaged vacuum fitting on the sensor or replace the MAP sensor; if no DTC 34, go to Step 3.
3. With the Map sensor harness disconnected, turn the ignition switch ON. Connect a multi-meter (DVOM) between the harness terminals **A** and **C**. If the reading is higher than 4 volts, you have a short to voltage in 432 circuit or a bad connection at the ECM; if lower than 4 volts, 814 circuit is open or you have a bad ECM connection.

1994–97 7.4L/8.2L AND 1998–01 454/502 Mag/8.2L MP/MPI W/O SCAN TOOL

1. With the ignition OFF install a vacuum gauge to the vacuum port on the lower front side of the plenum, start the engine and increase the idle to 1000 rpm. If the gauge reads steady at 14 in. Hg. or higher, go to Step 2; if it fluctuates or is lower, you have vacuum leak somewhere.

2. Turn off the engine and remove the gauge. Disconnect the sensor electrical lead and turn the ignition ON. Connect a multi-meter (DVOM) between harness terminals **A** and **C**. If the reading is over 4 volts, go to Step 4. If under 4 volts, go to Step 3.

3. Connect the meter between harness terminal **C** and an engine ground. If the reading is over 4 volts, check circuit 814 for an open or bad connections at the ECM. If both are OK, replace the ECM. If under 4 volts, check circuit 416 for an open or short to ground, or bad connections at the ECM. If both are OK, replace the ECM.

4. Connect the meter between harness terminals **B** and **C**. If the reading is over 4 volts, go to Step 6; if under 4 volts, go to Step 5.

5. Connect the meter between terminal **B** and an engine ground. If the reading is over 4 volts, circuit 432 is shorted to voltage. If under 4 volts, circuit 432 is open or there's a bad connection at the ECM. If both are OK, replace the ECM.

6. With the ignition OFF, connect a test light to the battery positive. Touch the probe to harness terminal **B**. If the light goes on, move to Step 7. If it doesn't light, replace the sensor.

7. Disconnect the terminals at the ECM and touch the probe to harness terminal **B** again. If the light goes on, circuit 432 is shorted to ground. If not, replace the ECM.

32068g72

Fig. 72 Pin locations for the MAP sensor

Altitude (F)	Voltage Range (V)
Below 1000	3.8-5.5
1,000-2,000	3.6-5.3
2,000-3,000	3.5-5.1
3,000-4,000	3.3-5.0
4,000-5,000	3.2-4.8
5,000-6,000	3.0-4.6
6,000-7,0C0	2.9-4.5
7,000-8,000	2.8-4.3
8,000-9,000	2.6-4.2
9,000-10,000	2.5-4.0

32068c73

Fig. 73 MAP sensor voltages—V8 engines

CODE 34

Manifold Absolute Pressure Sensor (MAP)

◆ **See Figure 72**

The MAP sensor responds to changes in pressure in the manifold (vacuum). This information is sent to the ECM as signal voltage and will vary from 1.0–2.0 volts at idle to about 4.0–5.0 volts at full throttle. The ECM compensates for a failing Map sensor by defaulting to a MAP value program that varies with rpm. Circuit 416 provides a reference value of about 5 volts to the MAP sensor, while 814 is the ground. Circuit 432 carries the signal to the ECM.

TBI AND 1998–01 MP/MPI ENGINES W/MEFI-2/3 W/SCAN TOOL

1. Connect a scan tool as per the manufacturer's instructions.

2. With the ignition switch OFF, install a vacuum gauge to the manifold. Start the engine an increase the idle to 1000 rpm with the engine in Neutral. If the reading is steady and above 14 in. Hg., go to Step 3; if the reading fluctuates or is below 14 in. Hg., look for and repair a vacuum leak.

3. With the engine idling, check the voltage reading on the scan tool. If sensor voltage is less than 1 volt, go to Step 4; if greater than 1 volt, check all sensor circuits for bad connections or scraped wires, if OK replace the MAP sensor.

4. With the ignition switch OFF, disconnect the sensor wiring harness. Connect a jumper wire between harness terminals **B** and **C**. Turn the ignition ON (engine OFF). If the tool reads less than 4 volts, go to Step 5; if the tool reads more than 4 volts, check for a leaking or damaged vacuum fitting on the sensor, otherwise replace the sensor.

5. With the ignition OFF, connect a multi-meter (DVOM) between the harness terminal **C** and a known good ground. Turn the ignition switch to the ON position. If sensor voltage is higher than 4 volts, 432 circuit is open or shorted to voltage or you have a bad connection at the ECM; if less than 4 volts, 416 circuit is open or shorted to ground, or you have a bad connection at the ECM. You should also the TP sensor circuit 416 for a short to ground.

TBI AND 1998–01 MP/MPI ENGINES W/MEFI-2/3 W/O SCAN TOOL

1. Connect a CodeMate Tester and switch it to the Normal mode. Turn the ignition switch OFF and disconnect the MAP sensor harness connector. Connect a jumper wire between the harness terminals **B** and **C**. Start the engine and increase the idle to 1000 rpm. Turn the ignition ON until the tester indicates a stored code. Turn the engine OFF, but leave the ignition switch in the ON and switch the tester to the Service mode and note the trouble code. If DTC 33, check the system circuits for bad connections or damaged wires, if OK replace the sensor; if no DTC 33, go to Step 2.

2. Remove the jumper wire and connect a multi-meter (DVOM) between the harness terminals **A** and **C**. If the reading is higher than 4 volts, 432 circuit is open or you have a short to voltage, or a bad connection at the ECM; if lower than 4 volts, 614 circuit is open or shorted to ground, or you have a bad ECM connection. You should also check the TP sensor circuit 416 for a short to ground.

CODE 41

Ignition Control Circuit (IC)

When the system is running on the ignition module or in cranking mode there is no voltage through circuit 424 and the module grounds circuit 423. The ECM is programmed for low voltage in this condition so if it detects voltage, Code 41 is set and it will not move into the IC mode.

TBI AND 1998–01 MP/MPI ENGINES w/MEFI-2/3 ECM

1. Clear the trouble code as detailed in this section. Start the engine and allow it to idle for two minutes or until the code sets itself. If DTC 41 sets, go to Step 2; if not, check the system harness and connections for a bad connection.

2. With the ignition OFF, disconnect the ECM harness connectors. Connect a multi-meter (DVOM), set it to Ohms and probe the ECM circuit 423 to ground. If the resistance is over 3000 ohms, go to Step 3; if under 3000 ohms, go to Step 4.

3. Reconnect the ECM, start the engine and allow it to idle for two minutes or until DTC 41 is set. If the code sets, move to Step 4; if not, check the system harness and wiring for a bad connection on circuit 423.

4. Look for circuit 423 to be open or a bad ECM connection. If you found either of these problems, fix it; if no problems were found, replace the IC module.

CODE 42

Ignition Control Circuit (IC)

When the idle speed reaches the necessary rpm for IC and voltage is applied to circuit 424, circuit 423 should no longer be grounded. If 424 is open or grounded the module will not switch to IC mode, 423 voltage will be low and code 42 will then be set.

TBI AND 1998–01 V6 AND 5.7L/6.2L/7.4L/8.1L V8 MP/MPI ENGINES

1. Clear the code as detailed in this section. Start the engine and allow it to idle for two minutes or until the code sets itself again. If DTC 42 sets itself again, go to Step 2; if not, circuit 424 is open or shorted to ground, or circuit 423 has a short.
2. With the ignition OFF, disconnect the ECM harnesses. Connect a multi-meter, set it to Ohms and then ground circuit 423. If resistance is over 3000 ohms, go to Step 3; if under 3000 ohms, circuit 423 has a short to ground or there is a bad connection at the ECM.
3. With the DVOM still connected and grounding 423, connect a test light to the positive battery cable and touch the probe to circuit 424. The resistance should switch from over 3000 ohms to under 1000 ohms. If it does, circuit 424 is open or shorted to ground, or circuit 423 has a short; if it doesn't drop, go to Step 4.
4. Touch the test light probe to circuit 424. If it lights, go to Step 5; if not, circuit 424 is open .
5. With the light still connected from Step 4, disconnect the 4-wire connector at the module. If the light goes on, circuit 424 is shorted to ground or there is a bad connection at the ECM; if not, replace the IC module.

1994–97 7.4L/8.2L AND 1998–01 454/502 Mag/8.2L MP/MPI ENGINES

1. Clear all codes as detailed in this section. Start the engine and allow it to idle for 1 minute or until the MIL comes on. Turn off the engine but leave the ignition ON. Switch the tester to Service mode and see if DTC 42 resets itself. If you get DTC 42, go to the next step; if no code, check the system circuits for bad connections or frayed wires.
2. Turn the ignition OFF and disconnect both ECM connectors. Connect a multi-meter, set it to Ohms and then ground circuit 423. If resistance is over 3000 ohms, go to the next step; if under 3000 ohms, circuit 423 is open or there is a bad connection at the ignition module. If both are OK, replace the module.
3. With the DVOM still connected and grounding 423, connect a test light to the positive battery cable and touch the probe to circuit 424. If the light comes on, go to Step 7; if it doesn't come on, go to the next step.
4. With the DVOM still connected and grounding 423, connect a test light to the positive battery cable and touch the probe to circuit 424. The resistance should switch from over 3000 ohms to under 1000 ohms. If it does, go to the next step; if it doesn't drop, go to Step 6.
5. Reconnect the ECM connectors, start the engine and allow it to idle for 1 minute or until the MIL comes on. If it comes on and shows DTC 42, replace the ECM. If it doesn't come on, check the system circuits for bad connections or frayed wires.
6. Disconnect the 4-wire connector at the distributor (with the meter still connected and circuit 423 still grounded); resistance should go high. If it did, circuit 424 is open or there is a bad connection at the ignition module. If both are OK, replace the module. If resistance didn't go to high, circuit 423 is shorted to ground.
7. With the light still connected, disconnect the 4-wire connector at the module. If the light goes on, circuit 424 is shorted to ground; if not, replace the IC module.

CODE 43

Knock Sensor (KS)

◆ **See Figure 74**

On models with MEFI-1/MEFI-2 ECM's, detonation or spark knock is sensed by the module which sends a voltage signal to the ECM. As the sensor detects knock, the voltage drops, signaling the ECM to begin retarding timing. The ECM will retard timing whenever knock is detected, and rpm and idle speed are above a specified level.

On models with MEFI-3 ECM's, detonation or spark knock is sensed by the sensor which sends a voltage signal to the ECM. As the sensor detects knock, the voltage increases, signaling the ECM to begin retarding timing. The ECM will retard timing whenever knock is detected, and rpm and idle speed are above a specified level. MEFI-3 ECM's are not equipped with a module.

Octane ratings that are to high may trip either of these codes.

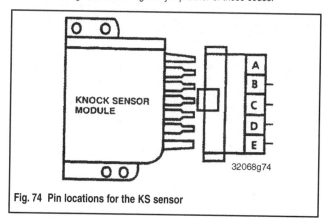

Fig. 74 Pin locations for the KS sensor

TBI AND 1998–01 V6 AND 5.7L/6.2L/7.4L/8.1L V8 MP/MPI ENGINES WITH MEFI-1/2 ECM

1. Disconnect the 5-wire KS module connector. Turn the ignition switch ON, connect a test light to ground and then touch the probe to the module harness terminal **B**. If the light comes on, go to Step 2; if it doesn't come on, circuit 439 is open or shorted to ground.
2. Connect the test light to battery positive and touch the probe to module harness terminal **D**. If the light comes on, go to Step 3; if it doesn't come on, circuit 486 is open and should be repaired.
3. With the ignition OFF, reconnect the module connector and disconnect the J1 connector at the ECM. Turn the ignition back ON and connect a multi-meter (DVOM) between circuit 485 and a known ground. If the voltage is 8–10 volts, go to Step 4; if not in the range, go to Step 5.
4. Allow the voltage to stabilize and connect a test light to battery positive. Touch the probe to circuit 496. If voltage changes from Step 3, check the system circuit for bad connections or frayed wires. Also make sure that the circuit 496 wire is not routed too close to any ignition wires; if voltage does not change, circuit 496 is open or shorted to ground.
5. Check to see if circuit 483 is open or shorted to ground. If there is a problem, fix it; if no problems are found, replace the KS module.

TBI AND 1998–01 MP/MPI ENGINES W/MEFI-3 ECM

1. Disconnect the J1 connector at the ECM. Connect a multi-meter (DVOM) and measure the resistance between circuit 496 and ground. If resistance is 3000–5000 ohms, check the system circuit for bad connections or frayed wires. Also make sure that the circuit 496 wire is not routed too close to any ignition wires; if voltage does not change, circuit 496 is open or shorted to ground; if not in the specified range, go to Step 2.
2. Disconnect the KS connector and measure the resistance between the sensor and ground. If resistance is 3000–5000 ohms, circuit 496 is open or shorted to ground; if resistance is not as specified, replace the KS sensor.

1994–97 7.4L/8.2L AND 454/502 Mag/8.2L MP/MPI ENGINES W/SCAN TOOL

1. Install the scan tool, start the engine and allow it to reach normal operating temperature. If the tool indicates a fixed knock retard value of more than **0** degrees, go to Step 2. If not, check circuit 496 for an open or short. If OK, replace the sensor.
2. With the module connector hooked up, disconnect the sensor harness connector. Connect a test light to the battery positive and start the engine. While holding the engine rpm steady at 2500, quickly touch the probe to harness terminal for circuit 496. If you experienced a noticeable drop in rpm, check for an open or short in circuit 496. If everything is OK, replace the sensor. If no drop in engine speed, go to Step 3.
3. Turn the ignition OFF, disconnect the rear (J2) connector at the ECM and turn the ignition back to ON. Connect a multi-meter (DVOM) between

harness connector terminal **C** and a good ground. If the meter shows 8–10 volts, go to Step 5. If not, go to Step 4.

4. Connect a test light to ground and disconnect the module harness connector. Touch the probe to terminal **B** on the module harness connector. If it lights, circuit 485 is open or shorted to ground. If OK, replace the module. If it doesn't light up, circuit 439 is either open or shorted to ground.

5. Allow the meter to stabilize and then touch the probe to circuit 485 on the sensor harness connector. If the voltage changes, replace the ECM. If no change, go to Step 6.

6. Disconnect the module harness connector and touch the probe to harness connector terminal **D**. If it lights, circuit 496 is either open or shorted to ground; if OK, replace the module. If it doesn't light up, circuit 486 is open to ground.

1994–97 7.4L/8.2L AND 454/502 Mag/8.2L MP/MPI ENGINES W/O SCAN TOOL

1. Disconnect the 5-wire KS module connector. Connect a multi-meter (DVOM) and measure the resistance between terminal **E** and ground. If between 3300 and 4500 ohms, go to Step 2. If not, circuit 496 is either open or shorted. If its OK, replace the sensor.

2. Reconnect the 5-wire connector at the module and disconnect the connector at the sensor. Connect a test light to battery positive and start the engine. Hold idle steady at 2500 rpm while touching the probe to harness connector terminal for circuit 496 a few times. If you detect a noticeable drop in rpm, check the sensor terminal contacts and, if they're OK, replace the sensor. If no drop in engine speed, go to Step 3.

3. Turn the ignition OFF and disconnect the rear (J2) connector at the ECM. Turn the ignition back ON and connect a multi-meter (DVOM) between harness connector terminal **C** and ground. If 8–10 volts are present, go to Step 4. If not, go to Step 6.

4. Allow voltage to stabilize and touch the test light probe to the harness connector terminal for circuit 496. If the voltage changes, replace the ECM. If no change, go to the next step.

5. Disconnect the module connector again and touch the probe to its terminal **D**. If it lights, circuit 496 is either open or shorted to ground. If OK, replace the module. If it doesn't light, look for an open ground on circuit 486.

6. Connect a test light to ground, disconnect the 5-wire connector at the module and touch the probe to terminal **B**. If the light goes on, circuit 485 is either open or shorted to ground. If OK, replace the KS module. If it doesn't go on, circuit 439 is open or grounded.

CODE 44

Knock Sensor (KS)

◆ See Figure 74

On models with MEFI-2 ECM's, detonation or spark knock is sensed by the module which sends a voltage signal to the ECM. As the sensor detects knock, the voltage drops, signaling the ECM to begin retarding timing. The ECM will retard timing whenever knock is detected, and rpm and idle speed are above a specified level.

On models with MEFI-3 ECM's, detonation or spark knock is sensed by the sensor which sends a voltage signal to the ECM. As the sensor detects knock, the voltage increases, signaling the ECM to begin retarding timing. The ECM will retard timing whenever knock is detected, and rpm and idle speed are above a specified level. MEFI-3 ECM's are not equipped with a module.

Octane ratings that are to high may trip either of these codes.

TBI AND 1998–01 MP/MPI ENGINES (EXC. 7.4L MPI) WITH MEFI-2 ECM

1. Disconnect the 5-wire KS module connector. Turn the ignition switch ON, connect a test light to ground and then touch the probe to the module harness terminal **B**. If the light comes on, go to Step 2; if it doesn't come on, circuit 439 is open or shorted to ground.

2. Connect the test light to battery positive and touch the probe to module harness terminal **D**. If the light comes on, go to Step 3; if it doesn't come on, circuit 486 is open and should be repaired.

3. With the ignition OFF, reconnect the module connector and disconnect the J1 connector at the ECM. Turn the ignition back ON and connect a multi-meter (DVOM) between circuit 485 and a known ground. If the voltage is 8–10 volts, go to Step 4; if not in the range, go to Step 5.

4. Allow the voltage to stabilize and connect a test light to battery positive. Touch the probe to circuit 496. If voltage changes from Step 3, check the system circuit for bad connections or frayed wires. Also make sure that the circuit 496 wire is not routed too close to any ignition wires; if voltage does not change, circuit 496 is open or shorted to ground.

5. Check to see if circuit 483 is open or shorted to ground. If there is a problem, fix it; if no problems are found, replace the KS module.

1998–01 7.4L/8.1L MPI ENGINES

1. Disconnect the 5-wire KS module connector. Connect a multi-meter (DVOM) and measure the resistance between harness terminal **E** and a good ground. If resistance is 3500–4700 ohms, go to the next step. If not, check circuit 496 for an open or short to ground; if no problem is found, go to Step 6.

2. Reconnect the module harness and then disconnect the sensor harness. Start the engine and hold the engine speed steady above 2500 rpm. Connect a test light to the battery positive and then touch the probe to harness terminal 496 a few times. If a noticeable drop in rpm occurs, check the sensor terminals contacts and listen for odd engine noises; if both are OK, replace the sensor. If no drop in engine speed occurs when touching the probe to the terminal, go to the next step.

3. Turn the ignition OFF and disconnect the J1 (front) connector at the ECM. Connect a multi-meter (DVOM) between harness terminal J1-1 (circuit 485) and a known ground. Voltage should be 8–10 volts. Allow the voltage to stabilize and then connect a test light to battery positive and touch the probe to the sensor harness terminal for circuit 496 a few times. If the voltage changes, check for bad connections at the ECM; if OK, replace the ECM. If the voltage doesn't change, go to the next step.

4. Disconnect the 5-wire connector at the KS module. Connect a test light to battery positive and touch the probe to module harness terminal **D**. If the light goes on, move to the next step. If it doesn't light, check for an open ground in circuit 486.

5. Now connect the test light to ground and touch the probe to module harness terminal **C**. If the light goes on, replace the KS module. If it doesn't go on, replace the KS module.

6. Disconnect the electrical leads at the knock sensors. Connect a multi-meter (DVOM) and measure the resistance between each sensor terminal and ground. If the meter shows resistance of 7600–8800 ohms, check for bad system connections or frayed wires. If not, replace the knock sensor.

TBI AND 1998–01 MP/MPI ENGINES W/MEFI-3 ECM

1. Disconnect the J1 connector at the ECM. Connect a multi-meter (DVOM) and measure the resistance between circuit 496 and ground. If resistance is 3000–5000 ohms, check the system circuit for bad connections or frayed wires. Also make sure that the circuit 496 wire is not routed too close to any ignition wires; if voltage does not change, circuit 496 is open or shorted to ground; if not in the specified range, go to Step 2.

2. Disconnect the KS connector and measure the resistance between the sensor and ground. If resistance is 3000–5000 ohms, circuit 496 is open or shorted to ground; if resistance is not as specified, replace the KS sensor.

CODE 45

Ignition Coil Driver

TBI AND 1998–01 ENGINES W/MEFI-3 ECM

1. With the ignition switch ON and the engine OFF, connect a multi-meter (DVOM) at the positive terminal on the coil. If greater than 12 volts, go to Step 2; if less than 12 volts, circuit 902 is open or shorted to ground.

2. With the ignition switch OFF, disconnect the J1 connector at the ECM. Check the continuity between the ECM circuit 121 and the negative terminal on the coil. If resistance is less than an ohm, go to Step 3; if more than an ohm, circuit 121 is open and should be repaired.

3. With the ignition switch OFF and the J1 terminal still disconnected, check the resistance between circuit 121 and ground. If resistance is less than an ohm, circuit 121 is shorted to ground; if more than an ohm, go to Step 4.

4. With the ignition switch OFF and the J1 terminal still disconnected, check the resistance between the positive and negative terminals on the coil. If less than 2 ohms, go to Step 5; if more than 2 ohms, replace the ignition coil.

5. With the ignition switch OFF and the J1 terminal still disconnected, remove the secondary ignition wire at the coil and check the resistance between the secondary coil tower and the negative coil terminal. Resistance should be between 5 ohms and 15 ohms. If it is, check the system circuits for bad connections or frayed wires; if it is not within the range, replace the ignition coil.

CODE 51

Electronic Control Module (ECM)

CALIBRATION MEMORY FAILURE

■ If DTC 51 has shown more than once but is intermittent, replace the ECM.

Turn the ignition switch ON and clear the code as detailed in this section. If the code resets itself, replace the ECM; if not, check all system circuits for bad connections and/or frayed wires.

CODE 52

Electronic Control Module (ECM)

EEPROM FAILURE

■ If DTC 52 has shown more than once but is intermittent, replace the ECM.

Clear the code as detailed in this section and Turn the ignition switch ON. If the code resets itself, replace the ECM; if not, check all system circuits for bad connections and/or frayed wires.

CODE 61

Fuel Pressure Sensor (FP)

◆ See Figure 75

The FP sensor sends a signal that changes as fuel pressure does. Signal voltage should always be 2.5–3.5 volts.

TBI AND 1998–01 MP/MPI ENGINES W/SCAN TOOL

1. Connect a scan tool as per the manufacturer's instructions and then turn the ignition switch ON. If sensor voltage is higher than 4 volts, go to Step 3; if lower than 4 volts, check the system circuit for bad connections or frayed wires. If sensor voltage is lower than 0.3 volts, go to Step 2; if not, check the system circuit for bad connections or frayed wires.
2. Turn the ignition OFF and disconnect the connector at the pressure sensor. Connect a jumper wire between terminals **B** and **C**, then turn the ignition ON. If the voltage reads higher than 4 volts, replace the sensor; if lower than 4 volts, go to Step 4.
3. Turn the ignition OFF and disconnect the connector at the sensor. Turn the ignition ON. If the reading is higher than 4 volts, check if circuit 475 is shorted to voltage; if less than 4 volts, go to Step 4.
4. Turn the ignition ON and connect a multi-meter (DVOM) between sensor harness terminals **A** and **B**. If the reading is over 4 volts, go to Step 5; if lower than 4 volts, go to Step 8.
5. Connect the meter between harness terminals **B** and **C**. If the reading is over 4 volts, go to Step 6; if lower than 4 volts, go to Step 7.
6. Turn the ignition OFF and connect a test light to the battery positive. Touch the probe to terminal **C**. If the light goes on, move to Step 9; if it does not light, replace the sensor.
7. Connect the multi-meter between harness terminal **C** and an engine ground. If the reading is over 4 volts, circuit 475 is shorted to voltage, if under 4 volts, circuit 475 is open or you have a bad connection at the ECM.
8. Now move the meter probe to terminal **B**. If over 4 volts, circuit 813 is open or you have a bad connection at the ECM; if under 4 volts, circuit 416 is open or shorted to ground, or you have a bad connection at the ECM.

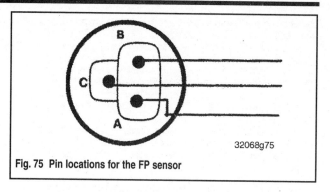

Fig. 75 Pin locations for the FP sensor

32068g75

9. Disconnect the ECM and then touch the test light probe to sensor terminal **C**. If it lights, circuit 475 is shorted to ground; if it doesn't light, circuit 475 is open or you have a bad ECM connection.

TBI AND 1998–01 MP/MPI ENGINES W/O SCAN TOOL

1. Turn the ignition ON and connect a multi-meter (DVOM) between sensor harness terminals **A** and **B**. If the reading is over 4 volts, go to Step 2; if lower than 4 volts, go to Step 4.
2. Connect the meter between harness terminals **B** and **C**. If the reading is over 4 volts, go to Step 3; if lower than 4 volts, go to Step 6.
3. Turn the ignition OFF and connect a test light to the battery positive. Touch the probe to terminal **C**. If the light goes on, move to Step 5; if it does not light, replace the sensor.
4. Now move the meter probe to terminal **B**. If over 4 volts, circuit 813 is open or you have a bad connection at the ECM; if under 4 volts, circuit 416 is open or shorted to ground, or you have a bad connection at the ECM.
5. Disconnect the ECM and then touch the test light probe to sensor terminal **C**. If it lights, circuit 475 is shorted to ground; if it doesn't light, circuit 475 is open or you have a bad ECM connection.

CODE 62

Fuel Pressure Sensor (FP)

◆ See Figure 75

The FP sensor sends a signal that changes as fuel pressure does. Signal voltage should always be 2.5–3.5 volts.

TBI AND 1998–01 MP/MPI ENGINES W/SCAN TOOL

1. Connect a scan tool as per the manufacturer's instructions. Turn the ignition OFF. If the sensor voltage is less than 0.3 volts, go to Step 2; if greater than 0.3 volts, check the system circuit for bad connections or frayed wires.
2. Turn the ignition OFF and disconnect the sensor harness. Connect a jumper wire between terminals **B** and **C**. Turn the ignition switch ON. If the tool shows voltage greater than 4 volts, check the system circuit for bad connections or frayed wires, if OK replace the sensor; if less than 4 volts, go to Step 3.
3. Turn the ignition OFF and connect a multi-meter (DVOM) between terminal **B** and a known ground. If voltage is greater than 4 volts, circuit 475 is open or shorted to ground, or you have a bad connection at the ECM; if less than 4 volts, circuit 416 is open or shorted to ground, or you have a bad connection at the ECM.

TBI AND 1998–01 MP/MPI ENGINES W/O SCAN TOOL

Connect a CodeMate tester and set it in the Normal mode. Turn the ignition OFF and disconnect the harness at the sensor. Connect a jumper wire between terminals **B** and **C**, then start the engine and let it idle for two minutes or until a stored code comes up. Turn the engine OFF but leave the ignition ON. Switch the tester to the Service mode. If DTC 61 shows, check the system circuit for bad connections or frayed wires, if OK replace the sensor; if no DTC 61, circuit 475 is open or shorted to ground, or you have a bad connection at the ECM

VACUUM DIAGRAMS

◆ See Figures 76, 77, 78, 79, 80, 81, 82, 83 and 84

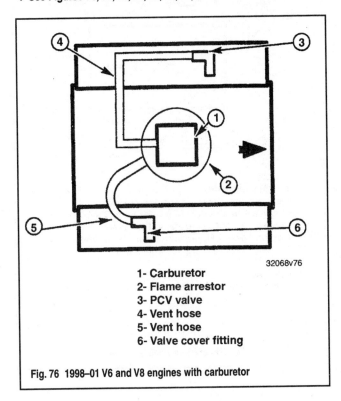

1- Carburetor
2- Flame arrestor
3- PCV valve
4- Vent hose
5- Vent hose
6- Valve cover fitting

32068v76

Fig. 76 1998–01 V6 and V8 engines with carburetor

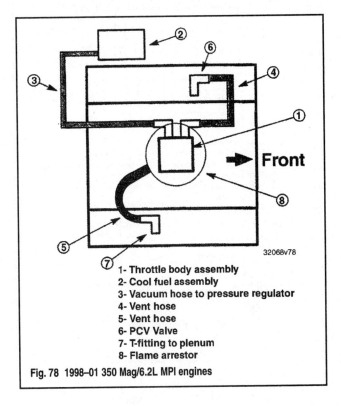

1- Throttle body assembly
2- Cool fuel assembly
3- Vacuum hose to pressure regulator
4- Vent hose
5- Vent hose
6- PCV Valve
7- T-fitting to plenum
8- Flame arrestor

32068v78

Fig. 78 1998–01 350 Mag/6.2L MPI engines

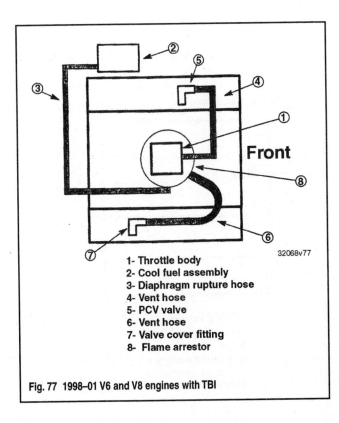

1- Throttle body
2- Cool fuel assembly
3- Diaphragm rupture hose
4- Vent hose
5- PCV valve
6- Vent hose
7- Valve cover fitting
8- Flame arrestor

32068v77

Fig. 77 1998–01 V6 and V8 engines with TBI

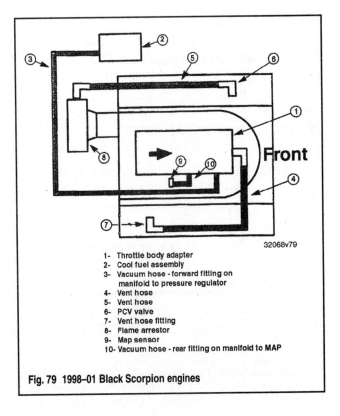

1- Throttle body adapter
2- Cool fuel assembly
3- Vacuum hose - forward fitting on manifold to pressure regulator
4- Vent hose
5- Vent hose
6- PCV valve
7- Vent hose fitting
8- Flame arrestor
9- Map sensor
10- Vacuum hose - rear fitting on manifold to MAP

32068v79

Fig. 79 1998–01 Black Scorpion engines

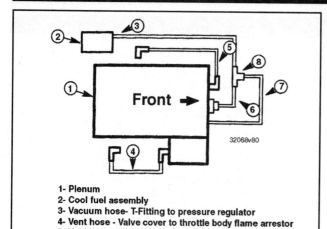

1- Plenum
2- Cool fuel assembly
3- Vacuum hose- T-Fitting to pressure regulator
4- Vent hose - Valve cover to throttle body flame arrestor
5- Vent hose - Front of plenum to PCV valve
6- Vacuum hose - Pressure damper to T-Fitting
7- Vacuum hose
8- T-Fitting

Fig. 80 1998–01 454/502/8.2L MPI engines w/MEFI-1 ECM

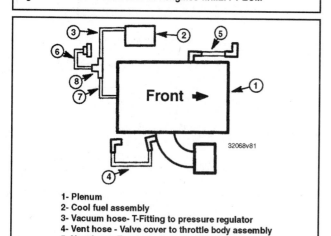

1- Plenum
2- Cool fuel assembly
3- Vacuum hose- T-Fitting to pressure regulator
4- Vent hose - Valve cover to throttle body assembly
5- Vent hose- Front of plenum to PCV valve
6- Vacuum hose- Fuel pressure damper to T-Fitting
7- Vacuum hose- T-Fitting to plenum
8- T-Fitting

Fig. 81 1998–01 7.4L MPI engines w/MEFI-2 ECM

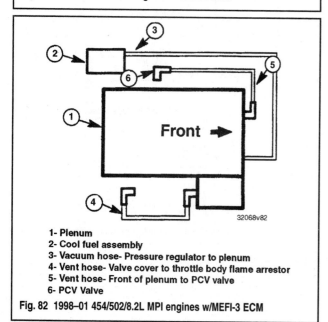

1- Plenum
2- Cool fuel assembly
3- Vacuum hose- Pressure regulator to plenum
4- Vent hose- Valve cover to throttle body flame arrestor
5- Vent hose- Front of plenum to PCV valve
6- PCV Valve

Fig. 82 1998–01 454/502/8.2L MPI engines w/MEFI-3 ECM

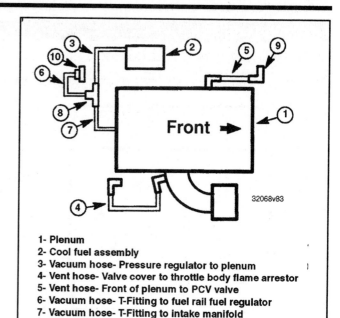

1- Plenum
2- Cool fuel assembly
3- Vacuum hose- Pressure regulator to plenum
4- Vent hose- Valve cover to throttle body flame arrestor
5- Vent hose- Front of plenum to PCV valve
6- Vacuum hose- T-Fitting to fuel rail fuel regulator
7- Vacuum hose- T-Fitting to intake manifold
8- T-Fitting
9- PCV Valve
10- GM fuel rail mounted pressure regulator not used

Fig. 83 1998–01 7.4L MPI engines w/MEFI-3 ECM

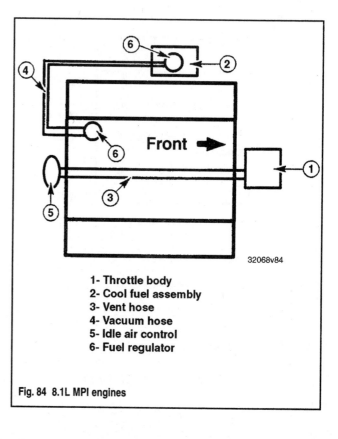

1- Throttle body
2- Cool fuel assembly
3- Vent hose
4- Vacuum hose
5- Idle air control
6- Fuel regulator

Fig. 84 8.1L MPI engines

WIRING SCHEMATICS

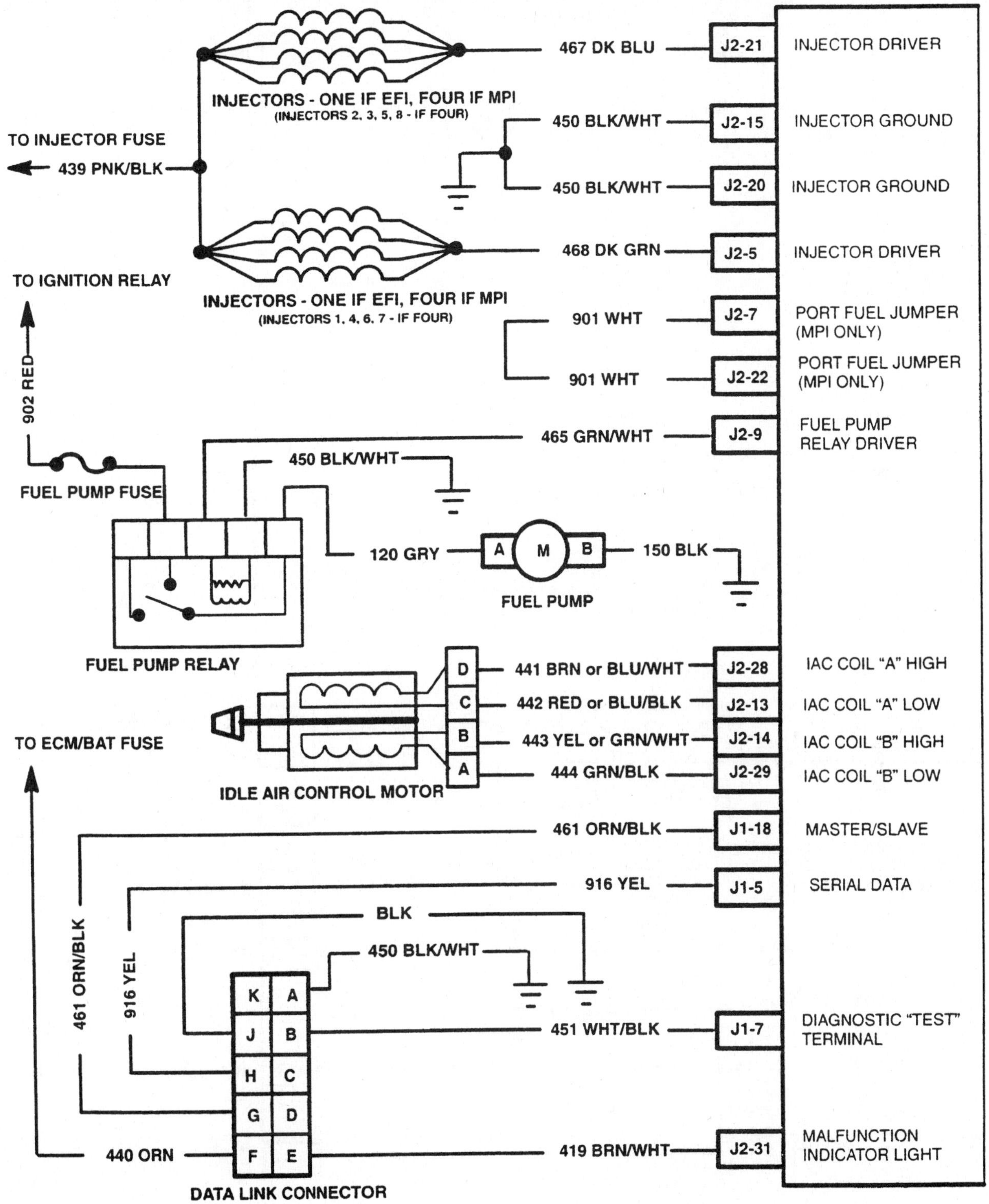

INJECTORS - ONE IF EFI, FOUR IF MPI
(INJECTORS 2, 3, 5, 8 - IF FOUR)

467 DK BLU — J2-21 — INJECTOR DRIVER

450 BLK/WHT — J2-15 — INJECTOR GROUND

450 BLK/WHT — J2-20 — INJECTOR GROUND

468 DK GRN — J2-5 — INJECTOR DRIVER

INJECTORS - ONE IF EFI, FOUR IF MPI
(INJECTORS 1, 4, 6, 7 - IF FOUR)

901 WHT — J2-7 — PORT FUEL JUMPER (MPI ONLY)

901 WHT — J2-22 — PORT FUEL JUMPER (MPI ONLY)

TO INJECTOR FUSE
439 PNK/BLK

TO IGNITION RELAY

902 RED

465 GRN/WHT — J2-9 — FUEL PUMP RELAY DRIVER

FUEL PUMP FUSE

450 BLK/WHT

120 GRY — A M B — 150 BLK

FUEL PUMP

FUEL PUMP RELAY

441 BRN or BLU/WHT — J2-28 — IAC COIL "A" HIGH

442 RED or BLU/BLK — J2-13 — IAC COIL "A" LOW

443 YEL or GRN/WHT — J2-14 — IAC COIL "B" HIGH

444 GRN/BLK — J2-29 — IAC COIL "B" LOW

TO ECM/BAT FUSE

IDLE AIR CONTROL MOTOR

461 ORN/BLK — J1-18 — MASTER/SLAVE

916 YEL — J1-5 — SERIAL DATA

461 ORN/BLK

916 YEL

BLK

450 BLK/WHT

451 WHT/BLK — J1-7 — DIAGNOSTIC "TEST" TERMINAL

440 ORN

DATA LINK CONNECTOR

419 BRN/WHT — J2-31 — MALFUNCTION INDICATOR LIGHT

Fig. 85 V6 and 1998-01 V8 engines, TBI/MPI, w/MEFI-1/2 ECM (1 of 4)

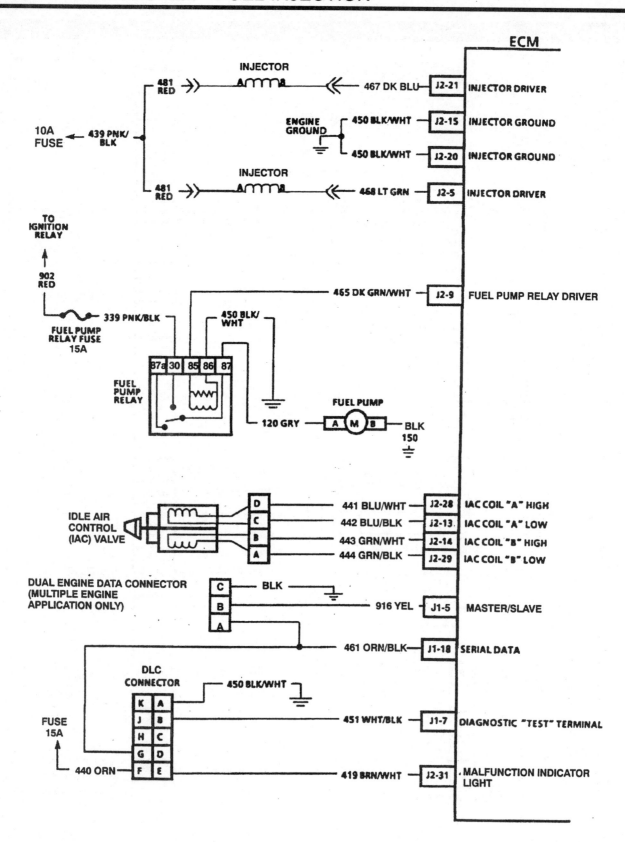

Fig. 86 1994-97 V8 TBI engines (1 of 4)

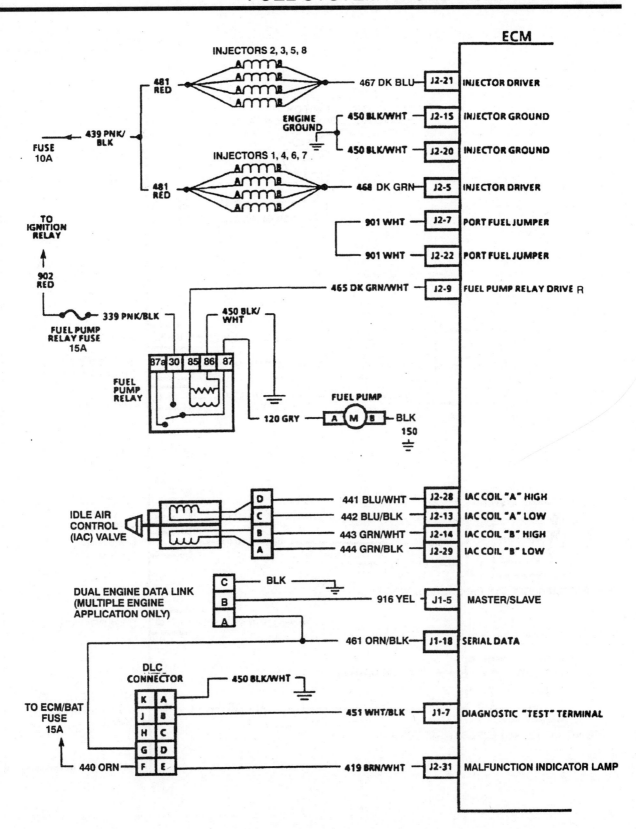

Fig. 87 1994-97 V8 MP/MPI engines, w/MEFI-1/2 ECM (1 of 4)

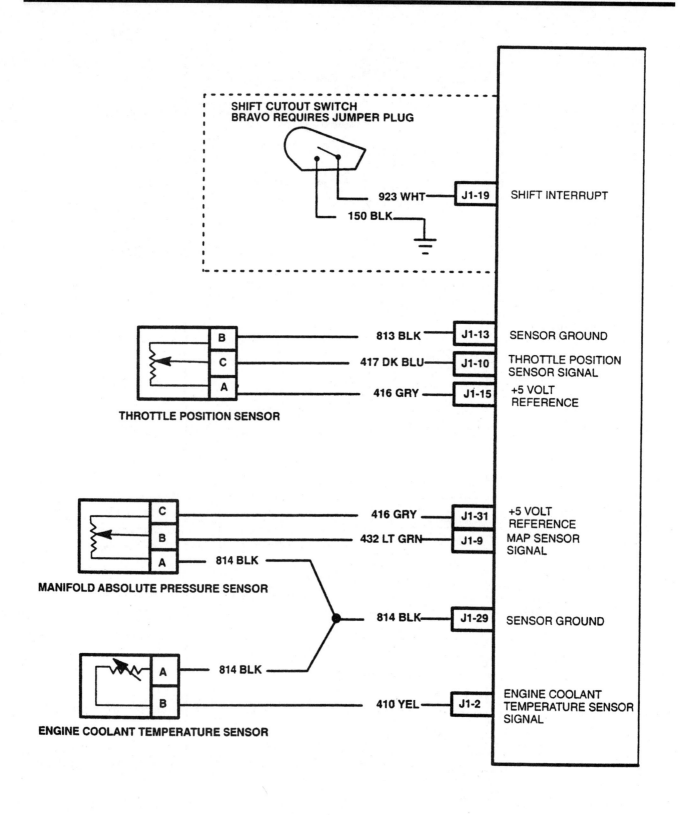

Fig. 88 V6/V8 TBI engines and 1998-01 5.7L/6.2L MP/MPI engines, w/MEFI-1/2 ECM (2 of 4)

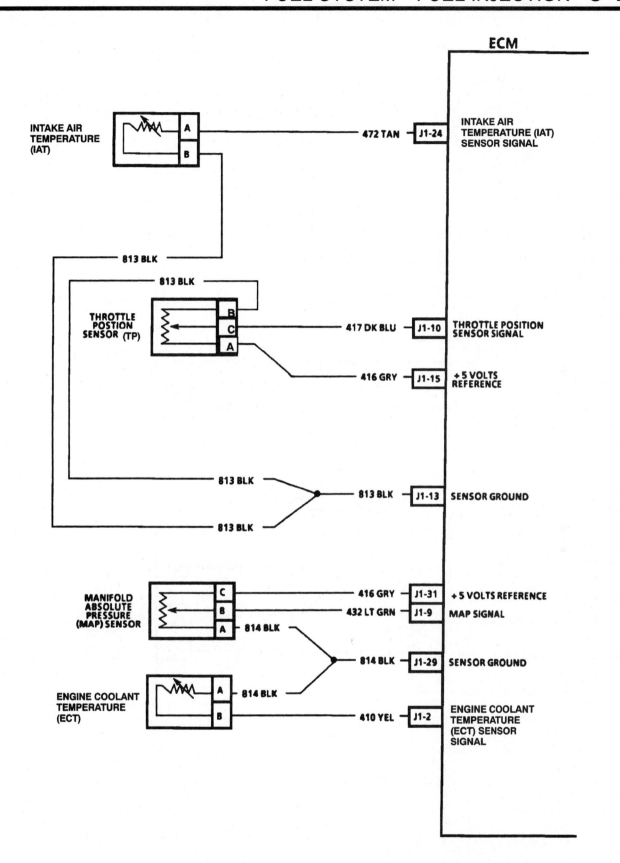

Fig. 89 1994-97 V8 MP/MPI engines and 1998-01 454/502/8.2L MPI engines-350 Mag engines may not use the IAT, w/MEFI-1/2 ECM (2 of 4)

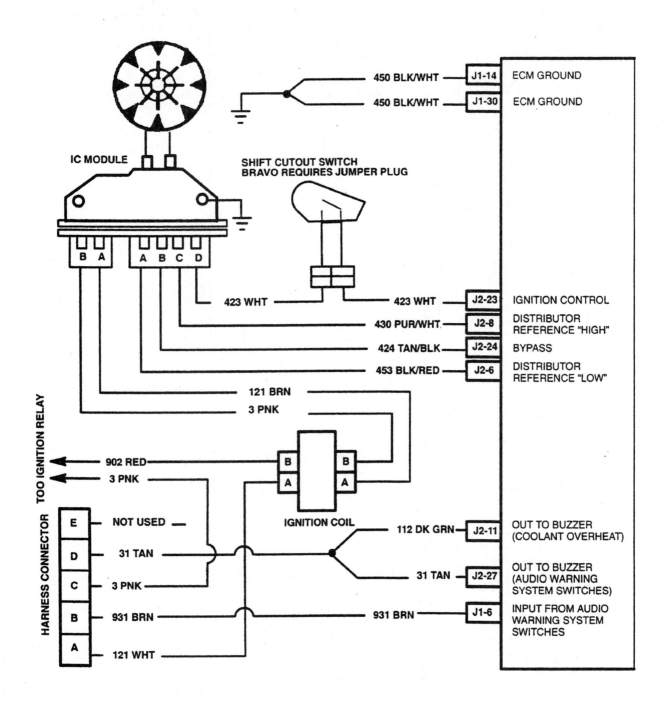

Fig. 90 V6 and 1998-01 V8 engines, TBI/MPI, w/MEFI-1/2 ECM (3 of 4)

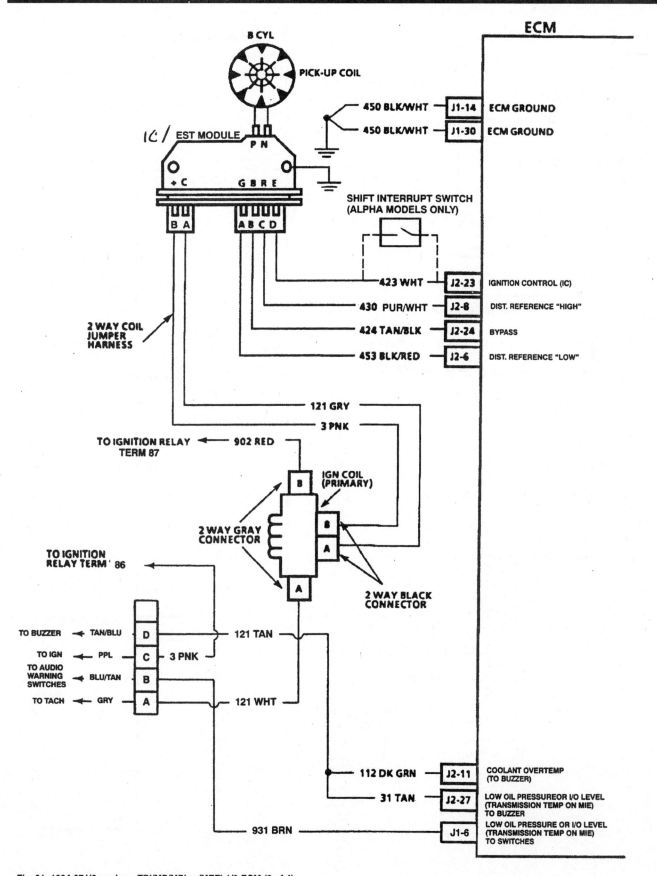

Fig. 91 1994-97 V8 engines, TBI/MP/MPI, w/MEFI-1/2 ECM (3 of 4)

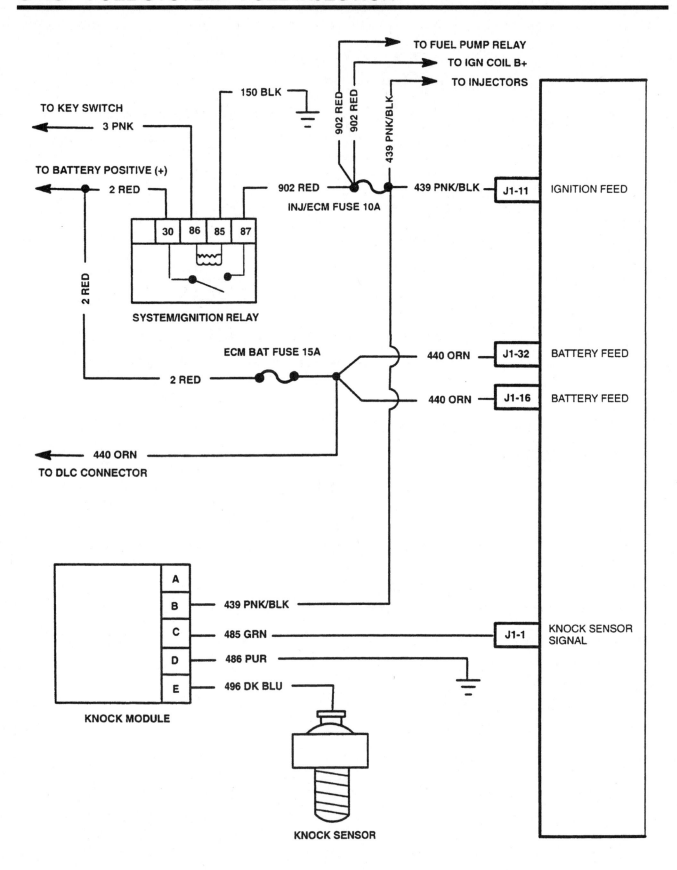

Fig. 92 V6 and 1998-01 V8 engines, TBI/MPI-1998-01 7.4L engine will have two knock sensors, w/MEFI-1/2 ECM (4 of 4)

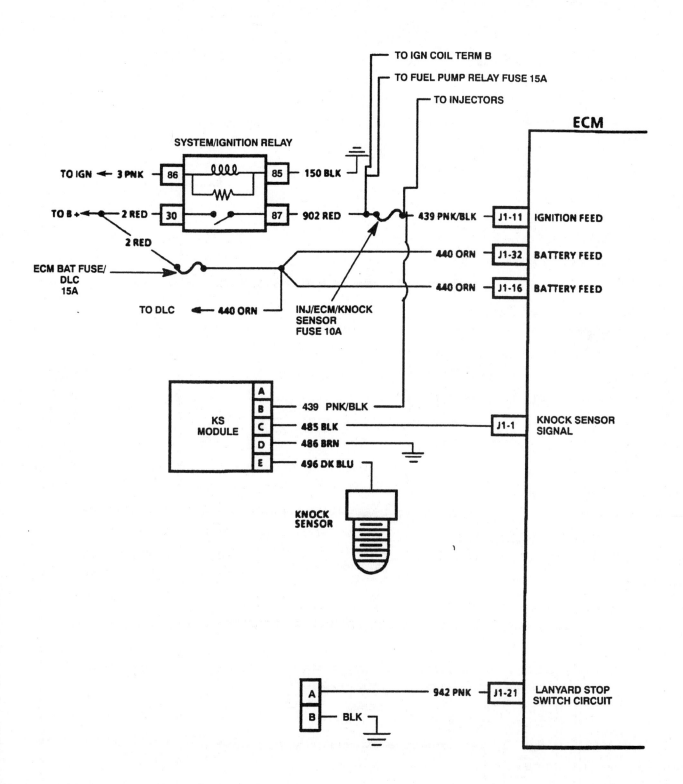

Fig. 93 1994-97 V8 engines, TBI/MP/MPI, w/MEFI-1/2 ECM (4 of 4)

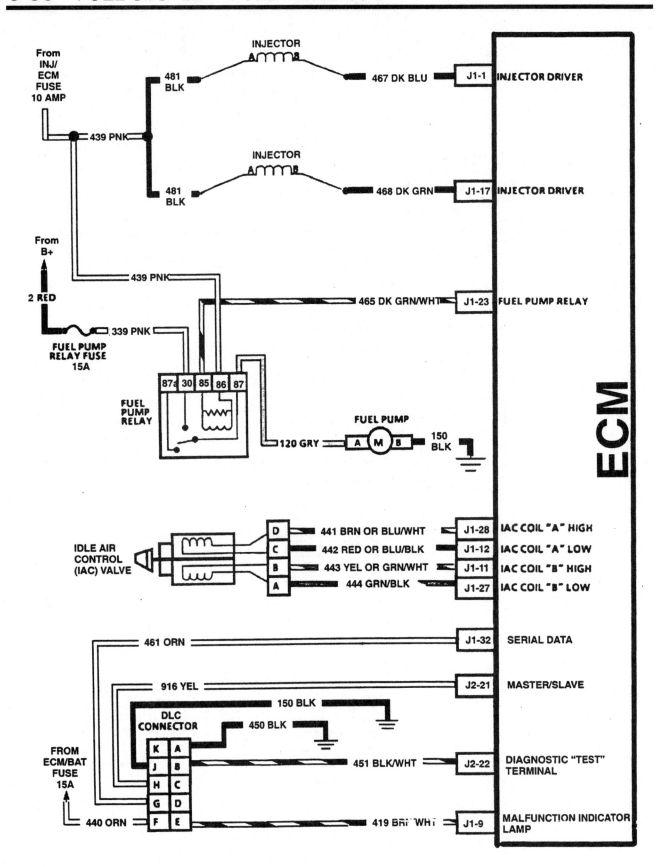

Fig. 94 V6 and 1998-01 V8 engines, TBI/MPI, w/MEFI-3 (1 of 4)

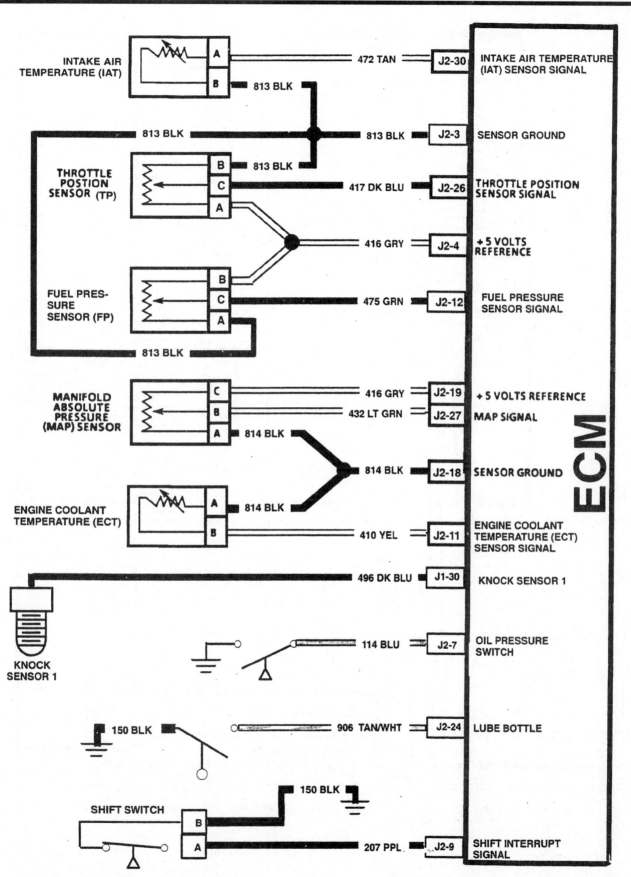

Fig. 95 V6 and 1998-01 V8 engines, TBI/MPI, w/MEFI-3 (2 of 4)

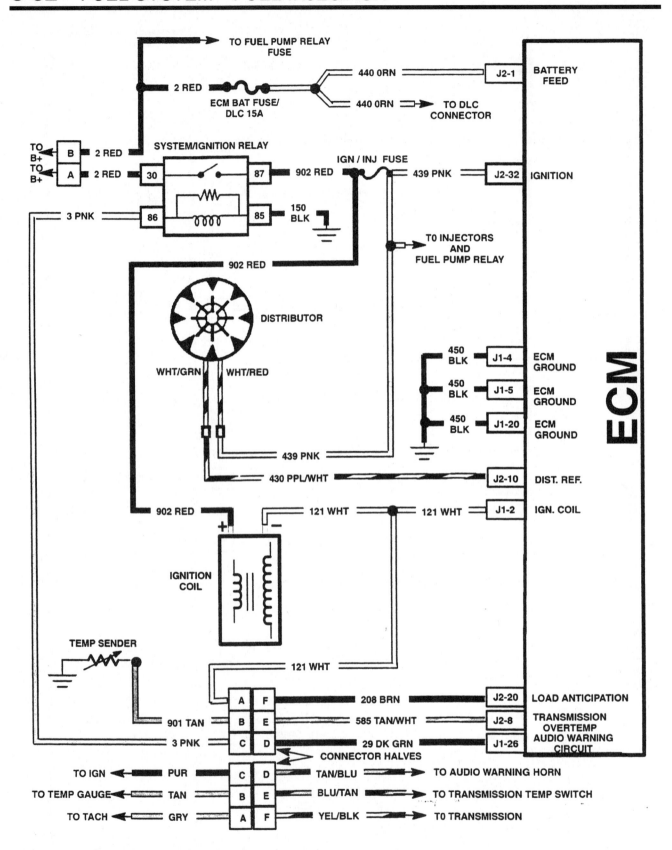

Fig. 96 V6 and 1998-01 V8 engines, TBI/MPI, w/MEFI-3, Mercury distributor (3 of 4)

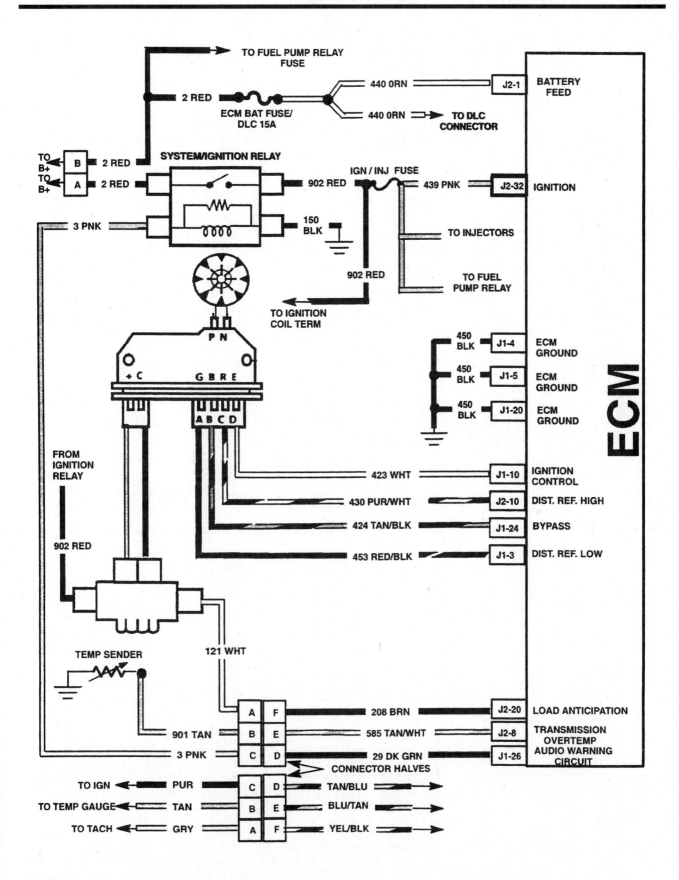

Fig. 97 V6 and 1998-01 V8 engines, TBI/MPI, w/MEFI-3, GM EST distributor (4 of 4)

Pin	Pin Function	Circuit (CKT) Number (#)	Wire Color	Normal Voltage Ignition ON	Normal Voltage Engine Running	Diagnostic Trouble Codes DTC(s)	Possible Symptoms
J1-1	Knock Sensor Signal	485	BLK	9.5V	9.5V	43	Poor Fuel Economy, Poor Performance Detonation
J1-2	ECT Signal	410	YEL	1.95V	1.95V	14	Poor Performance, Exhaust Odor, Rough Idle RPM Reduction
J1-3	Not Used	–	–	–	–	–	–
J1-4	Not Used	–	–	–	–	–	–
J1-5	Master/Slave	916	YEL	B+	B+	None	Lack Of Data From Other Engine (Dual Engine Only)
J1-6	Discrete Switch	931	BRN	–	–	None	Power Reduction Mode
J1-7	Diagnostic Test	451	WHT/BLK	B+	B+	None	Incorrect Idle, Poor Performance
J1-8	Not Used	–	–	–	–	–	–
J1-9	Map Signal	432	LT GRN	4.9V	1.46V	33	Poor Performance, Surge, Poor Fuel Economy, Exhaust Odor
J1-10	TP Signal	417	DK BLU	.62V	.62V	21	Poor Performance And Acceleration, Incorrect Idle
J1-11	Ignition Fused	439	PNK/BLK	B+	B+	None	No Start
J1-12	Not Used	–	–	–	–	–	–
J1-13	TP and IAT Ground	813	BLK	0	0	21,23	High Idle, Rough Idle, Poor Performance Exhaust Odor
J1-14	ECM Ground	450	BLK/WHT	0	0	None	No Start
J1-15	TP 5V Reference	416	GRY	5V	5V	21	Lack Of Power, Idle High
J1-16	Battery	440	ORN	B+	B+	None	No Start

32068c98

Fig. 98   J1 pin locations-1994-97 V8 MP/MPI engines

Pin	Pin Function	Circuit (CKT) Number (#)	Wire Color	Normal Voltage Ignition ON	Normal Voltage Engine Running	Diagnostic Trouble Codes DTC(s)	Possible Symptoms
J1-17	Not Used	–	–	–	–	–	–
J1-18	Serial Data	461	ORN/BLK	5V	5V	None	No Serial Data (NOTE 6)
J1-19	Not Used	–	–	–	–	–	–
J1-20	Not Used	–	–	–	–	–	–
J1-21	Lanyard Stop Switch	942	PNK	0	0	NONE	No Start
J1-22	Not Used	–	–	–	–	–	–
J1-23	Not Used	–	–	–	–	–	–
J1-24	IAT Sensor	472	TAN	5V		23	Poor Fuel Economy, Exhaust Odor
J1-25	Not Used	–	–	–	–	–	–
J1-26	Not Used	–	–	–	–	–	–
J1-27	Not Used	–	–	–	–	–	–
J1-28	Not Used	–	–	–	–	–	–
J1-29	MAP Ground	814	BLK	0	0	33	Lack Of Performance, Exhaust Odor, Stall
J1-30	ECM Ground	450	BLK/WHT	0	0	None	No Start
J1-31	MAP 5V Reference	416	GRY	5V	5V	33	Lack Of Power, Surge, Rough Idle, Exhaust Odor
J1-32	Battery	440	ORN	B+	B+	None	No Start

32068c99

Fig. 99   J1 pin locations-1994-97 V8 MP/MPI engines (cont'd)

Pin	Pin Function	Circuit (CKT) Number (#)	Wire Color	Normal Voltage		Diagnostic Trouble Codes DTC(s)	Possible Symptoms
				Ignition ON	Engine Running		
J2-1	Not Used	–	–	–	–	–	–
J2-2	Not Used	–	–	–	–	–	–
J2-3	Not Used	–	–	–	–	–	–
J2-4	Not Used	–	–	–	–	–	–
J2-5	Injector Driver	468	LT GRN	B+	B+	None	Rough Idle, Lack Of Power, Stall
J2-6	Ignition Control Ref. Low	463	BLK/RED	0	0	None	Poor Performance
J2-7	Port Fuel Jumper	901	WHT	–	–	–	–
J2-8	Ignition Control Ref. High	430	PUR/WHT	5V	1.6V	None	No Restart
J2-9	Fuel Pump Relay Driver	465	DK GRN/WHT	0	B+	None	No Start
J2-10	Not Used	–	–	–	–	–	–
J2-11	Coolant Over-temp	112	DK GRN	0	0	NONE	Power Reduction Mode or Improper Audio Warning
J2-12	Not Used	–	–	–	–	–	–
J2-13	IAC "A" Low	442	BLU/BLK	Not Usable	Not Usable	None	Rough Unstable or Incorrect Idle
J2-14	IAC "B" Low	443	GRN/WHT	Not Usable	Not Usable	None	Rough Unstable or Incorrect Idle
J2-15	Fuel Injector Ground	450	BLK/WHT	0	0	None	Rough Running, Lack Of Power, Poor Performance
J2-16	Not Used	–	–	–	–	–	–

Fig. 100 J2 pin locations-1994-97 V8 MP/MPI engines

32068c100

Pin	Pin Function	Circuit (CKT) Number (#)	Wire Color	Normal Voltage		Diagnostic Trouble Codes DTC(s)	Possible Symptoms
				Ignition ON	Engine Running		
J2-17	Not Used	–	–	–	–	–	–
J2-18	Not Used	–	–	–	–	–	–
J2-19	Not Used	–	–	–	–	–	–
J2-20	Fuel Injector Ground	450	BLK/WHT	0	0	None	Rough Running, Poor Idle, Lack Of Performance
J2-21	Injector Driver	467	DK BLU	B+	B+	None	Rough Idle, Lack Of Power, Stalling
J2-22	Port Fuel Jumper	901	WHT	–	–	–	–
J2-23	Ignition Control Signal	423	WHT	0	1.2V	42	Stall, Will Restart In Bypass Mode, Lack Of Power
J2-24	Ignition Control Bypass	424	TAN/BLK	0	4.5V	42	Lack Of Power, Fixed Timing
J2-25	Not Used	–	–	–	–	–	–
J2-26	Not Used	–	–	–	–	–	–
J2-27	Discrete Switch Signal	31	TAN	–	–	–	Power Reduction Mode
J2-28	IAC "A" High	441	BLU/WHT	Not Usable	Not Usable	None	Rough Unstable or Incorrect Idle
J2-29	IAC "B" Low	444	GRN/BLK	Not Usable	Not Usable	None	Rough Unstable or Incorrect Idle
J2-30	Not Used	–	–	–	–	–	–
J2-31	MIL Lamp	419	BRN/WHT	0	0	None	Lamp Inoperative
J2-32	Not Used	–	–	–	–	–	–

Fig. 101 J2 pin locations-1994-97 V8 MP/MPI engines (cont'd)

32068c101

Pin	Pin Function	CKT	Wire Color	Normal Voltage Ignition ON	Normal Voltage Engine Running	DTC	Possible Symptoms
J1-1	Knock Sensor	485	GRN	9.5V	9.5V	43	Poor Fuel Economy, Poor Performance Detonation
J1-2	ECT Signal	410	YEL	1.95V	1.95V	14	Poor Performance, Exhaust Odor, Rough Idle rpm Reduction
J1-4	Discrete Switch	931	BRN	-	-	None	Power Reduction Mode Alarm Activation
J1-5	Master/Slave	916	YEL	B+	B+	None	Lack Of Data From Other Engine (Dual Engine Only)
J1-6	Discrete Switch	931	BRN	-	-	None	Power Reduction Mode Alarm Activation
J1-7	Diagnostic Test	451	BLK/WHT	B+	B+	None	Incorrect Idle, Poor Performance
J1-9	Map Signal	432	LT GRN	4.9V	1.46V	33	Poor Performance, Surge, Poor Fuel Economy, Exhaust Odor
J1-10	TP Signal	417	DK BLU	.62V	.62V	21	Poor Performance And Acceleration, Incorrect Idle
J1-11	Ignition Fused	439	PNK/BLK	B+	B+	None	No Start
J1-13	Sensor Ground	813	BLK	0	0	21,23	High Idle, Rough Idle, Poor Performance Exhaust Odor
J1-14	ECM Ground	450	BLK/WHT	0	0	None	No Start
J1-15	TP 5V Power	416	GRY	5V	5V	21	Lack Of Power, Idle High
J1-16	Battery	440	ORN	B+	B+	None	No Start
J1-18	Serial Data	461	ORN/BLK	5V	5V	None	No Serial Data
J1-19	Shift Switch	923	WHT	0	0	None	Incorrect Idle
J1-21	Lanyard Stop	942	PNK	0	0	None	No Start
J1-29	MAP Ground	814	BLK	0	0	33	Lack Of Performance,Exhaust Odor, Stall
J1-30	ECM Ground	450	BLK/WHT	0	0	None	No Start
J1-31	MAP 5V Power	416	GRY	5V	5V	33	Lack Of Power, Surge, Rough Idle, Exhaust Odor
J1-32	Battery	440	ORN	B+	B+	None	No Start

32068c102

Fig. 102 J1 pin locations-V6 and V8 engines w/MEFI-1 ECM, exc. 1998-01 7.4L/8.2L engines

Pin	Pin Function	CKT	Wire Color	Normal Voltage Ignition ON	Normal Voltage Engine Running	DTC	Possible Symptoms
J2-5	Injector Driver	468	LT GRN	B+	B+	None	Rough Idle, Lack Of Power, Stall
J2-6	Ignition Control Ref. Low	463	RED/BLK	0	0	None	Poor Performance
J2-7	Not Used	-	-	-	-	None	Not Used
J2-8	Ignition Control Ref. High	430	PUR/WHT	5V	1.6V	None	No Restart
J2-9	Fuel Pump Relay Driver	465	DK GRN/WHT	0	B+	None	No Start
J2-11	Coolant Over temp.	112	DK GRN	0	0	None	Power Reduction Mode or Improper Audio Warning
J2-13	IAC "A" Low	442	RED	Not Usable	Not Usable	None	Rough Unstable or Incorrect Idle
J2-14	IAC "B" Low	443	YEL	Not Usable	Not Usable	None	Rough Unstable or Incorrect Idle
J2-15	Injector Ground	450	BLK/WHT	0	0	None	Rough Running, Lack Of Power, Poor Performance

32068c103

Fig. 103 J2 pin locations-V6 and V8 engines w/MEFI-1 ECM,

Pin	Pin Function	Circuit (CKT) Number (#)	Wire Color	Normal Voltage Ignition ON	Normal Voltage Engine Running	Diagnostic Trouble Codes DTC(s)	Possible Symptoms
J1-1	Knock Sensor Signal	485	BLK	9.5V	9.5V	43	Poor Fuel Economy, Poor Performance Detonation
J1-2	ECT Signal	410	YEL	1.95V	1.95V	14	Poor Performance, Exhaust Odor, Rough Idle RPM Reduction
J1-3	Not Used	-	-	-	-	-	-
J1-4	Not Used	-	-	-	-	-	-
J1-5	Master/Slave	916	YEL	B+	B+	None	Lack Of Data From Other Engine (Dual Engine Only)
J1-6	Discrete Switch	931	BRN	-	-	None	
J1-7	Diagnostic Test	451	WHT/BLK	B+	B+	None	Incorrect Idle, Poor Performance
J1-8	Not Used	-	-	-	-	-	
J1-9	Map Signal	432	LT GRN	4.9V	1.46V	33	Poor Performance, Surge, Poor Fuel Economy, Exhaust Odor
J1-10	TP Signal	417	DK BLU	.62V	.62V	21	Poor Performance And Acceleration, Incorrect Idle

Fig. 105 J1 pin locations-1998-01 454/502/8.2L engines w/MEFI-1 ECM

32068c105

Pin	Pin Function	CKT	Wire Color	Normal Voltage Ignition ON	Normal Voltage Engine Running	DTC	Possible Symptoms
J2-20	Fuel Injector Ground	450	BLK/WHT	0	0	None	Rough Running, Poor Idle, Lack Of Performance
J2-21	Injector Driver	467	DK BLU	B+	B+	None	Rough Idle, Lack Of Power, Stalling
J2-22	Not Used	-	-	-	-	-	Not Used
J2-23	Ignition Control Signal	423	WHT	0	1.2V	42	Stall, Will Restart In Bypass Mode, Lack Of Power
J2-24	Ignition Control Bypass	424	TAN/BLK	0	4.5V	42	Lack Of Power, Fixed Timing
J2-26	Discrete Switch	31	TAN	-	-	-	Audio Warning System Activation
J2-27	Discrete Switch	31	TAN	-	-	-	Audio Warning System Activation
J2-28	IAC "A" High	441	BRN	Not Usable	Not Usable	None	Rough, Unstable or Incorrect Idle
J2-29	IAC "B" Low	444	GRN/BLK	Not Usable	Not Usable	None	Rough, Unstable or Incorrect Idle
J2-31	MIL Lamp	419	BRN/WHT	0	0	None	Lamp Inoperative

32068c104

Fig. 104 J2 pin locations-V6 and V8 engines w/MEFI-1 ECM, exc. 1998-01 7.4L/8.2L engines (cont'd)

Pin	Pin Function	Circuit (CKT) Number (#)	Wire Color	Normal Voltage		Diagnostic Trouble Codes DTC(s)	Possible Symptoms
				Ignition ON	Engine Running		
J1-28	Not Used	–	–	–	–	–	–
J1-29	MAP Ground	814	BLK	0	0	33	Lack Of Performance, Exhaust Odor, Stall
J1-30	ECM Ground	450	BLK/WHT	0	0	None	No Start
J1-31	MAP 5V Reference	416	GRY	5V	5V	33	Lack Of Power, Surge, Rough Idle, Exhaust Odor
J1-32	Battery	440	ORN	B+	B+	None	No Start

Fig. 107 J1 pin locations-1998-01 454/502/8.2L engines w/MEFI-1 ECM (cont'd)

32068c107

Pin	Pin Function	Circuit (CKT) Number (#)	Wire Color	Normal Voltage		Diagnostic Trouble Codes DTC(s)	Possible Symptoms
				Ignition ON	Engine Running		
J1-11	Ignition Fused	439	PNK/BLK	B+	B+	None	No Start
J1-12	Not Used	–	–	–	–	–	–
J1-13	TP and IAT Ground	813	BLK	0	0	21,23	High Idle, Rough Idle, Poor Performance Exhaust Odor
J1-14	ECM Ground	450	BLK/WHT	0	0	None	No Start
J1-15	TP 5V Reference	416	GRY	5V	5V	21	Lack Of Power, Idle High
J1-16	Battery	440	ORN	B+	B+	None	No Start
J1-17	Not Used	–	–	–	–	–	–
J1-18	Serial Data	461	ORN/BLK	5V	5V	None	No Serial Data (NOTE 6)
J1-19	Not Used	–	–	–	–	–	–
J1-20	Not Used	–	–	–	–	–	–
J1-21	Lanyard Stop Switch	942	PNK	0	0	None	No Start
J1-22	Not Used	–	–	–	–	–	–
J1-23	Not Used	–	–	–	–	–	–
J1-24	IAT Sensor	472	TAN	5V		23	Poor Fuel Economy, Exhaust Odor
J1-25	Not Used	–	–	–	–	–	–
J1-26	Not Used	–	–	–	–	–	–
J1-27	Not Used	–	–	–	–	–	–

Fig. 106 J1 pin locations-1998-01 454/502/8.2L engines w/MEFI-1 ECM (cont'd)

32068c106

Pin	Pin Function	Circuit (CKT) Number (#)	Wire Color	Normal Voltage		Diagnostic Trouble Codes DTC(s)	Possible Symptoms
				Ignition ON	Engine Running		
J2-6	Ignition Control Ref. Low	463	BLK/RED	0	0	None	Poor Performance
J2-7	Port Fuel Jumper	901	WHT	–	–	None	–
J2-8	Ignition Control Ref. High	430	PUR/WHT	5V	1.6V	None	No Restart
J2-9	Fuel Pump Relay Driver	465	DK GRN/WHT	0	B+	None	No Start
J2-10	Not Used	–	–	–	–	–	–
J2-11	Coolant Over temp.	112	DK GRN	0	0	None	Improper Audio Warming
J2-12	Not Used	–	–	–	–	–	–
J2-13	IAC "A" Low	442	BLU/BLK	Not Usable	Not Usable	None	Rough Unstable or Incorrect Idle
J2-14	IAC "B" Low	443	GRN/WHT	Not Usable	Not Usable	None	Rough Unstable or Incorrect Idle
J2-15	Injector Ground	450	BLK/WHT	0	0	None	Rough Running, Lack Of Power, Poor Performance
J2-16	Not Used	–	–	–	–	–	–
J2-17	Not Used	–	–	–	–	–	–
J2-18	Not Used	–	–	–	–	–	–
J2-19	Not Used	–	–	–	–	–	–
J2-20	Fuel Injector Ground	450	BLK/WHT	0	0	None	Rough Running, Poor Idle, Lack Of Performance
J2-21	Injector Driver	467	DK BLU	B+	B+	None	Rough Idle, Lack Of Power, Stalling

32068c109

Fig. 109 J2 pin locations-1998-01 454/502/8.2L engines w/MEFI-1 ECM (cont'd)

Pin	Pin Function	Circuit (CKT) Number (#)	Wire Color	Normal Voltage		Diagnostic Trouble Codes DTC(s)	Possible Symptoms
				Ignition ON	Engine Running		
J2-1	Not Used	–	–	–	–	–	–
J2-2	Not Used	–	–	–	–	–	–
J2-3	Not Used	–	–	–	–	–	–
J2-4	Not Used	–	–	–	–	–	–
J2-5	Injector Driver	468	LT GRN	B+	B+	None	Rough Idle, Lack Of Power, Stall

32068c108

Fig. 108 J2 pin locations-1998-01 454/502/8.2L engines w/MEFI-1 ECM

Pin	Pin Function	Circuit (CKT) Number (#)	Wire Color	Normal Voltage		Diagnostic Trouble Codes DTC(s)	Possible Symptoms
				Ignition ON	Engine Running		
J1-1	Knock Sensor Signal	485	GRN	9.5V	9.5V	43	Poor Fuel Economy, Poor Performance Detonation
J1-2	ECT Signal	410	YEL	1.95V	1.95V	14	Poor Performance, Exhaust Odor, Rough Idle RPM Reduction
J1-3	Not Used	–	–	–	–	–	
J1-4	Discrete Switches	911	BRN	–	–	None	Audio Warning Activated (Low Oil Pressure/ Low I/O Fluid/Transmission Fluid/Overheat) Note: Earlier Models Use J1-4
J1-5	Master/Slave	916	YEL	B+	B+	None	Lack Of Data From Other Engine (Dual Engine Only)
J1-6	Discrete Switches	911	BRN	–	–	None	Audio Warning Activated (Low Oil Pressure/ Low I/O Fluid/Transmission Fluid/Overheat) Note: Later Models Use J1-6
J1-7	Diagnostic Test	451	BLK/WHT	B+	B+	None	Incorrect Idle, Poor Performance
J1-8	Not Used						

32068c111

Fig. 111 J1/J2 pin locations-1998-01 7.4L engines w/MEFI-2 ECM

Pin	Pin Function	Circuit (CKT) Number (#)	Wire Color	Normal Voltage		Diagnostic Trouble Codes DTC(s)	Possible Symptoms
				Ignition ON	Engine Running		
J2-22	Port Fuel Jumper	901	WHT	–	–	–	–
J2-23	Ignition Control Signal	423	WHT	0	1.2V	42	Stall, Will Restart In Bypass Mode, Lack Of Power
J2-24	Ignition Control Bypass	424	TAN/BLK	0	4.5V	42	Lack Of Power, Fixed Timing
J2-25	Not Used	–	–	–	–	–	–
J2-26	Not Used	–	–	–	–	–	–
J2-27	Discrete Switch Signal	31	TAN	–	–	–	Audio Warning System Activated
J2-28	IAC "A" High	441	BLU/WHT	Not Usable	Not Usable	None	Rough Unstable or incorrect Idle
J2-29	IAC "B" Low	444	GRN/BLK	Not Usable	Not Usable	None	Rough Unstable or incorrect Idle
J2-30	Not Used	–	–	–	–	–	–
J2-31	MIL Lamp	419	BRN/WHT	0	0	None	Lamp Inoperative
J2-32	Not Used	–	–	–	–	–	–

32068c110

Fig. 110 J2 pin locations-1998-01 454/502/8.2L engines w/MEFI-1 ECM (cont'd)

Pin	Pin Function	Circuit (CKT) Number (#)	Wire Color	Normal Voltage		Diagnostic Trouble Codes DTC(s)	Possible Symptoms
				Ignition ON	Engine Running		
J1-27	Not Used	-	-	-	-	-	-
J1-28	Not Used	-	-	-	-	-	-
J1-29	MAP Ground	814	BLK	0	0	33	Lack Of Performance, Exhaust Odor, Stall
J1-30	ECM Ground	450	BLK	0	0	None	No Start
J1-31	MAP 5V Reference	416E	GRY	5V	5V	33	Lack Of Power, Surge, Rough Idle, Exhaust Odor
J1-32	Battery	440	ORN	B+	B+	None	No Start
J2-1	Not Used	-	-	-	-	-	-
J2-2	Not Used	-	-	-	-	-	-
J2-3	Not Used	-	-	-	-	-	-
J2-4	Not Used	-	-	-	-	-	-
J2-5	Injector Driver	468	DRK GRN	B+	B+	None	Rough Idle, Lack Of Power, Stall
J2-6	Ignition Control Ref. Low	463	RED/BLK	0	0	None	Poor Performance
J2-7	Port Fuel Jumper	901	BLK	-	-	None	-
J2-8	Ignition Control Ref. High	430	PUR/WHT	5V	1.6V	None	No Restart
J2-9	Fuel Pump Relay Driver	465	DK GRN/WHT	0	B+	None	No Start
J2-10	Not Used	-	-	-	-	-	-
J2-11	Coolant Over temp.	112	DK GRN	0	0	NONE	Improper Audio Warning

32068c113

Fig. 113 J1/J2 pin locations-1998-01 7.4L engines w/MEFI-2 ECM (cont'd)

Pin	Pin Function	Circuit (CKT) Number (#)	Wire Color	Normal Voltage		Diagnostic Trouble Codes DTC(s)	Possible Symptoms
				Ignition ON	Engine Running		
J1-9	Map Signal	432	LT GRN	4.9V	1.46V	33	Poor Performance, Surge, Poor Fuel Economy, Exhaust Odor
J1-10	TP Signal	417	DK BLU	.62V	.62V	21	Poor Performance And Acceleration, Incorrect Idle
J1-11	Ignition Fused	439	PNK	B+	B+	None	No Start
J1-12	Not Used	-	-	-	-	-	-
J1-13	TP and IAT Ground	813	BLK	0	0	21,23	High Idle, Rough Idle, Poor Performance Exhaust Odor
J1-14	ECM Ground	450	BLK	0	0	None	No Start
J1-15	TP 5V Reference	416	GRY	5V	5V	21	Lack Of Power, Idle High
J1-16	Battery	440	ORN	B+	B+	None	No Start
J1-17	Not Used	-	-	-	-	-	-
J1-18	Serial Data	461	ORN	5V	5V	None	No Serial Data (NOTE 6)
J1-19	Not Used	-	-	-	-	-	-
J1-20	Not Used	-	-	-	-	-	-
J1-21	Not Used	-	-	-	-	-	-
J1-22	Not Used	-	-	-	-	-	-
J1-23	Not Used	-	-	-	-	-	-
J1-24	IAT Sensor	472	TAN	5V	-	23	Poor Fuel Economy, Exhaust Odor
J1-25	Not Used	-	-	-	-	-	-
J1-26	Not Used	-	-	-	-	-	-

32068c112

Fig. 112 J1/J2 pin locations-1998-01 7.4L engines w/MEFI-2 ECM (cont'd)

Pin	Pin Function	Circuit (CKT) Number (#)	Wire Color	Normal Voltage		Diagnostic Trouble Codes DTC(s)	Possible Symptoms
				Ignition ON	Engine Running		
J2-26	Discrete Switch Signal	31	DK GRN	–	–	–	Audio Warning System Activated Audio Warning to Buzzer
J2-27	Not Used	–	–	–	–	–	–
J2-28	IAC "A" High	441	LT BLU/WHT	Not Usable	Not Usable	None	Rough Unstable or Incorrect Idle
J2-29	IAC "B" Low	444	LT GRN/BLK	Not Usable	Not Usable	None	Rough Unstable or Incorrect Idle
J2-30	Not Used	–	–	–	–	–	–
J2-31	MIL Lamp	419	BRN/WHT	0	0	None	Lamp Inoperative
J2-32	Not Used	–	–	–	–	–	–

32068c115

Fig. 115 J1/J2 pin locations-1998-01 7.4L engines w/MEFI-2 ECM (cont'd)

Pin	Pin Function	Circuit (CKT) Number (#)	Wire Color	Normal Voltage		Diagnostic Trouble Codes DTC(s)	Possible Symptoms
				Ignition ON	Engine Running		
J2-12	Not Used	–	–	–	–	–	–
J2-13	IAC "A" Low	442	LT BLU/BLK	Not Usable	Not Usable	None	Rough Unstable or Incorrect Idle
J2-14	IAC "B" Low	443	LT GRN/WHT	Not Usable	Not Usable	None	Rough Unstable or Incorrect Idle
J2-15	Injector Ground	450	BLK	0	0	None	Rough Running, Lack Of Power, Poor Performance
J2-16	Not Used	–	–	–	–	–	–
J2-17	Not Used	–	–	–	–	–	–
J2-18	Not Used	–	–	–	–	–	–
J2-19	Not Used	–	–	–	–	–	–
J2-20	Fuel Injector Ground	450	BLK	0	0	None	Rough Running, Poor Idle, Lack Of Performance
J2-21	Injector Driver	467	DK BLU	B+	B+	None	Rough Idle, Lack Of Power, Stalling
J2-22	Port Fuel Jumper	901	BLK	–	–	–	–
J2-22	Port Fuel Jumper	901	BLK	–	–	–	–
J2-23	Ignition Control Signal	423	WHT	0	1.2V	42	Stall, Will Restart in Bypass Mode, Lack Of Power
J2-24	Ignition Control Bypass	424	TAN/BLK	0	4.5V	42	Lack Of Power, Fixed Timing
J2-25	Not Used	–	–	–	–	–	–

32068c114

Fig. 114 J1/J2 pin locations-1998-01 7.4L engines w/MEFI-2 ECM (cont'd)

Pin	Pin Function	CKT	Wire Color	Normal Voltage		DTC	Possible Symptoms
				Ignition ON	Engine Running		
J1-23	Fuel Pump Relay Driver	465	DK GRN/WHT	0	B+	None	No Start
J1-24	Ignition Control Bypass	424	TAN/BLK	0	4.5V	42	Lack Of Power, Fixed Timing
J1-26	Audio Warning Horn	29	DK GRN	–	–	None	–
J1-27	IAC "B" Low	444	GRN/BLK	Not Usable	Not Usable	None	Rough Unstable or Incorrect Idle
J1-28	IAC "A" High	441	BLU/WHT	Not Usable	Not Usable	None	Rough Unstable or Incorrect Idle
J1-30	Knock Sensor Signal	496	BLU	–	–	43, 44	Poor Fuel Economy, Poor Performance Detonation
J1-32	Serial Data	461	ORN	5V	5V	None	No Serial Data

32068c117

Fig. 117 J1 pin locations-V6 and V8 engines w/MEFI-3 ECM (cont'd)

Pin	Pin Function	CKT	Wire Color	Normal Voltage		DTC	Possible Symptoms
				Ignition ON	Engine Running		
J1-1	Injector Driver	467	DK BLU	B+	B+	None	Rough Idle, Lack Of Power, Stalling
J1-2	Ignition Coil	121	WHT	Not Usable	Not Usable	45	Rough Running, Poor Idle, Lack of Performance
J1-3	Ignition Control Ref. Low	453	RED/BLK	0	0	None	Poor Performance
J1-4	ECM Ground	450	BLK	0	0	None	No Start
J1-5	ECM Ground	450	BLK	0	0	None	No Start
J1-9	MIL Lamp	419	BRN/WHT	0	0	None	Lamp Inoperative
J1-10	Ignition Control Signal	423	WHT	0	1.2V	42	Stall, Will Restart In Bypass Mode, Lack Of Power
J1-11	IAC "B" Low	443	GRN/WHT	Not Usable	Not Usable	None	Rough Unstable or Incorrect Idle
J1-12	IAC "A" Low	442	BLU/BLK	Not Usable	Not Usable	None	Rough Unstable or Incorrect Idle
J1-17	Injector Driver	468	DK GRN	B+	B+	None	Rough Idle, Lack Of Power, Stall
J1-20	ECM Ground	450	BLK	0	0	None	Rough Running, Poor Idle, Lack Of Perfomance

32068c116

Fig. 116 J1 pin locations-V6 and V8 engines w/MEFI-3 ECM

Pin	Pin Function	CKT	Wire Color	Normal Voltage		DTC	Possible Symptoms
				Ignition ON	Engine Running		
J2-1	Battery	440	ORN	B+	B+	None	No Start
J2-3	Sensor Ground	813	BLK	0 (NOTE 5)	0 (NOTE 5)	21,23	High Idle, Rough Idle, Poor Performance Exhaust Odor
J2-4	TP 5V Power	416	GRY	5V	5V	21	Lack Of Power, Idle High
J2-7	Discrete Switch	114	BLU	-	-	None	-
J2-8	Discrete Switch	585	TAN/WHT	-	-	None	-
J2-9	Shift Switch	923	WHT	0	0	None	Incorrect Idle
J2-10	Ignition Control Ref. High	430	PUR/WHT	5V	1.6V	None	No Restart
J2-11	ECT Signal	410	YEL	1.95V (NOTE 2)	1.95V (NOTE 2)	14	Poor Performance, Exhaust Odor, Rough Idle rpm Reduction
J2-12	Fuel Pressure	475	GRN	3V	3V	61, 62	-
J2-18	MAP Ground	814	BLK	0 (NOTE 5)	0 (NOTE 5)	33	Lack Of Performance, Exhaust Odor, Stall
J2-19	MAP 5V Reference	416	GRY	5V	5V	33	Lack Of Power, Surge, Rough Idle, Exhaust Odor
J2-20	Discrete Switch	208	BRN	-	-	-	-

32068c118

Fig. 118 J2 pin locations-V6 and V8 engines w/MEFI-3 ECM

Pin	Pin Function	CKT	Wire Color	Normal Voltage		DTC	Possible Symptoms
				Ignition ON	Engine Running		
J2-21	Master/Slave	916	YEL	B+	B+	None	Lack Of Data From Other Engine (Dual Engine Only)
J2-22	Diagnostic Test	451	BLK/WHT	B+	B+	None	Incorrect Idle, Poor Performance
J2-24	Discrete Switch	906	TAN/WHT	-	-	None	-
J2-26	TP Signal	417	DK BLU	.62V (NOTE 4)	.62V (NOTE 4)	21	Poor Performance And Acceleration, Incorrect Idle
J2-27	Map Signal	432	LT GRN	4.9V	1.46V (NOTE 3)	33	Poor Performance, Surge, Poor Fuel Economy, Exhaust Odor
J2-30	iAT Sensor	472	TAN	5V	(NOTE 2)	23	Poor Fuel Economy, Exhaust Odor
J2-32	Ignition Fused	439	PNK	B+	B+	None	No Start

32068c119

Fig. 119 J2 pin locations-V6 and V8 engines w/MEFI-3 ECM (cont'd)

DIAGNOSTIC TROUBLE CODES

DTC Chart - V8
(MP/MPI w/MEFI-1 ECM)

Code	Description
14	Engine coolant temperature sensor
21	Throttle position sensor
23	Intake air temperature sensor
33	Manifold absolute pressure
42	Ignition control
43	Knock sensor
51	ECM calibration memory failure

32068c120

DTC Chart - V6 And V8
(TBI w/MEFI-1 ECM)

Code	Description
14	Engine coolant temperature sensor
21	Throttle position sensor
33	Manifold absolute pressure
42	Ignition control
43	Knock sensor
51	ECM calibration memory failure
52	ECM EEProm failure

32068c121

DTC Chart - V8
(TBI/MP/MPI w/MEFI-2 ECM)

Code	Description
14	Engine coolant temperature sensor (low temp. indicated)
15	Engine coolant temperature sensor (high temp. indicated)
21	Throttle position sensor (signal voltage high)
22	Throttle position sensor (signal voltage low)
23	Intake air temperature sensor (low temp. indicated)
25	Intake air temperature sensor (high temp. indicated)
33	Manifold absolute pressure (signal voltage high)
34	Manifold absolute pressure (signal voltage low)
41	Ignition control (open circuit)
42	Ignition control (gounded circuit, open or grounded bypass)
43	Knock sensor (continuous knock detected)
44	Knock sensor (no knock detected)
51	ECM calibration memory failure
52	ECM EEProm failure

32068c122

DTC Chart - V6 And V8
(TBI/MP/MPI w/MEFI-3 ECM exc. 7.4L/8.2L)

Code	Description
14	Engine coolant temperature sensor (low temp. indicated)
15	Engine coolant temperature sensor (high temp. indicated)
21	Throttle position sensor (signal voltage high)
22	Throttle position sensor (signal voltage low)
23	Intake air temperature (low temp. indicated)
25	Intake air temperature (high temp. indicated)
33	Manifold absolute pressure (signal voltage high)
34	Manifold absolute pressure (signal voltage low)
41	Ignition control (open circuit)
42	Ignition control (gounded circuit, open or grounded bypass)
43	Knock sensor (continuous knock detected)
44	Knock sensor (no knock detected)
45	Coil driver
51	ECM calibration memory failure
52	ECM EEProm failure
61	Fuel pressure high
62	Fuel pressure low

32068c123

DTC Chart - V8
(7.4L/8.2L MP/MPI w/MEFI-3 ECM)

Code	Description
14	Engine coolant temperature sensor (high temp. indicated)
15	Engine coolant temperature sensor (low temp. indicated)
21	Throttle position sensor (signal voltage high)
22	Throttle position sensor (signal voltage low)
23	Intake air temperature (high temp. indicated)
25	Intake air temperature (low temp. indicated)
33	Manifold absolute pressure (signal voltage high)
34	Manifold absolute pressure (signal voltage low)
41	Ignition control (open circuit)
42	Ignition control (gounded circuit, open or grounded bypass)
43	Knock sensor (continuous knock detected)
44	Knock sensor (no knock detected)
45	Coil driver
51	ECM calibration memory failure
61	Fuel pressure high
62	Fuel pressure low

32068c124

TORQUE SPECIFICATIONS

Component		inch lbs. (Nm)	ft. lbs. (Nm)
Adapter-to-manifold		-	30 (41)
Cool fuel assy.-to-engine block		-	30 (41)
Cool fuel retaining bracket		50 (6)	-
ECT sensor		①	①
Flame arrestor		50 (6)	-
Fuel lines			
	Inlet/return	-	23 (31)
	Connections	-	18 (24)
Fuel meter cover (TBI)		28 (3)	-
Fuel rail (MPI)		105 (12)	-
IAC valve		20 (2)	-
IAT sensor		①	①
Knock sensor		-	14 (19)
MAP sensor			
	Inlet/return	53 (6)	-
	Connections	44-62 (5-7)	-
Plenum-to-intake manifold			
	All, exc.	150 (17)	-
	1998-01 7.4L MPI	-	30 (41)
Pressure regulator-to-rail (MPI)		-	10 (12)
Pressure regulator-to-fuel cooler		53 (6)	-
Throttle body-to-adapter			
	1995-97	-	30 (40)
	1998-01	-	15 (19)
Throttle body-to-plenum			
	5.7L/6.2L	75 (8.5)	-
	1994-97 7.4L/8.2L	-	8 (11)
	1998-01 7.4L MPI	75 (8.5)	-
	1998-01 454/502 Mag/8.2L	165 (18.6)	-
	8.1L	89 (10)	-
TP sensor		20 (2)	-
VST-to-engine		105 (12)	-
VST cover-to-body		-	6 (8)

① Handtighten and then an additional 2 1/2 turns

9

COOLING SYSTEM

COOLING SYSTEMS

Description & Operation

All MerCruiser engines are cooled by means of one of two systems: an external-water, Seawater system; or a Closed system, which actually incorporates the Seawater system into a closed automotive-style anti-freeze system. All engines may come equipped with either of the two systems. Please refer to the flow diagrams at the end of this section.

SEAWATER SYSTEM

As implied by the name, this system utilizes water from outside the boat to cool the engine and certain related components. All versions of this system utilize two water pumps, hoses and a thermostat.

Models with an Alpha One stern drive unit use a pump mounted on top of the lower unit which draws water in through the drive unit and sends it on to an engine circulating pump attached to the front of the cylinder block—although different in construction, this pump is quite similar to a typical water pump you would find on your car or truck. From here the water is circulated through the engine block, cylinder heads and exhaust manifold(s); and then expelled back into the body of water where it originated.

Models with a Bravo stern drive or inboard use a belt-driven seawater pump mounted on the lower end of the engine to draw the water up through the drive unit and then on to the engine circulating pump just like on Alpha models. If you've performed any fuel system work, you'll recognize that 1994–97 7.4L/8.2L engines incorporate their mechanical fuel pump into the seawater pump.

CLOSED SYSTEM

This system is actually two systems in one—a closed freshwater system and a seawater system. All versions of this system utilize two water pumps, hoses, a thermostat and a heat exchanger.

In the closed portion of the system, a mixture of freshwater and anti-freeze is circulated through-out the engine block, cylinder head and a heat exchanger (similar to your car's radiator). On 1994–97 V6 and V8 engines, the exhaust manifolds are also cooled. The pressurized water/coolant never leaves the system and is thermostatically controlled.

Lacking a radiator and fan like an automobile, it is necessary to find another means of keeping the fluid in the closed system from boiling and this is where the second portion of this system comes into play. Seawater from outside the vessel is drawn in by means of a pick-up pump in the stern drive unit on Alpha models, or by a belt-driven engine-mounted seawater pump on all other models. Unlike true Seawater models as described previously, the seawater is pumped into the heat exchanger rather than through the engine. In fact, it bypasses the engine circulating pump altogether. Once in the exchanger, the seawater is routed through tubes surrounding the closed system tubes, thus cooling the fresh water/anti-freeze mix as it passes through the exchanger. The heated seawater is then routed through the exhaust manifold(s) and then back overboard. On 1994–97 V8 engines, the seawater bypasses the manifold and goes right to the exhaust elbow, and then goes overboard.

Troubleshooting

The following paragraphs list troubles encountered in the various portions of the system with accompanying probable causes of the problem.

The causes are given in a logical order of checking until the problem is corrected.

ENGINE OVERHEATS—SEAWATER SYSTEM

- Loose or broken circulating pump belt or pick-up water pump belt.
- Inaccurate temperature gauge or sender.
- An accessory or barnacles in front of water pickup causing turbulence.
- Defective water pick-up pump or seawater pump.
- Loose hose connections between pick-up and pump—sucking air.
- Pump fails to hold prime due to air leaks.
- Ice in water passages.
- Defective engine circulating pump.
- Defective thermostat.

ENGINES OVERHEATS—CLOSED SYSTEM, IN ADDITION TO ABOVE

- Closed cooling system reservoir level low.
- Plugged heat exchanger cores.
- Too much anti-freeze in closed system.
- Fresh water kit not installed properly.
- Exhaust elbow dump fittings "bottomed" on inner water jacket.

ENGINE OVERHEATS—MECHANICAL

- Incorrect ignition timing.
- Spark plug wires crossed.
- Lean air/fuel mixture.
- Pre-ignition—spark plugs are wrong heat range.
- Engine laboring—engine rpm below specifications at WOT.
- Poor lubrication.
- Water in cylinders—due to warped cylinder head.
- Distributor not functioning properly.
- Clogged exhaust elbows.
- Exhaust flappers stuck closed.
- Cabin hot water heater incorrectly connected to engine.

WATER IN CYLINDERS

- Backwash through exhaust system.
- Loose cylinder head bolts.
- Blown cylinder head gasket.
- Warped cylinder head.
- Cracked or corroded exhaust manifold.
- Cracked block in valve lifter area.
- Improper engine or exhaust hose installation.
- Cracked cylinder wall.

WATER IN OIL

- Backwash through exhaust system.
- Water seeping past piston rings from flooded combustion chamber.
- Thermostat stuck open or missing—condensation forms because engine is operating too cool.
- Cracked cylinder block.
- Intake manifold water passage leak.

MAINTENANCE AND TESTING

First, the most important words in this manual: The engine cannot be operated for even five seconds without water moving through the water pick-up/seawater pump or the pump impeller will be damaged. Therefore, never start the engine, even for testing purposes, without the boat being in the water, or provision having been made for water to pass through the seawater pick-up pump.

■ Marine thermostats are rated at 143 degrees–160 degrees. An automotive-type thermostat must never be used because of the higher temperature ratings. Such a high rating would cause the engine to run much hotter than normal.

Cooling System

The cooling system, seawater or closed, should be cleaned and flushed at least once every two years. Please refer to Section 3, Maintenance for detailed flushing procedures.

PRESSURE TEST

Closed System Only

✳✳ WARNING

Make sure the engine is cool before attempting to remove the cap. To be safe, always turn the cap 1/4 turn and allow any residual pressure to escape before removing it completely. Use a heavy rag or wear gloves.

1. Remove the pressure cap from the reservoir or exchanger.
2. Perform the pressure cap Testing procedure. If the cap is bad, this may be your problem. Replace the cap and ensure that the problem does not go away; if it persists, move to the next step.
3. Add coolant so that the level is within one inch of the bottom of the filler neck. Attach a cooling system pressure tester to the filler neck and pressurize the system to 17 psi on 3.0L and 1992–97 V6 and 5.0L/5.7L V8 engines; 20 psi on 1998–01 V6 and 5.0L/5.7L/6.2L V8 engines, and all 7.4L/8.2L V8 engines. On 8.1L engines, pressurize to 19 psi.
4. Watch the gauge for about 2 minutes. If the pressure remains steady, you're OK. If the pressure drops, move to the next step.
5. Maintain the specified pressure and check the entire closed system for any leaks—hoses, plugs, petcocks, pump seals, etc.). Also, listen very closely for any hissing or bubbling.
6. If you've still not found any leaks, test the heat exchanger as detailed in this section.
7. If the exchanger is OK then you most likely have an problem with loose head bolts, a bad head gasket or a warped cylinder head, not the cooling system.

DRAINING & FILLING

The cooling system should be drained, cleaned, and refilled each season although Mercury's recommendations are for every two years on normal anti-freeze systems and 5 years or 1000 hours on systems with Extended Life coolant. We think its cheap insurance to do it every season, but you certainly can't go wrong by following the factory's suggestion. The bow of the boat must be higher than the stern to properly drain the cooling system. If the bow is not higher than the stern, water will remain in the cylinder block and in the exhaust manifold. Insert a piece of wire into the drain holes, but not in the petcock to ensure sand, silt, or other foreign material is not blocking the drain opening.

If the engine is not completely drained for winter storage, trapped water can freeze and cause severe damage. The water in the oil cooler—if so equipped with power steering—must also be drained.

For complete storage or pre-season preparation procedures—see Section 3.

Seawater System

◆ See Figures 1, 2, 3, 4, 5, 6, 7, 8 and 9

■ Some 3.0L engines may be equipped with a single point drain system for the engine. Disconnect the two lines from their top fittings and bend them down so they are lower than their bottom connections. If they are difficult to lower, use the T-handle to persuade them. Reconnect the lines after draining.

1. If the boat is in the water, close the seacock to prevent water from entering the cooling system. If the boat is not equipped with a seacock, disconnect the seawater inlet line and plug it. Make sure you leave yourself a note by the ignition switch reminding yourself to turn the valve back on or connect the line, especially if the boat is going to sit for a period of time.
2. Position suitable containers under all drain plugs and hoses (if space permits) to catch any water being drained or else it will collect in the bilge.

3. Remove the drain plug on the side of the cylinder block (check both sides of the block, as most engines have them on each).
4. Remove the drain plug(s) on the exhaust manifold or elbow. Certain V6and V8 engines may not be equipped with a drain so simply disconnect the hose from the lower fitting.
5. On engines equipped with risers, remove the drain plug.
6. On models equipped with power steering, disconnect the lower seawater hose at the cooler—some models may actually have a drain plug on the bottom side of the cooler.
7. On engines with closed cooling, remove the drain plug from the bottom of the heat exchanger. If there is no plug, disconnect the seawater inlet hose at the bottom of the exchanger.
8. Disconnect and remove the lower end of all hoses and hold them as low as possible while draining. Reconnect when all water has stopped draining. On models with closed cooling, refer to the flow diagrams at the rear of this section so that you do not disconnect any hoses used for the closed portion of the system.
9. Using a stiff piece of wire, clean out all drain holes. Don't forget any drain holes in the drive unit!
10. Turn the ignition switch ON and crank the engine over a few times to force out any residual water in the block or seawater pump. DO NOT START THE ENGINE!
11. Coat the threads of all drain plugs with Perfect Seal and then reinstall them. Tighten the plugs and petcock securely.
12. If you are winterizing the boat and its going to sit for a few months, it's a good idea to remove the thermostat housing and fill the cylinder block and head with a mixture of water and anti-freeze. Remove the inlet line at the exhaust manifold and do the same. Make sure you drain all coolant into a suitable container prior to refilling the system with seawater next season.
13. If you closed the seacock or disconnected the water inlet line, make sure you open or reconnect it PRIOR to restarting the engine.

Closed System

3.0L AND 1992–97 V6 ENGINES

This system should be kept filled with the appropriate water and anti-freeze solution year-round; even if laying the boat up for the winter. Coolant should be changed at least every other year.

■ **Don't forget to drain the seawater system if winterizing the boat.**

1. Carefully remove the pressure cap from the system. Follow the precautions detailed in the Pressure Cap procedures in this section if the engine is hot.
2. Loosen and remove the petcock(s) or drain plug(s) on the cylinder block making sure you have a suitable container under the drains or else the coolant will flow into the bilge,.
3. Disconnect the coolant hose (the other is a seawater hose—refer to the flow diagrams) from the engine circulating pump at the heat exchanger. Loosen and remove the drain plug at the bottom of the exchanger.
4. Clean all drain holes with a stiff piece of wire.
5. If equipped with a coolant recovery or overflow tank, drain it also.
6. Coat the threads of all plugs with Perfect Seal and install them securely. Reconnect the coolant hose at the heat exchanger.
7. Pour the appropriate mix of water and anti-freeze into the reservoir filler neck until the level is within 1 in. of the bottom of the neck. Remember, many 1998 and later engines utilize Extended Life Coolant.
8. With the cap still off, start the engine and run it at 1500–1800 rpm. Add additional coolant until the level is within 1 in. of the bottom of the filler neck. If the boat is out of the water, don't forget to attach a flushing device to ensure proper flow of seawater.
9. After the engine has reached normal operating temperature, check that the level in the filler neck is correct; add coolant if necessary.
10. Install the pressure cap, check the coolant recovery reservoir and fill it to the FULL line.
11. If any leakage or overheating is noticed, shut the engine down immediately and look for the cause of the problem.

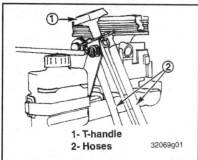

1- T-handle
2- Hoses

32069g01

Fig. 1 Certain models utilize a single point drain system for the engine—3.0L engines

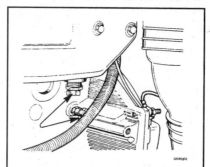

32069g02

Fig. 2 Loosen the drain plugs for the block and manifold—3.0L engines

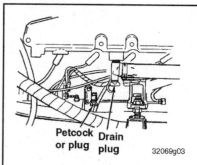

Petcock or plug   Drain plug

32069g03

Fig. 3 Loosen the drain plugs for the block and manifold—V6 and V8 engines

DRAIN PLUG

32069p04

Fig. 4 A good view of the cylinder block drain plug

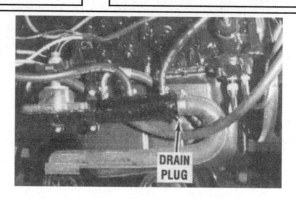

DRAIN PLUG

Fig. 7 Most power steering coolers have a drain plug

DRAIN PLUG

32069p05

Fig. 5 An even better view

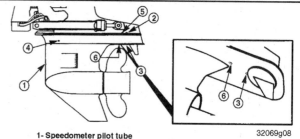

1- Speedometer pilot tube
2- Gear housing cavity drain hole
3- Trim tab cavity vent hole
4- Trim tab cavity drain passage
5- Gear housing water drain hole (One ea. -port and starboard)
6- Gear housing cavity vent hole

32069g08

Fig. 8 Drain holes on the drive unit should be cleaned with a stiff wire—Alpha

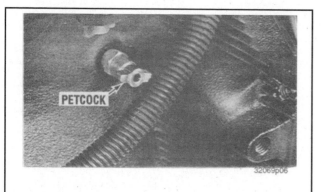

PETCOCK

32069p06

Fig. 6 Some engine will use a petcock in place of the plug

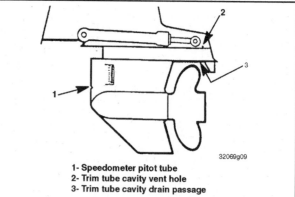

1- Speedometer pitot tube
2- Trim tube cavity vent hole
3- Trim tube cavity drain passage

32069g09

Fig. 9 Drain holes on the drive unit should be cleaned with a stiff wire—Bravo

1998–01 V6/V8 ENGINES EXCEPT 8.1L

◆ **See Figures 10 and 11**

This system should be kept filled with the appropriate water and anti-freeze solution year-round; even if laying the boat up for the winter. Coolant should be changed at least every other year, or every 5 years on models with Extended Life coolant systems.

■ **Don't forget to drain the seawater system if winterizing the boat.**

1. Carefully remove the pressure cap from the system. Follow the precautions detailed in the Pressure Cap procedures in this section if the engine is hot.
2. Loosen and remove the petcock(s) or drain plug(s) on the cylinder block making sure you have a suitable container under the drains or else the coolant will flow into the bilge,.
3. Disconnect the coolant hose (the other is a seawater hose—refer to the flow diagrams) from the engine circulating pump at the heat exchanger. Loosen and remove the drain plug at the bottom of the exchanger.
4. Clean all drain holes with a stiff piece of wire.
5. If equipped with a coolant recovery or overflow tank, drain it also.
6. Coat the threads of all plugs with Perfect Seal and install them securely. Reconnect the coolant hose at the heat exchanger.
7. Open the bleeder valve screw on top of the thermostat housing.
8. Pour the appropriate mix of water and anti-freeze into the reservoir filler neck until coolant comes out the bleeder valve. Close the bleeder and tighten it securely.
9. Continue adding coolant until the level comes into the filler neck and begins to flow into the coolant overflow container.
10. With the cap still off, start the engine and run it at idle. Add additional coolant until the level is within 1 in. of the bottom of the filler neck. If the boat is out of the water, don't forget to attach a flushing device to ensure proper flow of seawater.
11. After the engine has reached normal operating temperature and the thermostat is fully open, check that the level in the filler neck is correct; add coolant if necessary.
12. Install the pressure cap, check the coolant recovery reservoir and fill it to the FULL line.
13. If any leakage or overheating is noticed, shut the engine down immediately and look for the cause of the problem.

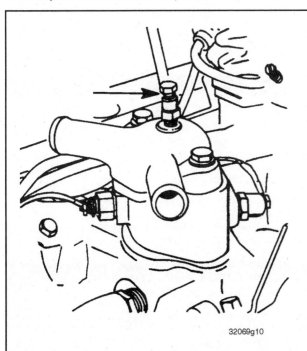

Fig. 10 There is a bleeder valve on the top of the thermostat housing

32069p11

Fig. 11 Most heat exchangers have a drain plug

1992–97 V8 ENGINES

◆ **See Figures 10 and 11**

This system should be kept filled with the appropriate water and anti-freeze solution year-round; even if laying the boat up for the winter. Coolant should be changed at least every other year.

■ **Don't forget to drain the seawater system if winterizing the boat.**

1. Carefully remove the pressure cap from the system. Follow the precautions detailed in the Pressure Cap procedures in this section if the engine is hot.
2. Loosen and remove the petcock(s) or drain plug(s) on the cylinder block making sure you have a suitable container under the drains or else the coolant will flow into the bilge,.
3. Disconnect the coolant hose (the other is a seawater hose—refer to the flow diagrams) from the engine circulating pump at the heat exchanger. Loosen and remove the drain plug at the bottom of the exchanger.
4. Clean all drain holes with a stiff piece of wire.
5. If equipped with a coolant recovery or overflow tank, drain it also.
6. Coat the threads of all plugs with Perfect Seal and install them securely. Reconnect the coolant hose at the heat exchanger.
7. Open the bleeder valve screw on top of the thermostat housing.
8. Pour the appropriate mix of water and anti-freeze into the reservoir filler neck until coolant comes out the bleeder valve. Close the bleeder and tighten it securely.
9. Continue adding coolant until the level comes to within 1 in. of the bottom of the filler neck.
10. With the cap still off, start the engine and run it at 1500–1800 rpm. Add additional coolant until the level is within 1 in. of the bottom of the filler neck. If the boat is out of the water, don't forget to attach a flushing device to ensure proper flow of seawater.
11. After the engine has reached normal operating temperature and the thermostat is fully open, check that the level in the filler neck is correct; add coolant if necessary.
12. Install the pressure cap, check the coolant recovery reservoir and fill it to the FULL line.
13. If any leakage or overheating is noticed, shut the engine down immediately and look for the cause of the problem.

8.1L ENGINES

This system should be kept filled with the appropriate water and anti-freeze solution year-round; even if laying the boat up for the winter. Coolant should be changed at least every other year, or every 5 years on models with Extended Life coolant systems.

■ **Don't forget to drain the seawater system if winterizing the boat.**

1. Carefully remove the pressure cap from the system. Follow the precautions detailed in the Pressure Cap procedures in this section if the engine is hot.
2. Loosen and remove the petcock(s) or drain plug(s) on the cylinder block making sure you have a suitable container under the drains or else the coolant will flow into the bilge,.
3. Disconnect the coolant hose (the other is a seawater hose—refer to the flow diagrams) from the engine circulating pump at the heat exchanger or crossover. Loosen and remove the drain plug at the bottom of the exchanger.

4. Clean all drain holes with a stiff piece of wire.

5. If equipped with a coolant recovery or overflow tank, drain it also.

6. Coat the threads of all plugs with Perfect Seal and install them securely. Reconnect the coolant hose at the heat exchanger and crossover.

7. Fill the reservoir with the appropriate mixture of water and anti-freeze until it reaches the FULL line.

8. Replace the cap and start the engine. Check for leaks.

Crossover

REMOVAL & INSTALLATION

8.1L Engines Only

1. Drain the cooling system.

2. Disconnect the coolant lines at the heat exchanger if you haven't already. Disconnect the hose leading to the overflow tank.

3. Loosen the heat exchanger mounting bolts and then loosen the hex bolt on the retaining strap.

4. Remove the retaining strap, take the mounting bolts out and then lift off the exchanger.

5. Remove the thermostat retainer and pull out the thermostat.

6. Tag and disconnect any electrical connections at the crossover.

7. Disconnect all coolant lines at the crossover.

8. Remove the two mounting bolts on each side of the crossover at the cylinder heads and remove the crossover.

To install:

9. Align the bypass and inlet hoses at the water pump with the crossover and place it into position.

10. Make sure all gaskets and lines are in alignment and then tighten the mounting bolts to 37 ft. lbs. (50 Nm).

11. Reconnect all electrical leads and coolant lines. Tighten the hose clamps securely.

12. Insert the thermostat and install the retainer.

13. Carefully position the heat exchanger and install the retaining strap. Do not tighten to tight or you could damage the holding slots.

14. Tighten the retainer bolts securely. Refill the cooling system, start the engine and check the system for leaks.

Drive Belts

For all removal and adjustment procedures, please refer to Section 3.

Heat Exchanger

CLEANING

Closed Cooling System Only

◆ See Figure 12

■ On some models it may be necessary to remove the exchanger prior to cleaning it. If you need to do this, make sure to refill the closed portion of the system again.

1. Remove the drain plug at the bottom of the exchanger making sure you have a suitable container underneath it. When the water has drained completely, coat the threads of the drain plug with Perfect Seal and tighten the plug securely.

2. Remove the end plate retaining bolts on each side of the exchanger and then pry off the plates, seals and gaskets. Carefully scrape any residual gasket material from the mating surfaces.

3. Clean each water passage in the unit with a small wire brush (carefully!) and then blow out each passage with compressed air if available.

4. Coat both side of new gaskets and seals with Perfect Seal. Install the gasket, seal and then end plate. Tighten the bolt to 190 inch lbs. (22 Nm) on 1992–97 engines; 36–72 inch lbs. (4–8 Nm) on 1998–01 engines except for the 8.1L where they should be tightened to 54 inch lbs. (6 Nm). Fill the system with coolant if necessary.

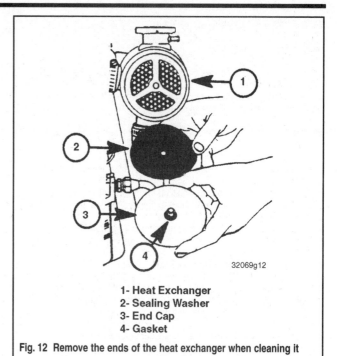

1- Heat Exchanger
2- Sealing Washer
3- End Cap
4- Gasket

Fig. 12 Remove the ends of the heat exchanger when cleaning it

TESTING

Closed Cooling System Only

1. An internal leak will cause coolant to mix with the seawater system when pressurized. Remove the seawater hose from the exchanger, but do not drain it. Install a cooling system tester and pressurize the circuit to 14–20 psi on 1992–97 engines; 16–20 psi on 1998–01 engines. If seawater begins to seep out of the fitting, there is a leak and the exchanger will require replacement.

2. To check for blockage, remove the end caps and inspect for any visible blockage. Remove the coolant hoses and inspect the tubes just inside the fittings. Although you can plug the coolant lines here, you may wish to drain the system instead. If any blockage is apparent, replace the exchanger.

Pressure Cap

TESTING

Closed System Only

◆ See Figures 13, 14 and 15

The pressure cap on the coolant reservoir is designed to maintain closed system pressure at approximately 14 psi when the engine is at normal operating temperature. Clean, inspect and pressure test the cap at the end of your first season and then every 100 hours (or once a year) thereafter.

✳✳ WARNING

Make sure the engine is cool before attempting to remove the cap. To be safe, always turn the cap 1/4 turn and allow any residual pressure to escape before removing it completely. Use a heavy rag or wear gloves.

1. Remove the cap and wash it thoroughly to remove any debris from the sealing surfaces.

2. Inspect the gasket and rubber seal for tears, cuts or cracks. If you notice anything, we recommend replacing the entire cap although you can replace the gasket on some older engines if you like.

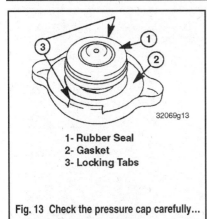

1- Rubber Seal
2- Gasket
3- Locking Tabs

Fig. 13 Check the pressure cap carefully...

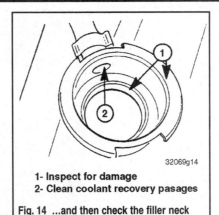

1- Inspect for damage
2- Clean coolant recovery pasages

Fig. 14 ...and then check the filler neck

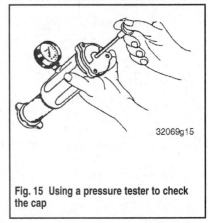

Fig. 15 Using a pressure tester to check the cap

3. Make sure that the locking tabs on the cap are not bent or damaged in any way. If so, replace the cap.

4. Use a cooling system pressure tester and, following the manufacturer's instructions, install the cap. Check that the cap relieves pressure at 14 psi (16 psi on 1992–97 7.4L/8.2L V8 engines and all 1998–01 V6/V8 engines) and holds pressure for 30 seconds without falling below 11 psi. Replace the cap if it fails.

5. Check the inside of the filler neck for debris; it should be completely smooth. Check that the lock flanges on the filler neck are not bent or damaged and then install the cap.

Sea Strainer

REMOVAL & INSTALLATION

② ◁ MODERATE

If performing the following while boat is in the water, close the seacock. If your boat is not equipped with a seacock, disconnect and plug the seawater inlet line to prevent water from entering the system.

■ **The engine should be OFF and cool.**

1. Disconnect the inlet and outlet hoses at the strainer.
2. Remove the mounting bolts and lift off the strainer. Clean the strainer and inspect it as detailed in Section 3.
3. Mount the strainer so that the arrow on the casing points toward the seawater pump and tighten the mounting bolts securely.
4. Install the inlet and outlet lines with two hose clamps on each side.

Seawater Pick-Up Pump

REMOVAL & INSTALLATION

② ◁ MODERATE

Except 1992–97 7.4L/8.2L V8 Engines

◆ **See Figures 16 and 17**

If performing the following while boat is in the water, close the seacock. If your boat is not equipped with a seacock, disconnect and plug the seawater inlet line to prevent water from entering the system.

■ **The engine should be OFF and cool.**

1. Disconnect the inlet and outlet hoses at the strainer.
2. If you plane on disassembling the pump, loosen the 4 pump pulley mounting bolts. Do not remove them yet.
3. Loosen the power steering pump bracket and move the pump until tension is relieved on the drive belt. Remove the power steering belt and then remove the seawater pump belt. Now you can remove the pulley bolts and lift off the pulley.
4. Loosen the pump bracket and brace mounting bolts at the cylinder block. Pull the bolts out and remove the pump assembly. Most models will come off with the brackets and brace attached although there are certain applications where you simply unbolt the pump from the bracket and brace. Either way, your goal is to eventually get the pump free of the bracket and brace, so you be the judge.

To Install:

5. Inspect the pump assembly for cracks, damage or other signs of wear.

6. Position the pump in the mounting bracket so that the hose fittings are vertical, with the outlet being on top of the inlet—this is very important, reversing the positions will cause overheating and damage to the engine.

7. Install the flat washer (spacer) between the mounting bracket bosses and then install the clamp screw and lockwasher, but do not tighten it yet.

8. Attach the brace to the bracket (if you actually removed the pump while attached) with the bolt and nut. Don't forget the lockwasher and don't tighten it yet.

9. Install the pulley with the mounting bolts fingertight.

10. Attach the entire assembly to the cylinder block, screwing in the bolts only fingertight.

11. Move the pump in the bracket until the pulley is in alignment with the crankshaft pulley and then tighten the clamp bolt to 20 ft. lbs. (27 Nm) and the pulley mounting bolts securely. Install the drive belt and adjust it as detailed in Section 3. Now you can tighten the remaining bolts. The brace bolt and the bracket-to-block bolts should be tightened to 30 ft. lbs. (41 Nm). The bracket-to-pump nuts should be tightened to 120 inch lbs. (4 Nm).

12. Reconnect the water hoses, outlet on top, and tighten the hose clamps securely.

13. Install the power steering belt and adjust it if equipped. Start the engine and check for leaks.

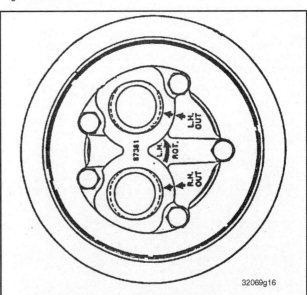

Fig. 16 When installing the pump in the mounting bracket, the outlet line MUST be on top—standard rotation engines...

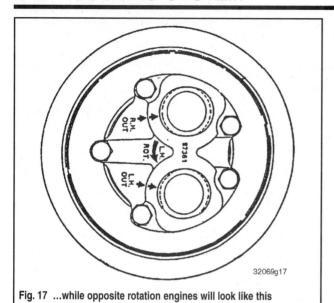
Fig. 17 ...while opposite rotation engines will look like this

1992–97 7.4L/8.2L V8 Engines

◆ See Figures 18, 19 and 20

If performing the following while boat is in the water, close the seacock. If your boat is not equipped with a seacock, disconnect and plug the seawater inlet line to prevent water from entering the system.

These engines have the fuel pump as part of the seawater pump.

■ The engine should be OFF and cool.

1. Disconnect the battery cables.
2. Loosen the idler pulley adjustment bolt and relieve tension on the drive belt. Remove the belt.
3. Disconnect the inlet and outlet lines at the pump. Make sure you have a suitable container handy.
4. Disconnect the two fuel lines at the fuel pump and plug them so they don't leak further.
5. Remove the two pump bracket-to-cylinder block mounting bolts. Remove the two pump-to-idler pulley bracket bolts and lift out the pump assembly.

To Install:

6. Inspect the pump assembly for cracks, damage or other signs of wear.
7. On Mark IV engines, position the pump so that the hose fittings are vertical, with the outlet being on top of the inlet; while on Generation V engines, the outlet fitting should be on the bottom—this is very important, reversing the positions will cause overheating and damage to the engine.

Fig. 18 Loosen the idler pulley nut before removing the drive belt

Fig. 19 Disconnect the inlet and outlet lines

Fig. 20 The water pump, fuel pump and idler are removed as an asssembly

Tighten the bracket and idler pulley bolts to 30 ft. lbs. (41 Nm).

8. Reconnect the fuel lines.
9. Install the drive belt and check the belt tension as detailed in Section 3. Tighten the idler locknut to 30 ft. lbs. (41 Nm).
10. Reconnect the water hoses and tighten the hose clamps securely.
11. Start the engine and check for leaks.

DISASSEMBLY & ASSEMBLY

Except 1992–97 7.4L/8.2L V8 Engines

◆ See Figures 21, 22, 23 and 24

1. Remove the pump.
2. Remove the 5 pump cover mounting bolts and their washers and lift the cover off the pump body.
3. Remove the gaskets and wear plate and then slide the pump body up and off of the shaft.
4. Pull out the rubber plug and then lift out the impeller.
5. Lift or scrape the gasket off the lower mating surface and pinch out the quad ring seal.
6. Install a Universal Puller Plate (# 91-37241) over the hub and then press it off the shaft.
7. Puncture the front oil seal with an awl and pry it out of the bearing housing.

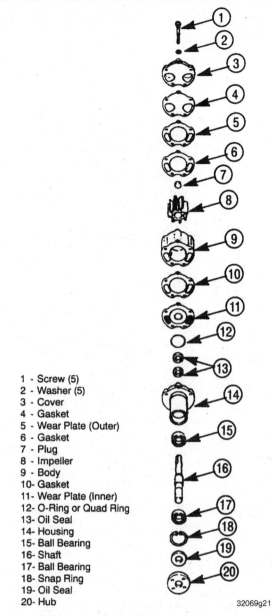

1 - Screw (5)
2 - Washer (5)
3 - Cover
4 - Gasket
5 - Wear Plate (Outer)
6 - Gasket
7 - Plug
8 - Impeller
9 - Body
10- Gasket
11- Wear Plate (Inner)
12- O-Ring or Quad Ring
13- Oil Seal
14- Housing
15- Ball Bearing
16- Shaft
17- Ball Bearing
18- Snap Ring
19- Oil Seal
20- Hub

32069g21

Fig. 21 Exploded view of the seawater pump

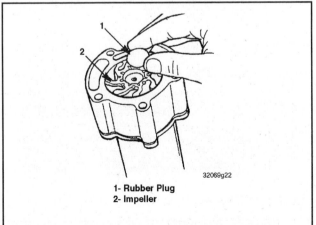

1- Rubber Plug
2- Impeller

32069g22

Fig. 22 Remove the rubber plug before lifting out the impeller

8. Using a pair of snap-ring pliers, remove the ring from the bore and then press the shaft and bearings out of the bore through the pulley end.

✳✳ CAUTION

The pump bearings are slip-fitted into the housing so do not use excessive force when removing them.

9. If the bearings need replacement, press them off the shaft with the Puller tool. Remember that if you remove them, you must replace them.
10. Press out the rear seals.
To assemble:
11. Coat two new rear housing seals with Loctite and then install them so that the lips face the impeller. Press in the lower seal until it bottoms and then the upper so that it is flush with the housing. Pack the cavity between the seals with Shell Alvania No. 2 grease (in a pinch, you can use Quicksilver 2-4-C).
12. Press the two bearings onto the shaft until they seat and then pack them with the above grease only. When installing the bearings, press on the inner race only in order to avoid damaging them. Slide the bearings and shaft into the housing and then pop in the snap-ring.
13. Coat a new front seal with Loctite (outer edge only) and press it into the housing, with the lip facing in, until it bottoms.
14. Coat the pump shaft with Quicksilver Special Lubricant 101 and then press the pulley hub onto the shaft until it's outer surface is 17/64 in. (6.6mm) from the outer end of the shaft.
15. Mount the bearing housing carefully into a vise with padded jaws so the flange end is pointing up. Coat the quad ring seal with 2-4-C and press it into the groove in the housing.
16. Position the wear plate onto the housing flange so the holes are in alignment. Coat both sides of a new gasket with Perfect Seal and lay it on top of the plate.
17. Position the impeller over the bore in the pump body and then turn it slowly in its normal direction of rotation while pressing down on it until it seats into the bore. Install the pump assembly onto the shaft so that all holes line up with those in the flange and gasket. Press in the rubber plug.
18. Coat both side of the two cover gaskets with Perfect Seal and position them onto the pump body so that all holes align correctly. The gasket with the single large hole goes down first, then the wear plate and finally the gasket with the two holes. Drop the cover into place and tighten the bolts to 120 inch lbs. (14 Nm).
19. Install the pump.

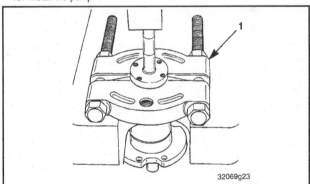

32069g23

Fig. 23 A special puller tool and an arbor press are needed to remove the hub and bearings

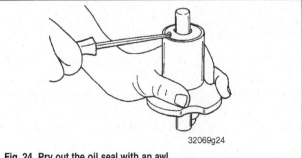

32069g24

Fig. 24 Pry out the oil seal with an awl

1992–97 7.4L/8.2L V8 Engines

◆ **See Figures 25, 26, 27, 28, 29, 30, 31, 32, 33, 34, 35, 36, 37 and 38**

1. Remove the pump assembly.
2. Paint a small line across the pump housing and the mounting bracket for reassembly.
3. Remove the 5 pump mounting bolts and their washers and lift the bracket off the pump body. Note that one bolt has a nut on the end.
4. Pull the pump end cover off of the housing.
5. Remove the gaskets and separator plate.
6. Pull out the rubber plug and then lift the housing off of the pump base.
7. Carefully press the impeller out of the housing.
8. Lift or scrape the gasket and end plate off the lower mating surface.
9. Remove the fill and drain screws from the side of the fuel pump along with their seal washers.

■ **It is normal to find a bit of metal particles on the magnetic drain screw. Large amounts of particles indicate abnormal wear.**

10. Loosen the two mounting bolts and remove the fuel pump from the pump housing.
11. Attach a two-armed puller to the pulley and remove it from the assembly.
12. Pinch out the quad ring seal.
13. Puncture the oil seal with an awl and pry it out of the bearing housing.
14. Using a pair of snap-ring pliers, remove the ring from the bore and then press the shaft and bearings out of the bore through the pulley end.
15. Install a Universal Puller Plate (# 91-37241) over the bearings and then press them off the shaft.
16. Slide the washer and slip ring off of the shaft.
17. Carefully pry out the inner and outer seals (oil and water).

To assemble:

18. Coat the outer edge of two new rear housing seals with Loctite and then install them so that the lip of the inner seal faces the oil pump and the lip of the outer seal faces the water pump. The inner oil seal should be pressed in until the edge is even with the edge of the bore; while the outer water seal should be flush with the face of the housing. Pack the cavity between the seals with Quicksilver Special Lubricant 101.
19. Lubricate all shaft components and bearings with Quicksilver High Performance Gear Lube.
20. Slide the slip ring and washer onto the shaft from the impeller side.
21. Press the bearings onto the shaft until they are fully seated.
22. Slowly lower the shaft assembly into the housing while rotating it slightly so that you don't roll the seal lips as it passes through them. Press it all the way in until seated—when you can see the snap-ring groove above the upper bearing, you'll know its all in position.
23. Install a new snap-ring.
24. Coat a new front seal with Loctite (outer edge only) and press it into the housing, with the lip facing in, until it bottoms.
25. Install the quad ring seal on the water pump end of the shaft.
26. Press the pulley back onto the fuel pump shaft so that its front edge is exactly 0.5 in. (13mm) from the end of the shaft. Rotate the pulley until the low spot on the pump cam is facing the pump lever. Coat a new gasket with Perfect Seal and install to the water pump. Tighten the mounting bolts to 25–28 ft. lbs. (34–38 Nm).
27. Install the drain screw and tighten to 18 inch lbs. (2 Nm). Fill the pump with Quicksilver High Performance Gear Lube until it reaches the bottom edge of the filler holes and then install the screw. Tighten to 18 inch lbs. (2 Nm).
28. Coat the impeller with soapy water. Position it over the bore in the pump body and then turn it slowly in its normal direction of rotation while pressing down on it until it seats into the bore.
29. Run the mounting bolts through the pump cover and then slide on the two gaskets and wear plate. The two-holed gasket goes on first, then the plate and finally the gasket with the single large hole.
30. Press in the rubber plug and then slide the entire cap assembly onto the housing. Slide the remaining gasket and wear plate over the ends of the protruding bolts. Swivel the pump shaft until the flat side is in approximate alignment with the flat on the inside of the impeller bore and then slide the entire assembly over the shaft, rotating it as you go until the bolts line up with their holes.

31. Slide the bracket over the assembly and thread in four of the five bolts. Install a lockwasher and nut to the fifth and then tighten them all to 10–15 ft. lbs. (14–20 Nm).
32. Install the pump.

Fig. 25 Matchmark the pump housing to the bracket

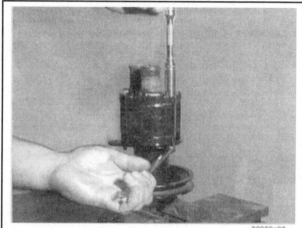

Fig. 26 Remove the five mounting bolts

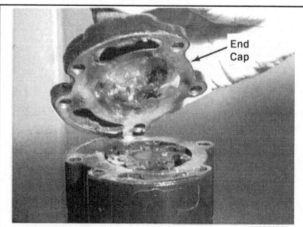

Fig. 27 Lift off the end cover

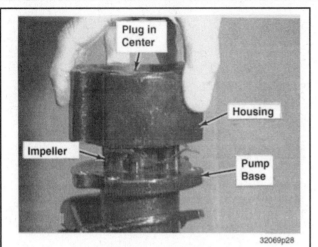

Fig. 28 Remove the plug and then lift off the housing

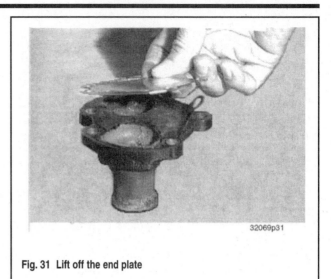

Fig. 31 Lift off the end plate

Fig. 29 A badly damaged impeller

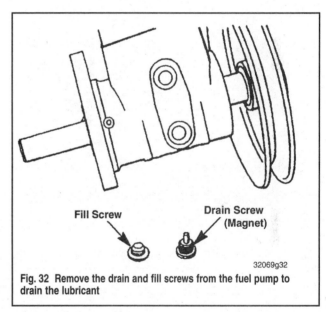

Fig. 32 Remove the drain and fill screws from the fuel pump to drain the lubricant

Fig. 30 Clean the old gasket material from the pump base

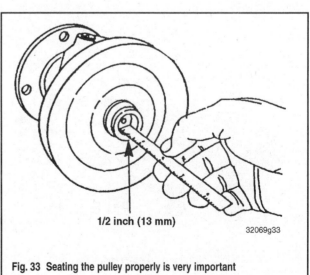

1/2 inch (13 mm)

Fig. 33 Seating the pulley properly is very important

Fig. 34 Install a new gasket on the pump base

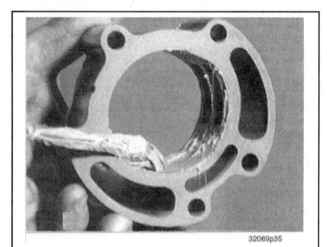
Fig. 35 Coat the inside of the housing with grease

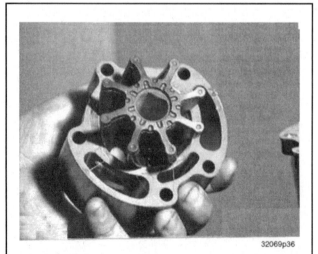
Fig. 36 Coat the impeller with soapy water and press it into the housing

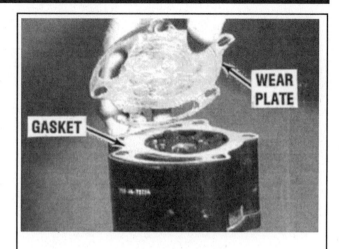
Fig. 37 Position the a new gasket and then install the wear plate

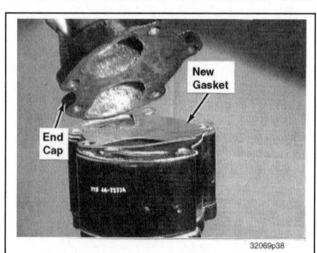
Fig. 38 Position the other gasket and install the cap

Stern Drive Seawater Pick-Up Pump

OEM ② MODERATE

REMOVAL & INSTALLATION

Alpha Models Only

◆ See Figures 39, 40 and 41

1. Remove the lower unit on the stern drive as detailed later in this manual.
2. Slide the water seal up and over the shaft.
3. Pull out the water tube coupling.
4. Loosen the pump mounting bolts and carefully slide the pump up and off the shaft. You may need to use a small prybar (or two) to loosen the pump from the mating surface.
5. Lift off the impeller and key and then pull off the face plate and gaskets.
   To install:
6. Inspect the water tube assembly for damage. There are two O-rings that you should pay special attention to. Inspect the impeller for wear on all edges of the blades, replace if any wear or damage is found. Check that the face plate and pump housing are free of grooves and are smooth.
7. Slide the small-holed gasket over the shaft so that all the holes line up. Do the same with the face plate and then position the large-holed gasket.

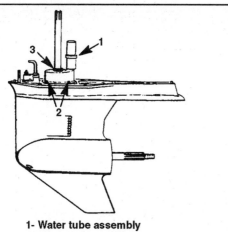

1- Water tube assembly
2- Water pump screws (4)
3- Water seal

32069g39

Fig. 39 Remove the water seal and water tube before removing the pump

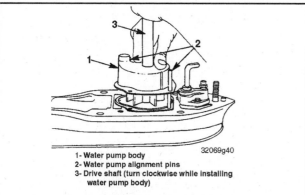

32069g40

1- Water pump body
2- Water pump alignment pins
3- Drive shaft (turn clockwise while installing water pump body)

Fig. 40 You will need special alignment pins to install the pump...

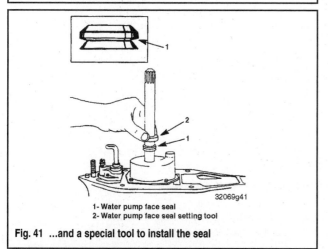

32069g41

1- Water pump face seal
2- Water pump face seal setting tool

Fig. 41 ...and a special tool to install the seal

8. Dab a little Quicksilver 2-4-C lubricant on the flat surface of the impeller key and position it into the keyway on the shaft.

9. Slide the impeller down the shaft and over the key. Make sure that the curl of the blades is facing counterclockwise. Never install the impeller with its blades oriented in the reverse direction.

10. Install the two pump locating pins (# 91-821571A1) through the gaskets and face plate and then slide the pump over the shaft. Carefully spin the shaft clockwise while pushing down on the pump housing until the pump slides over the impeller blades.

11. Install two mounting bolts and tighten them fingertight. Remove the locating pins and screw in the remaining two bolts. Tighten all of them to 60 inch lbs. (7.9 Nm).

12. Lightly coat the exposed shaft with Quicksilver 2-4-C. Slide the new pump seal about half way down the shaft and then slide on the Seal Setting tool (it comes with the seal kit) down the shaft until it is sitting on the seal. Press the seal the remainder of the way down with the tool and then remove the tool.

13. Coat the O-rings in the water tube with 2-4-C and press the assembly into the housing.

14. Install the lower unit

Thermostat

For all removal and testing procedures, please refer to Section 3.

Water (Engine) Circulating Pump

② MODERATE

REMOVAL & INSTALLATION

◆ **See Figure 42**

1. Drain all water/coolant from the cylinder block.
2. Loosen, but do not remove, the pump pulley mounting bolts.
3. Loosen the power steering pump and alternator bracket bolts and swivel them in until you are able to remove the belt(s). Different engines may have different systems, follow the belt back and loosen the appropriate components.
4. Now you can remove the pump pulley bolts along with the lockwashers and clamping ring. Pull off the pulley.
5. Disconnect the water hoses at the pump.
6. Disconnect the water hoses at the pump, remove the mounting bolts and lift the pump off of the block.

To install:

7. Carefully scrape any old gasket material off both mounting surfaces. Inspect the pump for blockage, cracks or any other damage. Inspect the impeller for cracks. Replace either if necessary.

8. Coat both sides of a new gasket with Perfect Seal and position it on the cylinder block. Coat the threads of the pump mounting bolts with Perfect Seal, install the pump and tighten the bolts to 30 ft. lbs. (41 Nm) on 1992–97 V6 and 5.0L/5.7L V8 engines, or 35 ft. lbs. (48 Nm) on 1998–01 V6 and 5.0L/5.7L/6.2L engines and all 7.4L/8.1L/8.2L engines.

9. Reconnect the water hoses and tighten the hose clamps securely.

10. Position the pump pulley and clamping ring on the boss. Screw the mounting bolts and lock washers in and tighten them securely.

11. Install the drive belt(s) and adjust them as detailed in Section 3. Start the engine and check the system for leaks.

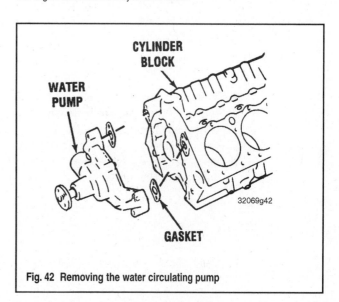

CYLINDER BLOCK

WATER PUMP

GASKET

32069g42

Fig. 42 Removing the water circulating pump

DRAIN DIAGRAMS

◆ See Figures 43, 44, 45, 46, 47, 48 and 49

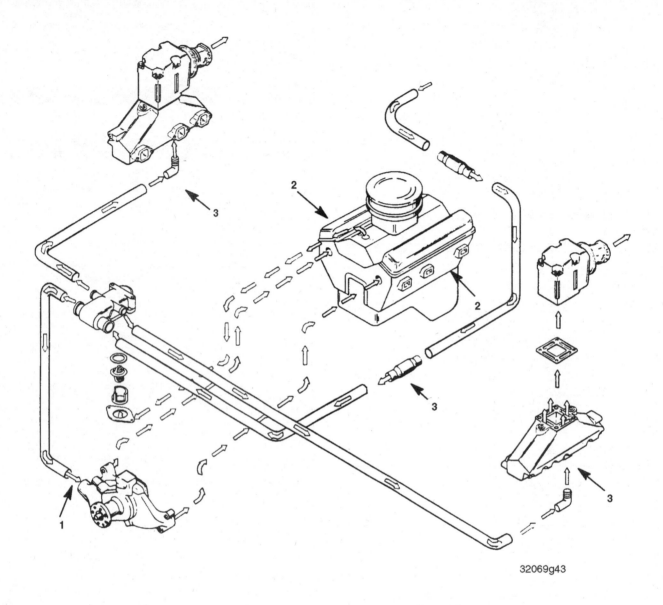

32069g43

1- Remove hoses (lift, lower or bend to completely drain).
2- Remove block plugs (repeatedly clean out holes using a stiff wire until entire system is drained).
3- Remove drain plugs from exhaust manifold. Drain elbows and fuel cooler or water tube (repeatedly clean out holes using a stiff wire until entire system is drained).

Fig. 43 1998–01 V6 engines with seawater system—Alpha

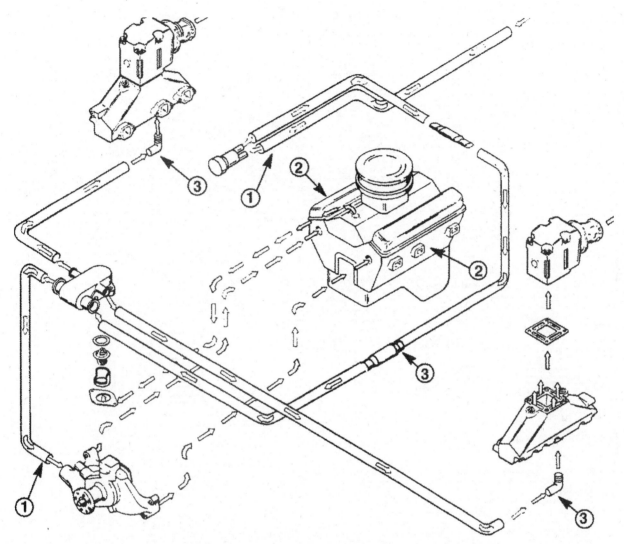

32069g44

1- Remove hoses (lift, lower or bend to completely drain).
2- Remove block plugs (repeatedly clean out holes using a stiff wire until entire system is drained).
3- Remove drain plugs from exhaust mainfold. Drain elbows and fuel cooler or water tube (repeatedly clean out holes using a stiff wire until entire system is drained).

Fig. 44 1998–01 V6 engines with seawater system—Bravo

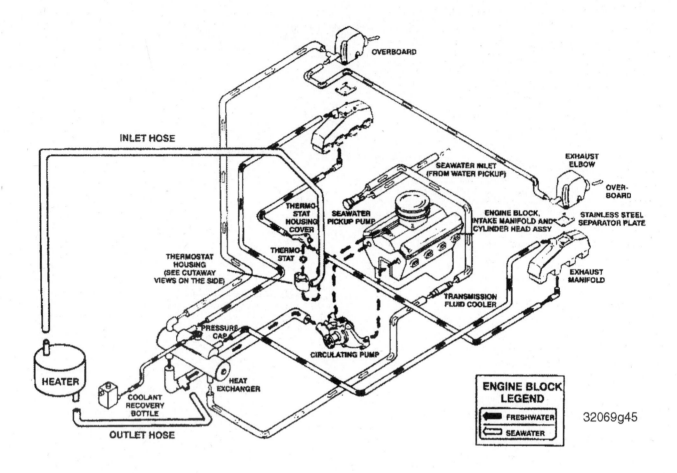

OVERBOARD

INLET HOSE

SEAWATER INLET
(FROM WATER PICKUP)

EXHAUST
ELBOW

OVER-
BOARD

THERMO-
STAT
HOUSING
COVER

SEAWATER
PICKUP PUMP

ENGINE BLOCK,
INTAKE MANIFOLD AND
CYLINDER HEAD ASSY

STAINLESS STEEL
SEPARATOR PLATE

THERMOSTAT
HOUSING
(SEE CUTAWAY
VIEWS ON THE SIDE)

THERMO-
STAT

EXHAUST
MANIFOLD

TRANSMISSION
FLUID COOLER

PRESSURE
CAP

CIRCULATING PUMP

HEATER

HEAT
EXCHANGER

COOLANT
RECOVERY
BOTTLE

OUTLET HOSE

ENGINE BLOCK LEGEND

FRESHWATER

SEAWATER

32069g45

Fig. 45 1992–97 5.0L/5.7L V8 engines with closed cooling system

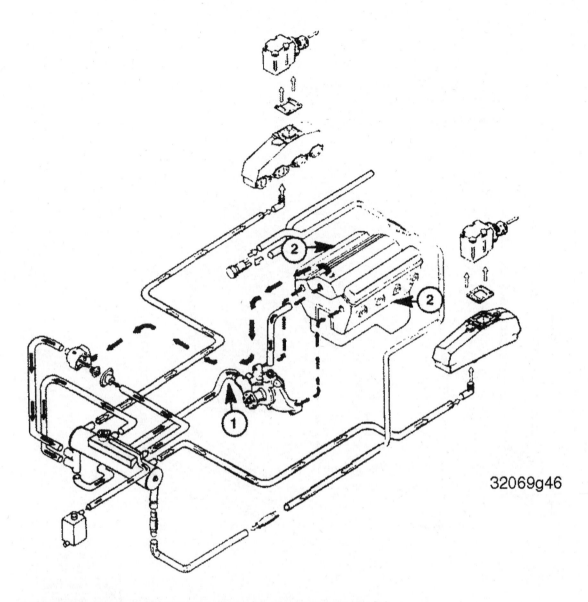

32069g46

1- Remove hoses (lift, lower or bend to completely drain).
2- Remove block plugs (repeatedly clean out holes using a stiff wire until
    entire system is drained).

Fig. 46 1998–01 5.0L/5.7L/6.2L V8 engines with closed cooling system

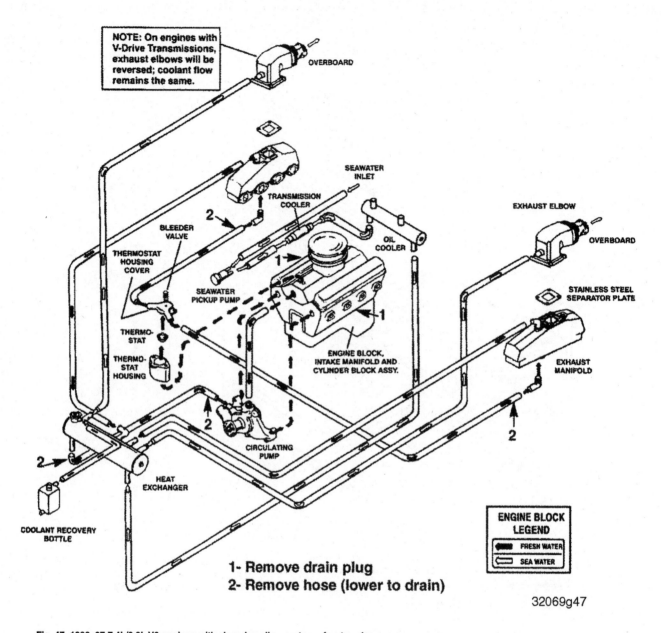

NOTE: On engines with V-Drive Transmissions, exhaust elbows will be reversed; coolant flow remains the same.

OVERBOARD

SEAWATER INLET

TRANSMISSION COOLER

EXHAUST ELBOW

OVERBOARD

OIL COOLER

BLEEDER VALVE

STAINLESS STEEL SEPARATOR PLATE

THERMOSTAT HOUSING COVER

SEAWATER PICKUP PUMP

THERMO-STAT

ENGINE BLOCK, INTAKE MANIFOLD AND CYLINDER BLOCK ASSY.

EXHAUST MANIFOLD

THERMO-STAT HOUSING

CIRCULATING PUMP

HEAT EXCHANGER

COOLANT RECOVERY BOTTLE

ENGINE BLOCK LEGEND
FRESH WATER
SEA WATER

1- Remove drain plug
2- Remove hose (lower to drain)

32069g47

Fig. 47 1992–97 7.4L/8.2L V8 engines with closed cooling system—front exchanger

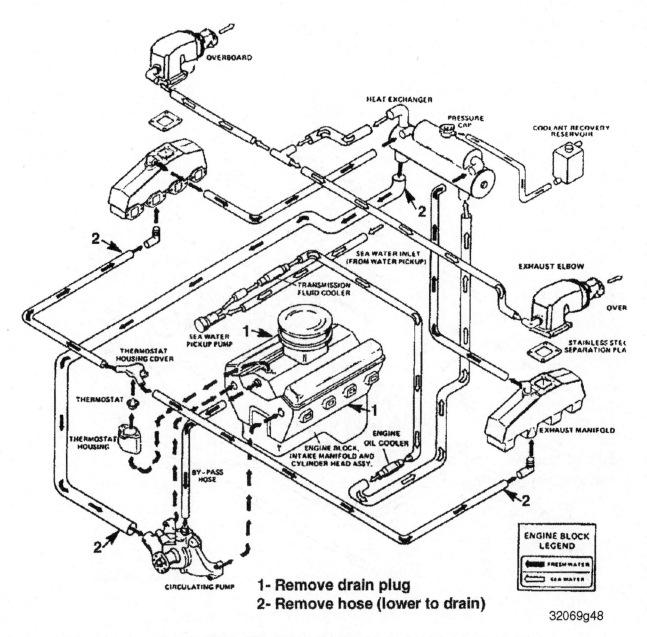

OVERBOARD

HEAT EXCHANGER

PRESSURE CAP

COOLANT RECOVERY RESERVOIR

2

SEA WATER INLET (FROM WATER PICKUP)

TRANSMISSION FLUID COOLER

EXHAUST ELBOW

OVER

2

SEA WATER PICKUP PUMP

STAINLESS STEEL SEPARATION PLA

1

THERMOSTAT HOUSING COVER

THERMOSTAT

ENGINE OIL COOLER

1

EXHAUST MANIFOLD

THERMOSTAT HOUSING

ENGINE BLOCK, INTAKE MANIFOLD AND CYLINDER HEAD ASSY.

BY-PASS HOSE

2

2

CIRCULATING PUMP

ENGINE BLOCK LEGEND

FRESH WATER

SEA WATER

1- **Remove drain plug**
2- **Remove hose (lower to drain)**

32069g48

Fig. 48 1992–97 7.4L/8.2L V8 engines with closed cooling system—rear exchanger

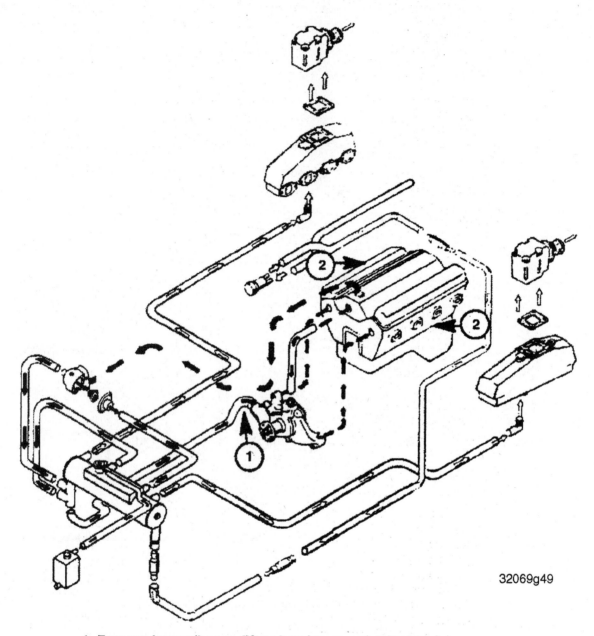

32069g49

1- Remove hoses (hoses, lift or bend to completely drain).
2- Remove block plugs (repeatedly clean out holes using a stiff wire until entire system is drained).

Fig. 49 1998–01 7.4L/8.2L V8 engines with closed cooling system

FLOW DIAGRAMS

◆ See Figures 50, 51, 52, 53, 54, 55, 56, 57, 58, 59, 60, 61. 62. 63, 64,
65, 66, 67, 68, 69, 70, 71, 72, 73, 74, 75, 76, 77 and 78

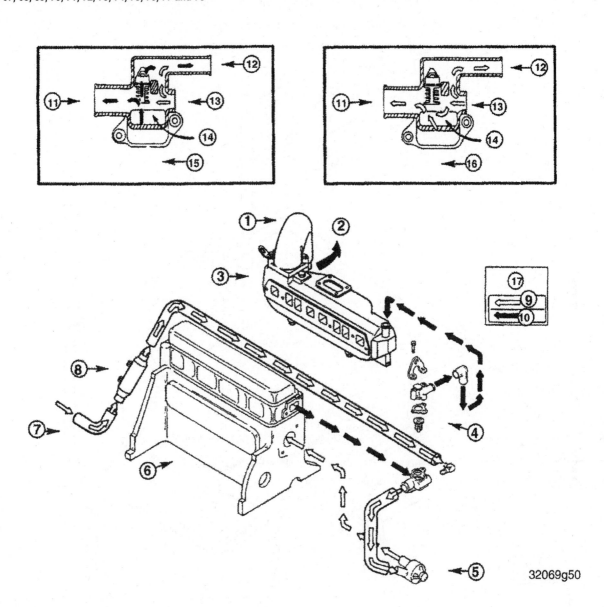

32069g50

Fig. 50 3.0L engines—seawater system

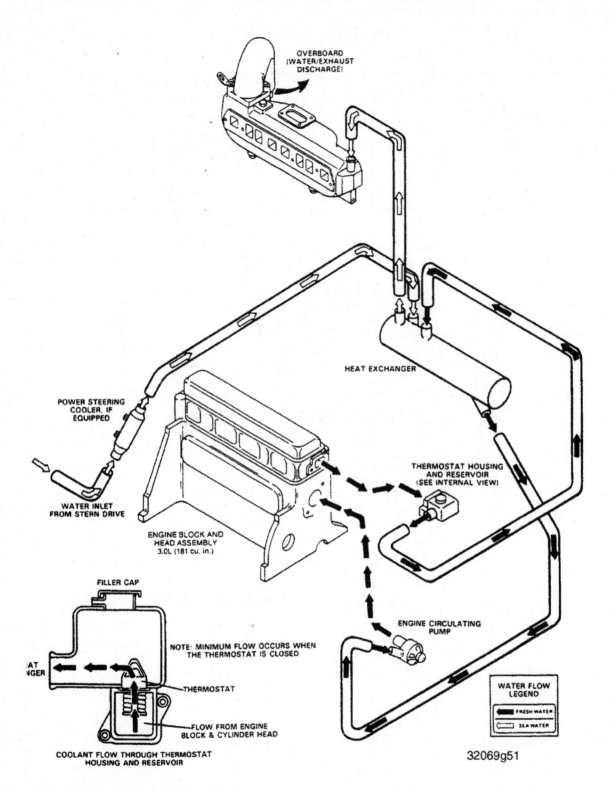

OVERBOARD
(WATER/EXHAUST
DISCHARGE)

HEAT EXCHANGER

POWER STEERING
COOLER, IF
EQUIPPED

THERMOSTAT HOUSING
AND RESERVOIR
(SEE INTERNAL VIEW)

WATER INLET
FROM STERN DRIVE

ENGINE BLOCK AND
HEAD ASSEMBLY
3.0L (181 cu. in.)

ENGINE CIRCULATING
PUMP

FILLER CAP

NOTE: MINIMUM FLOW OCCURS WHEN
THE THERMOSTAT IS CLOSED

AT
NGER

THERMOSTAT

FLOW FROM ENGINE
BLOCK & CYLINDER HEAD

COOLANT FLOW THROUGH THERMOSTAT
HOUSING AND RESERVOIR

WATER FLOW
LEGEND

FRESH WATER

SEA WATER

32069g51

Fig. 51 3.0L engine—closed system

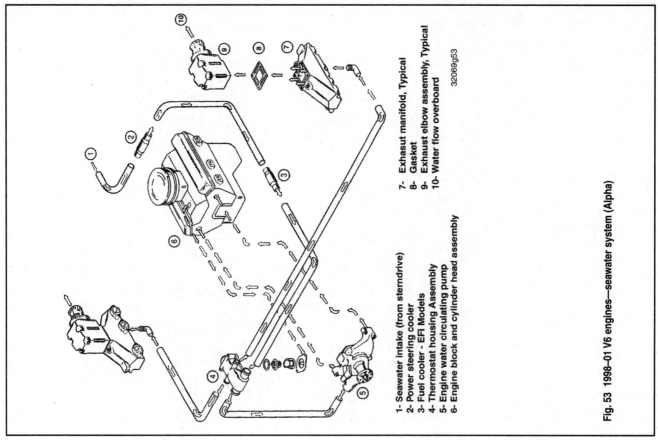

1- Seawater intake (from sterndrive)
2- Power steering cooler
3- Fuel cooler - EFI Models
4- Thermostat housing Assembly
5- Engine water circulating pump
6- Engine block and cylinder head assembly
7- Exhasut manifold, Typical
8- Gasket
9- Exhaust elbow assembly, Typical
10- Water flow overboard

32069g53

Fig. 53 1998-01 V6 engines—seawater system (Alpha)

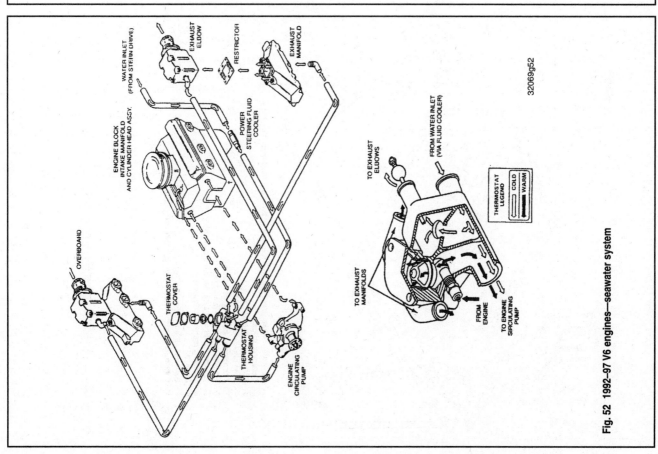

32069g52

Fig. 52 1992-97 V6 engines—seawater system

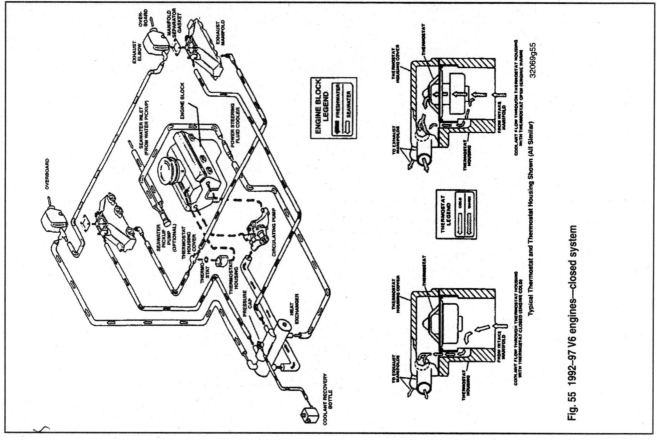

Fig. 55 1992–97 V6 engines—closed system

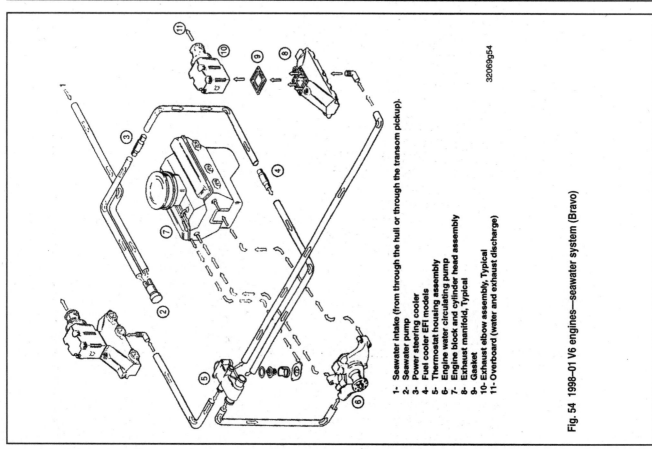

1- Seawater intake (from through the hull or through the transom pickup).
2- Seawater pump
3- Power steering cooler
4- Fuel cooler EFI models
5- Thermostat housing assembly
6- Engine water circulating pump
7- Engine block and cylinder head assembly
8- Exhaust manifold, Typical
9- Gasket
10- Exhaust elbow assembly, Typical
11- Overboard (water and exhaust discharge)

Fig. 54 1998–01 V6 engines—seawater system (Bravo)

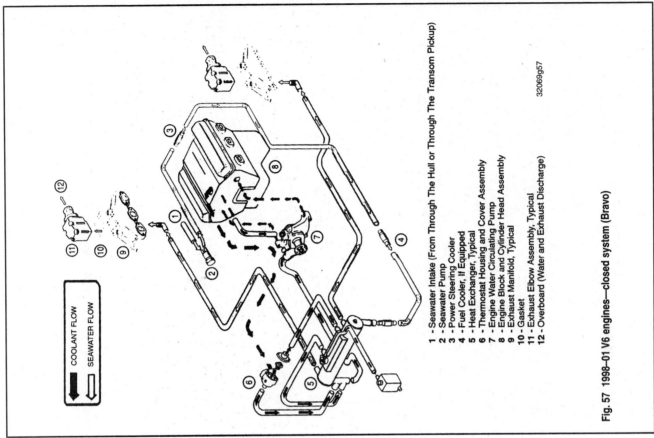

COOLANT FLOW
SEAWATER FLOW

1 - Seawater Intake (From Through The Hull or Through The Transom Pickup)
2 - Seawater Pump
3 - Power Steering Cooler
4 - Fuel Cooler, If Equipped
5 - Heat Exchanger, Typical
6 - Thermostat Housing and Cover Assembly
7 - Engine Water Circulating Pump
8 - Engine Block and Cylinder Head Assembly
9 - Exhaust Manifold, Typical
10 - Gasket
11 - Exhaust Elbow Assembly, Typical
12 - Overboard (Water and Exhaust Discharge)

32069g57

Fig. 57 1998–01 V6 engines—closed system (Bravo)

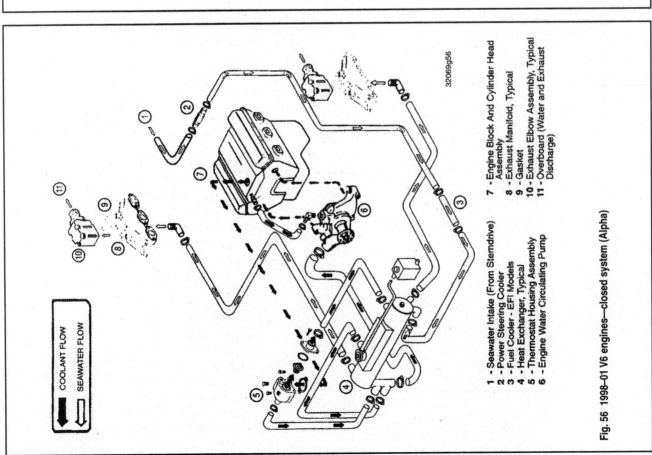

COOLANT FLOW
SEAWATER FLOW

1 - Seawater Intake (From Sterndrive)
2 - Power Steering Cooler
3 - Fuel Cooler - EFI Models
4 - Heat Exchanger, Typical
5 - Thermostat Housing Assembly
6 - Engine Water Circulating Pump

7 - Engine Block And Cylinder Head Assembly
8 - Exhaust Manifold, Typical
9 - Gasket
10 - Exhaust Elbow Assembly, Typical
11 - Overboard (Water and Exhaust Discharge)

32069g56

Fig. 56 1998–01 V6 engines—closed system (Alpha)

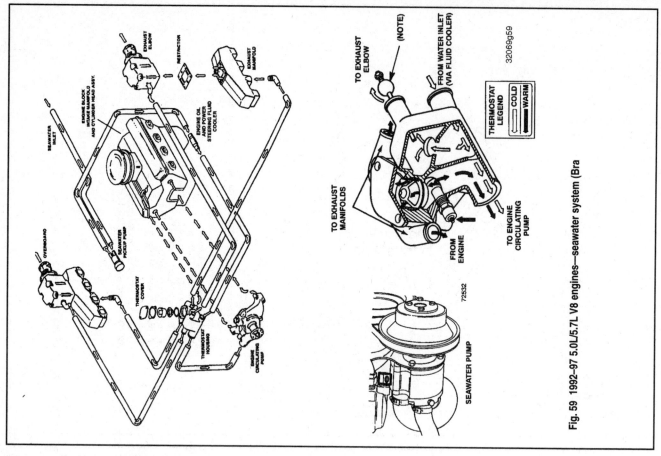

Fig. 59 1992–97 5.0L/5.7L V8 engines—seawater system (Bra

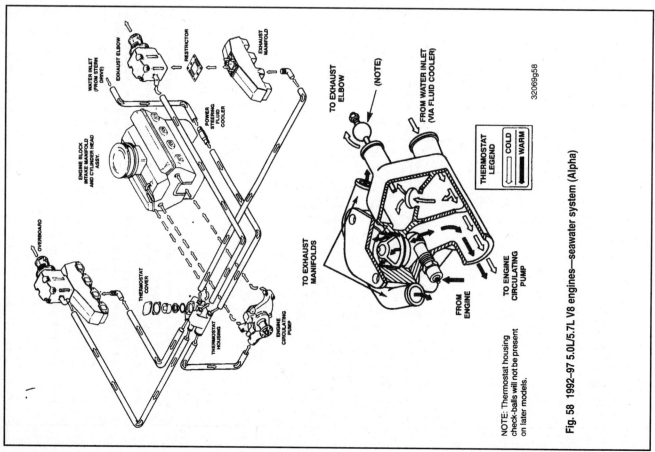

NOTE: Thermostat housing
check-balls will not be present
on later models.

Fig. 58 1992–97 5.0L/5.7L V8 engines—seawater system (Alpha)

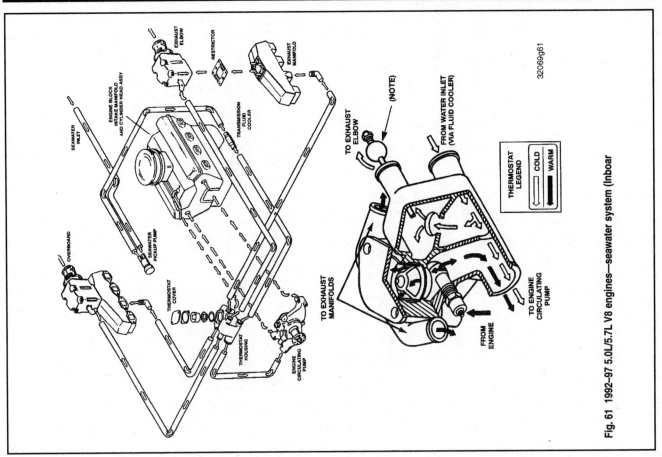

Fig. 61 1992–97 5.0L/5.7L V8 engines—seawater system (Inboar

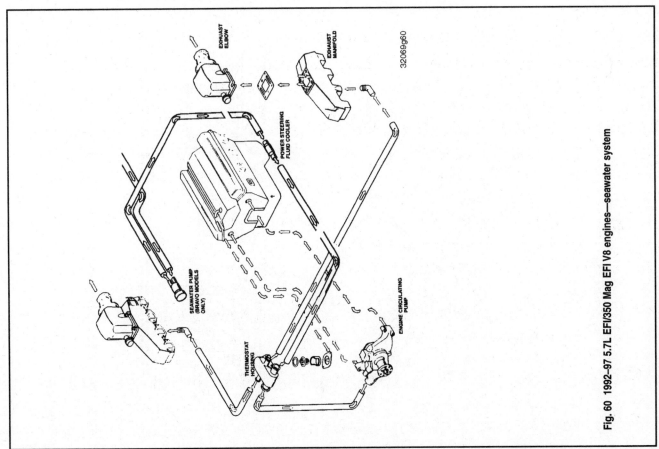

Fig. 60 1992–97 5.7L EFI/350 Mag EFI V8 engines—seawater system

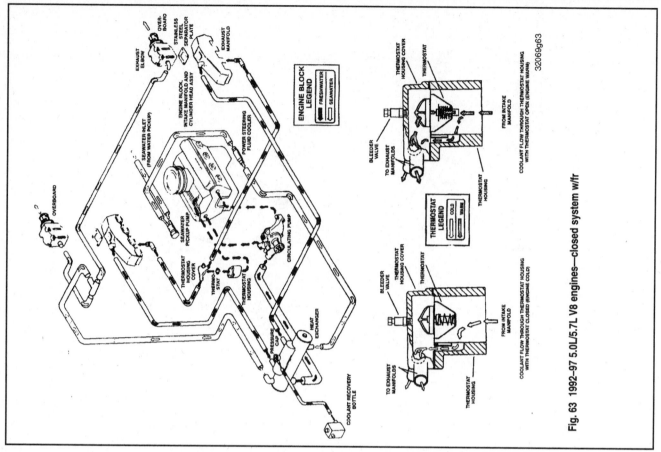

Fig. 63 1992–97 5.0L/5.7L V8 engines—closed system w/fr

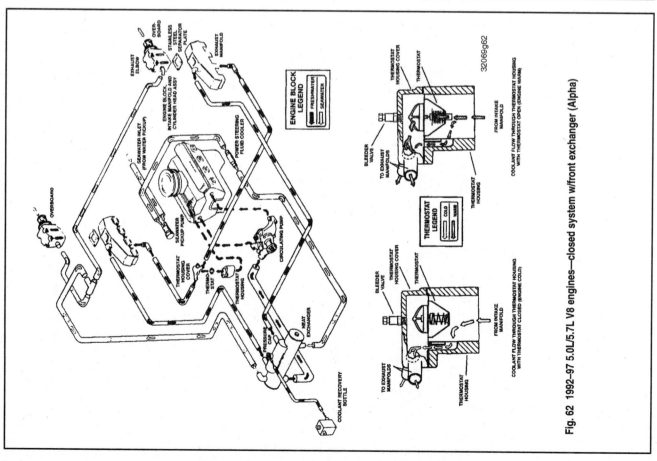

Fig. 62 1992–97 5.0L/5.7L V8 engines—closed system w/front exchanger (Alpha)

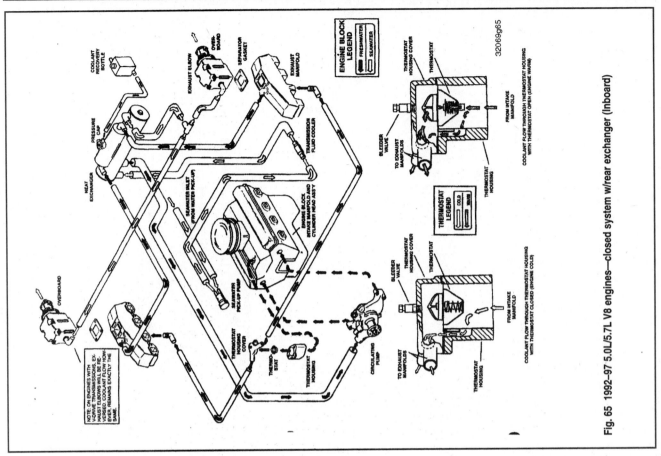

Fig. 65 1992-97 5.0L/5.7L V8 engines—closed system w/rear exchanger (Inboard)

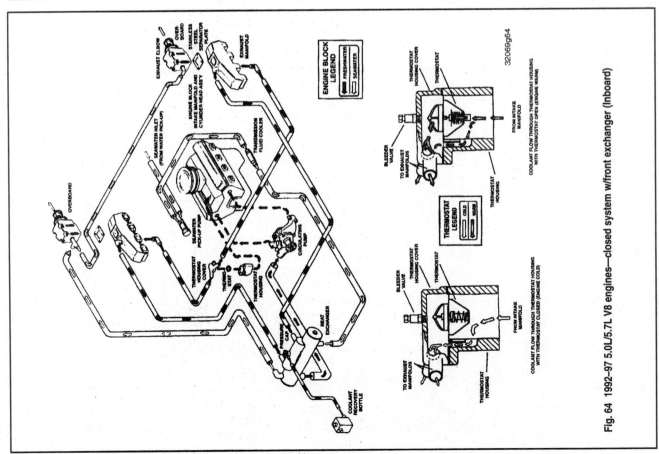

Fig. 64 1992-97 5.0L/5.7L V8 engines—closed system w/front exchanger (Inboard)

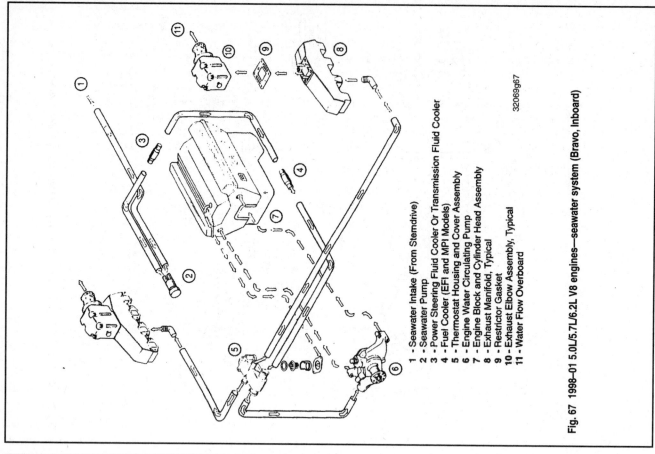

1 - Seawater Intake (From Sterndrive)
2 - Seawater Pump
3 - Power Steering Fluid Cooler Or Transmission Fluid Cooler
4 - Fuel Cooler (EFI and MPI Models)
5 - Thermostat Housing and Cover Assembly
6 - Engine Water Circulating Pump
7 - Engine Block and Cylinder Head Assembly
8 - Exhaust Manifold, Typical
9 - Restrictor Gasket
10 - Exhaust Elbow Assembly, Typical
11 - Water Flow Overboard

32069g67

Fig. 67 1998–01 5.0L/5.7L/6.2L V8 engines—seawater system (Bravo, Inboard)

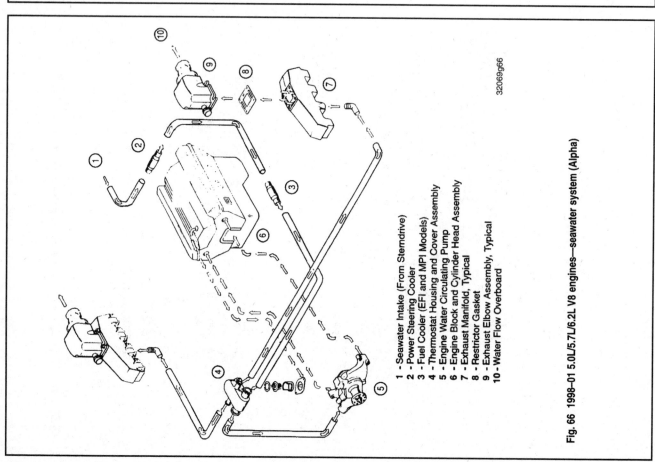

1 - Seawater Intake (From Sterndrive)
2 - Power Steering Cooler
3 - Fuel Cooler (EFI and MPI Models)
4 - Thermostat Housing and Cover Assembly
5 - Engine Water Circulating Pump
6 - Engine Block and Cylinder Head Assembly
7 - Exhaust Manifold, Typical
8 - Restrictor Gasket
9 - Exhaust Elbow Assembly, Typical
10 - Water Flow Overboard

32069g66

Fig. 66 1998–01 5.0L/5.7L/6.2L V8 engines—seawater system (Alpha)

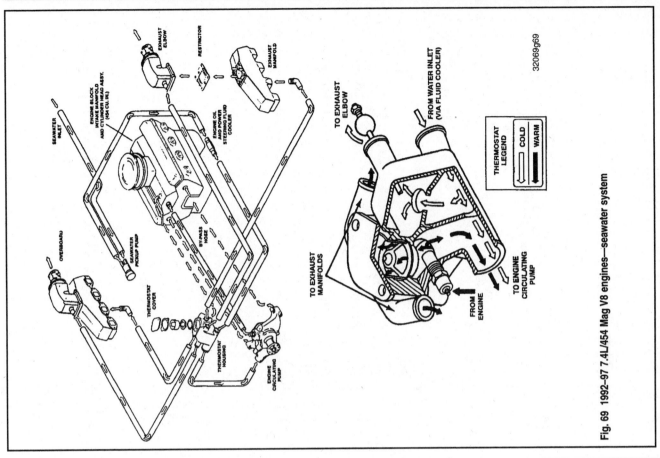

Fig. 69 1992–97 7.4L/454 Mag V8 engines—seawater system

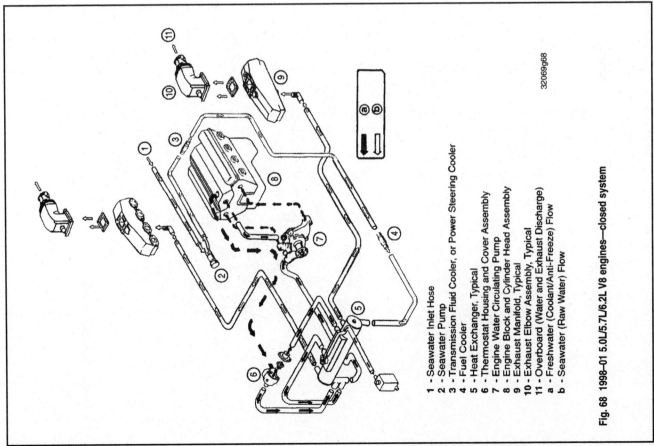

1 - Seawater Inlet Hose
2 - Seawater Pump
3 - Transmission Fluid Cooler, or Power Steering Cooler
4 - Fuel Cooler
5 - Heat Exchanger, Typical
6 - Thermostat Housing and Cover Assembly
7 - Engine Water Circulating Pump
8 - Engine Block and Cylinder Head Assembly
9 - Exhaust Manifold, Typical
10 - Exhaust Elbow Assembly, Typical
11 - Overboard (Water and Exhaust Discharge)
a - Freshwater (Coolant/Anti-Freeze) Flow
b - Seawater (Raw Water) Flow

Fig. 68 1998–01 5.0L/5.7L/6.2L V8 engines—closed system

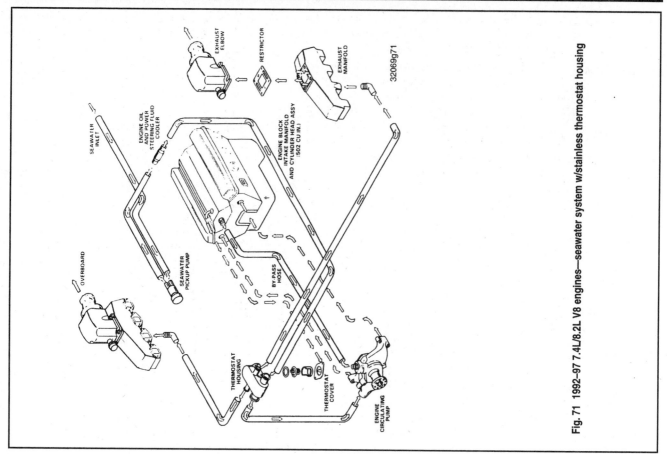

Fig. 71 1992–97 7.4L/8.2L V8 engines—seawater system w/stainless thermostat housing

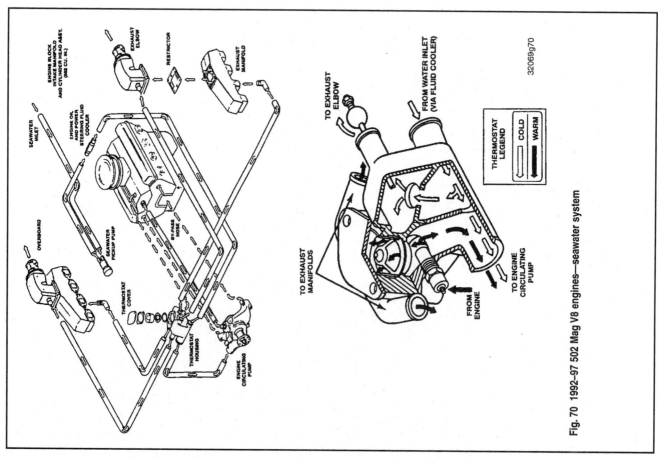

Fig. 70 1992–97 502 Mag V8 engines—seawater system

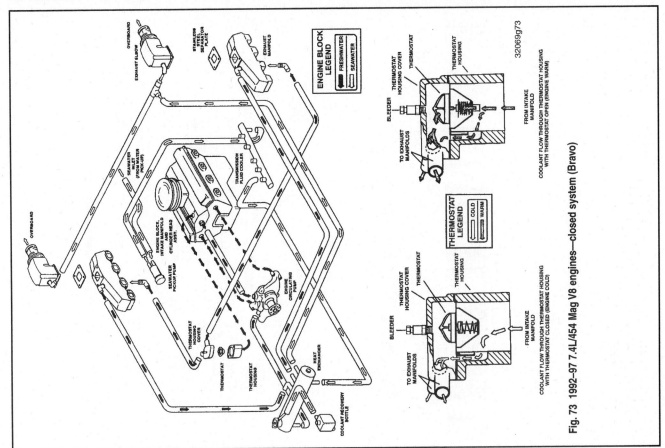

Fig. 73 1992–97 7.4L/454 Mag V8 engines—closed system (Bravo)

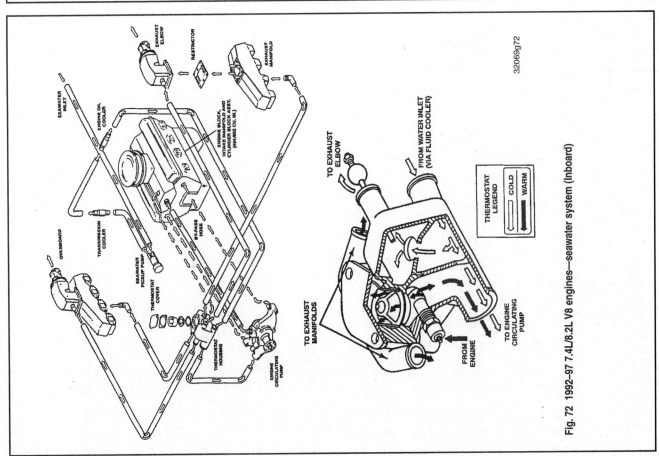

Fig. 72 1992–97 7.4L/8.2L V8 engines—seawater system (Inboard)

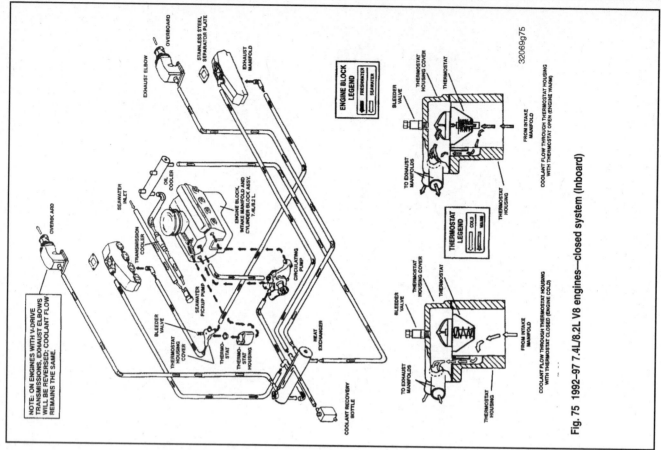

Fig. 75 1992-97 7.4L/8.2L V8 engines—closed system (Inboard)

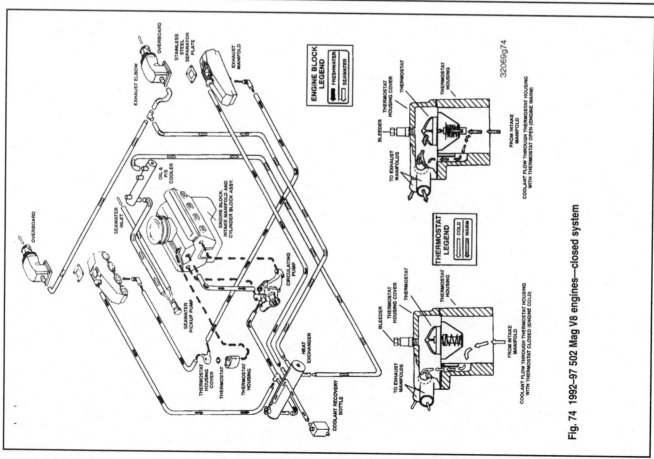

Fig. 74 1992-97 502 Mag V8 engines—closed system

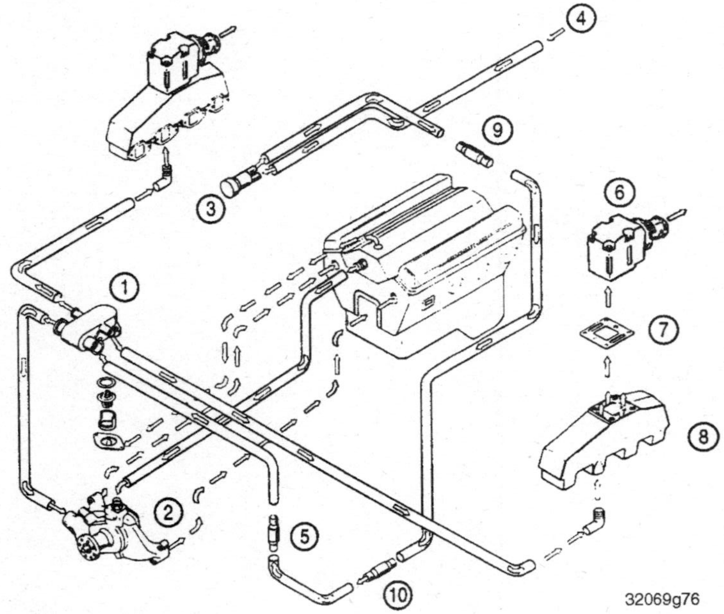

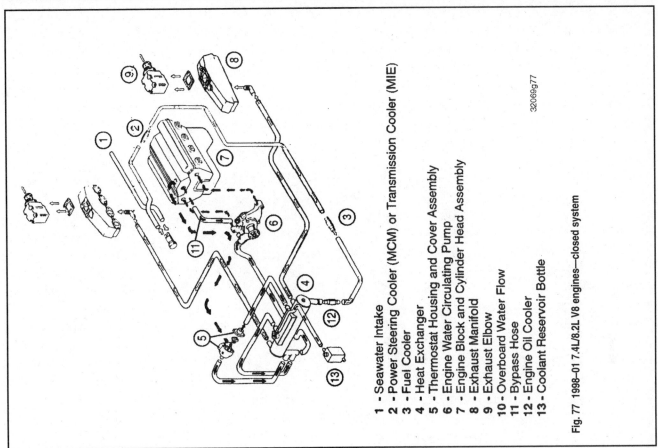

1 - Seawater Intake
2 - Power Steering Cooler (MCM) or Transmission Cooler (MIE)
3 - Fuel Cooler
4 - Heat Exchanger
5 - Thermostat Housing and Cover Assembly
6 - Engine Water Circulating Pump
7 - Engine Block and Cylinder Head Assembly
8 - Exhaust Manifold
9 - Exhaust Elbow
10 - Overboard Water Flow
11 - Bypass Hose
12 - Engine Oil Cooler
13 - Coolant Reservoir Bottle

32069g77

Fig. 77 1998–01 7.4L/8.2L V8 engines—closed system

32069g76

Fig. 76 1998–01 7.4L/8.2L V8 engines—seawater system

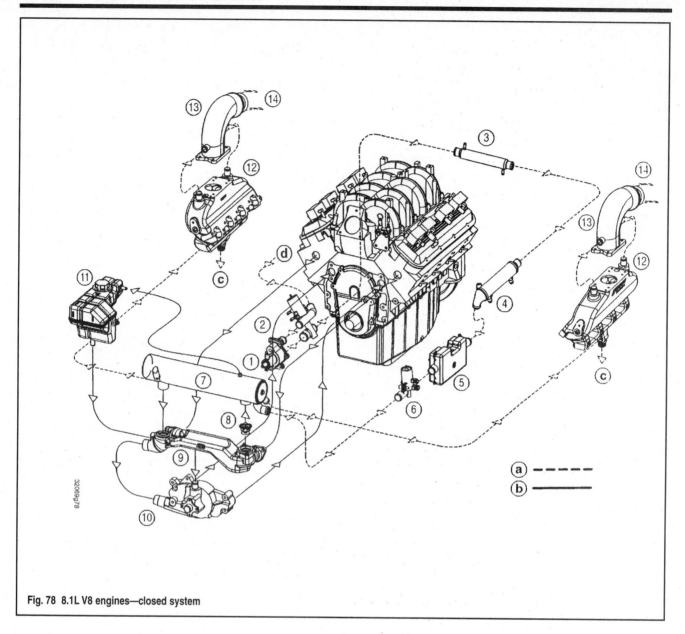

Fig. 78 8.1L V8 engines—closed system

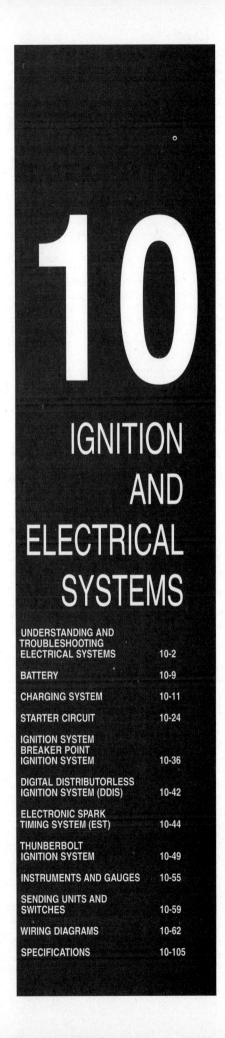

10

IGNITION
AND
ELECTRICAL
SYSTEMS

UNDERSTANDING AND TROUBLESHOOTING ELECTRICAL SYSTEMS

Basic Electrical Theory

◆ See Figure 1

For any 12 volt, negative ground, electrical system to operate, the electricity must travel in a complete circuit. This simply means that current (power) from the positive terminal (+) of the battery must eventually return to the negative terminal (–) of the battery. Along the way, this current will travel through wires, fuses, switches and components. If for any reason the flow of current through the circuit is interrupted, the component(s) fed by that circuit will cease to function properly.

Perhaps the easiest way to visualize a circuit is to think of connecting a light bulb (with two wires attached to it) to the battery—one wire attached to the negative (–) terminal of the battery and the other wire to the positive (+) terminal. With the two wires touching the battery terminals, the circuit would be complete and the light bulb would illuminate. Electricity would follow a path from the battery to the bulb and back to the battery. It's easy to see that with longer wires on our light bulb, it could be mounted anywhere. Further, one wire could be fitted with a switch so that the light could be turned on and off.

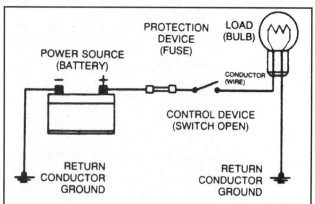

Fig. 1 This example illustrates a simple circuit. When the switch is closed, power from the positive (+) battery terminal flows through the fuse and the switch and then to the light bulb. The light illuminates and the circuit is completed through the ground wire back to the negative (–) battery terminal.

The normal marine circuit differs from this simple example in two ways. First, instead of having a return wire from each bulb to the battery, the current travels through a single ground wire, which handles all the grounds for a specific circuit. Secondly, most marine circuits contain multiple components, which receive power from a single circuit. This lessens the amount of wire needed to power components.

HOW ELECTRICITY WORKS: THE WATER ANALOGY

Electricity is the flow of electrons—the sub-atomic particles that constitute the outer shell of an atom. Electrons spin in an orbit around the center core of an atom. The center core is comprised of protons (positive charge) and neutrons (neutral charge). Electrons have a negative charge and balance out the positive charge of the protons. When an outside force causes the number of electrons to unbalance the charge of the protons, the electrons will split off the atom and look for another atom to balance out. If this imbalance is kept up, electrons will continue to move and an electrical flow will exist.

Many people have been taught electrical theory using an analogy with water. In a comparison with water flowing through a pipe, the electrons would be the water and the wire is the pipe.

The flow of electricity can be measured much like the flow of water through a pipe. The unit of measurement used is amps, frequently abbreviated as amps (a). You can compare amperage to the volume of water flowing through a pipe. When connected to a circuit, an ammeter will measure the actual amount of current flowing through the circuit. When relatively few electrons flow through a circuit, the amperage is low. When many electrons flow, the amperage is high.

Water pressure is measured in units such as pounds per square inch (psi), electrical pressure is measured in units called volts (v). When a voltmeter is connected to a circuit, it is measuring the electrical pressure. When electrical pressure is low, then voltage is considered to be low. When electrical pressure is high, then voltage is considered to be high.

The actual flow of electricity depends not only on voltage and amperage but also on the resistance of the circuit. The higher the resistance, the higher the force necessary to push the current through the circuit. The standard unit for measuring resistance is an ohm . Resistance in a circuit varies depending on the amount and type of components used in the circuit and the overall condition of the components and wires. If we assume that everything in our circuit is new, then, the main factors which determine resistance are:

- Material—some materials have more resistance than others. Those with high resistance are said to be insulators. Rubber materials (or rubber-like plastics) are some of the most common insulators used, as they have a very high resistance to electricity. Very low resistance materials are said to be conductors. Copper wire is among the best conductors. Silver is actually a superior conductor to copper and is used in some relay contacts but its high cost prohibits its use as common wiring. Most marine wiring is made of copper.
- Size—the larger the wire size being used, the less resistance the wire will have. This is why components that use large amounts of electricity usually have large wires supplying current to them.
- Length—for a given thickness of wire, the longer the wire, the greater the resistance. The shorter the wire, the less the resistance. When determining the proper wire for a circuit, both size and length must be considered to design a circuit that can handle the current needs of the component.
- Temperature—with many materials, the higher the temperature, the greater the resistance (positive temperature coefficient). Some materials exhibit the opposite trait of lower resistance with higher temperatures (negative temperature coefficient). These principles are used in many of the sensors on the engine.

OHM'S LAW

There is a direct relationship between current, voltage and resistance that can be summed up by a statement known as Ohm's law.
- Voltage (**E**) is equal to amperage (**I**) times resistance (**R**): $E = I \times R$
- Other forms of the formula are $R = E/I$ and $I = E/R$

In each of these formulas, **E** is the voltage in volts, **I** is the current in amps and **R** is the resistance in ohms. The basic point to remember is that as the resistance of a circuit goes up, the amount of current that flows in the circuit will go down, if voltage remains the same.

The amount of work that the electricity can perform is expressed as power. A unit of power is known as a watt (**W**). There is a direct relationship between power, voltage and current that can be summed up by the following formula:
- Power (**W**) is equal to amperage (**I**) times voltage (**E**): $W = I \times E$

■ This formula is only true for direct current (DC) circuits, The alternating current formula is a tad different but since the electrical circuits in most boats are DC type, we need not get into AC circuit theory.

Electrical Components

POWER SOURCE

◆ See Figure 2

Power is supplied to the boat by two devices: The battery and the alternator. The battery supplies electrical power during starting or during periods when the current demand of the boat's electrical system exceeds the output capacity of the alternator. The alternator supplies electrical current when the engine is running. The alternator does not just supply the current needs of the boat but it also recharges the battery.

In most modern boats, the battery is a lead/acid electrochemical device consisting of six 2 volt subsections (cells) connected in series, so that the unit is capable of producing approximately 12 volts of electrical pressure. Each subsection consists of a series of positive and negative plates held a short distance apart in a solution of sulfuric acid and water.

The two types of plates are of dissimilar metals, thus setting up a chemical reaction inside the battery case.

It is this reaction that produces current flow from the battery when its positive and negative terminals are connected to an electrical load. The alternator, restoring the battery to its original chemical state replaces the power removed from the battery.

The following is a brief description of how the system works:

The battery stores electricity and acts as a "sponge" for the whole system. It mops up generated
current until it's fully charged and it release s energy on demand.

Permanent magnets that create the moving magnetic field. If your engine has good spark, you can take it for granted that the magnets are in working order because the ignition and charging systems share the same magnets.

The alternator windings are the stationary coils of wire that the magnets rotate around. They produce the electrical charge. Simply put, the more windings in your stator, the greater the potential output in amps your charging system you'll have.

The rectifier consists of a series of diodes or electrical one-way valves. The rectifier overcomes one of the disadvantages of a current-generating system using permanent magnets and stator windings, which is that the current produced within the windings is alternating current (AC). You can't use AC to charge batteries. They accept only direct current (DC). So the rectifier is designed to convert AC current to a usable form of DC current simply called "rectified AC."

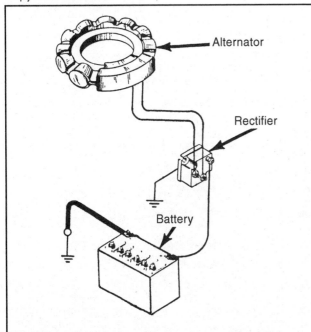

Fig. 2 Functional diagram of a typical charging circuit showing the relationship of the stator, solid-state rectifier and the battery

All engines covered here utilize a voltage regulator; either combined with the rectifier or standing alone. The regulator automatically reduces the output of generated current as the battery becomes fully charged.

GROUND

All boats use some sort of a ground return circuit. Direct ground components are grounded to an electrically conductive metal component through their mounting points. These electrically conductive metal components are then grounded to the battery.

All other components use some sort of ground wire which leads directly back to the battery. The electrical current runs through the ground wire and returns to the battery through the ground (–) cable. If you look, you'll see that the battery ground cable connects between the battery and a heavy gauge ground wire.

■ It should be noted that a good percentage of electrical problems can be traced to bad grounds.

PROTECTIVE DEVICES

◆ See Figure 3

It is possible for large surges of current to pass through the electrical system of your boat. If this surge of current were to reach components in the circuit, the surge could burn them out or severely damage them. Surges can also overload the wiring, causing the harness to get hot and melt the insulation. To prevent this, fuses, circuit breakers and/or fusible links are connected into the supply wires of the electrical system. These items are nothing more than a built-in weak spot in the system. When an abnormal amount of current flows through the system, these protective devices work as follows to protect the circuit:

● Fuse—when an excessive electrical current passes through a fuse, the fuse "blows" (the conductor melts) and opens the circuit, preventing the passage of current.

● Circuit Breaker—a circuit breaker is basically a self-repairing fuse. It will open the circuit in the same fashion as a fuse but when the surge subsides, the circuit breaker can be reset and does not need replacement.

● Fusible Link—a fusible link (fuse link or main link) is a short length of special, high temperature insulated wire that acts as a fuse. When an excessive electrical current passes through a fusible link, the thin gauge wire inside the link melts, creating an intentional open to protect the circuit. To repair the circuit, the link must be replaced. Some newer type fusible links are housed in plug-in modules, which are simply replaced like a fuse, while older type fusible links must be cut and spliced if they melt. Since this link is very early in the electrical path, it's the first place to look if nothing on the boat works, yet the battery seems to be charged and is properly connected.

✳✳ CAUTION

Always replace fuses, circuit breakers and fusible links with identically rated components. Under no circumstances should a component of higher or lower amperage rating be substituted.

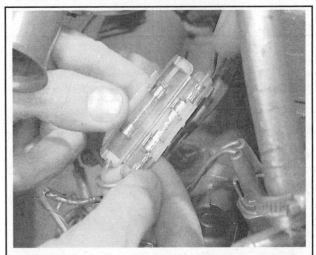

Fig. 3 Fuses protect the vessel's electrical system from abnormally high amounts of current flow

SWITCHES & RELAYS

◆ See Figure 4

Switches are used in electrical circuits to control the passage of current. The most common use is to open and close circuits between the battery and the various electric devices in the system. Switches are rated according to the amount of amperage they can handle. If a switch rated for the sufficient amperage is not used in a circuit, the switch could overload and cause damage.

Some electrical components which require a large amount of current to operate use a special switch called a relay. Since these circuits carry a large amount of current, the thickness of the wire in the circuit is also greater.

If this large wire were connected from the load to the control switch, the switch would have to carry the high amperage load and the space needed for wiring in the boat would be twice as big to accommodate the increased size of the wiring harness. To prevent these problems, a relay is used.

Relays are composed of a coil and a set of contacts. When the coil has a current passed though it, a magnetic field is formed and this field causes the contacts to move together, completing the circuit. Most relays are normally open, preventing current from passing through the circuit but they can take any electrical form depending on the job they are intended to do. Relays can be considered "remote control switches." They allow a smaller current to operate devices that require higher amperages. When a small current operates the coil, a larger current is allowed to pass by the contacts. Some common circuits that may use relays are horns, lights, starter, electric fuel pumps and other high draw circuits.

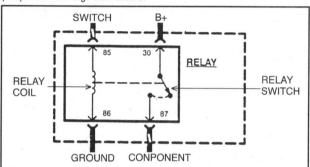

Fig. 4 Relays are composed of a coil and a switch. These two components are linked together so that when one operates, the other operates at the same time. The large wires in the circuit are connected from the battery to one side of the relay switch (B+) and from the opposite side of the relay switch to the load (component). Smaller wires are connected from the relay coil to the control switch for the circuit and from the opposite side of the relay coil to ground

LOAD

Every electrical circuit must include a "load" (something to use the electricity coming from the source). Without this load, the battery would attempt to deliver its entire power supply from one pole to another. This would result in a "short circuit" of the battery. All this electricity would take a short cut to ground and cause a great amount of damage to other components in the circuit by developing a tremendous amount of heat. This condition could develop sufficient heat to melt the insulation on all the surrounding wires and reduce a multiple wire cable to a lump of plastic and copper.

WIRING & HARNESSES

The average boat contains miles of wiring, with hundreds of individual connections. To protect the many wires from damage and to keep them from becoming a confusing tangle, they are organized into bundles, enclosed in plastic or taped together and called wiring harnesses. Different harnesses serve different parts of the boat. Individual wires are color coded to help trace them through a harness where sections are hidden from view.

Marine wiring can be either single strand wire, multi-strand wire or printed circuitry. Single strand wire has a solid metal core and is usually used inside such components as alternators, motors, relays and other devices. Multi-strand wire has a core made of many small strands of wire twisted together into a single conductor. Most of the wiring in a marine electrical system is made up of multi-strand wire, either as a single conductor or grouped together in a harness. All wiring is color coded on the insulator, either as a solid color or as a colored wire with an identification stripe. A printed circuit is a thin film of copper or other conductor that is printed on an insulator backing. Occasionally, a printed circuit is sandwiched between two sheets of plastic for more protection and flexibility. A complete printed circuit, consisting of conductors, insulating material and connectors is called a printed circuit board. Printed circuitry is used in place of individual wires or harnesses in places where space is limited, such as behind instrument panels.

Since marine electrical systems are very sensitive to changes in resistance, the selection of properly sized wires is critical when systems are repaired. A loose or corroded connection or a replacement wire that is too small for the circuit will add extra resistance and an additional voltage drop to the circuit.

The wire gauge number is an expression of the cross-section area of the conductor. Boats from countries that use the metric system will typically describe the wire size as its cross-sectional area in square millimeters. In this method, the larger the wire, the greater the number. Another common system for expressing wire size is the American Wire Gauge (AWG) system. As gauge number increases, area decreases and the wire becomes smaller. An 18 gauge wire is smaller than a 4 gauge wire. A wire with a higher gauge number will carry less current than a wire with a lower gauge number. Gauge wire size refers to the size of the strands of the conductor, not the size of the complete wire with insulator. It is possible, therefore, to have two wires of the same gauge with different diameters because one may have thicker insulation than the other.

It is essential to understand how a circuit works before trying to figure out why it doesn't. An electrical schematic shows the electrical current paths when a circuit is operating properly. Schematics break the entire electrical system down into individual circuits. In a schematic, usually no attempt is made to represent wiring and components as they physically appear on the boat, switches and other components are shown as simply as possible. Face views of harness connectors show the cavity or terminal locations in all multi-pin connectors to help locate test points.

CONNECTORS

◆ See Figures 5, 6, 7 and 8

Weatherproof connectors are most commonly used where the connector is exposed to the elements. Terminals are protected against moisture and dirt by sealing rings that provide a weather tight seal. All repairs require the use of a special terminal and the tool required to service it.

Unlike standard blade type terminals, these weatherproof terminals cannot be straightened once they are bent. Make certain that the connectors are properly seated and all of the sealing rings are in place when connecting leads.

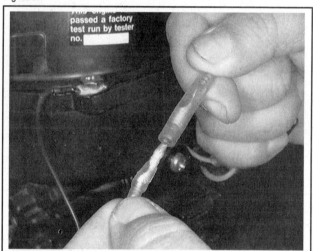

Fig. 5 Bullet connectors are some of the more common electrical connectors found on an engine

Fig. 6 A typical weatherproof electrical connector

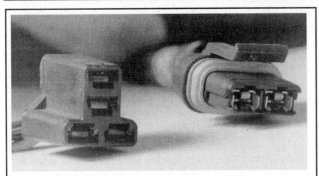

Fig. 7 Hard shell (left) and weatherproof (right) connectors have replaceable terminals

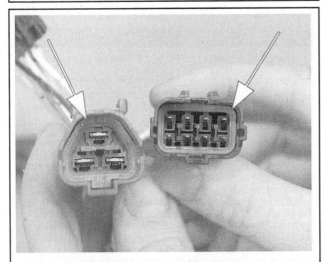

Fig. 8 The seals on weatherproof connectors must be kept in good condition to prevent the terminals from corroding

Test Equipment

Pinpointing the exact cause of trouble in an electrical circuit is most times accomplished by the use of special test equipment. The following sections describe different types of commonly used test equipment and briefly explain how to use them in diagnosis. In addition to the information covered below, the tool manufacturer's instruction manual (provided with most tools) should be read and clearly understood before attempting any test procedures.

JUMPER WIRES

◆ See Figure 9

✳✳ CAUTION °

Never use jumper wires made from a thinner gauge wire than the circuit being tested. If the jumper wire is of too small a gauge, it may overheat and possibly melt. Never use jumpers to bypass high resistance loads in a circuit. Bypassing resistances, in effect, creates a short circuit. This may, in turn, cause damage and fire. Jumper wires should only be used to bypass lengths of wire or to simulate switches.

Jumper wires are simple, yet extremely valuable, pieces of test equipment. They are basically test wires that are used to bypass sections of a circuit. Although jumper wires can be purchased, they are usually fabricated from lengths of standard marine wire and whatever type of connector (alligator clip, spade connector or pin connector) that is required for the particular application being tested. In cramped, hard-to-reach areas, it is advisable to have insulated boots over the jumper wire terminals in order to prevent accidental grounding.

It is also advisable to include a standard marine fuse in any jumper wire. This is commonly referred to as a "fused jumper". By inserting an in-line fuse holder between a set of test leads, a fused jumper wire can be used for bypassing open circuits. Use a 5-amp fuse to provide protection against voltage spikes.

Jumper wires are used primarily to locate open electrical circuits. If an electrical component fails to operate, connect the jumper wire between the component and a good ground. If the component operates only with the jumper installed, the ground circuit is open.

If the ground circuit is good but the component does not operate, the circuit between the power feed and component may be open. By moving the jumper wire successively back from the component toward the power source, you can isolate the area of the circuit where the open is located. When the component stops functioning or the power is cut off, the open is in the segment of wire between the jumper and the point previously tested.

You can sometimes connect the jumper wire directly from the battery to the "hot" terminal of the component but first make sure the component uses 12 volts in operation. Some electrical components, such as fuel injectors or sensors are designed to operate on about 4 to 5 volts and running 12 volts directly to these components will cause damage.

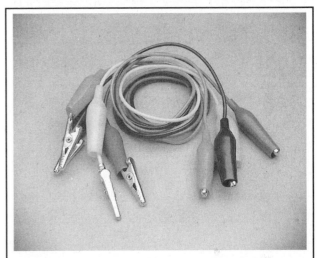

Fig. 9 Jumper wires are simple, yet extremely valuable, pieces of equipment

TEST LIGHTS

◆ See Figure 10

The test light is used to check circuits and components while electrical current is flowing through them. It is used for voltage and ground tests. To use a 12-volt test light, connect the ground clip to a good ground and probe wherever necessary with the pick. The test light will illuminate when voltage is detected. This does not necessarily mean that 12 volts (or any particular amount of voltage) is present, it only means that some voltage is present.

■ **It is advisable before using the test light to touch its ground clip and probe across the battery posts or terminals to make sure the light is operating properly.**

✳✳ WARNING

Do not use a test light to probe electronic ignition, spark plug or coil wires. Never use a pick-type test light to probe wiring on electronically controlled systems unless specifically instructed to do so. Any wire insulation that is pierced by the test light probe should be taped and sealed with silicone after testing.

Like the jumper wire, the 12-volt test light is used to isolate opens in circuits. But, whereas the jumper wire is used to bypass the open to operate the load, the 12-volt test light is used to locate the presence of voltage in a circuit. If the test light illuminates, there is power up to that point in the circuit, if the test light does not illuminate, there is an open circuit (no power). Move the test light in successive steps back toward the power source until the light in the handle illuminates. The open is between the probe and a point that was previously probed.

The self-powered test light is similar in design to the 12-volt test light but contains a 1.5-volt penlight battery in the handle. It is most often used in place of a multi-meter to check for open or short circuits when power is isolated from the circuit (continuity test).

The battery in a self-powered test light does not provide much current. A weak battery may not provide enough power to illuminate the test light even when a complete circuit is made (especially if there is high resistance in the circuit). Always make sure that the test battery is strong. To check the battery, briefly touch the ground clip to the probe, if the light glows brightly, the battery is strong enough for testing.

✳✳ WARNING

A self-powered test light should not be used on any electronically controlled system or component. The small amount of electricity transmitted by the test light is enough to damage many electronic marine components.

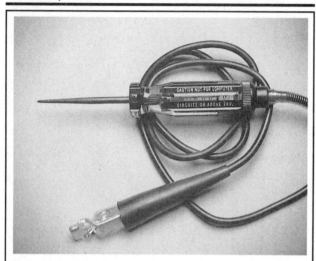

Fig. 10 A test light is used to detect the presence of voltage in a circuit

MULTI-METERS

◆ See Figure 11

Multi-meters are an extremely useful tool for troubleshooting electrical problems. They can be purchased in either analog or digital form and have a price range to suit any budget. A multi-meter is a voltmeter, ammeter and multi-meter (along with other features) combined into one instrument. It is often used when testing solid-state circuits because of its high input impedance (usually 10 mega-ohms or more). A brief description of the multi-meter main test functions follows:

• Voltmeter—the voltmeter is used to measure voltage at any point in a circuit or to measure the voltage drop across any part of a circuit. Voltmeters usually have various scales and a selector switch to allow the reading of different voltage ranges.

The voltmeter has a positive and a negative lead. To avoid damage to the meter, always connect the negative lead to the negative (–) side of the circuit (to ground or nearest the ground side of the circuit) and connect the positive lead to the positive (+) side of the circuit (to the power source or the nearest power source).

■ The negative voltmeter lead will always be Black and the positive voltmeter will always be some color other than Black (usually Red).

• Multi-meter—the multi-meter is designed to read resistance (measured in ohms) in a circuit or component. Most multi-meters will have a selector switch which permits the measurement of different ranges of resistance (usually the selector switch allows the multiplication of the meter reading by

10, 100, 1,000 and 10,000). Some multi-meters are "auto-ranging" which means the meter itself will determine which scale to use.

Since an internal battery powers the meters, the multi-meter can be used like a self-powered test light. When the multi-meter is connected, current from the multi-meter flows through the circuit or component being tested. Since the multi-meter's internal resistance and voltage are known values, the amount of current flow through the meter depends on the resistance of the circuit or component being tested.

The multi-meter can also be used to perform a continuity test for suspected open circuits. In using the meter for making continuity checks, do not be concerned with the actual resistance readings. Zero resistance (or any ohm reading) indicates continuity in the circuit. Infinite resistance indicates an opening in the circuit. A high resistance reading where there should be none indicates a problem in the circuit.

Checks for short circuits are made in the same manner as checks for open circuits, except that the circuit must be isolated from both power and normal ground. Infinite resistance indicates no continuity, while zero resistance indicates a dead short.

✳✳ WARNING

Never use a multi-meter to check the resistance of a component or wire while there is voltage applied to the circuit.

• Ammeter—an ammeter measures the amount of current flowing through a circuit in units called amps (amps). At normal operating voltage, most circuits have a characteristic amount of amps, called "current draw" which can be measured using an ammeter. By referring to a specified current draw rating, then measuring the amps and comparing the two values, one can determine what is happening within the circuit to aid in diagnosis.

For example, an open circuit will not allow any current to flow, so the ammeter reading will be zero. A damaged component or circuit will have an increased current draw, so the reading will be high.

The ammeter is always connected in series with the circuit being tested. All of the current that normally flows through the circuit must also flow through the ammeter, if there is any other path for the current to follow, the ammeter reading will not be accurate. The ammeter itself has very little resistance to current flow and, therefore, it will not affect the circuit but it will measure current draw only when the circuit is closed and electricity is flowing. Excessive current draw can blow fuses and drain the battery, while a Reduced current draw can cause motors to run slowly, lights to dim and other components to not operate properly.

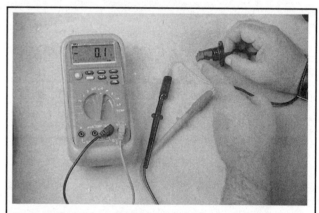

Fig. 11 Multi-meters are essential for diagnosing faulty wires, switches and other electrical components

Troubleshooting the Electrical System

When diagnosing any electrical problem organized troubleshooting is a must. The complexity of electrical systems on modern boats and their power plants demands that you approach any problem in a logical organized manner. There are certain troubleshooting techniques, which are standard:

• **Establish when the problem occurs**—Does the problem appear only under certain conditions? Were there any noises, odors or other unusual symptoms?

- **Check for obvious problems**—Problems such as broken wires and loose or dirty connections can cause major problems. Always check the obvious before assuming something complicated (or expensive) is the cause.

■ **Experience has shown that most problems tend to be the result of a fairly simple and obvious cause, such as loose or corroded connectors, bad grounds or damaged wire insulation, which causes a short. This makes careful visual inspection of components during testing essential to quick and accurate troubleshooting.**

- **Isolate the problem area**—Make some simple tests and observations, then eliminate the systems that are working properly. Test for problems systematically to determine the cause once the problem area is isolated. Are all the components functioning properly? Is there power going to electrical switches and motors. Performing careful, systematic checks will often turn up most causes on the first inspection, without wasting time checking components that have little or no relationship to the problem.
- **Verify all systems after repairs are completed**—Some causes can be traced to more than one component, so a careful verification of repair work is important in order to pick up additional malfunctions that may cause a problem to reappear or a different problem to arise. A blown fuse, for example, is a simple problem that may require more than another fuse to repair. If you don't look for a problem that caused a fuse to blow, a shorted wire (for example) may go undetected.

VOLTAGE

◆ **See Figure 12**

This test determines voltage available from the battery and should be the first step in any electrical troubleshooting procedure after visual inspection. Many electrical problems, especially on electronically controlled systems, can be caused by a low state of charge in the battery. Excessive corrosion at the battery cable terminals can cause poor contact that will prevent proper charging and full battery current flow.

1. Set the voltmeter selector switch to the 10-volt position.
2. Connect the multi-meter negative lead to the battery's negative (–) terminal and the positive lead to the battery's positive (+) terminal.
3. Turn the battery (ignition) switch **ON** to provide a load.
4. A well-charged battery should register over 12 volts. If the meter reads below 11.5 volts, the battery power may be insufficient to operate the electrical system properly.
5. Charge the battery and retest.

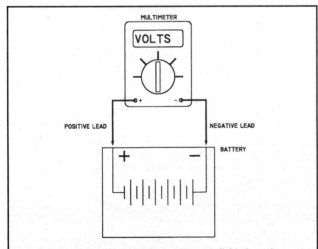

Fig. 12 The voltage test determines voltage available from the battery and should be the first step in any electrical troubleshooting procedure after visual inspection

VOLTAGE DROP

◆ **See Figure 13**

When current flows through a load, the voltage beyond the load drops. This voltage drop is due to the resistance created by the load and also by small resistances created by corrosion at the connectors and damaged insulation on the wires. The maximum allowable voltage drop under load is critical, especially if there is more than one load in the circuit, since all voltage drops are cumulative.

1. Set the voltmeter selector switch to the 10-volt position.
2. Connect the multi-meter negative lead to the battery negative (–) terminal or another good ground.
3. Touch the multi-meter positive lead to the battery's positive (+) terminal to determine battery voltage.
4. Operate the circuit and check the voltage prior to the first component (load).
5. There should be little or no voltage drop (from battery voltage) in the circuit prior to the first component. If a voltage drop exists, the wire or connectors in the circuit are suspect.
6. While operating the first component in the circuit, probe the ground-side of the component with the positive (+) meter lead and observe the voltage readings. A small voltage drop should be noticed. The resistance of the component causes this voltage drop.
7. Repeat the test for each component (load) down the circuit.
8. If a large voltage drop is noticed, the preceding component, wire or connector is suspect.

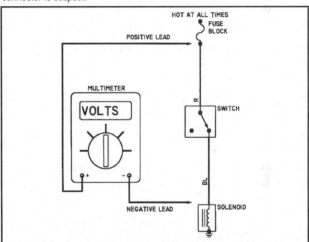

Fig. 13 Voltage drop is due to the resistance created by the load and also by small resistances created by corrosion at the connectors and damaged insulation on the wires

RESISTANCE

◆ **See Figures 14 and 15**

✳✳ WARNING

Never use an multi-meter with power applied to the circuit. The multi-meter is designed to operate on its own power supply. The normal 12-volt electrical system voltage can damage the meter!

1. Isolate the circuit from the boat's power source.
2. Ensure that the battery (ignition) switch is **OFF**.
3. Isolate at least one side of the circuit to be checked, in order to avoid reading parallel resistances. Parallel circuit resistances will always give a lower reading than the actual resistance of either of the branches.
4. Connect the meter leads to both sides of the circuit (wire or component) and read the actual resistance measured in ohms on the meter scale. Make sure the selector switch is set to the proper ohm scale for the circuit being tested, to avoid misreading the multi-meter test value.

5. Compare this reading to the resistance specification for the component or formulate the theoretical resistance using Ohms Law.

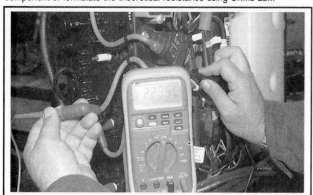

Fig. 14 Using a multi-meter to check resistance on the secondary side of the ignition coil

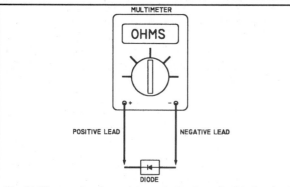

Fig. 15 When performing resistance tests, always isolate the circuit from power

OPEN CIRCUITS

◆ See Figures 16 and 17

This test already assumes the existence of an open in the circuit and it is used to help locate the open portion.
1. Isolate the circuit from power and ground.
2. Connect the self-powered test light or multi-meter ground clip to the ground-side of the circuit and probe sections of the circuit sequentially.
3. If the light is out or there is infinite resistance, the open is between the probe and the circuit ground.
4. If the light is on or the meter shows continuity, the open is between the probe and the end of the circuit toward the power source.

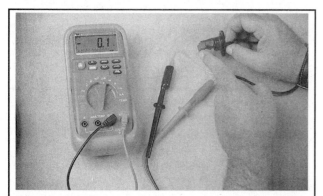

Fig. 16 The infinite display on this multi-meter (0.1) indicates that the circuit is open

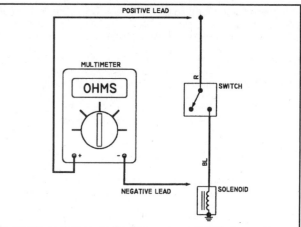

Fig. 17 The easiest way to illustrate an open circuit is to consider an example circuit with a switch. When the switch is turned OFF, power does not flow through the circuit to the load. Thus, the circuit is open

SHORT CIRCUITS

◆ See Figure 18

■ **Never use a self-powered test light to perform checks for opens or shorts when power is applied to the circuit under test. The test light can be damaged by outside power.**

1. Isolate the circuit from power and ground.
2. Connect the self-powered test light or multi-meter ground clip to a good ground and probe any easy-to-reach point in the circuit.
3. If the light comes on or there is continuity, there is a short somewhere in the circuit.
4. To isolate the short, probe a test point at either end of the isolated circuit (the light should be on or the meter should indicate continuity).
5. Leave the test light probe engaged and sequentially open connectors or switches, remove parts, etc. until the light goes out or continuity is broken.
6. When the light goes out, the short is between the last two circuit components, which were opened.

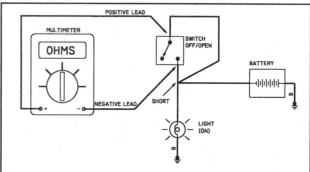

Fig. 18 In this illustration, the circuit between the battery and light should be open because the switch is turned OFF. However, battery voltage is reaching the light at the point of the short; this could possibly be caused by chaffed wires

Wire and Connector Repair

Almost anyone can replace damaged wires, as long as the proper tools and parts are available. Wire and terminals are available to fit almost any need. Even the specialized weatherproof, molded and hard shell connectors are now available from aftermarket suppliers.

Be sure the ends of all the wires are fitted with the proper terminal hardware and connectors. Wrapping a wire around a stud is never a permanent solution and will only cause trouble later. Replace wires one at a time to avoid confusion.

Always route wires in the same manner of the manufacturer.

When replacing connections, make absolutely certain that the connectors are certified for marine use. Automotive wire connectors may not meet United States Coast Guard (USCG) specifications.

■ **If connector repair is necessary, only attempt it if you have the proper tools. Weatherproof connectors require special tools to release the pins inside the connector. Attempting to repair these connectors with conventional hand tools will damage them.**

BATTERY

The battery is one of the most important parts of the electrical system. In addition to providing electrical power to start the engine, it also provides power for operation of the running lights, radio and electrical accessories.

Because of its job and the consequences (failure to perform in an emergency), the best advice is to purchase a well-known brand, with an extended warranty period, from a reputable dealer.

The usual warranty covers a pro-rated replacement policy, which means the purchaser is entitled to consideration for the time left on the warranty period if the battery should prove defective before the end of the warranty.

Many manufacturers have specifications on the size and type of battery to use for their engines. If in doubt as to how large a battery the boat requires, make a liberal estimate and then purchase the one with the next higher amp rating.

Please refer to Section 3 for all procedures on cleaning, testing, storage and maintenance.

BATTERY CONSTRUCTION

◆ **See Figure 19**

A battery consists of a number of positive and negative plates immersed in a solution of diluted sulfuric acid. The plates contain dissimilar active materials and are kept apart by separators. The plates are grouped into elements. Plate straps on top of each element connect all of the positive plates and all of the negative plates into groups.

The battery is divided into cells holding a number of the elements apart from the others. The entire arrangement is contained within a hard plastic case. The top is a one-piece cover and contains the filler caps for each cell. The terminal posts protrude through the top where the battery connections for the boat are made. Each of the cells is connected to its neighbor in a positive-to-negative manner with a heavy strap called the cell connector.

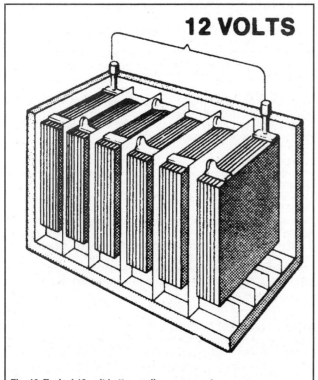

Fig. 19 Typical 12-volt battery cell arrangement

MARINE BATTERIES

◆ **See Figure 20**

Because marine batteries are required to perform under much more rigorous conditions than automotive batteries, they are constructed differently than those used in automobiles or trucks. Therefore, a marine battery should always be the No. 1 unit for the boat and other types of batteries used only in an emergency.

Marine batteries have a much heavier exterior case to withstand the violent pounding and shocks imposed on it as the boat moves through rough water and in extremely tight turns. The plates are thicker and each plate is securely anchored within the battery case to ensure extended life. The caps are spill proof to prevent acid from spilling into the bilge when the boat heels to one side in a tight turn or is moving through rough water. Because of these features, the marine battery will recover from a low charge condition and give satisfactory service over a much longer period of time than any type intended for automotive use.

✳✳ WARNING

Never use a Maintenance-free battery with an engine that is not voltage regulated. The charging system will continue to charge as long as the engine is running and it is possible that the electrolyte could boil out if periodic checks of the cell electrolyte level are not done.

Fig. 20 Cut-away look at a typical marine battery

BATTERY RATINGS

◆ **See Figure 21**

Three different methods are used to measure and indicate battery electrical capacity:
- Amp/hour rating
- Cold cranking performance
- Reserve capacity

The amp/hour rating of a battery refers to the battery's ability to provide a set amount of amps for a given amount of time under test conditions at a constant temperature. Therefore, if the battery is capable of supplying 4 amps of current for 20 consecutive hours, the battery is rated as an 80 amp/hour battery. The amp/hour rating is useful for some service operations, such as slow charging or battery testing.

Cold cranking performance is measured by cooling a fully charged battery to 0 degrees F (-17 degrees C) and then testing it for 30 seconds to determine the maximum current flow. In this manner the cold cranking amp rating is the number of amps available to be drawn from the battery before the voltage drops below 7.2 volts.

The illustration depicts the amount of power in watts available from a battery at different temperatures and the amount of power in watts required of the engine at the same temperature. It becomes quite obvious—the colder the climate, the more necessary for the battery to be fully charged.

Reserve capacity of a battery is considered the length of time, in minutes, at 80 degrees F (27 degrees C), a 25 amp current can be maintained before the voltage drops below 10.5 volts. This test is intended to provide an approximation of how long the engine, including electrical accessories, could operate satisfactorily if the stator assembly or lighting coil did not produce sufficient current. A typical rating is 100 minutes.

■ **If possible, the new battery should have a power rating equal to or higher than the unit it is replacing.**

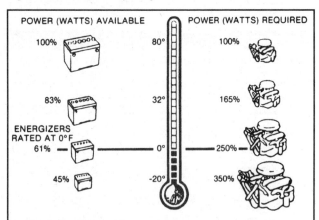

Fig. 21 Comparison of battery efficiency and engine demands at various temperatures

BATTERY LOCATION

◆ **See Figure 22**

Every battery installed in a boat must be secured in a well protected, ventilated area. If the battery area lacks adequate ventilation, hydrogen gas, which is given off during charging, is very explosive. This is especially true if the gas is concentrated and confined.

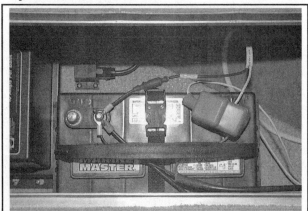

Fig. 22 A good example of a secured, well protected and ventilated battery area

DUAL BATTERY INSTALLATION

◆ **See Figures 23, 24, 25 and 26**

Three methods are available for utilizing a dual-battery hook-up.
1. A high-capacity switch can be used to connect the two batteries. The accompanying illustration details the connections for installation of such a switch. This type of switch installation has the advantage of being simple, inexpensive, and easy to mount and hookup. However, if the switch is forgotten in the closed position, it will let the convenience loads run down both batteries and the advantage of the dual installation is lost. However, the switch may be closed intentionally to take advantage of the extra capacity of the two batteries, or it may be temporarily closed to help start the engine under adverse conditions.
2. A relay, can be connected into the ignition circuit to enable both batteries to be automatically put in parallel for charging or to isolate them for ignition use during engine cranking and start. By connecting the relay coil to the ignition terminal of the ignition-starting switch, the relay will close during the start to aid the starting battery. If the second battery is allowed to run down, this arrangement can be a disadvantage since it will draw a load from the starting battery while cranking the engine. One way to avoid such a condition is to connect the relay coil to the ignition switch accessory terminal. When connected in this manner, while the engine is being cranked, the relay is open, but when the engine is running with the ignition switch in the normal position, the relay is closed, and the second battery is being charged at the same time as the starting battery.
3. A heavy duty switch installed as close to the batteries as possible can be connected between them. If such an arrangement is used it must meet the standards of the American Boat and Yacht Council, or the Fire Protection Standard for Motor Craft, N.F.P.A. No. 302.

BATTERY CHARGERS

◆ **See Figure 27**

Before using any battery charger, consult the manufacturer's instructions for its use. Battery chargers are electrical devices that change Alternating Current (AC) to a lower voltage of Direct Current (DC) that can be used to charge a marine battery. There are two types of battery chargers—manual and automatic.

A manual battery charger must be physically disconnected when the battery has come to a full charge. If not, the battery can be overcharged and possibly fail. Excess charging current at the end of the charging cycle will heat the electrolyte, resulting in loss of water and active material, substantially reducing battery life.

■ **As a rule, on manual chargers, when the ammeter on the charger registers half the rated amperage of the charger, the battery is fully charged. This can vary and it is recommended to use a hydrometer to accurately measure state of charge.**

Automatic battery chargers have an important advantage—they can be left connected (for instance, overnight) without the possibility of overcharging the battery. Automatic chargers are equipped with a sensing device to allow the battery charge to taper off to near zero as the battery becomes fully charged. When charging a low or completely discharged battery, the meter will read close to full rated output. If only partially discharged, the initial reading may be less than full rated output, as the charger responds to the condition of the battery. As the battery continues to charge, the sensing device monitors the state of charge and reduces the charging rate. As the rate of charge tapers to zero amps, the charger will continue to supply a few milliamps of current—just enough to maintain a charged condition.

BATTERY CABLES

Battery cables don't go bad very often but like anything else, they can wear out. If the cables on your boat are cracked, frayed or broken, they should be replaced.

When working on any electrical component, it is always a good idea to disconnect the negative (–) battery cable. This will prevent potential damage to many sensitive electrical components

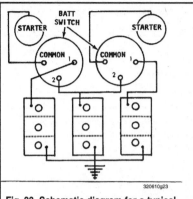

Fig. 23 Schematic diagram for a typical three battery, two engine hookup

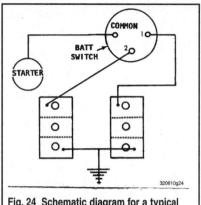

Fig. 24 Schematic diagram for a typical two battery, one engine hookup

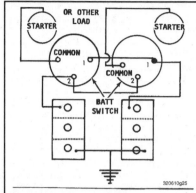

Fig. 25 Schematic diagram for a typical two battery, two engine hookup

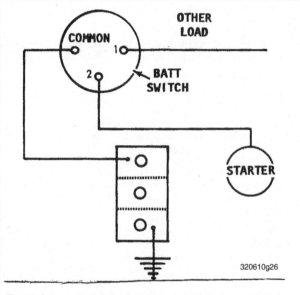

Fig. 26 Schematic diagram for a typical single battery, one engine hookup

Fig. 27 Automatic chargers, such as the Battery Tender® from Deltran, are equipped with a sensing device to allow the battery charge to taper off to near zero as the battery becomes fully charged

Always replace the battery cables with one of the same length or you will increase resistance and possibly cause hard starting. Smear the battery posts with a light film of dielectric grease or a battery terminal protectant-spray once you've installed the new cables. If you replace the cables one at a time, you won't mix them up.

■ **Any time you disconnect the battery cables, it is recommended that you disconnect the negative (–) battery cable first.**

This will prevent you from accidentally grounding the positive (+) terminal when disconnecting it, thereby preventing damage to the electrical system.

Before you disconnect the cable(s), first turn the ignition to the **OFF** position. This will prevent a draw on the battery that could cause arcing. When the battery cable(s) are reconnected (negative cable last), be sure to check all electrical accessories are all working correctly.

CHARGING SYSTEM

General Information

The charging system provides electrical power for operation of the vessel's ignition system, starting system and all electrical accessories. The battery serves as an electrical surge or storage tank, storing (in chemical form) the energy originally produced by the engine driven alternator. The system also provides a means of regulating output to protect the battery from being overcharged and to avoid excessive voltage to the accessories.

The storage battery is a chemical device incorporating parallel lead plates in a tank containing a sulfuric acid/water solution. Adjacent plates are slightly dissimilar, and the chemical reaction of the two dissimilar plates produces electrical energy when the battery is connected to a load such as the starter motor. The chemical reaction is reversible, so that when the alternator is producing a voltage (electrical pressure) greater than that produced by the battery, electricity is forced into the battery, and the battery is returned to its fully charged state.

Newer engines use alternating current alternators, because they are more efficient, can be rotated at higher speeds, and have fewer brush problems. In an alternator, the field usually rotates while all the current produced passes only through the stator winding. The brushes bear against continuous slip rings. This causes the current produced to periodically reverse the direction of its flow. Diodes (electrical one way valves) block the flow of current from traveling in the wrong direction. A series of diodes is wired together to permit the alternating flow of the stator to be rectified back to 12 volts DC for use by the vessel's electrical system.

The voltage regulating function is performed by a regulator. The regulator is often built into the alternator; this system is termed an integrated or internal regulator.

An alternator differs from a DC shunt generator in that the armature is stationary, and is called the stator, while the field rotates and is called the rotor. The higher current values in the alternator's stator are conducted to the external circuit through fixed leads and connections, rather than through a rotating commutator and brushes as in a DC generator. This eliminates a major point of maintenance.

The rotor assembly is supported in the drive end frame by a ball bearing and at the other end by a roller bearing. These bearings are lubricated during assembly and require no maintenance. There are six diodes in the end frame assembly. These diodes are electrical check valves that also change the alternating current developed within the stator windings to a Direct Current (DC) at the output (BAT) terminal. Three of these diodes are negative and are mounted flush with the end frame while the other three are positive and are mounted into a strip called a heat sink. The positive diodes are easily identified as the ones within small cavities or depressions.

The alternator charging system is a negative (–) ground system which consists of an alternator, a regulator, a charge indicator, a storage battery and wiring connecting the components, and fuse link wire.

The alternator is belt-driven from the engine. Energy is supplied from the alternator/regulator system to the rotating field through two brushes to two slip-rings. The slip-rings are mounted on the rotor shaft and are connected to the field coil. This energy supplied to the rotating field from the battery is called excitation current and is used to initially energize the field to begin the generation of electricity. Once the alternator starts to generate electricity, the excitation current comes from its own output rather than the battery.

The alternator produces power in the form of alternating current. The alternating current is rectified by 6 diodes into direct current. The direct current is used to charge the battery and power the rest of the electrical system.

When the ignition key is turned **ON**, current flows from the battery, through the charging system indicator light, to the voltage regulator, and to the alternator. When the engine is started, the alternator begins to produce current. As the alternator turns and produces current, the current is divided in two ways: part to the battery (to charge the battery and power the electrical components of the vessel), and part is returned to the alternator (to enable it to increase its output). In this situation, the alternator is receiving current from the battery and from itself. A voltage regulator is wired into the current supply to the alternator to prevent it from receiving too much current, which would cause it to put out too much current. Conversely, if the voltage regulator does not allow the alternator to receive enough current, the battery will not be fully charged and will eventually go dead.

The battery is connected to the alternator at all times, whether the ignition key is turned **ON** or not. If the battery were shorted to ground, the alternator would also be shorted. This would damage the alternator. To prevent this, a fuse link is installed in the wiring between the battery and the alternator. If the battery is shorted, the fuse link melts, protecting the alternator.

An alternator is better that a conventional, DC shunt generator because it is lighter and more compact, because it is designed to supply the battery and accessory circuits through a wide range of engine speeds, and because it eliminates the necessary maintenance of replacing brushes and servicing commutators.

General System Troubleshooting

The following symptoms and possible corrective actions will be helpful in restoring a faulty charging system to proper operation.

ALTERNATOR FAILS TO CHARGE

1. Drive belt loose or broken. Replace and/or adjust drive belt.
2. Corroded or loose wires or connection in the charging circuit. Inspect, clean, and tighten.
3. Open charging circuit. Trace and repair.

ALTERNATOR CHARGES LOW OR UNSTEADY

1. Drive belt loose or broken. Replace and/or adjust drive belt.
2. Battery charge too low. Charge or replace the battery.
3. High resistance at the battery terminals. Remove the cables, clean the connectors and battery posts, replace and tighten.
4. High resistance in the charging circuit. Trace and repair.
5. High resistance in the ground circuit. Trace and repair.

ALTERNATOR OUTPUT TOO HIGH BATTERY OVERCHARGED

1. Regulator base not grounded properly. Correct condition to make good ground.
2. Faulty ignition switch. Replace switch.

ALTERNATOR TOO NOISY

1. Worn, loose, or frayed drive belt. Replace belt and adjust properly.
2. Alternator mounting loose. Tighten all mounting hardware securely.
3. Worn alternator bearings. Replace.
4. Interference between rotor and stator leads or rectifier leads. Check and correct.
5. Rotor or fan damaged. Replace.
6. Open or shorted rectifier. Replace.
7. Open or shorted winding in stator. Replace.

AMMETER FLUCTUATES CONSTANTLY

1. High resistance connection in the alternator or voltage regulator circuit. Trace and repair.

Alternator

PRECAUTIONS

Several precautions must be observed when performing work on alternator equipment.
- If the battery is removed for any reason, make sure that it is reconnected with the correct polarity. Reversing the battery connections may result in damage to the one-way rectifiers.
- Never short across or ground any of the alternator terminals unless specifically mentioned in a particular test procedure.
- Never operate the alternator with the main circuit broken. Make sure that the battery, alternator, and regulator leads are not disconnected while the engine is running.
- Never attempt to polarize an alternator.
- When charging a battery that is installed in the vessel, disconnect the battery cables.
- When arc (electric) welding is to be performed on any part of the vessel, disconnect the negative battery cable and alternator leads.
- Never disconnect the battery cables while the engine is running

CHARGING SYSTEM INSPECTION

1. Make sure the battery connections are clean and tight and that the battery is in good condition and fully charged.
2. Check the drive belt for damage or looseness.
3. Check the wiring harness at the alternator. The harness connector should be tight and latched. Make sure that the output terminal of the alternator is connected to the vessel battery positive lead.
4. Verify that all charging system related fuses and electrical connections are tight and free of damage.
5. Check the mounting bolts for proper torque.

The alternator does not require period lubrication. The rotor shaft is mounted on bearings at the drive end and the slip ring end. Each bearing contains its own permanent grease supply.

TESTING

Many times the alternator is suspected of being defective when the battery is not receiving a charge and is constantly being depleted of its energy. Most of the time a heavily corroded wire terminal, broken wire or worn out battery is the actual problem. Perform the preliminary checks listed above to eliminate any problem areas in the charging circuitry before performing the output tests.

If the battery is constantly undercharged, verify all accessories are being switched off when the engine is not running. Check to see if a new accessory (fish finder, live bait tank, etc.) has been added which will place a heavy ampere draw on the battery when operating at low speeds. The battery may be drained when operating at slow speeds for long periods of time.

Check the physical condition and charge state of the battery. The battery should be 75% (1.230 specific gravity) of a full charge. If one or more cells in the battery are defective, the battery should be replaced.

Inspect the alternator system wiring for corroded or loose terminals, damaged or frayed wiring and/or loose wire harness connectors. Check the drive belt for physical condition and proper tension.

If the charging system has passed all the above visual checks, perform the output tests to determine if the alternator is defective.

The following tests require the use of a voltmeter/multi-meter capable of reading 0-20 volts DC. These tests will determine if the alternator and other components within the alternator circuit are in satisfactory working condition.

Mando And Delco

CURRENT OUTPUT

◆ See Figure 27a

This test requires the engine to be operated at approximately 2,000 rpm. Therefore, the boat should be launched and well secured to a dock prior to performing this test. *Do not* attempt to perform this test with only a flush devise attached to the stern drive. Such action could seriously damage the engine or the stern drive or both. Be sure the engine compartment is well ventilated and there are no gasoline vapors present before starting the engine.

1. Disconnect the negative battery cable.
2. Connect a multi-meter as per the manufacturer's instructions.
3. Disconnect the output lead (Orange wire) from the back of the alternator. Connect the meter in series between the output terminal and the Orange wire lead. Be sure the positive side of the meter is connected to the output terminal and the negative side of the meter is connected to the Orange output wire lead.
4. Connect the negative battery cable to the battery.
5. Remove the coil lead from the distributor cap and ground it to the engine block. Turn on all the accessories and crank the engine for 15-20 seconds to drain some of the energy from the battery. Shut off all accessories and reconnect the coil lead to the distributor cap. Start the engine and run the idle speed to 1,500-2,000 rpm. The meter should indicate 30 amps or greater.
6. If the reading is low, stop the engine and connect a jumper wire between the output terminal and the regulator terminal on the back of the alternator.
7. Repeat Steps 5 and 6.
8. If the reading is now 30 amps or greater, the diodes in the rectifier are damaged. Replace the alternator.
9. If the ammeter indication is still low with the jumper wire connected, perform the following Voltage Output Test to determine if the regulator or the alternator internal component is damaged.

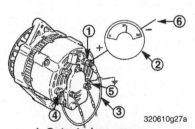

1- Output wire - orange
2- Ammeter (0-50 Amps)
3- Jumper lead
4- Regulator read
5- Ground
6- Output lead (orange)

Fig. 27a Testing the current output test

DIODE TRIO

◆ See Figure 28

1. Obtain an ohmmeter and set the switches to Rx1 scale. Connect one meter lead to the common side of the diode. Connect the other meter lead to the opposite end of the diode and note the meter reading.
2. Reverse the meter leads and again note the meter reading. The meter should indicate a high resistance (no movement) in one direction and a low resistance indication when the leads are reversed.
3. If both readings are high, indicating the diode is open, or if both readings are low indicating the diode is shorted, the rectifier assembly must be replaced. Repeat this step on the other two diodes.

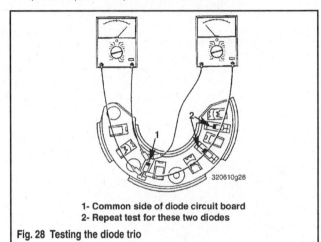

1- Common side of diode circuit board
2- Repeat test for these two diodes

Fig. 28 Testing the diode trio

EXCITATION CIRCUIT

◆ See Figure 29

1. On 1992-97 3.0L engines, remove the four mounting screws and disconnect the voltage regulator from the back of the alternator.
2. Connect a multi-meter as per the manufacturer's instructions.
3. Connect the positive lead to the alternator regulator terminal (Green wire) on 1992-97 3.0L engines, or to the tie strap terminals on all other engines. Connect the negative lead to the ground terminal on the alternator.
4. Turn the ignition switch ON. The meter should read 1.3-2.5 volts. If no reading is observed, there is an open in the excitation lead or circuit on the regulator.
5. Disconnect the Purple lead at the regulator and connect the positive lead to it. If the meter indicates battery voltage, replace the regulator. If there is no reading, check the excitation circuit for bad connections or frayed wiring. If the reading is 0.75-1.1 volts, the rotor field circuit is either shorted or grounded. Test the rotor. If the reading is 6-7 volts, the rotor field circuit is open. Inspect the rotor for worn brushes or dirty slip-rings. Replace the brushes if they are less than 1/4 in. (6mm) long. If both are OK, test the rotor.

OUTPUT CIRCUIT

◆ See Figure 30

1. Connect a multi-meter as per the manufacturer's instructions.
2. Connect the positive lead to the alternator output wire terminal (Orange). Connect the negative lead to the ground terminal on the alternator.
3. The meter should indicate approximate battery voltage (13.8-14.2 V). Now, wiggle the engine wire harness and look for any drop in the meter reading that would indicate a loose or broken wire. If no reading is obtained or the reading varies, check the alternator output circuit (Orange) to the cranking motor and Red wire from the starter motor to the battery for loose or corroded connection or possibly damaged wiring.
4. Disconnect the multi-meter leads from the alternator.

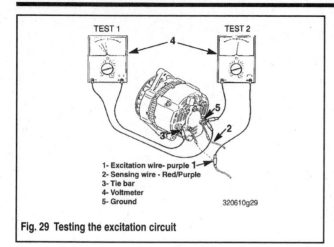

1- Excitation wire- purple
2- Sensing wire - Red/Purple
3- Tie bar
4- Voltmeter
5- Ground

320610g29

Fig. 29 Testing the excitation circuit

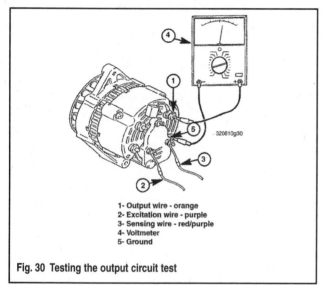

320610g30

1- Output wire - orange
2- Excitation wire - purple
3- Sensing wire - red/purple
4- Voltmeter
5- Ground

Fig. 30 Testing the output circuit test

RECTIFIER (-) NEGATIVE

◆ See Figure 31

1. The rectifier must be disconnected from the stator to perform the following test. Each rectifier must be tested to be sure it is not open or shorted. Check the three rectifiers using the following procedures.

2. Obtain an ohmmeter and set the switches to Rx1 scale. Connect one lead of the meter to the negative rectifier diode heat sink. Connect the other meter lead to one of the rectifier terminals and note the reading. Reverse the meter leads and again note the meter reading. The meter should indicate a high resistance (no movement) in one direction and a low resistance indication when the leads are reversed.

3. If both readings are high, indicating the rectifier is open, or if both readings are low, indicating the rectifier is shorted, the rectifier assembly must be replaced. Repeat this step on the other two rectifiers.

RECTIFIER (+) POSITIVE

◆ See Figure 32

The rectifier must be disconnected from the stator to perform the following test. Each rectifier must be tested to be sure it is not open or shorted. Check the three rectifiers using the following procedures.

1. Obtain an ohmmeter and set the switches to Rx1 scale. Connect one meter lead to the 1/4 in. stud on the positive rectifier heat sink. Connect the other meter lead to one of the rectifier diode terminals and note the reading.

2. Reverse the meter leads and again note the meter reading.

The meter should indicate a high resistance (no movement) in one direction and a low resistance indication when the leads are reversed.

3. If both readings are high, indicating the rectifier is open, or if both readings are low indicating the rectifier is shorted, the rectifier assembly must be replaced. Repeat this step on the other two rectifiers.

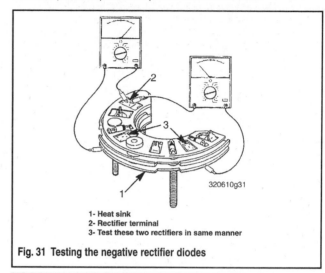

1- Heat sink
2- Rectifier terminal
3- Test these two rectifiers in same manner

Fig. 31 Testing the negative rectifier diodes

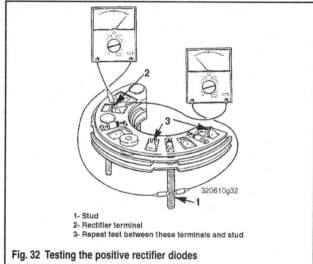

320610g32

1- Stud
2- Rectifier terminal
3- Repeat test between these terminals and stud

Fig. 32 Testing the positive rectifier diodes

ROTOR

◆ See Figure 33

1. Check the rotor windings for a short or ground by first connecting an ohmmeter between one brush slip ring and the shaft. The meter must indicate an open circuit on the R-100 scale. Next, check the rotor windings for continuity. Connect the ohmmeter between the two slip rings, and the reading should be between 4.2-5.5 ohms at a room temperature of 70ºF to 80ºF (22º-27ºC).

SENSING CIRCUIT

◆ See Figure 34

1. Connect a multi-meter as per the manufacturer's instructions.
2. Unplug the Red/Purple lead from the voltage regulator. Connect the positive lead to Red/Purple wire lead and the negative lead to the alternator ground terminal. The voltmeter should indicate approximately battery voltage. If battery voltage is not present, check the Red/Purple sensing lead for loose dirty or damaged connections.

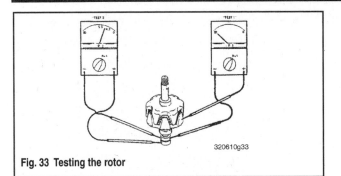

Fig. 33 Testing the rotor

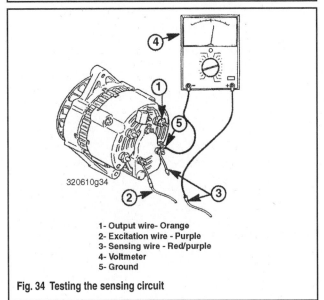

1- Output wire- Orange
2- Excitation wire - Purple
3- Sensing wire - Red/purple
4- Voltmeter
5- Ground

Fig. 34 Testing the sensing circuit

STATOR

◆ See Figure 35

1. The stator is not checked for shorts due to the very low resistance of the windings and the need for specialized test equipment. However, tests are made to check the stator for an open or grounded circuit. Use the following procedures to check the stator for possible defects.

2. Test the stator for ground circuits by obtaining an ohmmeter and setting the switches for Rx1 scale. Connect one meter lead to one of the stator leads. Make a good positive contact with the other meter lead to the stator frame.

3. The meter should indicate no continuity (no movement). If continuity exists, the stator is grounded and must be replaced.

4. Test the stator for open circuits by obtaining an ohmmeter and setting the switches for Rx1 scale. Connect one meter lead to one of the stator leads. Connect the other meter lead to another stator lead. The meter should indicate continuity.

5. Connect the meter leads to each pair of stator leads until all three stator leads have been checked.

6. If no continuity exists on any stator lead, the stator circuit is open and must be replaced.

7. If all of the tests check out satisfactorily, but the alternator still fails to meet its rated output, the rotor field windings may be shorting or grounding out, due to centrifugal force.

VOLTAGE OUTPUT

◆ See Figure 36

This test requires the engine to be operated at approximately 2,000 rpm. Therefore, the boat should be launched and well secured to a dock prior to

performing this test. Do not attempt to perform this test with only a flush devise attached to the stern drive. Such action could seriously damage the engine or the stern drive or both. Be sure the engine compartment is well ventilated and there are no gasoline vapors present before starting the engine.

1. Connect a multi-meter as per the manufacturer's instructions.
2. Connect the positive lead to the battery positive terminal and the negative lead to the battery negative terminal.

✳✳ CAUTION

Water must circulate through the lower unit to the engine any time the engine is run to prevent damage to the water pump mounted on the engine. Just a few seconds without water will damage the water pump.

3. Start the engine and operate it at a fast idle until the engine reaches normal operating temperature. Run the engine to 1500-2000 rpm while observing the meter. The meter should indicate between 13.9 and 14.7 volts DC.

4. If the reading is high, check for a loose or corroded regulator ground lead connection. If the connections are good and the Sensing Circuit Test was good, the regulator is at fault, replace the alternator.

5. Turn off the engine.

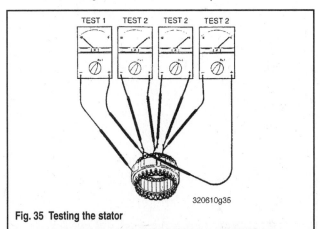

Fig. 35 Testing the stator

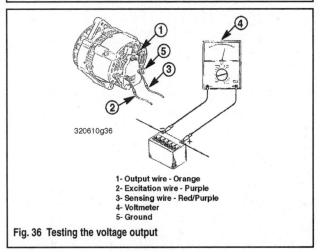

1- Output wire - Orange
2- Excitation wire - Purple
3- Sensing wire - Red/Purple
4- Voltmeter
5- Ground

Fig. 36 Testing the voltage output

Prestolite 55 Amp

DIODE TRIO

Obtain an ohmmeter and select the Rx1000 scale. Attach one ohmmeter lead to any one of the three small tabs. Attach the other meter lead to the "light" terminal (the terminal with the metal strap). Observe the meter reading, and then reverse the meter leads and again observe the reading. The ohmmeter should indicate continuity in one direction and an infinite resistance, when the meter leads are reversed. Test the other two small tabs in the same manner. If any one of the three tabs fail the test, the diode trio, as an assembly, must be replaced.

RECTIFIER BRIDGE NEGATIVE DIODE

Attach one ohmmeter lead to the rectifier bridge body. Attach the other meter lead to one of the three stator attachment terminals. As in the previous two tests, the ohmmeter should indicate continuity in one direction and an infinite resistance when the meter leads are reversed. Test the other two stator attaching terminals in the same manner. If any one of the three stator attaching terminals fail the test, the entire rectifier bridge assembly must be replaced.

RECTIFIER BRIDGE POSITIVE DIODE

Attach one ohmmeter lead to the positive terminal strap. Attach the other meter lead to one of the three stator attachment terminals. As in the previous test, the ohmmeter should indicate continuity in one direction and an infinite resistance when the meter leads are reversed. Test the other two stator attaching terminals in the same manner. If any one of the three stator attaching terminals fails the test, the entire rectifier bridge assembly must be replaced.

ROTOR GROUND TEST

◆ See Figure 37

Select the Rx1000 scale on the ohmmeter. Make contact with one meter lead to any one of the two slip rings. Make contact with the other meter lead to the rotor body. The meter should indicate infinity, or no continuity. Any other reading would indicate the rotor assembly is grounded and therefore must be replaced.

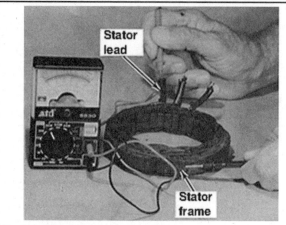

Fig. 37 Performing the rotor ground test

ROTOR OPEN/SHORT CIRCUIT

◆ See Figure 38

Select the Rx10 scale on the ohmmeter. Make contact with one meter lead to one slip ring, and the other slip ring with the second meter lead. The meter should indicate 3.5-4.7 ohms. A lower reading indicates shorted windings. A higher reading indicates an excessive resistance. An infinite reading indicates an open circuit. If the reading is not within specification, the rotor assembly must be replaced.

STATOR GROUND

◆ See Figure 39

Select the Rx1000 scale on the ohmmeter. Make contact with one meter lead to the stator frame. Make contact with the other meter lead to one of the stator leads. The meter must indicate infinity, or no continuity. Repeat this test for the other two stator leads. If any one lead fails the test, the stator assembly must be replaced.

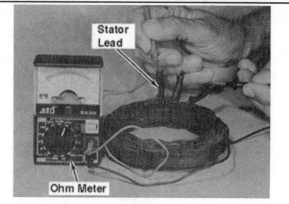

Fig. 38 Rotor open/short circuit test

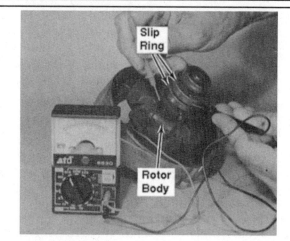

Fig. 39 Stator ground test

STATOR OPEN CIRCUIT

◆ See Figure 40

Select the lowest possible scale on the ohmmeter. Make contact with the meter leads to the first and second stator leads. The meter should register approximately 0.1 ohm. Repeat this test with the first and third stator leads, and then the second and third stator leads. In each test, the resistance value should be the same. If the results vary, the stator assembly must be replaced.

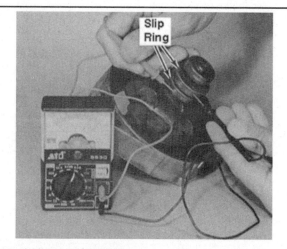

Fig. 40 Stator open circuit test

Prestolite 65 Amp

ALTERNATOR OUTPUT TEST

1. Connect a multi-meter as per the manufacturer's instructions.
2. Connect the positive meter lead to the output terminal (Orange wire) and the negative lead to the ground terminal. Start the engine and idle at approximately 1500 rpm with all accessories turned ON. Voltage should be 13.8-14.8.

CURRENT OUTPUT

◆ See Figure 27

This test requires the engine to be operated at approximately 2,000 rpm. Therefore, the boat should be launched and well secured to a dock prior to performing this test. *Do not* attempt to perform this test with only a flush devise attached to the stern drive. Such action could seriously damage the engine or the stern drive or both. Be sure the engine compartment is well ventilated and there are no gasoline vapors present before starting the engine.

1. Disconnect the negative battery cable.
2. Connect a multi-meter as per the manufacturer's instructions.
3. Disconnect the output lead (Orange wire) from the back of the alternator. Connect the meter in series between the output terminal and the Orange wire lead. Be sure the positive side of the meter is connected to the output terminal and the negative side of the meter is connected to the Orange output wire lead.
4. Connect the negative battery cable to the battery.
5. Remove the coil lead from the distributor cap and ground it to the engine block. Turn on all the accessories and crank the engine for 15-20 seconds to drain some of the energy from the battery. Shut off all accessories and reconnect the coil lead to the distributor cap. Start the engine and run the idle speed to 1,500-2,000 rpm. The meter should indicate 30 amps or greater.
6. If the reading is low, stop the engine and connect a jumper wire between the output terminal and the regulator terminal on the back of the alternator.
7. Repeat Steps 5 and 6.
8. If the reading is now 30 amps or greater, the diodes in the rectifier are damaged. Replace the alternator.
9. If the ammeter indication is still low with the jumper wire connected, perform the following Voltage Output Test to determine if the regulator or the alternator internal component is damaged.

HARNESS VOLTAGE TEST

1. Connect a multi-meter as per the manufacturer's instructions.
2. With the ignition switch OFF, unplug the sensor wire (Red/Purple) and check for voltage. If no voltage is found, check for an open in the harness.

OPEN DIODE TRIO TEST

1. Connect a multi-meter as per the manufacturer's instructions.
2. Check for battery voltage between the output terminal (Orange) and ground.
3. Connect the meter leads between the regulator terminal and ground; it should be 1.5-3.0 volts.
4. Now connect a jumper wire between the regulator and output terminals. Start the engine and allow it to idle. If voltage is present at the output terminal, replace the diode trio.

RECTIFIER (-) NEGATIVE

◆ See Figure 41

1. The rectifier must be disconnected from the stator to perform the following test. Each rectifier must be tested to be sure it is not open or shorted. Check the three rectifiers using the following procedures.

1. Obtain an ohmmeter and set the switches to Rx1 scale. Connect one lead of the meter to the negative rectifier diode heat sink. Connect the other meter lead to one of the rectifier terminals and note the reading. Reverse the meter leads and again note the meter reading. The meter should indicate a high resistance (no movement) in one direction and a low resistance indication when the leads are reversed.
2. If both readings are high, indicating the rectifier is open, or if both readings are low, indicating the rectifier is shorted, the rectifier assembly must be replaced. Repeat this step on the other two rectifiers.

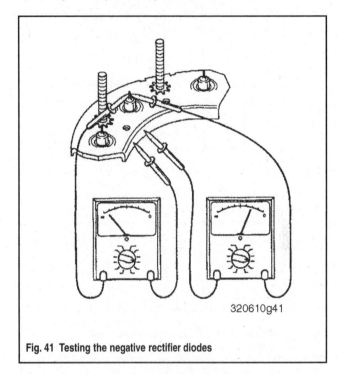

320610g41

Fig. 41 Testing the negative rectifier diodes

RECTIFIER (+) POSITIVE

◆ See Figures 32 and 42

1. The rectifier must be disconnected from the stator to perform the following test. Each rectifier must be tested to be sure it is not open or shorted. Check the three rectifiers using the following procedures.
2. Obtain an ohmmeter and set the switches to Rx1 scale. Connect one lead of the meter to the negative rectifier diode heat sink. Connect the other meter lead to one of the rectifier terminals and note the reading. Reverse the meter leads and again note the meter reading. The meter should indicate a high resistance (no movement) in one direction and a low resistance indication when the leads are reversed.
3. If both readings are high, indicating the rectifier is open, or if both readings are low, indicating the rectifier is shorted, the rectifier assembly must be replaced. Repeat this step on the other two rectifiers.
4. Check for continuity between the common side of the diodes and one of the terminals on the other side of the diode. Now reverse the leads and check it in the other direction. Continuity should be found in only one direction.

ROTOR

◆ See Figure 33

1. Check the rotor windings for a short or ground by first connecting an ohmmeter between one brush slip ring and the shaft. The meter must indicate an open circuit on the R-100 scale. Next, check the rotor windings for continuity. Connect the ohmmeter between the two slip rings, and the reading should be between 4.1-4.7 ohms at a room temperature of 70°F to 80°F (22°-27°C).

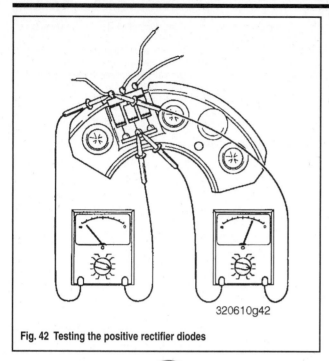

320610g42

Fig. 42 Testing the positive rectifier diodes

STATOR

◆ **See Figure 35**

1. The stator is not checked for shorts due to the very low resistance of the windings and the need for specialized test equipment. However, tests are made to check the stator for an open or grounded circuit. Use the following procedures to check the stator for possible defects.
2. Test the stator for ground circuits by obtaining an ohmmeter and setting the switches for Rx1 scale. Connect one meter lead to one of the stator leads. Make a good positive contact with the other meter lead to the stator frame.
3. The meter should indicate no continuity (no movement). If continuity exists, the stator is grounded and must be replaced.
4. Test the stator for open circuits by obtaining an ohmmeter and setting the switches for Rx1 scale. Connect one meter lead to one of the stator leads. Connect the other meter lead to another stator lead. The meter should indicate continuity.
5. Connect the meter leads to each pair of stator leads until all three stator leads have been checked.
6. If no continuity exists on any stator lead, the stator circuit is open and must be replaced.
7. If all of the tests check out satisfactorily, but the alternator still fails to meet its rated output, the rotor field windings may be shorting or grounding out, due to centrifugal force.

REMOVAL & INSTALLATION

1. Disconnect both the negative and positive battery cables from the battery terminal posts. On the back of the alternator, disconnect the Orange, Purple, and Red/Purple wire leads from the alternator terminals.
2. Loosen the alternator pivot and belt adjustment bolts. Pivot the alternator inward and remove the alternator drive belt or serpentine belt from the pulley.
3. Remove the belt adjustment bolt and brace on the alternator (drive belt models only). Remove the alternator pivot nut, bolt and washer. Lift the alternator free of the engine.
4. Install the alternator into the mounting bracket with the bolt, washers, spacer and nut. If your engine uses washers, make sure they go on either side of the spacer. Do not tighten the bolt yet.

5. Attach the brace to the alternator on models with drive belts.
6. Install the drive belt and adjust it as detailed in Section 3.
7. Reconnect the wiring and tighten the alternator-to-brace bolt to 192 inch lbs. (28 Nm), the alternator-to-bracket bolt to 35 ft. lbs. (48 Nm) and the brace/bracket-to-engine bolt to 30 ft. lbs. (41 Nm).

DISASSEMBLY

■ As expressed earlier in this manual, some times the equipment, small parts and experience required to repair the alternator can be more cost effective if the alternator is taken to a shop which specializes in alternator rebuilding. Be sure to inform the shop the alternator is for marine application, to ensure the terminal ends and covers are installed properly.

If a specialty shop is not available, use the following instructions and exploded diagram to disassemble the alternator.

Delco

There are no repair procedures or internal replacement parts for this model alternator. If it fails any of the preceding tests, replace the entire unit.

Mando

1992-97 3.0L ENGINES

◆ **See Figures 43, 44, 45 and 46**

1. Remove the alternator.
2. Place the alternator on a suitable clean work bench or mounted in a vise with the end frame facing up.
3. Disconnect the regulator leads. Remove the 4 mounting screws and pull the regulator away from the alternator until you can disconnect the field leads. Remove the regulator and its rubber seal.
4. Remove the 2 brush set screws and lift out the brush set.
5. Scribe a mark on both end frames and matching marks on the stator, as an aid to properly assemble the alternator frame.
6. Remove the three/four thru-bolts. Separate the drive end frame and rotor assembly from the stator by carefully inserting 2 screwdrivers no more than 1/16 inch into the slots on opposite sides of the stator frame. Pry the drive end frame and then the rear end frame from the stator. Never pry anywhere except at the slot or the castings will be damaged.
7. Place the rear end frame on the work bench with the stator pointing down. Remove the nuts, washers, insulators and condenser from the studs.
8. Turn the end frame over and remove the 3 Phillips head screws securing the rectifier assembly to the rear end frame.
9. Unsolder the diode trio at the insulated regulator terminal.
10. Place the tips of needle nose pliers onto the diode terminal between the solder joints and diode body. This action will serve as a heat sink, while soldering the stator terminals and prevent damaging the diodes. Using a soldering iron, unsolder the 3 stator leads and the 3 coated leads from the rectifier heat sink.

■ Perform the next step only if the pulley is damaged, front bearing is defective or if the pulley must be transferred to another alternator rotor shaft.

11. Place an oversized V-belt around the pulley and clamp the pulley in a vise. Make sure the vise is clamping on the V-belt only. Place a 7/8 inch socket or wrench over the pulley nut and remove the nut, lockwasher, pulley, fan spacer, and fan from the end of the rotor shaft. Slide the front end frame off the rotor shaft.
12. Remove the three Phillips head screws and washers securing the front bearing retaining plate to the front end frame and lift off the retaining plate. Place the end frame in an arbor press with the face pointing up. Use a suitable size mandrel and press the bearing out of the front end frame and discard the bearing.

1- Screws (3)
2- Nut (9)
3- Flat Washer (4)
4- Sensing wire (red/purple)
5- Excitation wire (purple)
6- Cover
7- Tie strap
8- Rubber gasket
9- Condensor
10- Insulator
11- Bolt (4)
12- End frame (rear)
13- Cap (2)
14- Brush / regulator assembly
15- Rectifier assembly
16- Flat washer
17- Screw
18- Stator
19- Rotor and slip ring
20- Retaining plate
21- Front bearing
22- End frame (front)
23- Screw (3)
24- Fan spacer
25- Fan
26- Pulley spacer
27- Pully
28- Lockwasher
29- Nut

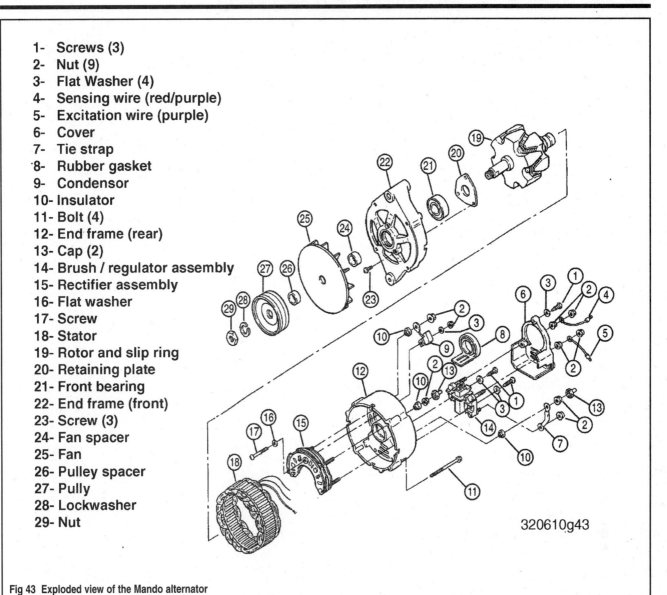

320610g43

Fig 43 Exploded view of the Mando alternator

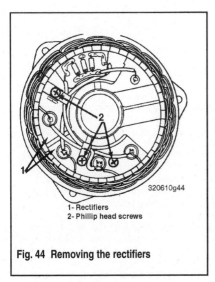

320610g44

1- Rectifiers
2- Phillip head screws

Fig. 44 Removing the rectifiers

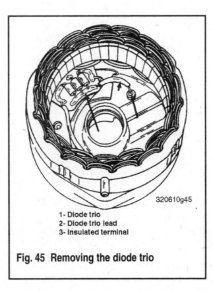

320610g45

1- Diode trio
2- Diode trio lead
3- Insulated terminal

Fig. 45 Removing the diode trio

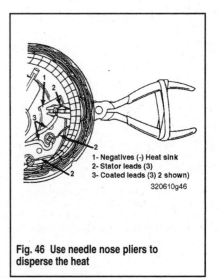

1- Negatives (-) Heat sink
2- Stator leads (3)
3- Coated leads (3) 2 shown)

320610g46

Fig. 46 Use needle nose pliers to disperse the heat

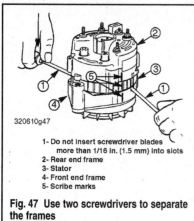

1- Do not insert screwdriver blades more than 1/16 in. (1.5 mm) into slots
2- Rear end frame
3- Stator
4- Front end frame
5- Scribe marks

320610g47

Fig. 47 Use two screwdrivers to separate the frames

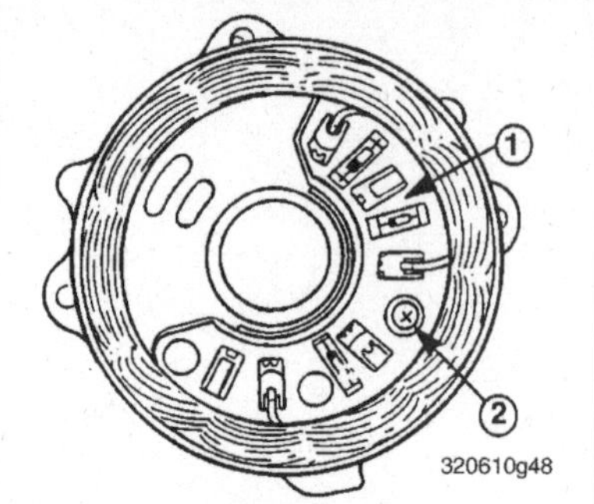

1- Rectifier assembly
2- Phillips head screw

320610g48

Fig. 48 A single screw holds the rectifier assembly

1- Heat sink
2- Stator leads

320610g49

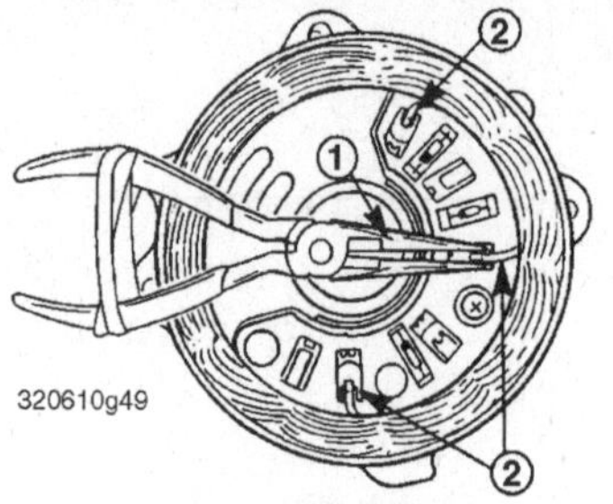

Fig. 49 Use needle nose pliers to dissipate heat when using a solder gun

ALL OTHERS

◆ **See Figures 43, 47, 48 and 49**

1. Remove the alternator.

2. Place the alternator on a suitable clean work bench or mounted in a vise with the end frame facing up.

3. Disconnect the regulator leads. Remove the 4 mounting screws and pull the regulator away from the alternator until you can disconnect the field leads. Remove the regulator and its rubber seal.

4. Remove the stud cover insulator, two nuts and the tie strap from the assembly.

5. Remove the 2 brush/regulator screws and lift out the assembly.

6. Scribe a mark on both end frames and matching marks on the stator, as an aid to properly assemble the alternator frame.

7. Remove the four thru-bolts. Separate the end frame from the stator by carefully inserting 2 screwdrivers no more than 1/16 inch into the slots on opposite sides of the stator frame. Pry the end frame and then the rear end frame from the stator. Never pry anywhere except at the slot or the castings will be damaged.

8. Place the rear end frame on the work bench with the stator pointing down. Remove the nuts, washers, insulators and condenser from the studs.

9. Turn the end frame over and remove the Phillips head screw securing the rectifier assembly to the end frame.

10. Insert screwdrivers into the slots again and separate the stator and rectifier.

11. Unsolder the three stator leads at the rectifier heat sink. Place the tips of needle nose pliers onto the diode terminal between the solder joints and diode body. This action will serve as a heat sink, while soldering the stator terminals and prevent damaging the diodes

■ **Perform the next step only if the pulley is damaged, front bearing is defective or if the pulley must be transferred to another alternator rotor shaft.**

12. Place an oversized V-belt around the pulley and clamp the pulley in a vise. Make sure the vise is clamping on the V-belt only. Place a 7/8 inch socket or wrench over the pulley nut and remove the nut, lockwasher, pulley, fan spacer, and fan from the end of the rotor shaft. Slide the front end frame off the rotor shaft.

13. Remove the three Phillips head screws and washers securing the front bearing retaining plate to the front end frame and lift off the retaining

plate. Place the end frame in an arbor press with the face pointing up. Use a suitable size mandrel and press the bearing out of the front end frame and discard the bearing.

Prestolite 55

◆ **See Figures 50 and 51**

1. Remove the alternator.
2. Run an oversize drive belt around the pulley and then clamp it into a vise so that the belt protects the pulley/alternator. Loosen the pulley nut.
3. Lift the assembly out of the vise and place it carefully on a clean workbench. Remove the pulley nut, pulley, fan, woodruff key and the spacer.
4. Remove the three upper nuts and the two screws in the center and pry off the rear cover.
5. Tag and disconnect the three stator wires on top of the rectifier bridge, remove the strap on the lower left side between the bridge and the brush holder, and then lift off the diode trio and bridge.
6. Remove the two nuts and lift off the brush holder.
7. Remove the regulator. Make sure you don't lose the two fiber washers!
8. Scribe (or paint) a mark across the front and rear housings, loosen the four screws and separate the pieces.
9. Separate the drive end frame and rotor assembly from the stator by carefully inserting 2 screwdrivers no more than 1/16 inch into the slots on opposite sides of the stator frame. Pry the drive end frame and then the rear end frame from the stator. Never pry anywhere except at the slot or the castings will be damaged.
10. Install the front housing in an arbor press and then press the rotor out of the front bearing.

Prestolite 65

◆ **See Figure 54**

1. Remove the alternator.
2. Run an oversize drive belt around the pulley and then clamp it into a vise so that the belt protects the pulley/alternator. Loosen the pulley nut.
3. Lift the assembly out of the vise and place it carefully on a clean workbench. Remove the pulley nut, pulley, fan, woodruff key and the spacer.
4. Tag and disconnect the regulator leads from the terminals on the end frame. Remove the four screws, pull the regulator away from the frame and disconnect the field lead. Remove the regulator.
5. Remove the two mounting screws and lift out the brush holder.
6. Scribe a mark on both end frames and matching marks on the stator, as an aid to properly assemble the alternator frame.
7. Remove the four thru-bolts. Separate the end frame from the stator by carefully inserting 2 screwdrivers no more than 1/16 inch into the slots on opposite sides of the stator frame. Pry the end frame and then the rear end frame from the stator. Never pry anywhere except at the slot or the castings will be damaged.
8. Support the front housing and press the rotor out with an arbor press.

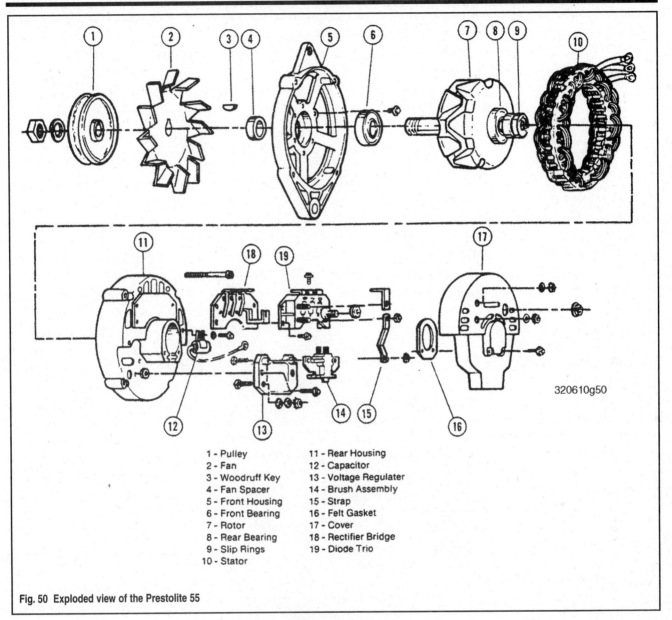

320610g50

1 - Pulley
2 - Fan
3 - Woodruff Key
4 - Fan Spacer
5 - Front Housing
6 - Front Bearing
7 - Rotor
8 - Rear Bearing
9 - Slip Rings
10 - Stator

11 - Rear Housing
12 - Capacitor
13 - Voltage Regulater
14 - Brush Assembly
15 - Strap
16 - Felt Gasket
17 - Cover
18 - Rectifier Bridge
19 - Diode Trio

Fig. 50 Exploded view of the Prestolite 55

320611p51

Fig. 51 Disconnect the three stator wires and the strap

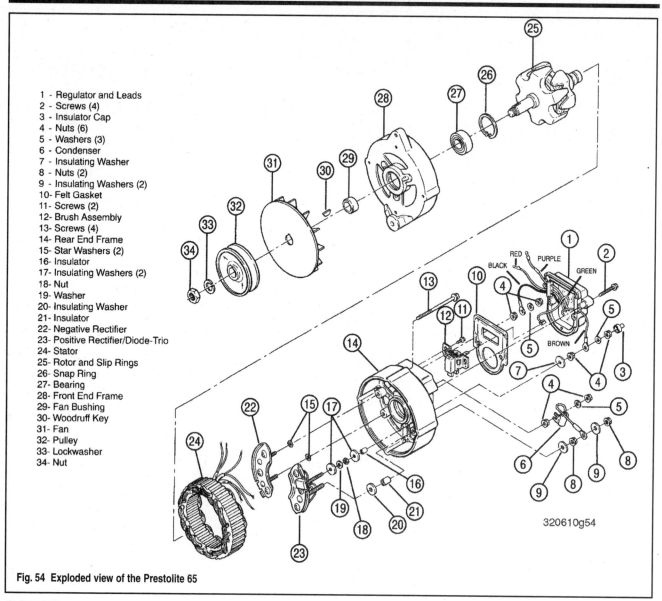

1 - Regulator and Leads
2 - Screws (4)
3 - Insulator Cap
4 - Nuts (6)
5 - Washers (3)
6 - Condenser
7 - Insulating Washer
8 - Nuts (2)
9 - Insulating Washers (2)
10- Felt Gasket
11- Screws (2)
12- Brush Assembly
13- Screws (4)
14- Rear End Frame
15- Star Washers (2)
16- Insulator
17- Insulating Washers (2)
18- Nut
19- Washer
20- Insulating Washer
21- Insulator
22- Negative Rectifier
23- Positive Rectifier/Diode-Trio
24- Stator
25- Rotor and Slip Rings
26- Snap Ring
27- Bearing
28- Front End Frame
29- Fan Bushing
30- Woodruff Key
31- Fan
32- Pulley
33- Lockwasher
34- Nut

320610g54

Fig. 54 Exploded view of the Prestolite 65

CLEANING

The following components are to be visually inspected for serviceability. Electrical checks and tests are detailed previously in this section.

Use a clean soft and wipe any dirt or debris off the parts. Do not clean electrical components with solvent, because such action may damage the item.

Brush/Regulator

Inspect the brush casing for cracks or damaged brush leads. Check the brush leads for poor solder connections and damaged leads. Check for broken springs and excessive brush wear. If the brushes are worn to less than a 1/4 inch (6.35mm) long the brush set must be replaced on the Mando; on the Prestolite, they should be no less than 3/16 inch (5mm).

Rotor

Inspect the end of the rotor shaft for stripped or damaged threads. Check the pole piece fingers for damage caused by worn bearings. Inspect the rear bearing for smooth quiet rolling action. If the threads cannot be repaired or the bearing is defective, the entire rotor assembly must be replaced.

Clean the rotor slip rings with 400 grit sand paper. Blow off any dust using compressed air and inspect the rotor slip rings for grooves, pits, flat

spots or out-of-round more than 0.002 in. (0.051mm). If any of the above conditions are found, the rotor assembly must be replaced.

Stator

Inspect the insulating enamel for heat discoloration because discoloration is an indication of a shorted diode or grounded winding. Check the stator for damaged insulation and wires from contact with the rotor. Replace the stator if any of the above defects are obvious.

End Frames

Inspect the end frames for cracks, distortion, stripped threads or worn bearing bore (bearing seized on shaft and spinning in bore). If any of the above damage is obvious the end frame must be replaced.

Fan and Pulley

Inspect the fan for bent or cracked fins, broken welds or worn mounting hole. Check the pulley sheaves for trueness, excessive wear, grooves, pits, nicks and corrosion. If the pulley sheave damage cannot be repaired with a file or wire brush, the pulley must be replaced or drive belt wear will be accelerated.

ASSEMBLY

Mando

1992–97 3.0L ENGINES

◆ **See Figures 43, 55 and 56**

1. Position a new bearing into the front end frame bearing bore. Position the front end frame into an arbor press. Using the appropriate size mandrel, press the bearing into the bore until the bearing contacfs the end frame. Place the bearing retainer plate over the bearing and secure with three Phillips head screws and washers.

2. Slide the threaded end of the rotor shaft through the front end frame. Slide the fan spacer, fan, pulley spacer, pulley, lockwasher and nut onto the end of the shaft. Place an oversize "V" belt around the pulley and tighten the pulley in a vise. Using a 7/8 inch socket, tighten the pulley nut to a torque value of 50 ft. lbs. (68 Nm).

■ **The insulating washers must be installed correctly or you will damage the alternator.**

3. Install the positive and negative rectifier heat sinks.

4. Position the rear end frame over the front and align the scribe marks made previously. Press the frames together and then install the screws, tightening them securely.

5. Position the brush set in the cavity on the back of the rear end frame and tighten the two mounting screws securely.

6. Position a new rubber seal on the regulator mating surface. Connect the field leads and the regulator lead to their respective terminals and then move the regulator into place. Tighten the 4 screws and the ground terminal nut securely.

7. Install the alternator.

ALL OTHER ENGINES

◆ **See Figure 43 and 57**

1. Position a new bearing into the front end frame bearing bore. Position the front end frame into an arbor press. Using the appropriate size mandrel, press the bearing into the bore until the bearing contacts the end frame. Place the bearing retainer plate over the bearing and secure with three Phillips head screws and washers.

2. Slide the threaded end of the rotor shaft through the front end frame. Slide the fan spacer, fan, pulley spacer, pulley, lockwasher and nut onto the end of the shaft. Place an oversize V-belt around the pulley and tighten the pulley in a vise. Using a 7/8 inch socket, tighten the pulley nut to 42 ft. lbs. (58 Nm).

■ **The insulating washers must be installed correctly or you will damage the alternator.**

3. Assemble the stator and rectifier and then solder the three leads. Position into the rear frame and install the mounting screw.

4. Position the rear end frame over the front and align the scribe marks made previously. Press the frames together and then install the screws, tightening them to 55 inch lbs. (5.5 Nm).

5. Push the brushes down so they are flush with the top of the holder and then insert a #54 (0.050 in./1mm) drill bit into the hole in the holder so they stay compressed. Position the brush/regulator in the cavity on the back of the rear end frame and tighten the two mounting screws to 42 inch lbs. (4.2 Nm). Pull out the drill bit so that the brushes release against the slip rings.

6. Install the tie strap between the two studs and tighten the nuts securely.

7. Install the cover. Install the screw, connect the two leads and then install the two nuts; tighten all three securely. Pop on the insulator caps.

8. Install the alternator.

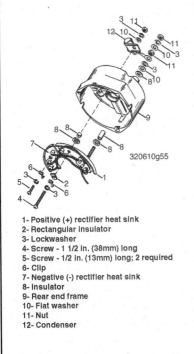

1- Positive (+) rectifier heat sink
2- Rectangular insulator
3- Lockwasher
4- Screw - 1 1/2 in. (38mm) long
5- Screw - 1/2 in. (13mm) long; 2 required
6- Clip
7- Negative (-) rectifier heat sink
8- Insulator
9- Rear end frame
10- Flat washer
11- Nut
12- Condenser

320610g55

Fig. 55 Reassembling the rear end frame properly is of the utmost importance

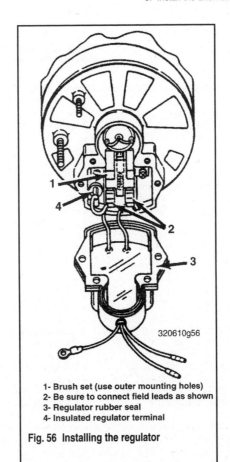

1- Brush set (use outer mounting holes)
2- Be sure to connect field leads as shown
3- Regulator rubber seal
4- Insulated regulator terminal

320610g56

Fig. 56 Installing the regulator

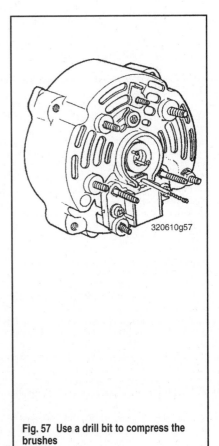

320610g57

Fig. 57 Use a drill bit to compress the brushes

Prestolite 55

1. Press the rotor into the front bearing with an arbor press. Make sure you only press on the bearing outer race.
2. Position the stator in one of the housings and then line up the alignment marks made earlier and connect the front and rear housings. Tighten the thru-bolts to 50–60 inch lbs. (6 Nm).
3. Lay the fiber washers over the regulator mounting bolt holes and install the regulator.
4. Install the brush holder over the regulator. Tighten the right nut finger tight and leave the left one loose.
5. Swab a little heat sink compound over the mounting surface and place the diode/rectifier bridge in position. Tighten the mounting screws and then connect the stator wires and the strap. Tighten all screws and nuts securely.
6. Connect the condenser lead to the output terminal and position the felt washer. Install the rear cover and tighten the screws and nuts securely.
7. Install the insulator and nut on the output terminal.
8. Install the spacer, woodruff key, fan and pulley and then tighten the nut finger tight. Place the assembly back in a vise with the drive belt around the pulley and tighten the nut to 35–50 ft. lbs. (48–68 Nm).

9. Install the alternator.

Prestolite 65

1. Press the rotor into the front bearing with an arbor press. Make sure you only press on the bearing outer race.
2. Position the stator in one of the housings and then line up the alignment marks made earlier and connect the front and rear housings. Tighten the thru-bolts to 50–60 inch lbs. (6 Nm).
3. Install the brush holder into the end frame and press it forward until the brushes are up against the slip ring. Install the two screws and tighten them securely
4. Position the felt regulator gasket and then attach the field lead to the brushes. Position the regulator and tighten the four screws securely. Reconnect the regulator leads to their terminals on the end frame.
5. Connect the condenser lead to the output terminal and position the felt washer.
6. Install the spacer, woodruff key, fan and pulley and then tighten the nut finger tight. Place the assembly back in a vise with the drive belt around the pulley and tighten the nut to 35–50 ft. lbs. (48–68 Nm).
7. Install the alternator.

STARTER CIRCUIT

Description and Operation

All engines utilize an electric starter motor coupled with a mechanical gear mesh between the starter motor and the flywheel, similar to the method used to crank an automobile engine.

As the name implies, the sole purpose of the starter motor circuit is to control operation of the starter motor to crank the engine until it is operating. The circuit includes a relay or magnetic switch (solenoid) to connect or disconnect the motor from the battery. The operator controls the switch with a key switch.

A neutral safety switch is installed into the circuit to permit operation of the starter motor only if the shift control lever is in neutral. This switch is a safety device to prevent accidental engine start when the engine is in gear.

The starter motor is a series-wound electric motor that draws a heavy current from the battery. It is designed to be used only for short periods of time to crank the engine for starting. To prevent overheating the motor, cranking should not be continued for more than 30-seconds without allowing the motor to cool for at least three minutes. Actually, this time can be spent in making preliminary checks to determine why the engine fails to start.

Power is transmitted from the starter motor to the flywheel through a Bendix drive. This drive has a pinion gear mounted on screw threads. When the motor is operated, the pinion gear moves upward and meshes with the teeth on the flywheel ring gear.

When the engine starts, the pinion gear is driven faster than the shaft and as a result, it screws out of mesh with the flywheel. A rubber cushion is built into the Bendix drive to absorb the shock when the pinion meshes with the flywheel ring gear. The parts of the drive must be properly assembled for efficient operation. If the screw shaft assembly is reversed, it will strike the splines and the rubber cushion will not absorb the shock.

The sound of the motor during cranking is a good indication of whether the starter motor is operating properly or not. Naturally, temperature conditions will affect the speed at which the starter motor is able to crank the engine. The speed of cranking a cold engine will be much slower than when cranking a warm engine. An experienced operator will learn to recognize the favorable sounds of the engine cranking under various conditions.

The job of the starter motor relay is to complete the circuit between the battery and starter motor. It does this by closing the starter circuit electromagnetically, when activated by the key switch. This is a completely sealed switch, which meets SAE standards for marine applications. Do not substitute an automotive-type relay for this application. It is not sealed and gasoline fumes can be ignited upon starting the engine. The relay consists of a coil winding, plunger, return spring, contact disc and four externally mounted terminals. The relay is installed in series with the positive battery cables mounted to the two larger terminals. The smaller terminals connect to the neutral switch and ground.

To activate the relay, the shift lever is placed in neutral, closing the neutral switch. Electricity coming through the ignition switch goes into the relay coil winding which creates a magnetic field. The electricity then goes on to ground in the engine. The magnetic field surrounds the plunger in the relay, which draws the disc contact into the two larger terminals. Upon contact of the terminals, the heavy amperage circuit to the starter motor is closed and activates the starter motor. When the key switch is released, the magnetic field is no longer supported and the magnetic field collapses. The return spring working on the plunger opens the disc contact, opening the circuit to the starter.

When the armature plate is out of position or the shift lever is moved into forward or reverse gear, the neutral switch is placed in the open position and the starter control circuit cannot be activated. This prevents the engine from starting while in gear.

Troubleshooting the Starting System

If the starter motor spins but fails to crank the engine, the cause is usually a corroded or gummy Bendix drive. The drive should be removed, cleaned and given an inspection.

1. Before wasting too much time troubleshooting the starter motor circuit, the following checks should be made. Many times, the problem will be corrected.

- Battery fully charged.
- Shift control lever in neutral.
- Main 20-amp fuse is good (not blown).
- All electrical connections clean and tight.
- Wiring in good condition, insulation not worn or frayed.

2. Starter motor cranks slowly or not at all.

- Faulty wiring connection
- Short-circuited lead wire
- Shift control not engaging neutral (not activating neutral start switch)
- Defective neutral start switch
- Starter motor not properly grounded
- Faulty contact point inside ignition switch
- Bad connections on negative battery cable to ground (at battery side and engine side)
- Bad connections on positive battery cable to magnetic switch terminal
- Open circuit in the coil of the magnetic switch (relay)
- Bad or run-down battery
- Excessively worn down starter motor brushes
- Burnt commutator in starter motor
- Brush spring tension slack
- Short circuit in starter motor armature

3. Starter motor keeps running.

- Melted contact plate inside the magnetic switch
- Poor ignition switch return action

4. Starter motor picks up speed, put pinion will not mesh with ring gear.

- Worn down teeth on clutch pinion
- Worn down teeth on flywheel ring gear

Starter Motor

DESCRIPTION & OPERATION

Delco-Remy Direct Drive

Delco-Remy direct drive starters consist of a set of field coils positioned over pole pieces, which are attached to the inside of a heavy iron frame. An armature, an over-running clutch drive mechanism, and a solenoid are included inside the iron frame.

The armature consists of a series of iron laminations placed over a steel shaft, a commutator, and the armature winding. The windings are heavy copper ribbons assembled into slots in the iron laminations. The ends of the windings are soldered or welded to the commutator bars. These bars are electrically insulated from each other and from the iron shaft.

An overrunning clutch drive arrangement is installed near one end of the starter shaft. This clutch drive assembly contains a pinion that is made to move along the shaft by means of a shift lever to engage the engine ring gear for cranking. The relationship between the pinion gear and the ring gear on the engine flywheel provides sufficient gear reduction to meet cranking requirement speed for starting.

The overrunning clutch drive has a shell and sleeve assembly, which is splined internally to match the spiral splines on the armature shaft. The pinion is located inside the shell. Spring-loaded rollers are also inside the shell and they are wedged against the pinion and a taper inside the shell. Some starters use helical springs and others use accordion type springs. Four rollers are used. A collar and spring, located over a sleeve completes the major parts of the clutch mechanism.

When the solenoid is energized and the shift lever operates, it moves the collar endwise along the shaft. The spring assists movement of the pinion into mesh with the ring gear on the flywheel. If the teeth on the pinion fail to mesh for just an instant with the teeth on the ring gear, the spring compresses until the solenoid switch is closed; current flows to the armature; the armature rotates; the spring is still pushing on the pinion; the pinion teeth mesh with the ring gear; and cranking begins.

Torque is transferred from the shell to the pinion by the rollers, which are wedged tightly between the pinion and the taper cut into the inside of the shell. When the engine starts, the ring gear drives the pinion faster than the armature; the rollers move away from the taper; the pinion overruns the shell; the return spring moves the shift lever back; the solenoid switch is opened; current is cutoff to the armature; the pinion moves out of mesh with the ring gear; and the cranking cycle is completed. The start switch should be opened immediately when the engine starts to prevent prolonged overrun.

Delco-Remy Gear Reduction

The Delco-Remy PG series gear reduction starter motors are small and light weight for the amount of work produced. These starters have small permanent magnets mounted inside the field frame. The placement and strength of these magnets differ between models. Therefore the field frames are not interchangeable.

The permanent magnets take the place of the large current-carrying iron core field-coil magnets previously used on the larger direct drive starter motors. The motor armature is supported on both ends by a permanently lubricated roller or ball bearing assembly to reduce the drag and friction created by the motor speed of approximately 7,000 rpm.

A planetary gear reduction unit is mated between the drive motor armature and the Bendix drive gear. This planetary gear drive results in an over all gear reduction of approximately 4:1. Through the gear reduction, the conversion of motor high speed, low torque is converted to a high torque, low speed gear drive output. The gear reduction results in a final engine cranking speed of approximately 1,750 rpm.

The starter is designed to operate under heavy loads and produce high power for only short periods of time. Therefore, never crank the engine for more than 30 seconds without allowing a minimum cooling off time of two minutes, before attempting to crank the engine again.

As with all marine installations, a safety switch is installed in the remote control shift box. This switch is designed to open the starter circuit to prevent the engine from starting, if the shift lever is in any position other than the neutral position.

The unit has a Bendix follow-thru type drive designed to overcome disengagement of the flywheel ring gear when engine speed exceeds cranking motor speed.

A helical cut shaft is designed to quickly engage and disengage the Bendix drive. An internal one-way clutch in the Bendix allows the motor to drive the Bendix. If the engine should start, and the flywheel begin to drive the Bendix, the one-way clutch will release and allow the drive to overrun and release the cranking motor armature. The helical splined shaft will then disengage the Bendix from the flywheel.

On the PG 200 and 250 series starters the solenoid switch, plunger, return spring, and shift lever are all permanently mounted inside the drive housing. If a failure occurs, the entire housing assembly must be replaced. On the 260-F1 and some 260 series, the solenoid and lever are replaceable components. Any other differences between these models are minor and will be identified in the text.

Mando

This starter motor, installed on certain later model 7.4L/8.2L V8 engines is not serviceable. No replacement parts are available. Replacement should always be with a Delco PG260 motor.

PRECAUTIONS

1. Always make sure that each battery cable is connected to the correct terminal on the battery.
2. Never disconnect the battery while the engine is running.
3. When using a battery charger or booster, always make sure that the positive battery cable on the charger is connected to the positive terminal on the battery. The same goes for the negative side.
4. Always make sure that both battery cables are disconnected prior to connecting a charger or booster.
5. Always make sure the battery is in good operating condition.
6. Always make sure the battery leads and terminals are clean.

TROUBLESHOOTING

◆ See Figures 58 and 59

Regardless of how or where the solenoid is mounted, the basic circuits of the starting system on all makes of cranking motors are the same and similar tests apply. In the following testing and troubleshooting procedures, the differences are noted.

✳✳ CAUTION

Always take time to vent the bilge when making any of the tests as a prevention against igniting any fumes accumulated in that area. As a further precaution, remove the high-tension wire from the center of the distributor cap and ground it securely to prevent sparks.

All cranking motor problems fall into one of three areas:
1. The cranking motor fails to rotate.
2. The cranking motor spins rapidly, but does not crank the engine.
3. The cranking motor cranks the engine, but too slowly to affect engine start.

The following paragraphs provide a logical sequence of tests designed to isolate a problem in the cranking system.

Battery

1. Turn on several of the cabin lights (or any accessories). Turn the ignition switch ON and note the effect on the brightness of the lights. With a properly functioning electrical system, the lights will dim slightly and the

starter will crank the engine at a reasonable rate. If the lights dim considerably and the engine does not turn over, one of several causes may be at fault.

2. If the lights go out completely, or dim considerably, the battery charge is low or almost dead. The obvious remedy is to charge the battery; switch over to a secondary battery if one is available; or to replace it with a known fully charged one.

3. If the starting relay clicks, sounding similar to a machine gun firing, the battery charge is too low to keep the starting relay engaged when the starter load is brought into the circuit.

4. If the starter spins without cranking the engine, the drive is broken. The starter will have to be removed for repairs.

5. If the lights do not dim, and the starter does not operate, then there is an open circuit. Proceed to the Cable Connection Test.

Cable Connection

1. If the starter fails to operate and the lights do not dim when the ignition switch is turned to start, the first area to check is the connections at the battery, starting relay, starter and neutral-safety switch.

2. First, remove the cables at the battery; clean the connectors and posts; replace the cables; and tighten them securely.

3. Now, try the starter. If it still fails to crank the engine, try moving the shift box selector lever from neutral to forward to determine if the neutral-safety switch is out of adjustment or the electrical connections need attention.

4. Sometimes, after working the shift lever back-and-forth and perhaps a bit sideways, the neutral-switch connections may be temporarily restored and the engine can be started. Disconnect the leads; clean the connectors and terminals on the switch; replace the leads; and tighten them securely at the first opportunity.

5. If the starter still fails to crank the engine, move on to the Solenoid Test.

Solenoid

◆ **See Figure 60**

1. The solenoid, commonly called the starter relay, is checked by directly bridging between the terminal from the battery (the large heavy one) to the terminal from the ignition switch.

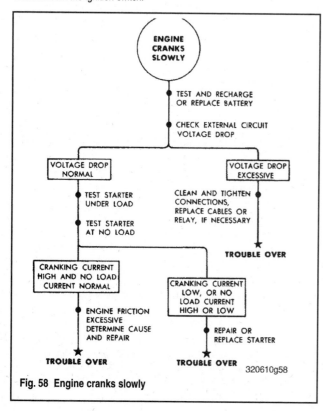

Fig. 58 Engine cranks slowly

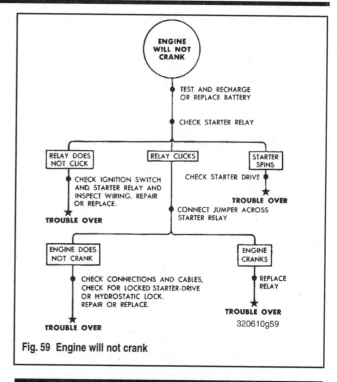

Fig. 59 Engine will not crank

※※ CAUTION

Take every precaution to ensure there is no gasoline fumes in the bilge before making these tests.

2. If a bilge blower is installed, operate it for at least five minutes to clear any fumes accumulated in the bilge.

3. Turn the ignition switch to the ON position. Now, connect a jumper wire between the battery lead terminal and the ignition lead terminal with a very heavy piece of wire. If the relay operates, the trouble is in the circuit to the ignition switch. If the starter motor still fails to operate, continue with the Current Draw Test.

Current Draw

◆ **See Figure 61**

Lay an amperage gauge on the cable between the battery and the starter motor. Attempt to crank the engine and note the current draw reading of the amperage gauge under load. The current draw should not exceed 190 amperes.

TESTING

Starter Circuit Resistance

If the starter motor turns very slowly or not at all, of if the solenoid fails to engage the starter with the flywheel, the cause may be excessive resistance in the starting circuit.

The following checks can be performed with the starter motor installed on the engine in the boat:

1. Test the battery and bring it up to a full charge, if necessary. Ground the distributor primary lead to prevent the engine from firing during the following checks.

2. Measure the voltage drop during cranking between the positive battery post and the battery lead terminal of the solenoid.

3. Measure the voltage drop during cranking between the battery lead terminal of the solenoid and the motor lead terminal of the solenoid.

4. Measure the voltage drop during cranking between the negative battery post and the cranking motor frame.

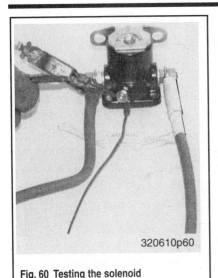

320610p60

Fig. 60 Testing the solenoid

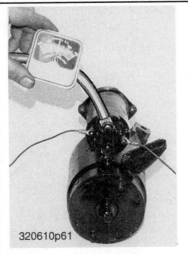

320610p61

Fig. 61 Testing the current draw

Fig. 62 Testing the field coils

If the voltage drop during any of the previous three tests is more than 0.2 volt, excessive resistance is indicated in the circuit being checked. Trace and correct the cause of the resistance.

If the solenoid fails to pull in, the problem may be due to excessive high voltage drop in the solenoid circuit. To check voltage drop in this circuit, measure the voltage drop during cranking, between the battery terminal of the solenoid and the switch terminal of the solenoid. If the voltage drop is more than 2.5 volts, the resistance is excessive in the solenoid circuit.

If the voltage drop is not more than 2.5 volts and the solenoid does not pull in, measure the voltage available at the switch terminal of the solenoid. The solenoid should pull in with 8.0 volts at temperatures up to 200º F. if it does not pull in, remove the starter and test the solenoid.

Starter

◆ **See Figures 62, 63, 64, 65, 66 and 67**

The following paragraphs provide a logical sequence of tests designed to isolate a defective part in the starter.

1. Remove the starter motor and then remove the commutator end frame.

2. Make contact with one probe of a test light on each end of the field coils connected in series. If the test light fails to come on, there is an open in the field coils and repair or replacement is required.

3. Disconnect the shunt coil or coil ground. Make contact with one probe of the test light on the connector strap and on the field frame with the other probe. If the test light comes on, the field coils are grounded and the defective coils must be repaired or replaced.

4. Disconnect the shunt coil grounds. Make contact with one probe of the test light on each end of the shunt coil, or coils. If the light fails to come on, the shunt coil is open and must be repaired or replaced.

5. Always replace the drive-end bushing during a starter overhaul.

6. True the commutator, if necessary, in a lathe. Never undercut the mica because the brushes are harder than the insulation. Check the armature for a short circuit by placing it on a growler and holding a hack saw blade over the armature core while the armature is rotated. If the saw blade vibrates, the armature is shorted. Clean between the commutator bars, and then check again on the growler. If the saw blade still vibrates, the armature must be replaced.

7. Connect one probe of the test light to the armature core or shaft and the other probe to the commutator. If the light comes on, the armature is grounded and must be replaced.

8. Wash the brush holder in solvent, and then blow it dry with compressed air. Use the test light to verify two of the brush holders are grounded and two are insulated.

9. The overrunning clutch is secured to the armature by a snap ring. This ring may be removed if the clutch requires replacement. Never wash an overrunning clutch in solvent or the lubricant will be dissolved; the clutch will fail; the engine will drive the cranking motor armature at high speed; the windings will be thrown out by centrifugal force; and the armature will be destroyed.

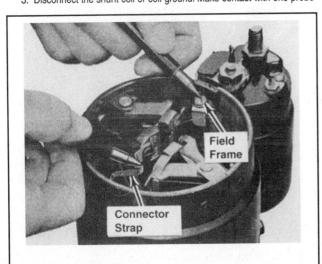

Fig. 63 Testing the field coils

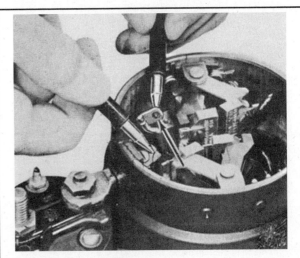

Fig. 64 Testing the shunt coil

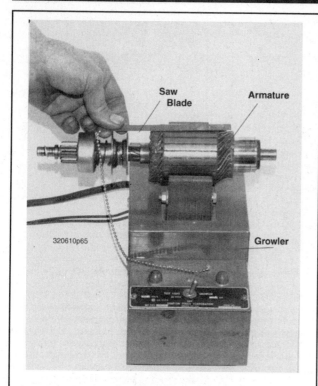

Fig. 65 Use a growler to test the armature

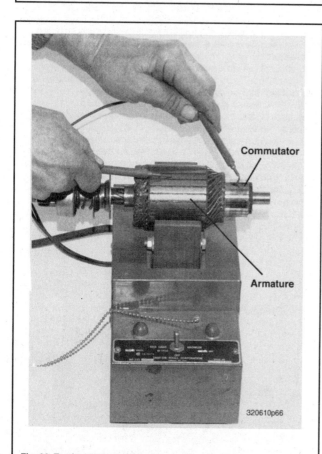

Fig, 66 Testing the armature

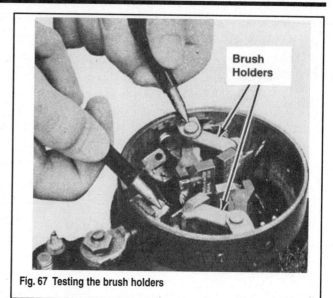

Fig. 67 Testing the brush holders

Pinion Clearance

◆ See Figure 68

Pinion clearance should always be checked after reassembly of the starter to insure proper adjustment.

1. Disconnect the field coil connector at the solenoid terminal and insulate it carefully.

2. Connect the positive side of the battery to the solenoid switch terminal and connect the negative side to the frame of the solenoid.

3. Connect a jumper wire to the solenoid motor terminal and quickly touch the other end to the body of the starter so the pinion moves into the cranking position.

4. Push the pinion back toward the commutator end until there is no slack. Measure the gap between the pinion and the retainer. If not 0.010–0.140 in. (0.25–3.5mm) on direct drive models or 0.010–0.160 in. (0.25–4.06mm) on gear reduction models, the solenoid linkage or shift lever yoke buttons are worn and require replacement; or the shift lever mechanism has been assembled improperly.

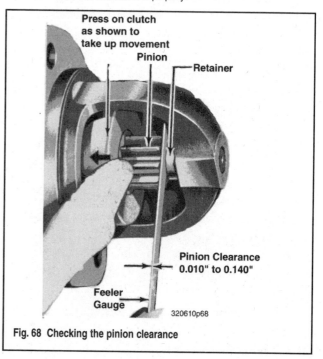

Fig. 68 Checking the pinion clearance

REMOVAL & INSTALLATION

1. Disconnect the battery cables.
2. Tag and disconnect all wires at the starter solenoid. Move them out of the way.
3. Loosen the two thru-bolts and then pull them out of the starter housing. Carefully lift out the starter. Certain versions of the PG260 starter may have three thru-bolts used for mounting the starter to the block.

■ **Be aware that many applications will utilize a shim(s) between the motor and cylinder block. Do not lose this shim!**

To install:

4. Position the shim (if equipped) and then install the starter. Slide in the thru-bolts and tighten them to 35 ft. lbs (48 Nm) on 3.0L direct drive starters and 50 ft. lbs. (68 Nm) on V8 direct drives and the Mando; or 30 ft. lbs. (41 Nm) on gear reduction starters..
5. Connect all wires to the solenoid and then connect the battery cables.

DISASSEMBLY & ASSEMBLY

■ **Before disassembling begins, read these few words first on a time and money saving idea. Most dealers do not find it economical to stock all the small parts or spend the time to service a starter motor. In most cases the labor cost alone will exceed the cost of a new or rebuilt unit.**

Throughout the country, specialty shops may be found specializing in and servicing starter motors, alternators and other electrical assemblies. A customer can usually obtain a new or rebuilt starter for a modest cost usually on the spot or same day service. Some marine dealers even stock new and rebuilt cranking motors ready for installation. This means the repair can be completed and the boat back in service before the end of the day.

If the decision is made to disassemble the unit, check with the marine dealer first for parts availability, order time, and cost compared to a new or rebuilt unit off the shelf. These units usually come with some type of warranty. Therefore, if it fails within the warranty period, additional cost is not involved.

Delco-Remy Direct Drive

◆ **See Figures 69, 70 and 71**

1. Remove the starter.
2. Remove the screw and washer retaining the motor field coil connector to the solenoid terminal.
3. Loosen and remove the thru-bolts. Separate the commutator end frame, with field frame and armature assemblies, from the drive housing.
4. Slide the thrust collar off of the armature shaft.
5. Locate a small piece of pipe (1/2 inch I.D) and slide it over the shaft. Tap it lightly with a hammer until you drive the retainer in toward the armature. Remove the snap-ring from its groove and then slide the retainer and armature assembly off of the shaft.
6. Remove the retaining pin from the brush support and holders and then remove them all after disconnecting the leads.

To assemble:

7. Clean all components thoroughly and inspect them. Perform the previous tests.
8. Install the brushes into the brush holders. Assemble the insulated and grounded brush holders together with the V-spring and position them as a unit on the support pin. Push the holders and springs to the bottom of the support, and then rotate the spring to engage the center of the V-spring in the support.
9. Attach the ground wire to the grounded brush and the field lead wire to the insulated brush.
10. Lubricate the drive end of the armature shaft with a thin coating of 10W motor oil or silicone lubricant. Slide the assist spring and clutch

assembly onto the armature shaft so that the pinion is facing outward. Slide the retainer onto the shaft with the cupped surface facing the end of the shaft (away from the pinion).

11. Stand the armature on its end on a wooden surface with the commutator down. Position the snap ring on the upper end of the shaft and hold it in place with a block of wood. Now, tap on the wood block with a hammer to force the snap ring over the end of the shaft. Slide the snap ring down into the groove. Assemble the thrust collar onto the shaft with the shoulder next to the snap ring.
12. Place the armature flat on the work bench, and then position the retainer and thrust collar next to the snap ring. Next, using two pair of pliers at the same time (one pair on each side of the shaft), grip the retainer and thrust collar and squeeze until the snap ring is forced into the retainer.
13. Lubricate the drive housing bushing with a thin coating of silicone lubricant. Make sure the thrust collar is in place against the snap ring and the retainer.
14. Squirt a few drops of 10W oil into the drive housing bushing and then slide the armature and clutch assembly into place in the drive housing while engaging the shift lever with the clutch.
15. Position the field frame over the armature and apply a thin coating of liquid neoprene (Gaco) between the frame and the solenoid case. Place the frame in position against the drive housing. Take care not to damage the brushes.
16. Apply a coating of silicone lubricant to the bushing in the commutator end frame. Place the leather brake washer onto the armature shaft and slide the commutator end frame onto the shaft. Connect the field coil connectors to the motor solenoid terminal.
17. Check the pinion clearance. Install the starter.

Delco-Remy Gear Reduction

PG200/250

◆ **See Figures 72, 72a, 72b, 73, 74, 74a, 74b, 74c and 74d**

1. Disconnect the wire lead from the brush block to the solenoid terminal. Place a scribe mark or a strip of tape on the drive housing, motor case and end plate. The tape or scribe marks will assist in the alignment of the drive housing and motor case during the assembling procedures.
2. Remove the two long thru-bolts on each side of the motor case. Slide the bolts out from the end plate. Pull the motor field frame, armature, and end frame, out and away from the drive housing.
3. Remove the two bolts securing the end frame to the brush holder. Lift off the end frame and slide the armature out from the field frame.
4. Lift off the shield from the rear housing.
5. Slide the planetary gear set, solenoid lever, grommet, and Bendix drive out from the drive housing. If the planetary gear set is stuck inside the housing, secure the drive housing in a vise with soft face jaws. Insert a screwdriver or blunt punch up through the drive housing opening. Tap evenly on each side of the sun gear until the gear is free from the housing. Now, slide the planetary gear set, solenoid lever, grommet, and Bendix drive free of the housing.
6. Slide the thrust collar washer off the end of the shaft. Tap on the edge of the collar and pry it off and away from the retaining ring. Spread the retaining ring open and slide it up and free of the shaft. Slide the Bendix drive and collar off the gear shaft. Slide the planet gears and shaft out from the sun gear.
7. Inspect the brush height before removing the ball bearing and brush holder from the commutator end of the armature shaft. If the brushes are worn down even with or below the brush holder, or if the brush leads are touching the guides, replace the brushes. Check the ball bearing on the end of the commutator shaft for rough spots, dragging or wear. If the bearing is defective, then it must be replaced. If the brushes and ball bearing are fit for further service, the armature should not be disassembled any further. Proceed to the Cleaning and Inspecting section. If any of the above components are defective, worn, or the armature is suspected of being defective, continue with the next step.

■ **The end frame ball bearing must be removed from the commutator shaft before removing the brush holder. If the bearing is removed it must be discarded because the bearing puller will damage this bearing during removal.**

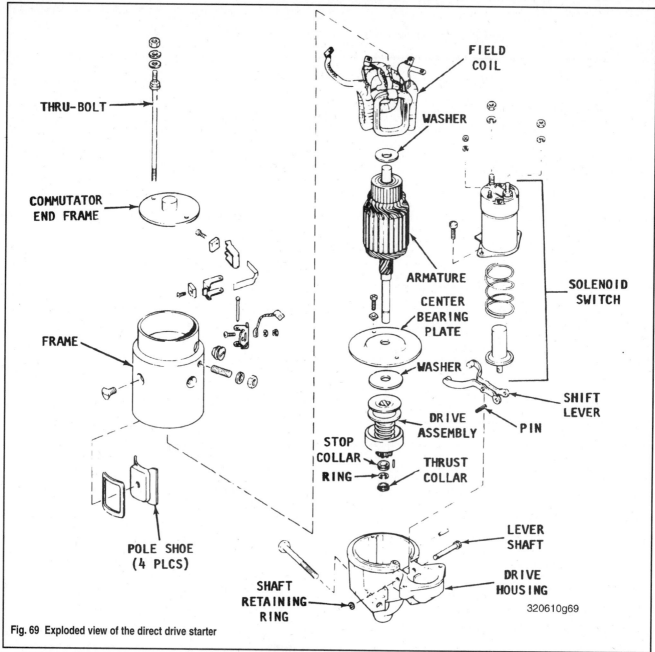

THRU-BOLT

COMMUTATOR
END FRAME

FRAME

POLE SHOE
(4 PLCS)

FIELD
COIL

WASHER

ARMATURE

CENTER
BEARING
PLATE

WASHER

DRIVE
ASSEMBLY

STOP
COLLAR
RING

THRUST
COLLAR

SOLENOID
SWITCH

SHIFT
LEVER

PIN

SHAFT
RETAINING
RING

LEVER
SHAFT

DRIVE
HOUSING

320610g69

Fig. 69 Exploded view of the direct drive starter

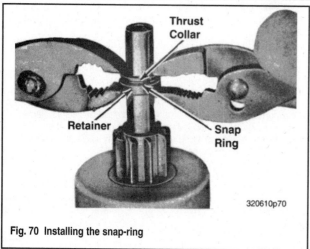

Thrust
Collar

Retainer

Snap
Ring

320610p70

Fig. 70 Installing the snap-ring

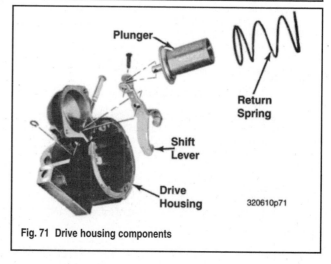

Plunger

Return
Spring

Shift
Lever

Drive
Housing

320610p71

Fig. 71 Drive housing components

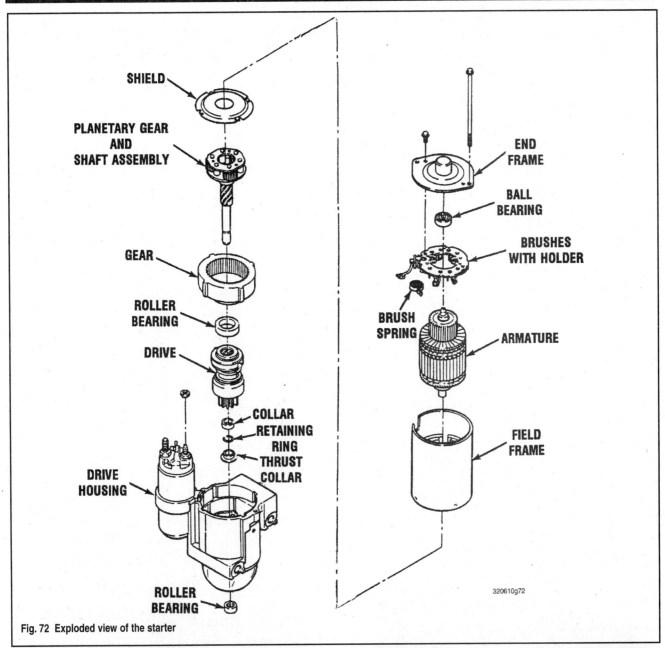

SHIELD

PLANETARY GEAR
AND
SHAFT ASSEMBLY

GEAR

ROLLER
BEARING

DRIVE

DRIVE
HOUSING

COLLAR
RETAINING
RING
THRUST
COLLAR

ROLLER
BEARING

END
FRAME

BALL
BEARING

BRUSHES
WITH HOLDER

BRUSH
SPRING

ARMATURE

FIELD
FRAME

320610g72

Fig. 72 Exploded view of the starter

Brush block
wire lead

Tape

Fig. 72a Place tape across the components

Motor and
Frame

Drive
housing

Fig. 72b Removing the motor and frame

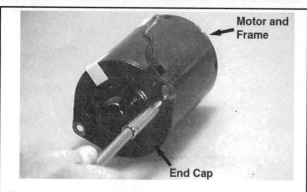

Fig. 73 Removing the end frame

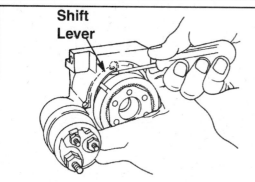

Fig. 74 Use a screwdriver to disconnect the lever from the drive unit

Fig. 74a Removing the planet gear

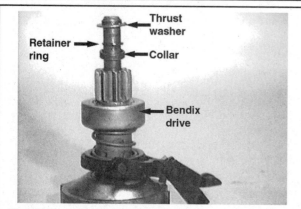

Fig. 74b Removing the Bendix drive

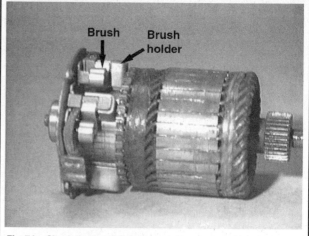

Fig. 74c Check the brush height

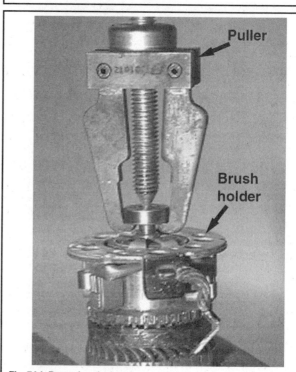

Fig. 74d Removing the bearing

8. Place a universal bearing puller around the end frame ball bearing. Set the unit in an arbor press and press the bearing off the armature shaft. Discard the ball bearing.

9. Slide the brush holder off the end of the armature commutator. Test the armature as detailed previously.

10. Place the sun gear face down on an arbor press. Using the appropriate size mandrel, press the sun gear roller bearing out the front face.

11. Place the drive end housing in an arbor press with the housing in a vertical position. Using the correct size driver, press the roller bearing out from the housing and discard the bearing.

To assemble:

◆ See Figures 70, 75, 76, 76a, 76b and 76c

12. Clean the armature, shaft and brush holder with a brush and compressed air. Do not use a grease dissolving solvent to clean electrical components, planetary gears or the Bendix drive. Solvent will damage wiring insulation and washout the lubricant in bearings and drive gears.

13. If you didn't remove them, verify the brush holder is not damaged and the brushes are held firmly against the commutator.

14. Inspect the armature commutator for grooves or out of round condition. If the commutator is worn it can be turned down on a lathe. Check the armature for shorts using a growler or test light as detailed previously.

15. Check the roller bearings for wear and rough spots. If there is any roughness or the bearing is stiff, replace the bearing.

16. Check the planetary gear set and the gear teeth for broken or missing teeth. Check to be sure the gears mesh and roll freely without binding or rough spots. If any of these conditions are found, replace the entire planetary gear set.

17. Check the one-way clutch in the Bendix drive. The gear teeth should turn freely in one direction (overrun) and must not slip in the opposite direction. Verify the drive spring is not broken, the drive teeth are not chipped or broken and the drive collar is not excessively worn. Replace the Bendix drive, if necessary.

18. Place a new sun gear bearing with the number side facing toward the gear teeth on the front of the sun gear. Using the appropriate size mandrel, press or drive the bearing into the front face of the sun gear until it is flush. Continue to drive or press the bearing until the bearing case is recessed 0.011–0.014 in. (0.28–0.38mm) into the face of the sun gear.

19. Turn the housing over in the arbor press with the end, or nose of the housing facing down. Place a new roller bearing into the drive housing bearing opening with the numbered end facing up. Using the correct size mandrel, press the new roller bearing into the drive housing. Continue pressing the bearing into the housing until the bearing is recessed 0.009–0.017 in. (0.25–0.45mm) into the housing.

20. Apply a coating of Quicksilver 2-4-C Marine Lubricant, or equivalent, to the sun gear, planetary gear and roller bearing. Slide the planet gear shaft through the sun gear. Rotate the gears until they are fully seated.

21. Invert the planet gear assembly and set it onto a clean workbench. Slide the Bendix drive gear onto the planet gear shaft with the drive gear facing up. Rotate the drive and index the pins in the drive with the helical grooves in the shaft. Slide the collar over the end of the shaft with the open side facing up. Place the retaining ring over the end of the shaft. Tap the ring onto the shaft with a wood block and hammer or spread the ring open with a pair of pliers.

22. Slide the retainer ring down and into the groove on the shaft. Squeeze the ring closed around the groove on the shaft.

23. Slide the thrust collar onto the shaft against the retainer ring on one side and the collar up against the retainer ring on the opposite side. Use a pair of pliers and squeeze the collar evenly around until the collar is fully seated over the retainer ring.

24. Spread the ends of the solenoid lever open and insert the tabs of the Bendix drive into the lever. Place a dab of grease on the thrust collar to hold it in place on the shaft. Insert the Bendix drive, planet gear set, and rubber grommet into the drive housing.

25. Insert the solenoid lever return spring and solenoid into the drive housing opening. Position the solenoid with the brush block terminal at the 6 o'clock position, or next to the rubber grommet. Secure the solenoid with three screws and lock washer.

26. Place the brush holder over the commutator end of the armature. Insert the brushes, one at-a-time into the holder, followed by the brush spring. Repeat this step for all four brushes. Be sure the brushes are pressing firmly against the commutator.

27. Set the armature into an arbor press and place a new ball bearing assembly onto the commutator end of the armature shaft. Set the numbered side of the bearing facing the armature. Place the correct size mandrel over the inner race of the bearing, and then press the bearing down onto the shaft until the inner race contacts the armature.

28. Mount the cranking motor in a vise with soft face jaws, as shown in the accompanying illustration.

29. Place the shield over the end of the planet gear assembly.

30. Slide the armature into the field frame assembly. Align the scribe marks or tape on the outside of the motor field frame assembly with the tape or marks on the drive housing. Guide the end of the armature into the center of the planetary gear set while aligning the marks on the motor case.

31. Position the end frame over the brush block and align the tape or scribe marks with the motor case. Guide the wire lead from the brush block and rubber grommet into place in the end frame. Secure the end frame to the brush block with two short bolts. Tighten the bolts securely.

32. Insert two long bolts down through the end frame. Align the bolts with the threaded openings in the drive housing. Tighten the bolts alternately and evenly until they are secure.

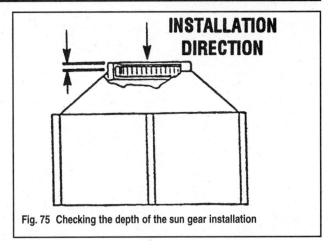

Fig. 75 Checking the depth of the sun gear installation

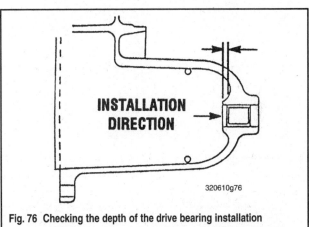

Fig. 76 Checking the depth of the drive bearing installation

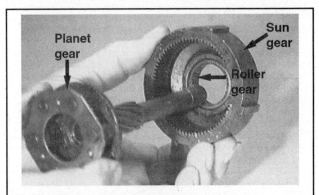

Fig. 76a Slide the planet gear through the sun gear

PG260

◆ See Figures 75, 76, 77 and 78

1. Remove the starter.

2. Disconnect the wire brush lead from the brush block to the solenoid terminal. Place a scribe mark or a strip of tape on the drive housing, field frame and end cap. The tape or scribe marks will assist in the alignment of the drive housing and field frame during the assembling procedures.

3. Remove the two long thru bolts on each side of the motor case. Slide the bolts out from the end cap.

4. Remove the two bolts securing the end cap to the brush holder. Separate the brush holder from the end cap. Pull the armature and field frame out from the drive housing. Separate the armature from the field frame.

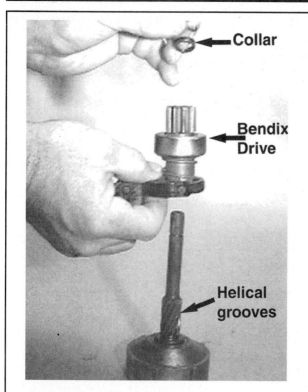

Fig. 76b Slide the Bendix onto the shaft

Fig. 76c Make sure the collar is seated over the retaining ring

■ **The armature will be held in the field frame by permanent magnets.**

5. Lift out the shield and washer from the open end of the drive housing. Remove the three screws securing the solenoid to the drive housing and pull the solenoid free of the housing.

6. Slide the planetary gear set, solenoid lever, grommet, and Bendix drive out from the drive housing.

7. Slide the thrust collar washer off the end of the shaft. Tap on the edge of the collar and pry it off and away from the retaining ring. Spread the retaining ring open and slide it up and free of the shaft. Slide the Bendix drive and collar off the gear shaft. Slide the planet gears and shaft out from the sun gear.

To assemble:

8. Clean the armature, shaft and brush holder with a brush and compressed air. Do not use a grease dissolving solvent to clean electrical components, planetary gears, or Bendix drive. Solvent will damage wiring insulation and washout the lubricant in bearings and drive gears.

9. Verify the brush holder is not damaged and the brushes are held firmly against the commutator.

10. Inspect the armature commutator for grooves or out of round condition. If the commutator is worn it can be turned down on a lathe. Check the armature for shorts using a growler or test light.

11. Check the roller bearings for wear and rough spots. If there is any roughness or the bearing is stiff, replace the bearing.

12. Check the planetary gear set and the gear teeth for broken or missing teeth. Check to be sure the gears mesh and roll freely without binding or rough spots. If any of these conditions are found, replace the entire planetary gear set.

13. Place the sun gear face down on an arbor press. Using the appropriate size mandrel, press the sun gear roller bearing out the front face. Place a new bearing with the number side facing toward the gear teeth on the front of the sun gear. Using the appropriate size mandrel, press or drive the bearing into the front face of the sun gear until it is flush. Continue to drive or press the bearing until the bearing case is recessed 0.011–0.014 in. (0.28–0.38mm) into the face of the sun gear.

14. Place the drive end housing in an arbor press with the housing in a vertical position. Using the correct size driver, press the roller bearing out from the housing and discard the bearing. Turn the housing over in the arbor press with the end, or nose of the housing facing down. Place a new roller bearing into the drive housing bearing opening with the numbered end facing up. Using the correct size mandrel, press the new roller bearing into the drive housing. Continue pressing the bearing into the housing until the bearing is recessed 0.009-0.017 in. (0.25–0.45mm) into the housing.

15. Apply a coating of Quicksilver 2-4-C Marine Lubricant, or equivalent, to the sun gear, planetary gear and roller bearing. Slide the planet gear shaft through the sun gear. Rotate the gears until they are fully seated.

16. Invert the planet gear assembly and set it onto a clean workbench. Slide the Bendix drive gear onto the planet gear shaft with the drive gear facing up. Rotate the drive and index the pins in the drive with the helical grooves in the shaft. Slide the collar over the end of the shaft with the open side facing up. Place the retaining ring over the end of the shaft. Tap the ring onto the shaft with a wood block and hammer or spread the ring open with a pair of pliers.

17. Slide the retainer ring down and into the groove on the shaft. Squeeze the ring closed around the groove on the shaft.

18. Slide the thrust collar onto the shaft against the retainer ring on one side and the collar up against the retainer ring on the opposite side. Use a pair of pliers and squeeze the collar evenly around until the collar is fully seated over the retainer ring.

19. Spread the ends of the solenoid lever open and insert the tabs of the Bendix drive into the lever. Place a dab of grease on the thrust collar to hold it in place on the shaft. Insert the Bendix drive, planet gear set, and rubber grommet into the drive housing.

20. Insert the solenoid lever return spring and solenoid into the drive housing opening. Position the solenoid with the brush block terminal at the 6 o'clock position, or next to the rubber grommet. Secure the solenoid with three screws and lock washer.

21. Place the shield over the end of the planet gear assembly. Place the large washer against the shield.

22. Slide the armature into the field frame assembly. Align the scribe marks or tape on the outside of the motor field frame assembly with the tape or marks on the drive housing. Guide the end of the armature into the center of the planetary gear set while aligning the marks on the motor case.

23. Position the end frame over the brush block and align the tape or scribe marks with the motor case. Guide the wire lead from the brush block and rubber grommet into place in the end frame. Secure the end frame to the brush block with two short bolts. Tighten the bolts securely.

24. Insert two long bolts down through the end frame. Align the bolts with the threaded openings in the drive housing. Tighten the bolts alternately and evenly until they are secure.

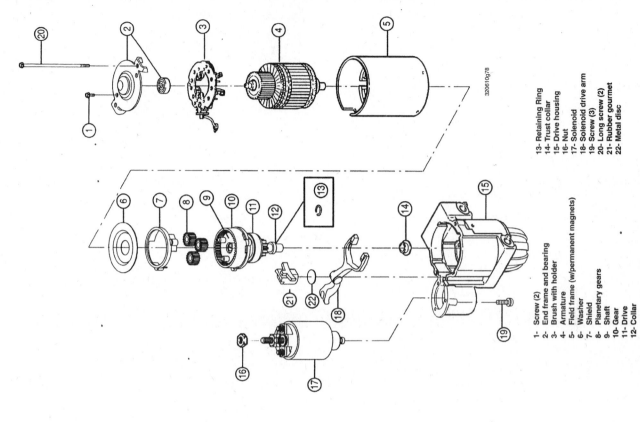

13- Retaining Ring
14- Trust collar
15- Drive housing
16- Nut
17- Solenoid
18- Solenoid drive arm
19- Screw (3)
20- Long screw (2)
21- Rubber gourmet
22- Metal disc

1- Screw (2)
2- End frame and bearing
3- Brush with holder
4- Armature
5- Field frame (w/permanent magnets)
6- Washer
7- Shield
8- Planetary gears
9- Shaft
10- Gear
11- Drive
12- Collar

Fig. 78 Exploded view of the Delco PG260-F1 starter

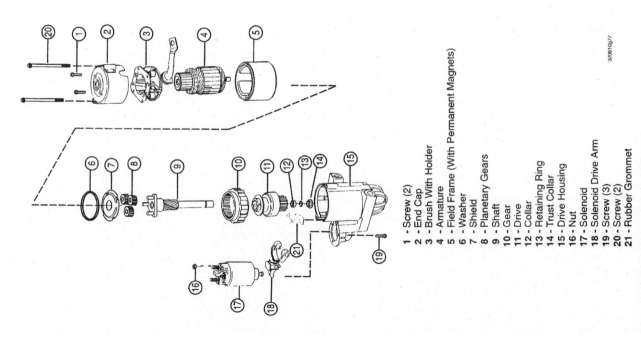

1 - Screw (2)
2 - End Cap
3 - Brush With Holder
4 - Armature
5 - Field Frame (With Permanent Magnets)
6 - Washer
7 - Shield
8 - Planetary Gears
9 - Shaft
10 - Gear
11 - Drive
12 - Collar
13 - Retaining Ring
14 - Trust Collar
15 - Drive Housing
16 - Nut
17 - Solenoid
18 - Solenoid Drive Arm
19 - Screw (3)
20 - Screw (2)
21 - Rubber Grommet

Fig. 77 Exploded view of the Delco PG260 starter

Solenoid

REMOVAL & INSTALLATION

Delco-Remy Direct Drive

◆ See Figure 79

1. Remove the starter motor.
2. Remove the screw/washer from the starter connector strap terminal.
3. Remove the two solenoid mounting screws, twist it clockwise until you can remove the flange key from the slot in the starter and then lift off the solenoid. Don't lose the return spring.

To install:

4. Slide the return spring over the plunger.
5. Position the solenoid at the starter and rotate it until the flange key engages the keyway.
6. Install the two mounting screws and tighten them securely.
7. Install the connector strap and then install the starter.

Delco-Remy Gear Reduction

PG200/250/260

The solenoid, along with the plunger, spring and shift lever are sealed,

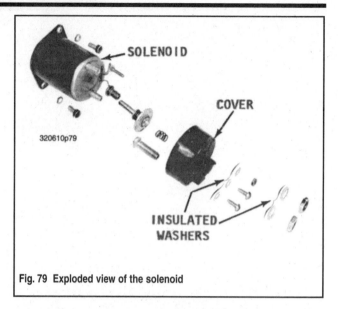

Fig. 79 Exploded view of the solenoid

and permanently mounted in the drive housing. Defective solenoids must be replaced with the entire housing. No repair is possible.

IGNITION SYSTEM

Mercruiser engine covered in this manual utilize one of five different ignition systems: conventional breaker points, Digital Distributorless Ignition System (DDIS), Electronic Spark Timing System (EST), Thunderbolt IV and Thunderbolt V.

3.0L inline engines may come with any of the first three systems, although breaker points were confined to only the earliest models covered here.

V6 and V8 engines will be equipped with either of the two Thunderbolt breakerless systems except late model Black Scorpion engines and 1998–01 7.4L/8.1L/8.2L V8s which utilize EST.

Descriptions of each system, and complete repair procedures follow.

BREAKER POINT IGNITION SYSTEM

■ **Please refer to Section 3 for breaker point, dwell and timing adjustment procedures.**

Description

The ignition system consists of a primary and a secondary circuit. The low-voltage current of the ignition system is carried by the primary circuit. Parts of the primary circuit include the ignition switch, ballast resistor, neutral-safety switch, primary winding of the ignition coil, contact points in the distributor, condenser, and the low-tension wiring.

The secondary circuit carries the high-voltage surges from the ignition coil that result in a high-voltage spark between the electrodes of each spark plug. The secondary circuit includes the secondary winding of the coil, coil-to-distributor high-tension lead, distributor rotor and cap, ignition cables, and the spark plugs.

When the contact points are closed and the ignition switch is on, current from the battery or from the alternator flows through the primary winding of the coil, through the contact points to ground. The current flowing through the primary winding of the coil creates a magnetic field around the coil windings and energy is stored in the coil. Now, when the contact points are opened by rotation of the distributor cam, the primary circuit is broken. The current attempts to surge across the gap as the points begin to open, but the condenser absorbs the current. In so doing, the condenser creates a sharp break in the current flow and a rapid collapse of the magnetic field in the coil. This sudden change in the strength of the magnetic field causes a voltage to be induced in each turn of the secondary windings in the coil.

The ratio of secondary windings to the primary windings in the coil increases the voltage to about 20,000 volts. This high voltage travels through a cable to the center of the distributor cap, through the rotor to an adjacent distributor cap contact point, and then on through one of the ignition wires to a spark plug.

When the high-voltage surge reaches the spark plug it jumps the gap between the insulated center electrode and the grounded side electrode.

This high voltage jump across the electrodes produces the energy required to ignite the compressed air/fuel mixture in the cylinder.

The entire electrical build-up, breakdown, and transfer of voltage is repeated as each lobe of the distributor cam passes the rubbing block on the contact breaker arm, causing the contact points to open and close. At high engine rpm operation, the number of times this sequence of actions takes place is staggering.

Beginning at the key switch, current flows to the ballast resistor and then to the positive side of the coil. When the resistor is cold its resistance is approximately one ohm. The resistance increases in proportion to the resistor's rise in temperature.

While the engine is operating at idle or slow speed, the cam on the distributor shaft revolves at a relatively slow rate. Therefore, the breaker points remain closed for a slightly longer period of time. Because the points remain closed longer, more current is allowed to flow and this current flow heats the ballast resistor and increases its resistance to cut down on current flow thereby reducing burning of the contact points.

During high rpm engine operation, the reduced current flow allows the resistor to cool enough to reduce resistance, thus increasing the current flow and effectiveness of the ignition system for high-speed performance.

The voltage drops about 25% during engine cranking due to the heavy current demands of the starter. These demands reduce the voltage available for the ignition system. In order to reduce the problem of less voltage, the ballast resistor is by-passed during cranking. This releases full battery voltage to the ignition system.

The shift cut-out switch is connected between the primary side of the ignition coil and ground. This switch is normally open. The function of this switch is to ground the ignition system during a shift to neutral. By grounding the ignition system, gear pressure is released and the shift is made much easier. Obviously, if the ignition system is grounded, the cylinders will not fire during this period. In actual practice only a few cylinders fail to fire and it is usually not noticeable. The shift cutout switch is mounted on the transom and is activated by the remote control shift cable. If this switch is not adjusted properly, or if it shorts out (is grounded) then the primary side of the ignition coil will be grounded and the engine will not start.

In order to obtain the maximum performance from the engine, the timing of the spark must vary to meet operating conditions. For idle, the spark advance should be as low as possible. During high-speed operation, the spark must occur sooner, to give the air/fuel mixture enough time to ignite, burn, and deliver power to the piston for the power stroke.

Manual setting and centrifugal advance are the two methods of obtaining the constantly changing demands of the engine. The manual setting is made at idle speed. This setting allows the contact points to open at a specified position of the piston in the same manner as with conventional ignition systems. Idle timing must be set below speeds of 900 rpm.

The ignition amplifier provides for ignition advance. After initial timing, the advance starts at 1000 rpm and increases to a maximum of 24 degrees advance between 3600 and 3800 rpm. If the key is left in the on position when the engine is at rest, the coil primary current will automatically turn off in a short time to protect the ignition coil.

The ignition amplifier will withstand a reverse battery connection for only about a minute.

Troubleshooting

Any problem in the ignition system must first be localized to the primary or secondary circuit before the defective part can be identified.

GENERAL TESTS

◆ See Figure 80

Disconnect the wire from the center of the distributor cap and hold it about 1/4 in. from a good ground. Turn the ignition switch to start, and crank the engine with the starter. If you observe a good spark, go to Test 5 (Secondary Circuit Test). If you do not have a good spark, go to Test 2 (Primary Circuit Test).

Fig. 80 Testing for spark

Primary Circuit Test

◆ See Figure 81

Remove the distributor cap; lift off the rotor; and then turn the crankshaft until the contact points close. Turn the ignition switch ON and open and close the contact points using a small screwdriver or a non-metallic object. Hold the high-tension coil wire about 1/4 in. from a good ground. If a good spark jump from the wire to the ground is observed, the primary circuit checks out. Proceed to Test 5, (Secondary Circuit Test). If there is no spark, proceed to Test 3, (Contact Point Test).

Contact Point Test

◆ See Figure 82

Remove the distributor cap and rotor. Rotate the crankshaft until the points are open, and then insert some type of insulator between the points. Now, hold the high-tension coil wire about 1/4 in. from a good ground, and at the same time move a small screwdriver up and down with the screwdriver shaft touching the moveable point and the tip making intermittent contact with the contact point base plate. In this manner, the screwdriver is being used for a set of contact points. If a spark is observed from the high-tension wire to ground, then the problem is in the contact points. Replace the set of points. If there is no spark from the high-tension wire to ground, the problem is either a defective coil or condenser. To test the condenser—see Test 4 (Condenser Test).

Condenser Test

◆ See Figure 82

Condensers seldom cause a problem. However, there is always the possibility one may short out and ground the primary circuit. Before testing the condenser, check to be sure one of the primary wires or connections inside the distributor has not shorted out to ground.

The most accurate method of testing a condenser is with an instrument manufactured for that purpose. However, seldom is one available, especially during an emergency. Therefore, the following procedure is outlined for emergency troubleshooting the condenser and the primary circuit insulation for a short.

First, remove the condenser from the system. Take care that the metallic case of the condenser does not touch any part of the distributor. Next, insert a piece of insulating material between the contact points. Now, move the blade of a small screwdriver up and down with the shaft of the screwdriver making contact with the movable contact point and the tip making and breaking contact with the contact point base plate. Observe for a low-tension spark between the tip of the screwdriver and the contact point base plate as you make and break the contact with the screwdriver tip. You should observe a spark during this test and it will prove the primary circuit complete through the neutral-safety switch, the primary side of the ignition coil, the ballast resistor, the shift cutout switch, and the primary wiring inside the distributor. If you have a spark, reconnect the condenser and again make the same test with the screwdriver. If you do not get a spark, either the condenser is defective and should be replaced, or the shift control switch should be adjusted or replaced.

If you were unable to get a spark with the condenser disconnected, it means no current is flowing to this point, or there is a short circuit to ground. Use a continuity tester and check each part in turn to ground in the same manner as you did at the movable contact point. If you get a spark, indicating current flow, at one terminal of the part, but not at the other, then you have isolated the defective unit.

Secondary Circuit Test

◆ See Figure 83

The secondary circuit cannot be tested using emergency troubleshooting procedures unless the primary circuit has been tested and proven satisfactory, or any problems discovered in the primary circuit have been corrected.

If the primary circuit tests are satisfactory, use the same procedures as outlined in Test 2, Primary Circuit Test, to check the secondary circuit.

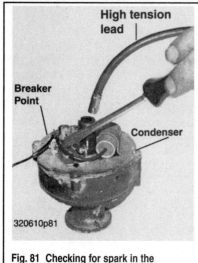

Fig. 81 Checking for spark in the primary circuit

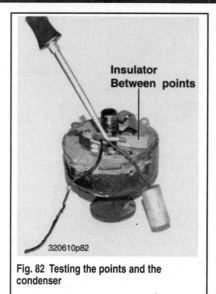

Fig. 82 Testing the points and the condenser

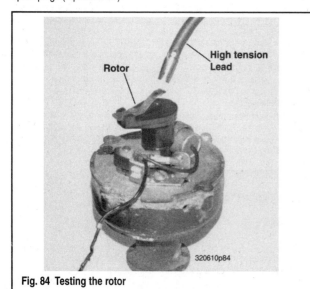

Fig. 83 Testing the secondary circuit

Hold the high-tension coil lead about 1/4 in. from a good ground and at the same time, use a small screwdriver to open and close the contact points. A spark at the high-tension lead proves the ignition coil is good. However, if the engine still fails to start and the problem has been traced to the ignition system, then the defective part or the problem must be in the secondary circuit.

The distributor cap, the rotor, high-tension leads, or the plugs may require attention or replacement. To test the rotor, go to Test 6 (Rotor Test). If you were unable to observe a spark during the secondary circuit test just described, the ignition coil is defective and must be replaced.

Rotor Test

◆ See Figure 84

With the distributor cap removed and the rotor in place on the distributor shaft, hold the high-tension coil lead about 1/4 in. from the rotor contact spring, and at the same time crank the engine with the ignition switch turned on. If a spark jumps to the rotor, it means the rotor is shorted to ground and must be replaced. If there is no spark to the rotor, it means the insulation is good and the problem is either in the distributor cap (check it for cracks), in the high-tension leads (check for poor insulation or replace it), or in the spark plugs (replace them).

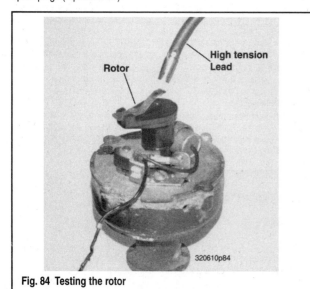

Fig. 84 Testing the rotor

IGNITION VOLTAGE TESTS

◆ See Figure 85

Many times hard starting and misfiring problems are caused by defective or corroded connections. Such a condition can lower the available voltage to the ignition coil. Therefore, make voltage tests at critical points to isolate such a problem. Move the voltmeter test probes from point-to-point in the following order.

Test 1—Voltage Loss Across Entire Ignition Circuit

◆ See Figure 86

Connect a voltmeter between the battery side of the ignition coil and the positive post of the battery, as shown in the Test 1 illustration. Crank the engine until the contact points are closed. Turn the ignition switch to ON. The voltage loss should not exceed 3.2 volts. This figure allows for a 0.2 loss across each of the connections in the circuit, plus a calibrated 2.4 volt drop through the ballast resistor. If the total voltage loss exceeds 3.2 volts, then it will be necessary to isolate the corroded connection in the circuit of the key resistor, the wiring, or at the battery.

TEST 2—Cranking System

◆ See Figure 87

Disconnect the high tension wire and ground it securely to minimize the danger of sparks and a possible fire. Next, connect a voltmeter between the battery side of the ignition coil and ground, as shown in Test 2 illustration. Now crank the engine and check the voltage. A normal system should have a reading of 8.0 volts. If the voltage is lower, the battery is not fully charged or the starter is drawing too much current.

Test 3—Contact Points and Condenser

◆ See Figure 88

Measure the voltage between the distributor primary terminal and ground, as shown in the illustration. Crank the engine until the contact points are closed. Turn the ignition switch to the ON position. The voltage reading must be less than 0.2 volts. A higher reading indicates the contact points are oxidized and must be replaced. To check the condenser, crank the engine until the contact points are open, and then take a voltage reading. If the reading is not equal to the battery voltage, the condenser is shorted to ground. Check the condenser installation or replace the condenser.

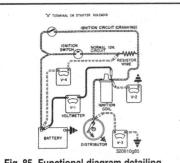

Fig. 85 Functional diagram detailing different voltmeter hook-ups

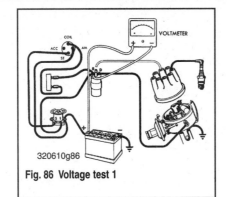

Fig. 86 Voltage test 1

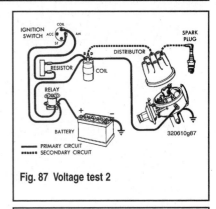

Fig. 87 Voltage test 2

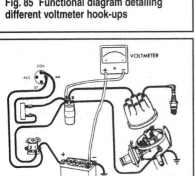

Fig. 88 Voltage test 3

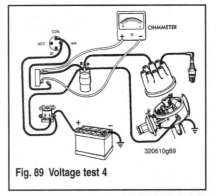

Fig. 89 Voltage test 4

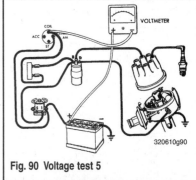

Fig. 90 Voltage test 5

Test 4—Primary Resistor

◆ See Figure 89

Disconnect the battery wire at the primary resistor to prevent damage to the ohmmeter. Connect an ohmmeter across the terminals of the resistor, as shown in the illustration. The specified resistance is between 1.3 and 1.4 ohms. If the reading does not fall within this range, replace the resistor.

Test 5—Voltage Loss In The Ignition Switch, Ammeter, Or Battery Cable

◆ See Figure 90

Crank the engine until the points are closed. Connect the voltmeter to the battery post (not the cable terminal) and to the load side of the ignition switch, as shown in the illustration. Now, turn the ignition switch to the ON position and note the voltage reading. The meter reading should not be more than 0.8 volts. A 0.2 volt drop across each of the connections is permitted.

If the voltage drop is more than 0.8 volts, move the test probe to the "hot" side of the ammeter. If the reading is 0.4 volts, the ignition switch is satisfactory. Once the corroded connection has been located, remove the nut, clean the wire terminal and connector, and then tighten the connection securely.

Test 6—Distributor Condition

The condition of the distributor can be quickly and conveniently checked with a timing light.

Under normal timing light procedures the trigger wire from the timing light is attached to the spark plug wire of the No. 1 cylinder. In this test, connect the trigger wire to the fourth cylinder in the firing order of an in-line engine or to the fifth cylinder in the firing order of a V8 engine.

The timing mark and the pointer should align in the same position as it did with the number one cylinder. If there is a variation of a few degrees, the distributor shaft bushings or cam lobes may be worn and the condition will have to be corrected.

Before setting the timing, make sure the point dwell is correct. Take care to aim the timing light straight at the mark. Sighting from an angle may cause an error of two or three degrees.

Distributor Cap and Rotor

REMOVAL & INSTALLATION

◆ See Figured 91 and 92

■ It is not necessary to remove the spark plug and ignition coil leads to remove the cap. If you do remove them, be sure to carefully mark each lead and tag it for reinstallation in the proper cap socket.

1. Loosen the two cap retaining screws and carefully lift off the cap. It never hurts to paint a small mark across the cap and distributor body to aid in installation.
2. Paint a small mark on the distributor body at the end of the rotor and then pull the rotor up and off the shaft.
3. If you have removed the electrical leads from the cap, clean it in warm, soapy water and allow it to dry completely. Check the cap for cracks or other damage. Marine caps should have brass contacts—if your's has aluminum contacts, replace it as someone has previously installed an automotive cap.
4. Check the rotor for cracks and other damage.
5. Install the rotor onto the shaft so that the end is pointing to the mark on the distributor base made earlier.
6. Install the cap onto the distributor. There should be locating tabs cut into the distributor housing to aid installation; you've also got the mark you made previously. Tighten the retaining screws securely.

Distributor

REMOVAL

■ It is not necessary to remove the spark plug leads to remove the cap. If you do remove them, be sure to carefully mark each lead and tag it for reinstallation in the proper cap socket.

1. Disconnect the distributor primary lead at the ignition coil terminal and then remove the cap and rotor.

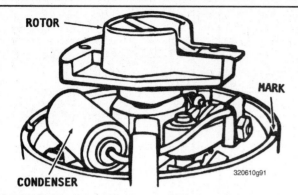

Fig. 91 Paint a small mark on the distributor housing for rotor alignment

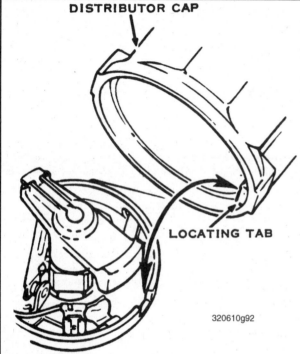

Fig. 92 Most models will have locating tabs cut into the housing to aid in installing the cap

2. Scribe or paint a small alignment mark on the distributor body at the rotor notch on the distributor shaft. Scribe or paint a small mark across the distributor housing and cylinder block.

3. Loosen and remove the distributor clamp bolt and then slowly lift the distributor out of the cylinder block.

INSTALLATION

Engine Not Disturbed

1. Install the rotor onto the shaft so that the marks made earlier align.
2. Install a new gasket on the housing.
3. Grasp the rotor and spin the shaft about 1/8 of a turn, counter clockwise, past the alignment mark. Position the distributor over the hole in the block so that the alignment marks match and then work it down into the hole until it engages completely with the oil pump shaft.

■ **It is OK to wiggle the rotor slightly during installation so that the distributor gear meshes properly with the oil pump. It is imperative**

though that the rotor ends up pointing to the alignment mark made during removal when installation is complete.

2. Install the clamp bolt and tighten it to 20 ft. lbs. (27 Nm).
3. Check the point gap and dwell as detailed in Section 3 and then install the cap and coil lead.
4. Check the ignition timing.

Engine Disturbed

1. Set the No. 1 piston to TDC. Remove the No. 1 spark plug, put your thumb over the hole and slowly crank the engine until you feel compression on the No. 1 cylinder. Crank the engine a bit more until the pointer lines up with the mark on the scale.
2. Install the rotor onto the shaft so that the marks made earlier align.
3. Install a new gasket on the housing.
4. Grasp the rotor and spin the shaft about 1/8 of a turn, counter clockwise, past the alignment mark. Position the distributor over the hole in the block so that the alignment marks match and then work it down into the hole until it engages completely with the oil pump shaft.

■ **It is OK to wiggle the rotor slightly during installation so that the distributor gear meshes properly with the oil pump. It is imperative though that the rotor ends up pointing to the alignment mark made during removal when installation is complete.**

5. Install the clamp bolt and tighten it finger-tight. Rotate the distributor carefully until the breaker points just begin to open and then tighten the clamp bolt to 20 ft. lbs. (27 Nm).
6. Install the distributor cap into position and confirm that the rotor points to the terminal for the No. 1 spark plug lead. If it does, go to the next step; if not, repeat the installation procedure.
7. Check the point gap and dwell as detailed in Section 3 and then install the cap and coil lead.
8. Check the ignition timing.

DISASSEMBLY & ASSEMBLY

◆ **See Figure 93**

1. Remove the distributor.
2. Disconnect the primary wire and the condenser lead from the breaker point assembly terminal. Remove the breaker point assembly by removing the two attaching screws. Remove the condenser attaching screw and the condenser.
3. Pull the primary lead through the opening in the housing. Remove the two breaker plate attaching screws, and then remove the breaker plate.
4. Identify one of the distributor weight springs and its bracket with a mark. Mark one of the weights and its pivot pin. Carefully unhook and remove the weight springs. Pull the lubricating wick out of the cam assembly. Remove the cam assembly by first removing the retainer, and then lifting the assembly off the distributor shaft.
5. Remove the thrust washer installed only on counterclockwise rotating engines.
6. Remove the weight retainers, and then remove the weights.
7. Remove the distributor cap clamps. Scribe a mark on the gear and a matching mark on the distributor shaft as an aid in locating the pin holes during assembly. Place the distributor shaft in a V-block, and then use a drift punch to remove the roll pin. Remove the gear from the shaft. Remove the shaft collar roll pin.

To assemble:

8. Never wash the distributor cap, rotor, condenser, or breaker plate assembly of a distributor in any type of cleaning solvent. Such compounds may damage the insulation of these parts or, in the case of the breaker plate assembly, saturate the lubricating felt.
9. Check the shaft for wear and fit in the distributor body bushings. If either the shaft or the bushings are worn, replace the shaft and distributor body as an assembly. Use a set of V-blocks and check the shaft alignment with a dial gauge. If the run-out is more than 0.002 in., the shaft and body must be replaced.
10. Inspect the breaker plate assembly for damage and replace it if there are signs of excessive wear.

11. Check to be sure the advance weights fit free on their pins and do not have any burrs or signs of excessive wear. Check the cam fit on the end of the shaft. The cam should not fit loose but it should still be free without binding.

✳✳ WARNING

Marine distributors have a corrosion resistant coating applied to the return spring on top of the breaker plate and on the two small weight springs under the plate. For this reason, as well as other good reasons, automotive parts—point sets, springs, or other distributor parts—should never be used as a replacement.

12. Always replace the points with a new set during a distributor overhaul. The condenser seldom gives trouble, but good shop practice a few years ago called for a new condenser with a new set of points. Some point sets still include a condenser in the package. If you have paid for a new condenser, you might as well install it and be free of concern over that part.

13. Lubricate the distributor shaft with crankcase oil, and then slide it into the distributor body. Slide the collar onto the shaft; align the holes in the collar with the hole in the shaft; and then install a new pin.

14. Install the distributor cap clamps. Left-hand rotating engine distributor assemblies have an additional thrust washer between the collar and the base. Use a feeler gauge between the collar and the distributor base to check the shaft end-play. The end-play should be between 0.024–0.035 in.

15. Install the gear onto the shaft with the marks, on the gear and the shaft made during disassembly, aligned. The holes through the gear and the shaft should be aligned after the gear is installed. Install the gear roll pin.

16. Fill the grooves in the weight pivot pins with distributor cam lubricant. Position the weights in the distributor with the weight you identified with a mark during disassembly matched with the marked pivot pin. Secure the weights in place with the retainers. Slide the thrust washer onto the shaft. Fill the upper distributor shaft grooves with distributor cam lubricant.

17. Install the cam assembly with the marked spring bracket near the marked spring bracket on the stop plate. If a new cam assembly is being installed, take care to be sure the cam is installed with the hypalon-covered stop in the correct cam plate control slot. The proper slot can be determined by measuring the length of the slot used on the old cam' and then using the corresponding slot on the new cam. Some new cams will have the size of the slot stamped in degrees near the slot. If the wrong slot is used, the maximum advance will not be correct.

18. Coat the distributor cam lobes with a light film of distributor cam lubricant. Install the retainer and wick. Use a few drops of SAE 10W engine oil on the wick. Install the weight springs, with the spring and bracket marked during disassembly, matched.

19. Place the breaker plate in position, and then secure with the attached screws.

20. Push the primary wire through the opening in the distributor. Place the breaker point assembly and the condenser in position and secure them in place with the attaching screws. Connect the primary wire and the condenser lead to the breaker point primary terminal.

21. Adjust the point gap and dwell as outlined in the following procedures, then install the distributor.

Ignition Coil

REMOVAL & INSTALLATION

1. Disconnect the negative battery cable.
2. Disconnect the high tension lead coming from the distributor.
3. Tag and disconnect the coil electrical leads.
4. Remove the coil mounting bolts/screws and lift out the coil.
5. Install the coil and tighten the mounting hardware securely.
6. Connect the electrical leads and the high tension lead.
7. Connect the battery cable.

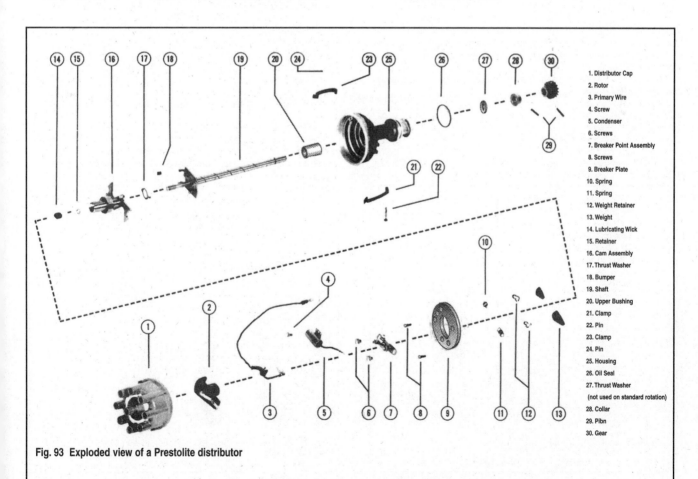

1. Distributor Cap
2. Rotor
3. Primary Wire
4. Screw
5. Condenser
6. Screws
7. Breaker Point Assembly
8. Screws
9. Breaker Plate
10. Spring
11. Spring
12. Weight Retainer
13. Weight
14. Lubricating Wick
15. Retainer
16. Cam Assembly
17. Thrust Washer
18. Bumper
19. Shaft
20. Upper Bushing
21. Clamp
22. Pin
23. Clamp
24. Pin
25. Housing
26. Oil Seal
27. Thrust Washer
    (not used on standard rotation)
28. Collar
29. Pibn
30. Gear

Fig. 93 Exploded view of a Prestolite distributor

DIGITAL DISTRIBUTORLESS IGNITION SYSTEM (DDIS)

Description

◆ See Figure 94

In 1992, a Digital Distributorless Ignition System (DDIS), was installed as standard equipment on the 3.0 LX 4-cylinder engine. This system was also made available as "optional" equipment for the 3.0L engine. As the name suggests, this ignition system does not have a distributor.

Where conventional systems have a distributor installed in the cylinder block, engines with this system have a motion sensor located where the conventional distributor would be.

The system consists of the following components:
- Two ignition coils
- An ignition amplifier
- A motion sensor
- A shift cut-out (interrupter) switch

Two ignition coils (combined into one unit) are mounted on the block. Each coil simultaneously provides the spark for two cylinders. As explained later, the spark to one cylinder is wasted. As a result, this system is sometimes referred to as the "waste spark" method of spark distribution. The unit has four high voltage towers and the coils are fired on both the compression and exhaust strokes.

The ignition amplifier is a solid-state "black box" type unit, also mounted on the engine block. The amplifier is connected to the ignition coil, motion sensor, and the shift cut-out switch. The amplifier controls ignition coil output, determines spark advance, and limits engine speed to a maximum of 5000 rpm.

The motion sensor is located next to the oil filter. The lower portion of the motion sensor resembles a conventional distributor. The spiral gear on the sensor shaft is driven by a matching gear on the camshaft in the block. The lower end of the shaft is slotted to drive the oil pump. The upper portion of the motion sensor is entirely different from a conventional distributor. A close-tolerance machined reluctor disc is mounted on top of the sensor shaft and rotates at engine speed. Six "high points" are machined into the reluctor disc. Four of the spaces between the "high points" are quite large, and two spaces are small, as shown in the illustration. A crankshaft position sensor is mounted to one side of the reluctor disc. This sensor is connected to the ignition amplifier.

The shift cut-out switch and its operation is described in complete detail previously in the Breaker Points section.

The reluctor disc at the upper end of the motion sensor shaft rotates at engine speed. Six "high points" are machined into the reluctor disc. Four of the spaces in between the "high points" are quite large and are spaced approximately 90º apart. The small spaces are 180º—opposite each other. As each space passes the crankshaft position sensor, an induced voltage pulse is created. The sensor monitors crankshaft position and engine speed. The sensor sends signals to the ignition amplifier.

The ignition amplifier compares time intervals between pulses. As a small space sweeps past the crank position sensor, the ignition amplifier sends a signal to energize the first ignition coil and fires one of the two cylinder pair. After the crankshaft has rotated through 180º, the second small space sweeps past the crankshaft position sensor. The ignition amplifier again sends a signal to energize the second ignition coil and fires the other cylinder pair.

Troubleshooting

◆ See Figure 95

PRECAUTIONS

- Make sure engine compartment is well-ventilated and no fuel vapors are present.
- Never touch or disconnect any ignition system component while the engine is running.
- Never reverse the battery cable connections.
- Never disconnect the battery cables while the engine is running.
- All tests should be performed at room temperature—68ºF (20ºC), with the engine cold.

IGNITION AMPLIFIER

1. Unplug the electrical connector running from the amplifier to the ignition coil.
2. Turn the ignition switch to the ON position and touch the multi-meter probe to terminal **A** (while grounding the other black probe) on the amplifier-side connector. The meter should read battery voltage.
3. Disconnect the engine harness from the amplifier and now touch the red probe to terminal **A**, you should read battery voltage once again.
4. If the readings in Steps 2 and 3 differ by more than 0.5 volt, replace the ignition amplifier.

IGNITION COIL

1. With the ignition switch in the OFF position, unplug the coil-to-amplifier connector.
2. Connect a multi-meter and measure the resistance between terminals **A** and **B** on the coil-side connector. Now measure the resistance between terminals **A** and **C**. Primary side resistance for both tests should be 1.9–2.5 ohms.
3. Tag and disconnect the spark plug leads at the coil towers and measure the resistance between towers **1** and **4** and then between towers **2** and **3**. When looking at the coil with the electrical connector at the bottom, towers **1** and **4** are on the left side (4 is on top) and towers **2** and **3** are on

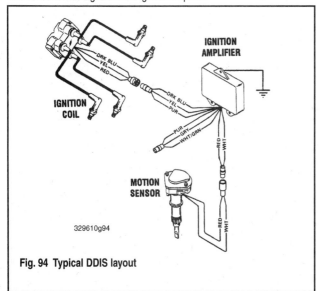

Fig. 94 Typical DDIS layout

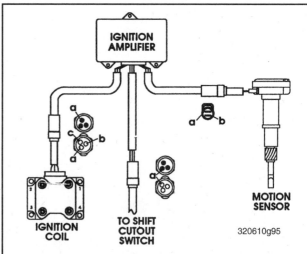

Fig. 95 Connector terminal identification to be used while testing the DDIS system

the right side (3 is on top). Secondary coil resistance for both tests should be 11,300–15,500 ohms.

4. With the connector in hand, now check the resistance between terminal **A** and an engine ground. Now do the same with each coil tower. Minimum coil insulation resistance should be 10 megohms.

MOTION SENSOR

■ **The motion sensor is set at the factory and not adjustable. Never remove the sensor cover unless you intend to replace it**

With the ignition switch in the OFF position, unplug the sensor-to-amplifier connector. Measure the resistance between terminals **A** and **B** on the sensor-side connector. If resistance is not 140–180 ohms, replace the motion sensor.

SHIFT INTERRUPTOR SWITCH

Unplug the switch-to-amplifier connection and check continuity between terminal **A** on the switch-side connector and an engine ground. The switch is normally open, but when shifting out of forward or reverse it should momentarily close and then reopen. If not, replace the switch.

Motion Sensor

REMOVAL & INSTALLATION

◆ **See Figures 96 and 97**

1. Locate the sensor near the oil filter and then unplug the connector.
2. Connect a remote starter and bump the engine over until the timing pointer is at TDC. You can use a friend and the ignition key to do this, but it's much easier with the remote starter.
3. Loosen and remove the clamp bolt and then carefully pull out the motion sensor. We recommend NOY turning over the engine while the sensor is out.

To install:

4. Remove the cover of the new motion sensor; or the old one if you removed it without needing to replace it for some reason. Check to be sure the timing mark has not moved from the TDC position. If necessary, crank the engine to align the marks. The No. 1 piston may be at the top of the compression or exhaust stroke—TDC—it makes no difference with the DDIS system—because of the "twin" firing.
5. Align the hole—not the roll pin hole—at the base of the gear with the notch on the lower housing. Install a new gasket. Lower the sensor into the block. If the slotted end of the sensor fails to index with the oil pump, use a long slotted screwdriver to realign the oil pump.
6. The sensor is correctly installed when the small drain hole in the housing faces directly away from the engine and the dimple on the reluctor aligns with the center of the crankshaft position sensor. Secure the sensor in place with the clamp and bolt. Tighten the bolt to 20 ft lb (27Nm). Install the cover over the motion sensor and tighten the bolts securely.

Ignition Amplifier

REMOVAL & INSTALLATION

1. Disconnect the negative battery cable.
2. Tag and unplug the three electrical connectors.
3. Remove the three mounting bolts and lift out the amplifier.
4. Install the amplifier and tighten the mounting bolts securely.
5. Reconnect the electrical leads and connect the battery cable.

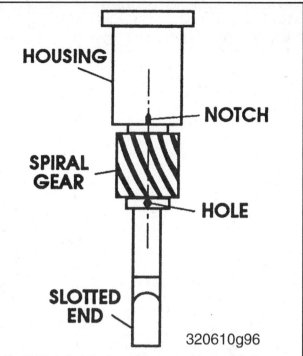

Fig. 96 The hole on the base of the gear and the notch in the housing must be aligned before inserting the sensor into the cylinder block

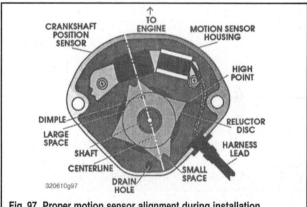

Fig. 97 Proper motion sensor alignment during installation

Ignition Coil

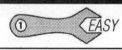

REMOVAL & INSTALLATION

1. Disconnect the negative battery cable.
2. Disconnect the four high tension lead coming from the spark plugs. Make sure you tag them properly for installation purposes.
3. Tag and disconnect the coil electrical leads from the ignition amplifier.
4. Remove the four coil mounting bolts and lift out the coil.
5. Install the coil and tighten the mounting hardware securely.
6. Connect the electrical lead and the four high tension leads.
7. Connect the battery cable.

ELECTRONIC SPARK TIMING SYSTEM (EST)

Description

Basic components of this system include a pointless distributor, a remote ignition coil, an electronic ignition module and a pick-up coil. The distributor does not contain breaker points, a condenser or a centrifugal advance arrangement. As the name of the system implies, spark advance and ignition timing are handled electronically.

The EST uses a magnetic pulse generator and electronic module for the primary circuit current. The secondary circuit is the same as with any standard ignition system with an ignition coil, high tension leads and spark plugs.

A trigger "wheel" is mounted near the upper end of the distributor shaft. This "wheel" has "teeth"—the number corresponding to the number of engine cylinders. An ignition module is mounted close to the wheel. This module contains a sensor. The "wheel" and the ignition module replace the breaker points and condenser on other distributors. Spark advance is controlled by the module.

The EST system contains two separate circuits—a primary circuit—and a secondary circuit. The primary circuit handles the low voltage current supplied by the battery or by the alternator. Major components of this circuit are the primary winding of the ignition coil, the electronic control module including the sensor and the low tension wiring. The secondary circuit handles the high voltage surges produced by the ignition coil. This high voltage is routed to each spark plug through the high tension leads. Components of this circuit include the secondary winding of the ignition coil, the high tension leads between the coil and the distributor and between the distributor, the high tension leads to each spark plug, the distributor rotor, and the distributor cap.

This EST system operates on the principle of a metal detector. The detector is the sensor in the module and the detected metal is the "tooth" in the trigger "wheel".

As the "wheel", near the upper end of the distributor shaft, rotates; the sensor detects the metal "tooth"; the oscillator in the ignition module is at a low level; the transistor is off; and primary current does not flow. If this were a breaker point system—the points would be open.

After the "tooth" of the "wheel" passes the sensor; the oscillator is at a high level; the transistor is on; and current flows in the primary winding. With a breaker point system—the points would be closed.

The high voltage surge developed in the secondary winding of the ignition coil is transferred to the proper spark plug through the cable; the center of the distributor cap; the distributor rotor; to the proper cap contact point when the rotor aligns with the point; and finally through the high tension lead to the spark plug in the cylinder.

This sequence is repeated as each "tooth" on the trigger "wheel" passes the sensor.

Troubleshooting

◆ See Figures 98, 99 and 100

✳✳ WARNING

This Delco EST breakerless ignition system requires the use of a jumper wire—MerCruiser P/N 91-818812A1—or one may be fabricated using a six inch piece of 16AWG wire and a male bullet terminal on each end. This jumper is required to shunt (turn off), the electronic spark advance function in the module while setting initial timing.

✳✳ CAUTION

Always ground the high tension of the coil any time it is disconnected from the distributor. Failure to disconnect the lead could result in fire or an explosion, if gas vapors are present. Reason: A high voltage discharge in the secondary circuit of the coil may occur when the ignition switch is turned to the on and off positions. This high voltage discharge may occur even if the engine is not cranked.

Before spending too much time and money attempting to trace a problem to the ignition system, a compression check of the cylinders should be made. If the cylinders do not have adequate compression, troubleshooting and attempted service of the ignition or fuel system will fail to give the desired results of satisfactory engine performance.

VISUAL CHECKS

Check the coil tower for carbon tracking. Check the terminals for secure connections. Verify the polarity is correct. Check the coil nipple to be sure it is sealed and insulated. If flashover should occur at this location, the engine will fail to start.

Check the distributor cap for carbon tracking. Clean the cap if it is dirty. Moisture and dirt make a good path for flashover. If a carbon track has started, the cap must be replaced. Check the rotor for carbon tracking and cleanliness. Again, if carbon track has started, the rotor must be replaced.

Check the high tension leads to the spark plugs for burning, cracking and any type of deterioration. Check the spark plug for fouling, proper gap and possible cracked insulator.

IGNITION COIL

◆ See Figure 101

Connect a multi-meter and set it to the Rx100 scale.

✳✳ CAUTION

Disconnect both connectors from the coil before performing the following tests.

1. Check for short to ground: Connect one meter lead to the frame as a ground. Make contact with the other meter lead to the Purple wire terminal (**B**). The meter should indicate an infinite reading. If the meter indicates less than an infinite reading, the ignition coil has a short to ground. The coil must be replaced.

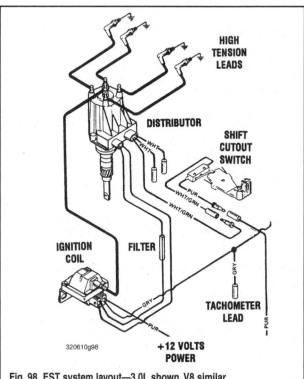

Fig. 98 EST system layout—3.0L shown, V8 similar

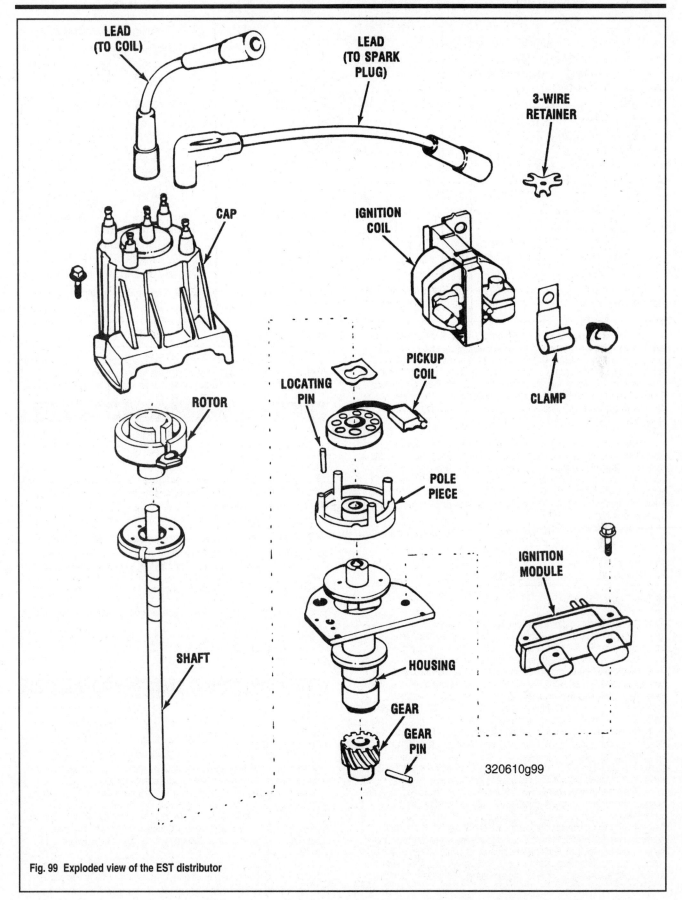

Fig. 99 Exploded view of the EST distributor

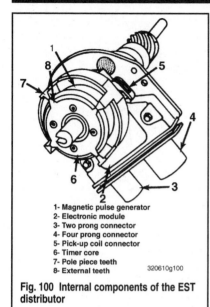

1- Magnetic pulse generator
2- Electronic module
3- Two prong connector
4- Four prong connector
5- Pick-up coil connector
6- Timer core
7- Pole piece teeth
8- External teeth

320610g100

Fig. 100 Internal components of the EST distributor

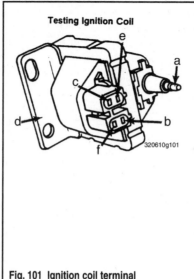

Testing Ignition Coil

320610g101

Fig. 101 Ignition coil terminal identification

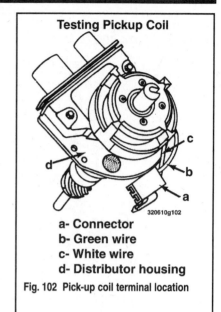

Testing Pickup Coil

320610g102

a- **Connector**
b- **Green wire**
c- **White wire**
d- **Distributor housing**

Fig. 102 Pick-up coil terminal location

2. Check for open or short in the Secondary circuit. Connect one meter lead to the Purple wire terminal (**B**). Make contact with the other meter lead to the high tension terminal (**A**). The meter should indicate 7800–8800 ohms. If the meter indicates less than 7800 ohms, the circuit has a short. If the meter indicates more than 8800 ohms, the circuit probably has an open. Either condition demands the coil must be replaced.

3. Connect one meter lead to the Purple wire terminal (**B**) and the other to terminal **C**. Reading should be approximately 0.4 ohms; if not, replace the coil.

4. Check for open or short in the Primary circuit: Set the ohmmeter to the Rx1 scale. Connect one meter lead to the Purple wire terminal (**B**). Make contact with the other meter lead to the Gray wire terminal (**F**) from the tach lead. The meter should indicate 0.35–0.45 ohms. If the meter indicates higher ohms, the primary circuit probably has an open. If the meter reading is low, the primary circuit has a short. Either condition requires that the coil must be replaced.

5. Install the black coil connector (distributor harness) first and then install the gray connector (engine harness).

PICK-UP COIL

◆ **See Figure 102**

1. Connect a multi-meter as per the manufacturers instructions.
2. Remove the screws securing the distributor cap, and then remove the cap and the rotor. Release the locking tab and unplug the pickup coil connector. On almost all models these two wires should be white and green.
3. Check for short to ground: Set the ohmmeter to the Rx10 scale. Connect one meter lead to the distributor body. Make contact with the other meter lead to the white wire terminal (**C**) of the pick-up coil. The meter should indicate an infinite reading. A reading of less than infinite, indicates the coil has a shorted circuit. The coil must be replaced.
4. Check for short to ground: Set the ohmmeter to the Rx10 scale. Connect one meter lead to the distributor body. Make contact with the other meter lead to the green wire terminal (**B**) of the pick-up coil. The meter should indicate an infinite reading. A reading of less than infinite, indicates the coil has a shorted circuit. The coil must be replaced.
5. Check for open or shorted coil: Set the ohmmeter to the Rx100 scale. Connect one meter lead to the green wire pickup coil terminal (**B**). Make contact with the other meter lead to the white wire coil terminal (**C**). The meter should have a constant indication of 500–1500 ohms. If the meter indicates more than 1500 ohms, the coil probably has an open circuit and the coil must be replaced. If the meter indicates less than 500 ohms the coil probably has a shorted circuit and the coil must be replaced. Make sure that you bend the leads slightly while testing to determine any intermittent open circuits.

IGNITION MODULE

The ignition module in the distributor either produces a spark or it fails to produce an adequate spark. If all other tests have been performed with satisfactory results and the problem still exists—replace the ignition module.

Distributor Cap

REMOVAL & INSTALLATION

① EASY

■ **It is not necessary to remove the spark plug leads at the cap when removing it. If you do remove them, be sure to tag each of them to ensure correct installation.**

1. Disconnect the battery cables.
2. Remove the distributor cap mounting screws and lift off the cap. Some caps may have a vent connection; if yours does, unplug the hose.
3. Clean the cap with warm soapy water—if you have not removed the spark plug wires, you must, obviously, do it at this time. Blow it dry with compressed air; or if not available, make sure it air dries completely prior to installation.
4. Check the cap contacts for excessive burning, wear or corrosion. Check the cap for cracks or wear.
5. Install the cap by fitting the tab into the notch in the distributor housing and tighten the screws securely. Reconnect the plug wires if disconnected.

Distributor

REMOVAL

② MODERATE

1. Begin by disconnecting the high tension leads from the distributor. Take time to identify and tag each lead as an aid during installation to ensure they are properly installed for correct cylinder firing.
2. Lift the locking tabs and disconnect the 2-and 4-terminal connectors. Rotate the crankshaft until No. 1 cylinder is in the firing position—both valves for No. 1 are closed and the timing mark on the harmonic balancer is aligned with the 0° mark on the grid attached to the front chain cover.
3. Remove the two screws securing the distributor cap in place, and then remove the cap.
4. Now, note the position of the rotor tip. Take time to make a reference mark on the distributor housing to enable the rotor and the housing to be properly aligned during installation.

5. Make a mark on the distributor base and a matching mark on the engine as an aid during installation to ensure the distributor will be installed back in its original position.

■ **Take care to prevent the crankshaft from being rotated—even slightly—while the distributor is out of the block. If the crankshaft should be rotated—follow the procedures listed under Timing Out Of Synchronization later in this section.**

6. Remove the distributor clamp bolt and lift the distributor straight up and clear of the engine. The gasket should be discarded.

INSTALLATION

Engine Not Disturbed

1. Perform the following procedures if the distributor and engine were marked prior to removal and there is no evidence to indicate that the crankshaft was rotated while the distributor was removed. If the distributor was not marked, as instructed in the removal procedures, or if the crankshaft was rotated—even slightly—while the distributor was out, proceed to the Engine Disturbed section.

2. On the 3.0L engine, install the rotor and rotate it approximately 1/8 turn counterclockwise from the reference mark made prior to disassembling. Rotor rotation at this time is necessary because as the distributor shaft gear indexes with the camshaft gear, the shaft will rotate clockwise about 1/8 turn—the rotor will then be back where it should be for the No. 1 cylinder to fire.

3. Position a new distributor gasket in place on the engine. A reference mark should have been made on the distributor and the block prior to removal, as instructed in the removal procedures. Install the distributor into the block with the mark on the distributor roughly aligned with the mark on the block. Push the distributor fully into the block until the housing is seated.

✳✳ WARNING

If necessary on the 3.0L engine, rotate the rotor slightly to permit the gear on the lower end of the distributor shaft to index with the gear on the camshaft. The "spade" on the lower end of the distributor shaft must engage the slot for the oil pump. Failure to index would result in no engine oil circulation—disaster. The distributor shaft collar should now be fully seated on the block. However, once the distributor is in place the rotor reference marks and the distributor housing marks should both be aligned. On the V8, if the distributor won't seat properly, remove it and insert a screwdriver into the hole to turn the oil pump driveshaft slightly to assist proper alignment.

4. Secure the distributor in place with the hold down clamp and bolt. Tighten the bolt to 20 ft. lbs (27 Nm) on the 3.0L and 30 ft. lbs. (40 Nm) on the V8.

5. Connect the 2-and 3-wire leads into the distributor.

6. Install the cap and tighten the screws securely. The cap can only be installed properly—one way. Lubricate the sockets in the distributor cap with Quicksilver 2-4-C lubricant, or the equivalent. Install the spark plug high tension leads (if they were removed) using the identification made during removal to ensure proper cylinder firing.

7. Check the ignition timing.

Engine Disturbed

The following procedures are to be performed if reference marks were not made for the rotor, the distributor housing and the engine block or if the crankshaft was rotated while the distributor was out of the block.

1. Rotate the crankshaft until the No. 1 cylinder is ready to fire—both valves are closed and the timing mark on the harmonic balancer is aligned with the 0° mark on the grid attached to the timing cover.

2. Position a new gasket in place on the engine block. Install the distributor into the block.

■ **On the 3.0L, if it is not possible to fully seat the distributor in place on the block, press down lightly on the distributor housing and at the same time rotate the rotor slightly. This action will permit the gear on**

the lower end of the distributor shaft to index with the camshaft gear. The "spade" on the lower end of the distributor shaft should engage the slot for the oil pump. The distributor shaft collar should now be fully seated on the block. On the V8, if the distributor won't seat properly, remove it and insert a screwdriver into the hole to turn the oil pimp driveshaft slightly to assist proper alignment.

3. Once the distributor is fully seated, install the clamp and bolt, but leave it just loose enough to permit rotating the distributor with strong hand pressure. At this point, the rotor must be in position to fire No. 1 cylinder.

4. Just place the cap in position on the distributor. Scribe a mark on the distributor housing aligned with the No. 1 spark plug terminal. Now, remove the cap and verify the rotor is aligned with the mark just made for the No. 1 spark plug terminal. If the rotor does not align with the mark, rotate the distributor housing—with difficulty, because the clamp was not tightened—right or left to align the rotor with the mark for the No. 1 terminal. If it is not possible to align the rotor with the No. 1 mark, the distributor must be removed the entire procedure started over again.

5. Tighten the distributor clamp bolt to 20 ft. lbs (27 Nm) on the 3.0L and 30 ft. lbs. (40 Nm) on the V8.

6. Check all high tension leads and connect them in the proper sequence for correct cylinder firing.

7. Connect the 2-wire engine harness connector to the distributor.

8. Check the ignition timing.

DISASSEMBLY

◆ **See Figures 103. 104, 105, 106, 107, 108, 109 and 110**

Remove the distributor from the engine, as outlined in the previous section. The following procedures pick up the work after the distributor has been removed and is on a suitable work surface.

Fig. 103 A good view of the EST distributor

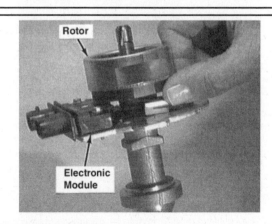

Fig. 104 Lift off the rotor

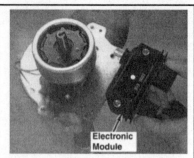

Fig. 105 Removing the ignition module

Fig. 106 Driving out the roll pin

Fig. 107 Close up of the pin and the gear

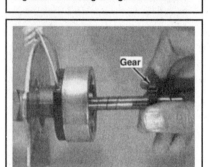

Fig 108 Removing the gear

Fig. 109 Pull the shaft up through the coil

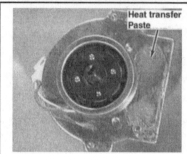

Fig. 110 Coat the plate with grease

1. Pull the rotor free from the upper end of the distributor shaft.

2. Disconnect the leads from the ignition module, and then pry the retainer free. Remove the pick-up coil.

3. Disconnect the ignition module lead. Remove the two screws securing the module to the distributor, and then remove the module. It may be necessary to carefully pry the module loose from the distributor.

4. Make a mark on the drive tang at the lower end of the distributor shaft and another mark on the collar of the gear aligned with the first mark as an aid to installing the gear back in its original position. After the marks have been made, drive the roll pin free of the gear with a small (4.5mm) punch.

5. Slide the gear free of the shaft, and then separate the shaft by pulling it up through the top of the pick-up coil.

6. Pry off the retainer on the end of the housing securing the shield, coil and pole piece. Slide them all off and discard the retainer.

CLEANING AND INSPECTING

Obtain a clean cloth and wipe the distributor cap clean. Inspect the cap for chips, cracks and any sign of a carbon path. If such a path is discovered, the cap must be replaced because such a path would permit high tension leakage.

Clean loose corrosion from the terminal segments inside the cap. Do not use emery cloth or sandpaper. If the segments are deeply grooved, a new cap should be installed.

Inspect the terminal sockets for corrosion. Clean the sockets using a stiff wire brush to loosen the corrosion. After the sockets are clean, lubricate them with Quicksilver 2-4-C lubricant, or equivalent.

Inspect the rotor for cracks. Check to be sure the tip of the rotor is not badly burned. Such a condition demands the rotor be replaced.

Inspect the trigger wheel for any sign of contact with the sensor. If the distributor has been removed from the block, make an attempt to check the distributor shaft for excessive wear between the shaft and the bushings in the housing.

ASSEMBLY

MODERATE

The following procedures pickup the work after disassembled parts have been cleaned and replacement items have been obtained and are on hand.

1. Align the pole piece with the housing and slide it onto the end of the housing. Align the tab on the coil with the hole in the base of the housing.

2. Install the shield and a new retainer over the end of the housing. Position a 5/8 in. socket over the retainer and lightly tap it into place over the end of the housing—make sure it is seated firmly into the groove.

3. Coat the distributor shaft with oil and slide into and through the center of the housing. Spin it several times to lubricate it and the housing.

4. Slide the star washer onto the shaft so that the teeth are pointing up toward the top of the housing. Slide the flat washer and the gear onto the shaft so that the alignment marks made earlier line up. Insert the roll pin and drive it down until it is flush.

5. Make sure the surface of the distributor plate is clean and then spread an even coat of Thermal Conductive grease onto the bottom of the module. Position the module on the plate and tighten the two screws securely; clean any excess grease from the plate and module.

6. Connect the lead from the coil to the back of the module and lock it in place.

7. Align the rotor with the notch in the distributor shaft, and then press it to secure it in place.

Ignition Coil

REMOVAL & INSTALLATION

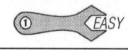

EASY

1. Disconnect the negative battery cable.
2. Disconnect the high tension lead coming from the distributor.

3. Tag and disconnect the coil electrical leads (two).
4. Remove the two coil mounting bolts and lift out the coil with the bracket attached.

5. Install the coil and tighten the mounting hardware securely.
6. Connect the electrical leads and the high tension lead.

7. Connect the battery cable.

THUNDERBOLT IGNITION SYSTEM

Description

ELECTRONIC IGNITION

Engine performance and efficiency are, to a large degree, governed by how fine the engine is tuned to factory specifications, as determined by the designers. The service work outlined in this section must be performed in the sequence given and to the specifications listed.

The ignition system still consists of both primary and secondary ignition circuits. The primary circuit carries the low voltages of the ignition system while the secondary circuit carries the high voltages. Components of the low voltage or primary circuit include the battery, ignition switch, distributor, pick-up coil or pulse generator, and a control module. The secondary circuit components are the ignition coil, distributor cap, rotor, high tension spark plug wires, and spark plugs.

In an electronic ignition system, the cams and points in a conventional point distributor have been replaced by a magnetic pulse generator, sensing coil, and electronic module mounted either internally inside the distributor or externally. A dry type or oil filled ignition coil provides the high-energy voltage to the spark plugs.

The distributor does not contain centrifugal weights or a vacuum diaphragm for timing advancement. This function is now controlled by the distributor electronic module or Engine Control Module (ECM) on fuel injected models. The module contains all the electronics for system operation.

The electronic module is sealed in an epoxy based plastic, therefore it is non-serviceable. The plastic protects the electronics within the module from moisture, extreme temperatures, shock and vibration. Major components of the electronic ignition system are the distributor, electronics module, primary coil, high tension leads and spark plugs.

Pulse Generator or Pick-Up Coil

The pulse generator is mounted on the driven shaft of the distributor and rotates with the distributor rotor. The pulse generator contains an equally spaced number of teeth, in relation to the number of engine cylinders—eight for eight cylinder engine). When the teeth of the pulse generator are exactly aligned with the teeth on the sensor coil, a voltage pulse is created and sent to the electronic module.

Sensing Coil

The sensing coil is stationary and mounted around the pulse generator/pickup coil. The sensing coil has the same number of teeth as the pulse generator and the teeth are equally spaced around the inside of the sensing coil. A fine wire coil is connected to each tooth. As the pulse generator rotates, a voltage is induced in the sensing coil. When the teeth of both the pulse generator and the sensing coil are exactly aligned, the voltage within the coil goes to "0". This voltage drop is detected by the electronic module as a pulse. The pulse signal turns a transistor ON and OFF within the electronic module.

Electronic Module

The electronic module houses a number of integral circuits made up of transistors, diodes, and capacitors. In addition, there is a current limiting circuit that regulates the primary current in the ignition coil to a maximum of 5.5 amperes.

When the teeth of both the pulse generator and sensing coil are aligned, a pulsed voltage signal is sent to the module. A transistorized circuit within the module will turn OFF and stop the flow of current to the primary windings of the ignition coil. This action is the same as points opening in a conventional point ignition circuit. When the teeth of the sensing coil and pulse generator are no longer aligned, the voltage through the sensing coil to the electronic module will turn ON the transistorized circuit, again allowing current flow to the primary windings of the ignition coil. This action is the same as points closed in a conventional point circuit. Each time the transistorized circuit turns OFF the flow of current through the coil ceases. As the field in the coil collapses, a high voltage is generated. This voltage is then sent from the coil, through the secondary high voltage cable, to the distributor cap. At this time the end of the rotor should be pointing to the

corresponding terminal in the distributor cap. The high voltage jumps across the rotor to the terminal in the cap and flows down the high tension lead to the spark plug. The voltage leaps across the tip of the spark plug creating a spark and igniting the compressed gasses.

Normal ignition timing advancement is controlled by the electronic module for carburetor models or ECM on fuel injected models. As the pickup coil pulse rate changes with engine speed, the timing is advanced up to a total of 27 degrees.

The electronic module is mounted on a plate directly under the pulse generator/pickup coil. Due to the extreme heat generated within the module, the air gap between the module and mounting plate must be filled with a silicone grease or heat transfer paste. Failure to fill this void will result in insufficient heat transfer from the module to the mounting plate, resulting in failure of the electronic module.

Primary Coil

The primary coil develops the high voltage required to jump the gap between the spark plug electrodes. The ratio of the secondary windings to the primary windings in the coil increases the voltage to about 35 kVA.

On HEI ignition systems, the coil is a dry-type rather than an oil filled coil that provides the normal cooling of the coil. The large mounting bracket and exposed surface provides the necessary heat sink to cool a dry type coil.

High Tension Leads

The high tension leads are of a carbon fiber core, radio suppression, 7mm or 8mm AWG wire. These wires are designed and constructed to handle the increased voltages of modern high energy ignition systems and reduce the possibility of a leakage path.

THUNDERBOLT IV AND V

◆ See Figure 111

The Thunderbolt IV is an electronic transistorized High Energy Ignition (HEI) system. This system is quite different from the standard point-type ignition systems, discussed in the previous paragraphs. This ignition system was used on all V6 and V8 engines from 1992–1994.

In 1995 a revised and improved version of this ignition system was introduced as the Thunderbolt V. These improvements made the Thunderbolt ignition system compatible with the new Electronic Fuel Injection systems used on MerCruiser engines. Both the Thunderbolt IV and V ignition systems operate identically except the Thunderbolt V system has a Knock Control Module to control ignition timing and prevent spark knock.

The system does not utilize breaker points or a mechanical advance mechanism. It was designed to be almost maintenance free without the requirement for periodic adjustment.

The Thunderbolt IV and V ignition systems look very much like a conventional type distributor shaft, housing, cap, and rotor. The ignition coil is mounted externally in close proximity to the distributor. The ignition amplifier or module is mounted on the exhaust elbow for early models or on the side of the distributor housing on later models.

The Thunderbolt V ignition system is easily identifiable because the distributor has a new ignition control module mounted on the side of the distributor and is known as the Digital Electronic Spark Advance (D.E.S.A.). A second module for spark Knock Control is mounted on top of the D.E.S.A. module.

Internally, both distributors are identical. A sensor and sensor wheel are mounted near the top of the distributor shaft under a conventional-type rotor. The ignition sensor performs similar to a switch, when subjected to a magnetic field. The distributor electrical operation uses a Hall effect vane switch assembly, causing the ignition coil to be switched ON and OFF by the ignition module on a Thunderbolt IV system and by the ECM on a Thunderbolt V ignition system. The sensor has a Hall switch on one side and a permanent magnet on the opposite side. The rotary sensor wheel is made of a ferrous metal and triggers the on and off signal. When a window opening of the sensor wheel cup is in front of the magnet and the Hall effect switch, a magnetic flux field is completed from the magnet through the Hall effect switch and back to the magnet.

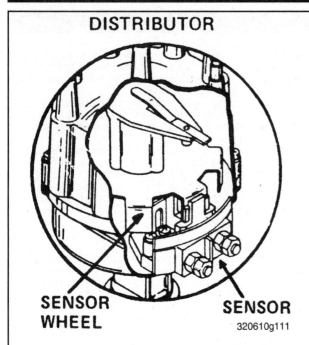

DISTRIBUTOR

SENSOR WHEEL

SENSOR

320610g111

Fig. 111 Thunderbolt systems use a sensor and wheel in place of points

As the sensor wheel cup rotates and closes the window opening, the flux lines are shunted through the vane and back to the magnet. During this time, a voltage is produced as the vane passes through the window opening. When the vane clears the opening, the window edge causes the signal to go to 0 volts. This pulsed signal controls the firing of the ignition coil on the Thunderbolt IV system.

The pulsed signal provides a "REF HI" and "REF LO" signal to the ECM on Thunderbolt V ignition systems. These two signals from the distributor to the ECM provide a precise indication of engine speed to the ECM for spark and timing control. Two additional lines called "IC" and "BYPASS" between the distributor and the ECM are used to sense and control ignition module functions.

Ignition timing advancement on the Thunderbolt V system is normally controlled by the ECM. Should a failure occur or a fault be detected within the ECM, the ignition module will take control of the ignition timing causing the engine to operate at a reduced power output.

Initial timing on the Thunderbolt IV system is adjusted by rotating the distributor housing in the same manner as with a conventional ignition system. Idle timing must be set below speeds of 900 rpm.

To change the base timing on engines equipped with the Thunderbolt V ignition system and ECM, requires the use of a scan tool or code tool. Once the tool is connected into the DTC connector and set to on, the ECM will shift into Ignition Control Mode and the distributor ignition module will shift to base timing. The initial timing can now be adjusted by rotating the distributor, as with other ignition systems.

The Thunderbolt IV ignition module provides for ignition advance. After initial timing, the advance starts at approximately 1000 rpm and increases to a maximum of 14° advance (plus initial setting of 8° for a total of 22°), between 1100 and 2500 rpm. If the key is left in the on position when the engine is at rest, the coil primary current will automatically turn off in a short time to protect the ignition coil.

The ignition module will withstand a reverse battery connection for only about a minute, before it is severely damaged.

The ignition coil appears to be a standard size and shape. However, the HEI coil has a special winding and core. If a standard coil is used, the ignition module will not be damaged, but it will supply a low output and will overheat.

Standard tachometers and most synchronizers monitoring ignition impulses at the negative terminal of the coil will operate satisfactorily with the Thunderbolt IV and V ignition systems.

Troubleshooting

The only equipment needed to troubleshoot the Thunderbolt IV or V ignition systems are an analog or digital voltmeter (multi-meter) and a spark gap tester.

✳✳ WARNING

Check to be sure the engine compartment is well ventilated and free of any gasoline vapors before starting any of the following tests. A spark will be generated creating a potential fire hazard if fuel vapors are present.

1. Check to be sure the battery is up to full charge. If not, correct the condition by charging the battery or making a substitution. Check to be sure the distributor, and clamp bolt are tight. If loose, rotate the distributor and check the timing.

2. Check all terminal connections on the distributor, ignition module and ignition coil. Verify the Gray Tachometer lead from the coil (-) side, to the tachometer on the control panel is not shorted to ground. Temporarily disconnect the tachometer lead if it is suspected of having a short.

3. Set the ignition key switch to the RUN position. Connect a multi-meter and set the switches for 12V DC reading. Attach the Red meter lead to the positive terminal of the ignition coil and the Black meter lead to a good engine ground. The meter should indicate 12 volts. If no voltage is present, check the engine and instrument wiring harness, battery cables, and the key switch for damaged wiring or unplugged harness connectors. If 12 volts is indicated on the meter and a Thunderbolt IV ignition system, is being serviced—proceed to Step 7. If 12 volts is indicated on the meter and a Thunderbolt V ignition system is being serviced—proceed to Step 5.

4. Connect the Red meter lead to the White/Red terminal on the distributor. Connect the Black meter lead to a good engine ground. If the meter indicates 1–12 volts, proceed to Step 6. If no voltage is indicated continue with this step. Unplug the White/Red bullet connector from the distributor terminal. Connect the Red meter lead to the ignition module side of the White/Red lead. Connect the Black meter lead to a good engine ground. If there is a voltage indication, replace the ignition sensor in the distributor. If there is no voltage, check the module wire harness for damage wiring and repair or replace the harness. If the wiring is good, then replace the defective ignition module and repeat this test. If 12 volts is present, proceed to Step 6.

5. Disconnect the White/Red wire lead from the distributor bullet connector. Connect the Red meter lead to the ignition module side of the White/Red lead. Connect the Black meter lead to a good engine ground. The meter should indicate 12 volts. If no voltage is present, replace the ignition module. If 12 volts is present, proceed to Step 6.

6. Connect the White/Red bullet connectors at the distributor, if previously disconnected. Remove the high tension lead from the distributor to the coil. Insert a spark gap tester from the coil tower to ground. Disconnect the White/Green lead from the distributor terminal. Turn the ignition switch to the RUN position. Now, with a rapid motion, strike the White/Green lead coming from the distributor 2–3 times per second against a good engine ground. If there is spark at the coil, replace the defective ignition sensor in the distributor. If there is no spark at the coil, proceed with Step 7.

7. Substitute a new or known good ignition coil and repeat Step 6. If there is now spark at the coil, install a new ignition coil. If there is no spark at the coil, replace the defective ignition module.

KNOCK CONTROL MODULE

The knock control module contains the circuits that monitor AC voltage signal output from the knock sensor. If no spark knock is present, the knock control module supplies 8–10 volts to the ignition control module and normal timing advancement will occur. However, if spark knock is detected by the sensor, the knock control module will remove the 8–10 volt signal to the ignition control module and the ignition timing will then be retarded to eliminate the spark knock.

Inspect the routing of the Blue wire lead from the knock sensor to the module. If this wire is too close to a high tension lead, the knock module may interpret this interference as a knock signal, which could result in a false retard timing.

1. Obtain a multi-meter and an un-powered test light. The test light should be low power and use less than 300mA of current draw. Set the switches on the meter to auto range or 12 volt DC reading.

✳✳ WARNING

Water must circulate through the lower unit to the engine any time the engine is run to prevent damage to the water pump mounted on the engine. Just a few seconds without water will damage the water pump.

2. Start the engine and allow it to warm to normal operating temperature.

3. Connect the Red lead from the DVOM to the Purple/White bullet connector between the ignition control and knock control modules. Connect the Black lead from the DVOM to a good engine ground. The meter should indicate 8–10 volts. If there is no voltage, move the Red meter lead to terminal <B>B</B> on the knock control module connector plug, or probe the Purple wire to the connector. The meter should indicate 12 volts power to the knock control module. If the meter indicates 12 volts to the module, but no output from the module when connected to the Purple/White lead, then the knock control module is probably defective. Continue with the remainder of the test to eliminate all probable failures before replacing the knock control module.

4. Connect the Red meter lead back to the Purple/White bullet connector. If there is no voltage indicated on the Purple lead, troubleshoot the engine harness for an open circuit or damaged wiring.

5. Disconnect the Blue wire connector from the knock sensor on the side of the engine block. Connect the test light to a 12 volt power source. Move the throttle remote control until engine speed is slightly above 1500 rpm. Now, simulate an AC voltage signal from the knock sensor by tapping on the knock sensor harness terminal with the test light while observing the meter. If a voltage drop is observed, the wiring from the sensor to the knock control module is good but the knock sensor could be defective. If no voltage drop is observed, check the Blue wire from the sensor to the knock module for a short or open circuit.

6. Connect the Blue wire terminal onto the knock sensor. Using a small hammer, tap rapidly on the side of the engine block near the knock sensor. If a voltage drop is observed on the meter, the knock sensor and wiring is functioning properly. If no voltage drop is observed on the meter, then the knock sensor is defective and should be replaced.

7. Place the remote control throttle in the idle position. Shut down the engine and disconnect the DVOM meter and test light.

PRECAUTIONS

The following safety precautions should always be practiced when servicing or testing any portion of the ignition system.

1. Check to be sure the engine compartment is well ventilated. After the engine hood has been removed, allow some time for any existing fumes to dissipate and be replaced with fresh air. Service work should not be started if any gasoline vapors are present to prevent the possibility of fire.

2. Work slowly and deliberately—thinking of the next movement to made before proceeding with the work.

3. Take extra precautions to keep hands, feet and clothing completely clear of any moving part.

4. Be especially careful not to touch or disconnect any ignition component while the engine is operating.

5. Never reverse the battery connections for any reason. The electrical and ignition systems comprise a negative (-) ground.

6. Never disconnect either end of the battery cables while the engine is operating.

■ **A careful inspection should be made of the wiring during troubleshooting and before expensive parts are purchased for replacement. The manufacturer uses specially designed features and environmental protection extensively to protect ignition contacts. The following areas could be considered likely causes of an electrical or ignition problem:**

7. Damaged contacts caused by improper engagement.

8. The mating surfaces of contacts contaminated by corrosion, sealer or other foreign material.

9. Terminals not fully seated.

10. Microscopic damage or holes in an electrical lead allowing water to enter and cause corrosion, or actual circuit failure.

Distributor Cap

REMOVAL & INSTALLATION

◆ **See Figures 112, 113 and 114**

■ **It is not necessary to remove the spark plug leads at the cap when removing it. If you do remove them, be sure to tag each of them to ensure correct installation.**

1. Disconnect the battery cables.

2. Remove the four distributor cap mounting screws and lift off the cap. Some caps may have a vent connection; if yours does, unplug the hose.

3. Clean the cap with warm soapy water—if you have not removed the spark plug wires, you must, obviously, do it at this time. Blow it dry with compressed air; or if not available, make sure it air dries completely prior to installation.

4. Check the cap contacts for excessive burning, wear or corrosion. Check the cap for cracks or wear.

5. Install the cap by fitting the tab into the notch in the distributor housing and tighten the screws securely. Reconnect the plug wires if disconnected.

Rotor/Sensor Wheel

REMOVAL & INSTALLATION

◆ **See Figures 115, 116 and 117**

1. Remove the distributor cap.

■ **The rotor and sensor wheel are mounted to the distributor shaft with Loctite; you may wish to carefully use a heat gun or hair drier before prying the assembly off of the shaft**

2. Position two small flat blade screwdrivers underneath the sensor wheel so the end of the blade is against the distributor shaft. The screwdrivers should be 180º apart. Press down on each screwdriver (using the housing as a fulcrum) and pry the assembly up and off the shaft..

3. Inspect the locating key in found in the ramp at the bottom of the splined hole in the assembly. It should be about 1/8 in. (3mm) wide and should not be shaved or burred in any way.

4. Check the rotor for burns, corrosion, cracks and carbon tracks. Replace the rotor if necessary—just remove the three small hex bolts securing it to the sensor wheel. When installing the new rotor to the wheel, make sure that the pin on the rotor fits into the locating hole on the wheel and then tighten the three bolts securely.

5. Check that the carbon brush tang gap between it and the rotor is 1/4 in. (6.4mm). It can be adjusted by carefully bending it until the correct gap is achieved.

6. Squeeze two drops of Loctite 271 onto the locating key in the rotor and then squeeze two more drops into the keyway on the upper edge of the distributor shaft and then immediately install the rotor assembly onto the shaft. Ensure that the key is positioned fully in the keyway and then press the assembly completely down with the palm of your hand until it seats fully.

■ **Let the Loctite cure overnight before operating the engine. It is a good idea to allow it to cure with the distributor upside down so no sealant can run into the housing bushing. Yes, this means you should remove the distributor.**

7. Install the distributor cap and connect the battery cables.

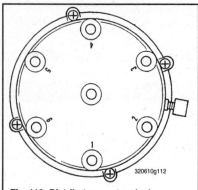

Fig. 112 Distributor cap terminal locations—V6

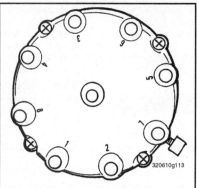

Fig. 113 Distributor cap terminal locations—V8 (left hand rotation)

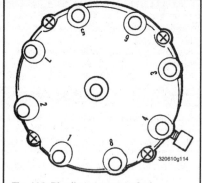

Fig. 114 Distributor cap terminal locations—V8 (right hand rotation)

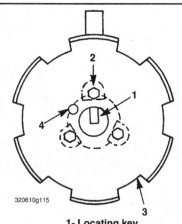

1- Locating key
2- Bolts (hex head)
3- Sensor Wheel
4- Locating pin

Fig. 115 Check the sensor wheel indexing by laying it upside down (teeth pointing up) on top of this illustration so that the bolt holes and locating pin hole are aligned. All teeth should be in alignment with those in the illustration if indexed correctly—V6

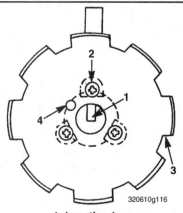

1- Locating key
2- Screws
3- Sensor wheel
4- Locating pin

Fig. 116 Check the sensor wheel indexing by laying it upside down (teeth pointing up) on top of this illustration so that the bolt holes and locating pin hole are aligned. All teeth should be in alignment with those in the illustration if indexed correctly—V8

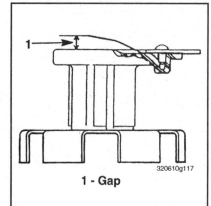

1 - Gap

Fig. 117 Adjusting the carbon brush gap

Sensor

REMOVAL & INSTALLATION

② MODERATE

◆ See Figure 117a

1. Disconnect the battery cables, remove the distributor cap and the rotor/sensor wheel assembly.
2. Remove the two mounting screws and lift the sensor out of the housing.
3. Inspect the two jumper leads hanging from the inside edge of the sensor with a magnifying glass. If either is cracked, replace the sensor.
4. Install the sensor into the distributor housing and tighten the two screws securely. Install the rotor assembly and cap. Connect the battery cables.

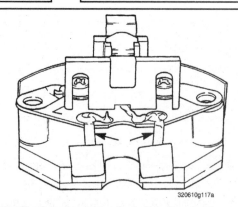

Fig. 117a Check the two jumper leads. Note the mounting screw holes on each side of the sensor

Ignition Module

REMOVAL & INSTALLATION

◆ **See Figure 118**

1. Disconnect the battery cables.
2. Unsnap the wiring harness at the module.
3. Although a few early models had the module mounted on the exhaust elbow, almost all others have it mounted onto the side of the distributor housing. Remove the mounting bolts and pull off the module.
4. Coat the seating base of the module with thermalconductive grease and install the module. Tighten the mounting screws to 10 inch lbs. (1.1 Nm).
5. Reconnect the wiring harness and the battery cables.

Distributor

REMOVAL

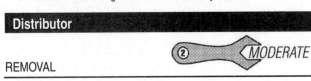

◆ **See Figures 119, 120, 121 and 122**

1. Disconnect the battery cables.
2. Disconnect the high tension leads from the distributor cap and remove the cap. Take time to identify each lead as an aid during the installation to ensure they are properly installed for correct cylinder firing. Also, it's a good idea to paint a little mark on the housing even with there the No. 1 plug wire terminal is. Although this is not necessary, it's a good idea anyway
3. If the distributor has no module(s) mounted on the side, disconnect the wires from the terminals. If the distributor has an ignition or knock module, or both, lift the locking tabs on the module wire connector and disconnect the wire harness from the distributor.
4. Rotate the crankshaft until the No. 1 cylinder is in the firing position—both valves for the No. 1 cylinder are closed and the timing mark on the harmonic balancer is aligned with the 0° mark on the timing grid attached to the front cover. Now, note the position of the rotor tip—it should be pointing toward the spot where the No. 1 cylinder terminal was in the distributor cap.
5. Make a mark on the distributor base and a matching mark on the engine block as an aid during installation to ensure the distributor will be installed back into its original position.

■ **Take care to prevent the crankshaft from rotating—even slightly—while the distributor is out of the block. If the crankshaft should be**

Fig. 118 Some early models had the module mounted on the exhaust elbow

rotated—follow the procedures listed under Engine Disturbed following the assembly instructions.

6. Loosen and remove the distributor clamp bolt and clamp. Lift the distributor straight up and clear of the engine. Remove the gasket from the base of the distributor and discard the gasket.

INSTALLATION

Engine Not Disturbed

1. Install a new gasket over the distributor shaft.
2. Turn the rotor approximately 1/8 of a turn counterclockwise past the mark painted on the housing and then install the distributor slowly into the cylinder block. You may have to wiggle the rotor slightly until it meshes with the camshaft gear to get the unit fully seated. The key is that the rotor ends up pointing to the mark on the housing made during removal.
3. Install the clamp and clamp bolt and tighten it to 20 ft. lbs. (27 Nm).
4. Connect all harness leads and install the cap. Connect the spark plug leads if removed.
5. Connect the battery cables.

Engine Disturbed

1. Set the No. 1 piston to TDC. Remove the No. 1 spark plug, put your thumb over the hole and slowly crank the engine until you feel compression on the No. 1 cylinder. Crank the engine a bit more until the pointer lines up with the mark on the scale.

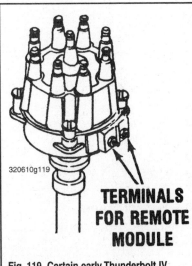

Fig. 119 Certain early Thunderbolt IV engines had the ignition module remotely mounted

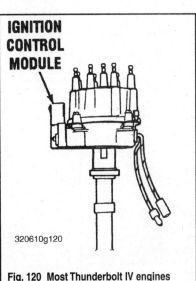

Fig. 120 Most Thunderbolt IV engines had the module mounted on the distributor housing.

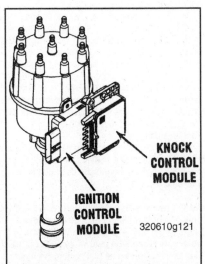

Fig. 121 Thunderbolt V engines had both an ignition module and a knock control module mounted on the housing

Fig. 122 Loosen the clamp bolt to remove the distributor

2. Install a new gasket on the housing.

3. Grasp the rotor and spin the shaft about 1/8 of a turn, counterclockwise, past the alignment mark. Position the distributor over the hole in the block so that the alignment marks match and then work it down into the hole until it engages completely with the camshaft gear.

■ **It is OK to wiggle the rotor slightly during installation so that the distributor gear meshes properly with the oil pump. It is imperative though that the rotor ends up pointing to the alignment mark made during removal when installation is complete.**

4. Install the clamp bolt and tighten it finger-tight. Rotate the distributor carefully until the breaker points just begin to open and then tighten the clamp bolt to 20 ft. lbs. (27 Nm).

5. Install the distributor cap into position and confirm that the rotor points to the terminal for the No. 1 spark plug lead. If it does, go to the next step; if not, repeat the installation procedure.

6. Install the cap and coil/plug leads.

7. Check the ignition timing.

DISASSEMBLY & ASSEMBLY ② MODERATE

◆ **See Figures 123 and 124**

1. Remove the distributor, cap, rotor/sensor wheel assembly and sensor as detailed previously.

2. Carefully clamp the distributor in a vise and drive out the roll pin at the base of the shaft with a small drift.

3. Slide the gear and washer off the end of the distributor shaft. Check for side play between the distributor shaft and the distributor housing bushings by pulling and pushing the shaft sideways in the distributor housing. If the shaft moves sideways in the housing in excess of 0.002 in. (0.05 mm), the bushings are worn and the distributor should be replaced.

4. Pull the distributor shaft out of the distributor housing from the top. Check the distributor shaft for being trueness by placing it in a pair of V-blocks. Place a dial indicator gauge against the shaft and rotate the distributor shaft. If the shaft maximum run-out exceeds 0.002 in. (0.5mm), the shaft is bent and must be replaced.

To assemble:

5. Apply a coating of engine oil to the distributor driveshaft. Install the E-clip onto the shaft, if it was removed during disassembling. Slide the shaft down through the center of the distributor housing until the clip contacts the distributor housing.

■ **If a new driven gear is being installed, the new gear will come with only one hole drilled in it. Slide the new gear onto the driveshaft, aligning the hole in the gear with the hole in the distributor shaft. This hole is offset and the gear will only fit on the shaft in one direction. Use the two holes as a pilot guide and insert a 3/16 in. carbide tipped drill bit through the hole in the gear and shaft. Now, drill the second hole through the other side of the gear.**

6. Slide the washer and driven gear onto the distributor shaft aligning the hole in the gear with the hole in the driveshaft. This hole is offset and the gear will only fit in one direction onto the shaft. With the gear and shaft holes aligned, insert the roll pin. Using a hammer and blunt end punch, tap the roll pin until it is flush with the gear surface.

7. Install the sensor, rotor/wheel, cap and distributor as detailed previously.

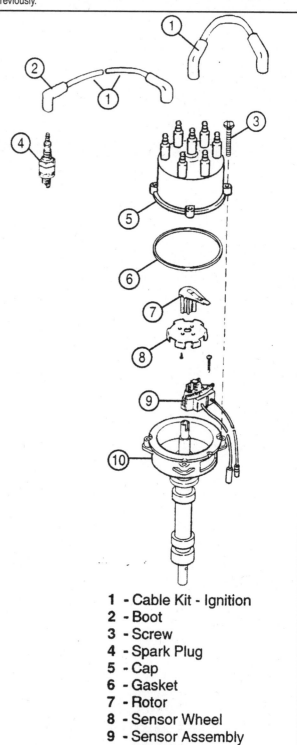

1 - Cable Kit - Ignition
2 - Boot
3 - Screw
4 - Spark Plug
5 - Cap
6 - Gasket
7 - Rotor
8 - Sensor Wheel
9 - Sensor Assembly
10 - Housing

Fig. 123 Exploded view of the Thunderbolt V distributor

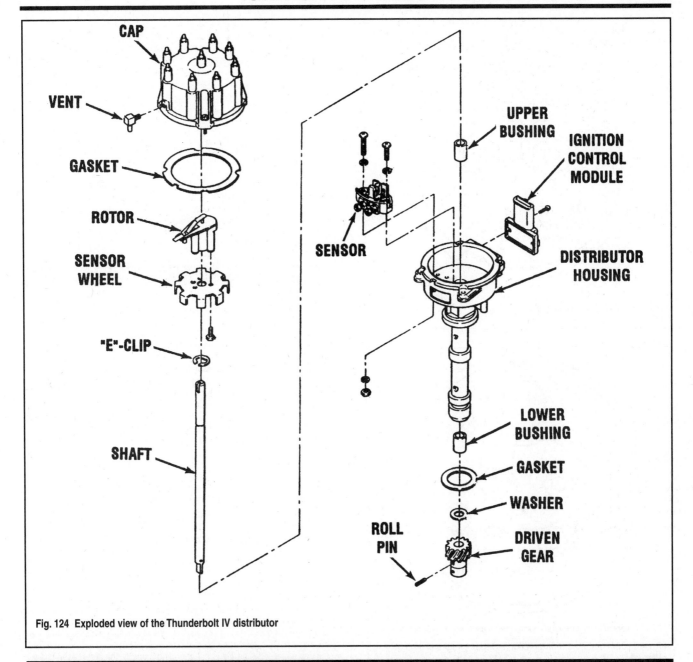

Fig. 124 Exploded view of the Thunderbolt IV distributor

INSTRUMENTS AND GAUGES

Oil and Temperature Gauges

TROUBLESHOOTING

◆ See Figures 125 and 126

The body of oil and temperature gauges must be grounded and they must be supplied with 12 volts. Many gauges have a terminal on the mounting bracket for attaching a ground wire. A tang from the mounting bracket makes contact with the gauge. Check to be sure the tang does make good contact with the gauge.

Ground the wire to the sending unit and the needle of the gauge should move to the full right position indicating the gauge is in serviceable condition.

Check the sender unit for a defective temperature warning system.

If a problem arises on a boat equipped with water, temperature, and oil pressure lights, the first area to check is the light assembly for loose wires or burned-out bulbs.

When the ignition key is turned ON, the light assembly is supplied with 12 volts and grounded through the sending unit mounted on the engine. When the sending unit makes contact because the water temperature is too hot or the oil pressure is too low, the circuit to ground is completed and the lamp should light.

To check the bulb: turn the ignition switch ON. Disconnect the wire at the sending unit, and then ground the wire. The lamp on the dash should light. If it does not light, check for a burned-out bulb or a break in the wiring to the light.

■ Certain Commodore and International Series gauges can supply power to the bulb via the ignition switch or and alternate lighting switch. These gauges can be identified by the lack of a contact strip on the left side of the rear of the gauge.

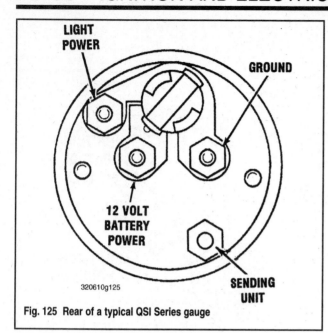

LIGHT
POWER

GROUND

12 VOLT
BATTERY
POWER

SENDING
UNIT

320610g125

Fig. 125 Rear of a typical QSI Series gauge

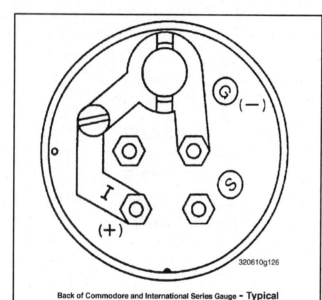

(—)

(+)

320610g126

Back of Commodore and International Series Gauge - **Typical**

Fig. 126 Commodore and International Series gauges are slightly different

TESTING

1. With the ignition switch OFF, remove the wire to the sending unit (lower right, looking at the back of the gauge).
2. Turn the ignition switch to RUN and confirm that the gauge needle is seated on the post at the left side; or all the way to the left of the scale if there is no post.
3. Turn the ignition switch OFF and connect a jumper wire between the ground terminal (upper right from the back of gauge) and the sending unit terminal.
4. Turn the ignition switch to the RUN position and confirm that the gauge needle is seated on the post at the right side or all the way to the right of the scale if there is no post.
5. Replace the gauge if anything fails.

REMOVAL & INSTALLATION

1. Disconnect the battery cables.
2. Tag and disconnect the electrical leads at the back of the gauge.
3. Disconnect the light socket, remove the holding strap and lift out the gauge.
To install:
4. Position the gauge into the mounting hole, install the strap and tighten the nuts securely.

■ **Be careful not to tighten the holding strap too tightly or you risk distorting the gauge casing.**

5. Connect the ground wire and then connect the remaining leads.
6. Install the light socket and then coat all terminal connections with liquid neoprene or equivilent.
7. Connect the battery cables.

Fuel Gauge

HOOKUP

The Boating Industry Association recommends the following color coding be used on all fuel gauge installations:
- Black—for all grounded current-carrying conductors.
- Pink—insulated wire from the fuel gauge sending unit to the gauge.
- Red—insulated wire for a connection from the positive side of the battery to any electrical equipment.
1. Connect one end of a pink insulated wire to the terminal on the gauge marked tank and the other end to the terminal on top of the tank unit.
2. Connect one end of a black wire to the terminal on the fuel gauge marked IGN and the other end to the ignition switch.
3. Connect one end of a second black wire to the fuel gauge terminal marked GRD and the other end to a good ground. It is important for the fuel gauge case to have a good common ground with the tank unit. Aboard an all-metal boat, this ground wire is not necessary. However, if the dashboard is insulated, or made of wood or plastic, a wire must be run from the gauge ground terminal to one of the bolts securing the sending unit in the fuel tank, and then from there to the negative side of the battery.

TROUBLESHOOTING

In order for the fuel gauge to operate properly the sending unit and the receiving unit must be of the same type and preferably of the same make. The following symptoms and possible corrective actions will be helpful in restoring a faulty fuel gauge circuit to proper operation.

If you suspect the gauge is not operating properly, the first area to check is all electrical connections from one end to the other. Be sure they are clean and tight. Next, check the common ground wire between the negative side of the battery, the fuel tank, and the gauge on the dash.

If all wires and connections in the circuit are in good condition, check the sending unit.

If the pointer does not move from the empty position one of four faults could be to blame:
1. The dash receiving unit is not properly grounded.
2. No voltage at the dash receiving unit.
3. Negative meter connections are on a positive grounded system.
4. Positive meter connections are on a negative grounded system.

If the pointer fails to move from the full position, the problem could be one of three faults.
1. The tank sending unit is not properly grounded.
2. Improper connection between the tank sending unit and the receiving unit on the dash.
3. The wire from the gauge to the ignition switch is connected at the wrong terminal.

If the pointer remains at the 3/4 full mark, it indicates a six-volt gauge is installed in a 12-volt system.

If the pointer remains at about 3/8 full, it indicates a 12-volt gauge is installed in a six-volt system.

Erratic Fuel Gauge Readings

Inspect all of the wiring in the circuit for possible damage to the insulation or conductor. Carefully check:
1. Ground connections at the receiving unit on the dash.
2. Harness connector to the dash unit.
3. Body harness connector to the chassis harness.
4. Ground connection from the fuel tank to the trunk floor pan.
5. Feed wire connection at the tank sending unit.

Gauge Always Reads Full—When Ignition Switch Is In ON Position

1. Check the electrical connections at the receiving unit on the dash; the body harness connector to chassis harness connector; and the tank unit connector in the tank.
2. Make a continuity check of the ground wire from the tank to the tank floor pan.
3. Connect a known good tank unit to the tank feed wire and the ground lead. Raise and lower the float and observe the receiving unit on the dash. If the dash unit follows the arm movement, replace the tank sending unit.

Gauge Always Reads Empty—When Ignition Switch Is ON

Disconnect the tank unit feed wire and do not allow the wire terminal to ground. The gauge on the dash should read full.

If Gauge Reads Empty

1. Connect a spare control unit into the control unit harness connector and ground the unit. If the spare unit reads full, the original unit is shorted and must be replaced.
2. A reading of empty indicates a short in the harness between the tank sending unit and the gauge on the control panel.

If Gauge Reads Full

1. Connect a known good tank sending unit to the tank feed wire and the ground lead.
2. Raise and lower the float while observing the gauge on the control panel. If the control panel gauge follows movement of the float, replace the tank sending unit.

Gauge Never Reads Full

This test requires shop test equipment.
1. Disconnect the feed wire to the tank unit and connect the wire to ground thru a variable resistor or thru a spare tank unit.
2. Observe the control panel gauge reading. The reading should be full when resistance is increased to about 90 ohms. This resistance would simulate a full tank.
3. If the check indicates the control panel gauge is operating properly, the trouble is either in the tank sending unit rheostat being shorter, or the float is binding. The arm could be bent, or the tank may be deformed. Inspect and correct the problem.

TESTING

1. With the ignition switch OFF, remove the wire to the sending unit (lower right, looking at the back of the gauge).
2. Turn the ignition switch to RUN and confirm that the gauge needle is seated on the post at the left side; or all the way to the left of the scale if there is no post.
3. Turn the ignition switch OFF and connect a jumper wire between the ground terminal (upper right from the back of gauge) and the sending unit terminal.

4. Turn the ignition switch to the RUN position and confirm that the gauge needle is seated on the post at the right side or all the way to the right of the scale if there is no post.
5. Replace the gauge if anything fails.

REMOVAL & INSTALLATION

1. Disconnect the battery cables.
2. Tag and disconnect the electrical leads at the back of the gauge.
3. Disconnect the light socket, remove the holding strap and lift out the gauge.

To install:
4. Position the gauge into the mounting hole, install the strap and tighten the nuts securely.

■ **Be careful not to tighten the holding strap too tightly or you risk distorting the gauge casing.**

5. Connect the ground wire and then connect the remaining leads.
6. Install the light socket and then coat all terminal connections with liquid neoprene or equivilent.
7. Connect the battery cables.

Battery Gauge

TESTING

1. Make sure that the battery is fully charged and then turn the ignition switch to the RUN position. The gauge should read battery voltage, otherwise replace it.
2. If the gauge is out of the dash, turn the ignition switch OFF and connect a jumper wire between the ground terminal (upper right looking at the back of the gauge) and the negative battery terminal. Connect another jumper between the power terminal (lower left looking at the back of the gauge) and the positive battery terminal. The gauge should register battery voltage, otherwise replace it.

REMOVAL & INSTALLATION

1. Disconnect the battery cables.
2. Tag and disconnect the electrical leads at the back of the gauge.
3. Disconnect the light socket, remove the holding strap and lift out the gauge.

To install:
4. Position the gauge into the mounting hole, install the strap and tighten the nuts securely.

■ **Be careful not to tighten the holding strap too tightly or you risk distorting the gauge casing.**

5. Connect the ground wire and then connect the remaining leads.
6. Install the light socket and then coat all terminal connections with liquid neoprene or equivilent.
7. Connect the battery cables.

CruiseLog

TESTING

◆ See Figure 127

Turn the ignition switch to the RUN position. The run indicator dial should be turning; if not, replace the gauge.

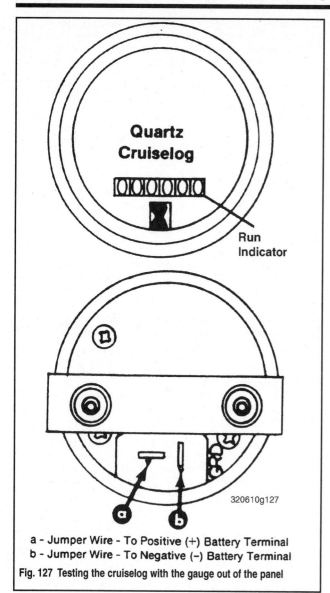

a - Jumper Wire - To Positive (+) Battery Terminal
b - Jumper Wire - To Negative (–) Battery Terminal

Fig. 127 Testing the cruiselog with the gauge out of the panel

If the gauge is out of the dash, connect a jumper wire between terminal **A** and the positive battery cable. Connect another jumper between terminal **B** and the negative battery cable. The run indicator dial should be turning; if not, replace the gauge.

REMOVAL & INSTALLATION

1. Disconnect the battery cables.
2. Tag and disconnect the electrical leads at the back of the gauge.
3. Disconnect the light socket, remove the holding strap and lift out the gauge.
To install:
4. Position the gauge into the mounting hole, install the strap and tighten the nuts securely.

■ **Be careful not to tighten the holding strap too tightly or you risk distorting the gauge casing.**

5. Connect the ground wire and then connect the remaining leads.
6. Install the light socket and then coat all terminal connections with liquid neoprene or equivilent.
7. Connect the battery cables.

Speedometer

TESTING

■ **An air compressor is necessary for this procedure. It is imperative that its pressure gauge is extremely accurate.**

Disconnect the hose at the back of the gauge and then connect a line from an air compressor. Apply 5.3 psi of air pressure to the speedometer and check that the gauge reads 20 mph (+/- 1 mph). Now apply 27.8 psi pressure and check that the gauge reads 45 mph (+/- 1 mph). If either of the readings is not as specified, replace the speedometer.

REMOVAL & INSTALLATION

1. Disconnect the battery cables.
2. Tag and disconnect the electrical leads and the hose at the back of the gauge.
3. Disconnect the light socket, remove the holding strap and lift out the gauge.
To install:
4. Position the gauge into the mounting hole, install the strap and tighten the nuts securely.

■ **Be careful not to tighten the holding strap too tightly or you risk distorting the gauge casing.**

5. Connect the ground wire and then connect the remaining leads. Connect the hose.
6. Install the light socket and then coat all terminal connections with liquid neoprene or equivilent.
7. Connect the battery cables.

Tachometer

TESTING

Connect a tachometer as per the manufacturer's instructions. Start the engine and check a few different engine speeds on the boat's tachometer against the service tach. If using a 6000 rpm service tach, variations of +/- 150 rpm are acceptable; if using an 8000 rpm tach, variations of +/- 200 rpm are acceptable. Replace the tachometer if not within specifications.

REMOVAL & INSTALLATION

1. Disconnect the battery cables.
2. Tag and disconnect the electrical leads at the back of the gauge.
3. Disconnect the light socket, remove the holding strap and lift out the gauge.
To install:
3. Position the gauge into the mounting hole, install the strap and tighten the nuts securely.

■ **Be careful not to tighten the holding strap too tightly or you risk distorting the gauge casing.**

4. Connect the ground wire and then connect the remaining leads. .
5. Install the light socket and then coat all terminal connections with liquid neoprene or equivilent.
6. Connect the battery cables.

SENDING UNITS AND SWITCHES

Oil Pressure Sending Unit

The oil pressure sending unit is located on the starboard side of the engine block, above the starter motor.

TESTING

◆ **See Figure 128**

■ **This test should be performed only after confirming that the oil pressure gauge is operating properly.**

Connect a multi-meter as per the manufacturer's instructions. Connect the positive lead of the meter to the sender terminal and the negative lead to the hex nut on the back of the sender housing. With the engine not running, the meter should register continuity as shown in the accompanying chart.

Start the engine and check the sender at the different oil pressure readings as shown in the chart. If any readings vary, replace the unit.

■ **Dual station units can be identified by 353-AM stamped on the hex nut of the sender.**

Oil Pressure (psi)	Meter Reading (ohms)	
	Single	Dual
0	227-257	114-129
20	142-163	71-81
40	92-114	46-57
80	9-49	5-25

320610g128

Fig. 128 Oil pressure sender unit test specifications

Water Temperature Sender

The water temperature sending unit is located in the thermostat housing.

TESTING

3.0L and V6 Engines

There are no *accurate* tests for the temperature sending unit. If the gauge is operating properly and the sending unit is suspect, replace it.

V8 Engines

1. Disconnect the electrical lead and remove the switch from the thermostat housing.
2. Connect a multi-meter to the switch with the positive lead on the terminal and the negative lead on the hex. Carefully immerse it a pan of water. Heat the water and observe that the readings are as follows:
   2a. 140ºF (60ºC): 121–147 ohms
   2b. 194ºF (90ºC): 47–55 ohms
   2c. 212ºF (100ºC): 36–41 ohms
3. Turn the heat off and allow the water to cool, checking the readings again as it cools.

Fuel Tank Sending Unit

TESTING

Flange Type

◆ **See Figure 129**

1. Disconnect the electrical lead at the sender and loosen the mounting screw with the ground wire attached; remove the ground wire.
2. Remove the remaining mounting screws and lift the sender out of the fuel tank—be careful not to damage the float arm.
3. Connect a multi-meter as per the manufacturer's instructions. Connect the positive lead to the terminal on the sender and the negative lead to the flange on the housing. Move the float arm to the full position (arm is horizontal) and check that the meter reads 30 ohms (+/- 5).
4. Move the float arm to the empty position (arm is vertical) and check that the meter reads 240 ohms (+/- 5).
5. Replace the sender unit if it is outside either of the specifications.

Capsule Type

◆ **See Figure 130**

1. Disconnect the electrical lead at the sender unit.
2. Remove the two mounting screws and carefully lift the unit from the fuel tank.
3. Position a large magnet under the bottom side of the unit.
4. Connect a multi-meter as per the manufacturer's instructions. Connect the positive lead to the sender terminal and the negative lead to the metal portion of the unit housing. Rotate the magnet counterclockwise until the gauge reads empty. The meter should show 240 ohms (+/- 5).
5. Rotate the magnet the other way until the gauge reads FULL. The meter should read 30 ohms (+/- 5).
6. If the unit is not within specifications on either test, replace it.

Ignition Switch

TESTING

■ **Before performing this test, please ensure that all fuses and the starter motor are in good condition.**

1. Disconnect the battery cables and connect a multi-meter as per the manufacturer's instructions.
2. With the switch in the OFF position, test for continuity across all of the terminals on the back of the switch—there should be none.
3. Move the ignition switch to the RUN position. Check for continuity between terminals **A** and **B** and between terminals **B** and **I**. There should be no continuity between terminal **C** and any of the other three terminals.
4. Move the ignition switch to START. There should be continuity between terminals **A** and **B**, **B** and **I** and between terminals **B** and **C**.
5. If the switch fails any of the tests, unsolder the wire connections and remove the actual switch and repeat Steps 2–4. There should be no continuity between any of the terminals with the switch removed. If the switch passes this time, you've got a bad wiring harness.

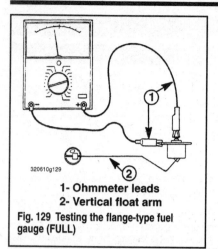

1- Ohmmeter leads
2- Vertical float arm

Fig. 129 Testing the flange-type fuel gauge (FULL)

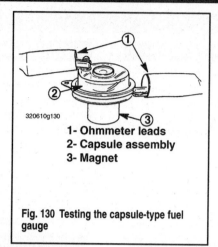

1- Ohmmeter leads
2- Capsule assembly
3- Magnet

Fig. 130 Testing the capsule-type fuel gauge

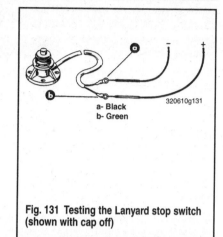

a- Black
b- Green

Fig. 131 Testing the Lanyard stop switch (shown with cap off)

Emergency (Lanyard) Stop Switch

TESTING

Button Style

◆ See Figure 131

1. Remove the electrical leads from their connections at the switch. Make sure that you tape the white lead back against the wiring harness with at least two turns of quality electrical tape.
2. Connect a multi-meter as per the manufacturer's instructions.
3. With the switch cap on, check for continuity between the Black and Green leads—there should be none.
4. Now pop the cap off, there should be continuity between the two leads.
5. Replace the switch if it fails either test.

Toggle Style

Disconnect the switch leads and connect a multi-meter across each lead. There should be continuity with the switch in the RUN position and no continuity with it in the OFF position.

■ The Black/Yellow lead is not used and should be taped back against the harness with a few wraps of electrical tape.

Remote Control Mounted

Disconnect the switch leads and connect a multi-meter across each lead (Black/Yellow on EFI; Purple on Carb). There should be continuity with the switch lanyard connected and no continuity with it disconnected.

Start/Stop Switch Panel

TESTING

◆ See Figure 132

1. Disconnect the battery cables and connect a multi-meter as per the manufacturer's instructions.
2. Check for continuity between the start switch terminals—there should be none. Press the button and you should see continuity.
3. Now check the stop switch in the same manner. If either switch fails either test, replace the panel.

Audio Warning System

TESTING

Old Style

◆ See Figure 133

1. Turn the ignition switch to the RUN position but do not start the engine. The horn should sound within 7–14 seconds.
2. If the horn didn't go make any sound, connect jumper wires between terminal **A** and the positive battery terminal, and a jumper between terminal **B** and a good ground. If the horn doesn't sound, replace the system. If the horn does sound, go to the next step.
3. Now connect a jumper wire from the end of lead **C** to the positive battery terminal, and a jumper from the end of lead **D** (tan/blue wire) to a good ground. If the horn sounds within 7–14 seconds, you've got a problem with the tan/blue wire running back to the engine (or some other switch).

New Style

◆ See Figure 134

1. Turn the ignition switch to the RUN position. The buzzer should sound and then turn off.
2. If the buzzer does not sound, disconnect the Tan/Blue wire at the back of the system and touch it to a known ground. If the buzzer sounds, you have a problem in the wire

Oil Pressure Switch

The oil pressure sending unit is located on the starboard side of the engine block, above the starter motor.

TESTING

364-AF

◆ See Figure 135

■ This test should be performed only after confirming that the oil pressure gauge is operating properly.

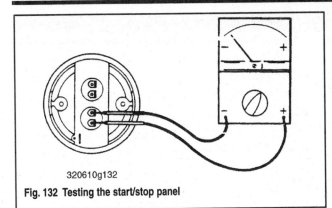

320610g132

Fig. 132 Testing the start/stop panel

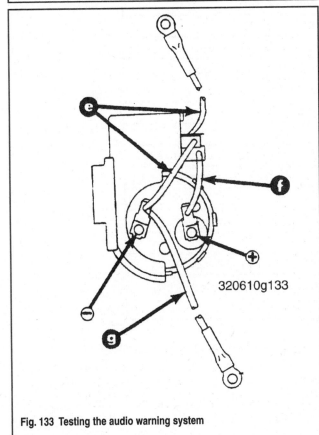

320610g133

Fig. 133 Testing the audio warning system

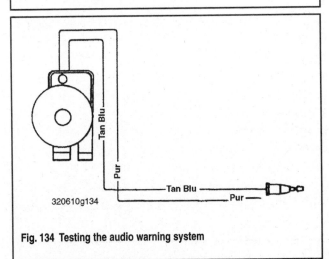

320610g134

Fig. 134 Testing the audio warning system

Connect a multi-meter as per the manufacturer's instructions. Connect the positive lead of the meter to the sender terminal and the negative lead to the hex nut on the back of the sender housing. With the engine running and the oil pressure above 6 psi, the meter should register no continuity. If it does, replace the switch.

■ **Dual station units can be identified by 353-AM stamped on the hex nut of the sender.**

Stern Drive Gear Lube Monitor Switch

With the fluid level correct this switch should normally be open (no continuity). When the level falls below a prescribed point the switch will close and sound the warning buzzer. If the buzzer sounds and the level is OK, disconnect the harness wires—if the sound stops then the switch will require replacement.

Transmission Temperature Switch

These switches are normally open until the engine reaches a certain temperature, at which time they closes. These switches are only used on inboard engines.

TESTING

1. Disconnect the electrical lead and remove the switch from the transmission.
2. Connect a multi-meter to the switch (each terminal on the top) and check that there is no continuity. If there is, replace the switch; if not, proceed to the next step.
3. With the meter still attached, carefully immerse the switch in a pan of water. Heat the water and verify that the switch closes (continuity present) at 220º–240ºF (104º–116ºC).
4. Turn off the heating element and allow the container to cool. The switch should open again when it reaches 180º–200ºF (82º–93ºC.
5. Replace the switch if it fails either test.

Water Temperature Switch

This switches is normally open until the engine reaches a certain temperature, at which time it closes.

TESTING

1. Disconnect the electrical lead and remove the switch from the thermostat housing.
2. Connect a multi-meter to the switch and carefully immerse it a pan of water. Heat the water and verify that the switch closes at 190º–200ºF (88º–93ºC) on 48592 switches (red sleeve); or at 215º–225ºF (102º–107ºC) on 87-86080 switches (black sleeve).
3. Turn off the heating element and allow the container to cool. The switch should open again when it reaches 150º–170ºF (66º–77ºC) on 48952 switches; or at 175º–195ºF (79º–91ºC) on 87-86080 switches (black sleeve).
4. Replace the switch if it fails either test.

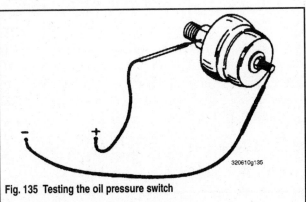

320610g135

Fig. 135 Testing the oil pressure switch

WIRING DIAGRAMS

Wire Color	Where Used
Black	All Grounds
Brown	Reference Electrode - MerCathode
Orange	Anode Electrode - MerCathode
Lt. Blue/White	Trim - "Up" Switch
Gray	Tachometer Signal
Green/White	Trim - "Down" Switch
Tan	Water Temp. Sender to Gauge
Lt. Blue	Oil Pressure Sender (10 Gauge)
Pink	Fuel Gauge Sender to Gauge
Brown/White	Trim Sender to Trim Gauge
Purple/White	Trim - "Trailer" Switch
Red	Unprotected Wires from Battery
Red/Purple	Protected (Fused) Wires from Battery
Red/Purple	Protected (+12V) to Trim Panel
Orange	Alternator Output
Purple/Yellow	Ballast Bypass
Purple	Ignition Switch (+12V)
Yellow	Starter Solenoid to Starter Motor
Yellow/Red	Starter Switch to Starter Solenoid to Neutral Start Switch

COLOR CODE

BLK • Black
BLU • Blue
BRN • Brown
GRY • Gray
ORN • Orange
PNK • Pink
PUR • Purple
RED • Red
TAN • Tan
WHT • White
YEL • Yellow
LIT • Light
DRK • Dark

Fig. 136 Typical wiring colors

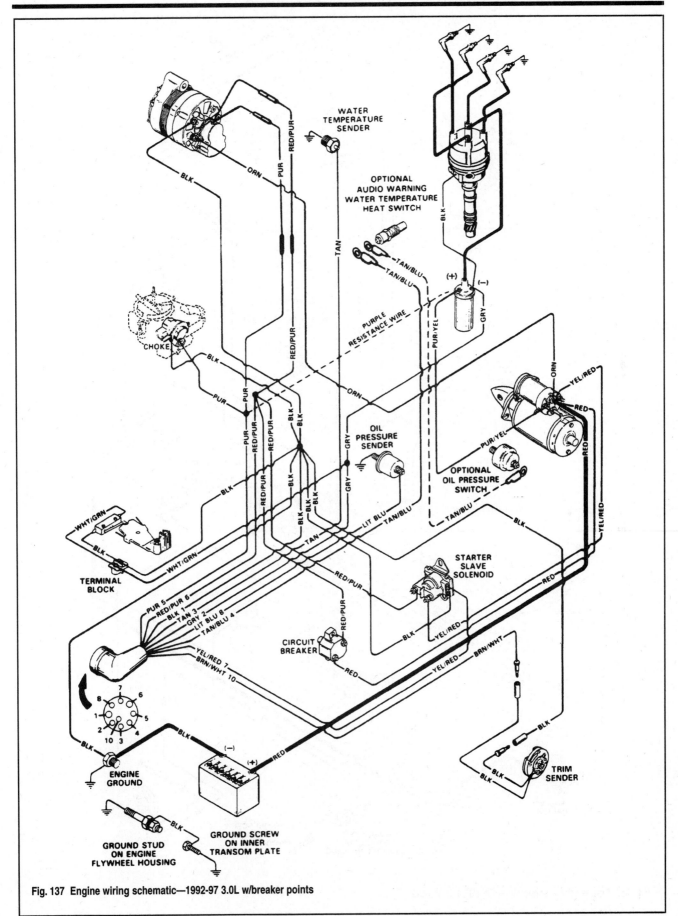

Fig. 137 Engine wiring schematic—1992-97 3.0L w/breaker points

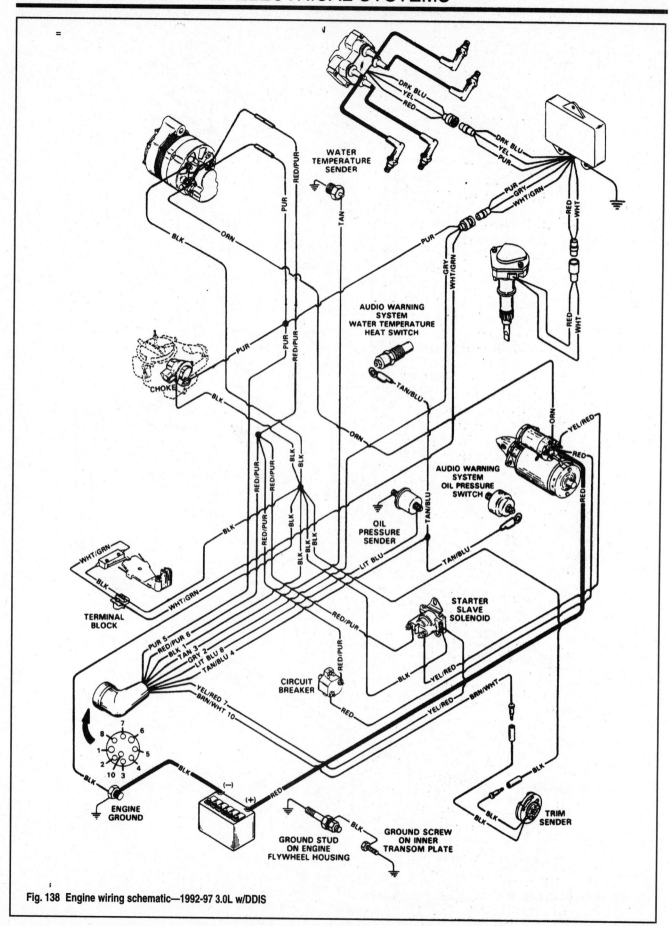

Fig. 138 Engine wiring schematic—1992-97 3.0L w/DDIS

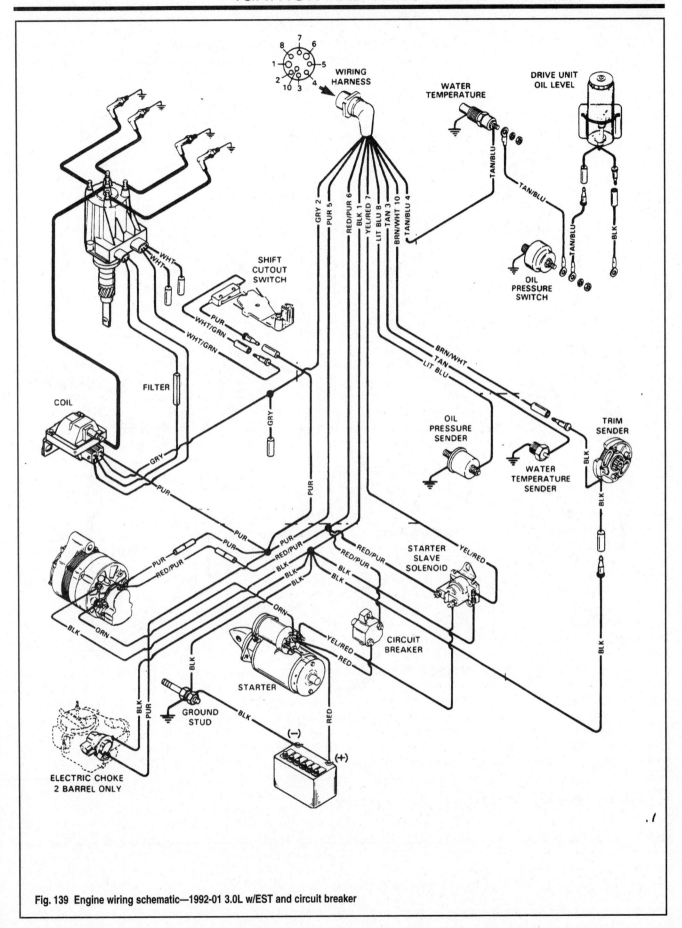

Fig. 139 Engine wiring schematic—1992-01 3.0L w/EST and circuit breaker

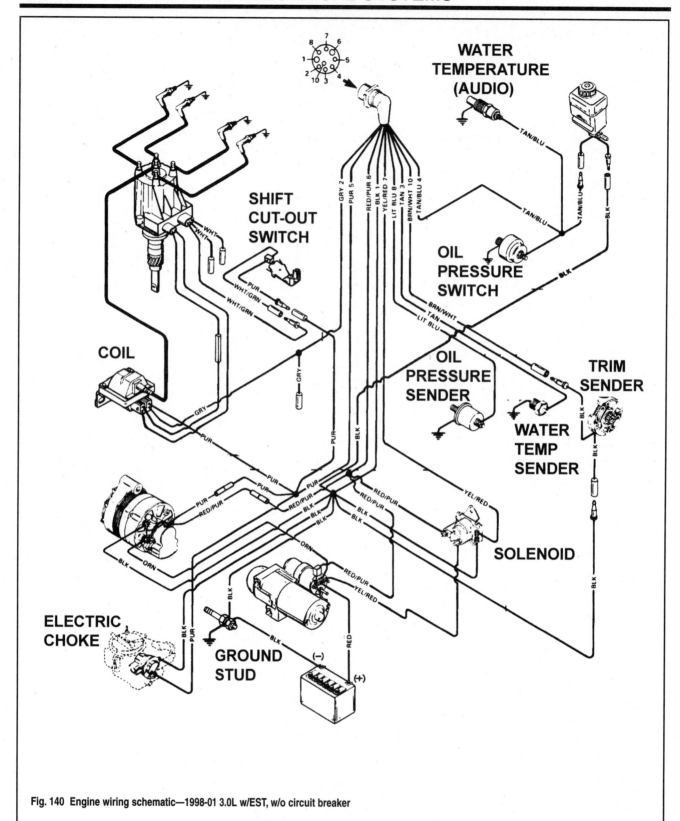

Fig. 140 Engine wiring schematic—1998-01 3.0L w/EST, w/o circuit breaker

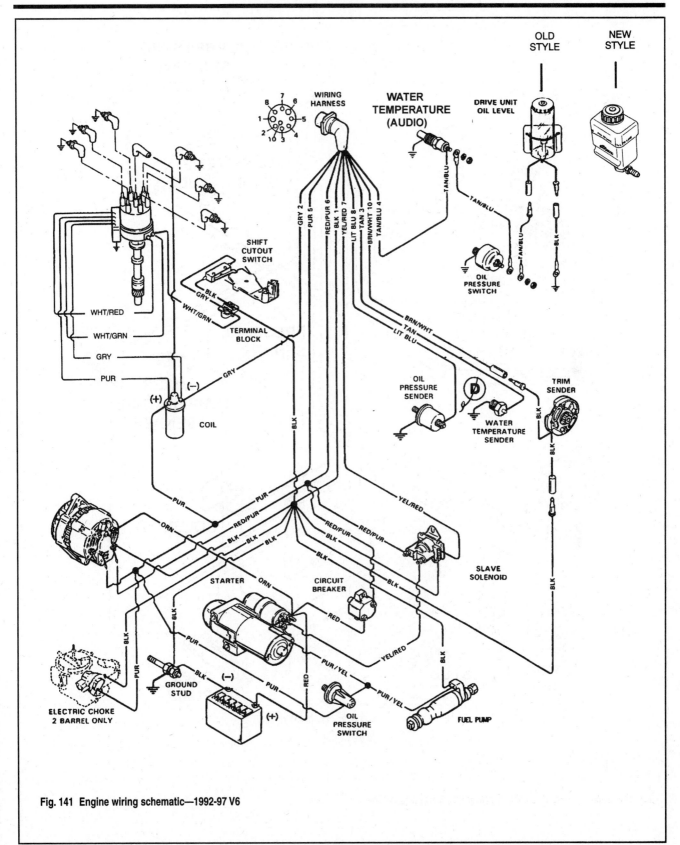

Fig. 141 Engine wiring schematic—1992-97 V6

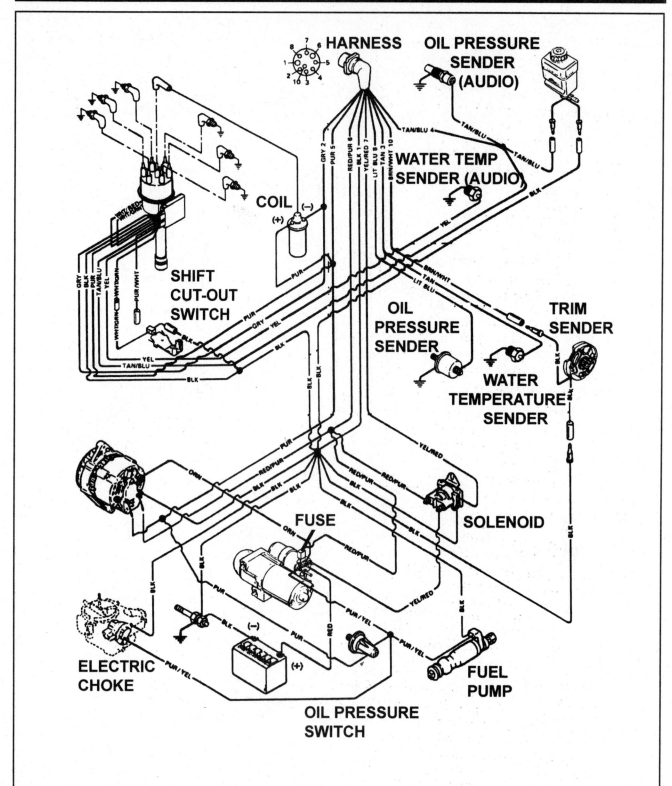

Fig. 142 Engine wiring schematic—1998-01 V6 w/carburetor, Alpha

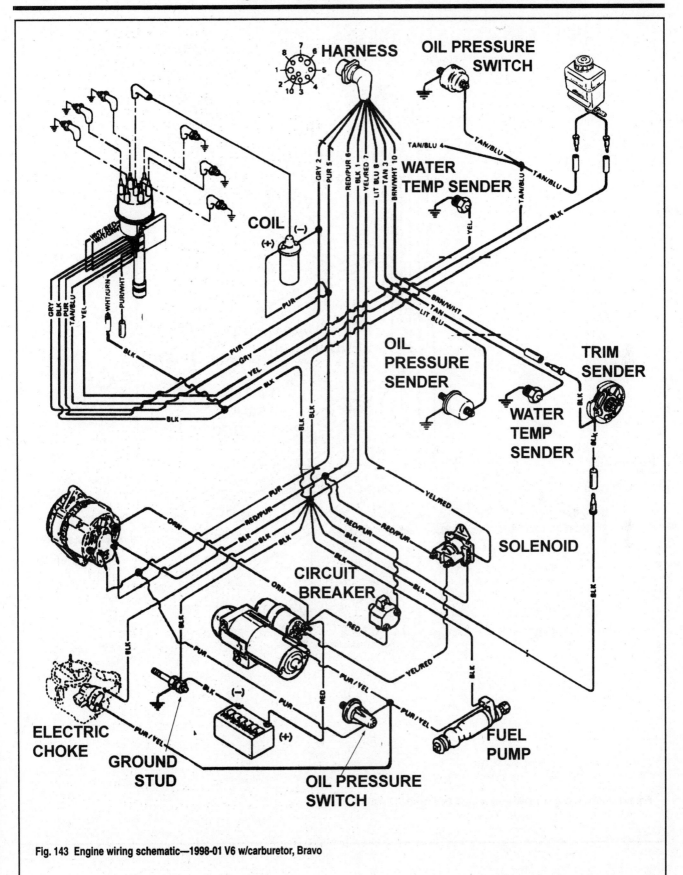

Fig. 143 Engine wiring schematic—1998-01 V6 w/carburetor, Bravo

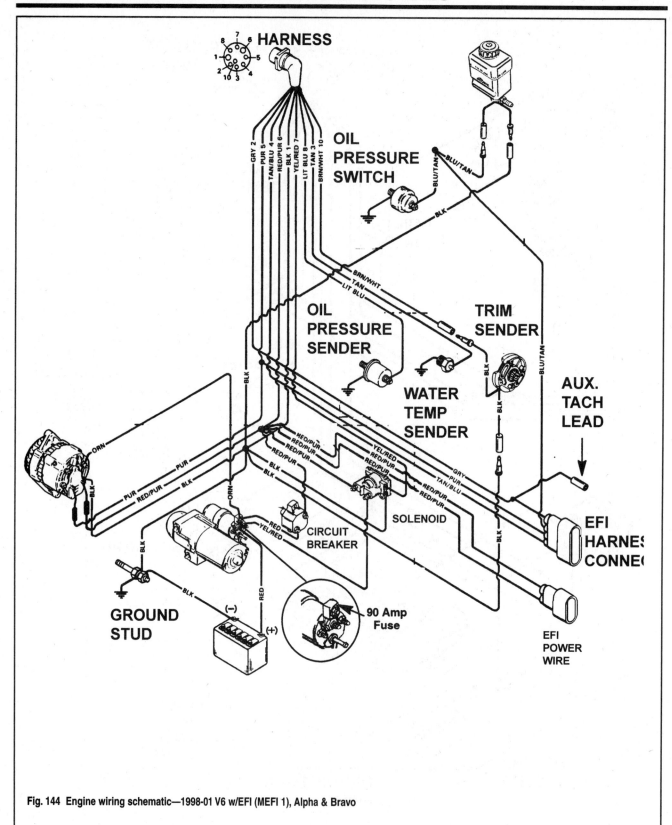

Fig. 144 Engine wiring schematic—1998-01 V6 w/EFI (MEFI 1), Alpha & Bravo

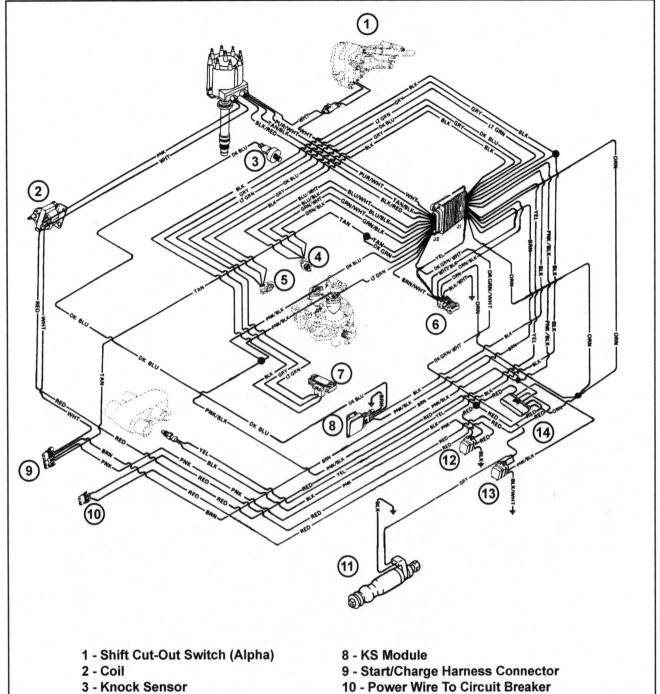

1 - **Shift Cut-Out Switch (Alpha)**
2 - **Coil**
3 - **Knock Sensor**
4 - **Idle Air Control Valve (IAC)**
5 - **Throttle Position Sensor (TP)**
6 - **Data Link Connector (DLC)**
7 - **MAP Sensor**
8 - **KS Module**
9 - **Start/Charge Harness Connector**
10 - **Power Wire To Circuit Breaker**
11 - **Fuel Pump**
12 - **Ignition System Relay**
13 - **Fuel Pump Relay**
14 - **Fuse**

Fig. 145 Fuel/ignition systems wiring schematic—1990-01 V6 w/EFI (MEFI 1), Alpha & Bravo

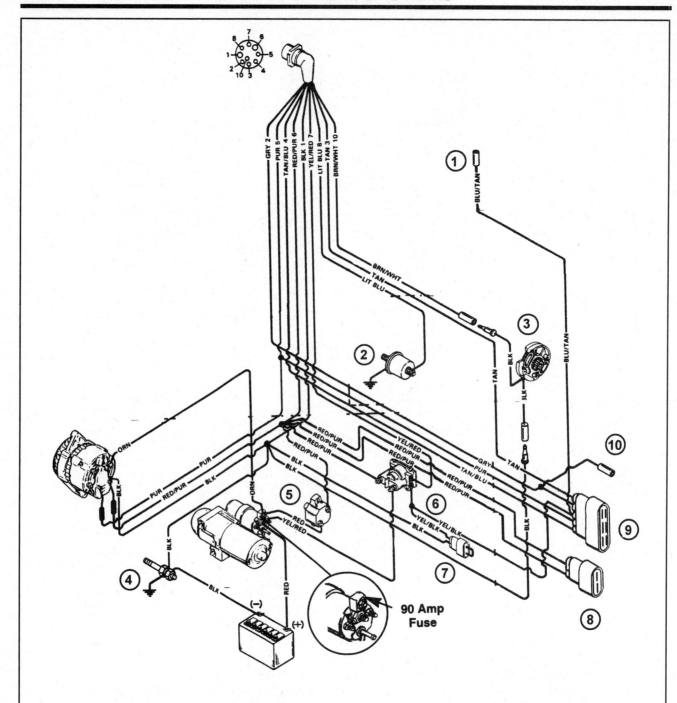

1 - Oil Pressure Switch (Audio)
2 - Oil Pressure Sender
3 - Trim Sender
4 - Ground Stud
5 - Circuit Breaker

6 - Starter Solenoid
7 - Jumper Wire Connection
8 - EFI Power Wire
9 - EFI Harness Connector
10 - Auxiliary Tach Lead

Fig. 146 Engine wiring schematic—1998-01 V6 w/EFI (MEFI 3)

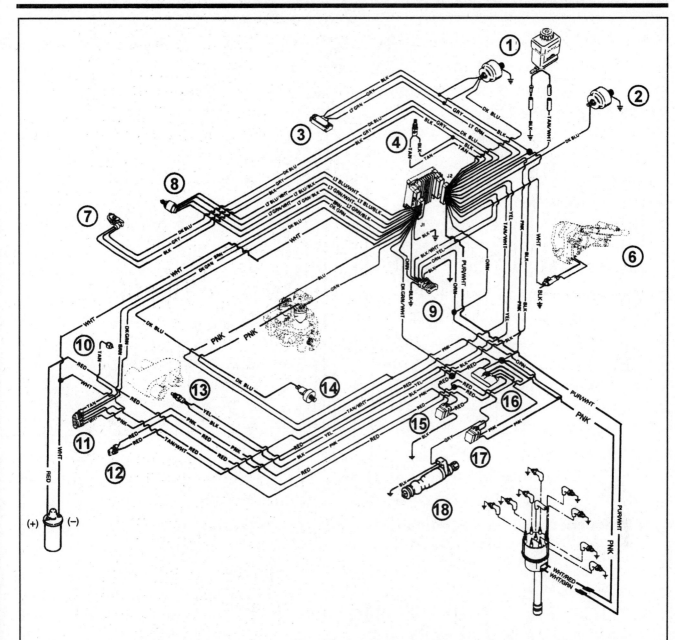

1 - Fuel Pressure Switch
2 - Oil Pressure Sensor
3 - MAP Sensor
4 - MAT Sensor
6 - Shift Plate (Alpha)
7 - TP Sensor
8 - IAC Valve
9 - Date Link Connector (DLC)
10 - Water Temperature Sender

11 - Start/Charge Harness Connector
12 - Power Wire To Circuit Breaker
13 - ECT Sensor
14 - Knock Sensor
15 - Ignition System Relay
16 - Fuses
17 - Fuel Pump Relay
18 - Fuel Pump

Fig. 147 Fuel/ignition systems wiring schematic—1990-01 V6 w/EFI (MEFI 3)

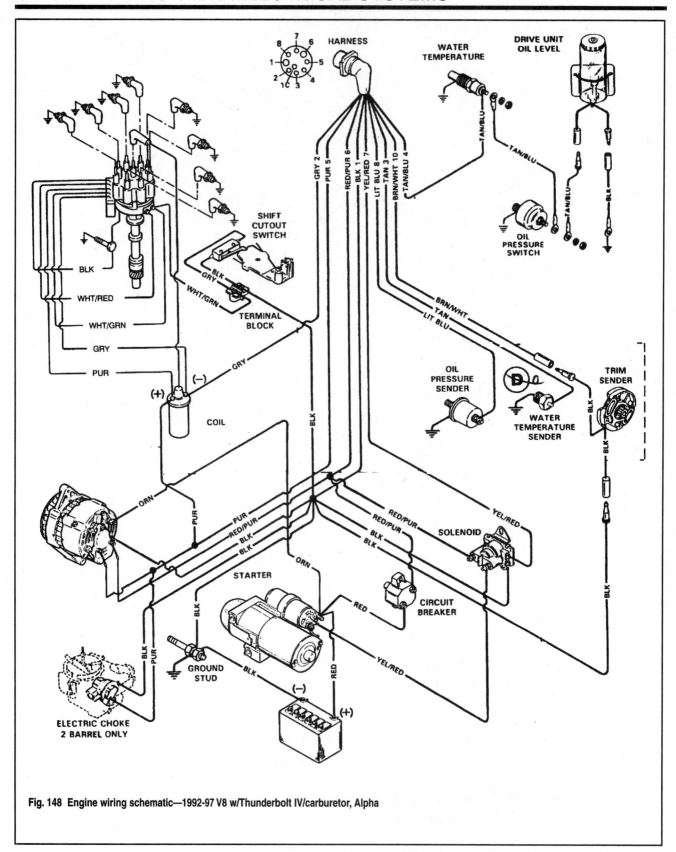

Fig. 148 Engine wiring schematic—1992-97 V8 w/Thunderbolt IV/carburetor, Alpha

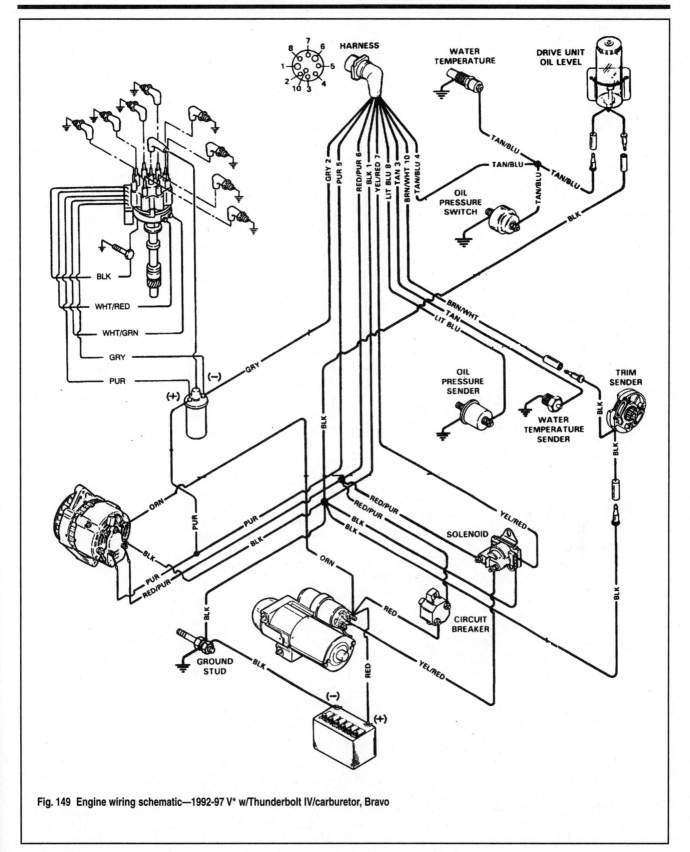

Fig. 149 Engine wiring schematic—1992-97 V* w/Thunderbolt IV/carburetor, Bravo

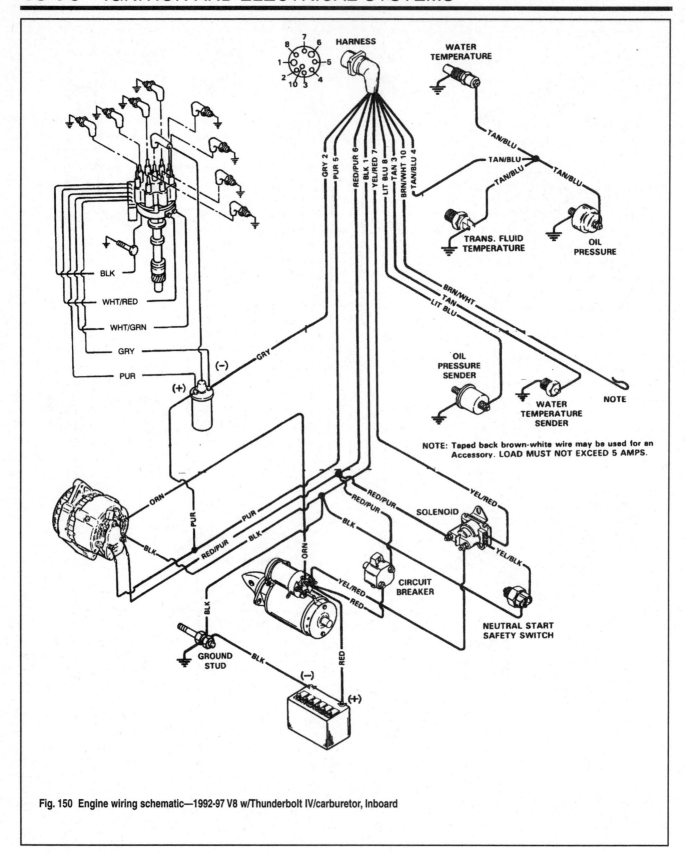

Fig. 150 Engine wiring schematic—1992-97 V8 w/Thunderbolt IV/carburetor, Inboard

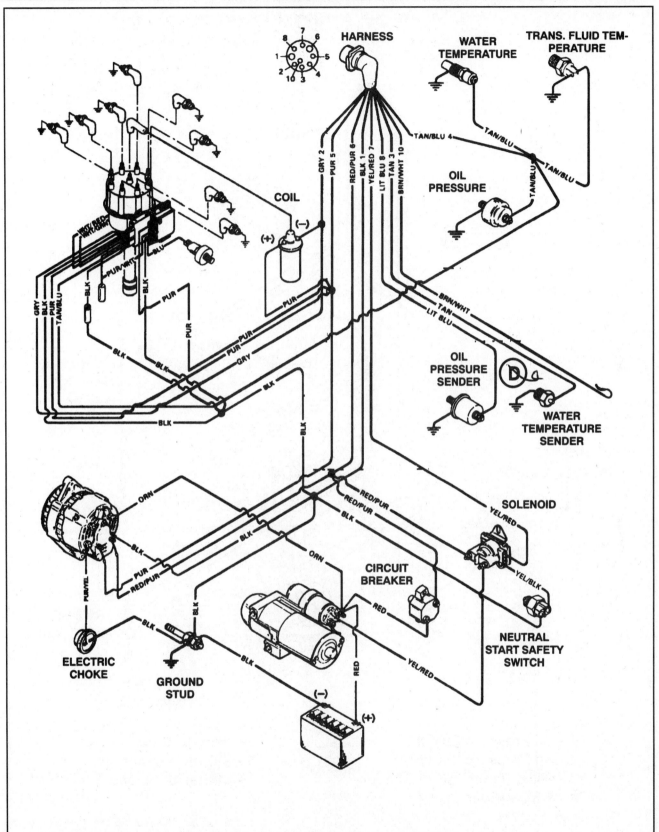

Fig. 151 Engine wiring schematic—1992-97 5.7L V8 w/Thunderbolt V/carburetor (7.4L/8.2L similar)

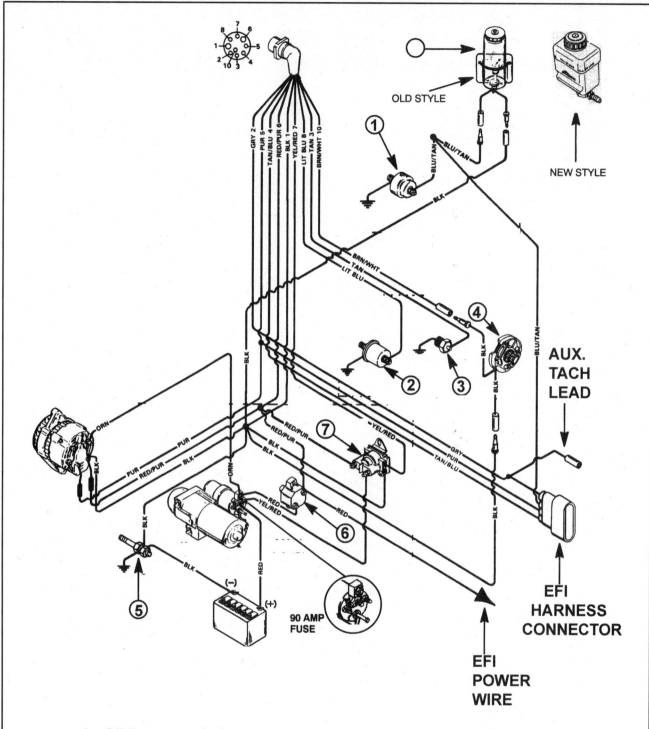

1 - Oil Pressure Switch
2 - Oil Pressure Sender
3 - Water Temperature Sender
4 - Trim Sender

5 - Ground Stud
6 - Circuit Breaker
7 - Starter Solenoid

Fig. 152 Engine wiring schematic—1992-97 V8 w/EFI, Alpha & Bravo

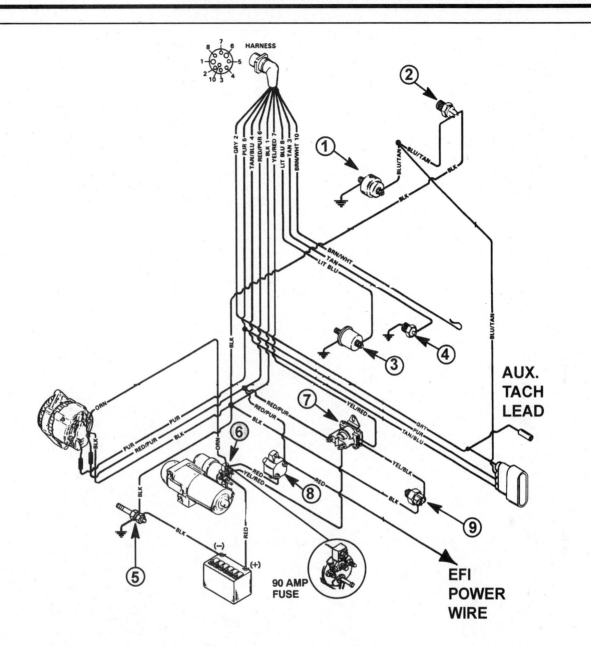

1 - Oil Pressure Switch
2 - Trans. Fluid Temp. Switch
3 - Oil Pressure Sender
4 - Water Temperature Sender
5 - Ground Stud

6 - Fuse
7 - Starter Solenoid
8 - Circuit Breaker
9 - Neutral Safety Switch

Fig. 153 Engine wiring schematic—1992-97 V8 w/EFI, Inboard

1 - VST
2 - Coil
3 - TP Sensor
4 - AIC Valve
5 - Knock Sensor
6 - MAP Sensor
7 - Start/Charge Harness Connector
8 - Power Wire To Circuit Breaker
9 - ECT Sensor

10 - KS Module
11 - Data Link Connector (DLC)
12 - Lanyard Stop Switch Harness
13 - Ignition System Relay
14 - Fuel Pump Relay
15 - 10 Amp Fuse
16 - 15 Amp Fuse (Fuel Pump)
17 - 15 Amp Fuse (ECM/DLC/Bat)
18 - Dual Eng. DLC Connector

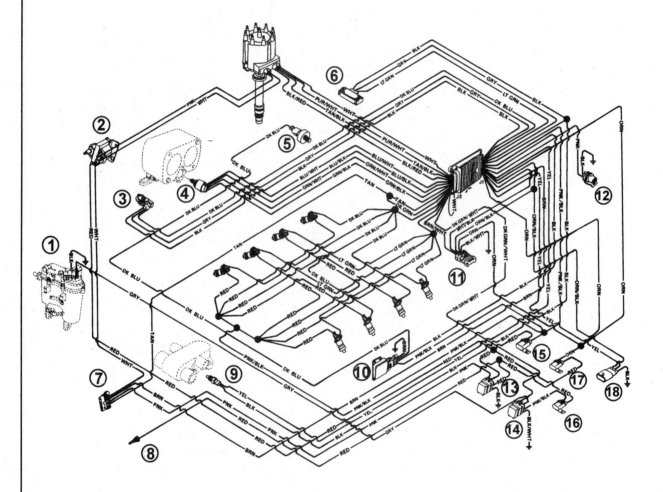

Fig. 154 Fuel/ignition systems wiring schematic—1992-97 5.7L V8 w/MPI (7.4L/8.2L similar)

1 - Coil
2 - Knock Sensor
3 - Shift Cut-out Switch (Alpha)
4 - Lanyard Stop Switch Connector
5 - Data Link Connector (DLC)
6 - IAC Valve
7 - TP Sensor
8 - VST
9 - Start/Charge Harness Connector
10 - Power Wire To Circuit Breaker
11 - ECT Sensor
12 - MAP Sensor
13 - KS Module
14 - Ignition System Relay
15 - Fuse (ECM/Ign. Feed/KS/ Injectors)
16 - Fuse (ECM/DLC/Battery)
17 - Dual Engine Data Link Connector
18 - Fuel Pump Fuse
19 - Fuel Pump Relay

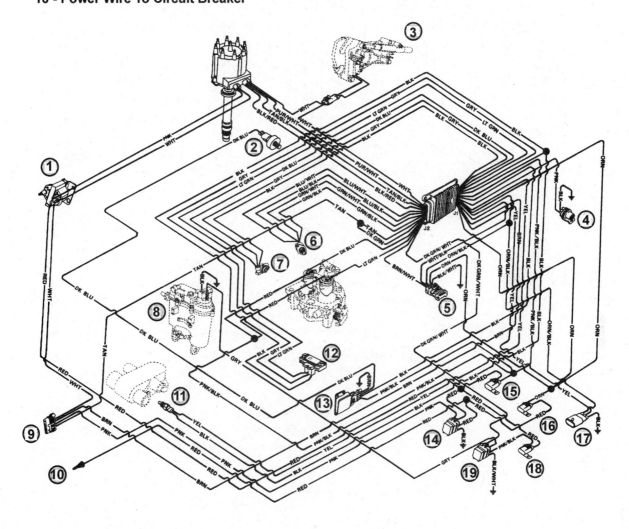

Fig. 155 Fuel/ignition systems wiring schematic—1992-97 V8 w/TBI

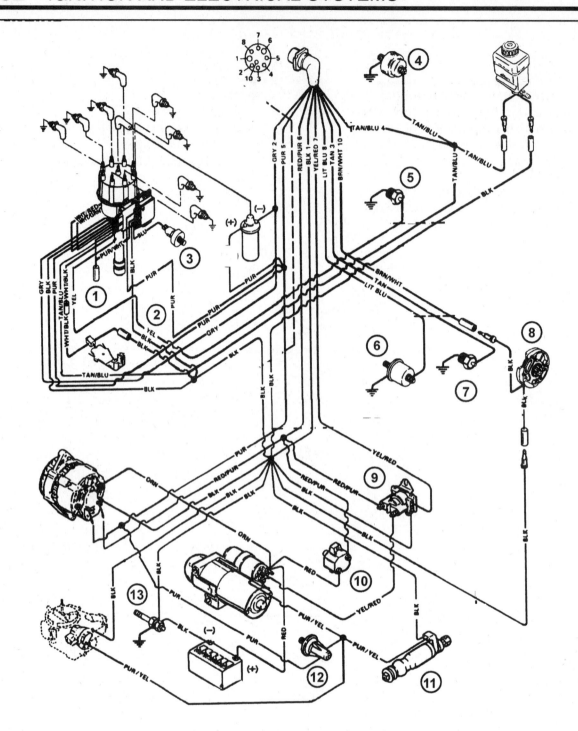

1 - Timing Lead
2 - Shift Interupt Switch (Alpha)
3 - Knock Sensor (5.7L)
4 - Oil Pressure Switch
5 - Water Temperature Switch
6 - Oil Pressure Sender
7 - Water Temperature Sender

8 - Trim Sender
9 - Starter Solenoid
10 - Circuit Breaker (5.7L)
11 - Fuel Pump
12 - Oil Pressure Switch
13 - Ground Stud

Fig. 156 Engine wiring schematic—1998-01 V8 w/carburetor, Alpha & Bravo

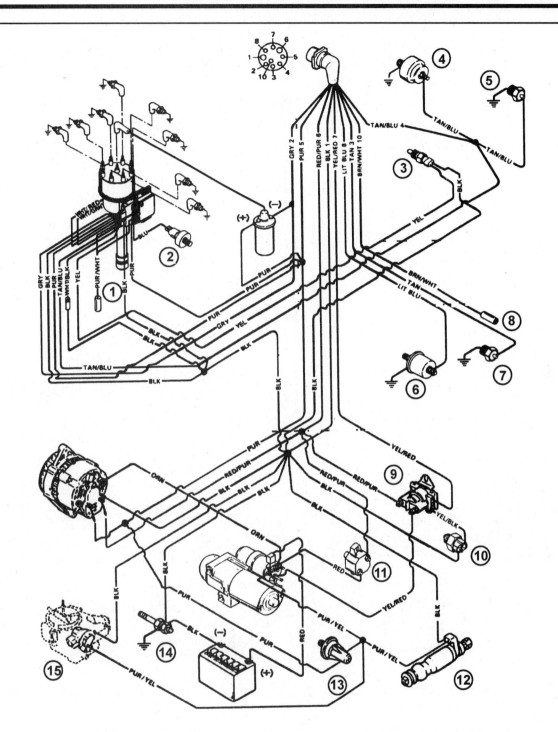

1 - Timing Lead
2 - Knock Sensor
3 - Water Temperature Switch
4 - Oil Pressure Switch
5 - Trans. Temperature Switch
6 - Oil Pressure Sender
7 - Water Temperature Sender
8 - For Accessory

9 - Starter Solenoid
10 - Neutral Safety Switch
11 - Circuit Breaker
12 - Fuel Pump
13 - Oil Pressure Switch
14 - Ground Stud
15 - Electric Choke

Fig. 157 Engine wiring schematic—1998-01 5.7L V8 w/carburetor, Inboard

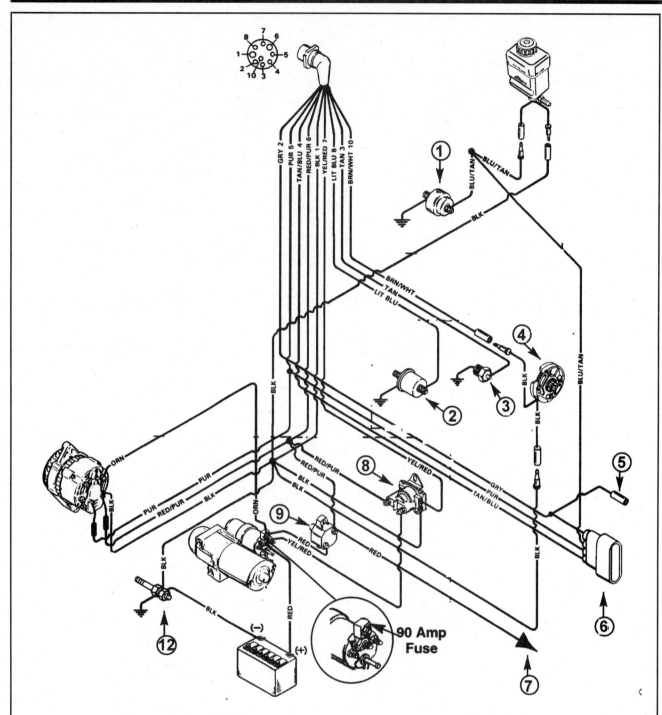

1 - Oil Pressure Switch
2 - Oil Pressure Sender
3 - Water Temperature Sender
4 - Trim Sender
5 - Auxiliary Tach Lead

6 - EFI System Harness Connector
7 - Power Wire To EFI System
8 - Starter Solenoid
9 - Circuit Breaker
10 - Ground Stud

Fig. 158 Engine wiring schematic—1998-01 V8 w/EFI (MEFI 1/2)

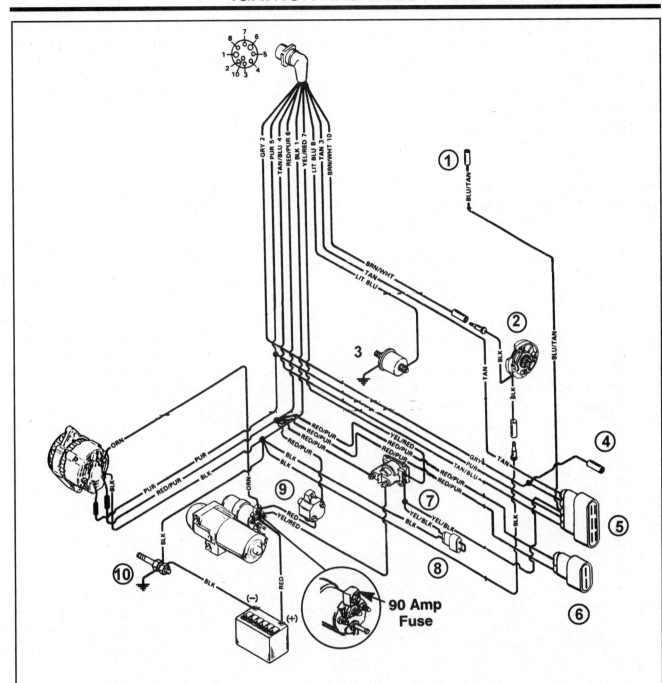

1 - Oil Pressure Switch
2 - Trim Sender
3 - Oil Pressure Sender
4 - Auxiliary Tach Lead
5 - EFI Harness Connector

6 - EFI System Power Wire
7 - Starter Solenoid
8 - Jumper Wire (exc. 7.4L/454/502 Mag)
9 - Circuit Breaker
10 - Ground Stud

Fig. 159 Engine wiring schematic—1998-01 V8 w/EFI (MEFI 3)

1 - Fuel Pump
2 - Throttle Body
3 - Distributor
4 - Coil
5 - Electronic Spark Control (KS) Module
6 - Data Link Connector (DLC)
7 - Manifold Absolute Pressure (MAP) Sensor
8 - Knock Sensor
9 - Idle Air Control (IAC)
10 - Throttle Position (TP) Sensor
11 - Engine Coolant Temperature (ECT) Sensor

12 - Electronic Control Module (ECM)
13 - Fuel Pump Relay
14 - Ignition/System Relay
15 - Fuse (15 Amp) Fuel Pump
16 - Fuse (15 Amp) ECM/DLC/Battery
17 - Fuse (10 Amp) ECM/Injector/Ignition/Knock Module
18 - Harness Connector To Starting/Charging Harness
19 - Positive (+) Power Wire To Engine Circuit Breaker

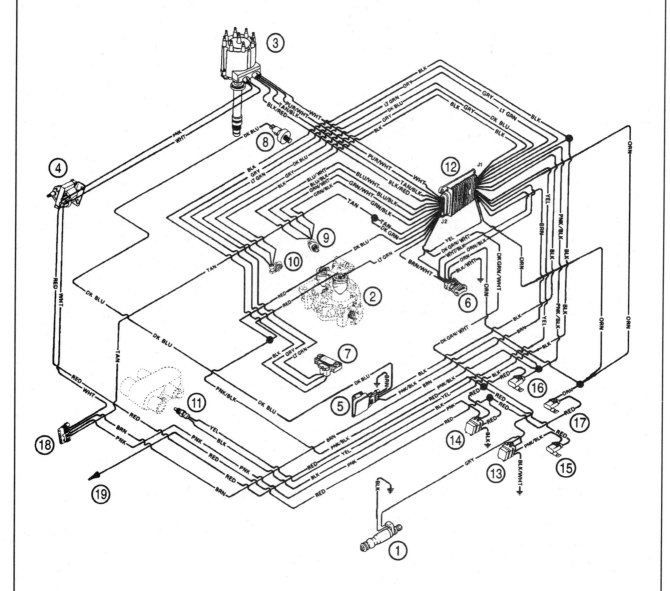

Fig. 160 Fuel/ignition systems wiring schematic—1998-01 V8 w/TBI (MEFI 1), Alpha

1 - Fuel Pump
2 - Throttle Body
3 - Distributor
4 - Coil
5 - Manifold Air Temperature (MAT) Sensor
6 - Data Link Connector (DLC)
7 - Manifold Absolute Pressure (MAP) Sensor
8 - Knock Sensor
9 - Idle Air Control (IAC)
10 - Throttle Position (TP) Sensor
11 - Engine Coolant Temperature (ECT) Sensor
12 - Electronic Control Module (ECM)
13 - Water Temperature Sender

14 - Fuel Pump Relay
15 - Ignition/System Relay
16 - Fuses (15 Amp) Fuel Pump, (15 Amp) ECM/DLC/Battery, (10 Amp) ECM/Injector/Ignition/Knock Module
17 - Oil Pressure Sensor
18 - Harness Connector To Starting/Charging Harness
19 - Positive (+) Power Wire To Engine Circuit Breaker
20 - Shift Plate
21 - Gear Lube Monitor
22 - Fuel Pressure Switch

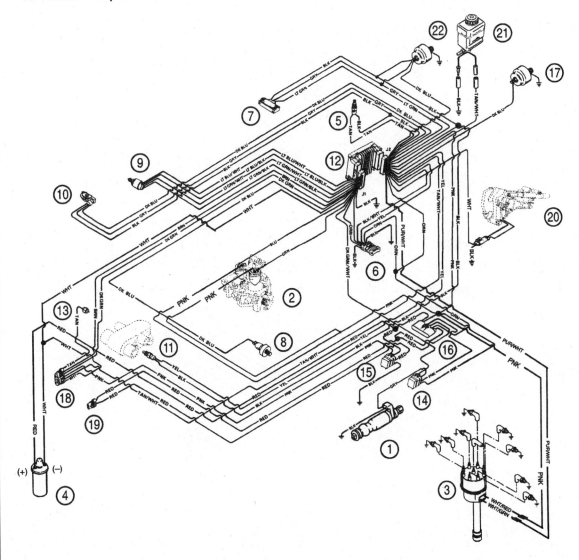

Fig. 161 Fuel/ignition systems wiring schematic—1998-01 V8 w/TBI (MEFI 3)

1 - Fuel Pump
2 - Distributor
3 - Coil
4 - Electronic Spark Control (KS) Module
5 - Data Link Connector (DLC)
6 - Manifold Absolute Pressure (MAP) Sensor
7 - Knock Sensor
8 - Idle Air Control (IAC)
9 - Throttle Position (TP) Sensor
10 - Engine Coolant Temperature (ECT) Sensor

11 - Electronic Control Module (ECM)
12 - Fuel Pump Relay
13 - Ignition/System Relay
14 - Fuse (15 Amp) Fuel Pump
15 - Fuse (15 Amp) ECM/DLC/Battery
16 - Fuse (10 Amp) ECM/Injector/Ignition/Knock Module
17 - Harness Connector To Starting/Charging Harness
18 - Positive (+) Power Wire To Engine Circuit Breaker

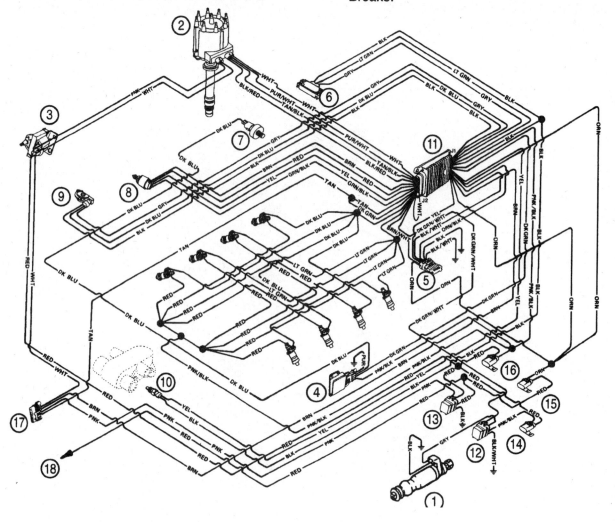

Fig 162 Fuel/ignition systems wiring schematic—1998-01 5.7L V8 w/MPI (MEFI 1/2), Bravo

1 - Fuel Pump
2 - Distributor
3 - Coil
4 - Knock Sensor (KS) Module
5 - Data Link Connector (DLC)
6 - Manifold Absolute Pressure (MAP) Sensor
7 - Idle Air Control (IAC)
8 - Throttle Position (TP) Sensor
9 - Engine Coolant Temperature (ECT) Sensor
10 - Electronic Control Module (ECM)
11 - Fuel Pump Relay
12 - Ignition/System Relay

13 - Fuse (15 Amp) Fuel Pump, Fuse (15 Amp) ECM/DLC/Battery,Fuse (10 Amp) ECM/Injector/Ignition/Knock Module
14 - Harness Connector To Starting/Charging Harness
15 - Positive (+) Power Wire To Engine Circuit Breaker
16 - Shift Plate (Not used on Ski models)
17 - Oil Pressure (Audio Warning System)
18 - Gear Lube Bottle (Not used on Ski models)
19 - Fuel Pressure Switch
20 - Water Temperature Sender

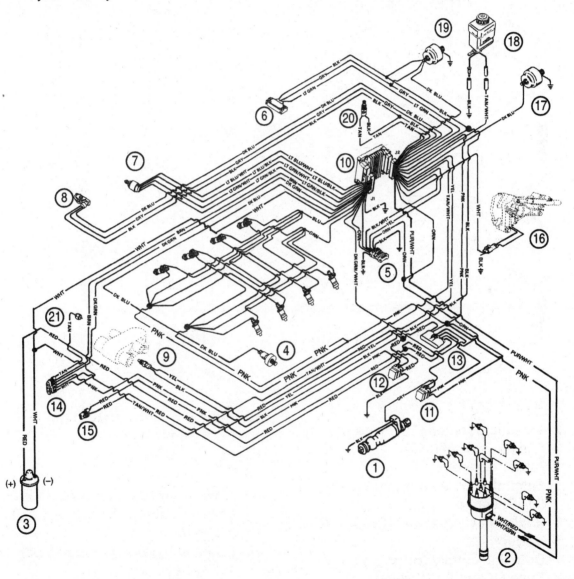

Fig. 163 Fuel/ignition systems wiring schematic—1998-01 5.7L/6.2L V8 w/MPI (MEFI 3)

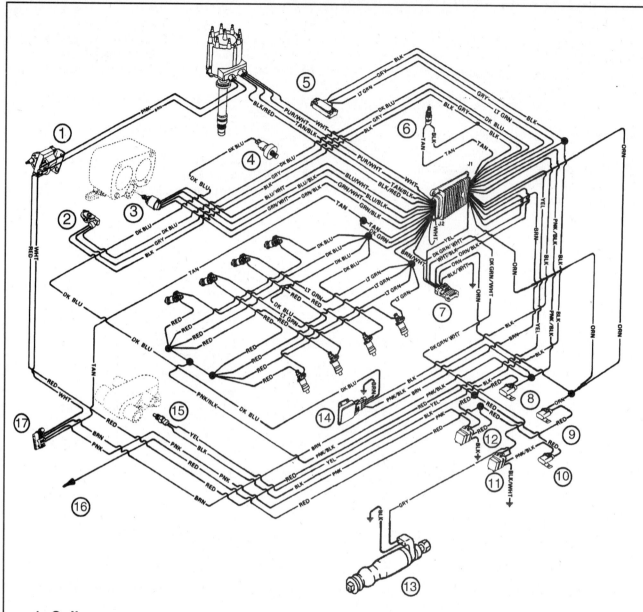

1- Coil
2- Throttle position (TP) Sensor
3- Idle air control (IAC)
4- Knock sensor
5- Manifold absolute pressure (MAP) sensor
6- Intake air temperature (IAT) sensor
7- Data link connector (DLC)
8- Fuse (10 amp) ECM/Injector/ Ignition/knock module
9- Fuse (15 amp) ECM/DCL/Battery

10- Fuse (15 amp) Fuel pump
11- Fuel pump relay
12- Ignition/system relay
13- Fuel pump
14- Electronic spark control module
15- Engine coolant temperature (ECT) sensor
16- Positive (+) power wire to engine circuit breaker
17- Harness connector to Starting/ Charging harness

Fig. 164 Fuel/ignition systems wiring schematic—1998-01 454/502 Mag/8.1L/8.2L V8 w/MPI (MEFI 1)

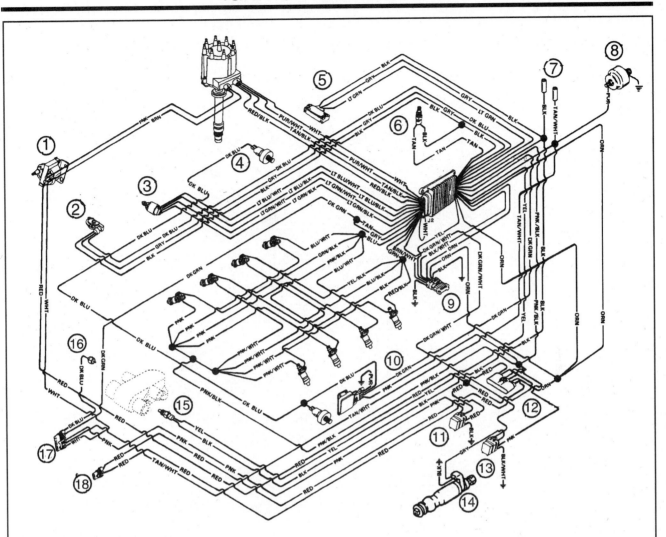

1- Coil
2- Throttle position (TP) sensor
3- Idle air control (IAC)
4- Knock sensor
5- Manifold absolute pressure (MAP) sensor
6- Intake Air Temperature (IAT) sensor
7- Not used
8- Water temperature sender (gauge)
9- Data link connector (DLC)

10- Electronic spark control module
11- Ignition/system relay
12- Fuse (15 amp) fuel pump fuse (15 amp) ECM/DLC/Battery fuse (10 amp) ECM/injector/ignition/knock module
13- Fuel pump relay
14- Fuel pump
15- Engine coolant temperature (ECT) sensor
16- Oil pressure - Audio warning switch
17- Harness connector to starting/charging harness
18- Positive (+) power wire to engine circuit breaker

Fig. 165 Fuel/ignition systems wiring schematic—1998-01 7.4L V8 w/MPI (MEFI 2)

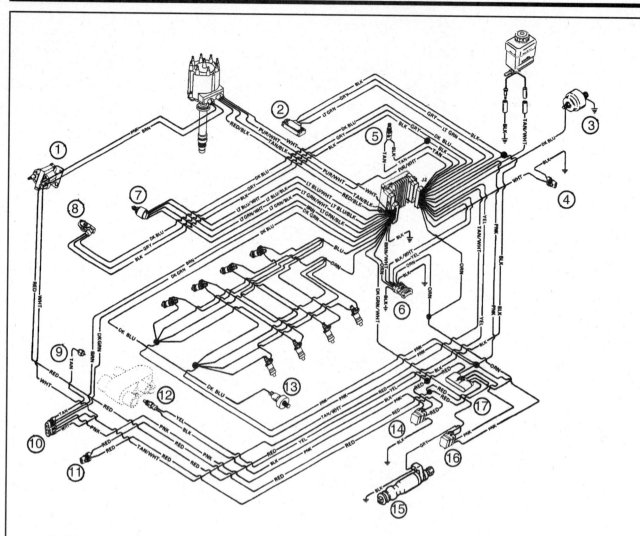

1- Coil
2- Manifold absolute pressure (MAP) sensor
3- Oil pressure (audio warning system)
4- Load anticipation circuit
5- Electronic control module (ECM)
6- Data link connector (DLC)
7- Idle air control (IAC)
8- Throttle position (TP) sensor
9- Water temperature sender
10- Harness connector to starting/charging harness

11- Positive (+) power wire to engine circuit breaker
12- Engine coolant temperature (ECT) sensor
13- Knock sensor (KS) module
14- Ignition/System relay
15- Fuel pump
16- Fuel pump relay
17- Fuses (15 amp) fuel pump, (15 amp) ECM/DLC/Battery, (10 amp)ECM/ Injector/ Ignition/knock module

Fig. 166 Fuel/ignition systems wiring schematic—1998-01 454/502 Mag/8.1L/8.2L V8 w/MPI (MEFI 3)

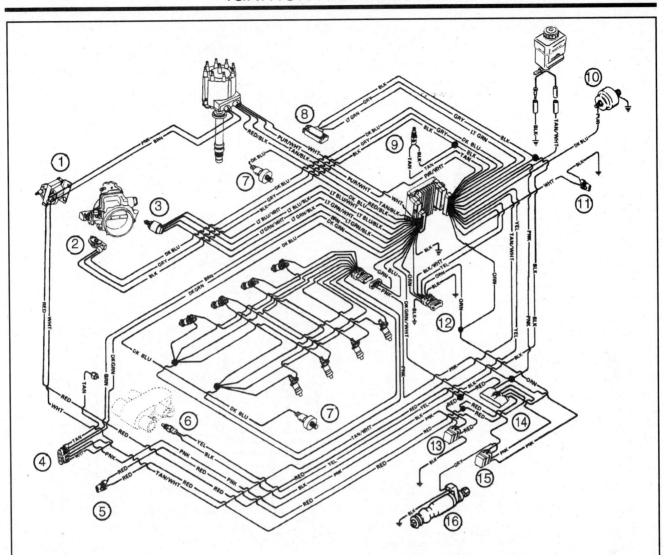

1- Coil
2- Throttle position (TP) sensor
3- Idle air control (IAC)
4- Harness connector to starting/
   charging harness
5- Positive (+) power wire to engine
   circuit breaker
6- Engine coolant temperature (ECT) sensor
7- Knock sensor
8- Manifold absolute pressure (MAP)
   sensor
9- Manifold air temperature (MAT) sensor
10- Oil pressure - audio warning switch

11- Load anticipation circuit (not used on
    MCM)
12- Data link connector (DLC)
13- Ignition/System relay
14- Fuse (15 amp) Fuel pump/ECM/DLC/
    Battery (10 amp) ECM/Injector/
    Ignition/knock module
15- Fuel pump relay
16- Fuel pump

Fig. 167 Fuel/ignition systems wiring schematic—1998-01 7.4L V8 w/MPI (MEFI 3)

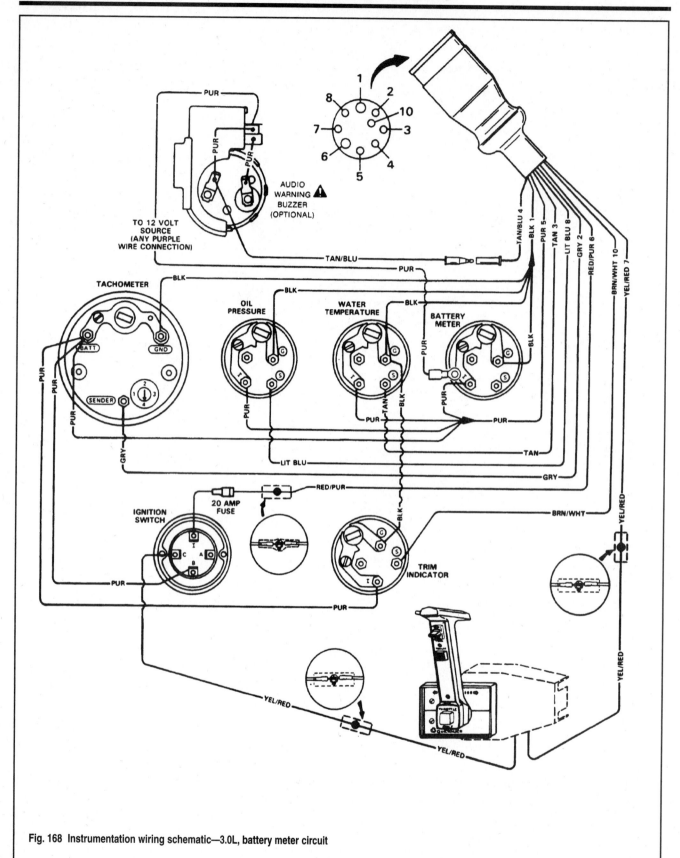

Fig. 168 Instrumentation wiring schematic—3.0L, battery meter circuit

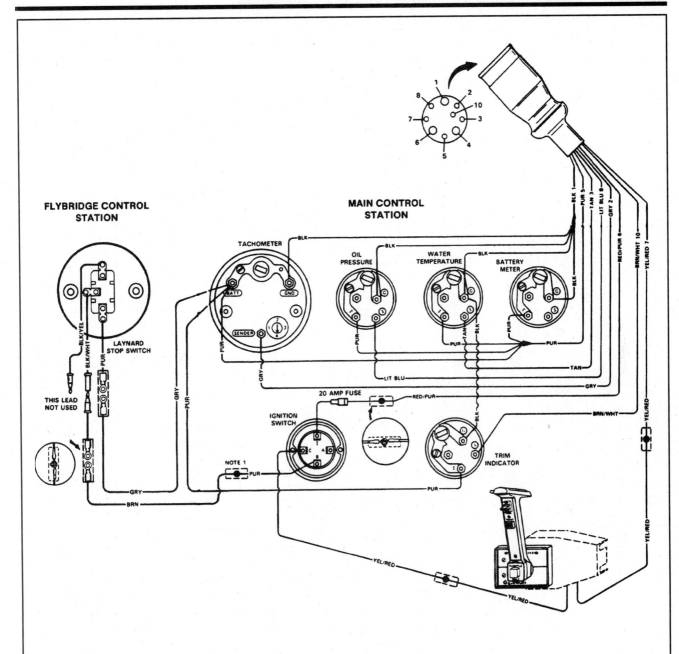

Fig. 169 Instrumentation wiring schematic—1992-97 3.0L, lanyard stop circuit (second station)

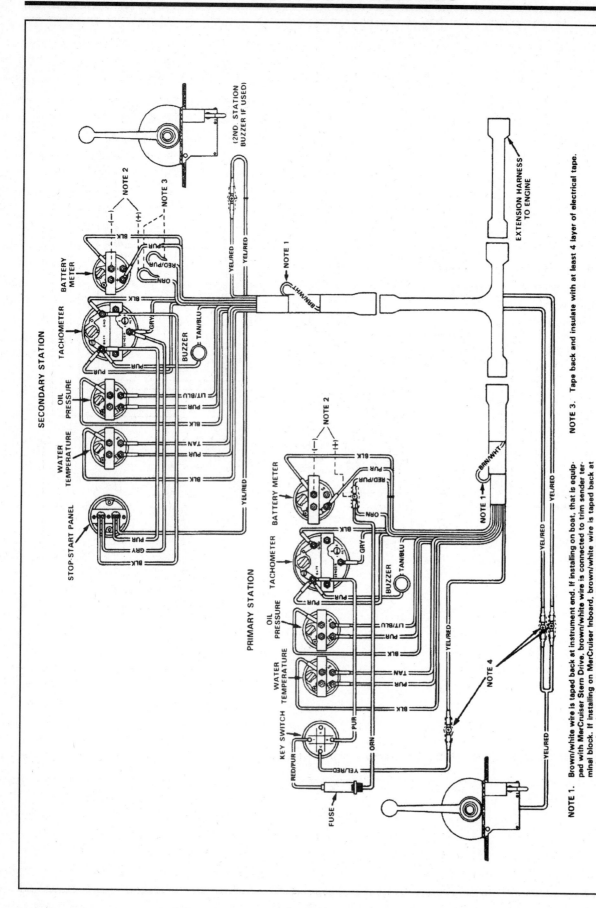

NOTE 1. Brown/white wire is taped back at instrument end. If installing on boat, that is equipped with MerCruiser Stern Drive, brown/white wire is connected to trim sender terminal block. If installing on MerCruiser Inboard, brown/white wire is taped back at engine end, or it may be used for an accessory (limit 5 amps.).

NOTE 2. An accessory fuse panel may be connected at this location. The combined current draw of the primary station and secondary station MUST NOT exceed 35 amps.

NOTE 3. Tape back and insulate with at least 4 layer of electrical tape.

Fig. 170 Instrument harness wiring schematic—3.0L w/breaker points, dual station, single neutral safety switch, dual switches similar

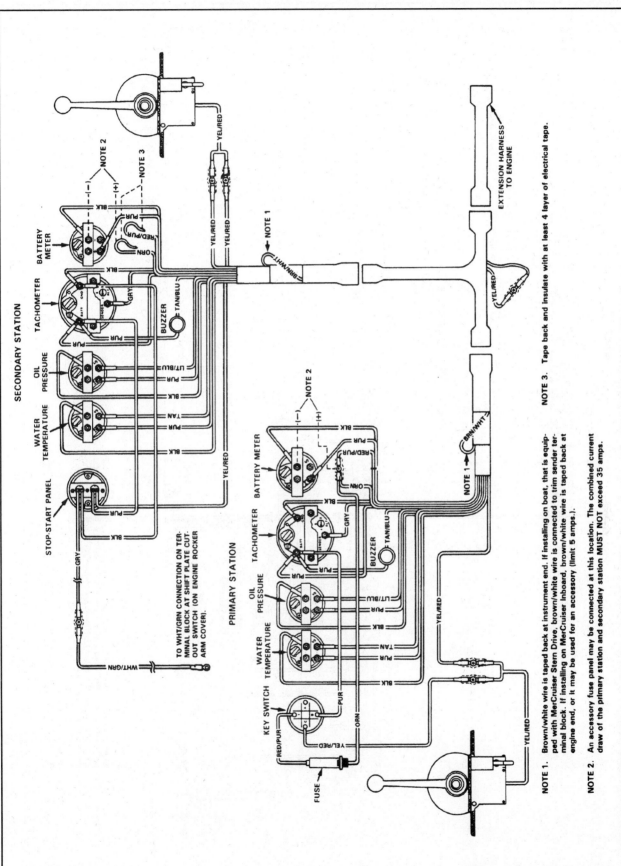

NOTE 1. Brown/white wire is taped back at instrument end. If installing on boat, that is equipped with MerCruiser Stern Drive, brown/white wire is connected to trim sender terminal block. If installing on MerCruiser Inboard, brown/white wire is taped back at engine end, or it may be used for an accessory (limit 5 amps.).

NOTE 2. An accessory fuse panel may be connected at this location. The combined current draw of the primary station and secondary station MUST NOT exceed 35 amps.

NOTE 3. Tape back and insulate with at least 4 layer of electrical tape.

Fig. 171 Instrument harness wiring schematic—3.0L w/DDIS or EST, dual station, two neutral safety switches, single switch similar

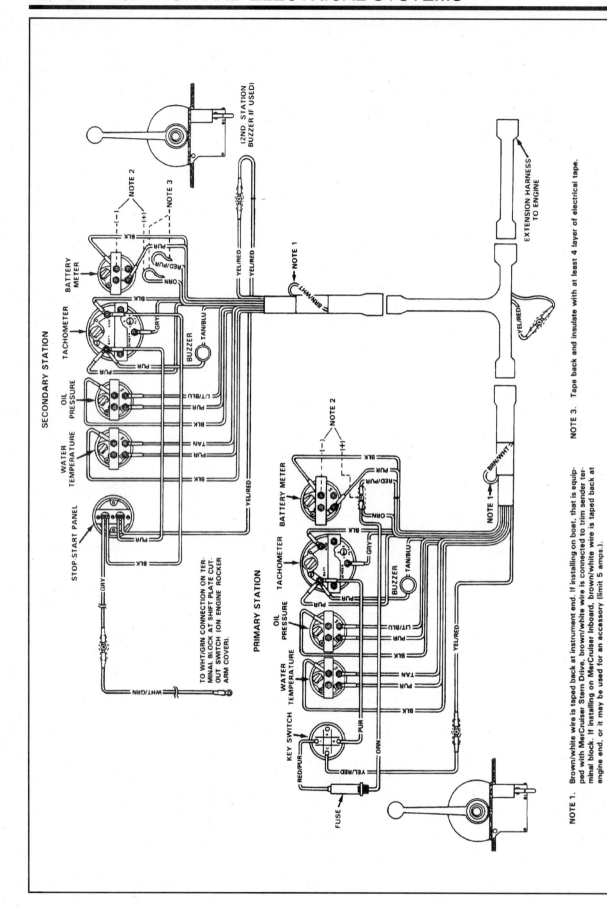

NOTE 1. Brown/white wire is taped back at instrument end. If installing on boat, that is equipped with MerCruiser Stern Drive, brown/white wire is connected to trim sender terminal block. If installing on MerCruiser Inboard, brown/white wire is taped back at engine end, or it may be used for an accessory (limit 5 amps.).

NOTE 2. An accessory fuse panel may be connected at this location. The combined current draw of the primary station and secondary station MUST NOT exceed 35 amps.

NOTE 3. Tape back and insulate with at least 4 layer of electrical tape.

Fig. 172 Instrument harness wiring schematic—3.0L w/DDIS or EST, neutral safety switch in harness

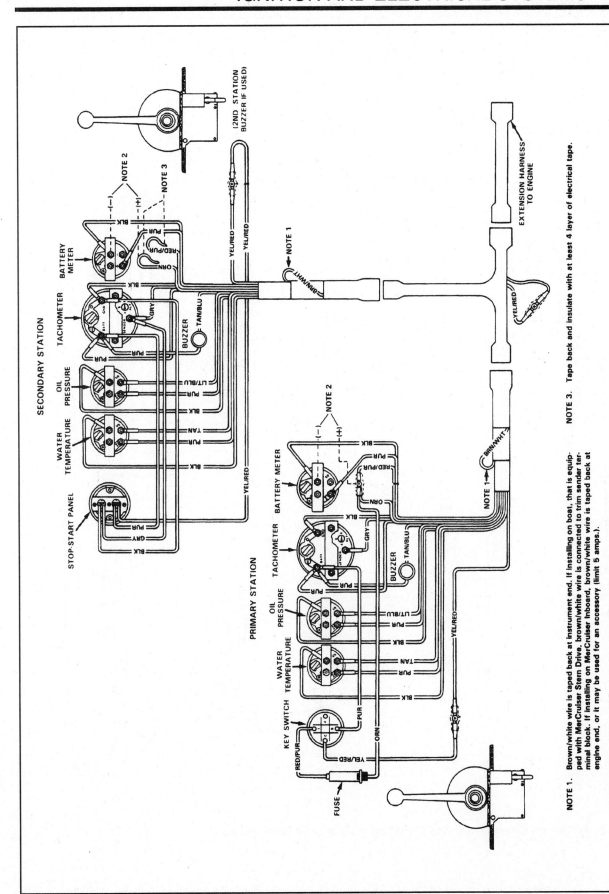

Fig. 173 Instrument harness wiring schematic—3.0L w/breaker points, neutral safety switch on transmission (inboard)

NOTE 1.   Brown/white wire is taped back at instrument end. If installing on boat, that is equipped with MerCruiser Stern Drive, brown/white wire is connected to trim sender terminal block. If installing on MerCruiser Inboard, brown/white wire is taped back at engine end, or it may be used for an accessory (limit 5 amps.).

NOTE 2.   An accessory fuse panel may be connected at this location. The combined current draw of the primary station and secondary station MUST NOT exceed 35 amps.

NOTE 3.   Tape back and insulate with at least 4 layer of electrical tape.

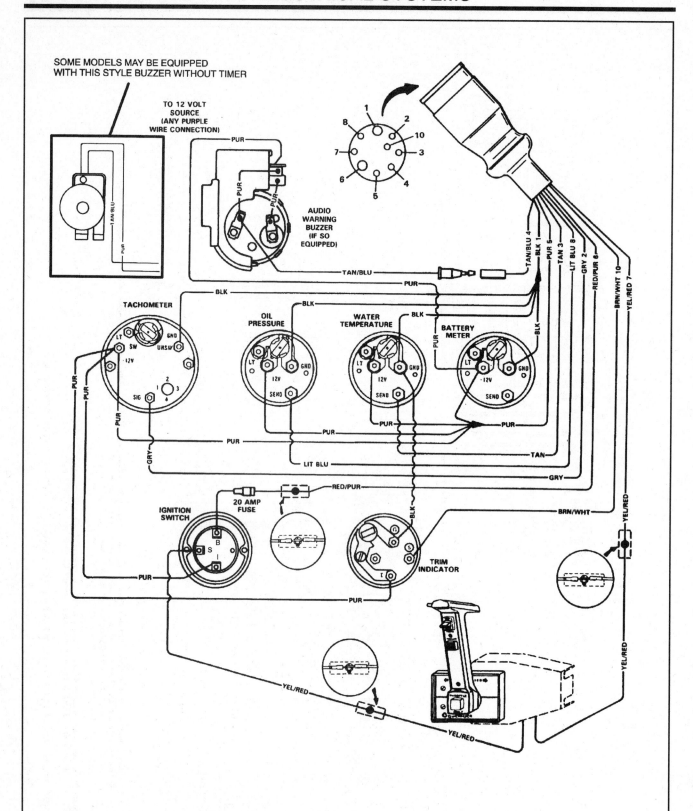

Fig. 174 Instrument harness wiring schematic—1992-97 V6/V8, single station (w/Quicksilver gauges)

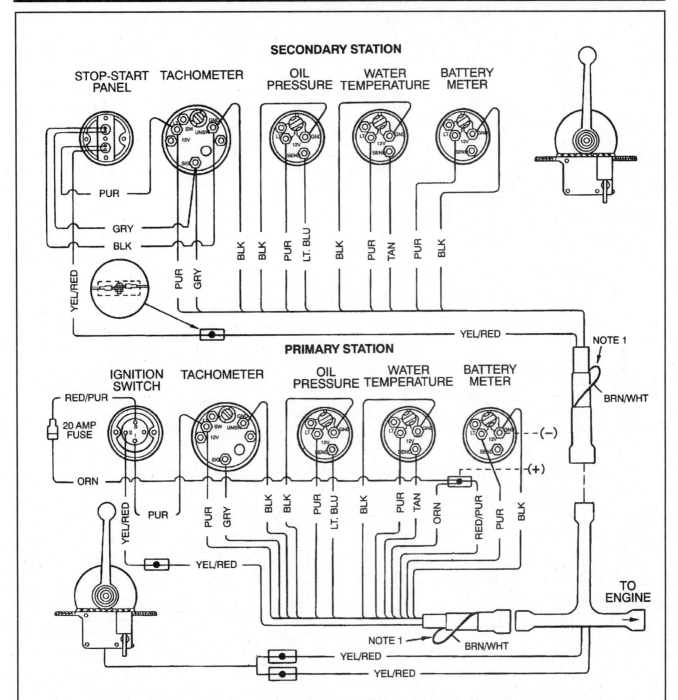

NOTE 1: Brown/white wire is taped back at instrument end. If installing on boat that is equipped with MerCruiser Stern Drive, brown/white wire is connected to trim sender terminal block. If installing on Mercruiser Inboard, brown/white wire is taped back at engine end, or it may be used for an accessory (limit 5 amps).

Fig. 175 Instrument harness wiring schematic—1992-97 V6/V8, dual station, single neutral safety switch

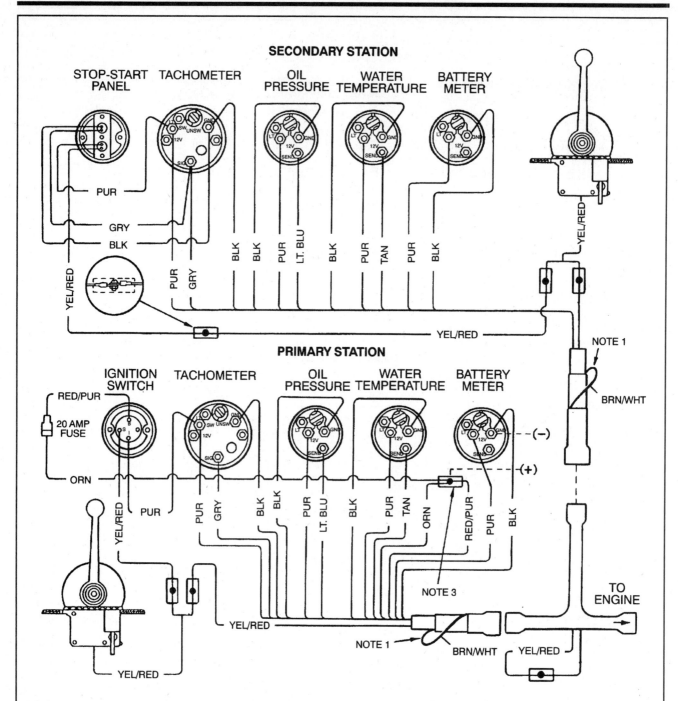

NOTE 1: Brown/white wire is taped back at instrument end. If installing on boat that is equipped with MerCruiser Stern Drive, brown/white wire is connected to trim sender terminal block. If installing on Mercruiser Inboard, brown/white wire is taped back at engine end, or it may be used for an accessory (limit 5 amps).

Fig. 176 Instrument harness wiring schematic—1992-97 V6/V8, dual station, two neutral safety switches

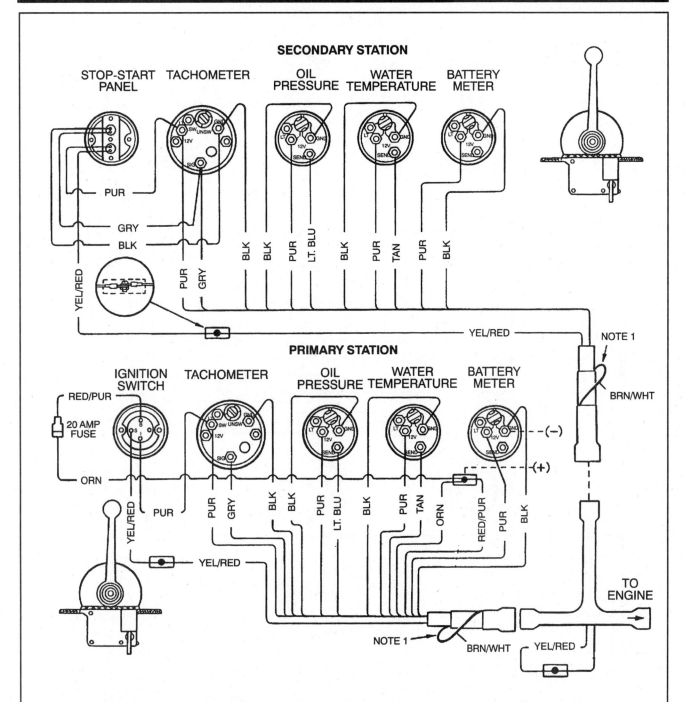

NOTE 1: Brown/white wire is taped back at instrument end. If installing on boat that is equipped with MerCruiser Stern Drive, brown/white wire is connected to trim sender terminal block. If installing on Mercruiser Inboard, brown/white wire is taped back at engine end, or it may be used for an accessory (limit 5 amps).

Fig. 177 Instrument harness wiring schematic—1992-97 V6/V8, dual station, single neutral safety switch (in harness)

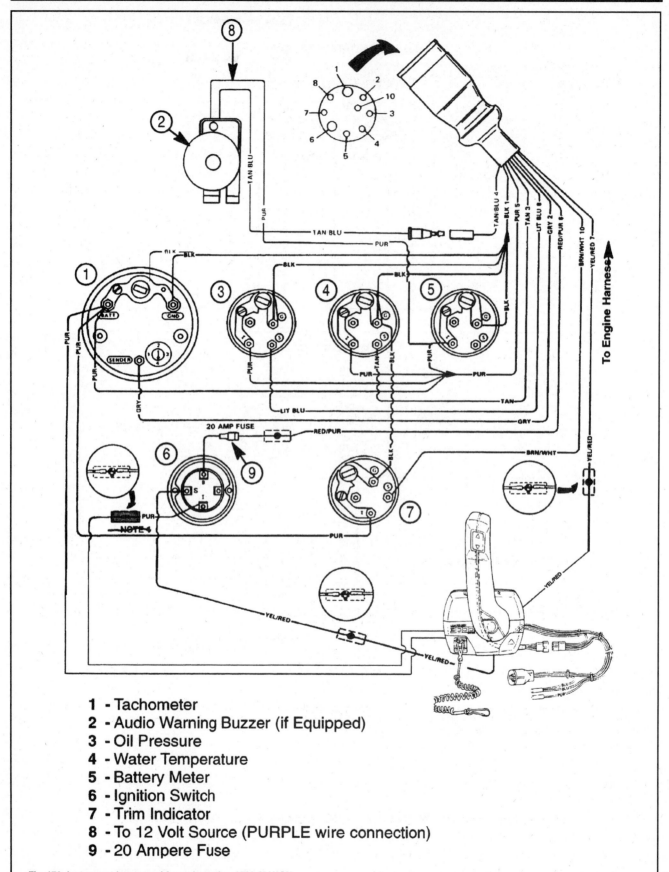

1 - Tachometer
2 - Audio Warning Buzzer (if Equipped)
3 - Oil Pressure
4 - Water Temperature
5 - Battery Meter
6 - Ignition Switch
7 - Trim Indicator
8 - To 12 Volt Source (PURPLE wire connection)
9 - 20 Ampere Fuse

Fig. 178 Instrument harness wiring schematic—1998-01 V6/V8

ALTERNATOR SPECIFICATIONS

Engine	Year	Alternator
3.0L	1992-97	Mando 55
		Prestolite 55
	1998-01	Mando 55
		Mando 65
4.3L	1992-97	Mando 55
	1998-01	Mando 55
		Mando 65
		Delco 65
5.0L/5.7L/6.2L	1992-97	Mando 55
		Mando 65
	1998-01	Mando 55
		Mando 65
		Delco 65
7.4L/8.1L/8.2L	1992-97	Mando 55
		Mando 65
		Prestolite 65
	1998-01	Mando 65
		Delco 65

Alternator	Output (Amps)		(Volts)	Excitation Circuit (Volts)	Condenser Capacity	Minimum Brush Length In. (mm)
Delco	65		13.8-14.8	NA	NA	NA
Mando	55					
		3.0L	13.8-14.8	1.5-3.0	0.5 MFD	3/16 (5)
		All others	13.9-14.7	1.3-2.5	0.5 MFD	1/4 (6)
	65		13.9-14.7	1.3-2.5	0.5 MFD	1/4 (6)
Prestolite	55		13.8-14.8	1.5-3.0	0.5 MFD	3/16 (5)
	65		13.8-14.8	1.5-3.0	0.5 MFD	3/16 (5)

STARTER SPECIFICATIONS

Engine	Year	Model	Type
3.0L	1992-97	14MT	Direct Drive
		PG200	Gear Reduction
	1998-01	PG260	Gear Reduction
		PG260F1	Gear Reduction
4.3L	1992-97	PG200	Gear Reduction
	1998-01	PG260	Gear Reduction
		PG260F1	Gear Reduction
5.0L/5.7L/6.2L	1992-97	PG200	Gear Reduction
		PG250	Gear Reduction
	1998-01	14MT	Direct Drive
		PG260F1	Gear Reduction
7.4L/8.1L/8.2L	1992-97	PG200	Gear Reduction
		PG250	Gear Reduction
		PG260	Gear Reduction
		14MT	Direct Drive
	1998-01	PG260	Gear Reduction
		PG260F1	Gear Reduction
		14MT	Direct Drive
		Mando	Gear Reduction

Starter				No Load			Brush Spring Tension Oz. (g)	Pinion Clearance In. (mm)	Commutator End Frame Gap In. (mm)	Gear Bearing Depth In. (mm)	Housing Bearing Depth In. (mm)
Brand	Type	Volts	Amps	Amps	rpm	rpm					
Delco	14MT	10.6	70	120	5400	10800	56-105 (1588-2976)	0.010-0.140 (0.25-3.5)	0.025 Max (0.6 Max)	NA	NA
	PG200	10.6	60	90	3000	3300	83-104 (2353-2948)	0.010-0.160 (0.25-4.06)	0.015 Max (0.4 Max)	0.011-0.014 (0.28-0.38)	0.009-0.017 (0.4 Max)
	PG250	10.6	60	95	2750	3250	0.011-0.014 (0.28-0.38)	0.010-0.160 (0.25-4.06)	NA	0.011-0.014 (0.28-0.38)	0.009-0.017 (0.4 Max)
	PG260	10.6	60	95	2750	3250	0.011-0.014 (0.28-0.38)	0.010-0.160 (0.25-4.06)	NA	0.011-0.014 (0.28-0.38)	0.009-0.017 (0.4 Max)
	PG260F1	11.5	40	90	3200	4800	NA	0.009-0.160 (0.23-4.06)	NA	Flush	0.009-0.017 (0.4 Max)
Mando		11	NA	90	2800	2900	83-104 (2353-2948)	0.010-0.160 (0.25-4.06)	NA	NA	NA

11

DRIVE SYSTEMS ALPHA

STERN DRIVE UNIT – ALPHA

Description

◆ **See Figures 1 and 2**

That which we refer to as the stern drive is actually a number of individual components attached and working together to transfer the power of the engine into a viable propulsion system for your boat. All stern drive units can be broken down into their components assemblies.

The transom assembly, consisting of an inner transom plate, a gimbal housing and gimbal plate, and a bell housing is just what it sounds like—the unit attached to the transom of the vessel. The inner transom plate is, obviously, attached to the inner side of the transom and actually makes up the rear engine mounts.

On the other side of the transom, and attached to the transom plate, are the gimbal housing, gimbal plate (or ring) and bell housing. The bell housing is attached to the gimbal plate via roller bearings and is what allows for the up and down (trim) movement of the stern drive unit itself. The gimbal plate is also attached to the gimbal housing via roller bearings and is what allows for side-to-side movement of the unit, or, steering.

The stern drive unit, or at least that thing that is most visible when viewing the stern of the boat, is made of two component assemblies: the driveshaft housing and the gear housing.

The driveshaft housing, frequently called the upper gear housing or simply the upper unit is attached to the bell housing at the top and the gear housing at the bottom. Power from the engine, brought through the transom assembly via the driveshaft is transferred to a vertical shaft leading to the lower unit by means of a set of drive and driven gears.

The gear housing, or lower unit, is attached to the bottom of the driveshaft housing. Power, or propulsion, comes through the vertical shaft from the upper unit, is transferred to the propeller shaft via a pinion gear and causes the propeller to rotate.

Output power from the engine is connected to the stern drive through a horizontal driveshaft. A coupler is bolted to the flywheel and has a splined hub in the center. The end of the horizontal driveshaft indexes with and slides into the center of the hub. Power from the engine is then transmitted through the horizontal driveshaft to a pinion gear set where power direction is changed from horizontal to vertical.

The upper driven gear is pressed onto the outside diameter of the upper driveshaft. The upper driveshaft is splined on the lower end. When the upper gear housing is mated to the lower unit, the end of the lower unit driveshaft indexes into the splined end of the upper driveshaft. Engine power is then transferred down into the lower gear unit.

The horizontal and vertical driveshafts are both mechanically connected. Therefore, anytime the engine is operating, the horizontal and vertical driveshafts are constantly rotating with engine rpm. A double yoke universal joint assembly in the horizontal driveshaft allows the stern drive to be raised or lowered to a required trim/tilt position (within limits), while the engine is operating.

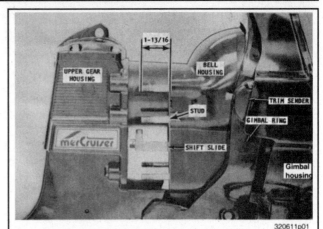

Fig. 1 A good look at the entire drive system

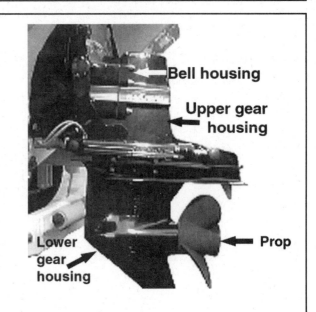

Fig. 2 Stern drive unit components

General Information

The following are a list of potential drive unit problems and their possible causes:

GEAR HOUSING NOISE

1. Metal particles in the unit oil supply.
2. Propeller incorrectly installed.
3. Propeller or propeller shaft bent.
4. Incorrect drive gear shimming—gear housing back-lash or pinion gear height.
5. Worn or damaged gears or bearings.

DRIVESHAFT HOUSING NOISE

1. Steering lever may be contacting the transom cut-out edges when turning.
2. Flywheel housing on the engine coming in contact with the inner transom plate or the exhaust pipe.
3. Bad propeller.
4. U-joint cross and bearing assembly O-rings of incorrect size or installed wrong.
5. U-joint cross and bearing assemblies have excessive side-play.
6. Bearing caps on the U-joint are coming in contact with the center socket or the driveshaft housing bearing retainer.
7. Rough or scored U-joint cross and bearings.
8. Worn or missing O-rings on the U-joint shaft rattling against the gimbal bearing.
9. Worn or damaged splines on the driveshafts and/or couplers.
10. Incorrect engine alignment.
11. Damaged, rough, worn or loose gimbal bearing.
12. Gimbal bearing seated incorrectly.
13. Incorrect clearance between the gimbal housing and plate.
14. Bell housing and gimbal housing incorrectly aligned.
15. Weak or flexing transom.
16. Weak, missing or mis-aligned rear engine mount.

DRIVE UNIT WILL NOT SLIDE INTO BELL HOUSING

1. U-joint and engine coupler splines not aligned.
2. Unit not in Forward gear.
3. Incorrectly aligned shift shaft coupler.
4. Engine out of alignment.
5. Incorrectly installed gimbal bearing.
6. Damaged or worn splines on the driveshaft or coupler.

DRIVE UNIT WILL NOT SHIFT—SHIFT HANDLE MOVES

1. Shift cables not adjusted properly.
2. Shift cables not connected.
3. Inner wire on cable broken.
4. Gear housing crank installed incorrectly.

DRIVE UNIT WILL NOT SHIFT—SHIFT HANDLE DOES NOT MOVE

1. Remote control box/assembly installed incorrectly.
2. Broken, worn or damaged linkage.
3. Shift shaft or lever stuck.
4. Controls or cables installed or adjusted incorrectly.

HARD SHIFTING

1. Shift cables out of adjustment.
2. Shift cut-out switch broken or improperly adjusted.
3. Shift cable too short, or too long.
4. Corroded cables.
5. Shift shaft bushings corroded or damaged.
6. Shift crank or clutch actuating spool worn or damaged.
7. Shaft shaft damaged, worn or broken.
8. Shift cable attaching nuts too tight.

JUMPS OUT OF GEAR

1. Incorrectly adjusted shift cables.
2. Worn or damaged clutch or gears.

Stern Drive Unit

REMOVAL & INSTALLATION

◆ **See Figures 3, 4, 5, 6, 7, 8, 9, 10, 11 and 12**

1. Drain the stern drive unit oil as detailed in Section 3. Check the drained lubricant carefully to determine if it contains any water or metal particles. Rub some of the lubricant between your fingers. Any metal particles in the lubricant will thus be evident. Do not be mislead if the color of the lubricant is metal colored. This is not a harmful condition and is caused by the lubricant used in the unit the first time after manufacture.

2. Shift the unit into Forward gear. On engines with left hand (counter) rotation, move the lever to Reverse gear.

3. Remove the plastic cap over the anchor pin connecting the trim cylinders to the aft end of the upper unit. Remove the e-rings and washers and pry out the bushings. Disconnect the cylinder arms from the unit and carefully move them back against the transom.

■ **Although not imperative, we highly recommend the use of an engine hoist or other lifting device when removing the unit.**

4. Install a lifting device if so desired.

5. Raise the stern drive to the full UP position. Reach in on top of the anti-ventilation plate at the transom end and grasp the top half of the plastic fitting for the speedometer. Twist the fitting 1/4 turn counterclockwise and pull up—separating the upper fitting half from the lower half. Move the cable out of the way.

6. Remove the six 5/8 in. nuts and washers securing the stern drive to the bell housing. There are only five washers as the lower starboard bolt (looking toward the bow) uses a permanent ground plate instead of a washer.

7. Secure the lifting chains so they are taught and then remove the stern drive unit by pulling it straight back and free of the bell housing. Remove and discard the gasket.

■ If the stern drive unit is difficult to remove, the driveshaft splines may be frozen in the engine coupler or the shaft may be frozen onto the gimbal bearing. One of two methods may be used to break the stern drive loose from the coupler. The first involves disconnecting the engine mounts and then moving the engine forward slightly to enable the driveshaft to be pryed from the splines. The second method is to remove the top cover plate and then drive off the yoke nut in front of the drive gear with a punch and hammer. The stern drive can then be removed, leaving the frozen shaft in the bell housing.

8. If you were unable to remove the drive unit with the driveshaft attached, remove the bell housing as detailed later in this section and then cut or remove the exhaust and U-joint bellows.

9. Remove the snap-ring securing the gimbal bearing with a pair of tru-arc pliers—yes the U-joints are in the way, but be patient and you'll get it out.

10. Use a slide hammer and pull out the shaft assembly. If it is frozen to the bearing, or the bearing is frozen, apply a little heat from a torch while working the slide hammer. See the Transom Assembly procedures later in this section for complete details on the gimbal bearing.

To install:

The engine must be properly aligned or the female splines in the engine coupler and the male splines of the driveshaft will be destroyed after a short time of engine operation.

If troubleshooting indicates the coupler is damaged and requires replacement, remove the engine and replace the coupler. The coupler is simply secured to the flywheel with attaching bolts. If the necessary tools are not available for proper engine alignment, the boat should be taken to an authorized marine dealer.

11. Install a new O-ring in the water passageway between the upper unit and the bell housings and the rubber sealing ring inside the bell housing. Install a new gasket on the face of the bell housing.

12. Coat all six studs and their threads with Quicksilver 2-3-C marine lubricant.

13. Check the bell housing bearing to be sure it turns freely without binding or rough spots.

14. Oil the shaft to prevent damage to the seals, and then slide two new O-ring seals over the splines of the universal joint shaft, and into the grooves provided.

15. Coat the splines of the driveshaft and the driveshaft housing pilot, with Multi-purpose Lubricant. As an aid to installation, apply a light coating of Multi-purpose Lubricant to the inside surfaces of the bell housing bore.

16. Apply a coating of Multi-purpose Lubricant onto the shift lever coupler, and then align the slot straight by moving the lever to the right. Lubricate the slot and the reverse lock roller.

17. Check to be sure the stern drive unit is in Forward gear. Rotate the propeller shaft to align the intermediate shift shaft coupling with the upper shift shaft slot. Bad News: If the coupler of the intermediate shift shaft in the driveshaft housing and the locating slot of the shift lever in the bell housing are not aligned properly, the bell housing and shift shaft couplers will be damaged when the drive unit is installed. Remember that on engines with left hand rotation, the unit must be in Reverse.

18. Coat the U-joint O-rings and gimbal housing ball bearing with Multi-purpose Lubricant, to assist installation. Install the drive unit into the bell housing bore. If the driveshaft splines do not index with the splines of the engine coupler, slide the propeller onto the propeller shaft, and then slowly rotate the shaft counterclockwise until the drive unit can be pushed completely into position. While performing this installation maneuver, make sure the shifting slide assembly does not turn out of position. If the stern drive will not move completely into place, check underneath and observe if the shift cam is properly aligned in the groove on the stern drive.

19. If you r boat utilizes a gear lube monitor, be sure that the housing gasket has a hole where the passage is.

20. Secure the driveshaft housing to the bell housing with the six elastic-stop nuts with flat washers. Tighten the nuts alternately and evenly to a torque value of 50 ft.lbs. (68 Nm). If you removed the top cover because the driveshaft was stuck, tighten the screws to 17–23 ft. lbs. (23–31 Nm).

21. Connect the speedometer cable in the opposite manner that you disconnected it—turn the fitting clockwise this time.

22. Install the trim cylinders.

23. Fill the unit with oil.

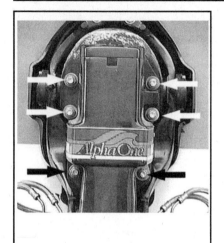

Fig. 3 There are six mounting bolts...

Fig. 4 ...remove them in a criss-cross pattern

Fig. 5 You can try it this way, but we recommend a lifting hoist

Fig. 6 Removing the yoke nut with a punch

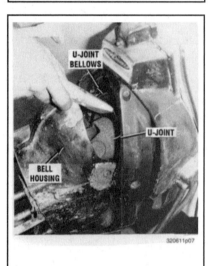

Fig. 7 Cut back the bellows after removing the bell housing

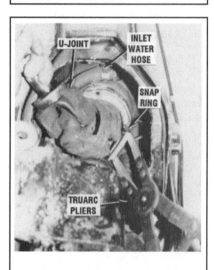

Fig. 8 Use the pliers to remove the gimbal bearing snap-ring

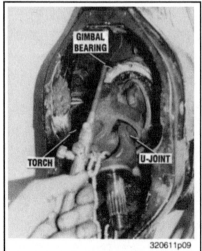

Fig. 9 You will probably need a torch to free up the frozen bearing

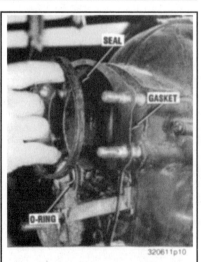

Fig. 10 Don't forget the O-ring and seal when installing the gasket

Fig. 11 Make sure two install two new O-rings

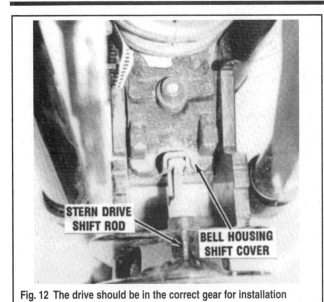

Fig. 12 The drive should be in the correct gear for installation

Trim Cylinder

REMOVAL & INSTALLATION

MODERATE

◆ See Figures 13 and 14

1. Raise the stern drive to the full UP position. Reach in on top of the anti-ventilation plate at the transom end and grasp the top half of the plastic

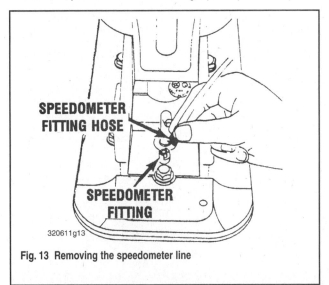

Fig. 13 Removing the speedometer line

fitting for the speedometer. Twist the fitting 1/4 turn counterclockwise and pull up—separating the upper fitting half from the lower half.

✳✳ CAUTION

Be sure the stern drive is in the full DOWN position before proceeding with this procedure. If the trim/tilt cylinders are disconnected with the stern drive in the up position, the stern drive will swing down suddenly causing severe damage to the stern drive and possibly personal injury.

2. Using a screwdriver, pry off the plastic caps on each end of the trim cylinders where they attach to the upper gear housing. Remove the E-clip from each side of the aft anchor pin and lift off the flat washers.
3. Grasp the cylinder and pull it off the end of the anchor pin, or use a blunt end punch and tap the anchor pin out from the stern drive upper gear housing. Remove the bushings—four total—from the trim/tilt ends and place it with the rest of the attaching hardware to guard against possible loss. Position the trim/tilt cylinders away from the work area and secure them with a piece of rope.

To install:
4. Place a bushing inside each cylinder end cap, if it was removed or fell off during disassembling. Insert the anchor pin through the aft end of the upper gear housing.
5. Now, slide each cylinder end cap onto the anchor pin. Install the second bushing into the end cap followed by a flat washer. Secure the trim/tilt cylinder to the anchor pin with the E-lip inserted into the inboard groove on the anchor pin. The outboard grooves in the anchor pin are for the plastic caps. Install the plastic cap over the end of the anchor pin and push in until it locks.
6. Raise the stern drive to the full UP position. Grasp the speedometer fitting and insert it into the fitting in the lower half on top of the anti-ventilation plate. Push down and twist the fitting 1/4 turn clockwise to lock it in place.

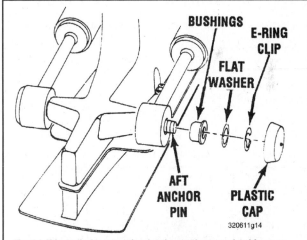

Fig. 14 Trim cylinder mounting hardware; there are bushings on the inside of each arm also

GEAR HOUSING (LOWER UNIT)

Description

The upper gear housing transfers engine power to the lower unit through a driveshaft. Splines on the forward end of the drive shaft mate with matching splines of a coupler attached to the engine flywheel. The aft end of the driveshaft is connected to a universal joint and the upper gear assembly. The pinion gear and drive gear changes the direction of the power train from horizontal to vertical. The power is transferred from the upper gear housing down to the lower unit through a vertical driveshaft. A water pump installed in the lower unit is constantly driven by the vertical driveshaft. This pump supplies cooling water directly to the engine, or indirectly through a heat exchanger.

The forward and reverse gears are contained within the lower unit. A sliding clutch, actuated by the shift linkage, engages a forward gear and creates a direct coupling for transmitting power through the pinion gear and forward gear to the propeller shaft. The reverse gear utilizes the same mechanics as the forward gear, except that the clutch is coupled with the reverse gear.

Gear Housing (Lower Unit)

REMOVAL & INSTALLATION ② ◁ MODERATE

◆ **See Figures 15, 16, 17, 18, 19, 20 and 21**

The lower unit may be separated and removed from the upper gear housing while the stern drive remains attached to the boat or after the stern drive has been removed.

1. If you plan on disassembling the unit, remove the propeller from the lower gear housing as detailed later in this section and also in Section 3.

2. If the unit is still on the boat, drain the drive oil as detailed in Section 3. Allow the lubricant to drain into the container. As the lubricant drains, check the color. Dark brown to black indicates normal old lubricant. A chalky white to cream color indicates the presence of water. The presence of any water in the gear lubricant is bad news. The unit must be completely disassembled, inspected, the cause of the problem determined and corrected. Close attention should be given to the back-to-back seals on the propeller shaft, driveshaft and the bearing carrier O-ring. Examine the magnet on the end of the fill/drain plug for evidence of metal particles. The presence of tiny small metal dust like shavings indicates normal wear of the gears, bearings and shafts within the lower unit. Large metal chips or heavy fillings indicate extensive internal damage is taking place and the lower unit must be completely disassembled and inspected. All worn and/or damaged components must be replaced.

3. If the stern drive is installed on the boat, place the remote control shift lever in the Forward gear position.

4. Place an alignment mark on the trim tab trailing edge and the anti-cavitation plate as an aid during assembling. Remove the plastic plug from the top of the anti-cavitation plate. Remove the bolt directly above the trim tab, lift off the trim tab. Remove the four hex nuts and washers (two on each side), from the sides of the anti-cavitation plate. Disconnect the speed indicator fitting on the leading edge of the anti-cavitation plate by pushing down on the fitting and turning it 1/4 turn counterclockwise. Remove the nut and washer in front of the fitting. Remove the bolt up inside the cavity for the trim tab.

5. Separate the two housings by tapping downward on the lower unit with a soft face mallet and a wooden block. In some cases the housing may be difficult to separate because the driveshaft may be corroded and "frozen" in the upper gear housing vertical shaft. Lower the gear housing away from the upper gear housing.

To install:

6. Verify the remote control shift handle is in the Forward gear position for standard right hand stern drives. The 90º shift lever should be facing forward and in the center of the lower unit. Rotate the propeller shaft clockwise to verify gear engagement.

■ **If the unit being serviced is a counter-rotation—left hand drive— the gear shift lever should be in the reverse gear position.**

7. Verify the trim tab bolt is in place in the aft section of the lower unit.

8. Check to be sure the O-ring is in place on the passageway for the gear oil.

9. Apply a coating of Quicksilver 2-4-C Lubricant, or equivalent, to the splines on the end of the driveshaft.

10. Some lower unit housings may have an aluminum dam behind the water pump. If the aluminum dam is corroded or damaged it must be replaced with a rubber filler unit. If the aluminum dam is serviceable, check the drain hole on the starboard side of the dam to verify it is not plugged. If this drain hole is plugged, severe damage to the gear housing will result.

11. Check to be sure the plastic guide tube is installed in the water pump housing. Verify the water pump face seal is positioned on top of the water pump cover. This seal prevents exhaust gases from entering the water cooling system.

If the unit has an aluminum dam and it has not been replaced with the rubber plug, place a bead of Permatex Ultra Blue Silicone Sealant, or equivalent, across the top of the dam. If a rubber dam is installed, no sealant is required.

12. Now, raise the lower unit (or lower the upper unit) to mate with the upper gear housing, and at the same time check to be sure the water tube starts into the water pump guide tube. Verify the 90º end of the shift shaft indexes with the slotted arm in the upper gear housing.

13. It may be necessary to slowly rotate the propeller shaft clockwise until the splines on the upper and lower drive shafts index.

14. Screw the nut onto the stud at the front of the unit. Install the aft screw into the forward hole in the trim tab well.

15. Install the two port and two starboard bolts and their nuts and washers. Tighten the four 5/8 in. nuts alternately, and a little at-a-time, until the lower unit is tight against the upper gear housing.

16. Now tighten the front nut and the four side nuts to 35 ft. lbs. (47.5 Nm). Tighten the aft screw to 28 ft. lbs. (41 Nm).

17. Install the trim tab with the marks made during disassembly aligned. Tighten the screw(s) to 23 ft. lbs. (31 Nm).

■ **The trim tab performs two very important jobs, one of which you may not realize. First, the tab compensates for steering torque. If the boat continually seems to move to port or starboard while the helmsman is on a straight course, the trim tab can be adjusted to the side of the pull. The tab also prevents electrolysis from damaging expensive parts. The tab is not expensive; it should show signs of electrolytic action; and should always be replaced after some of the material has been eaten away or is pitted. Now, if the tab shows no signs of electrolytic action after the boat has been in use over a period of time, the grounding should be checked to ensure more expensive parts are not being damaged. Install the trim tab plastic cover plug.**

18. Install the propeller if it was removed and refill the drive unit with oil.

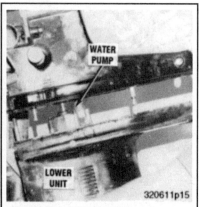

Fig. 15 Separating the upper and lower units

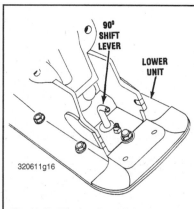

Fig. 16 Positioning the shift lever

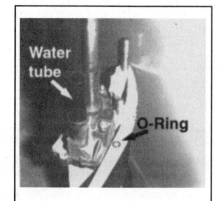

Fig. 17 A view of the water pump and tube, don't forget the O-ring

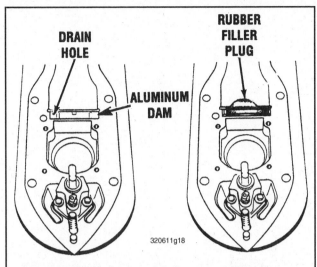

Fig 18 Some early model Alpha One sterndrives were equipped with an aluminum dam. Later models replace the aluminum dam with a rubber filler plug. Aluminum dams require a bead of silicone sealant, as explained in the text.

DISASSEMBLY

◆ **See Figures 22, 23, 24, 25, 26, 27 28, 29, 30, 31 and 32**

The following procedures assume that the lower unit, water pump, seal carrier and propeller have already been removed.

1. To remove the reverse bearing carrier, first straighten the locking tab on the tab washer. Loosen the cover nut turning it counterclockwise with special wrench P/N C-91-61069. Remove the cover nut and tab washer. Remove the aligning key.

■ **If the retainer is corroded in place, drill 4 holes in the nut and then break the retainer into pieces with a chisel. Pry out the pieces.**

2. Pull out the reverse gear bearing carrier assembly using a bearing carrier puller. The thrust hub for the propeller, which was removed previously, can be used to keep the puller jaws in place.

■ **If the reverse gear bearing carrier is corroded in place, apply heat to the housing to loosen the corrosion. While the housing is still hot, use a bearing carrier puller to remove the assembly do not overheat the housing or it will become distorted and useless.**

3. Reach in and remove the thrust washer-to-gear housing shim material. Tag and wire them together, as an aid during assembling.

4. Remove the driveshaft pinion nut by sliding the pinion nut holding tool, P/N 91-61067A2 over the propeller shaft. Turn the driveshaft counterclockwise with the special splined wrench adapter P/N C-91-56775 and a break-over handle. If the pinion nut holding tool is not available, a substitute can be made from a box-end wrench of the proper size for the nut. Grind the wrench down on both sides until it will clear the clutch dog and the pinion gear. Place the wrench on the nut and proceed with disassembling. Remove the nut.

5. Remove the outer race retaining nut in order for the driveshaft to clear the housing. To remove the nut, obtain driveshaft bearing retainer nut tool P/N 91-43506. Slide the tool down the driveshaft until the tool is indexed over the nut. Now, use a breaker bar on the special tool and loosen the nut. Slide the special tool free of the driveshaft, and then back the nut off and clear of the driveshaft.

6. Clamp the driveshaft in a vise equipped with brass soft jaws. If soft jaws are not available, clamp the driveshaft on a cast surface, away from the machined splines. Use a block of wood to protect the housing from damage when hammering on it. Drive the housing downward by striking the block of wood with a hammer. Once the driveshaft is free of the pinion gear, slide the housing clear of the driveshaft. When the driveshaft is removed from the housing, 18 needle bearings in the pinion gear bearing may fall out into the propeller end of the housing. Reach inside the propeller end of the housing and retrieve the 18 loose needle bearings, the pinion gear, washer and nut.

7. On the top of the housing where the driveshaft bearing was positioned, reach inside the bearing bore and remove the shim material for the bearing. Identify and save the shim material for later measurement and count.

8. Move the propeller shaft to the port side of the lower unit to pass by the shift cam. Pull out the propeller shaft assembly and forward gear.

9. Remove the two bolts securing the shift shaft bushing to the lower unit housing. Grasp the shift shaft bushing and pull it straight up and out of the lower unit housing. Rotate the lower unit housing or reach inside through the propeller shaft opening and retrieve the shift crank from the gear housing.

■ **The next step is not required, unless the bearings are to be replaced. Any time the bearings are replaced, they must be adjusted properly with shim material or their service life will be drastically reduced.**

10. Using a slide hammer, remove the propeller shaft tapered roller bearing cup. Tag and wire the shim material pieces from behind the bearing cup together, as an aid during assembly.

Fig. 19 Make sure that the water tube slides into the guide

Fig. 20 Tightening the port attaching nuts

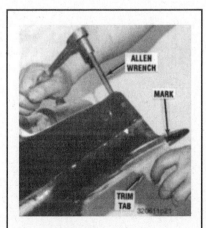

Fig. 21 Installing the trim tab

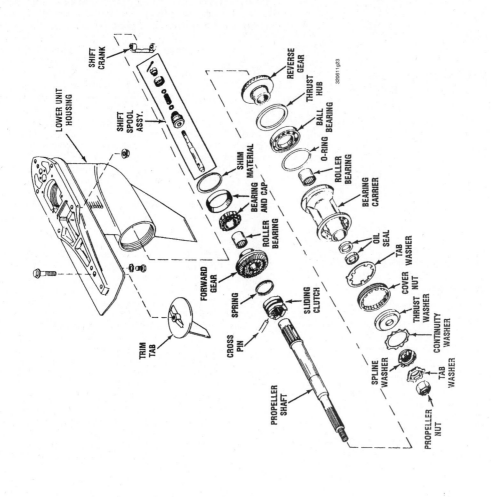

Fig. 23 Exploded view of the propeller shaft

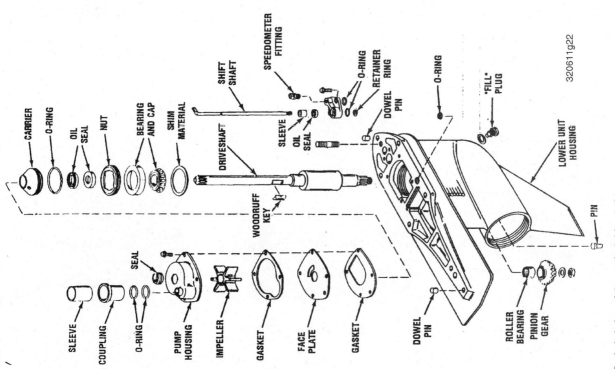

Fig. 22 Exploded view of the vertical driveshaft

Fig. 24 Bend out the tab on the washer

Fig. 25 If the nut is corroded, drill some holes and then break it with a chisel

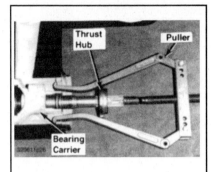

Fig. 26 Pulling out the bearing carrier

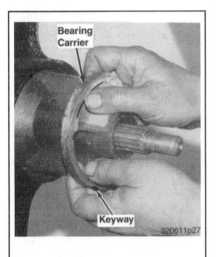

Fig. 27 Removing the bearing carrier after disconnecting the puller

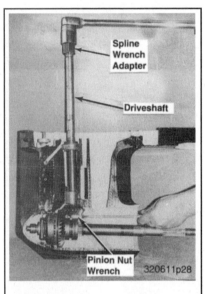

Fig. 28 Removing the driveshaft pinion nut

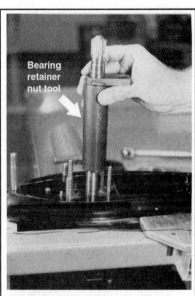

Fig. 29 Bearing nut retainer tool in position

Fig 30 Removing the driveshaft

Fig. 31 Removing the propeller shaft

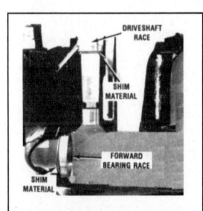

Fig. 32 Shim material locations

Bearing Carrier And Reverse Gear

◆ **See Figures 33, 34, 35 and 36**

1. Clamp the bearing carrier in a vise with soft jaws. Be sure to clamp on the bearing carrier at the reinforcement ribs. Place the jaws of a universal slide hammer puller down through the center of the reverse gear and bearing. Pull the reverse gear, thrust hub, and bearing out of the bearing carrier. Check the bearing for roughness, catches, and overheating. If the bearing is removed from the reverse gear, it must be replaced with a new item. If the bearing is found to be defective, press the bearing from the reverse gear using a universal bearing plate P/N 91-37241 and an arbor press.

2. Remove the two propeller shaft oil seals from the bearing carrier. Press the needle bearing free of the carrier, using an arbor press.

To assemble:

3. Wash all parts in solvent and blow them dry with compressed air. Remove any corrosion from the outside surface of the bearing carrier. Remove any seal and gasket material from all mating surfaces. Blow all water, oil passageways, and screw holes clean with compressed air. After all parts are clean and dry, apply a light coating of gear lubricant to the bright surfaces of all gears, bearings, and shaft, as a prevention against rusting.

4. Inspect the gears for any sign of excessive wear, nicks, or broken teeth. Check all splines to be sure they have not been damaged. Inspect the bearing sets for pits, grooves, or uneven wear. If the bearing carrier is worn or damaged, it may be replaced with a new assembly at the same time the bearings and seals are installed.

5. Install the new bearing into the bearing carrier with the numbers on the bearing facing the installation tool, as shown.

6. Press two new oil seals into the front of the bearing carrier. These seals are installed back-to-back. Seat the first one with the lip facing in to prevent the lubricant from escaping. Seat the second seal with the lip facing out to prevent water from entering.

7. Take time to obtain the proper mandrel for the ball bearing to ensure the force will be applied to the inner race of the bearing. Press the reverse gear ball bearing and thrust washer into place with the washer facing the reverse gear. Turn the bearing carrier over with the seal side down and then press the reverse gear and bearing assembly into the carrier. Always use an adapter to protect the reverse gear. Set the assembly aside for later installation.

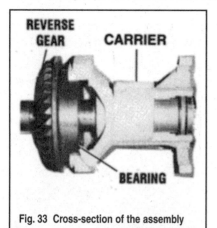

Fig. 33 Cross-section of the assembly

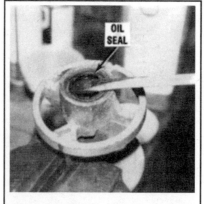

Fig. 34 Removing the oil seal

Fig. 35 Installing the roller bearing

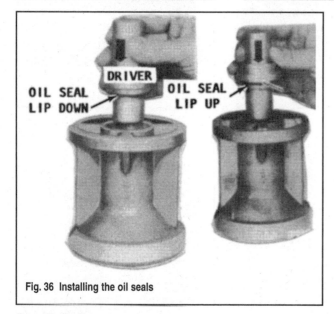

Fig. 36 Installing the oil seals

Propeller Shaft

◆ **See Figures 37, 38 and 39**

1. With a small screwdriver, unwind the spring from around the sliding clutch. Take care not to damage the spring by bending it out of shape. Remove the cross-pin. Slide the shift-spool and the shift-actuating shaft out. Remove the forward gear-and-bearing assembly.

2. Remove the cotter pin and nut, and then pull the spool off the shift-actuating shaft. Remove the bearings from the forward gear by first splitting the needle bearing inside the gear with a chisel and hammer. Remove the bearings. Use a universal puller plate to remove the tapered roller bearing. Press the gear from the bearing.

To assemble:

3. Wash all parts in solvent and blow them dry with compressed air. Remove any seal and gasket material from all mating surfaces. Blow all water, oil passageways, and screw holes clean with compressed air. After all parts are clean and dry, apply a light coating of gear lubricant to the bright surfaces of all gears, bearings, and shaft, as a prevention against rusting.

4. Inspect the gears for any sign of excessive wear, nicks, or broken teeth. Check all splines to be sure they have not been damaged. Inspect the bearing sets for pits, grooves, or uneven wear.

5. Use a file, and clean the grooves inside the propeller hub. Inspect the propeller and remove any nicks and burrs with a file. Take care not to remove any more material than is absolutely necessary.

6. Inspect the propeller for cracks, damage, or bent condition. Roll the propeller shaft on a flat surface and check for a bent condition. If the shaft is bent over 0.005 in. (0.13mm), it must be replaced.

7. Check the clutch "dog" and gears for worn lugs. If the lugs are worn, the clutch "dog" or gear must be replaced. If the 17 lugs show wear, check the shift cable between the inside of the boat and the stern drive.

8. Assemble the parts of the shift clutch actuating shaft. Place the first washer, spring, and second washer into the brass shift actuating spool, and then thread on the retainer. Tighten the retainer securely.

9. Slide the spool assembly over the clutch actuating shaft and thread the adjusting sleeve on fingertight. Now, back off the adjusting sleeve until the cotter key slides through the first hole.

10. Use a suitable adapter and press the tapered roller bearing onto the forward gear. The force must be applied only against the center race. Seat the bearing against the shoulder of the forward gear. Use the proper adapter and press the forward gear needle bearing into the gear from the numbered

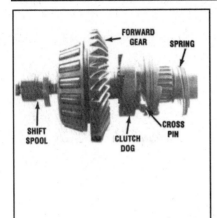

Fig. 37 A good view of the forward gear assembly

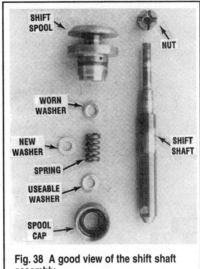

Fig. 38 A good view of the shift shaft assembly

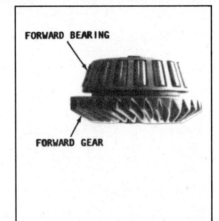

Fig. 39 Press the forward roller bearing into the forward gear

side. Seat the bearing against the inner gear shoulder.

11. Place the sliding clutch on the propeller shaft, with the grooves in the clutch toward the reverse gear. The clutch on early models do not have grooves. Instead, the clutch has a copper coating on only one side. The face without the copper coating must face toward the reverse gear. Install the forward gear-and-bearing assembly, with the gear facing the sliding clutch. Insert the clutch actuating shaft assembly into the hole in the forward end of the propeller shaft. Rotate the shaft assembly until the hole in the shaft is aligned with the holes in the sliding clutch and the propeller shaft. Insert the cross-pin through the sliding clutch, propeller shaft, and actuating shaft. Install the cross-pin retainer spring by winding it through the sliding clutch in the clutch retainer groove. Do not overstretch the spring. Set the assembly aside for later installation.

CLEANING AND INSPECTING

②◁MODERATE

1. Wash all parts in solvent and blow them dry with compressed air. Remove all traces of seal and gasket material from all mating surfaces. Blow all water, oil passageways and screw holes clean with compressed air. After cleaning, apply a light coating of gear lubricant to the bright surfaces of all gears, bearings, and shafts as a prevention against rusting and corrosion. Check the shift crank to be sure it is not bent.
2. Inspect the water pump impeller plate for wear and corrosion. Replace any worn, corroded, or damaged parts. Use a fine file to remove burrs. Replace all O-rings, gaskets, and seals to ensure satisfactory service from the unit. Clean the corrosion from inside the housing where the bearing carrier was removed.
3. Check to be sure the water intake is clean and free of any foreign material.
4. Inspect the gear case, housings, and covers inside and out for cracks. Check carefully around screw and shaft holes. Check for burrs around machined faces and holes. Check for stripped threads in screw holes and traces of gasket material remaining on mating surfaces.
5. Check O-ring seal grooves for sharp edges, which could cut a new seal. Check all oil holes.
6. Inspect the bearing surfaces of the shafts, splines, and keyways for wear and burrs. Look for evidence of an inner bearing race turning on the shaft. Check for damaged threads. Measure the run-out on all shafts to detect any bent condition. If possible, check the shafts in a lathe for out-of-roundness.
7. Inspect the gear teeth and shaft holes for wear and burrs. Hold the center race of each bearing and turn the outer race. The bearing must turn freely without binding or evidence of rough spots. Never spin a ball bearing with compressed air or it will be ruined. Inspect the outside diameter of the outer races and the inside diameter of the inner races for evidence of turning in the housing or on a shaft. Deep discoloration and scores are evidence of overheating.

8. Inspect the thrust washers for wear and distortion. Measure the washers for uniform thickness and flatness.
9. Inspect the forward and reverse gears for distortion, burrs, and cracks. Check for any sign of discoloration, which means the gears are running too hot.
10. Replace all seals, O-rings, and gaskets to ensure maximum service after the work is completed.

ASSEMBLY

OEM ④ ◁SKILLED

◆ See Figures 40, 41, 42, 43. 44. 45, 46, 47, 48, 49, 50, 51 and 52

✳✳ CAUTION

The driveshaft bearing has a slight taper. Therefore, if the driveshaft roller bearing is reversed during installation, the driveshaft will fail. The only way the bearing can be installed properly, is with the numbered end of the bearing case facing up and away from the pinion gear and toward the anti-cavitation plate.

1. Coat the driveshaft roller bearing bore in the housing with a thin layer of Multi-Purpose Lubricant. Use Bearing Installation and Removal Kit P/N 91-31229A5, and install the driveshaft roller bearing, pay attention to the previous Caution. Pull the bearing up until it bottoms on the lower unit housing shoulder.
2. If the forward gear, forward gear bearing and bearing cup were removed from the lower unit housing, and none of these components are being replaced with new items, install the same amount of shim material behind the bearing cup which was removed during disassembling. Use the same thickness and quantity of new shim material because the old shim material will have been distorted and give a false thickness.
3. If a new forward gear, or bearing and bearing cup are installed, place a nominal thickness of 0.020 in. (0.51mm) shim(s) material between the bearing cup and the housing.
Lubricate the bearing cup bore with a light coating of gear lubricant. Place the shim material into the forward gear bore in the housing. Slide the bearing cup into the housing and aligned in the bore. Place the driver cup special tool P/N 91-36577 onto the bearing cup. Using the bearing carrier as a pilot and the propeller shaft as a driver, tap the end of the propeller shaft until the forward gear bearing cup is seated in the housing bore.
4. Pick up the shift actuating crank. Now, reach all the way into the lower unit to the forward end and install the crank on the locating pin. Be sure the shift cam is facing toward the port side of the housing.
5. Apply a coating of 2-4-C marine lubricant to the O-rings and oil seal for the shift shaft assembly. Place the O-rings into the grooves in the bottom of the shift shaft bushing. Slide the oil seal over the end of the shaft and against the top of the shift shaft bushing. Ensure the E-clip is seated into the groove on the shift shaft.

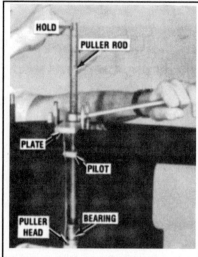

Fig. 40 Use the special tool to install the driveshaft roller bearing

Fig. 41 The shift cam should be facing the port side

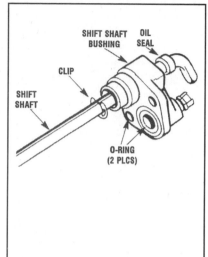

Fig. 42 Assemble the shift shaft...

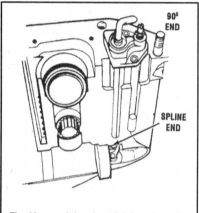

Fig. 43 ...and then install it into the unit

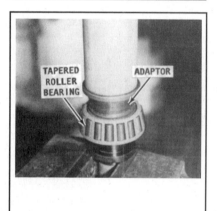

Fig. 44 Pressing the bearing onto the driveshaft

Fig. 45 Use the tool to tighten the drive-shaft retaining nut

6. Lower the shift shaft assembly into the lower unit housing. When the male splines of the shift shaft index with the female splines of the shift crank, be sure the 90° end of the shift shaft is facing forward.

7. If the pinion gear bearing cage was not removed, but the needle bearings fell out upon disassembly, coat the 18 needle bearing with Quicksilver Needle Bearing Assembly Lubricant, or equivalent. Place the 18 needle bearings into the bearing cage. Do not mix the needle bearings with needle bearings from another bearing and never use needle bearings from a new bearing in a used bearing cage.

■ Two methods are available to properly shim the lower unit. The first method involves shimming and checking the backlash before the unit is assembled. The second method is the factory approved procedure which involves assembling the unit; taking the required clearance measurements; and then disassembling the unit if necessary, in order to change the number of shims installed. This second method requires the use of special tools obtained only from a MerCruiser dealer, while the alternate procedure can be accomplished using standard tools or modified standard tools.

■ Three steps are necessary to properly shim the lower unit. First the pinion gear must be shimmed to the correct depth. Secondly, the forward gear must be shimmed to the pinion gear for the proper backlash. And finally, the reverse gear is shimmed to the pinion gear with the correct backlash.

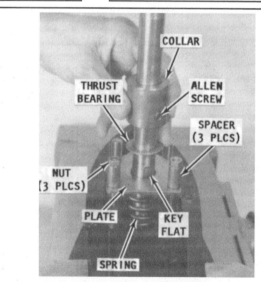

Fig. 46 Installing the pinion gear preload tool

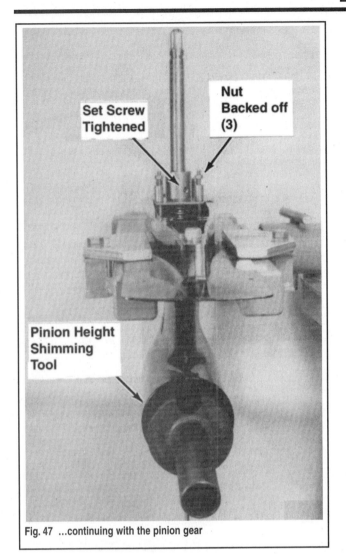

Fig. 47 ...continuing with the pinion gear

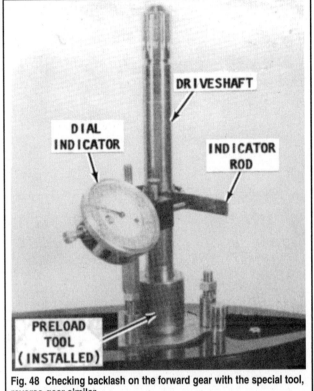

Fig. 48 Checking backlash on the forward gear with the special tool, reverse gear similar

4. Secure the bearing in place by tightening the retaining nut to a torque value of 100 ft. lbs. (130 Nm). Install the propeller shaft. Push in on the propeller shaft and at the same time, pull up on the driveshaft, then rotate the driveshaft clockwise from 25–30 turns. This action will establish a forward and pinion gear wear pattern.

CHECKING FORWARD/REVERSE GEAR BACKLASH

1. Push hard on the propeller shaft to exert pressure on the forward gear and to keep it from rotating. Now, while exerting this force on the propeller shaft, turn the driveshaft back-and-forth with a light pressure. The movement of the driveshaft is the approximate backlash of the forward gear. The amount of "play" should be 0.017–0.0028 in. (0.43–0.71mm).

2. A dial indicator is necessary to accurately determine the amount of backlash. If the backlash is too tight, remove shim material from behind the forward gear bearing race. If the backlash to too great, add shim material behind the forward gear bearing race. As a general guide—for 0.001 in. of backlash, add or remove 0.001 in. of shim material; adding shims will reduce backlash, removing shims will increase backlash.

3. Hold onto the driveshaft and remove the pinion nut, pinion gear, and the forward gear assembly. Examine the wear pattern on the forward gear

Shimming W/O Special Tools

◆ See Figures 53, 54, 55 and 56

1. Paint the forward and pinion gear teeth with Dyken machine dye, or an equivalent substance.

2. Assemble the forward bearing onto the forward gear, and then lower the assembly into the lower unit.

3. Slide the outer driveshaft bearing race and retainer nut down the driveshaft over the bearing. Insert the driveshaft into the lower unit with the outer bearing race seating in the cavity of the lower unit.

Fig. 49 Cut-away of the lower unit showing the clutch dog engaged with the forward gear...

Fig. 50 ...and the reverse gear

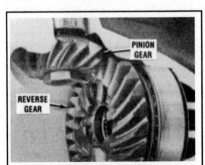

Fig. 51 The pinion gear is properly engaged with the teeth of the reverse gear

Fig. 52 A good shot of the shim material locations on the propeller shaft

and on the pinion gear. The pattern must be within 1/16 in. of the bottom of the gear cut onto the tooth. Also, the wear pattern should start about 1/8 in. in on the tooth and end about 1/8 in. from the end of the tooth. To change the wear pattern, the pinion gear must be raised or lowered by adding or removing shim material. If the backlash was satisfactory, but it is necessary to change the shim material from under the driveshaft bearing, the backlash should be checked a second time. Once the pinion depth is established for the forward gear, or backlash, it cannot be changed for reverse gear. The reverse gear will be checked only for proper backlash.

4. Install the thrust washer onto the reverse gear, and then install the assembly into the bearing carrier. Install the same amount of reverse gear shim material into the lower unit housing as was removed and noted during disassembling. If the amount was not recorded; if the shim material was lost; or if a new assembly is being installed, begin with 0.015 in. (0.38mm) shim material. Coat the inside surface of the lower unit with oil as an aid to installing the bearing carrier.

■ Do not install the O-ring onto the bearing carrier, at this time. Without the O-ring, the task of installing and removing the carrier will be much easier. The O-ring will be installed during final assembling.

5. Slide the bearing carrier into place. Install the tab washer and retainer nut. Tighten the nut to 210 ft. lbs. (285Nm). Take care not to jam the teeth of the gears together while tightening the retainer nut, by continually moving the driveshaft to ensure the proper amount of backlash is being maintained. Insert a long-blade screwdriver through the propeller shaft hole in the bearing carrier. Exert a force on the screwdriver to prevent the reverse gear from rotating.

6. At the same time the force is being exerted on the reverse gear with the screwdriver, pull up on the driveshaft and at the same time rotate the driveshaft back-and-forth to determine the amount of backlash.

The movement of the driveshaft is the approximate backlash of the reverse gear. This amount of "play" should be 0.040–0.060 in. (1.02–1.52mm). A dial indicator is necessary to accurately determine the amount of backlash. If the backlash to too tight, add shim material to the reverse gear. If the backlash to too great, remove shim material from the reverse gear. As a general guide—for 0.001 in. of backlash, add or remove 0.001 in. of shim material. Adding shims will reduce backlash while removing shims will increase backlash.

7. Remove the retainer nut, tab washer, and bearing carrier. Remove the pinion nut, pinion gear and driveshaft. Discard the old pinion nut because it has lost its locking ability. Always use a new pinion nut during final assembling.

ASSEMBLY (continued)

8. Install the assembled propeller shaft into the lower unit housing. This is accomplished by tilting the propeller end of the shaft toward the oil fill hole side of the lower unit housing to allow the actuating shaft-and spool to engage the shift crank. Straighten the propeller shaft and operate the shift shaft. The sliding clutch should not move unless the shift crank and spool are moved.

9. If the tapered roller bearing was pressed off the driveshaft, install the bearing onto the driveshaft. Insert the driveshaft into a suitable adapter, against the bearing shoulder on the shaft, and then press against the inner bearing race to seat the bearing. Use a wrench adapter on the upper end of the driveshaft to protect the splines.

10. Lower the driveshaft pinion gear into the cavity. Move the propeller shaft slightly to allow the pinion gear to drop into position and mesh with the forward gear. Secure the lower unit housing in the horizontal position. Install the driveshaft, with the threaded end through the two bearings and the pinion gear. Coat the driveshaft locknut threads with a small mount of Loctite.

■ A thin washer is used to more evenly distribute the load of the nut against the pinion gear. When the washer is new, both sides are flat. However, after use, one side becomes slightly concave, the other side slightly raised. When installing a used washer, the raised side must face toward the pinion gear.

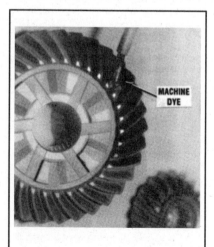

Fig. 53 Paint the gear teeth with dye

Fig. 54 Slide the forward bearing and gear into the unit

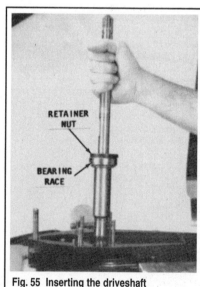

Fig. 55 Inserting the driveshaft

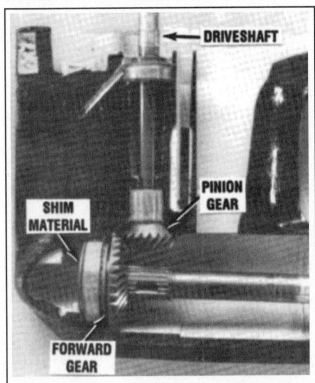

Fig. 56 Press down hard on the propeller shaft while measuring backlash

11. After the driveshaft is in place, slide the outer bearing race and retainer nut down the driveshaft and over the bearing. Place the nut in the pinion nut holding tool P/N C-91-61067A2. If a pinion nut holding tool is not available, a substitute can be made from a box-end wrench, as described in the disassembling procedures. Place the nut in the wrench and proceed with the installation.

12. Slide the tool into position under the driveshaft, and then thread the driveshaft into position, using driveshaft adapter P/N C-91-56775, over the splines.

13. Secure the retaining nut by tightening it with special tool P/N 91-43506 to a torque value of 100 ft. lbs. (135 Nm).

14. Obtain special pre-load tool P/N 91-14311A2. This tool consists of a spring which will seat on top of the driveshaft retaining nut; a plate which will fit over the driveshaft; three water pump studs; a thrust bearing to ride on top of the plate; a collar with an Allen head screw that slides down the driveshaft to bear on a flat on the driveshaft; and three nuts for the water pump studs.

15. Proceed with the pre-load tool as follows: Install the spring, plate, spacers, and nuts of the tool, as described and shown. One of the nuts takes a 1/2 in. wrench and other two a 7/16 in. wrench. Now, tighten the nuts alternately and evenly—moving the plate down the driveshaft as the spring is compressed. Continue tightening the nuts until the 1/2 in. nut bottoms out (all the threads on the stud are used). Check to be sure the plate is fairly level with the surface of the lower unit.

16. Slide the thrust bearing down the driveshaft and onto the surface of the plate. The thrust bearing may be installed with either side facing down. Check to be sure the Allen set screw in the collar is backed out to allow the collar to slide down the driveshaft. Move the collar down the driveshaft with the Allen screw end going down first, as shown and positioned to allow the Allen set screw to bear onto the water pump impeller key flat.

17. Tighten the Allen set screw securely, and then back-off the three nuts on the water pump studs. The driveshaft has now been forced upward placing an upward load on the driveshaft bearing.

18. Insert pinion gear shimming tool, C-91-56048, over the propeller shaft and into the housing. The design of this tool will allow the flat portion of the front of the gauge to clear the pinion gear. After the tool is inserted all the way past the pinion gear, turn the tool until the flat on the tool is away from the pinion gear. Now, with the tool held in this position, insert a 0.025 in.

(0.64mm) feeler gauge between the bottom face of the pinion gear and the rounded part of the gauge. Most feeler gauges are not long enough to make this measurement properly. However, a suitable gauge can be made by grinding the teeth off a hacksaw blade and then using it to take the clearance measurement. If the clearance is over 0.025 in., remove shims from under the upper bearing cup. If the clearance is under 0.025 in., add shims under the bearing cup. Leave the preload shimming tool installed for forward and reverse gear shimming procedures.

19. Install the carrier shims, which were tagged and wired together during disassembly. Install a new O-ring seal on the carrier between the thrust washer and the carrier housing. Coat the surfaces of the O-ring seal and carrier with anti-corrosion grease. Insert the carrier assembly.

20. Insert the bearing carrier-to-gear housing alignment key. If the housing has two notches, use the top notch. Install the bearing carrier retainer tabbed washer, with the V-tab aligned with the V-notch on the bearing carrier.

21. Thread the gear housing cover (retainer nut) into place by hand to prevent cross-threading. Tighten it a couple of turns by hand, and then use bearing carrier wrench P/N C-91-61069 and tighten the cover to 210 ft. lbs. (285 Nm). Do not secure the tabbed washer at this time. The tab is not secured until after the forward and reverse gear backlash has been adjusted.

■ **The driveshaft pre-load tool will remain installed for both the forward and reverse gear shimming procedures.**

22. Install bearing carrier puller with the arms of the puller on the carrier and the center bolt on the end of the propeller shaft. Tighten the puller center bolt to a torque value of 45 inch lbs. (5 Nm). This action places a pre-load on the forward gear, pushing the forward gear into the forward gear bearing. Rotate the driveshaft about three full revolutions, and then recheck the torque value on the center puller bolt.

23. Attach the backlash dial indicator rod P/N 91-53459 to the driveshaft. Position the dial indicator shaft to the line marked "I" on the dial indicator rod. Tighten the indicator rod to the driveshaft and then rotate the driveshaft while observing the dial indicator. Rotate the drive shaft until the dial indicator needle makes one full revolution to "0" and then stop rotating the shaft.

24. Move the driveshaft back-and-forth, and observe movement of the dial indicator. The propeller shaft must remain stationary. Note and record the dial indicator reading. Loosen the dial indicator rod and reposition the driveshaft 90°, and then tighten the rod again with the dial indicator resting on "0". Repeat this step until four backlash readings have been obtained. Add all four readings and then divide the total by 4 to determine the average backlash. The average forward gear backlash should be 0.017–0.028 in. (0.43–0.71mm).

25. Remove the bearing carrier puller from the lower unit. Regardless of the backlash measurements, if they are correct or incorrect, perform the reverse gear backlash check before making any shim adjustments on the forward gear.

27. The reverse gear backlash adjustment outlined in this step should be checked before correcting the forward gear backlash. In this sequence, the changing of shim material may be accomplished in one disassembly job.

28. Slide the pinion nut adapter tool P/N 91-611067A2 onto the propeller shaft with the larger end of the pinion adapter toward the bearing carrier. Install the special washer P/N 12-54048 over the end of the exposed threads of the propeller shaft. Thread the propeller nut onto the shaft, and then tighten the nut to a torque value of 45 inch lbs. (5.0 Nm). This action will place a preload onto the reverse gear—pushing it into the reverse gear bearing.

29. Attach the backlash dial indicator rod 91-53459 to the driveshaft. Position the dial indicator shaft to the line marked "I" on the dial indicator rod. Tighten the indicator rod to the driveshaft and then rotate the driveshaft while observing the dial indicator. Rotate the drive shaft until the dial indicator needle makes one full revolution to "0" and then stop rotating the shaft.

30. Move the driveshaft back-and-forth, and observe movement of the dial indicator. The propeller shaft must remain stationary. Note and record the dial indicator reading. Loosen the dial indicator rod and reposition the driveshaft 90°, and tighten the rod again with the dial indicator resting on "0". Repeat this step until a total of four backlash readings have been obtained. Add all four readings and then divide the total by 4 to determine the average backlash. The average reverse gear backlash should be 0.040–0.060 in. (1.02–1.52mm).

31. Remove the propeller nut, special washer, and pinion nut adapter tool P/N 91-611067A2 from the propeller shaft.

32. If the average backlash for the Forward Gear Check is too much, add shim material behind the forward gear bearing race. If the backlash is too small, remove shim material from behind the forward gear bearing race. For each 0.001 in. of backlash, add or remove 0.001 in. shim material. Adding shims will reduce backlash, removing shims will increase backlash.

33. If the average backlash for the Reverse Gear Check is too much, remove shim material from in front of the bearing carrier thrust washer. If the backlash is too small, add shim material in front of the bearing carrier thrust washer. For each 0.001 in. of backlash, add or remove 0.001 in. shim material. Adding shims will reduce backlash, removing shims will increase backlash.

34. After correcting the forward gear backlash by changing the thickness of the shim material behind the forward gear cup, or the reverse gear backlash by changing the thickness of shim material between the reverse gear and the reverse gear bearing carrier assembly, proceed with the next step.

35. Remove the bearing carrier assembly and clean off all lubricant used during assembling. After the assembly is clean, install a new O-ring onto the carrier. Apply a liberal coating of Perfect Seal Sealant, or equivalent, to the outer diameter of the bearing carrier which contacts the lower unit housing. Do not allow any sealant to enter the ball bearings or the reverse gear. Coat the threads of the bearing carrier nut with sealant and install the nut. Using Bearing Carrier Nut special tool P/N 91-61069, tighten the bearing carrier retainer nut to a torque value of 210 ft. lbs. (285 Nm). Bend one of the tabs on the locking washer into one of the slots in the cover. Install the water pump per the procedures outlined earlier in this section.

Propeller

REMOVAL & INSTALLATION

  MODERATE

◆ See Figures 57, 58, 59, 60, 61, 62, 63 and 64

✳✳ CAUTION

If the drive unit or the lower unit is not disconnected or removed from the boat it is imperative that you disconnect the battery cables and remove the ignition key prior to performing this procedure.

1. Bend the locking tabs on the tab washer out and away from the grooves in the splined washer. Never pry on the end of the propeller hub because any small distortion will affect propeller performance or possibly break the hub off the propeller. Place a block of wood between one blade of the propeller and the anti-cavitation plate to prevent the shaft from rotating. Place the correct size socket and breaker bar over the propeller nut and remove the nut from the propeller shaft. Slide the tab washer, splined washer, continuity washer, propeller, and thrust hub (washer) off of the propeller shaft.

2. If the propeller is frozen to the shaft, heat must be applied to the shaft so as to melt out the rubber inside the hub. Using heat will destroy the hub, but there is no other way. As heat is applied, the rubber will expand and the propeller will actually be blown from the shaft. Therefore, stand clear to avoid personal injury.

3. Use a knife and cut the hub off the inner sleeve.

4. The sleeve can be removed by cutting it off with a hacksaw or cold chisel, or it can be removed with a puller. Again, if the sleeve is frozen, it may be necessary to apply heat.

5. Remove the thrust hub from the propeller shaft.

To install:

6. Check to be sure the splines on the propeller shaft are in good condition. If there is any visible sign of corrosion on the shaft, the corrosion must be removed before assembly is completed. Verify the threaded end of the propeller shaft is clean and the threads are in good condition.

7. Verify the splined steel hub inside the propeller is free of corrosion. Check to be sure the rubber hub is tight and secured to the propeller inner hub. Verify there is no sign of the rubber hub spinning on the inside of the propeller.

8. Slide the propeller thrust washer onto the shaft with the smaller side of the washer facing out—to the rear, to the propeller. Coat the propeller shaft and splines with one of the following lubricants in the order of preference; Quicksilver Special Lubricant 101, 2-4-C Lubricant or Perfect Seal.

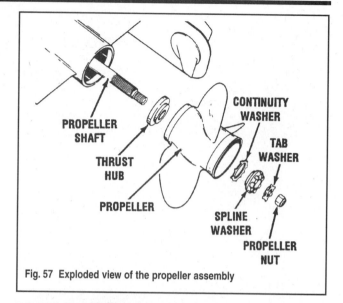

Fig. 57 Exploded view of the propeller assembly

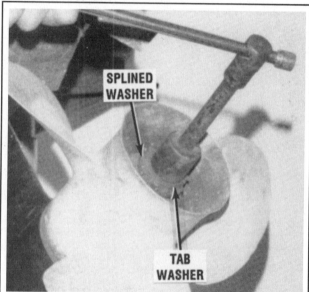

Fig. 58 Bend back the tabs on the washer before loosening the propeller nut

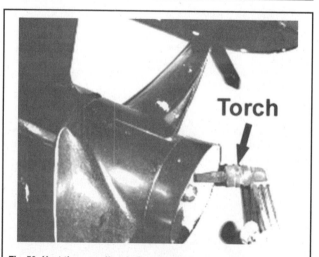

Fig. 59 Heat the propeller shaft to loosen the frozen hub

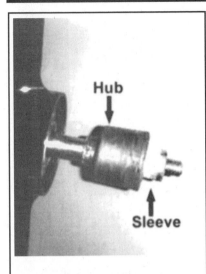

Fig. 60 The rubber hub rides on the sleeve, inside the propeller flange

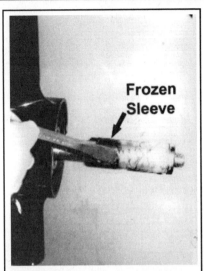

Fig. 61 Break off the sleeve after heating the shaft again…

Fig. 62 …and then remove the thrust washer

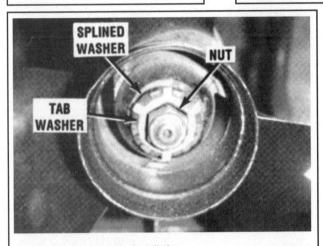

Fig. 63 A good look at the installation

Fig. 64 Don't forget to bend those tabs over

9. Slide the propeller onto the shaft followed by the continuity washer, splined washer, tab washer and then the propeller nut, as indicated in the accompanying illustration. These components must be assembled in the correct order.

10. Position a block of wood between one of the propeller blades and the anti-cavitation plate to keep the propeller from rotating. Tighten the propeller nut to a torque value of 55 ft. lbs. (75 Nm). Continue tightening the nut until three tabs on the tab washer align with the grooves in the splined washer. There should be a minimum of two threads showing on the end of the propeller nut after it has been tightened to the proper torque value.

11. Bend three of the tab washer tabs into the grooves of the splined washer using a blunt end punch and hammer. The tabs will prevent the propeller nut from backing off.

Water Pump/Seal Carrier

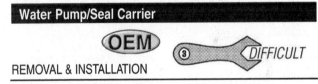

REMOVAL & INSTALLATION

◆ See Figures 65, 66, 67, 68, 69, 70, 71 and 72

1. Remove the lower unit.
2. Pull the water tube and coupling (water) seal out of the water pump body. Slide the water seal up and free of the driveshaft. Discard the seal.
3. Remove the bolts securing the water pump body, and then slide the body up and free of the driveshaft. It may be necessary to use a few small prybars—one on each mounting flange—to gently persuade the pump body to "break" loose from the plate.
4. Slide the impeller up and free of the driveshaft. Remove the Woodruff key from the cutout in the driveshaft.
5. Remove the gasket from the under side of the body. It is possible this gasket will remain on the plate when the body is removed.
6. Slide the plate and gasket up and free of the driveshaft.
7. Use two small prybars—one on each side, as shown—to pry the oil seal carrier free from the gear housing. Remove the carrier.
8. If the oil seals are damaged, they may be pried out of the carrier with a screwdriver after the carrier has been clamped in a vise. Be sure replacements have been obtained and are at hand, because removing the seals will destroy their sealing qualities.

To install:

✳✳ WARNING

The water pump impeller must be in very good condition for satisfactory service. The pump performs an extremely important function by supplying sufficient water to properly cool the stern drive and the engine. Therefore, good shop practice dictates—replace the water pump impeller whenever the unit is disassembled.

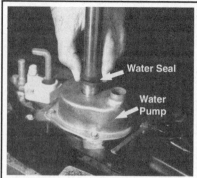

Fig. 65 Slide the water seal up and off the shaft

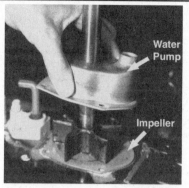

Fig. 66 Lift off the pump housing...

Fig. 67...and then remove the old gasket

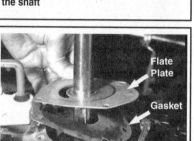

Fig. 68 Lift off the face plate and gasket...

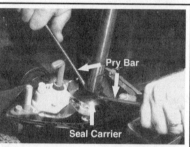

Fig. 69 ...and then pry off the seal carrier

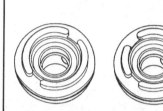

Fig. 70 A new style oil seal on the right and an old style oil seal on the left. The new style seal has back to back seals

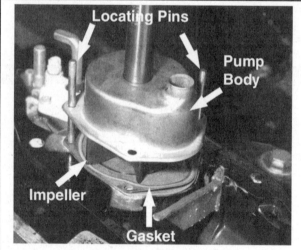

Fig. 71 Use alignment pins when installing the pump...

Fig. 72 ...and then tighten the bolts

9. Inspect the water tube coupling for wear or damage. Its always a good idea to replace the two O-rings.

10. Inspect the impeller for any wear or damage. Replace as necessary.

11. If the seals in the seal carrier were removed, apply a thin coating of Quicksilver Perfect Seal to the oil seal bore, and then press the seals in back-to-back with the small seal going in first, with the lip facing down—away from seal driver (P/N 91-817569). The seal is properly installed when the driver bottoms against the carrier. Do not press further or the carrier may be damaged. Press the large seal in with the lip facing up, until the driver bottoms on the carrier. This places the seals in the carrier back-to-back. Install a new O-ring around the perimeter of the carrier.

12. Apply a light coating of Quicksilver 2-4-C Marine Lubricant, or equivalent, to the lips of the oil seals and to the O-ring. Slide the carrier down the driveshaft and into the lower unit opening. Do not use a hammer to seat the carrier. Use only hand pressure.

13. Slide the small hole gasket down the driveshaft, followed by the face plate and the large hole gasket. Holes in the gaskets and plate will only align with each other and the holes in the lower unit one way. If the holes do not align one or more of the items is upside down. Correct the situation by turning one or more of the items over.

14. Apply just a "dab" of grease to the Wood-ruff key, and then place it in the driveshaft keyway. Slide the impeller down the driveshaft and onto the face plate with the cutout in the impeller indexed over the Woodruff key. If an old impeller with a "set" to the blades is being installed, face the curl of the blades in a counterclockwise direction. If the direction is reversed, premature impeller failure will surely occur.

15. Slide the pump body down the driveshaft and just to the top of the impeller. Keeping the gasket, impeller, and body mounting holes aligned is not an easy task. However, if a couple of pins are inserted down through just two opposite holes, as shown, the holes will stay aligned while the body is worked down over the impeller. The pins can be old drill bits, small diameter bolts, rod, whatever is handy. Exert some downward pressure on the pump body and at the same time rotate the driveshaft clockwise and the impeller blades will "set" in the proper direction.

16. Start a couple of the body mounting bolts, and then remove the two pins. Install the remaining mounting bolts, and tighten them all to 60 inch lbs. (7.9 Nm).

■ **A special water pump seal seating tool is required to properly seat the seal on top of the water pump case. This tool is only available from MerCruiser in a kit, P/N 26-81657A2.**

DRIVESHAFT HOUSING (UPPER UNIT)

Description

Output power from the engine is connected to the stern drive through a horizontal driveshaft. A coupler is bolted to the flywheel and has a splined hub in the center. The end of the horizontal driveshaft indexes with and slides into the center of the hub. Power from the engine is then transmitted through the horizontal driveshaft to a pinion gear set where power direction is changed from horizontal to vertical.

The upper driven gear is pressed onto the outside diameter of the upper driveshaft. The upper driveshaft is splined on the lower end. When the upper gear housing is mated to the lower unit, the end of the lower unit driveshaft indexes into the splined end of the upper driveshaft. Engine power is then transferred down into the lower gear unit.

The horizontal and vertical driveshafts are both mechanically connected. Therefore, anytime the engine is operating, the horizontal and vertical driveshafts are constantly rotating with engine rpm. A double yoke universal joint assembly in the horizontal driveshaft allows the stern drive to be raised or lowered to a required trim/tilt position (within limits), while the engine is operating.

Gear Ratio Identification

The gear ratio for the stern drive must be known to permit the proper special tool selection to be made in order to properly position the gears in the housing. The gear ratio for all Alpha stern drives is identified in two places—a decal is affixed to the port side of the upper gear housing. A number such as 1.50R is followed by the serial number.

The second location of the gear ratio is on the universal joint splined yoke. The identification mark here will be a letter such as "F".

This letter represents the gear ratio in the upper gear housing. Use the accompanying chart to determine the gear ratio when the letter is known.

Gear Ratio	
Letter	Ratio
B	1.98:1
C	1.65:1
D	1.84:1
F	1.50:1
M	1.50:1 Mag

If the gears inside the housing have previously been changed from the factory markings on the housing, another method may be used to determine the gear ratio. This method is the least desired because the unit has to be disassembled before the ratio is known. With the gears removed from the housing count the number of teeth on the drive and driven gears. Compare the gear teeth number count to the chart below for the gear ratio of the unit being serviced.

Gear Ratio		
Drive Gear	Driven Gear	Gear Ratio
20	24	1.98:1
17	19	1.65:1
24	24	1.84:1
20	22	1.50:1

17. Apply a light coating of lubricant to the driveshaft. Slide the water pump face seal down the driveshaft until it is one inch above the pump body. Obtain special seal seating tool from the kit identified above. Slide the tool down the driveshaft and onto the seal. Push the seal down with the tool until the tool makes contact with the pump body. If the tool is not available, a large washer with an inside diameter slightly larger than the driveshaft can be used. Push the seal down evenly onto the water pump body until the seal face makes light contact with the pump body.

Driveshaft Housing (Upper Unit)

REMOVAL & INSTALLATION

◆ **See Figure 15**

■ **Please refer to the preceding Lower Unit removal section for illustrations of the two unit separation process.**

1. Remove the sterndrive unit.
2. Place an alignment mark on the trim tab trailing edge and the anti-cavitation plate as an aid during assembling. Remove the plastic plug from the top of the anti-cavitation plate. Remove the bolt directly above the trim tab, lift off the trim tab. Remove the four hex nuts and washers (two on each side), from the sides of the anti-cavitation plate. Disconnect the speed indicator fitting on the leading edge of the anti-cavitation plate by pushing down on the fitting and turning it 1/4 turn counterclockwise. Remove the nut and washer in front of the fitting. Remove the bolt up inside the cavity for the trim tab.
3. Separate the two housings (upper and lower) by tapping downward on the lower unit with a soft face mallet and a wooden block. In some cases the housing may be difficult to separate because the vertical driveshaft may be corroded and "frozen" in the upper gear housing vertical shaft. Lower the gear housing away from the upper gear housing.

To install:

4. Verify the remote control shift handle is in the Forward gear position for standard right hand stern drives. The 90° shift lever should be facing forward and in the center of the lower unit. Rotate the propeller shaft clockwise to verify gear engagement.

■ **If the unit being serviced is a counter-rotation—left hand drive—the gear shift lever should be in the reverse gear position.**

5. Verify the trim tab bolt is in place in the aft section of the lower unit.
6. Check to be sure the O-ring is in place on the passageway for the gear oil.
7. Apply a coating of Quicksilver 2-4-C Lubricant, or equivalent, to the splines on the end of the vertical driveshaft.
8. Some lower unit housings may have an aluminum dam behind the water pump. If the aluminum dam is corroded or damaged it must be replaced with a rubber filler unit. If the aluminum dam is serviceable, check the drain hole on the starboard side of the dam to verify it is not plugged. If this drain hole is plugged, severe damage to the gear housing will result.
9. Check to be sure the plastic guide tube is installed in the water pump housing. Verify the water pump face seal is positioned on top of the water pump cover. This seal prevents exhaust gases from entering the water cooling system.

If the unit has an aluminum dam and it has not been replaced with the rubber plug, place a bead of Permatex Ultra Blue Silicone Sealant, or equivalent, across the top of the dam. If a rubber dam is installed, no sealant is required.

10. Now, raise the lower unit (or lower the upper unit) to mate with the upper gear housing, and at the same time check to be sure the water tube starts into the water pump guide tube. Verify the 90° end of the shift shaft indexes with the slotted arm in the upper gear housing.
11. It may be necessary to slowly rotate the propeller shaft clockwise until the splines on the upper and lower drive shafts index.
12. Screw the nut onto the stud at the front of the unit. Install the aft screw into the forward hole in the trim tab well.

13. Install the two port and two starboard bolts and their nuts and washers. Tighten the four 5/8 in. nuts alternately, and a little at-a-time, until the lower unit is tight against the upper gear housing.

14. Now tighten the front nut and the four side nuts to 35 ft. lbs. (47.5 Nm). Tighten the aft screw to 28 ft. lbs. (41 Nm).

15. Install the trim tab with the marks made during disassembly aligned. Tighten the screw(s) to 23 ft. lbs. (31 Nm).

■ The trim tab performs two very important jobs, one of which you may not realize. First, the tab compensates for steering torque. If the boat continually seems to move to port or starboard while the helmsman is on a straight course, the trim tab can be adjusted to the side of the pull. The tab also prevents electrolysis from damaging expensive parts. The tab is not expensive; it should show signs of electrolytic action; and should always be replaced after some of the material has been eaten away or is pitted. Now, if the tab shows no signs of electrolytic action after the boat has been in use over a period of time, the grounding should be checked to ensure more expensive parts are not being damaged. Install the trim tab plastic cover plug.

16. Install the propeller if it was removed and refill the drive unit with oil.

ASSEMBLY & DISASSEMBLY

◆ See Figures 75 and 76

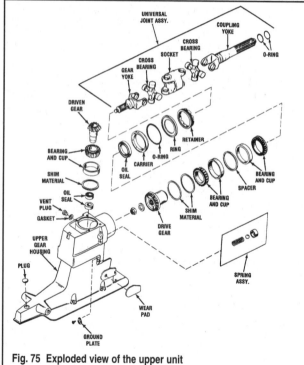

Fig. 75 Exploded view of the upper unit

U-Joint/Drive Gear Assembly

◆ See Figures 77, 78, 79, 80, 81, 83, 84, 85, 86, 87, 88, 89, 90 and 91

Clamp the upper gear housing into a large shop vise or place the unit into a shop holding fixture designed for the upper gear housing.

1. Remove the dipstick and gasket from the center of the top cover. Next, remove the four bolts securing the cover to the housing. Lift off the top cover. If the cover is difficult to remove, insert a small pry bar into the cut-out

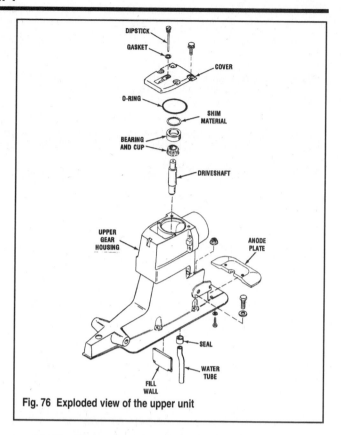

Fig. 76 Exploded view of the upper unit

on each side of the housing and pry up on the cover. Check the O-ring around the bearing race inside the cover for damage and deterioration.

2. Scribe a mark on the bearing retainer and a matching mark on the housing, as an aid during assembling. Loosen the U-joint roller bearing retainer with special wrench P/N 91-17256.

3. Work the tool *counterclockwise* to remove the retainer. Pull the U-joint/shaft assembly from the housing. Leave the tool attached to the retainer

4. Reach inside the housing bore and remove the shim material. Save and tag the shim material, as an aid during assembling.

5. Clean the assembly in solvent and allow it to dry thoroughly. Inspect the drive (pinion) gear and the driven gear for broken teeth, pitting, etc. Make sure all teeth are worn evenly

6. Slide the two small O-rings off the end of the coupling yoke and then clamp the retainer tool in a vise. Insert the assembly, gear side facing up, into the tool.

7. Slide a breaker bar through the joint so that it is levered against the side of the vise. Remove the locknut and washer from the end of the gear yoke and pull off the driven gear/bearing. Set it aside for now.

8. If the drive or driven gears are found to be defective, both gears must be replaced as a set. Obtain a Universal Bearing Puller Plate P/N 91-37241 or equivalent tool. Place the tool under the tapered bearing and tighten the plate jaws under the tapered bearing. Set the assembly into an arbor press and press the tapered bearing off the drive gear. Lift off the spacer and then the bearing race. Repeat this step and press the remaining tapered roller bearing off the driven gear. Remove the shim material from the gear and measure the thickness, for assembling.

9. Inspect the large O-ring around the oil seal carrier and replace it if its worn. Inspect the joint retainer for cracks or other damage, replacing if anything is found. Do the same with the thrust ring.

10. Inspect the oil seal and carrier for defects or damage. If there is a problem with the carrier it must be replaced with the seal as a unit. If the seal is bad, press it out with a punch and hammer. Use a seal driver tool (#91-36577) to press a new seal back into the retainer.

11. Inspect the splines on both yokes for wear or cracking, replacing anything if a problem is found.

12. Inspect the joints themselves for wear, knocking or too much side play. If problems are found, separate the joints.

13. Drive the eight C-rings off the U-joint bearings with a punch and hammer. Remove the two O-rings from the yoke shaft.

Fig. 77 Loosen the retainer to remove the joint assembly...

Fig. 78 ...and then pull out the shims

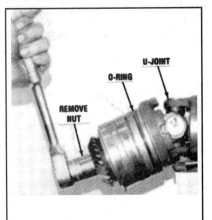

Fig. 79 After clamping the assembly in a vise, remove the nut...

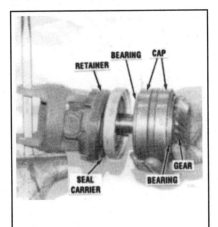

Fig. 80 ...and then slide off the drive gear/bearing assembly

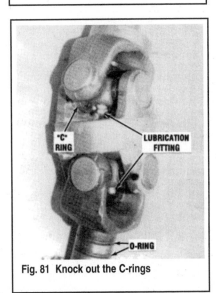

Fig. 81 Knock out the C-rings

Fig. 83 Install the shim and cup into the top cover

14. Obtain a suitable adaptor to support the U-joint yoke. Press one bearing until the opposite bearing is pressed into the adaptor. Remove the two loose bearings. Rotate the U-joint assembly 180°. Use the adaptor and press on the bearing cross-member and remove the bearings. Remove the cross-member from the yoke. Press the other four bearings out in the same manner, as described in the previous step and in the first part of this step.

To assemble:

15. Lubricate the U-joint bearing cups with a liberal amount of Quicksilver 2-4-C lubricant. Place the lubricated cups in the yoke and start them on the bearing crossmembers. Press the bearings through the yoke onto the crossmembers, as shown.

16. Drive the bearing cup retaining C-rings into place with a hammer and punch. Lubricate and install the other sets of bearings in the same manner.

17. Lubricate all bearings, cups, and gears with Quicksilver High Performance Gear lube or an equivalent, before assembling. Set the drive gear into an arbor press with the gear facing down. Place the tapered roller bearing onto the drive gear with the tapered side facing up. Using driver tool P/N 91-90774, press the bearing down onto the gear until the aft side of the bearing makes contact with the gear.

18. Place the bearing cup over the tapered roller bearing followed by the large spacer.

19. Set the second tapered bearing cup onto the large spacer ring with the tapered side facing up.

Place the tapered roller bearing onto the shaft. Obtain a suitable size mandrel and slowly press the tapered roller bearing down into the bearing cup until the roller barely makes contact with the bearing cup—then stop.

The bearing will be drawn closer into the cup when the bearing preload is set.

20. Press a new oil seal into the oil seal carrier, with the lip of the seal facing the concave side of the carrier, until the seal is flush with the carrier. Coat the lip of the seal with Multi-Purpose lubricant, or equivalent. Install the oil seal carrier, with the lip of the oil seal facing down, toward the drive gear. The seal prevents lubricant from working its way out of the upper housing and onto the U-joints.

21. Slide the retainer ring, thrust washer, O-ring, oil seal carrier, and drive gear bearing assembly onto the U-joint yoke shaft. Install the washer and nut—fingertight.

22. Clamp the unit in a vise with soft face jaws or clamp the retainer tool in a vise and set the U-joint into the retainer tool, as shown. Insert a breaker bar into the lower U-joint to prevent the yoke from rotating.

23. Place a screw type hose clamp around the bearing cup and spacer on the drive gear bearing assembly to hold the spacer and bearing cup aligned with the bearing. Tighten the nut on the end of the drive gear yoke until the preload on the bearing begins to increase slightly. Remove the hose clamp from around the bearing. Pull the bar out of the lower U-joint to permit the shaft to rotate.

24. Attach a socket and an inch-lb torque wrench to the nut on the end of the drive gear yoke. Hold the bearings stationary and at the same time, rotate the torque wrench and nut two or more turns. Observe the torque wrench indication for a bearing preload torque of 6–10 inch lbs. (0.7–1.1Nm). If the bearing preload torque is under specifications, repeat the procedure, tightening the nut slightly each time until the required bearing preload is obtained.

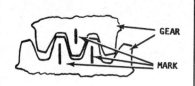

Fig. 84 Most drive and driven gears will have matchmarks on them

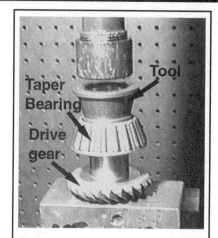

Fig. 85 Press the taper bearing onto the drive gear...

Fig. 86 ...and then slide on the bearing cup and the large spacer

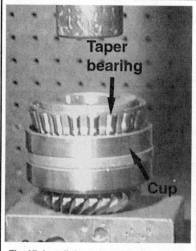

Fig. 87 Install the second bearing cup...

Fig. 88 ...and then install the oil seal and carrier

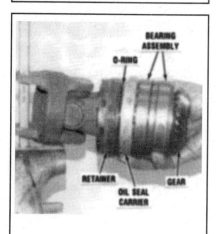

Fig. 89 Slide the gear assembly onto the yoke...

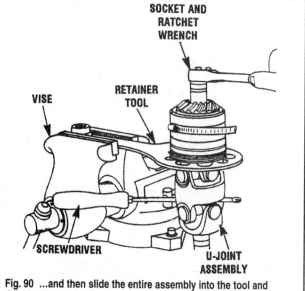

Fig. 90 ...and then slide the entire assembly into the tool and tighten the nut

Fig. 91 Use a torque wrench to adjust the preload

■ If the bearing preload torque exceeds the specified limit of 10 in. lbs, the bearings must be removed from the gear, inspected and then reassembled, as previously described in this section. If the bearing is damaged it must be replaced with a new bearing and cup. Failure to follow these procedures will place excessive loads on the bearing and gear, resulting in early bearing failure and possible gear damage.

25. Install the same number of bearing-to-driveshaft housing shims in the top cover that were removed and tagged during disassembling.

26. Most MerCruiser drive and driven gears have matching marks. These marks must be aligned when the gears are meshed in the gear housing. However, on some models, the gears do not have these marks. In this case, the gears may be installed with any gear mesh. Position the driven and drive gears to mesh properly before installing the U-joint shaft assembly.

27. Install the O-ring seal. Insert the assembled shaft, with the gears properly meshed into the upper gear housing. Tighten the bearing retainer nut to 200 ft. lbs. (271 Nm). It is very difficult to tighten the retainer nut to the proper torque value. An alternate method is to tighten the nut securely and then to bring it around to the mark you made during disassembling. From this point tighten it another 1/4 in. past the mark.

28. Install the top cover and tighten the screws to 17–23 ft. lbs. (23–31 Nm).

Driven Gear/Driveshaft Assembly

◆ See Figure 92, 93, 94, 95, 96, 97, 98, 99, 100, 101 and 102

1. Withdraw the driven gear assembly straight up and out of the housing bore.

2. If the driven gear large bearing race is in good condition, the shim material under the race providing bearing clearance is also in good condition—does not have to be changed—skip this step and proceed to Step 5.

3. Check the driven gear taper bearing race for evidence of the race spinning inside the housing bore. Look for pits, grooves, scoring and discoloration of the bearing race from overheating and contamination. If any of these conditions exist, the bearing race and shim material must be replaced.

4. Place slide hammer tool P/N 91-34569A into the housing and expand the jaws out under the bearing race. Using the slide hammer, pull the bearing race out of the housing bore. Discard the bearing race.

5. Reach inside the housing and remove the shim material under the taper bearing race. Save and tag the shim material, as an aid during assembling.

■ The top cover should only be disassembled if the bearing cup is damaged, or the shim material under the cup needs to be changed for gear location or bearing pre-load. These procedures are identified and given in detail later in this section.

6. Inspect the bearing cup in the top cover for signs of the cup spinning inside the bore of the cover. Look for pits, grooves, scoring and discoloration on the bearing cup from overheating and contamination. If any of these conditions exist, the cover, bearing cup, shims, and small bearing on the end of the upper drive gear assembly must all be replaced as one unit.

7. Pull the bearing cup out of the top cover. Reach inside the top cover and remove the shim material under the bearing cup. Save and tag the shim material, as an aid during assembling.

8. Remove the water tube and seal from inside the housing. Discard the seal. If the water tube is damaged, replace the tube.

9. Insert a long shank common screwdriver down through the driveshaft bore in the housing. Tap the two oil seals free of the driveshaft bore. Discard the seals.

10. Clean the driven gear and tapered bearings in solvent and blow them dry with clean compressed air. Do not spin the bearing with compressed air because such action will damage the bearing. Visually examine the gear for pitting, chipped, broken or damaged teeth. Examine both the large and small taper bearings on the shaft. Check for smooth rolling action without any rough spots or dragging. Check to be sure the bearing race inside the top cover and the gear housing are free of pits, grooves, scores, uneven wear and discoloration from overheating. If any of the above damage is found on the bearings or gear, replace the bearing or gear.

11. Place the Universal Puller plate P/N 91-37241 between the large tapered roller bearing and the driven gear. Tighten the puller plate securely between the driven gear and the large tapered bearing. Press the large tapered bearing free of the driveshaft.

12. Position the puller plate under the small tapered bearing, and then press the bearing free of the shaft.

13. Removal of the gear from the driveshaft is not necessary unless it is damaged and unfit for further service. If the gear is damaged, position the gear under a suitable size mandrel and press the gear free of the driveshaft.

To assemble:

14. Press the driveshaft into the driven gear using an arbor press and suitable mandrel until the shoulder on the shaft seats against the gear collar.

15. Position the tapered roller bearing over the end of the shaft with the number side of the bearing facing towards the gear teeth. Using a suitable size mandrel, press the bearing onto the shaft until the inner race of the bearing contacts the shoulder on the shaft. Remove the unit from the arbor press.

16. Inspect the housing, and cover inside and out for cracks. Check carefully around screw and shaft holes. Inspect machined faces and holes for burrs. Verify all old gasket material has been removed. Check O-ring seal grooves for sharp edges, which could cut a new seal. Inspect all oil passages to ensure they are clear and free of obstruction.

17. Inspect the bearing surfaces of the shafts, splines, and keyways for wear and burrs. Check for evidence of an inner bearing race rotating on the shaft. Measure the run-out on all shafts to detect any bent condition. If possible, check the shafts in a lathe for out-of-roundness.

18. Inspect the gear teeth and shaft holes for wear and burrs. Hold the center race of each bearing and turn the outer race. The bearing must turn freely without any evidence of binding or rough spots. Never spin a bearing with compressed air or the bearing will be ruined. Inspect the balls and rollers for pitting and flat spots. Inspect the outside diameter of the outer race and the inside diameter of the inner races for evidence of turning in the housing or on a shaft. Dark discoloration and/or deep scores are evidence the bearing has overheated.

19. Coat the rubber grommet for the water pick-up tube with a small amount of Perfect Seal. Install the grommet into the housing. Coat the end of the water pickup tube with Quicksilver 2-4-C Marine Lubricant or equivalent and slide the water pick-up tube into the water pick-up.

20. Coat the outside metal case of the lower driveshaft oil seal with a small amount of Loctite. Drive the seal into the cavity with the lip of the seal facing the driven gear. Continue to drive the seal into place until it is flush with the housing.

21. To ease the installation of the upper driveshaft, lubricate the lip of the seal with Multi-Purpose Lubricant. This oil seal prevents exhaust gases and water from entering the splined area of the driveshaft. Do not install the upper driveshaft oil seal until the upper driveshaft bearing preload has been measured.

22. The remainder of the components are installed while the bearing is being preloaded and measurements are taken during the assembling procedures.

Preload And Shim Adjustments

The following procedures cover preload and shim adjustments for the upper gear housing. The sequence of instructions are divided into three sections and they all must be performed in the order given to ensure efficient and long life operation of the upper gear housing.

UPPER DRIVESHAFT BEARING PRELOAD

◆ See Figures 103, 104 and 105

1. If the shim material thickness is known, install the shim material removed during disassembling, under the bearing race. If the shim thickness is unknown, then begin with a nominal thickness of 0.015 in. (0.038mm).

2. Place the shim material into the bore of the gear housing. Coat the bore in the housing for the bearing race with Quicksilver High Performance Gear lube or equivalent. Now, slide the bearing race down into the bore of the housing. Place the correct size mandrel (# 91-33493) onto the bearing

Fig. 92 Lift out the driven gear assembly...

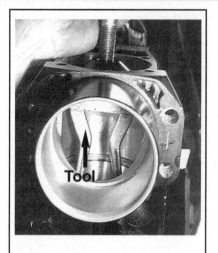

Fig. 93 ...and then pull the bearing race out with a slide hammer

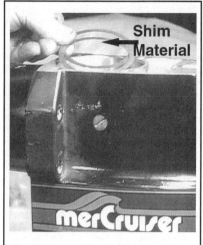

Fig. 94 Removing the shims from the bore

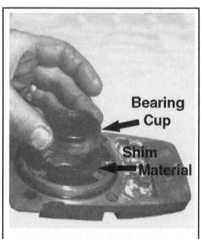

Fig 95 Remove the bearing cup from the cover and lift out the shims

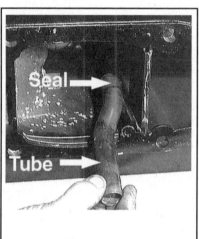

Fig. 96 Remove the water tube and seal...

Fig. 97 ...and then pry out the two oil seals

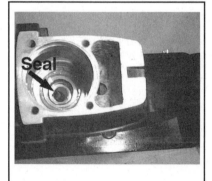

Fig. 98 Drive the lower seal into the case

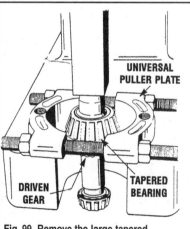

Fig. 99 Remove the large tapered bearing...

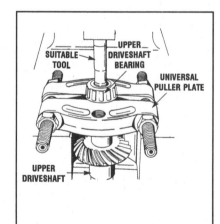

Fig. 100 ...and then remove the small one

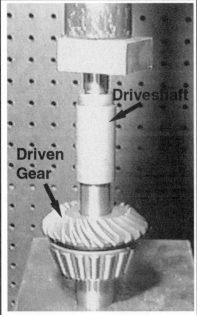

Fig. 101 Press the driveshaft into the driven gear...

Fig. 102 ...and then press on the bearing

Fig. 103 Add shims under the bearing race in the top cover

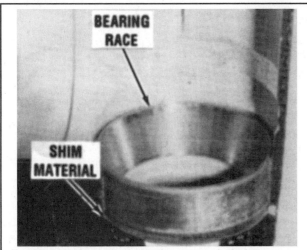

Fig. 104 Add shims under the bearing cup race in the bore

Fig. 105 Insert an old shaft into the unit from the bottom

race and drive the race into the bore until it contacts the shim material on the shoulder in the housing.

3. If the upper oil seal is in place, it must be removed before replacing the driveshaft. This seal must be removed in order to obtain an accurate bearing preload measurement. The friction a seal exerts on the shaft would give a false measurement.

4. Coat the gear, bearings, seals, and O-rings with Quicksilver High Performance Gear lube or equivalent. Install the upper driveshaft into the gear housing.

5. Install the same amount of shim material under the bearing race of the top cover as was removed during disassembling. Press in the top tapered bearing race. Install the top cover temporarily, without the O-ring seal in place. A gasket is not used under the top cover. Now, tighten the four bolts, securing the top cover in place, to a torque value of 17–23 ft. lbs. (23–31 Nm).

6. With the unit turned upside down, insert an old gear housing driveshaft, with a pinion gear retaining nut in place, into the upper driveshaft splines. Determine the effort required to turn the upper driveshaft. This may be done with a torque wrench calibrated in inch pounds. The torque value

for new or used bearings should be as follows;
- Used Bearings—3.0–7.5 inch lbs (0.3–0.8Nm)
- New Bearings—6.0–15.0 inch lbs (0.7–1.7Nm)

7. If the reading is too high, remove shim material from under the bearing cup in the top cover. If the torque value is less than the minimum listed, add shim material under the bearing cup. If the shim material pack under the top cover is changed, the bolts securing the cover is place must be tightened again to the proper torque value.

8. Repeat this procedure until the required bearing preload is obtained.

DRIVEN GEAR SHIMMING

◆ **See Figures 106 and 107**

These instructions pickup the work after the driveshaft bearing preload has been properly adjusted, as described in the previous section. The measurements in this section must be done slowly and precisely, to avoid problems surely to develop if inaccurate measurements are taken.

1. Begin by obtaining a shimming gauge tool (#91-60526), and then install the tool into the drive gear cavity, with the proper tool position (X, Y or Z), according to the table below.

2. Align the gauge tool with at least two full teeth of the driven gear, making sure that one full tooth is on either side of the tool centerline.

3. Measure the clearance between the gear face and the shimming gauge. This clearance should be 0.025 in. (0.64mm). Most feeler gauges do not have a blade of sufficient length to make this clearance measurement. A hacksaw blade is usually about 0.023 in. thick. Therefore, if the teeth of the blade are ground off, it may then be used to check the clearance. It would be best to check the thickness of the hacksaw blade with a micrometer before grinding the teeth down.

4. Rotate the shimming tool slightly each way to obtain a slight drag on the feeler gauge when it is aligned with one outside tooth. Without moving the shimming tool, remove the gauge and insert it between the shimming tool and the other outside tooth. This procedure is necessary to align the face of the gauge parallel to the driven gear teeth.

5. Accurately determine the clearance. If the measured clearance is less than 0.025 in. (0.64mm), subtract the measured clearance from the specified 0.025 in. (0.64mm), and then remove shim material of that thickness from under the driven gear tapered roller bearing race. Add an equal amount of shim material under the upper driveshaft tapered roller bearing race to maintain the previously corrected bearing preload.

6. Check the clearance a second time. If the clearance is correct, coat the outside metal case of the upper driveshaft oil seal with a thin layer of Loctite. Apply a thin coating of Multi-Purpose Lubricant to the lip of the seal to ease installation of the upper driveshaft.

■ It was necessary to remove this seal when measuring the driveshaft bearing preload because of the extra drag a seal places on a rotating shaft. This seal prevents lubricants from working from the driveshaft housing into the exhaust chamber.

Driven Gear Shim Tool Position	
Letter	Ratio
Y	1.98:1
X	1.65:1
Y	1.84:1
Z	1.50:1

Fig. 107 Use a feeler gauge and the special tool to shim the driven gear

DRIVE (PINION) GEAR SHIMMING

◆ **See Figures 108 and 109**

This procedure must be followed precisely as described in order to obtain a correct reading of the clearance between the drive gear and the shimming tool. The measurement is needed to ensure the shimming tool face is parallel with the face of the gear.

1. Insert shimming tool (#91-60523) into the driveshaft housing top cover cavity. Align the proper tool face (X, Y or Z) with two full teeth of the drive gear. One full tooth should be on each side of the tool centerline. Insert a 0.025 in. feeler gauge blade between one of the outside teeth and the shimming tool. Rotate the tool to obtain a slight drag on the feeler gauge blade, and then, without moving the shimming tool, remove the gauge and insert it between the tool and the other outside tooth. If the clearance is greater than the gauge thickness, repeat the measuring procedure with a thicker gauge until the same clearance is obtained between both outside gear teeth and the tool. If the clearance is less, use a thinner thickness gauge blade and repeat the measuring procedures described. Now, if the measurement between the tool and the gear is less than 0.025 in., subtract the reading obtained from 0.025 in. and add shims of that thickness between the U-joint tapered roller bearing race and the driveshaft housing. If the final measured clearance is more than 0.025 in. subtract 0.025 in. from the measurement and remove shims of that thickness. Check the clearance a second time after making changes to the shims. Remove the shimming tool.

2. After the shimming procedure is complete, position a new O-ring seal on the upper cover. Apply a coating of Quicksilver Perfect Seal, or equivalent, to the area shown on the cover. Install the cover and tighten the four bolts to 17–23 ft. lbs. (23–31 Nm).

The upper driveshaft housing assembly is now completely rebuilt and ready to be assembled to the lower unit, if the lower unit does not require service.

Drive Gear Shim Tool Position	
Letter	Ratio
Y	1.98:1
Y	1.65:1
Y	1.84:1
Z	1.50:1

Fig. 109 Use a feeler gauge and the special tool to shim the drive gear

TRANSOM ASSEMBLY

Description

The stern drive unit, consisting of the upper gear housing and the lower unit, is attached to the bell housing. The bell housing is secured to the gimbal ring by two roller bearings. These bearings permit up-and-down trim movement. The gimbal ring is mounted to the gimbal housing by two roller bearings. This second set of roller bearings permits steering movement to port and starboard. The gimbal housing is mounted on the transom of the boat and attached to an inner transom plate. This plate contains the rear engine mounts and attachments for the steering and shift cables.

The gimbal housing is mounted on the transom of the boat and attached to an inner transom plate. This plate contains the rear engine mounts and attachments for the steering and shift cables. The bell housing, gimbal ring, gimbal housing and inner plate all make up the transom assembly.

The bell housing has extended flanges to connect the exhaust, universal and shift cable bellows, and water hoses between the stern drive unit and the gimbal housing.

The bell housing is held in place by the gimbal ring.

Gimbal housing/transom plate service can generally be performed without removing the gimbal housing. However, in those cases when a part of the housing has been broken, or if the inner transom plate must be removed, the gimbal housing must be removed before the transom plate.

Before the gimbal housing can be removed, several tasks involving considerable work and time must be performed in the following order:

Stern drive removed.
Engine removed.
Bell housing and gimbal ring removed.

Gimbal Bearing

REMOVAL

◆ See Figures 110, 111, 112 and 113

■ **The gimbal bearing and carrier are a matched set and must be replaced as an assembly; further, the tolerance ring must be replaced whenever the bearing is replaced.**

1. Remove the stern drive unit. After the stern drive unit has been removed, the gimbal housing bearing can be checked by reaching through the U-joint bellows attached to the bell housing and rotating the inner bearing race. There should be no evidence of binding or rough spots. Check the race for side play by pulling and pushing on the race. If there is any sign of roughness, binding, or excessive side play, the bearing should be replaced. A special puller is required to remove the gimbal housing bearing. This puller is designed to establish alignment from the face of the bell housing. Never remove the gimbal bearing unless it is to be replaced, because it will be damaged during removal.

2. Assemble the plates of special tool (#91-29310). Position the plates between the top and middle studs located on the bell housing. Use a 3-jaw puller from the Slide Hammer Puller Set (#91-34569A1). If the bearing assembly is tight, tap the end of the Puller Shaft (#91-31229), with a mallet while attempting to turn the nut (#11-24156).

3. Remove the tolerance ring from the carrier.

4. Reach into the housing and remove the grease seal.

To install:

5. Clean all metal parts in solvent and dry them with compressed air. Never spin ball bearings with compressed air, because such action will ruin the bearing.

6. Inspect the bellows carefully for cracks, cuts, and punctures. Verify the bellows are still flexible. If the condition of the bellows is doubtful, replace them. In most cases, if the gimbal bearing is damaged due to water, the water has entered through, or around, the bellows.

The gimbal bearing has been replaced by a new style bearing. These are easily identified because the old style bearing carrier is black (top), and the new style is silver (bottom). The notches must face forward when the bearing is installed.

7. The gimbal housing bearing can only be installed with the bell housing in place in order to establish an alignment reference, so if you have removed the housing install it now.

8. Lubricate the outside of a new gimbal bearing carrier assembly with Multi-Purpose Lubricant, or equivalent.

9. Install the tolerance ring over the carrier. The opening in the tolerance band must align with the lubrication opening in the bearing carrier before the carrier is installed. The opening must also align with the lubrication opening in the gimbal housing after the carrier is installed.

■ **Observe the notches (could be called "cutouts), in the bearing carrier, indicated in the accompanying illustration. These notches must face forward when the carrier is installed.**

10. Install the assembled bearing carrier using the tools indicated in the cutaway line drawing labeled for Step 2. Insert the Driver Head (#91-32325), through the Mandrel (#91-30366), and into the inside diameter of the new bearing. Align the bearing with the gimbal housing by positioning Plate (#91-29310) between the top and middle studs on the bell housing. Positioning the plate as described will ensure the driver rod will remain at right angles to the bearing carrier bore.

11. Now, drive the bearing into the gimbal housing cavity with a hammer. Lubricate the bearing using only Quicksilver Multi-Purpose Lubricant through the grease fitting. To lubricate the bearing properly, pump 40 full strokes, to deliver about one full ounce of lubricant.

12. Install the stern drive.

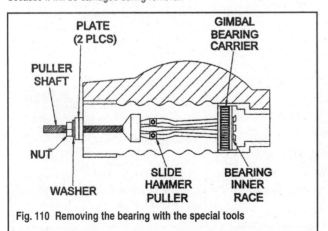

Fig. 110 Removing the bearing with the special tools

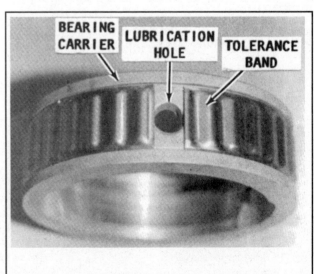

Fig. 111 It is important to install the tolerance ring correctly

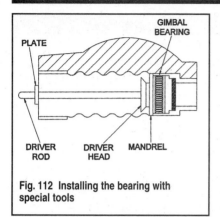

Fig. 112 Installing the bearing with special tools

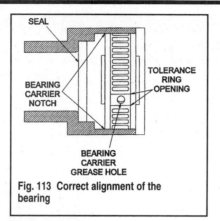

Fig. 113 Correct alignment of the bearing

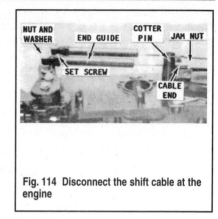

Fig. 114 Disconnect the shift cable at the engine

Shift Cable

OEM ② MODERATE

REMOVAL & INSTALLATION

◆ See Figures 114, 115, 116, 117, 118, 119, 120 and 121

1. Disconnect the stern drive unit shift cable from the shift plate mounted on the engine. This is accomplished by removing the nut, washer, and cotter pin.

2. Loosen the set screws from the cable end guide, and then remove the end guide. If only the inner cable is to be replaced, burrs made by the set screws must be removed from the core wire, to prevent the inner lining of the shift cable from being damaged when the wire is removed. Loosen the jam nut on the metal cable end, and then turn the metal end out of the cable.

3. Pull the threaded support tube from the end of the inner core wire.

4. Remove the protective wrapping from the shift cable in the area where the cable passes through the transom. Remove the stern drive.

5. If the inner wire is to replaced, hold the cable slide and pull the inner wire out of the shift cable. Now, remove the safety wire on the end of the cable slide. Remove the Allen head screw and cable slide from the inner shift wire.

6. Obtain special shift cable removal and installation tool (#91-12037). Remove the shift cable bellows clamp at the small end of the bellows. Pull the shift cable through the bellows. Using the special tool, completely loosen the locking nut and remove the cable from the bell housing. The locking nut is located on the aft (stud) end of the bell housing.

To install:

7. Inspect the shift bellows for cracks, cuts, and punctures. Verify the bellows are still flexible. If there is the least doubt about the condition of the bellows, they should be replaced. If the bellows are defective and leak, water will enter the boat.

8. Inspect the cable locking nut threads for any damage such as cross-threading and check for any indication of the locking nut separating from the outer casing. Check the length of the shift cable for kinks, cuts or chafing and the inner core wire for signs of unraveling.

■ The manufacturer strongly recommends that the small shift cable bellows clamp be the crimp type, not a worm type clamp and not a tie wrap. To crimp such a clamp, a special tool is required. However, a pair of common pliers may be easily and quickly modified to do the job. To customize a standard pair of pliers to crimp the bellows clamp, first tack weld a 3/4 in. nut to the pliers with the gripping surfaces of the pliers contacting two opposite sides of the nut. Clamp the pliers in a vice and drill out most of the threads using a 1/2 in. drill bit. With the pliers still in the vise, cut the nut in half, leaving an equal amount of the nut on each side of the pliers, as shown in the accompanying illustration.

9. Apply a coating of Perfect Seal to the locking nut threads. Secure the shift cable to the bell housing and tighten the nut securely using special cable tool. No more than two threads on the retainer should be visible.

10. Install the bellows clamp on the small end of the shift bellows with the end of the bellows 2 in. from the cable locking nut. Use the modified pliers, described in the previous Note, and squeeze the bellows clamp securely around the bellows. After the clamp is properly installed, water will not be able to enter the bellows and find its way into the boat.

11. Install the inner shift wire through the cable slide from the stern drive side. Install the Allen head screw securely, and then back it off about 1/8 turn. This adjustment will permit the inner wire to rotate freely. Install the safety wire, twist it tightly and cut off any excess.

12. Coat the inner shift wire with light weight oil, and then feed it into the shift cable until the cable slide enters the bell housing. Now, from inside the boat, carefully pull on the inner wire and be sure it is fully extended.

13. Install the protective wrapping around the cable.

14. Install the threaded tube on the shift cable end and tighten it until it just bottoms; now tighten the jam nut against the cable end.

15. Install the stern drive.

16. Push in on the drive unit shift cable while spinning the propeller counterclockwise until it stops. Place a bungee cord over the prop to keep the clutch engaged.

17. Install the cable end guide over the core wire and then insert the core wire through the anchor. Tighten the screws securely.

18. Rotate the barrel on the cable threads until the distance between the center of the barrel and the hole in the end of the guide is 6 in. (153mm).

19. Adjust the cable.

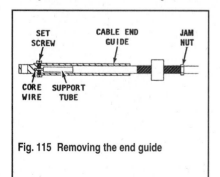

Fig. 115 Removing the end guide

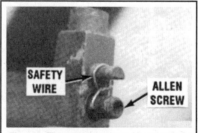

Fig. 116 The screw on the cable slide will be safety wired

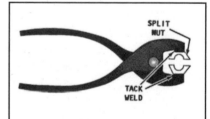

Fig. 116 Fabricate a crimping tool with a pair of pliers and a nut

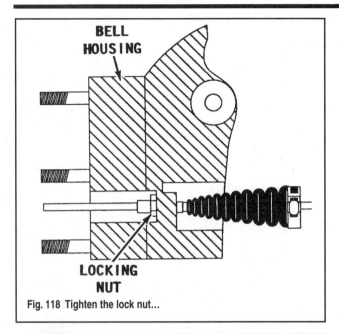

Fig. 118 Tighten the lock nut...

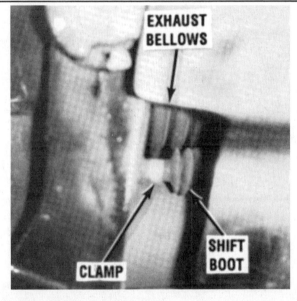

Fig. 119 ...and then install the clamp

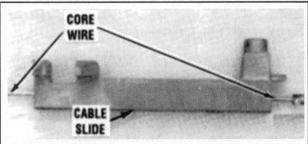

Fig. 120 Pull the wire through the slide from the stern drive side

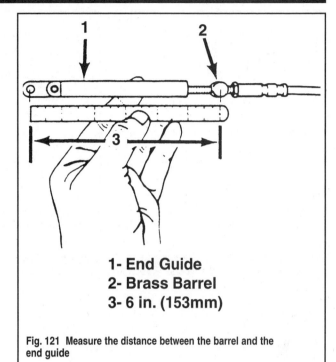

1- End Guide
2- Brass Barrel
3- 6 in. (153mm)

Fig. 121 Measure the distance between the barrel and the end guide

ADJUSTMENT

◆ **See Figures 122, 123 and 124**

1. The brass barrel should be in the retainer with a cotter pin holding it in position. Make sure that the end guide has washers on both sides and that the locknut has been backed out from bottom 1/4–1/2 turns. The adjustable stud that the end guide rides on should be at the bottom of the slot.

2. On a right hand rotation drive, move the remote control lever to full throttle in Forward. On a left hand rotation unit, move the lever to full throttle in Reverse.

3. Push in on the drive unit cable while having someone rotate the propeller counterclockwise until it stops; this will ensure clutch engagement. Maintain a light pressure on the cable so there is no slack at all. If the shift cut-out switch moves, you are exerting too much pressure.

4. Now, pull lightly on the remote control cable end guide and adjust the barrel until it lines up with the stud. The hole in the end guide should still be lined up with the clevis hole. You should continue to maintain light pressure on the drive cable during this.

5. Once the barrel hole and the stud are aligned, turn the barrel 4 complete turns away (to the right as you look at the assembly) from the end guide.

6. Install the cable to the clevis and stud.

7. On a RH rotation drive, move the remote control lever to full throttle in Reverse. On a LH rotation unit, move the lever to full throttle in Forward. At the same time, have someone rotate the propeller clockwise until it stops.

8. Check the shift cut-out switch lever position—the roller should be centered in the detent. If not, check that the adjustable stud is at the bottom of the slot, check that the remote control cable output is 3 +/- 1/8 in. If both are OK, check that the drive unit cable is not crushed or kinked.

9. Now that the remote control cable is adjusted properly, reinstall the cable and shift assist assembly (if equipped) and secure all hardware as shown in the illustration. If the shift assembly attaching points will not align, push or pull on the assembly until they fit. Do not attempt to readjust the shift cable.

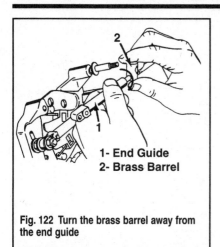

1- End Guide
2- Brass Barrel

Fig. 122 Turn the brass barrel away from the end guide

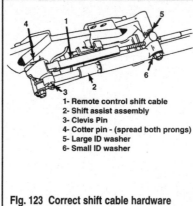

1- Remote control shift cable
2- Shift assist assembly
3- Clevis Pin
4- Cotter pin - (spread both prongs)
5- Large ID washer
6- Small ID washer

Fig. 123 Correct shift cable hardware locations—w/shift assist

1- Pin
2- Cotter Pin
3- Washer (Large ID)
4- Washer (Small ID)
5- Cable end guide
6- Shift lever
7- Lock nut

Fig. 124 Correct shift cable hardware locations—w/o shift assist

Shift Cut-Out Switch

ADJUSTMENT

MODERATE

◆ See Figures 125 and 126

1. Disconnect the switch wire (White/Green) at the terminal block.
2. Connect an ohmmeter as per the manufacturer's instructions. Connect the positive lead to the White/Green wire and the negative lead to the switch's black wire at the terminal block. Set the meter to the RX1 scale.
3. Slowly move the switch roller off of its seat. The circuit should close (showing continuity) when the roller has been moved 1/8 in. (3mm). If the switch closes before being moved as specified, bend the roller away from its seat. If it closes after the specification, bend it toward the seat.
4. Reconnect the wires at the block and cover them with Liquid Neoprene.

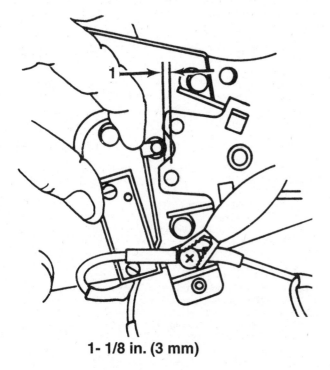

1- 1/8 in. (3 mm)

Fig. 125 Move the roller of its seat until the circuit closes

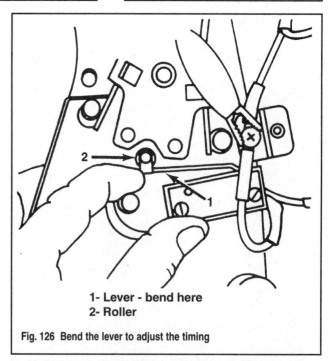

1- Lever - bend here
2- Roller

Fig. 126 Bend the lever to adjust the timing

Exhaust Bellows

OEM   DIFFICULT

REMOVAL & INSTALLATION

◆ See Figures 127, 128, 129 and 130

■ If the stern drive has been removed, simply undo the bellows and replace them. The following procedure is with the drive still attached and may prove difficult.

1. Tilt the stern drive slightly up and move it to the port side. Access to the aft clamp on the exhaust bellows is from the bottom of the bell housing. Remove the clamp. Move the stern drive to the center position. Access to the forward clamp on the exhaust bellows is through the access hole on the port side of the gimbal housing. Insert a screwdriver through the access hole and remove the forward exhaust bellows clamp.
2. Pull the exhaust bellows from the gimbal housing and bell housing exhaust flanges. It may be necessary to exert considerable force to pull the bellows loose because of the adhesive used during installation. Raising the stern drive to the full up position, will be helpful.

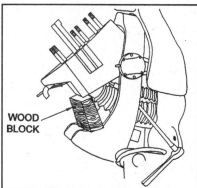

Fig. 127 With the stern drive removed from the boat, a 2 x 4 wood support under the bell housing will be helpful to disconnect the clamps on the exhaust bellows

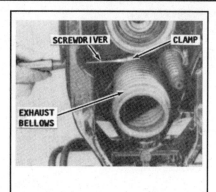

Fig. 128 An access hole in the housing provide a way to get your screwdriver on the clamp screw

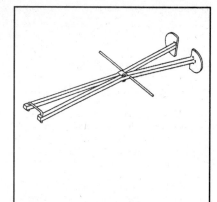

Fig. 129 A bellows expander is necessary to get the bellows back onto the flange

To install:

3. Clean the bellows mounting flanges of the upper bell housing with a wire brush or sandpaper, and then wipe the surface clean with lacquer thinner to remove any old glue.

4. If new bellows are being installed (and if you've removed them, you should install new one), be sure to clean the powder residue from the bellows surface with warm soapy water and dry thoroughly. The powdery substance is a mold release agent and the adhesive will not bond if this powder is left on the sealing surfaces.

5. Check the flanges and housing for cracks, nicks, or corrosion. Clean the bellows clamps thoroughly to ensure a good ground. New bellows have a clip on each end of the bellows to ground the clamp. Check the clamps for cracks or nicks. Replace the clamp if in doubt as to their condition. Always use stainless steel clamps for satisfactory service.

✳✳ WARNING

Bellows Adhesive must be used to ensure a satisfactory installation. This adhesive is extremely toxic and flammable. Therefore, make every effort to ensure adequate ventilation in the work area during its use. Work in the outdoors, if at all possible. Vapors from the adhesive may cause flash fire or ignite explosively.

Keep the adhesive away from heat, sparks, and open flame. Observe no smoking. Extinguish all flames and pilot lights in the area. Turn off stoves, heaters, electric motors, and all other possible sources of ignition while using the adhesive and until there is no doubt but what all vapors have left the area. Close the container immediately after use.

The adhesive is harmful or fatal if swallowed. Avoid prolonged contact with the skin or breathing of the vapors. If swallowed, do not induce vomiting. Call a physician immediately. Keep the adhesive out-of-reach of children.

6. Apply a coating of Bellows Adhesive to the inside of each end of the bellow and around the mounting flanges. Allow the adhesive to dry approximately 10 minutes, or until the material is no longer tacky. Install the clamp on one end of the bellows, and then install it to the gimbal housing flange so that the clamp screw is accessible through the access hole. Tighten the clamp screw to 35 inch lbs. (4 Nm).

7. Now the hard part. Install the bellow clamp on the other end of the bellows.

■ A special bellows tool is almost a necessity to "pull" the bellows over the flange of the stern drive. Therefore, obtain expander tool (#91-45497A1) and place the tool into the first bellows convolution. Pull on the tool until the tool touches the flange on the bell housing—the hose starts to slip onto the flange—then release the tool. The accompanying cross-section line drawing illustrates the position of the tool in the first convolution of the bellows.

8. Move the tool into the third bellows convolution and pull the bellows onto the bell housing flange. Tighten the hose clamp to a torque value of 35 in lbs (4Nm).

9. Start the engine and check for leaks.

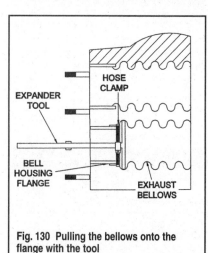

Fig. 130 Pulling the bellows onto the flange with the tool

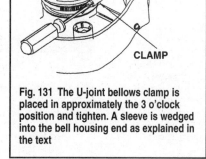

Fig. 131 The U-joint bellows clamp is placed in approximately the 3 o'clock position and tighten. A sleeve is wedged into the bell housing end as explained in the text

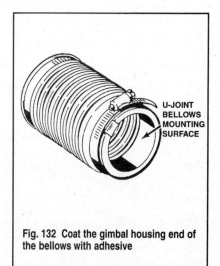

Fig. 132 Coat the gimbal housing end of the bellows with adhesive

U-Joint Bellows

REMOVAL & INSTALLATION   ③ ⟨DIFFICULT⟩

◆ **See Figures 131, 132, 133 and 134**

■ Although not necessary, we recommend removing the stern drive. The bell housing does not have to be removed in order to replace the U-joint bellows. However, this is not an easy task. Therefore, work slowly and carefully. If the job proves too difficult, the decision may be made to remove the exhaust bellows and the water intake hose first.

1. Remove the single hose clamp securing the bellows to the gimbal housing using a long shank screwdriver. Work the bellows off the gimbal housing flange. Considerable force may be required to remove the bellows due to the adhesive used during installation.

2. There is no hose clamp securing the other end of the bellows to the bell housing. Instead, a sleeve fits tightly inside the bellows to hold it against the bell housing flange. The sleeve must be pried free of the bellows. Obtain a can of aerosol cleaner such as "Gunk" or WD-40. Spray the cleaner around the edge of the sleeve to help loosen it. Pry the sleeve free using a thin blade screwdriver. Push the bellows back from the bell housing flange until it is free. No adhesive is used at this end of the bellows, therefore, just push.

To install:

3. Remove all bellow adhesive from the inside diameter of the bellows, if the bellows are to be reused. Inspect the bellows for cracks, cuts, punctures and to be sure it is still flexible. If the least bit of doubt exists concerning the condition of the bellow, install new bellows.

4. Clean the bellow mounting flanges of the bell housing with a wire brush or sandpaper, and then wipe the surface clean with lacquer thinner.

5. Check the flanges and housing for cracks, nicks, or corrosion. Clean the bellow clamps thoroughly. Check the clamps for cracks or nicks. Replace the clamps if in doubt as to their condition. Always use stainless steel clamps as a replacement.

✳✳ WARNING

Bellows Adhesive must be used to ensure a satisfactory installation. This adhesive is extremely toxic and flammable. Therefore, make every effort to ensure adequate ventilation in the work area during its use. Work in the outdoors, if at all possible. Vapors from the adhesive may cause flash fire or ignite explosively.

Keep the adhesive away from heat, sparks, and open flame. Observe no smoking. Extinguish all flames and pilot lights in the area. Turn off stoves, heaters, electric motors, and all other possible sources of ignition while using the adhesive and until there is no doubt but what all vapors have left the area. Close the container immediately after use. The adhesive is harmful or fatal if swallowed. Avoid prolonged contact with the skin or breathing of the vapors. If swallowed, do not induce vomiting. Call a physician immediately. Keep the adhesive out-of-reach of children.

6. Apply a coating of Bellows Adhesive to the inside diameter of the gimbal housing bellow end. Do not apply the adhesive to the bell housing end. Allow the adhesive to dry approximately 10 minutes, or until the material is no longer tacky. Install a grounding clip over the forward edge of the bellows with the shorter side of the clip on the inside of the bellows.

7. Find the word TOP embossed on the bellows. This word must face up after installation. Be sure to position the bead on the inner diameter of the bellows in the groove on the gimbal housing flange. Install the U-joint bellows over the flange and position the clamp tightening screw at the 3 o'clock position. Tighten the screw 35 inch lbs. (4 Nm). Don't forget the ground clip.

8. Apply Locquic Primer "T" to the internal threads of the bell housing. Apply the primer to the external hinge pin threads. Allow the primer at both locations to dry. Apply Loctite 271 to the threads of the bell housing, and then install the hinge pins. Tighten the hinge pins to a torque value of 95 ft. lbs (129 Nm).

9. Position the U-joint bellows on the bell housing. Ensure the bell housing flange is indexed in the second groove from the end of the bellows.

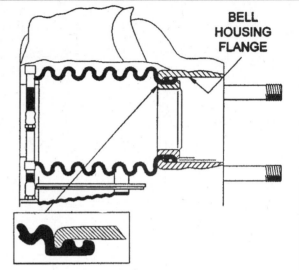

Fig. 133 Make sure that the bell housing flange rests in the second groove on the bellows

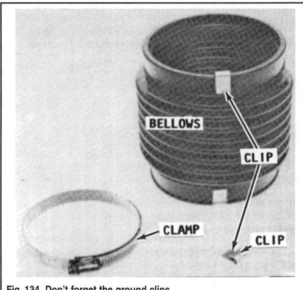

Fig. 134 Don't forget the ground clips

10. Obtain Sleeve Installation Tool (#91-818162). Lubricate the outside surface of the sleeve with soapy water or engine cleaner, and then install the sleeve tool. Use a suitable driving rod and move the sleeve in place on the bell housing.

11. If the water hose and the exhaust bellows were removed, install them to the gimbal housing.

12. Replace the stern drive.

13. Start the engine and check the completed work.

Bell Housing

REMOVAL & INSTALLATION   ③ ⟨DIFFICULT⟩

◆ **See Figures 135, 136, 137, 138 and 139**

1. Remove the stern drive as detailed previously.

2. Loosen the mounting screws, bend back the retainer clip and remove the trim position sender from the starboard side and the trim limit switch from the port side. Let them dangle temporarily.

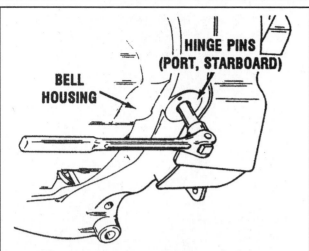

Fig. 135 The hinge pins are removed and installed with a special tool. Without this tool, removal of the hinge pins will almost be an impossible task

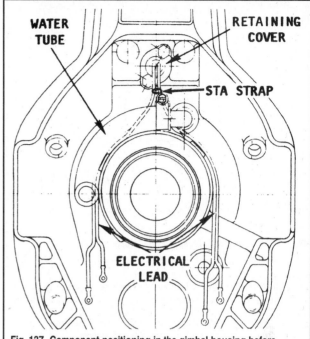

Fig. 137 Component positioning in the gimbal housing before installing the bell housing

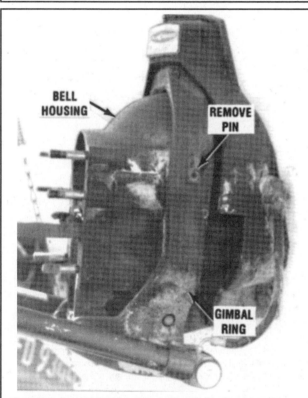

Fig. 136 Swivel the housing to one side in order to remove the opposite side's hinge pin

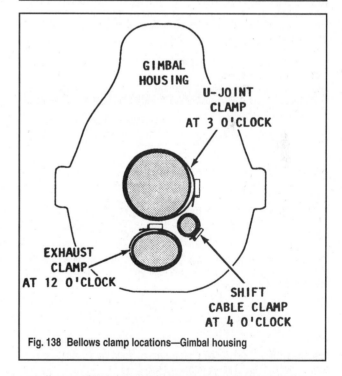

Fig. 138 Bellows clamp locations—Gimbal housing

3. Remove the shift cable.

4. Remove the two screws and lift off the water tube cover and the grommet underneath it. Push the tube through the gimbal housing.

5. Insert a sleeve removal tool (#91-818169) into the U-joint bellows hole, tighten the nut and remove the sleeve.

6. Remove the exhaust bellows.

7. Pop out the speedometer tubing clip at the bottom of the housing.

8. Attach a hinge pin tool (#91-78310) to a socket wrench. Swivel the housing to the port side and remove the starboard hinge pin. Now swivel it over to starboard and remove the port hinge pin. Carefully remove the housing.

9. If you need to replace the trim limit or position switches, loosen the screw and remove the retaining clamp. Disconnect the wire leads at the engine harness and pull them through the hole in the gimbal housing.

To install:

10. Install the bellows to the gimbal housing.

11. Lubricate the end of the shift cable with 2-4-C lubricant and insert the shaft cable into the shift cable bellows.

12. Insert the water tube through the gimbal housing. Install a new grommet and then install the plate cover.

13. If you disconnected the trim switches, route the trim limit switch and trim position sender leads and install a sta-strap around both sets of leads

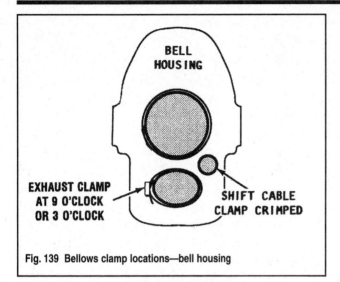

Fig. 139 Bellows clamp locations—bell housing

approximately 5" (12cm) from the retaining cover.

14. Bring the bell housing together with the gimbal housing. Push evenly on the perimeter of the bell housing until the U-joint bellows slides into the flange on the gimbal housing. Check to be sure the exhaust bellows aligns with the flange on the bell housing. Refer to the appropriate procedures in this section for pertinent details on final bellows connections.

15. Coat the hinge pin threads (inside the housing and on the bolts) with Locquic Primer "T" and allow the material to dry. Next, apply a coating of Loctite 271 to the housing threads. Install and tighten the hinge pins to 95 ft. lbs. (129 Nm).

16. Check final bellows installation against the individual procedures given earlier.

17. Install and adjust the shift cable.

18. Install the trim switches on either side of the gimbal ring

19. Install the stern drive.

Gimbal Ring

REMOVAL & INSTALLATION

This procedure can be completed with or without the engine and transom plate being removed.

Engine/Transom Assembly Installed

◆ See Figures 140, 141, 142, 143, 144 and 145

■ Unless a previous owner has already drilled and access hole in the gimbal housing, you will need to drill (and plug) your own hole in order to remove the steering lever.

1. Remove the stern drive and the bell housing.

2. Pump a liberal amount of grease into the fitting at the top of the gimbal housing.

3. You'll need Access Plug Kit (#22-88847A-1) from your dealer. Use the template provided and position it on the housing using the dimple in the front of the housing—it will be under the decal.

4. Use a center punch and the template to mark the location of the holes on each side of the gimbal housing and then drill a 1/4 in. (6mm) hole through each side of the housing.

5. Using a 1-1/8 in. hole saw with a 1/4 in. (6mm) pilot rod, drill a hole in each side using the previous holes as a guide. Make sure the drill is perpendicular .to the housing and do not rush this.

Fig. 140 Use a socket on the port side when removing the steering lever bolt

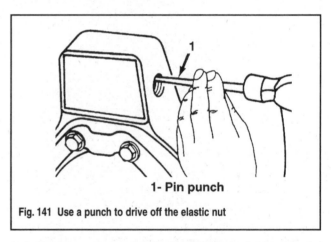

1- Pin punch

Fig. 141 Use a punch to drive off the elastic nut

6. Use a piece of tape and mark a 1 in. #180 pipe tap (hardware store) about 1-1/8 in. from the end. Coat it with grease and then thread it into the hole.

7. Insert a socket wrench into the port access hole and an open end into the starboard hole. Loosen and remove the steering lever clamping bolt. Now insert a punch through the starboard hole and unthread the elastic locknut. You may have to move the steering wheel a few time to jockey the lever and nuts into position.

8. Disconnect the trim cylinders at the gimbal ring and carefully set them aside—they should already have been disconnected on the other end when you removed the stern drive.

9. Disconnect the continuity wire and then remove the lower swivel cotter pin. Drive out the lower swivel pin and remove the anti-gauling washer.

10. Loosen the two upper gimbal ring bolts and then remove the upper swivel shaft and large washer. You may need a slide hammer if it is frozen. Lift out the gimbal ring. Look for the steering lever and its hardware and remove them along with the ring.

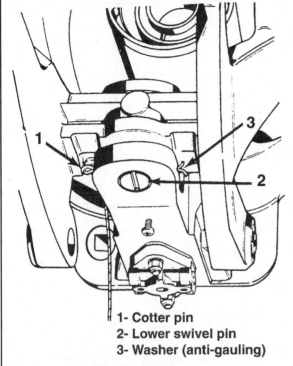

1- Cotter pin
2- Lower swivel pin
3- Washer (anti-gauling)

Fig. 142 Removing the lower swivel pin

Fig. 143 You may need a slide hammer to remove the upper swivel shaft

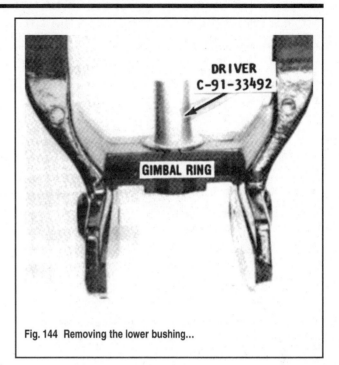

Fig. 144 Removing the lower bushing...

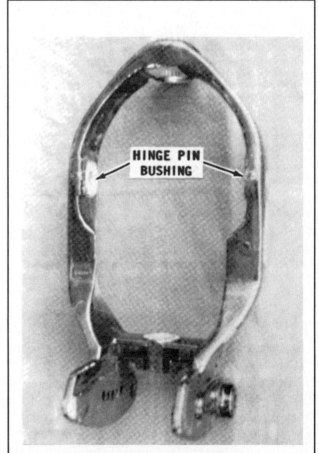

Fig. 145 ...and then remove the hinge pin bushings

11. Remove the gimbal ring lower shaft bushing, using a suitable driver. Remove the gimbal ring upper swivel shaft bushing and single oil seal, using a slide hammer with a two jaw puller attachment.

12. Remove the gimbal ring hinge pins, using a Hinge Pin tool (#91-78310) to remove the pins, if they are threaded into the housing.

13. Obtain a suitable driver and drive out both hinge pin bushings.

To install:

14. Clean all metal parts in solvent and blow them dry with compressed air. Never, spin ball bearings with compressed air, because such action will ruin the bearings.

15. Remove all bellow adhesive from the inside diameter of the bellows, if the bellows are to be reused. Inspect the bellows for cracks, cuts, punctures and to be sure they are still flexible. If there is the least bit of doubt concerning their condition, do not hesitate to install new bellows. Clean the bellow mounting flanges of the bell housing with a wire brush or sandpaper and then wipe the surface clean with lacquer thinner.

16. Check the shift cables for cuts or damage caused by a cable being pinched or bent too short.

17. Inspect the water hose for cracks, cuts, punctures, or worn spots. Inspect the shift shaft oil seal for tears, wear, or any roughness. Check the shift shaft and shift shaft bushings for wear.

18. Inspect the O-ring and rubber gasket for cuts, nicks, hardness, or cracks.

19. Inspect the surface of the lower swivel pin in the area where the needle bearing rides. Any pitting, grooves, or uneven wear is cause to replace the bearing and swivel pin. Inspect the lip surface of the oil seals for wear, tears, and roughness.

20. Obtain an ohmmeter and test the trim limit switch. Set the switches of the ohmmeter for continuity reading and connect the leads of the ohmmeter to the leads of the trim limit switch. Rotate the outer case while holding the inner rotor and observe the ohmmeter. The meter should indicate continuity through most of the movement and then no continuity. If the meter indicates continuity all the time or no continuity, the switch is defective and must be replaced.

20. Inspect the long steering lever retaining bolt. Any grooves found on the bolt are a result of friction against the upper swivel shaft. If grooves are discovered, both the steering lever and the bolt must be replaced.

21. Obtain Resiweld Sealer. Apply a coating of this sealer to the outer diameter of the lower bushing. Using a suitable driver, tap the bushing into place.

22. To install the small upper swivel shaft bushing, first obtain a Bearing and Seal Driver tool (#91-43578). Next, place the bushing on the tool, and then use a hammer and tap the bushing into place. Install the larger bushing in a similar manner.

23. Pack the single oil seal lip with Multi-Purpose lubricant and place the seal on the driver with the lip facing the smaller diameter of the tool. Install the oil seal into the gimbal ring until the seal seats against the larger bushing.

24. Check the condition of the synthane washers on the gimbal ring. If they are worn, or damaged, remove and discard the washers. Clean the washer mounting surface free of all adhesives, dirt, grease and oil. Peel the backing from a new washer and place it on the mounting surface, and then press the washer down firmly. The adhesive backing will hold the washer in place.

25. Apply a coating of Resiweld to the outer diameter of both hinge pin bushings. Using a suitable driver, tap the bushings into place on the port and starboard sides of the gimbal ring.

26. Apply a coat of 2-4-C Lubricant onto the surface of the lower swivel pin. Insert the switch harness, if equipped, through the gimbal housing, and then secure it in place with the screws. Place the gimbal ring in position in the gimbal housing. Place a washer between the gimbal ring and the housing. Secure the ring in place with the lower swivel pin and cotter pin.

27. Install the long bolt and nut into the steering lever. A washer is not used at this location. Tighten the nut until it is just "snug". If a new swivel shaft elastic nut is to be used, thread the nut onto the swivel shaft until it is tight, and then remove it. This action cuts the threads into the new nut (which should come with the access plug kit).

28. Check to be sure the gimbal ring rests squarely within the gimbal housing. Before installing the swivel shaft, check for a flat area machined onto the shaft splines. If a flat exists, the flat must face forward after installation. Place the larger washer on top of the ring, followed by the steering lever, then the smaller washer and finally, the large nut. Hold these components aligned together while the upper swivel shaft is passed through the gimbal ring and these parts. Start the large nut onto the shaft threads.

29. Tighten the bolt on the steering lever to 60 ft. lbs. (81 Nm). Use a punch through one of the side access holes to tighten the nut, until a clearance of 0.002–0.010 in. (0.05–0.25mm) exists between the lower swivel pin washer and the gimbal housing.

30. Use a rawhide mallet and strike the top of the gimbal ring flanges a couple times each to seat both pins. Recheck the clearance and adjust the large nut (using the punch and hammer method), as necessary to maintain the required clearance.

31. Tighten the two gimbal ring bolts to 55 ft. lbs. (74 Nm). Apply a coating of Perfect Seal to the threads or the sealing surface of the plugs and close the two large openings drilled earlier on the sides of the gimbal housing for access to the steering lever.

32. Install the trim cylinders by first attaching them in place with the forward anchor pin. Next, install the two hydraulic hoses to the connector. Finally, install the connector to the gimbal housing.

33. Connect the steering link rod to the steering lever. Tighten the castellated nut to 24 inch lbs. (3 Nm), and then back it off slightly and insert the cotter pin. If the assembly is secured with a Nylok nut instead of the cotter pin, tighten the nut to 5 ft. lbs. (7 Nm).

34. Install the bell housing and stern drive.

Engine/Transom Assembly Removed

◆ **See Figures 146 and 147**

1. Remove the stern drive and the bell housing.
2. Loosen the two upper gimbal ring screws.
3. Reach in through the transom and loosen the steering lever clamp bolt.
4. Reach in through the transom and remove the elastic locknut on top of the steering lever with a 1-1/16. in. wrench.
5. Disconnect the trim cylinders at the gimbal ring and carefully set them aside—they should already have been disconnected on the other end when you removed the stern drive.
6. Disconnect the continuity wire and then remove the lower swivel cotter pin. Drive out the lower swivel pin and remove the anti-gauling washer.
7. Remove the upper swivel shaft and large washer. You may need a slide hammer if it is frozen. Lift out the gimbal ring. Look for the steering lever and its hardware and remove them along with the ring.
8. Remove the gimbal ring lower shaft bushing, using a suitable driver. Remove the gimbal ring upper swivel shaft bushing and single oil seal, using a slide hammer with a two jaw puller attachment.
9. Remove the gimbal ring hinge pins, using a Hinge Pin tool (#91-78310) to remove the pins, if they are threaded into the housing.
10. Obtain a suitable driver and drive out both hinge pin bushings.

To install:

11. Clean all metal parts in solvent and blow them dry with compressed air. Never, spin ball bearings with compressed air, because such action will ruin the bearings.

12. Remove all bellow adhesive from the inside diameter of the bellows, if the bellows are to be reused. Inspect the bellows for cracks, cuts, punctures and to be sure they are still flexible. If there is the least bit of doubt concerning their condition, do not hesitate to install new bellows. Clean the bellow mounting flanges of the bell housing with a wire brush or sandpaper and then wipe the surface clean with lacquer thinner.

13. Check the shift cables for cuts or damage caused by a cable being pinched or bent too short.

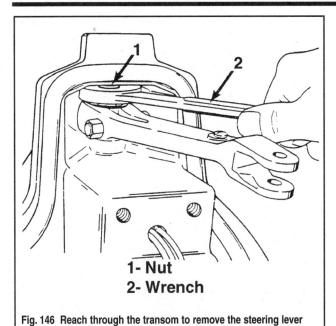

1- Nut
2- Wrench

Fig. 146 Reach through the transom to remove the steering lever

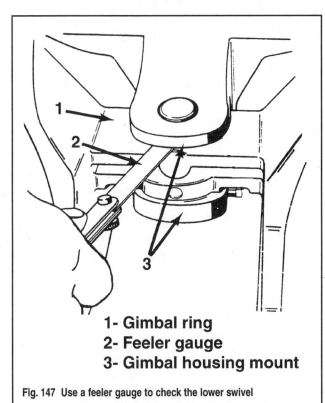

1- Gimbal ring
2- Feeler gauge
3- Gimbal housing mount

Fig. 147 Use a feeler gauge to check the lower swivel

14. Inspect the water hose for cracks, cuts, punctures, or worn spots. Inspect the shift shaft oil seal for tears, wear, or any roughness. Check the shift shaft and shift shaft bushings for wear.

15. Inspect the O-ring and rubber gasket for cuts, nicks, hardness, or cracks.

16. Inspect the surface of the lower swivel pin in the area where the needle bearing rides. Any pitting, grooves, or uneven wear is cause to replace the bearing and swivel pin. Inspect the lip surface of the oil seals for wear, tears, and roughness.

17. Obtain an ohmmeter and test the trim limit switch. Set the switches of the ohmmeter for continuity reading and connect the leads of the ohmmeter to the leads of the trim limit switch. Rotate the outer case while holding the inner rotor and observe the ohmmeter. The meter should indicate continuity through most of the movement and then no continuity. If the meter indicates continuity all the time or no continuity, the switch is defective and must be replaced.

18. Inspect the long steering lever retaining bolt. Any grooves found on the bolt are a result of friction against the upper swivel shaft. If grooves are discovered, both the steering lever and the bolt must be replaced.

19. Obtain Resiweld Sealer. Apply a coating of this sealer to the outer diameter of the lower bushing. Using a suitable driver, tap the bushing into place.

20. To install the small upper swivel shaft bushing, first obtain a Bearing and Seal Driver tool (#91-43578). Next, place the bushing on the tool, and then use a hammer and tap the bushing into place. Install the larger bushing in a similar manner.

21. Pack the single oil seal lip with Multi-Purpose lubricant and place the seal on the driver with the lip facing the smaller diameter of the tool. Install the oil seal into the gimbal ring until the seal seats against the larger bushing.

22. Check the condition of the synthane washers on the gimbal ring. If they are worn, or damaged, remove and discard the washers. Clean the washer mounting surface free of all adhesives, dirt, grease and oil. Peel the backing from a new washer and place it on the mounting surface, and then press the washer down firmly. The adhesive backing will hold the washer in place.

23. Apply a coating of Resiweld to the outer diameter of both hinge pin bushings. Using a suitable driver, tap the bushings into place on the port and starboard sides of the gimbal ring.

24. Apply a coat of 2-4-C Lubricant onto the surface of the lower swivel pin. Insert the switch harness, if equipped, through the gimbal housing, and then secure it in place with the screws. Place the gimbal ring in position in the gimbal housing. Place a washer between the gimbal ring and the housing. Secure the ring in place with the lower swivel pin and cotter pin.

25. Install the long bolt and nut into the steering lever. A washer is not used at this location. Tighten the nut until it is just "snug". If a new swivel shaft elastic nut is to be used, thread the nut onto the swivel shaft until it is tight, and then remove it. This action cuts the threads into the new nut (which should come with the access plug kit).

26. Check to be sure the gimbal ring rests squarely within the gimbal housing. Before installing the swivel shaft, check for a flat area machined onto the shaft splines. If a flat exists, the flat must face forward after installation. Place the larger washer on top of the ring, followed by the steering lever, then the smaller washer and finally, the large nut. Hold these components aligned together while the upper swivel shaft is passed through the gimbal ring and these parts. Start the large nut onto the shaft threads.

27. Tighten the bolt on the steering lever to 60 ft. lbs. (81 Nm). Now tighten it further until a clearance of 0.002–0.010 in. (0.05–0.25mm) exists between the lower swivel pin washer and the gimbal housing.

28. Use a rawhide mallet and strike the top of the gimbal ring flanges a couple times each to seat both pins. Recheck the clearance and adjust the large nut (using the punch and hammer method), as necessary to maintain the required clearance.

29. Tighten the two gimbal ring bolts to 55 ft. lbs. (74 Nm).

30. Install the trim cylinders by first attaching them in place with the forward anchor pin. Next, install the two hydraulic hoses to the connector. Finally, install the connector to the gimbal housing.

31. Connect the steering link rod to the steering lever. Tighten the castellated nut to 24 inch lbs. (3 Nm), and then back it off slightly and insert the cotter pin. If the assembly is secured with a Nylok nut instead of the cotter pin, tighten the nut to 5 ft. lbs. (7 Nm).

32. Install the bell housing and stern drive.

Gimbal Housing/Transom Plate

REMOVAL & INSTALLATION   ③ DIFFICULT

◆ **See Figures 148, 149 and 150**

1. Remove the stern drive, bell housing, gimbal ring and bushing.
2. Remove the engine.
3. If equipped with power steering, disconnect the clevis from the steering lever and then disconnect the cable at the clevis. Loosen the pivot bolts and then remove the power steering unit.
4. If equipped with manual steering, disconnect the steering cable at the lever. Loosen the pivot bolts and remove the steering swivel ring.
5. If equipped with a Gear Lube Monitor, disconnect the hose at the transom plate.
6. Disconnect the speedometer tube.
7. Tag and disconnect the MerCathode wires—see Section 3 for an illustration.
8. Tag and disconnect the trim switch wires at the engine and pull them through the assembly.
9. Disconnect the hydraulic lines at the plate and plug them securely. Carefully move them out of the way.
10. Remove the exhaust pipe at the plate.
11. Disconnect the ground wire at the steering lever.
12. Loosen the six nuts and two bolts that secure the inner plate to the transom. Support the gimbal housing and then remove the retainers and lift out the transom plate and gimbal housing.

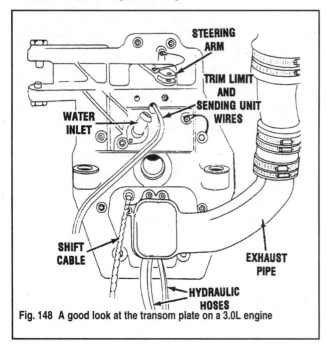

Fig. 148 A good look at the transom plate on a 3.0L engine

To install:

13. Rotate the inside race of the gimbal housing bearing and check for rough or any sign of binding. Push and pull on the inner race to check the bearing for side play. The bearing should move only a slight amount. Any excessive movement is just cause to replace the bearing.
14. Check the upper swivel shaft roller bearing and bushing, by inspecting the area of the swivel shaft, where the bearing and bushing ride. Any pits, grooves, or uneven wear is cause to replace the bearing, bushing, or swivel shaft.

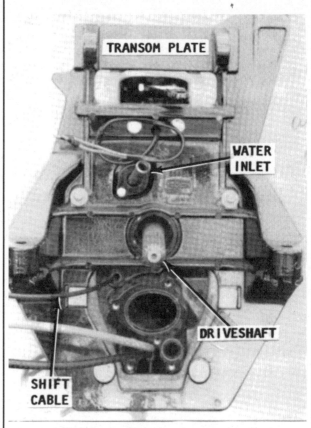

Fig. 149 Another good look with the exhaust pipe(s) removed

15. Check the transom plate for pitting or worn housing. Clean any old sealer from the back side of the plate. Clean the exhaust outlet. Use a tap, and "chase" the threads of the screw holes where the exhaust elbow is mounted.
16. Secure the water tube to the water hose and tighten the hose clamp securely. Install the water hose and tube through the gimbal housing. Route the trim limit switch and trim position sender leads down over the installed tube.
17. Secure both sets of leads with a sta-strap 5 in. (12cm) away from their retaining bracket. Tilt the bell housing up and secure the water hose over the flange on the bell housing. Tighten the hose clamp over the tube. Install the grommet and water tube cover and connect the engine water inlet hose to the water tube.
18. Insert the shift cable, the trim limit switch lead, and the hydraulic hose through the opening in the transom, and then move the gimbal housing into position. The hydraulic hose must be positioned on the port side of the exhaust elbow on in-line engines and on the starboard side for V6 or V8 engines.

■ **Do not hold onto the trim limit switch wires to support the gimbal housing. The trim limit switch or the switch leads will be damaged, if the gimbal housing is supported by the switch leads while fastening the inner transom plate.**

19. Hold the gimbal housing in position, and at the same time insert the shift cable and hydraulic hose through the large opening; insert the trim limit switch leads through the center hole of the three small holes in the inner transom plate; and then set the plate in position.

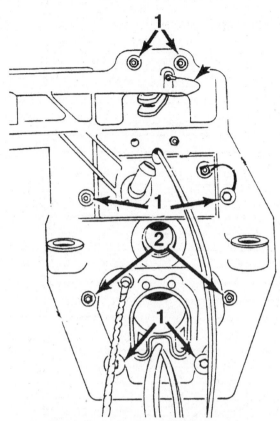

1- Flat washers and locknuts

2- Long screws, lock washers and square flat washers

Fig. 150 Remove these nuts and bolts to pull of the plate

20. Thread the two short cap screws, with lockwashers, through the top two holes of the inner transom plate and into the gimbal housing. Insert the two special anode head bolt assemblies, with rubber seals between the bolt and gimbal housing, through the bottom two holes of the gimbal housing, the boat transom, and the inner transom plate.

21. Insert a flat washer and elastic stop nut on each bolt but do not tighten them at this time. Install a flat washer and elastic stop nut onto each of the two studs protruding through the inner transom plate.

22. Do not attempt to drive the cap screws through the transom, because the threads in the gimbal housing will be damaged. Install the two long cap screws; with a flat washer and lockwasher, through the remaining holes in the inner transom plate and into the gimbal housing. The square flat washer must be against the transom plate.

23. Now, tighten the transom plate cap screws and elastic stop nuts evenly. Work from the center up, and then down. Tighten the screws and nuts to 20–25 ft. lbs. (27–34 Nm).

24. On 3.0L engines, check to be sure the mating surfaces on the exhaust elbow and the gimbal housing are clean. Place a new O-ring seal into the groove in the gimbal housing opening. Position lockwashers on the hex-head screws, and then install the exhaust elbow to the gimbal housing. Tighten the screws to 20–25 ft. lbs. (27–34 Nm), and at the same time check to be sure the O-ring remains properly seated in the groove. Slide hose clamps onto the rubber exhaust tube, and then install the tube onto the exhaust elbow. Tighten the hose clamps securely.

25. On V6/V8 engines, check to be sure the mating surfaces on the exhaust separator and the gimbal housing are clean. Place new O-ring seals into the grooves in the gimbal housing. Install the exhaust separator, with the exhaust elbows and exhaust bellows attached, to the gimbal housing. Thread the four hex-head cap screws and the one Allen-head cap screw, with lockwashers, into the gimbal housing. Tighten the screws to 20–25 ft. lbs. (27–34 Nm) and at the same time check to be sure the O-ring seals remain properly seated in the grooves.

■ **Move quickly while working with the hydraulic hoses to prevent spilling any more hydraulic fluid than is necessary. Follow the sequence and take care to route the hoses properly, or the hoses may be damaged and the system become inoperative.**

26. Connect the hydraulic lines and tighten the connections to 70–150 inch lbs. (9–17 Nm).

27. Install the steering gear and tighten the pivot bolts to 25 ft. lbs. (34 Nm). Bend the washer tabs up and against the bolt head.

28. Connect the Gear Lube line, speedometer tube and MerCathode wires.

29. Install the engine.

30. Install the bell housing and stern drive.

31. Bleed the hydraulic system. Adjust the shift cable.

TORQUE SPECIFICATIONS - ALPHA

Component		ft. lbs.	inch lbs.	Nm
Bellows hose clamps			35	4
Exhaust pipe-to-gimbal housing		20-25		27-34
Lower unit				
Bearing carrier retainer		210		285
Driveshaft retainer		100		136
Lower-to-upper unit				
	Nuts	35		47
	Screw	28		38
Pinion gear nut		70		95
Shift shaft bushing			60	6.8
Trim tab		23		31
Water pump			60	6.8
Gimbal ring		55		74
Gimbal ring-to-bell housing	Hinge pins	95		129
Propeller nut		55 Min. ①		75 Min. ①
Steering cable coupler				
	Nut	35		48
	Locking plate screw		60-72	7-8
Steering lever		60		81
Steering sys. Pivot bolts		25		24
Stern drive-to-bell housing		50		68
Transom assembly		20-25		27-34
Trim limit switch			90-100	10-11
Trim limit switch			90-100	10-11
Upper unit				
Top cover screws		17-23		23-31
U-joint drive gear nut		①		①
U-joint retainer nut		①		①
U-joint bearing preload				
	New		6-10	0.7-1.0
	Used		3-7.5	0.3-0.8
Upper shaft bearing preload				
	New		6-15	0.7-1.7
	Used		3-7.5	0.3-0.8

① See text

12

DRIVE
SYSTEMS
BRAVO
BLACKHAWK

STERN DRIVE UNIT — BRAVO/BLACKHAWK

Description

That which we refer to as the stern drive is actually a number of individual components attached and working together to transfer the power of the engine into a viable propulsion system for your boat. All stern drive units can be broken down into their components assemblies.

The transom assembly, consisting of an inner transom plate, a gimbal housing and gimbal plate, and a bell housing is just what it sounds like—the unit attached to the transom of the vessel. The inner transom plate is, obviously, attached to the inner side of the transom and actually makes up the rear engine mounts.

On the other side of the transom, and attached to the transom plate, are the gimbal housing, gimbal plate (or ring) and bell housing. The bell housing is attached to the gimbal plate via roller bearings and is what allows for the up and down (trim) movement of the stern drive unit itself. The gimbal plate is also attached to the gimbal housing via roller bearings and is what allows for side-to-side movement of the unit, or, steering.

The stern drive unit, or at least that thing that is most visible when viewing the stern of the boat, is made of two component assemblies: the driveshaft housing and the gear housing.

The driveshaft housing, frequently called the upper gear housing or simply the upper unit is attached to the bell housing at the top and the gear housing at the bottom. Power from the engine, brought through the transom assembly via the driveshaft is transferred to a vertical shaft leading to the lower unit by means of a set of drive and driven gears.

The gear housing, or lower unit, is attached to the bottom of the driveshaft housing. Power, or propulsion, comes through the vertical shaft from the upper unit, is transferred to the propeller shaft(s) via a pinion gear and causes the propeller(s) to rotate.

Output power from the engine is connected to the stern drive through a horizontal driveshaft. A coupler is bolted to the flywheel and has a splined hub in the center. The end of the horizontal driveshaft indexes with and slides into the center of the hub. Power from the engine is then transmitted through the horizontal driveshaft to a pinion gear set where power direction is changed from horizontal to vertical.

The upper driven gear is pressed onto the outside diameter of the upper driveshaft. The upper driveshaft is splined on the lower end. When the upper gear housing is mated to the lower unit, the end of the lower unit driveshaft indexes into the splined end of the upper driveshaft. Engine power is then transferred down into the lower gear unit.

All shifting is accomplished inside the upper gear housing. The shift cable actuates a yoke and cam assembly via linkage. The yoke rides in the center of a clutch spool between two brass collars of the clutch gears. The contact surfaces of the collars are not perfectly horizontal. By pivoting left or right, the yoke is able to block one of the clutch gears and permit the clutch shaft to rotate in only one direction.

The lower splines of the short clutch shaft are coupled with the upper splines of the driveshaft.

The pinion gear and single driven gears are contained within the lower unit. Unlike many other stern drive units, the lower unit is free of any shifting mechanisms and water pump components.

The horizontal and vertical driveshafts are both mechanically connected. Therefore, anytime the engine is operating, the horizontal and vertical driveshafts are constantly rotating with engine rpm. A double yoke universal joint assembly in the horizontal driveshaft allows the stern drive to be raised or lowered to a required trim/tilt position (within limits), while the engine is operating.

Troubleshooting

Troubleshooting must be done before the unit is removed from the boat to permit isolating the problem to one area.

1. Check the propeller and the rubber hub for shredding. If the propeller has been subjected to many strikes with underwater objects, it could slip on its hub. If the hub appears to be damaged, replace it with a new hub. Replacement of the hub must be done by a propeller rebuilding shop equipped with the proper tools and experience for such work.

2. **Shift Mechanism Check:** Verify the ignition switch is in the off position to prevent possible injury, should the engine start. Shift the unit into Reverse and at the same time have an assistant turn the propeller shaft to ensure the clutch is fully engaged. If the shift handle is hard to move, the

trouble may be in the stern drive unit, transom shift cable, remote control cable, or the shift box.

3. **Isolate the Problem:** Disconnect the remote control cable at the ransom plate, by first removing the two nuts, and then lifting off the remote control shift cable. Operate the shift lever. If shifting is still hard, the problem is in the shift cable or control box. If the shifting feels normal with the remote control cable disconnected, the problem must be in the stern drive. To verify the problem is in the stern drive unit, have an assistant turn the propeller and at the same time move the shift cable between the transom plate and stern drive back-and-forth. Determine if the clutch engages properly. Most of hard shifting problems are caused because this cable is not moved during the off-season.

Water entering the cable, especially salt water, will cause rapid corrosion and hard shifting. If the cable moves freely, then reconnect the control cables at the transom plate.

4. **Stern Drive Noise Check:** First, a word about stern drive noise. When the stern drive is positioned for dead-ahead operation, the U-joints will make very little noise. However, as the stern drive is moved to port or starboard, the noise will increase, due to the working of the U-joints. This test and procedure is for abnormal noises. Attach a Flush-Test device to the stern drive and turn on the water. Start the engine and shift into gear.

✳✳ CAUTION

Water must circulate through the lower unit to the engine any time the engine is run to prevent damage to the water pump mounted on the engine. Just a few seconds without water will damage the water pump.

Operate the engine at idle speed. Turn the stern drive slightly to port and then slightly to starboard, and at the same time, listen at the upper gear housing. An unusual noise is an indication the U-joints are worn or the gimbal housing bearing is defective. If the bearing is defective, see Chapter 12— Bell Housing. Shut down the engine. Turn off the water and disconnect the Flush-Test device.

5. **Tilt/Trim System Check:** Operate the control buttons on the control panel or on the shift control lever. Check to determine if the stern drive unit moves to the full up and full down position. If the drive unit fails to move, the U-joints may be "frozen" due to corrosion from water entering through the U-joint bellow; lack of lubrication; or from non-operation over a long period of time. If the stern drive does not move properly, the trim/tilt mechanism may need service, see the section on Trim/Tilt.

6. **Noise in Upper Gear Housing:**
- Oil level low.
- Worn U-joints.
- Worn bearings.
- Incorrect thrust washer placement.
- Gear timing incorrect.
- Engine not aligned properly.
- Worn engine coupler.
- Transom too light for stern drive.
- Worn, damaged, or loose internal parts.

Stern Drive Unit

REMOVAL & INSTALLATION

◆ See Figures 1, 2, 3, 4, 5, 6 and 7

1. Drain the stern drive unit oil as detailed in the Maintenance section. Check the drained lubricant carefully to determine if it contains any water or metal particles. Rub some of the lubricant between your fingers. Any metal particles in the lubricant will thus be evident. Do not be mislead if the color of the lubricant is metal colored. This is not a harmful condition and is caused by the lubricant used in the unit the first time after manufacture.

2. Shift the unit into Neutral and then move it to the full UP position.

3. On Bravo models, reach around to the rear of the anti-ventilation plate and move the speedometer hose fitting release lever over and up. Pull out the hose and move it out of the way.

4. Move the drive unit back to the full DOWN position. Remove the plastic cap over the anchor pin connecting the trim cylinders to the aft end of

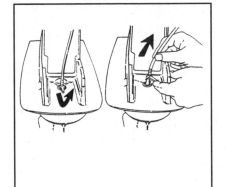

Fig. 1 Move the speedometer lever counterclockwise to release the tube

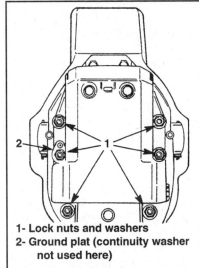

1- Lock nuts and washers
2- Ground plat (continuity washer not used here)

Fig. 2 Remove the six locknuts

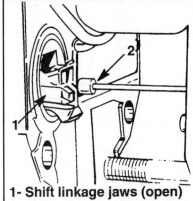

1- Shift linkage jaws (open)
2- Shift cable end (released from jaws)

Fig. 4 Make sure the shift cable comes out of the assembly jaws

the upper unit. Remove the nuts and washers and pry out the bushings. Disconnect the cylinder arms from the unit and carefully move them back against the transom. It's a good idea to tie them up and out of the way to avoid damage during drive removal.

■ **Although not imperative, we highly recommend the use of an engine hoist or other lifting device when removing the unit.**

5. Install a lifting device if so desired.
6. Remove the six 5/8 in. nuts and washers securing the stern drive to the bell housing. There are only five washers as the center, port bolt (looking toward the bow) uses a permanent ground plate instead of a continuity washer.
7. Secure the lifting chains so they are taught and then remove the stern drive unit by pulling it straight back and free of the bell housing. Remove and discard the gasket. Make sure that the jaws on the shift cable linkage open properly and release the cable end as you are pulling the unit back and out.

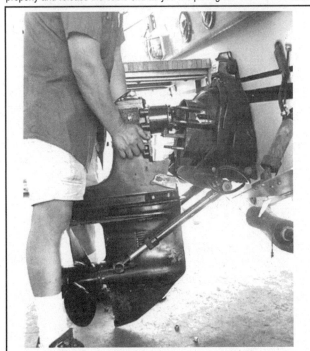

Fig. 3 You can try it this way, but we recommend a lifting hoist

To install:

The engine must be properly aligned or the female splines in the engine coupler and the male splines of the driveshaft will be destroyed after a short time of engine operation.

If troubleshooting indicates the coupler is damaged and requires replacement, remove the engine and replace the coupler. The coupler is simply secured to the flywheel with attaching bolts. If the necessary tools are not available for proper engine alignment, the boat should be taken to an authorized marine dealer.

8. Ensure that the remote control shift lever on the drive is in Neutral.
9. Install new O-rings into the two passageways just under the driveshaft on the forward side of the housing. Install a new gasket on the face of the bell housing.
10. Coat all six studs and their threads with Quicksilver 2-3-C marine lubricant.
11. Check the bell housing bearing to be sure it turns freely without binding or rough spots.
12. Oil the shaft with coupler grease to prevent damage to the seals, and then slide three new O-ring seals over the splines of the universal joint shaft, and into the grooves provided.
13. Coat the splines of the driveshaft and the driveshaft housing pilot, with Multi-purpose Lubricant. As an aid to installation, apply a light coating of Multi-purpose Lubricant to the inside surfaces of the bell housing bore.
14. Check the U-joint bellows for cracks wear or other damage. Replace it if there is any doubt as to its condition whatsoever as it acts as a seal between the bell housing and the upper unit..
15. Check the driveshaft bellows for cracks wear or other damage. Replace it if there is any doubt as to its condition whatsoever.
16. Grab the lower lip of the shift linkage assembly with needle nose pliers and pull it out until the jaws open all the way. Coat the underside of the assembly with Special Lubricant 101. While moving the drive unit into position with the bell housing, make sure that the shift cable enters the jaws and drops down into the slot in the lower lip. Once it is in the slot it should force the entire assembly back into the upper unit as you move the unit further into position, thus closing the jaws. You may want to guide the cable into the jaws by hand to make sure this goes smoothly.
17. Position the drive unit and slowly move it into the bell housing. The trim cylinders should be untied and positioned so they are pointing straight backwards. Make sure the driveshaft slips into the bell housing bore properly, through the gimbal bearing and into the engine coupler. You may have to wiggle the propeller slightly in order to get the splines to line up correctly.
18. Apply a coating of Multi-purpose Lubricant onto the shift lever coupler, and then align the slot straight by moving the lever to the right. Lubricate the slot and the reverse lock roller.
19. Secure the driveshaft housing to the bell housing with the six elastic-stop nuts with flat washers. Tighten the nuts alternately and evenly to a torque value of 50 ft. lbs. (68 Nm). If you removed the top cover, tighten the

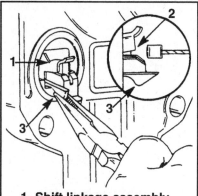

1- **Shift linkage assembly**
2- **Jaws - open**
3- **Underside of lower lip**

Fig. 5 Pull the linkage assembly out until the jaws open all the way

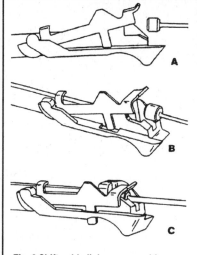

Fig. 6 Shift cable linkage assembly

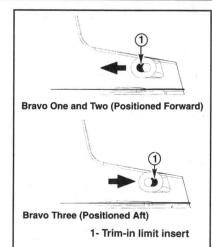

Bravo One and Two (Positioned Forward)

Bravo Three (Positioned Aft)

1- **Trim-in limit insert**

Fig. 7 It is very important that the trim-in insert is positioned correctly on 1998–01 models

screws to 20 ft. lbs. (27 Nm) on Bravo models, or 13–15 ft. lbs. (18–20 Nm) on Blackhawks.
20. Install the trim cylinders as detailed in the next section.

■ 1999 and later models utilize a Trim-In Limit insert to prevent over trimming at high speeds on Bravo III units. On Bravo I/II models the insert must be positioned all the way forward, while on Bravo III models the insert must be positioned all the way to the rear of the cut-out. This must take place before installing the trim cylinder anchor pin.

21. Connect the speedometer cable in the opposite manner that you disconnected it— turn the release lever clockwise this time (Bravo only).
22. Fill the unit with oil and adjust the shift cable.

Trim Cylinder

REMOVAL & INSTALLATION

② MODERATE

◆ See Figures 7, 8 and 9

■ 1999 and later models utilize a Trim-In Limit insert to prevent over trimming at high speeds on Bravo III units. On Bravo I/II models the insert must be positioned all the way forward, while on Bravo III models the insert must be positioned all the way to the rear of the cut-out.

✳✳ CAUTION

Be sure the stern drive is in the full DOWN position before proceeding with this procedure. If the trim/tilt cylinders are disconnected with the stern drive in the up position, the stern drive will swing down suddenly causing severe damage to the stern drive and possibly personal injury.

1. Remove the end caps on each end of the trim cylinders where they attach to the upper gear housing. Remove the nut from each side of the anchor pin and lift off the flat washers.
2. Grasp the cylinder and pull it off the end of the anchor pin, or use a blunt end punch and tap the anchor pin out from the stern drive upper gear housing. Remove the bushings — four total — from the trim/tilt ends and place it with the rest of the attaching hardware to guard against possible loss.
3. Carefully loosen the fitting nut and disconnect the UP hydraulic line from the cylinder. Plug the hole in the cylinder and the line quickly and move the line out of the way. Do the same with the DOWN hydraulic line, except at the connector on the gimbal housing.

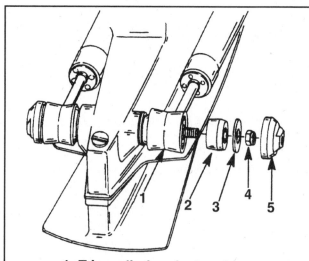

1- **Trim cylinder pivot ends**
2- **Rubber bushing**
3- **Small ID flat washer**
4- **Lock nut**
5- **Trim cylinder cap**

Fig. 8 Upper unit end of the trim cylinder

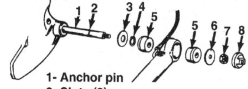

1- **Anchor pin**
2- **Slots (2)**
3- **Flat washer (large ID) (2)**
4- **Snap rings (2)**
5- **Rubber bushings (4)**
6- **Flat washer (small ID) (2)**
7- **Lock nut (2)**
8- **Plastic cap (2)**

Fig. 9 Trim cylinder mounting hardware at the transom assembly

4. Remove the end caps on each end of the trim cylinders where they attach to the transom assembly. Remove the nut from each side of the forward anchor pin and lift off the flat washers.

5. Grasp the cylinder and pull it off the end of the anchor pin. Remove the bushings—four total—from the trim/tilt ends and place it with the rest of the attaching hardware to guard against possible loss. Remove the snap ring and the remaining washer.

To install:

6. Install the washer and snap ring onto the forward anchor pin.

7. Coat the bushings with warm soapy water and press them into each side of the cylinder.

8. Press the cylinder onto the pin, slide on the washer and hand tighten the locknut after coating the threads with 2-4-C lubricant

9. Position a large washer over the aft anchor pin and then slide on a bushing so the small side is facing outward.

10. Now, slide each cylinder end onto the anchor pin. Install the second bushing into the cylinder followed by a flat washer. Turn the retaining nut on hand tight.

11. Tighten all four retaining nuts until the nuts and washers bottom out against the shoulder of the anchor pin—no more, no less! Install the end caps.

12. Reconnect the hydraulic lines and bleed the system.

GEAR HOUSING (LOWER UNIT)

Description

The upper gear housing transfers engine power to the lower unit through a driveshaft. Splines on the forward end of the drive shaft mate with matching splines of a coupler attached to the engine flywheel. The aft end of the driveshaft is connected to a universal joint and the upper gear assembly. The pinion gear and drive gear changes the direction of the power train from horizontal to vertical. The power is transferred from the upper gear housing down to the lower unit through a vertical driveshaft. A water pump installed in the lower unit is constantly driven by the vertical driveshaft. This pump supplies cooling water directly to the engine, or indirectly through a heat exchanger.

The forward and reverse gears are contained within the lower unit. A sliding clutch, actuated by the shift linkage (all in the upper unit), engages a forward gear and creates a direct coupling for transmitting power through the pinion gear and forward gear to the propeller shaft. The reverse gear utilizes the same mechanics as the forward gear.

Gear Housing (Lower Unit) Bravo I/II

REMOVAL & INSTALLATION   ③ DIFFICULT

◆ **See Figures 10 and 11**

The lower unit may be separated and removed from the upper gear housing while the stern drive remains attached to the boat or after the stern drive has been removed.

1. Remove the propeller(s) from the lower gear housing as detailed later in this section.

2. If the unit is still on the boat, drain the drive oil as detailed in the Maintenance section. Allow the lubricant to drain into the container. As the lubricant drains, check the color. Dark brown to black indicates normal old lubricant. A chalky white to cream color indicates the presence of water. The presence of any water in the gear lubricant is bad news. The unit must be completely disassembled, inspected, the cause of the problem determined and corrected. Close attention should be given to the back-to-back seals on the propeller shaft, driveshaft and the bearing carrier O-ring. Examine the magnet on the end of the fill/drain plug for evidence of metal particles. The presence of tiny small metal dust like shavings indicates normal wear of the gears, bearings and shafts within the lower unit. Large metal chips or heavy fillings indicate extensive internal damage is taking place and the lower unit must be completely disassembled and inspected. All worn and/or damaged components must be replaced.

3. If working on a Bravo I, place an alignment mark on the trim tab trailing edge and the anti-cavitation plate as an aid during assembling. Remove the plastic plug from the top of the anti-cavitation plate. Remove the 1/2 in. bolt directly above the trim tab, lift off the trim tab. You will need a long extension to your socket wrench for this one.

4. Remove the six hex nuts and washers (three on each side), from the sides of the anti-cavitation plate. Remove the single bolt hidden inside the trim tab cavity on the Bravo I; on the Bravo II this bolt is in a recess, just forward of the trim tab.

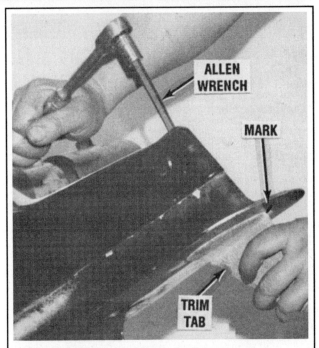

Fig. 10 Removing the trim tab bolt will require a socket extension—Bravo I

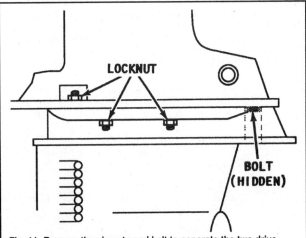

Fig. 11 Remove the six nuts and bolt to separate the two drive halves

5. Separate the two housings by tapping downward on the lower unit with a soft face mallet and a wooden block. In some cases the housing may be difficult to separate because the driveshaft may be corroded and "frozen" in the upper gear housing vertical shaft. Lower the gear housing away from the upper gear housing.

To install:

6. Coat a new O-ring with 3-M Adhesive and install it on the upper face of the lower unit at the water passage opening. No gasket is used between the two housings. All sealing is accomplished with two O-rings: The water passageway O-ring and the O-ring installed around the driveshaft spacer. The spacer has two vertical holes drilled through it to allow oil to pass through the spacer and lubricate not only the double roller bearing assembly on the driveshaft but the entire lower unit. Oil passes freely through the double bearing arrangement. Therefore, the oil passageway O-ring is perhaps the most crucial component in a Bravo stern drive.

✳✳ WARNING

In the Bravo Stern Drive, no locating pins are utilized in aligning the lower unit with the upper gearcase housing. Therefore, if the bolt holes show serious signs of elongation, or the securing bolts show signs of stress or "necking", the lower unit, the upper gearcase housing and all securing bolts should be replaced. This is not as drastic or expensive as it sounds. All internal components are removed and only the housing castings are replaced.

7. Install the coupler and retainer onto the upper end of the driveshaft. Raise the lower unit to mate with the upper gearcase housing. The lower end of the short clutch shaft must slide into the coupler at the upper end of the lower driveshaft. It may be necessary to slowly rotate the propeller shaft counterclockwise until the splines index.

8. Screw the aft bolt into the hole in the trim tab cavity on Bravo I or into the hole just forward of the trim tab on Bravo II..

9. Install the three port and three starboard nuts and their washers. Tighten the six nuts alternately, and a little at-a-time, until the lower unit is tight against the upper gear housing.

10. Now tighten the rear bolt and the six side nuts to 35 ft. lbs. (47.5 Nm).

11. Install the trim tab (Bravo I) with the marks made during disassembly aligned. Tighten the screw(s) to 23 ft. lbs. (31 Nm).

■ **The trim tab performs two very important jobs, one of which you may not realize. First, the tab compensates for steering torque. If the boat continually seems to move to port or starboard while the helmsman is on a straight course, the trim tab can be adjusted to the side of the pull. The tab also prevents electrolysis from damaging expensive parts. The tab is not expensive; it should show signs of electrolytic action; and should always be replaced after some of the material has been eaten away or is pitted. Now, if the tab shows no signs of electrolytic action after the boat has been in use over a period of time, the grounding should be checked to ensure more expensive parts are not being damaged. Install the trim tab plastic cover plug.**

12. Install the propeller(s) if it was removed and refill the drive unit with oil.

DISASSEMBLY OEM ③ DIFFICULT

◆ **See Figures 12, 13, 14, 15, 16, 17, 18, 19, 20, 21 and 22**

The following procedures assume that the lower unit and propeller have already been removed.

1. Remove the trim tab if working on a Bravo II.

2. Install a clamp plate over the driveshaft and tighten the nuts onto the studs securely.

3. To remove the bearing carrier, first straighten the locking tabs on the tab washer. Loosen the bearing carrier retainer by turning it counterclockwise with special wrench (#91-61069—Bravo I or #91-17257—Bravo II). Remove the retainer and tab washer.

4. Pull out the bearing carrier assembly using a bearing carrier puller (#91-90339—Bravo I; or #91-34569A1—Bravo II). The thrust hub for the propeller, which was removed previously, can be used to keep the puller jaws in place. Be careful not to lose the locating key from the side of the carrier housing on Bravo II units.

■ **If the reverse gear bearing carrier is corroded in place, apply heat to the housing to loosen the corrosion. While the housing is still hot, use a bearing carrier puller to remove the assembly do not overheat the housing or it will become distorted and useless.**

6. Reach in and remove the O-ring, washer and load ring. Tag and wire them together, as an aid during assembling.

7. Grab the end of the propeller shaft and pull it out of the housing (leaving the driven gear still in the housing).

8. Temporarily reinstall the bearing retainer by hand and then install a driveshaft adapter tool (#91-61077) on the top of the vertical shaft. Insert a long breaker bar (with socket) into the propeller shaft bore and remove the pinion (drive) gear nut. Lift out the nut and washer. Remember that on 1999 and later Bravo I models the pinion gear is attached with a screw instead of a nut.

9. Remove the clamp plate installed previously. Lift the O-ring, spacer, shim material and tab washer up and off of the driveshaft.

10. Grab the top of the driveshaft and pull it out of the housing. Pull out the pinion gear.

11. Reach into the propeller shaft bore and pull out the driven gear and bearing/cup.

■ **The next step is not required, unless the bearings are to be replaced. Any time the bearings are replaced, they must be adjusted properly with shim material or their service life will be drastically reduced.**

13. Using a slide hammer with a two-armed puller, remove the driveshaft tapered roller bearing cup. Tag and wire the shim material pieces from underneath the bearing cup together, as an aid during assembly.

14. Remove the needle bearings from the driveshaft needle bearing race unless you intend to remove the pinion bearing.

15. Pull the O-ring out of the water passage on top of the housing.

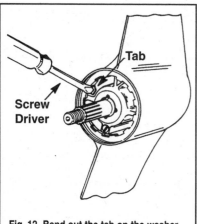

Fig. 12 Bend out the tab on the washer

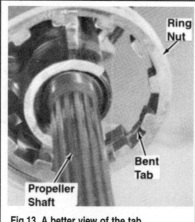

Fig.13 A better view of the tab

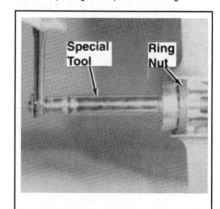

Fig. 14 Use a special tool to remove the retainer ring nut

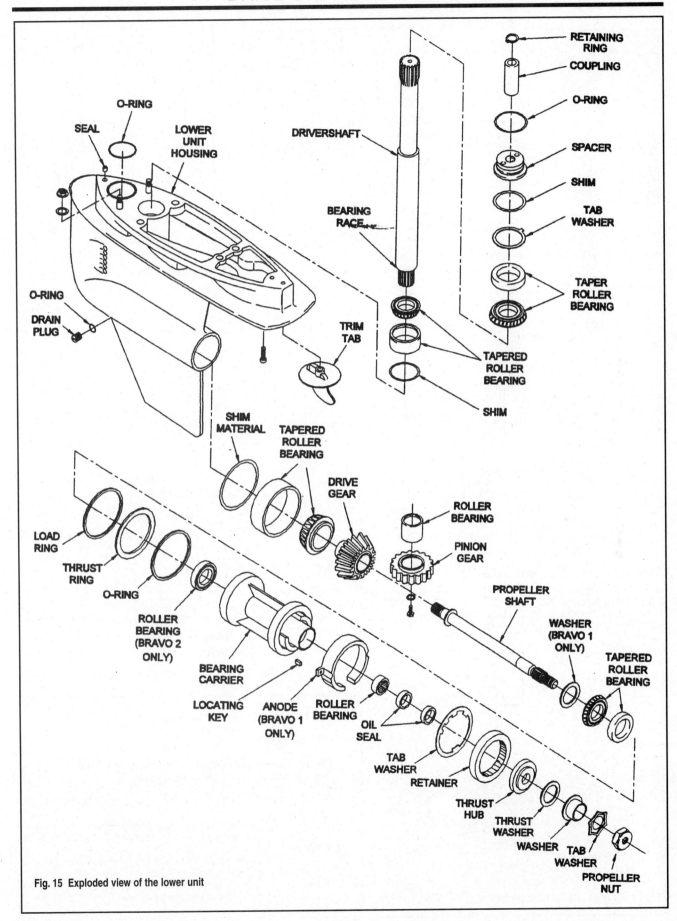

Fig. 15 Exploded view of the lower unit

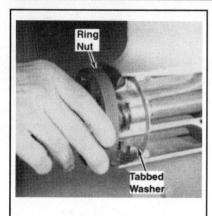

Fig. 16 Spin out the nut and remove the washer

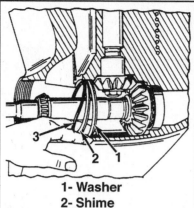

1- Washer
2- Shime
3- O-ring

Fig. 17 Reach in and pull out the O-ring, washer and load ring

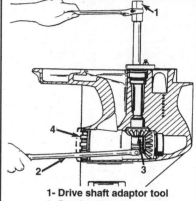

1- Drive shaft adaptor tool
2- Breaker bar and socket
3- Pinion gear
4- Bearing carrier retainer

Fig. 18 Remove the pinion nut...

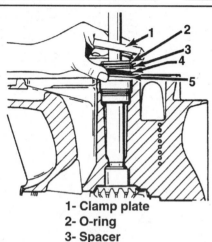

1- Clamp plate
2- O-ring
3- Spacer
4- Shim(s)
5- Tab washer

Fig. 19 ...lift off the rings, shims and spacers

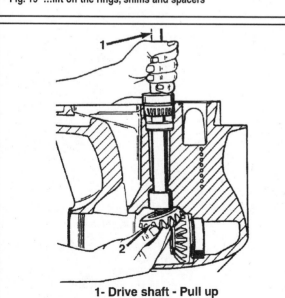

1- Drive shaft - Pull up
2- Pinion gear

Fig. 20 And then pull out the driveshaft

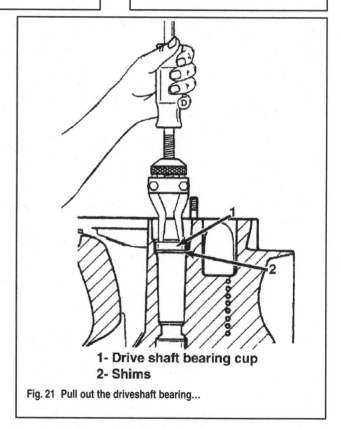

1- Drive shaft bearing cup
2- Shims

Fig. 21 Pull out the driveshaft bearing...

Driveshaft And Propeller Shaft Bearings

◆ See Figure 23

■ Bearing assemblies will be damaged on removal; always replace the roller bearings if they have been removed.

  1. Install a Universal Puller Plate (#91-37241) to the shaft to support the bearing.
  2. Install the assembly into and arbor press and press off the bearing.
  3. Repeat this procedure for the second tapered roller bearing on the driveshaft.
  4. Install a suitable mandrel and press the bearings onto the shaft. When installing the bearings on the driveshaft, first press the smaller bearing onto the shaft so that the smaller O.D. is facing the pinion end of the shaft and then press on the larger bearing making sure the larger O.D. faces the pinion end.

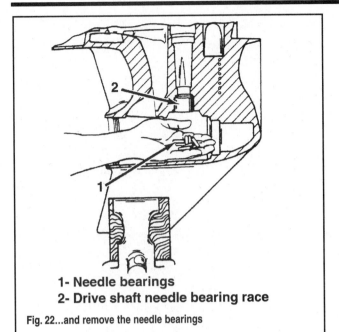

1- Needle bearings
2- Drive shaft needle bearing race

Fig. 22...and remove the needle bearings

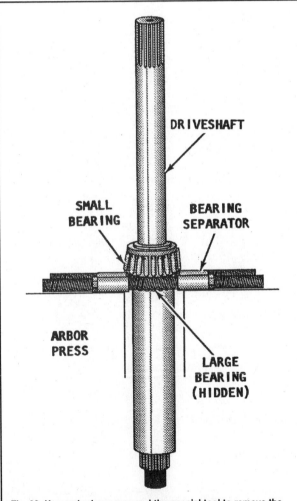

Fig. 23 Use and arbor press and the special tool to remove the roller bearings on either of the shafts

Bearing Carrier

◆ See Figures 24, 25 and 26

■ The oil seals or the bearing cup may be replace individually. It is not necessary to remove one in order to remove the other.

1. Clamp the bearing carrier in a vise with soft jaws. Be sure to clamp on the bearing carrier at the reinforcement ribs. Place the jaws of a universal slide hammer puller down through the center of the carrier until the teeth on the puller are through the bearing cup. Pull cup out of the bearing carrier.
2. Remove the two propeller shaft oil seals from the bearing carrier with a hammer and punch.

To assemble:

3. Wash all parts in solvent and blow them dry with compressed air. Remove any corrosion from the outside surface of the bearing carrier. Remove any seal and gasket material from all mating surfaces. Blow all water, oil passageways, and screw holes clean with compressed air. After all parts are clean and dry, apply a light coating of gear lubricant to the bright surfaces of all gears, bearings, and shaft, as prevention against rusting.
4. Install the new bearing into the bearing carrier with the bearing surface facing inward.
5. Coat the outer edges of two new seals with Loctite and then press two new oil seals into the front of the bearing carrier. These seals are installed back-to-back. Seat the first one with the lip facing outward. Seat the second seal with the lip facing inward. Fill the area between the two seals with 2-4-C lubricant.

Driven Gear Bearing

◆ See Figures 27 and 28

1. Install a Universal Puller Plate between the driven gear and the roller bearing.
2. Sit the assembly on two pieces of wood and press the bearing off with an arbor press.
3. Sit the gear on a block of wood with the teeth toward the wood.
4. Coat the inside of a new bearing with gear lube and position over the gear.
5. Place a suitable mandrel (an old bearing race works well) over the new bearing race and then press it into the gear.

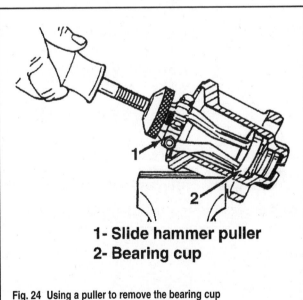

1- Slide hammer puller
2- Bearing cup

Fig. 24 Using a puller to remove the bearing cup

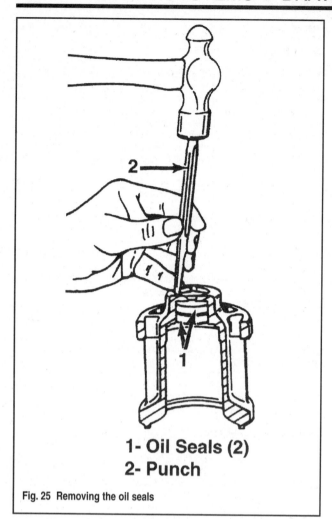

1- Oil Seals (2)
2- Punch

Fig. 25 Removing the oil seals

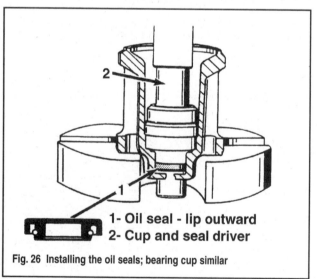

1- Oil seal - lip outward
2- Cup and seal driver

Fig. 26 Installing the oil seals; bearing cup similar

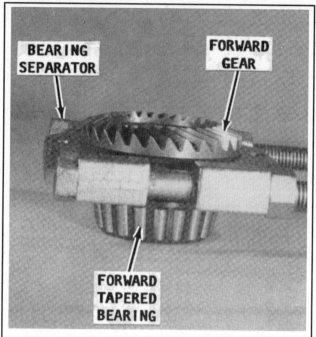

Fig. 27 Use the special tool to separate the bearing from the driven (forward) gear

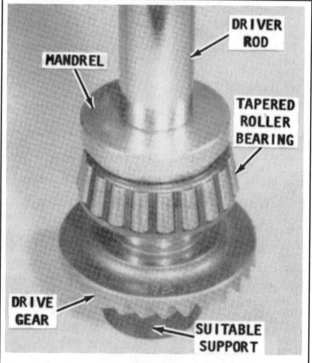

Fig. 28 Press the forward roller bearing into the forward gear

CLEANING AND INSPECTING

1. Wash all parts in solvent and blow them dry with compressed air. Remove all traces of seal and gasket material from all mating surfaces. Blow all water, oil passageways and screw holes clean with compressed air. After cleaning, apply a light coating of gear lubricant to the bright surfaces of all gears, bearings, and shafts as a prevention against rusting and corrosion. Check the shift crank to be sure it is not bent.

2. Inspect the water pump impeller plate for wear and corrosion. Replace any worn, corroded, or damaged parts. Use a fine file to remove burrs. Replace all O-rings, gaskets, and seals to ensure satisfactory service from the unit. Clean the corrosion from inside the housing where the bearing carrier was removed.

3. Check to be sure the water intake is clean and free of any foreign material.

4. Inspect the gear case, housings, and covers inside and out for cracks. Check carefully around screw and shaft holes. Check for burrs around machined faces and holes. Check for stripped threads in screw holes and traces of gasket material remaining on mating surfaces.

5. Check O-ring seal grooves for sharp edges, which could cut a new seal. Check all oil holes.

6. Inspect the bearing surfaces of the shafts, splines, and keyways for wear and burrs. Look for evidence of an inner bearing race turning on the shaft. Check for damaged threads. Measure the run-out on all shafts to detect any bent condition. If possible, check the shafts in a lathe for out-of-roundness.

7. Inspect the gear teeth and shaft holes for wear and burrs. Hold the center race of each bearing and turn the outer race. The bearing must turn freely without binding or evidence of rough spots. Never spin a ball bearing with compressed air or it will be ruined. Inspect the outside diameter of the outer races and the inside diameter of the inner races for evidence of turning in the housing or on a shaft. Deep discoloration and scores are evidence of overheating.

8. Inspect the thrust washers for wear and distortion. Measure the washers for uniform thickness and flatness.

9. Inspect the gears for distortion, burrs, and cracks. Check for any sign of discoloration, which means the gears are running too hot.

10. Replace all seals, O-rings, and gaskets to ensure maximum service after the work is completed.

ASSEMBLY

◆ **See Figures 29, 30, 31, 32 33, 34, 35, 36 and 37**

1. Lubricate all gears, splines and bearings with gear lube so as to get accurate pre-load readings.

2. Install a new O-ring into the water passage with 3-M adhesive.

3. Coat the needle bearings with bearing assembly lubricant and install them into the lower driveshaft bearing casing.

4. Position the driven gear and bearing into the propeller bore.

5. Drop the original shims into the driveshaft bore and drive the lower

cup into position. If the original shims were damaged or misplaced, install new ones with a thickness of 0.050 in. (1.27mm).

6. Slide the driveshaft into the bore and install the upper bearing cup, tab washer, original shims, spacer and the O-ring over the shaft. If the original shims were damaged or misplaced, install new ones with a thickness of 0.050 in. (1.27mm).

7. Install the clamp plate back onto the top of the gear housing.

8. Coat the threads of a new pinion nut with Loctite 271. Move the pinion gear into the bore and onto the bottom of the driveshaft, install the washer and then screw the nut on finger-tight. Install the driveshaft adaptor tool again with a torque wrench, make sure that the retainer is still screwed in to protect the threads and then tighten the pinion nut to 100 ft. lbs. (135 Nm); on 1999 and later Bravo I models, tighten the screw to 45 ft. lbs. (61 Nm). Remove the retainer.

9. With the driveshaft adapter still installed, check the rolling pre-load with a dial-type torque wrench. It should be 3–5 inch lbs. (0.3–0.6 Nm).

10. Insert a shimming tool (#91-42840—Bravo I; or #91-96512—Bravo II) into the propeller shaft bore and check the clearance between the tool and the gear with a feeler gauge. Check this in three places, 120° apart. Clearance must be 0.025 in. (0.635mm). If less than spec, add shims of the appropriate thickness under the lower bearing cup to bring it into specification. If greater than spec, remove shim material from under the bearing cup. Recheck the clearance again until you're sure it is correct.

■ **Any thickness of shims added, must be subtracted from the shim thickness at the upper bearing cup.**

11. Remove the clamp plate. Lift the O-ring, spacer and shims off of the driveshaft. Measure the distance between the top of the gear housing and the tab washer with a micrometer. Measure the thickness of the spacer from the top surface to the bottom surface. Correct shim thickness can be determined by subtracting the second measurement from the first and adding 0.001 in. Reinstall the correct shim(s), spacer and O-ring.

12. Recheck the pre-load.

13. Install the bearing carrier retaining nut and tighten it until resistance to the propeller shaft rotation can felt— you have pre-loaded the bearings. Install the load ring, thrust ring and O-ring into the unit. Install the bearing carrier with the tab washer. Reinstall the bearing carrier retainer and tighten it until resistance to the propeller shaft rotation can be felt.

14. Install a dial indicator adaptor (#91-83155), backlash indicator rod (#91-53459) and dial indicator onto the driveshaft so that the dial rod is aligned with the **II** on the indicator rod. Rotate the driveshaft back and forth slightly without turning the propeller shaft; the dial indicator should read 0.009–0.015 in. (0.23–0.38mm). If backlash is less than specified, remove shims from behind the driven gear bearing cup; if greater, add shims. Continue this process until determining that backlash is correct.

15. Remove the bearing carrier, O-ring, thrust washer and load ring. Reinstall everything using a new load ring. Lubricate the space between the carrier oil seals with 2-4-C lubricant and coat the rounds on the carrier with Perfect Seal. On the Bravo II, insert the locating key with a pair of needle nose pliers.

16. Install the tab washer, lubricate the threads of the retainer with Special Lubricant 101, and tighten it until resistance can just be felt. Install the retainer wrench and tighten to the final torque a little at a time, checking

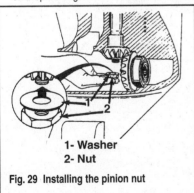

1- Washer
2- Nut

Fig. 29 Installing the pinion nut

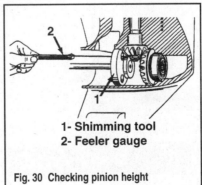

1- Shimming tool
2- Feeler gauge

Fig. 30 Checking pinion height

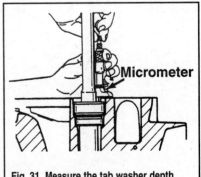

Micrometer

Fig. 31 Measure the tab washer depth with a micrometer

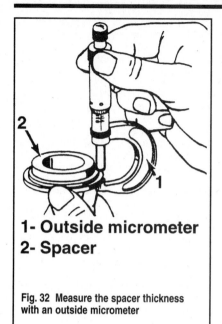

1- Outside micrometer
2- Spacer

Fig. 32 Measure the spacer thickness with an outside micrometer

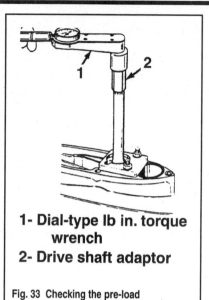

1- Dial-type lb in. torque wrench
2- Drive shaft adaptor

Fig. 33 Checking the pre-load

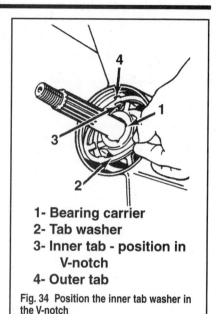

1- Bearing carrier
2- Tab washer
3- Inner tab - position in V-notch
4- Outer tab

Fig. 34 Position the inner tab washer in the V-notch

overall housing preload each time with a torque wrench on the propeller shaft. When preload is 8–12 inch lbs. (0.9–1.4 Nm) for new bearings, or 5–8 inch lbs. (0.6–0.9 Nm) on used bearings, bend one tab from the washer into the retainer and then bend the remaining tabs down into the housing.

17. Install the trim tab on the Bravo II units and tighten it to

Gear Housing (Lower Unit) — Bravo III

REMOVAL & INSTALLATION

③ DIFFICULT

◆ **See Figures 38, 39, 40 and 41**

The lower unit may be separated and removed from the upper gear housing while the stern drive remains attached to the boat or after the stern drive has been removed.

1. Remove the propeller(s) from the lower gear housing as detailed later in this section.
2. If the unit is still on the boat, drain the drive oil as detailed in the Maintenance section. Allow the lubricant to drain into the container. As the lubricant drains, check the color. Dark brown to black indicates normal old

lubricant. A chalky white to cream color indicates the presence of water. The presence of any water in the gear lubricant is bad news. The unit must be completely disassembled, inspected, the cause of the problem determined and corrected. Close attention should be given to the back-to-back seals on the propeller shaft, driveshaft and the bearing carrier O-ring. Examine the magnet on the end of the fill/drain plug for evidence of metal particles. The presence of tiny small metal dust like shavings indicates normal wear of the gears, bearings and shafts within the lower unit. Large metal chips or heavy fillings indicate extensive internal damage is taking place and the lower unit must be completely disassembled and inspected. All worn and/or damaged components must be replaced.

3. Remove the plastic plug from the top of the gear housing. Remove the bolt directly above the anode, lift off the anode. You may need an extension to your socket wrench for this one.
4. On 1992–98 models, look inside the cavity under the anode and remove the bolt holding the splash plate extension. Remove the plate. On 1999–01 models, look inside the cavity and remove the mounting bolt—there is no extension plate.
5. Remove the six hex nuts and washers (three on each side), from the sides of the housing.
6. Separate the two housings by tapping downward on the lower unit with a soft face mallet and a wooden block. In some cases the housing may be difficult to separate because the driveshaft may be corroded and "frozen"

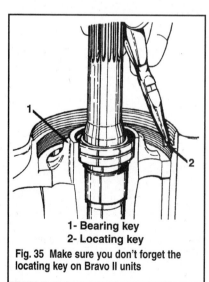

1- Bearing key
2- Locating key

Fig. 35 Make sure you don't forget the locating key on Bravo II units

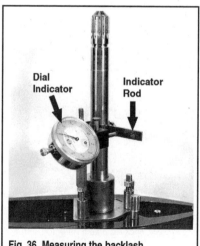

Dial Indicator   Indicator Rod

Fig. 36 Measuring the backlash

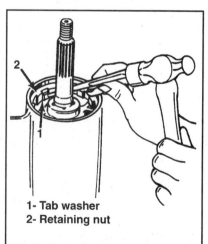

1- Tab washer
2- Retaining nut

Fig. 37 Bend in the tabs once the pre-load is within specifications

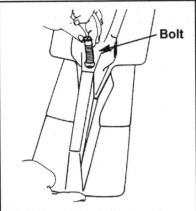

Fig. 38 Pop out the plastic plug and remove the anode mounting bolt

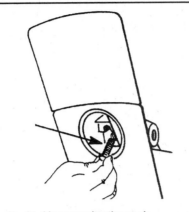

Fig. 39 After removing the anode, remove the bolt in the cavity...

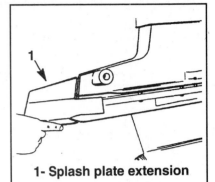

1- Splash plate extension

Fig. 40 ...and then pull out the splash plate extension

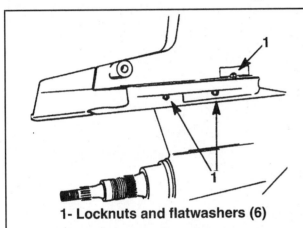

1- Locknuts and flatwashers (6)

Fig. 41 Remove the six nuts and bolt to separate the two drive halves

in the upper gear housing vertical shaft. Lower the gear housing away from the upper gear housing and place it in a holding device.

To install:

7. Install the coupler and retainer onto the upper end of the driveshaft. Raise the lower unit to mate with the upper gearcase housing. The lower end of the short clutch shaft must slide into the coupler at the upper end of the lower driveshaft. It may be necessary to slowly rotate the propeller shaft counterclockwise until the splines index.

8. Slide the splash plate extension into the slot in the driveshaft housing and tighten the bolt to 35 ft. lbs. (47.5 Nm) on 1992–98 models. On 1998 and later models, simply install the mounting bolt and tighten it..

9. Install the anode into the extension and tighten the bolt to 23 ft. lbs. (31 Nm).

10. Install the three port and three starboard nuts and their washers. Tighten the six nuts alternately, and a little at-a-time, until the lower unit is tight against the upper gear housing.

11. Now tighten the rear bolt and the six side nuts to 35 ft. lbs. (47.5 Nm).

12. Install the propeller(s) if it was removed and refill the drive unit with oil.

DISASSEMBLY

◆ **See Figures 19, 20, 21, 22, 42, 43, 44, 45, 45a, 45b, 45c, 45d, 45e, 45f 45g, 45h, 45j, 45k and 45l**

The following procedures assume that the lower unit and propeller have already been removed.

1. Install a clamp plate over the driveshaft and tighten the nuts onto the studs securely.

2. Install a bearing carrier removal tool (#91-805374-1) over the propeller shaft until it seats onto the leading edge of the carrier.

3. Heat the end of the housing around the carrier (with a gun or lamp, no torches) and then turn the carrier off the propeller shaft with a breaker bar. Remember that this is a left hand thread so you will lossen it by turning it clockwise—opposite of what you're used to.

4. Slide a bearing retainer tool (#91-805302) over the propeller shaft and into the bore until it fits into the bearing retainer. Install the carrier removal tool over the retainer tool and loosen the retainer by turning the tools clockwise also. Unscrew the retainer and remove it from the bore.

5. Grab the inner and outer propeller shaft assembly and pull it out of the housing bore, leaving the front driven gear in the bore. Reach back into the housing bore and remove the shim(s).

6. Install a driveshaft adapter tool (#91-61077) on the top of the vertical shaft (with 1-1/4 in. socket). Insert a long breaker bar (with 9/16 in. socket) into the propeller shaft bore and remove the pinion (drive) gear bolt. Lift out the bolt and washer.

7. Remove the clamp plate installed previously. Lift the O-ring, spacer, shim material and tab washer up and off of the driveshaft.

8. Grab the top of the driveshaft and pull it out of the housing.

9. Reach into the propeller shaft bore and pull out the pinion gear.

10. Reach into the propeller shaft bore and pull out the driven gear and bearing/cup.

■ **The next step is not required, unless the bearings are to be replaced. Any time the bearings are replaced, they must be adjusted properly with shim material or their service life will be drastically reduced.**

11. Using a slide hammer with a two-armed puller, remove the driveshaft tapered roller bearing cup. Tag and wire the shim material pieces from underneath the bearing cup together, as an aid during assembly.

12. Remove the needle bearings from the driveshaft needle bearing race unless you intend to remove the pinion bearing.

13. Pull the O-ring out of the water passage on top of the housing.

Driveshaft

◆ **See Figure 23**

■ **Bearing assemblies will be damaged on removal; always replace the roller bearings if they have been removed.**

1. Install a Universal Puller Plate (#91-37241) to the shaft to support the bearing.

2. Install the assembly into and arbor press and press off the bearing.

3. Repeat this procedure for the second tapered roller bearing on the driveshaft.

4. Install a suitable mandrel and press the bearings onto the shaft. When installing the bearings on the driveshaft, first press the smaller bearing onto the shaft so that the smaller O.D. is facing the pinion end of the shaft and then press on the larger bearing making sure the larger O.D. faces the pinion end.

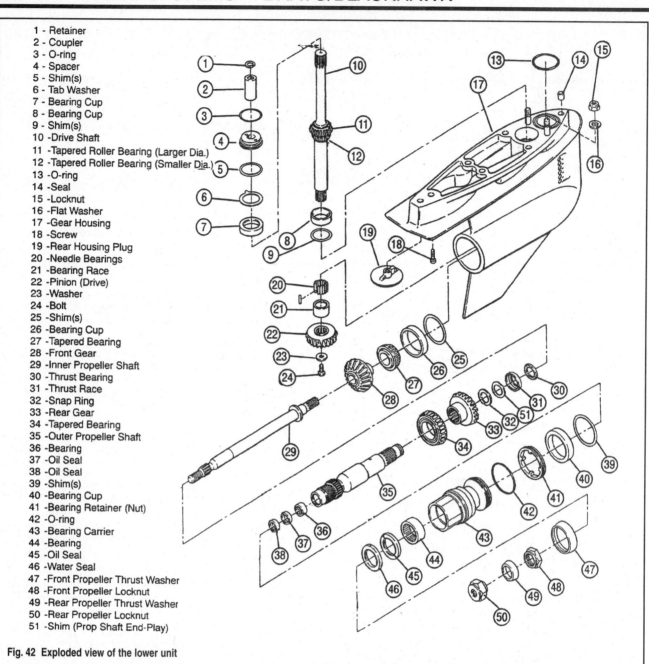

1 - Retainer
2 - Coupler
3 - O-ring
4 - Spacer
5 - Shim(s)
6 - Tab Washer
7 - Bearing Cup
8 - Bearing Cup
9 - Shim(s)
10 -Drive Shaft
11 -Tapered Roller Bearing (Larger Dia.)
12 -Tapered Roller Bearing (Smaller Dia.)
13 -O-ring
14 -Seal
15 -Locknut
16 -Flat Washer
17 -Gear Housing
18 -Screw
19 -Rear Housing Plug
20 -Needle Bearings
21 -Bearing Race
22 -Pinion (Drive)
23 -Washer
24 -Bolt
25 -Shim(s)
26 -Bearing Cup
27 -Tapered Bearing
28 -Front Gear
29 -Inner Propeller Shaft
30 -Thrust Bearing
31 -Thrust Race
32 -Snap Ring
33 -Rear Gear
34 -Tapered Bearing
35 -Outer Propeller Shaft
36 -Bearing
37 -Oil Seal
38 -Oil Seal
39 -Shim(s)
40 -Bearing Cup
41 -Bearing Retainer (Nut)
42 -O-ring
43 -Bearing Carrier
44 -Bearing
45 -Oil Seal
46 -Water Seal
47 -Front Propeller Thrust Washer
48 -Front Propeller Locknut
49 -Rear Propeller Thrust Washer
50 -Rear Propeller Locknut
51 -Shim (Prop Shaft End-Play)

Fig. 42 Exploded view of the lower unit

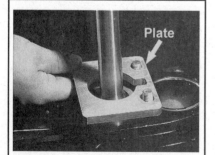

Fig. 43 Install the clamp plate over the shaft

Fig. 44 Unscrew the bearing carrier with the special tool...

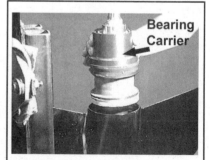

Fig. 45 ...and remove it from the housing

Fig. 45a Slide the retainer nut tool over the propeller shaft...

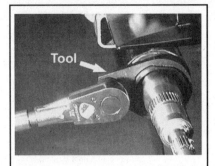

Fig. 45b ...and then install the carrier tool to remove the nut

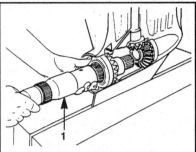

1- Inner and Outer Propeller Shaft Assembly

Fig. 45c Pull out the propeller shaft assembly...

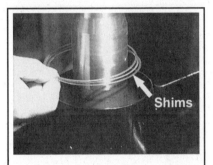

Fig. 45d ...and then pull out the shims

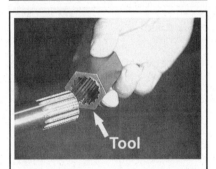

Fig. 45e Install and adaptor tool to the top of the shaft...

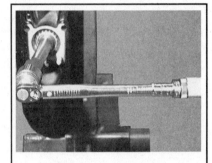

Fig. 45f ...and then loosen the pinion nut

Fig. 45g Pull out the driveshaft

Fig. 45h Pull out the pinion gear

Fig. 45i Pull out the forward gear and bearing

Fig 45j Pull out the bearing race...

Fig. 45k ...and the shim material

Fig. 45l Remove the shim material from the vertical bore

Bearing Carrier

◆ **See Figures 46**

■ The oil seals or the bearing cup may be replace individually. It is not necessary to remove one in order to remove the other.

1. Clamp the bearing carrier in a vise with soft jaws. Be sure to clamp on the bearing carrier at the reinforcement ribs. Place the jaws of a universal slide hammer puller down through the center of the carrier until the teeth on the puller are through the bearing cup. Pull cup out of the bearing carrier.
2. Remove the two propeller shaft oil seals from the bearing carrier with a puller also.

To assemble:

3. Wash all parts in solvent and blow them dry with compressed air. Remove any corrosion from the outside surface of the bearing carrier. Remove any seal and gasket material from all mating surfaces. Blow all water, oil passageways, and screw holes clean with compressed air. After all parts are clean and dry, apply a light coating of gear lubricant to the bright surfaces of all gears, bearings, and shaft, as a prevention against rusting.
4. Install the new bearing into the bearing carrier with the bearing surface facing inward.
5. Coat the outer edges of two new seals with Loctite and then press two new oil seals into the front of the bearing carrier. These seals are installed back-to-back. Seat the first one with the lip facing outward. Seat the second seal with the lip facing inward. Fill the area between the two seals with Special Lubricant 101.

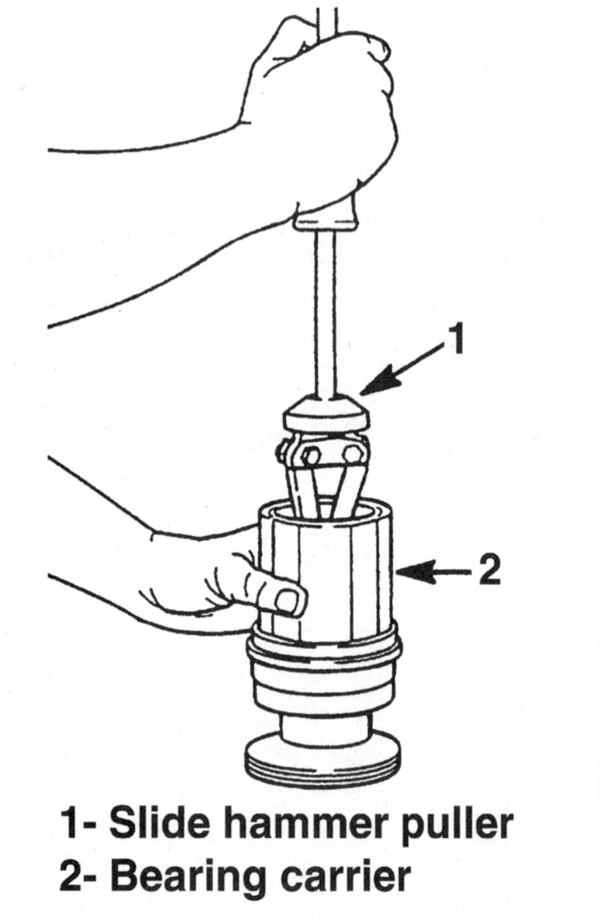

Driven Gear Bearing

◆ **See Figures 27 and 28**

1. Install a Universal Puller Plate between the driven gear and the roller bearing.
2. Sit the assembly on two pieces of wood and press the bearing off with an arbor press.
3. Sit the gear on a block of wood with the teeth toward the wood.
4. Coat the inside of a new bearing with gear lube and position over the gear.
5. Place a suitable mandrel (an old bearing race works well) over the new bearing race and then press it into the gear.

Propeller Shaft

◆ **See Figures 47, 48, 49, 50 and 50a**

1. Grasp the outer shaft and pull the inner shaft and thrust ring out of the assembly.
2. Remove the bearing cup from the outer shaft.
3. Position a punch on the thrust cap so it is lined up with the gap in the snap-ring and then press off the thrust cap. Remove the shim.
4. Using snap-ring pliers, remove the snap-ring from the end of the shaft and then slide off the driven gear and bearing assembly.
5. Insert a slide hammer with a three-armed puller attachment into the outer shaft and pull out the bearing and two oil seals.

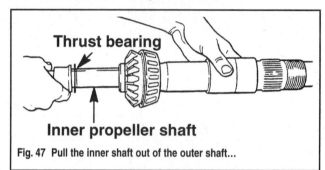

Thrust bearing

Inner propeller shaft

Fig. 47 Pull the inner shaft out of the outer shaft...

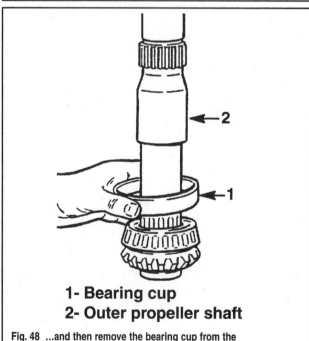

1- Slide hammer puller
2- Bearing carrier

Fig. 46 Removing the oil seals or the bearing with a puller

1- Bearing cup
2- Outer propeller shaft

Fig. 48 ...and then remove the bearing cup from the outer shaft

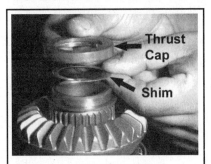

Fig. 49 Remove the thrust cap and shim...

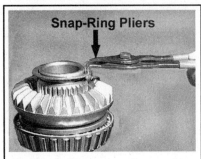

Fig. 50 ...and then remove the snap-ring

Fig. 50a ...and gear/bearing assembly

To assemble:

6. Press a new bearing into the outer shaft with the appropriate installation tool.

7. Coat the edges of the two new oil seals with Loctite 271 and press the seals into the outer shaft so the seal lips face away from each other. The inner seal lips should be facing inward and the outer seal lips should be facing out. Press the seals in until the tool bottoms out against the shaft and then lubricate the lips and the area between the seals with Special Lubricant 101.

8. Press the driven gear and bearing onto the end of the outer shaft and install a new snap-ring.

9. Coat the the inside ring of the thrust cap lightly with Loctite 242 and then position the shim into the cap. Position the cap and shim so it is squarely over the shaft and then press it on. If the original shim material was damaged or misplaced, use a 0.020 in shim pack as the starting point for installation and then adjust it after checking end-play.

10. Do not install the inner shaft at this time, see the Assembly procedure for this.

SPLINE LASH CHECK

◆ **See Figure 51**

■ **This test is necessary to complete the final backlash checks on final reassembly of the unit.**

1. Position the inner propeller shaft on two V-blocks.
2. Install a dial indicator rod on the shaft just behind the shaft splines.
3. Slide the front driven gear assembly onto the splines and then install a dial indicator so its tip lines up with the mark on the indicator rod.
4. Grasp the gear assembly and wiggle it back and forth a bit while observing the indicator reading.
5. Repeat this procedure for the outer shaft and record both reading for use later on in assembly.

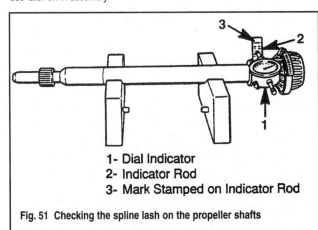

1- Dial Indicator
2- Indicator Rod
3- Mark Stamped on Indicator Rod

Fig. 51 Checking the spline lash on the propeller shafts

CLEANING AND INSPECTING

1. Wash all parts in solvent and blow them dry with compressed air. Remove all traces of seal and gasket material from all mating surfaces. Blow all water, oil passageways and screw holes clean with compressed air. After cleaning, apply a light coating of gear lubricant to the bright surfaces of all gears, bearings, and shafts as a prevention against rusting and corrosion. Check the shift crank to be sure it is not bent.

2. Inspect the water pump impeller plate for wear and corrosion. Replace any worn, corroded, or damaged parts. Use a fine file to remove burrs. Replace all O-rings, gaskets, and seals to ensure satisfactory service from the unit. Clean the corrosion from inside the housing where the bearing carrier was removed.

3. Check to be sure the water intake is clean and free of any foreign material.

4. Inspect the gear case, housings, and covers inside and out for cracks. Check carefully around screw and shaft holes. Check for burrs around machined faces and holes. Check for stripped threads in screw holes and traces of gasket material remaining on mating surfaces.

5. Check O-ring seal grooves for sharp edges, which could cut a new seal. Check all oil holes.

6. Inspect the bearing surfaces of the shafts, splines, and keyways for wear and burrs. Look for evidence of an inner bearing race turning on the shaft. Check for damaged threads. Measure the run-out on all shafts to detect any bent condition. If possible, check the shafts in a lathe for out-of-roundness.

7. Inspect the gear teeth and shaft holes for wear and burrs. Hold the center race of each bearing and turn the outer race. The bearing must turn freely without binding or evidence of rough spots. Never spin a ball bearing with compressed air or it will be ruined. Inspect the outside diameter of the outer races and the inside diameter of the inner races for evidence of turning in the housing or on a shaft. Deep discoloration and scores are evidence of overheating.

8. Inspect the thrust washers for wear and distortion. Measure the washers for uniform thickness and flatness.

9. Inspect the gears for distortion, burrs, and cracks. Check for any sign of discoloration, which means the gears are running too hot.

10. Replace all seals, O-rings, and gaskets to ensure maximum service after the work is completed.

ASSEMBLY

◆ **See Figures 29, 30, 32, 51a, 51b, 51c, 51d, 51e, 52, 53, 54, 55 and 56**

1. Lubricate all gears, splines and bearings with gear lube so as to get accurate pre-load readings.
2. Install a new O-ring into the water passage with 3-M adhesive.
3. Coat the needle bearings with bearing assembly lubricant and install them into the lower driveshaft bearing casing.

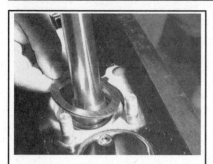

Fig. 51a Slide the tab washer over the shaft...

Fig. 51b ...so it fits into the cutout

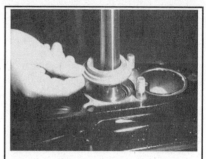

Fig. 51c Slide the spacer over the shaft

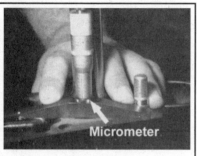

Fig. 51d Using a micrometer

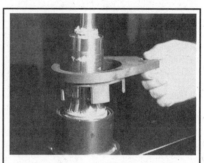

Fig. 51e Install the bearing carrier tool

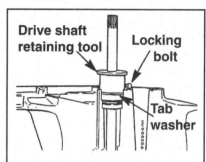

Fig. 53 To check the front driven gear backlash, install a driveshaft retainer...

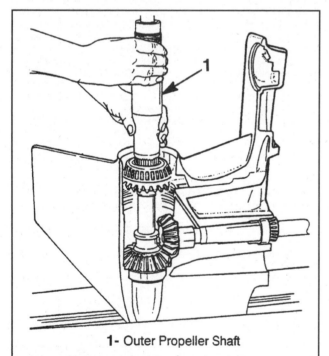

1- Outer Propeller Shaft

Fig. 52 Installing the outer shaft over the inner shaft

4. Position the driven gear and bearing into the propeller bore.

5. Drop the original shims into the driveshaft bore and drive the lower cup into position. If the original shims were damaged or misplaced, install new ones with a thickness of 0.050 in. (1.27mm).

6. Slide the driveshaft into the bore and install the upper bearing cup, tab washer, original shims, spacer and the O-ring over the shaft. If the original shims were damaged or misplaced, install new ones with a thickness of 0.050 in. (1.27mm).

7. Install the pinion gear onto the driveshaft, but leave the nut and washer off.

8. To determine the shim thickness required for the driveshaft bearing pre-load, measure the distance between the top of the gear housing and the tab washer with a micrometer. Measure the thickness of the spacer from the top surface to the bottom surface. Proper shim thickness can be determined by subtracting the second measurement from the first and adding 0.001 in.

9. Install the correct shim(s), spacer and O-ring.

10. Install the clamp plate back onto the top of the gear housing.

11. With the driveshaft adapter still installed, check the rolling pre-load with a dial-type torque wrench. It should be 3–10 inch lbs. (0.3–1.1 Nm).

12. Install the pinion bolt and washer and tighten it to 35 ft. lbs. (48 Nm) on 1992–98 models; 45 ft. lbs. (61 Nm) on 1999 and later models.

13. Insert a shimming tool (#91-805462) into the propeller shaft bore and check the clearance between the tool and the gear with a feeler gauge. Check this in three places, 120° apart. Clearance must be 0.023–0.028 in. (0.575–0.700mm). If less than spec, add shims of the appropriate thickness under the lower bearing cup to bring it into specification. If greater than spec, remove shim material from under the bearing cup. Recheck the clearance again until you're sure it is correct.

■ On 1999 and later models, use tool position 15:19 for units with 1.50:1 or 1.36:1 gear ratios; use tool position 16:27 for all other gear ratios.

■ Any thickness of shims added, must be subtracted from the shim thickness at the upper bearing cup.

14. If you haven't already done so, coat the inside ring of the thrust cap lightly with Loctite 242 and then position the shim into the cap. Position the cap and shim so it is squarely over the shaft and then press it on. If the original shim material was damaged or misplaced, use a 0.020 in shim pack as the starting point fro installation and then adjust it after checking end-play.

15. Install the inner propeller shaft and thrust bearing into the housing and then slide on the outer shaft.

16. Install a new shim pack of the same thickness as that removed, or start with a nominal thickness of 0.050 in. (1.3mm).

17. Install the bearing cup over the shaft assembly and seat it against the shim pack.

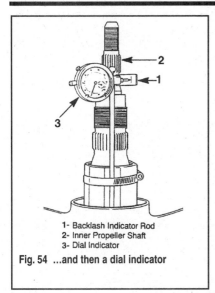

1- Backlash Indicator Rod
2- Inner Propeller Shaft
3- Dial Indicator

Fig. 54 ...and then a dial indicator

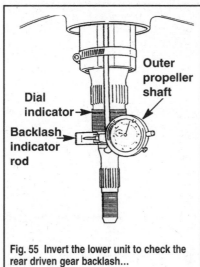
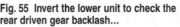

Fig. 55 Invert the lower unit to check the rear driven gear backlash...

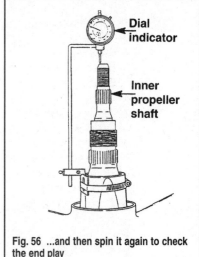

Fig. 56 ...and then spin it again to check the end play

18. Coat the threads of the bearing retainer with lubricant and screw it onto the shaft assembly. Install the retainer tool and tighten it to 200 ft. lbs. (271 Nm). Remember this is a left hand thread, so tighten it counterclockwise.

19. Coat the threads of the bearing carrier and the O-ring surface with Special Lubricant 101. Coat the tapered surface of the carrier with Perfect Seal and then screw the carrier onto the shaft and into the housing. Install the special tool and tighten the carrier to 150 ft. lbs. (203 Nm)—remember this is counterclockwise also!

20. Remove the clamp plate, O-ring, spacer and shims again and install a driveshaft retaining tool (# 91-805381) over the shaft and onto the tab washer. Do not tighten the lock bolt yet.

21. Press down lightly on the retainer while wiggling the driveshaft to seat the bearings in the cups and then tighten the lock bolt. Now move the entire unit so the propeller shaft is facing up and install a backlash indicator rod to the end of the inner propeller shaft. Mount a dial indicator to the bearing carrier with a hose clamp and position it so the tip is at the mark on the indicator rod. Wiggle the inner shaft back and forth slightly and record the reading. Subtract the reading observed in the Spline Lash Check detailed previously to arrive at your total gear backlash. It should be 0.012–0.016 in. (0.3–0.4mm). If the reading is too high, disassemble everything and add shims behind the front driven gear bearing cup; if too low, remove shims. Recheck front driven gear again backlash after reassembly.

22. Spin the unit 180º so that the propeller shaft is now facing down. Loosen the lock bolt on the driveshaft retainer, wiggle the propeller shaft a few times to seat the bearings and then press down on the retainer again and tighten the lock bolt. With the dial indicator and rod still installed from the previous step, wiggle the outer propeller shaft back and forth while observing the indicator. Record the reading. Subtract the reading observed in the Spline Lash Check detailed previously to arrive at your total gear backlash. It should be 0.012–0.016 in. (0.3–0.4mm). If the reading is too high, disassemble everything and remove shims from in front of the rear driven gear bearing cup; if too low, add shims. Recheck rear driven backlash again after reassembly.

23. Swivel the unit back around so the propeller shaft is again pointing upward and install the dial indicator as illustrated. Move the inner propeller shaft up and down while observing the indicator. End play should be 0.001–0.050 in. (0.025–1.27mm). If you reading is outside of this range, add or subtract the appropriate number of thrust race shims to the outer propeller shaft until the reading comes into range.

24. Remove the retainer tool and reinstall the O-ring, spacer and shims.

Gear Housing (Lower Unit) — Blackhawk

REMOVAL & INSTALLATION

Blackhawk models utilize a one-piece case housing both upper and lower gear components. Drive unit removal is sufficient—it is not possible to split the cases as on other Bravo models.

DISASSEMBLY

◆ **See Figures 44, 45, 57, 58 and 59**

1. Remove the stern drive as detailed previously and install it in a suitable holding device.

2. Remove all upper end components as detailed in the Upper Gear Housing section. Although it is not possible to split the drive unit, the upper driveshaft and clutch components are all covered in the section with the other drives.

3. Remove the propeller.

4. Using a small bar and a hammer bend the tabs of the tab washer away from the bearing carrier retainer. Install a bearing carrier retainer tool (#91-805377) so it fits into the retainer, attach a breaker bar and loosen the retainer. Remember this is a left hand thread so turn it clockwise to remove it.

5. Grab the tab washer with pliers and pull it off. Remove the O-ring.

6. Position a bearing carrier tool (#91-805374) over the carrier , attach a breaker bar and loosen the carrier. When you remove the carrier, keep the O-ring seal with it for assembly later.

7. Grab the propeller shaft and remove the inner and outer shaft assembly by pulling them straight back and out of the housing. Reach into the bore and remove the shims.

8. Install a bearing adaptor tool (#91-805846) with a breaker bar from the top end and remove the adaptor through the clutch assembly bore. Lift out the spline coupling underneath the adaptor.

9. Spin the bearing retainer back into the propeller bore temporarily to protect the threads. Attach a driveshaft adaptor tool through the upper bore and onto the top of the vertical driveshaft while inserting a long-handled socket wrench into the propeller shaft bore. Remove the pinion gear bolt. Remove the driveshft retaining nut.

10. Attach a slide hammer with a 3/16-16 thread to the upper end of the driveshaft and carefully remove the driveshaft out the top. Make sure you hold the pinion gear from the bottom side during removal so that the needle bearings don't drop out.

11. Reach in and remove the shims from the driveshaft bore.

12. Now you can reach in and remove the needle bearings from the pinion bearing race. While you've got your hand in the bore, remove the forward gear and bearing assembly.

13. Refer to the following procedures for further disassembly of the removed components.

Driveshaft

Please refer to the preceding Bravo III procedure.

Bearing Carrier

Please refer to the preceding Bravo III procedure.

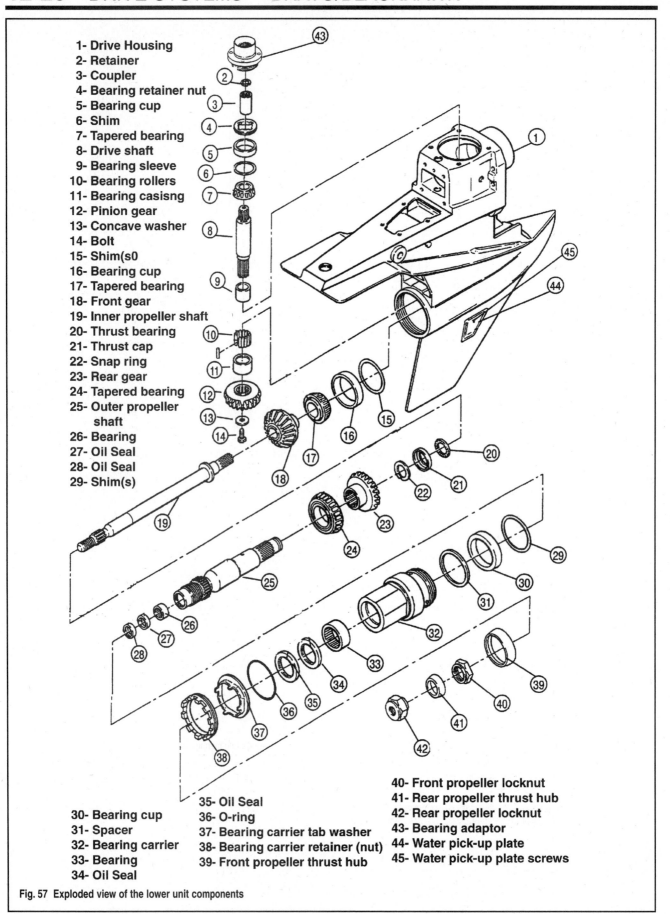

1- Drive Housing
2- Retainer
3- Coupler
4- Bearing retainer nut
5- Bearing cup
6- Shim
7- Tapered bearing
8- Drive shaft
9- Bearing sleeve
10- Bearing rollers
11- Bearing casisng
12- Pinion gear
13- Concave washer
14- Bolt
15- Shim(s0
16- Bearing cup
17- Tapered bearing
18- Front gear
19- Inner propeller shaft
20- Thrust bearing
21- Thrust cap
22- Snap ring
23- Rear gear
24- Tapered bearing
25- Outer propeller
    shaft
26- Bearing
27- Oil Seal
28- Oil Seal
29- Shim(s)

30- Bearing cup
31- Spacer
32- Bearing carrier
33- Bearing
34- Oil Seal

35- Oil Seal
36- O-ring
37- Bearing carrier tab washer
38- Bearing carrier retainer (nut)
39- Front propeller thrust hub

40- Front propeller locknut
41- Rear propeller thrust hub
42- Rear propeller locknut
43- Bearing adaptor
44- Water pick-up plate
45- Water pick-up plate screws

Fig. 57 Exploded view of the lower unit components

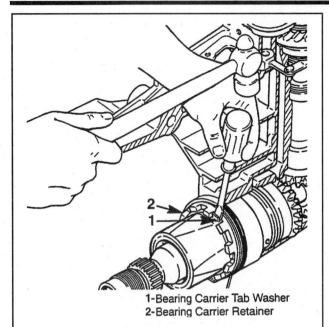

1-Bearing Carrier Tab Washer
2-Bearing Carrier Retainer

Fig. 58 Bend out the tab washer

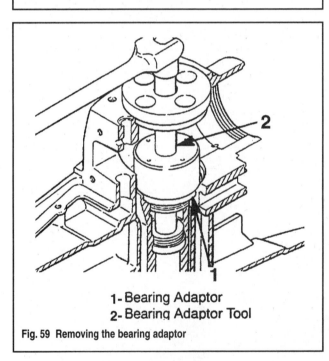

1- Bearing Adaptor
2- Bearing Adaptor Tool

Fig. 59 Removing the bearing adaptor

Driven Gear Bearing

Please refer to the preceding Bravo III procedure.

Propeller Shaft

Please refer to the preceding Bravo III procedure.

SPLINE LASH CHECK

Please refer to the preceding Bravo III procedure.

CLEANING AND INSPECTING

Please refer to the preceding Bravo III procedure.

ASSEMBLY

◆ **See Figures 52, 55 and 56**

1. Reinstall the lower bearing needles into the race with Assembly Lubricant so they stay in place.
2. Push the forward gear/bearing assembly into position in the propeller shaft bore.
3. Slide the driveshaft into the housing from the top. It's a good idea to hold the pinion gear against the lower bearing race incase shaft installation knocks out any needles.
4. Install the original shims back into the upper bore. If they were misplaced or damaged, use new ones of 0.040 in. (1mm) thickness as your starting point.
5. Using the special tool, install the outer bearing race and then the retaining nut. Tighten the nut to 100 ft. lbs. (135 Nm).
6. While the adapter tool is still in place, coat the threads of the pinion bolt with Loctite 271 and install the bolt. Tighten it to 37 ft. lbs. (48 Nm). Remember to remove the retainer after you've tightened the bolt.
7. Install a pre-load tool (#91-805458, 91-805473, 91-805466) onto the upper end of the driveshaft and tighten the nut on the tool to 3–10 inch lbs. (0.3–1.1 Nm) so that you draw the shaft up into the bearing taper. Now install a shimming tool (#91-805462) into the propeller shaft bore and measure the clearance between the tool and the pinion gear. The measurement should be 0.023–0.028 in. (0.575–0.0700mm) and you should check it at three different locations on the tool, 120° apart. If you find the clearance to be less than specifications, add shims under the tapered roller bearing cup. If greater than spec, remove shims until the pinion height is correct. Recheck the clearance again after adjustment and re-torque the nut to 100 ft. lbs. (136 Nm).
8. Install the inner propeller shaft and thrust bearing into the housing and then slide on the outer shaft.
9. Install a new shim pack of the same thickness as that removed, or start with a nominal thickness of 0.060 in. (1.5mm).
10. Install the bearing cup and spacer over the shaft assembly and seat it against the shim pack.
11. Coat the threads of the bearing carrier with lubricant and screw it into the housing over the shaft assembly. Install the special tool and tighten it to 250 ft. lbs. (339 Nm).
12. Install the O-ring onto the carrier and then install the tab washer to hold it in place.
13. Coat the threads of the bearing carrier retainer with Special Lubricant 101 and then screw it into the housing. Install the special tool and tighten the carrier to 150 ft. lbs. (203 Nm)—remember this is counterclockwise!
14. Now move the entire unit so the propeller shaft is facing up and install a backlash indicator rod to the end of the inner propeller shaft. Mount a dial indicator to the bearing carrier with a hose clamp and position it so the tip is at the mark on the indicator rod. Wiggle the inner shaft back and forth slightly and record the reading. Subtract the reading observed in the Spline Lash Check detailed previously to arrive at your total gear backlash. It should be 0.008–0.012 in. (0.2–0.3mm). If the reading is too high, disassemble everything and add shims behind the front driven gear bearing cup; if too low, remove shims. Recheck front driven gear again backlash after reassembly.
15. Spin the unit 180° so that the propeller shaft is now facing down. Loosen the nut on the driveshaft, wiggle the propeller shaft a few times to seat the bearings and then tighten the nut. With the dial indicator and rod still installed from the previous step, wiggle the outer propeller shaft back and forth while observing the indicator. Record the reading. Subtract the reading observed in the Spline Lash Check detailed previously to arrive at your total gear backlash. It should be 0.008–0.012 in. (0.2–0.3mm). If the reading is too high, disassemble everything and remove shims from in front of the rear driven gear bearing cup; if too low, add shims. Recheck rear driven backlash again after reassembly.
16. Swivel the unit back around so the propeller shaft is again pointing upward and install the dial indicator as illustrated. Move the inner propeller shaft up and down while observing the indicator. End play should be 0.014–0.082 in. (0.356–2.08mm). If you reading is outside of this range, add or subtract the appropriate number of thrust race shims to the outer

propeller shaft until the reading comes into range.

17. Bend the tabs on the tab washer into the slots on the retainer.

18. Remove the tool from the driveshaft housing and install the spline coupling. Coat the first few threads of the bearing adaptor with Loctite 271 and install it using the tool. Tighten it to 150 ft. lbs. (203 Nm).

19. Install all upper end components removed and then install the entire unit to the boat.

Propeller(s)

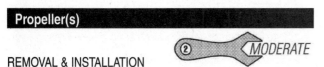

REMOVAL & INSTALLATION

Bravo I/II

◆ See Figures 60, 61, 62, 63, 64, 65, 66 and 67

✳✳ CAUTION

If the drive unit or the lower unit is not disconnected or removed from the boat it is imperative that you disconnect the battery cables and remove the ignition key prior to performing this procedure.

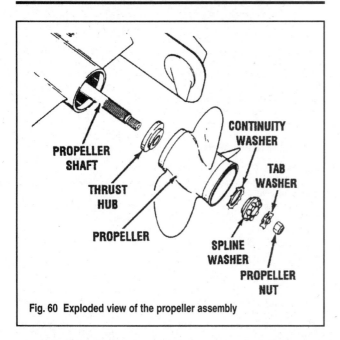

Fig. 60 Exploded view of the propeller assembly

1. Bend the locking tabs on the tab washer out and away from the grooves in the splined washer. Never pry on the end of the propeller hub because any small distortion will affect propeller performance or possibly break the hub off the propeller. Place a block of wood between one blade of the propeller and the anti-cavitation plate to prevent the shaft from rotating. Place the correct size socket and breaker bar over the propeller nut and remove the nut from the propeller shaft. Slide the tab washer, splined washer, continuity washer, propeller, and thrust hub (washer) off of the propeller shaft.

2. If the propeller is frozen to the shaft, heat must be applied to the shaft so as to melt out the rubber inside the hub. Using heat will destroy the hub, but there is no other way. As heat is applied, the rubber will expand and the propeller will actually be blown from the shaft. Therefore, stand clear to avoid personal injury.

3. Use a knife and cut the hub off the inner sleeve.

4. The sleeve can be removed by cutting it off with a hacksaw or cold chisel, or it can be removed with a puller. Again, if the sleeve is frozen, it may be necessary to apply heat.

5. Remove the thrust hub from the propeller shaft.

To install:

6. Check to be sure the splines on the propeller shaft are in good condition. If there is any visible sign of corrosion on the shaft, the corrosion must be removed before assembly is completed. Verify the threaded end of the propeller shaft is clean and the threads are in good condition.

7. Verify the splined steel hub inside the propeller is free of corrosion. Check to be sure the rubber hub is tight and secured to the propeller inner hub. Verify there is no sign of the rubber hub spinning on the inside of the propeller.

8. Slide the propeller thrust washer onto the shaft with the smaller side of the washer facing out— to the rear, to the propeller. Coat the propeller shaft and splines with one of the following lubricants in the order of preference; Quicksilver Special Lubricant 101, 2-4-C Lubricant or Perfect Seal.

9. Slide the propeller onto the shaft followed by the continuity washer, splined washer, tab washer and then the propeller nut, as indicated in the accompanying illustration. These components must be assembled in the correct order.

10. Position a block of wood between one of the propeller blades and the anti-cavitation plate to keep the propeller from rotating. Tighten the propeller nut to a torque value of 55 ft. lbs. (75 Nm). Continue tightening the nut until three tabs on the tab washer align with the grooves in the splined washer. There should be a minimum of two threads showing on the end of the propeller nut after it has been tightened to the proper torque value. On 1999–01 Bravo II models, simply tighten the nut to 60 ft. lbs. (82 Nm).

11. Bend three of the tab washer tabs into the grooves of the splined washer using a blunt end punch and hammer. The tabs will prevent the propeller nut from backing off.

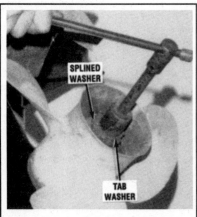

Fig. 61 Bend back the tabs on the washer before loosening the propeller nut

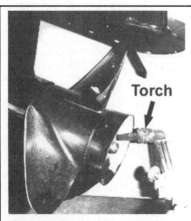

Fig. 62 Heat the propeller shaft to loosen the frozen hub

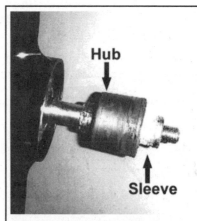

Fig. 63 The rubber hub rides on the sleeve, inside the propeller flange

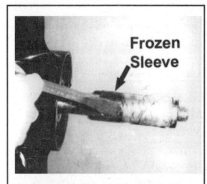

Fig. 64 Break off the sleeve after heating the shaft again...

Fig. 65 ...and then remove the thrust washer

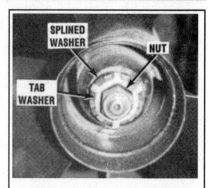

Fig. 66 A good look at the installation

Fig. 67 Don't forget to bend those tabs over

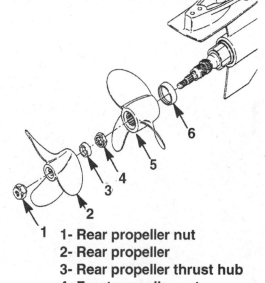

1- **Rear propeller nut**
2- **Rear propeller**
3- **Rear propeller thrust hub**
4- **Front propeller nut**
5- **Front propeller**
6- **Front propeller thrust hub**

Fig. 68 Exploded view of the propeller assembly

Bravo III/Blackhawk

◆ See Figures 68, 69, 70, 71 and 72

✳✳ CAUTION

If the drive unit or the lower unit is not disconnected or removed from the boat it is imperative that you disconnect the battery cables and remove the ignition key prior to performing this procedure.

1. Remove the aft propeller nut, and then remove the aft propeller. The aft propeller thrust hub will probably come off the shaft with the propeller, or it may remain on the shaft. Remove the collar from the propeller or the shaft.

2. Obtain Inner Propeller Nut tool (#91-805457). Move the tool into place over the inner propeller nut and remove the nut.

3. Carefully remove the inner propeller. The inner propeller centering ring may come off with the propeller or remain on the shaft. Remove the ring from the propeller or from the shaft.
To install:

4. Check to be sure the splines and the threaded end of both propeller shafts are clean and in good condition. Apply a liberal coating of Quicksilver Marine Lubricant 2-4-C to both the inner and outer propeller shaft.

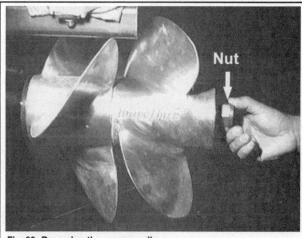

Fig. 69 Removing the rear propeller

Fig. 70 Install the special tool...

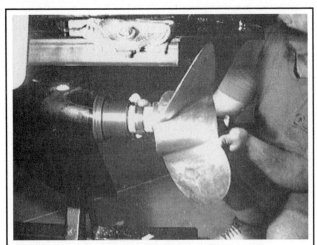

Fig. 71 ...and then remove the forward propeller

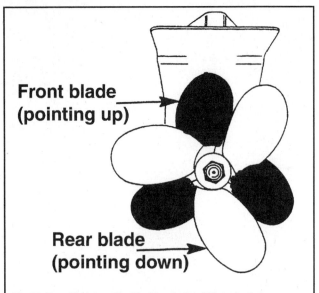

Front blade (pointing up)

Rear blade (pointing down)

Fig. 72 Prop blades on the Blackhawk should be oriented as shown

■ When installing the props on the Blackhawk, one blade of the inner one should be pointing straight UP (at 12:00); while one blade of the outer one should be pointing straight DOWN (at 6:00).

5. Slide the forward thrust hub onto the outer propeller shaft with the tapered side of the hub facing the aft end of the shaft. Slide the forward propeller onto the outer propeller shaft with the splines in the propeller indexed with the splines of the propeller shaft.

6. Install the forward propeller nut. Tighten the nut to 100 ft. lbs. (136 Nm) using the special tool.

7. Slide the rear thrust hub onto the inner propeller shaft with the tapered side of the hub facing the aft end of the propeller shaft. Install the aft propeller onto the inner propeller shaft with the splines in the propeller indexed with the splines on the shaft.

8. Install the aft propeller nut and tighten the nut to 60 ft. lbs. (81 Nm).

9. Recheck the torque on both propeller nuts after 20 hours of operation.

DRIVESHAFT HOUSING (UPPER UNIT)

Description

Output power from the engine is connected to the stern drive through a horizontal driveshaft. A coupler is bolted to the flywheel and has a splined hub in the center. The end of the horizontal driveshaft indexes with and slides into the center of the hub. Power from the engine is then transmitted through the horizontal driveshaft to a pinion gear set where power direction is changed from horizontal to vertical.

The upper driven gear is pressed onto the outside diameter of the upper driveshaft. The upper driveshaft is splined on the lower end. When the upper gear housing is mated to the lower unit, the end of the lower unit driveshaft (or clutch assembly) indexes into the splined end of the upper driveshaft. Engine power is then transferred down into the lower gear unit.

The horizontal and vertical driveshafts are both mechanically connected. Therefore, anytime the engine is operating, the horizontal and vertical driveshafts are constantly rotating with engine rpm. A double yoke universal joint assembly in the horizontal driveshaft allows the stern drive to be raised or lowered to a required trim/tilt position (within limits), while the engine is operating.

Gear Ratio Information

The gear ratio for the stern drive must be known to permit the proper special tool selection to be made in order to properly position the gears in the housing. The gear ratio for all stern drives is identified in two places—a decal is affixed to the port side of the upper gear housing. A number such as 1.50R is followed by the serial number.

The second location of the gear ratio is on the universal joint splined yoke. The identification mark here will be a letter such as "**C**".

This letter represents the gear ratio in the upper gear housing. Use the accompanying chart to determine the gear ratio when the letter is known.

Gear Ratio		
Model	**Letter**	**Ratio**
Bravo I	C	1.65:1
	F	1.50:1
	H	1.36:1
Bravo II	C	2.20:1
	F	2.00:1
	H	1.81:1
	T	1.65:1
Bravo III	B	2.00:1
	C	1.65:1
	D	1.81:1 ①
	F	1.50:1
	G	1.81:1 ②
	K	2.20:1
	N	2.43:1
	P	1.36:1

① Diesel
② Gasoline

If the gears inside the housing have previously been changed from the factory markings on the housing, another method may be used to determine the gear ratio. This method is the least desired because the unit has to be disassembled before the ratio is known. With the gears removed from the housing count the number of teeth on the drive and driven gears. Compare the gear teeth number count to the chart below for the gear ratio of the unit being serviced.

	Gear Ratio		
Model	Drive Gear	Driven Gear	Gear Ratio
Bravo I	23	30	1.65:1
	27	32	1.50:1
	27	29	1.36:1
Bravo II	23	30	2.20:1
	27	32	2.00:1
	27	29	1.81:1
	27	32	1.65:1
Bravo III			
1992-98	16	27	2.20:1
	16	27	2.00:1
	16	27	1.81:1
	18	25	1.65:1
	15	19	1.50:1
1999-01	23	30	2.43:1
	23	30	2.20:1
	27	32	2.00:1
	27	29	1.81:1
	27	32	1.65:1
	27	32	1.50:1
	27	29	1.36:1

Driveshaft Housing (Upper Unit)

OEM ③ DIFFICULT

REMOVAL & INSTALLATION

Bravo I/II

◆ See Figures 10 and 11

■ Please refer to the preceding Lower Unit removal section for illustrations of the two unit separation process.

1. Remove the stern drive unit.
2. If working on a Bravo I, place an alignment mark on the trim tab trailing edge and the anti-cavitation plate as an aid during assembling. Remove the plastic plug from the top of the anti-cavitation plate. Remove the 1/2 in. bolt directly above the trim tab, lift off the trim tab. You will need a long extension to your socket wrench for this one.
3. Remove the six hex nuts and washers (three on each side), from the sides of the anti-cavitation plate. Remove the single bolt hidden inside the trim tab cavity on the Bravo I; on the Bravo II this bolt is in a recess, just forward of the trim tab.
4. Separate the two housings by tapping downward on the lower unit with a soft face mallet and a wooden block. In some cases the housing may be difficult to separate because the driveshaft may be corroded and "frozen" in the upper gear housing vertical shaft. Lower the gear housing away from the upper gear housing (or lift the upper unit away from the lower!).
To install:
5. Coat a new O-ring with 3-M Adhesive and install it on the upper face of the lower unit at the water passage opening. No gasket is used between the two housings. All sealing is accomplished with two O-rings: The water passageway O-ring and the O-ring installed around the driveshaft spacer.

The spacer has two vertical holes drilled through it to allow oil to pass through the spacer and lubricate not only the double roller bearing assembly on the driveshaft but the entire lower unit. Oil passes freely through the double bearing arrangement. Therefore, the oil passageway O-ring is perhaps the most crucial component in a Bravo stern drive.

✳✳ WARNING

In the Bravo Stern Drive, no locating pins are utilized in aligning the lower unit with the upper gearcase housing. Therefore, if the bolt holes show serious signs of elongation, or the securing bolts show signs of stress or "necking", the lower unit, the upper gearcase housing and all securing bolts should be replaced. This is not as drastic or expensive as it sounds. All internal components are removed and only the housing castings are replaced.

6. Install the coupler and retainer onto the upper end of the driveshaft. Raise the lower unit to mate with the upper gearcase housing. The lower end of the short clutch shaft must slide into the coupler at the upper end of the lower driveshaft. It may be necessary to slowly rotate the propeller shaft counterclockwise until the splines index.
7. Screw the aft bolt into the hole in the trim tab cavity on Bravo I or into the hole just forward of the trim tab on Bravo II..
8. Install the three port and three starboard nuts and their washers. Tighten the six nuts alternately, and a little at-a-time, until the lower unit is tight against the upper gear housing.
9. Now tighten the rear bolt and the six side nuts to 35 ft. lbs. (47.5 Nm).
10. Install the trim tab (Bravo I) with the marks made during disassembly aligned. Tighten the screw(s) to 23 ft. lbs. (31 Nm).

■ The trim tab performs two very important jobs, one of which you may not realize. First, the tab compensates for steering torque. If the boat continually seems to move to port or starboard while the helmsman is on a straight course, the trim tab can be adjusted to the side of the pull. The tab also prevents electrolysis from damaging expensive parts. The tab is not expensive; it should show signs of electrolytic action; and should always be replaced after some of the material has been eaten away or is pitted. Now, if the tab shows no signs of electrolytic action after the boat has been in use over a period of time, the grounding should be checked to ensure more expensive parts are not being damaged. Install the trim tab plastic cover plug.

11. Install the propeller if it was removed and refill the drive unit with oil.

Bravo III

◆ See Figures 38, 39, 40 and 41

■ Please refer to the preceding Lower Unit removal section for illustrations of the two unit separation process.

1. Remove the stern drive unit.
2. Remove the plastic plug from the top of the gear housing. Remove the bolt directly above the anode, lift off the anode. You may need an extension to your socket wrench for this one.
3. On 1992–98 models, look inside the cavity under the anode and remove the bolt holding the splash plate extension. Remove the plate. On 1999–01 models, look inside the cavity and remove the mounting bolt— there is no extension plate.
4. Remove the six hex nuts and washers (three on each side), from the sides of the housing.
5. Separate the two housings by tapping downward on the lower unit with a soft face mallet and a wooden block. In some cases the housing may be difficult to separate because the driveshaft may be corroded and "frozen" in the upper gear housing vertical shaft. Lower the gear housing away from the upper gear housing.
To install:
6. Install the coupler and retainer onto the upper end of the driveshaft. Raise the lower unit to mate with the upper gearcase housing. The lower end of the short clutch shaft must slide into the coupler at the upper end of the lower driveshaft. It may be necessary to slowly rotate the propeller shaft counterclockwise until the splines index.

7. Slide the splash plate extension into the slot in the driveshaft housing and tighten the bolt to 35 ft. lbs. (47.5 Nm) on 1992–98 models.

8. Install the anode into the extension and tighten the bolt to 23 ft. lbs. (31 Nm).

9. Install the three port and three starboard nuts and their washers. Tighten the six nuts alternately, and a little at-a-time, until the lower unit is tight against the upper gear housing.

10. Now tighten the rear bolt and the six side nuts to 35 ft. lbs. (47.5 Nm).

11. Install the propeller(s) if it was removed and refill the drive unit with oil.

Blackhawk

Blackhawk models utilize a one-piece case housing both upper and lower gear components. Drive unit removal is sufficient—it is not possible to split the cases as on other Bravo models.

ASSEMBLY & DISASSEMBLY

◆ **See Figures 75, 76, 77, 78, 79, 80, 81, 82, 83, 84, 85, 86 and 87**

1. Make sure that the drive unit is in the Neutral detent position.

2. Loosen the three rear cover bolts evenly and remove the cover (the Blackhawk uses two Allen screws). Unscrew the vent screw on the forward starboard side of the housing (if equipped).

3. On the Blackhawk, remove the shifter assembly cover plate (4 screws) and the detent cover plate (4 screws also).

4. Insert an Allen wrench into the shift linkage cap screw, loosen it and remove the screw. Now insert and shift handle tool (#91-17302) into the linkage cap screw hole to hold it steady and then remove the shift cam cap screw with an Allen wrench.

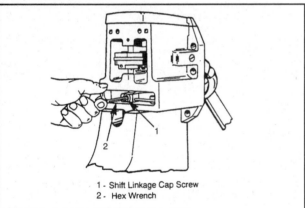

1 - Shift Linkage Cap Screw
2 - Hex Wrench

Fig. 75 Loosen the shift linkage cap screw

5. Remove the dipstick (if equipped). Loosen and remove the four top cover mounting bolts and lift off the cover. There may be a small Allen screw at the forward edge of the cover— leave it there. Throw away the O-ring.

6. Screw the shift handle tool onto the top of the shifter shaft from the top of the housing and then pull the shaft up and out of the housing.

7. Grab the shift linkage assembly from the rear, turn it so the link bar rotates about 1/4 turn clockwise and pull the entire assembly out of the housing. You may have to wiggle it a little as you're pulling it out. This can be tricky, but just examine the situation and you'll get it out. In fact, unless something is broken, there's really no reason to remove it in the first place other than to inspect the components.

8. Reach in and grab the shifter yoke and cam assembly.

9. Install a U-joint retainer wrench (#91-17256) over the yoke shaft and joints so that it indexes with the notches in the retaining ring nut. Loosen the ring nut and then pull the shaft assembly straight out of the housing. Do not twist or turn it in any manner, just pull it straight back.

10. Lift the upper thrust race and bearing out of the top of the housing. Put them aside and keep them together in the exact position that they were removed.

11. Reach in and carefully lift the clutch assembly straight up and out of the housing. Reach into the bore and lift out the lower thrust bearing and race. Store them as you did the upper set.

12. Inspect and disassemble the individual components as detailed in the following procedures.

To assemble:

13. There are two sets of numbers stamped on the rear of the housing, above and below the shifter cavity that designate the upper and lower thrust bearing race thickness. You guessed it…the top number is for the upper race and the lower number is for the lower one! These are guides to what the actual thickness should be—"94" equates to 0.094 in., etc. Measure the thickness of your races to ensure that they are within specification. If using the original races, make sure they are installed exactly as they were removed; do not flip them so the contact sides are different.

14. Position the correct lower race into the clutch bore Coat the lower face of the bottom clutch assembly gear and the thrust bearing with 60% mixture of High Performance Gear lube and 40% Special Lubricant 101. Stick the bearing onto the bottom of the gear and insert the assembly into the housing bore.

15. Repeat this procedure with the top gear and the upper bearing and attach the bearing to the gear set. Position the race on top of the bearing.

16. Each side of each clutch gear has an index mark and a plus (+) and minus (-) sign, align the upper and lower gears so that the index marks line up AND there is a plus (+) on one and a minus (-) on the other. It doesn't matter whether the plus or minus sign is on the top or bottom, just that they are different. There should never be two plus (+) signs or two minus (-) signs on top of each other. You will also notice that there is an index mark stamped above and below the shifter cavity (in between the number discussed earlier) — make sure the aligned index marks on the two clutch gears are in line with each of these.

17. Coat the threads of the retaining ring nut with Special Lubricant 101 and carefully guide the driveshaft assembly back into the housing. Tighten the ring nut to 200 ft. lbs. (271 Nm) with the special tool. Check the index

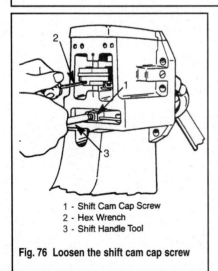

1 - Shift Cam Cap Screw
2 - Hex Wrench
3 - Shift Handle Tool

Fig. 76 Loosen the shift cam cap screw

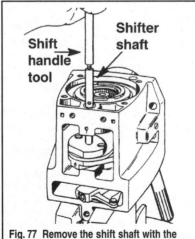

Shift handle tool

Shifter shaft

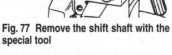

Fig. 77 Remove the shift shaft with the special tool

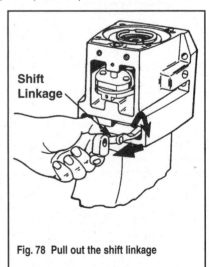

Shift Linkage

Fig. 78 Pull out the shift linkage

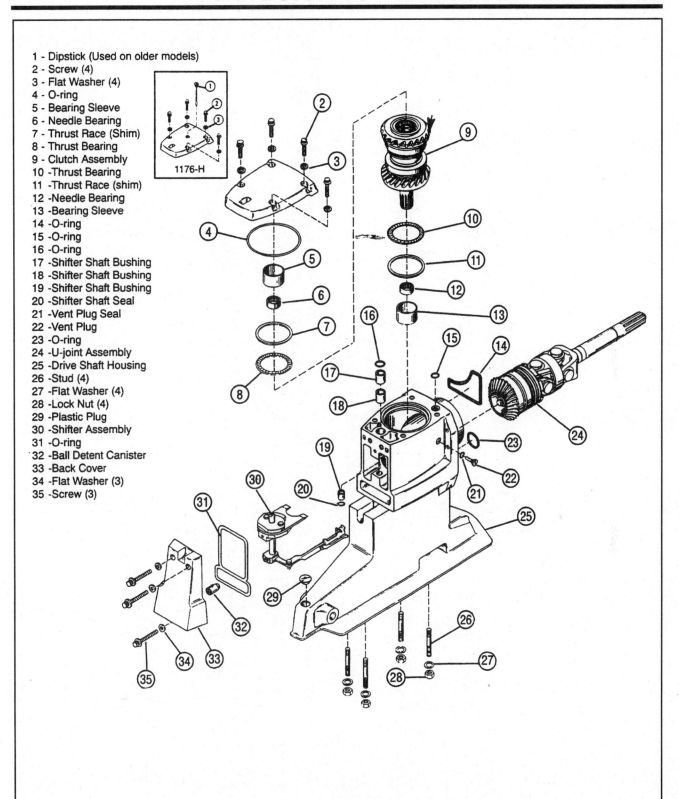

1 - Dipstick (Used on older models)
2 - Screw (4)
3 - Flat Washer (4)
4 - O-ring
5 - Bearing Sleeve
6 - Needle Bearing
7 - Thrust Race (Shim)
8 - Thrust Bearing
9 - Clutch Assembly
10 -Thrust Bearing
11 -Thrust Race (shim)
12 -Needle Bearing
13 -Bearing Sleeve
14 -O-ring
15 -O-ring
16 -O-ring
17 -Shifter Shaft Bushing
18 -Shifter Shaft Bushing
19 -Shifter Shaft Bushing
20 -Shifter Shaft Seal
21 -Vent Plug Seal
22 -Vent Plug
23 -O-ring
24 -U-joint Assembly
25 -Drive Shaft Housing
26 -Stud (4)
27 -Flat Washer (4)
28 -Lock Nut (4)
29 -Plastic Plug
30 -Shifter Assembly
31 -O-ring
32 -Ball Detent Canister
33 -Back Cover
34 -Flat Washer (3)
35 -Screw (3)

1176-H

Fig. 79 Exploded view of the upper unit— Bravo

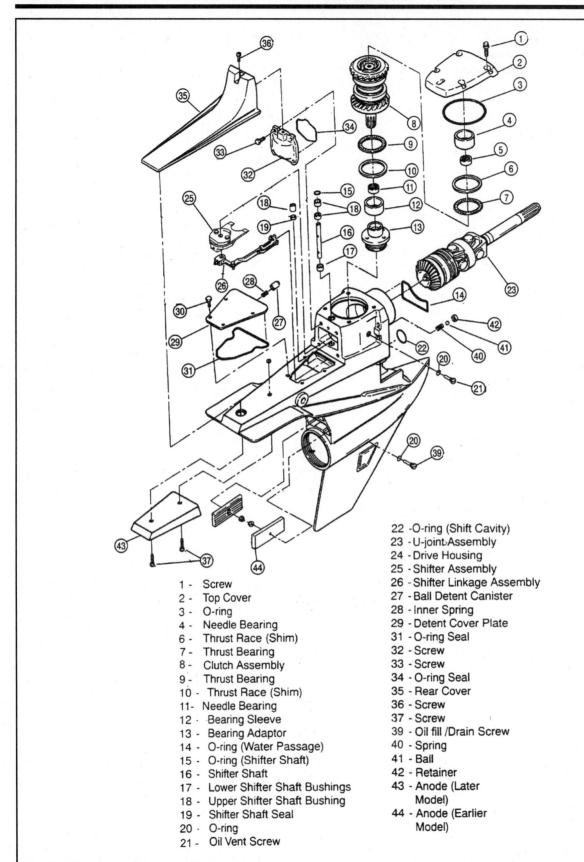

1 - Screw
2 - Top Cover
3 - O-ring
4 - Needle Bearing
6 - Thrust Race (Shim)
7 - Thrust Bearing
8 - Clutch Assembly
9 - Thrust Bearing
10 - Thrust Race (Shim)
11- Needle Bearing
12 - Bearing Sleeve
13 - Bearing Adaptor
14 - O-ring (Water Passage)
15 - O-ring (Shifter Shaft)
16 - Shifter Shaft
17 - Lower Shifter Shaft Bushings
18 - Upper Shifter Shaft Bushing
19 - Shifter Shaft Seal
20 - O-ring
21 - Oil Vent Screw

22 - O-ring (Shift Cavity)
23 - U-joint Assembly
24 - Drive Housing
25 - Shifter Assembly
26 - Shifter Linkage Assembly
27 - Ball Detent Canister
28 - Inner Spring
29 - Detent Cover Plate
31 - O-ring Seal
32 - Screw
33 - Screw
34 - O-ring Seal
35 - Rear Cover
36 - Screw
37 - Screw
39 - Oil fill /Drain Screw
40 - Spring
41 - Ball
42 - Retainer
43 - Anode (Later
      Model)
44 - Anode (Earlier
      Model)

Fig. 80 Exploded view of the upper unit— Blackhawk

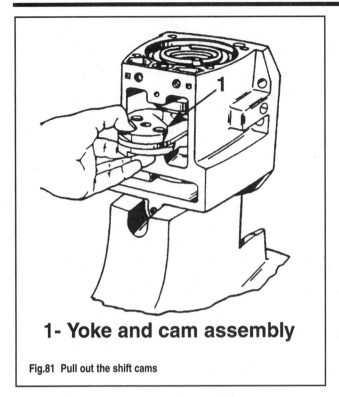

1- Yoke and cam assembly

Fig.81 Pull out the shift cams

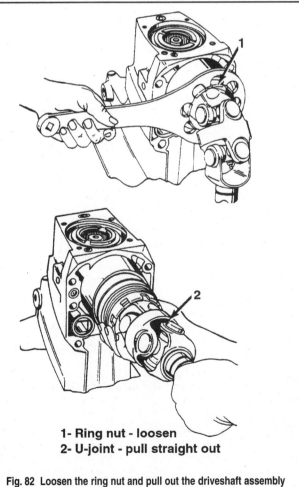

1- Ring nut - loosen
2- U-joint - pull straight out

Fig. 82 Loosen the ring nut and pull out the driveshaft assembly

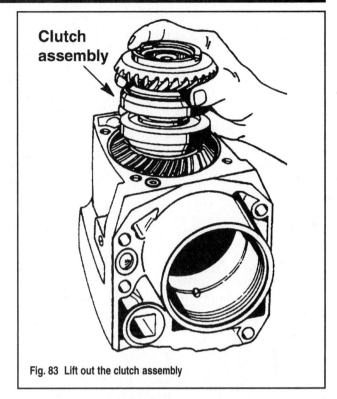

Clutch assembly

Fig. 83 Lift out the clutch assembly

marks on the rear of the housing and the clutch gears are still in alignment; if not, remove the shaft assembly and try it again.

18. Insert the shift cam assembly into the cavity in the rear of the housing so that the boss is on the bottom side, facing the bottom of the housing.

19. Slide the shifter linkage assembly into the cavity while twisting it side-to-side. Once in, rotate it 1/4 turn counterclockwise.

20. Reinstall the shift handle tool to the shift shaft and press the shaft down through the housing and shifter assemblies. Now take the shift handle tool and insert it through the shift linkage and into the shift shaft so you can swivel the shaft until the hole aligns with the one in the lower shift cam. Coat the 1st few threads of the cam cap screw with Loctite 271 and screw it into the cam. Tighten the screw to 100–120 inch lbs. (12–13 Nm).

21. Remove the tool and then install the linkage cap screw in the same way and to the same torque.

22. Move the linkage into the Neutral detent position.

23. Coat two new O-rings with 3-M adhesive and install them into the shift linkage and water passages in the front of the housing.

24. Coat two new O-rings with 3-M adhesive and install them into the shift shaft and oil passages in the top of the housing.

25. On the Blackhawk, install shift assembly cover and detent plates. Tighten the screws to 13–15 inch lbs. (1.5–1.7 Nm).

26. Coat a new rear cover O-ring with adhesive and install it into the groove in the cover. Do the same with the top cover.

27. Install the top cover and tighten the bolts, in a criss-cross pattern, to 18–22 ft. lbs. (25–29 Nm). Repeat this procedure for the rear cover; but on the Blackhawk tighten the screws to 13–15 inch lbs. (1.5–1.7 Nm).

28. Connect the upper and lower units (except Blackhawk) and install the stern drive.

U-Joint/Drive (Pinion) Gear Assembly

◆ See Figures 88, 89, 90 and 92

1. With the ring nut tool still attached to the retainer, mount the assembly in a vise using the tool to clamp down on.

2. Remove the nut and washer from the end of the pinion gear assembly and then carefully slide the assembly off of the inner yoke. Remove the sealing ring, O-ring, beveled washer, oil seal and carrier and the ring nut.

3. If gear is in good shape but the bearings are in need of replacement, both bearings must be replaced as a set.

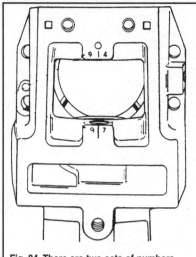
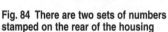

Fig. 84 There are two sets of numbers stamped on the rear of the housing

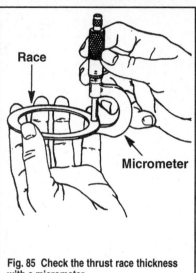

Fig. 85 Check the thrust race thickness with a micrometer

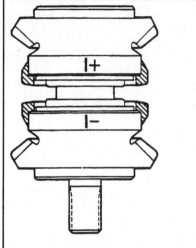

Fig. 86 The plus (+) sign and minus (-) signs must always be opposite each other

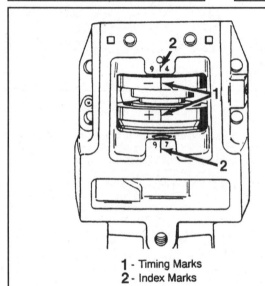

1 - Timing Marks
2 - Index Marks

Fig. 87 All four index marks must line up

Obtain a Universal Bearing Puller Plate (#91-37241) or equivalent tool. Place the tool under the tapered bearing and tighten the plate jaws under the tapered bearing. Set the assembly into an arbor press and press the tapered bearing off the r. Lift off the spacer and then the bearing cup. Repeat this step and press the remaining tapered roller bearing off the gear.

4. Clean the assembly in solvent and allow it to dry thoroughly. Inspect the drive (pinion) gear and the driven gear for broken teeth, pitting, etc. Make sure all teeth are worn evenly

5. Slide the three small O-rings off the end of the coupling yoke.

6. Inspect the large O-ring around the oil seal carrier and replace it if its worn. Inspect the joint retainer for cracks or other damage, replacing if anything is found. Do the same with the thrust ring.

7. Inspect the oil seal and carrier for defects or damage. If there is a problem with the carrier it must be replaced with the seal as a unit. If the seal is bad, press it out with a punch and hammer. Use a seal driver tool (#91-36577) to press a new seal back into the retainer.

8. Inspect the splines on both yokes for wear or cracking, replacing anything if a problem is found.

9. Inspect the joints themselves for wear, knocking or too much side play. If problems are found, separate the joints.

10. Drive the eight C-rings off the U-joint bearings with a punch and hammer

11. Obtain a suitable adaptor to support the U-joint yoke. Press one bearing until the opposite bearing is pressed into the adaptor. Remove the two loose bearings. Rotate the U-joint assembly 180°. Use the adaptor and press on the bearing cross-member and remove the bearings. Remove the cross-member from the yoke. Press the other four bearings out in the same manner, as described in the previous step and in the first part of this step.

To assemble:

12. Lubricate the U-joint bearing cups with a liberal amount of Quicksilver 2-4-C lubricant. Place the lubricated cups in the yoke and start them on the bearing cross-members. Press the bearings through the yoke onto the cross-members, as shown.

13. Drive the bearing cup retaining C-rings into place with a hammer and punch. Lubricate and install the other sets of bearings in the same manner.

14. Lubricate all bearings, cups, and gears with Quicksilver High Performance Gear lube or an equivalent, before assembling. Set the drive gear into an arbor press with the gear facing down. Place the tapered roller bearing onto the drive gear with the tapered side facing up. Using driver tool P/N 91-90774, press the bearing down onto the gear until the aft side of the bearing makes contact with the gear.

15. Place the bearing cup over the tapered roller bearing followed by the large spacer.

16. Set the second tapered bearing cup onto the large spacer ring with the tapered side facing up.

17. Place the tapered roller bearing onto the shaft. Obtain a suitable size mandrel and slowly press the tapered roller bearing down into the bearing cup until the roller barely makes contact with the bearing cup—then stop. The bearing will be drawn closer into the cup when the bearing preload is set.

18. Install the ring nut with the threaded end facing the end of the short yoke.

19. Press a new oil seal into the oil seal carrier, with the lip of the seal facing the concave side of the carrier, until the seal is flush with the carrier. Coat the lip of the seal with Multi-Purpose lubricant, or equivalent. Install the oil seal carrier, with the lip of the oil seal facing down, toward the drive gear. The seal prevents lubricant from working its way out of the upper housing and onto the U-joints.

20. Slide the gear/bearing assembly on the shaft and install the washer and nut. Tighten the nut until the washer just comes into contact with the gear.

21. Temporarily install the assembly into the housing (without the clutch assembly) and tighten the ring nut finger-tight.

22. Swivel the housing in the holding fixture until the shaft assembly is hanging straight down. Reach in through the shift cavity and tighten the pinion nut 1/16 of a turn at a time until the pre-load is 6–10 inch lbs. (0.7–1.1 Nm). If the nut becomes too tight before reaching the proper pre-load figure, loosen it a few turns and then give it a few light hits with a bar and hammer. Now attempt to set the pre-load again.

23. Remove the assembly from the housing until final assem

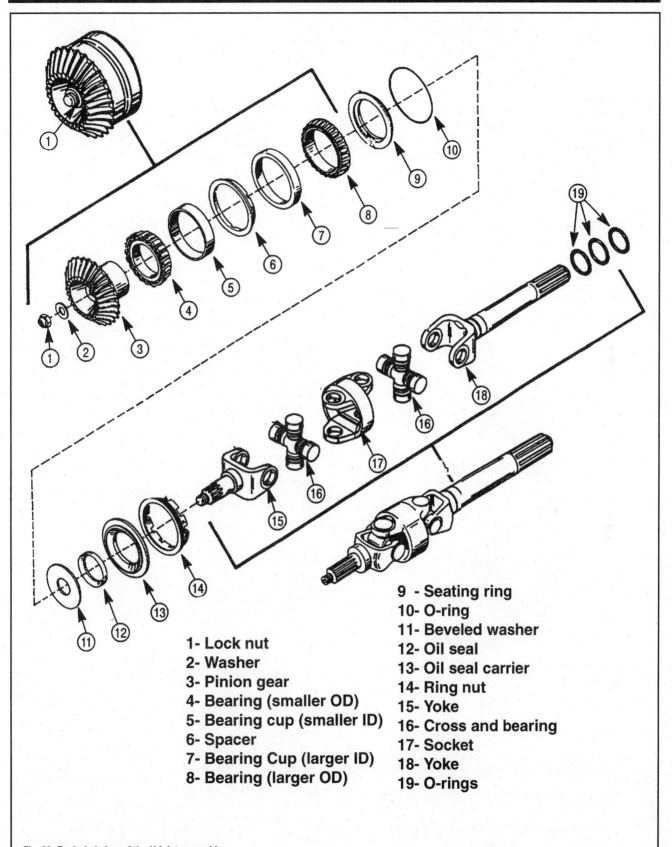

Fig. 88 Exploded view of the U-joint assembly

1- Lock nut
2- Washer
3- Pinion gear
4- Bearing (smaller OD)
5- Bearing cup (smaller ID)
6- Spacer
7- Bearing Cup (larger ID)
8- Bearing (larger OD)

9 - Seating ring
10- O-ring
11- Beveled washer
12- Oil seal
13- Oil seal carrier
14- Ring nut
15- Yoke
16- Cross and bearing
17- Socket
18- Yoke
19- O-rings

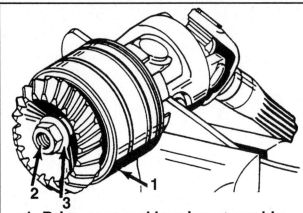

1- Drive gear and bearing assembly
2- Nut
3- Washer

Fig. 89 After clamping the assembly in a vise, remove the nut and then slide off the drive gear/bearing assembly

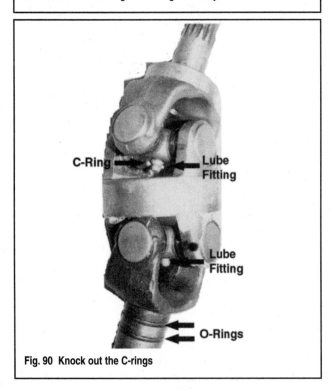

Fig. 90 Knock out the C-rings

Clutch Assembly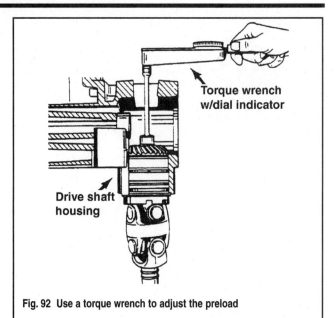

◆ See Figures 93, 94, 95, 96 and 97

1. Obtain holding fixture (#91-17301A1). Slide the clutch shaft splined end into the holding fixture, or clamp the splined end in a vise equipped with soft jaws.
2. Push down on the upper gear and upper thrust collar until the collar just clears the two keepers. Remove the two keepers from the groove in the clutch shaft.
3. Slide the thrust collar and gear free of the shaft.

Fig. 92 Use a torque wrench to adjust the preload

■ If either the gear or the internal needle bearing is unfit for further service, both are replaced as a set. Both gears, as removed, are not servicable, even though it appears a circlip holds the needle bearing within the gear. The gear and the needle bearing are purchased together under the same part number. Therefore, separating them would be a pointless task.

4. Lift out the upper thrust bearing, thrust washer and garter spring. Stack these items closely together if they are to be reused after inspection. Identify the stack as the upper set.
5. Rotate the clutch spool counterclockwise, up over the spiral worm gear on the clutch shaft and remove it.
6. Remove the lower garter spring, thrust washer and thrust bearing. Stack these items closely together, if they are to be reused after inspection. Identify the stack as the lower set.
7. Lift the lower gear from the clutch shaft.
To assemble:
8. Apply a good grade of gear oil to the splines. Slide the lower gear assembly down over the splines with the brass collar facing upward. The gear assembly will rotate clockwise as it follows the curved splines going down the shaft and will seat against the lower collar.
9. Slide the lower thrust bearing down over the shaft, followed by the lower thrust washer, with the washer lip facing downward. Stretch the lower garter spring over the thrust washer. The garter spring will not lie flat. However if it does, the spring has lost too much tension and needs to be replaced.
10. Lower the clutch spool down onto the shaft. The spool will rotate clockwise as it follows the curved splines downward to rest against the lower garter spring.
11. Slide the upper garter spring, upper thrust washer, with the washer lip facing upwards and finally the upper thrust bearing over the shaft. At this stage, these items will not appear to be seated. Fear not, for this is a normal condition.
12. Position the upper gear over the shaft with the brass collar facing downward. Install the upper thrust collar, with the taper (smaller diameter), facing upward.
13. Snap the gear down sharply to spread and seat both garter springs around each end of the clutch spool. This action will expose the upper groove in the shaft. Hold the gear down and insert the two keepers into the groove in the shaft. Release pressure on the gear allowing the upper thrust collar and gear to snap up and retain the split ring in the groove.

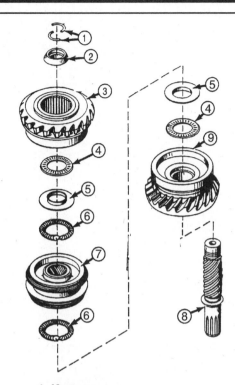

1- Keepers
2- Thrust collar
3- Upper gear assembly
4- Thrust Bearing
5- Thrust race
6- Garter spring
7- Clutch
8- Later style shaft (bottom thrust collar is partof the shaft, no keepers used)
9- Lower gear assembly

Fig. 93 Exploded view of the clutch assembly

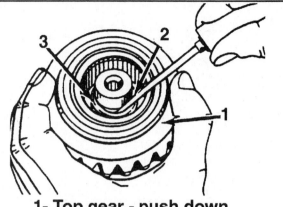

1- Top gear - push down
2- Thrust collar - push down
3- Keepers (2) - Remove

Fig. 94 Press the gear down and remove the keepers...

Top Cover Bearing And Sleeve

◆ See Figures 98, 99, 100, 101, 102 and 103

1. Remove the top cover and place it upside down on a clean surface.
2. Position puller jaws (#91-90777A1 & 91-90778) around the sleeve.
3. Position the puller guide (#91-90774) over the jaws and install the bolt supplied with the tool.
4. Install a driver guide (#92-90244) over the puller and then turn the bolt clockwise to pull off the sleeve.
5. Clamp the cover in a vise (carefully!) and pull out the bearing with a slide hammer.
To install:
6. Install a driver head (#91-90773) onto the puller guide and then position the new sleeve against the edge of the driver.
7. Move the sleeve and puller into place on the top cover and slide on the guide. Tap the bolt lightly with a mallet until the tool bottoms out.
8. Now install the bearing onto the tool and repeat the previous step.
9. Remove the tools, install a new O-ring and install the cover. Tighten the four bolts to 20 ft. lbs. (27 Nm).

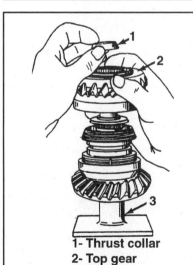

1- Thrust collar
2- Top gear
3- Stand

Fig. 95 ...and then lift off the collar and gear

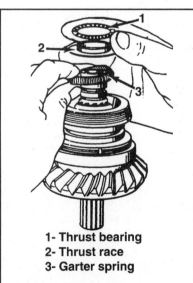

1- Thrust bearing
2- Thrust race
3- Garter spring

Fig. 96 Remove the bearing, race and spring...

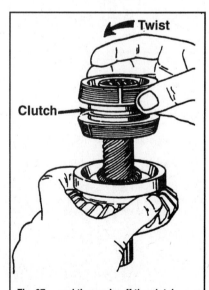

Fig. 97 ...and then spin off the clutch

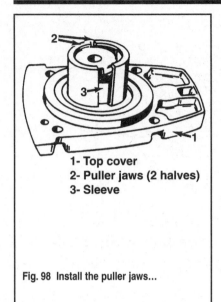

1- Top cover
2- Puller jaws (2 halves)
3- Sleeve

Fig. 98 Install the puller jaws...

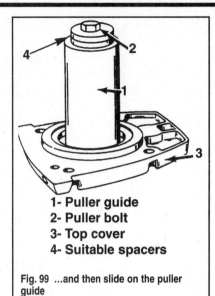

1- Puller guide
2- Puller bolt
3- Top cover
4- Suitable spacers

Fig. 99 ...and then slide on the puller guide

Fig. 100 Slide the driver guide over the puller and turn the bolt

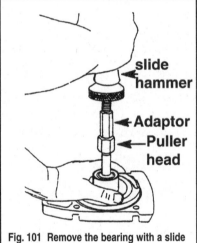

Fig. 101 Remove the bearing with a slide hammer

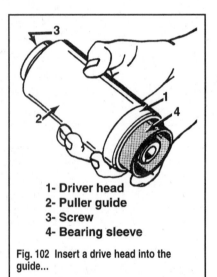

1- Driver head
2- Puller guide
3- Screw
4- Bearing sleeve

Fig. 102 Insert a drive head into the guide...

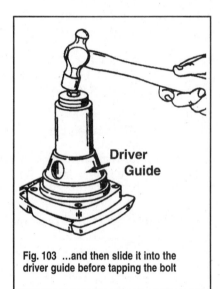

Fig. 103 ...and then slide it into the driver guide before tapping the bolt

Drive Housing Bearing And Sleeve

◆ See Figures 104, 105, 106, 107 and 108

1. Position puller jaws (#91-90777A1 & 91-90778) around the sleeve in the housing bore.
2. Position the puller guide (#91-90774) over the jaws and install the bolt supplied with the tool.
3. Install a driver guide (#92-90244) over the puller and then turn the

bolt clockwise to pull off the sleeve.
4. Position a suitable mandrel over the bearing in the clutch bore. Place a long bar over the mandrel and drive the bearing down into the oil cavity
To install:
5. Install a driver head (#91-90773) onto the puller guide and then position the new sleeve against the edge of the driver.
6. Move the sleeve and puller into place in the housing bore and slide on the guide. Tap the bolt lightly with a mallet until the tool bottoms out.
7. Now install the bearing onto the tool and repeat the previous step.
8. Remove the tools.

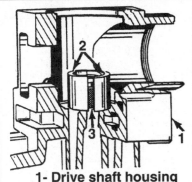

1- Drive shaft housing
2- Pull jaws (2 halves)
3- Sleeve

Fig. 104 Install the puller jaws...

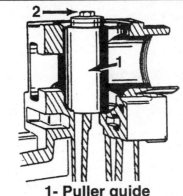

1- Puller guide
2- Puller bolt

Fig. 105 ...and then slide on the puller guide

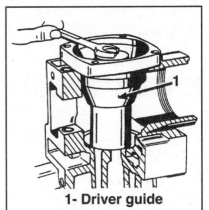

1- Driver guide

Fig. 106 Slide the driver guide over the puller and turn the bolt

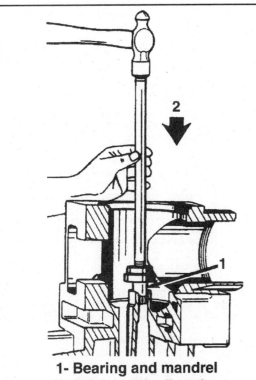

1- Bearing and mandrel
2- Drive in this direction

Fig. 107 Drive the bearing into the oil cavity

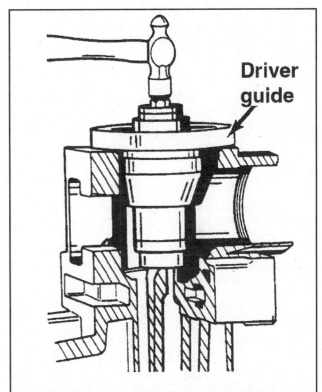

Fig. 108 Slide puller and head into the driver guide before tapping the bolt

TRANSOM ASSEMBLY

Description

The stern drive unit, consisting of the upper gear housing and the lower unit, is attached to the bell housing. The bell housing is secured to the gimbal ring by two roller bearings. These bearings permit up-and-down trim movement. The gimbal ring is mounted to the gimbal housing by two roller bearings. This second set of roller bearings permits steering movement to port and starboard. The gimbal housing is mounted on the transom of the boat and attached to an inner transom plate. This plate contains the rear engine mounts and attachments for the steering and shift cables.

The gimbal housing is mounted on the transom of the boat and attached to an inner transom plate. This plate contains the rear engine mounts and attachments for the steering and shift cables. The bell housing, gimbal ring, gimbal housing and inner plate all make up the transom assembly.

The bell housing has extended flanges to connect the exhaust, universal and shift cable bellows, and water hoses between the stern drive unit and the gimbal housing.

The bell housing is held in place by the gimbal ring.

Gimbal housing/transom plate service can generally be performed without removing the gimbal housing. However, in those cases when a part of the housing has been broken, or if the inner transom plate must be removed, the gimbal housing must be removed before the transom plate.

Before the gimbal housing can be removed, several tasks involving considerable work and time must be performed in the following order:
Stern drive removed.
Engine removed.
Bell housing and gimbal ring removed.

Gimbal Bearing

REMOVAL

◆ **See Figures 109, 110, 111 and 112**

■ **The gimbal bearing and carrier are a matched set and must be replaced as an assembly; further, the tolerance ring must be replaced whenever the bearing is replaced.**

1. Remove the stern drive unit. After the stern drive unit has been removed, the gimbal housing bearing can be checked by reaching through the U-joint bellows attached to the bell housing and rotating the inner bearing race. There should be no evidence of binding or rough spots. Check the race for side play by pulling and pushing on the race. If there is any sign of roughness, binding, or excessive side play, the bearing should be replaced. A special puller is required to remove the gimbal housing bearing. This puller is designed to establish alignment from the face of the bell housing. Never remove the gimbal bearing unless it is to be replaced, because it will be damaged during removal.
2. Assemble the plates of special tool (#91-29310). Position the plates between the top and middle studs located on the bell housing. Use a 3-jaw puller from the Slide Hammer Puller Set (#91-34569A1). If the bearing assembly is tight, tap the end of the Puller Shaft (#91-31229), with a mallet while attempting to turn the nut (#11-24156).
3. Remove the tolerance ring from the carrier.
4. Reach into the housing and remove the grease seal.
To install:
5. Clean all metal parts in solvent and dry them with compressed air. Never spin ball bearings with compressed air, because such action will ruin the bearing.
6. Inspect the bellows carefully for cracks, cuts, and punctures. Verify the bellows are still flexible. If the condition of the bellows is doubtful, replace them. In most cases, if the gimbal bearing is damaged due to water, the water has entered through, or around, the bellows.
The gimbal bearing has been replaced by a new style bearing. These are easily identified because the old style bearing carrier is black (top), and the new style is silver (bottom). The notches must face forward when the bearing is installed.
7. The gimbal housing bearing can only be installed with the bell housing in place in order to establish an alignment reference, so if you have removed the housing install it now.
8. Lubricate the outside of a new gimbal bearing carrier assembly with Multi-Purpose Lubricant, or equivalent.
9. Install the tolerance ring over the carrier. The opening in the tolerance band must align with the lubrication opening in the bearing carrier before the carrier is installed. The opening must also align with the lubrication opening in the gimbal housing after the carrier is installed.

■ **Observe the notches (could be called "cutouts), in the bearing carrier, indicated in the accompanying illustration. These notches must face forward when the carrier is installed.**

10. Install the assembled bearing carrier using the tools indicated in the cutaway line drawing labeled for Step 2. Insert the Driver Head (#91-32325), through the Mandrel (#91-30366), and into the inside diameter of the new bearing. Align the bearing with the gimbal housing by positioning Plate (#91-29310) between the top and middle studs on the bell housing. Positioning the plate as described will ensure the driver rod will remain at right angles to the bearing carrier bore.
11. Now, drive the bearing into the gimbal housing cavity with a hammer. Lubricate the bearing using only Quicksilver Multi-Purpose Lubricant through the grease fitting. To lubricate the bearing properly, pump 40 full strokes, to deliver about one full ounce of lubricant.
12. Install the stern drive.

Shift Cable

REMOVAL & INSTALLATION

◆ **See Figures 113, 114 and 115**

1. Disconnect the stern drive unit shift cable from the shift plate mounted on the engine. This is accomplished by removing the nut, washer, and cotter pin.
2. Loosen the set screws from the cable end guide, and then remove the end guide. If only the inner cable is to be replaced, burrs made by the set screws must be removed from the core wire, to prevent the inner lining of the shift cable from being damaged when the wire is removed. Loosen the jam nut on the metal cable end, and then turn the metal end out of the cable.
3. Pull the threaded support tube from the end of the inner core wire.
4. If the inner wire is to replaced, hold the cable slide and pull the inner wire out of the shift cable.
5. Hold the shift cable nut on the transom side of the housing and remove the retaining nut on the other side.
6. Remove the protective wrapping from the shift cable in the area where the cable passes through the transom.
7. Obtain special shift cable removal and installation tool (#91-12037). Remove the shift cable bellows clamp at the small end of the bellows. Pull the shift cable through the bellows.
To install:
8. Inspect the shift bellows for cracks, cuts, and punctures. Verify the bellows are still flexible. If there is the least doubt about the condition of the bellows, they should be replaced. If the bellows are defective and leak, water will enter the boat.
9. Inspect the cable locking nut threads for any damage such as cross-threading and check for any indication of the locking nut separating from the outer casing. Check the length of the shift cable for kinks, cuts or chafing and the inner core wire for signs of unraveling.

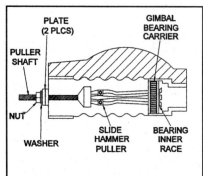

Fig. 109 Removing the bearing with the special tools

Fig. 110 It is important to install the tolerance ring correctly

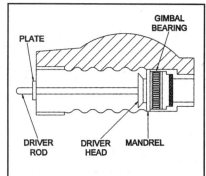

Fig. 111 Installing the bearing with special tools

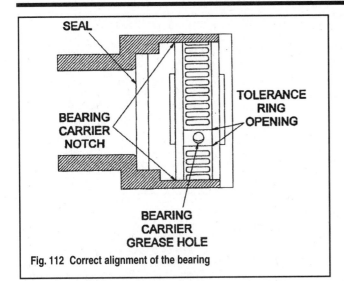

Fig. 112 Correct alignment of the bearing

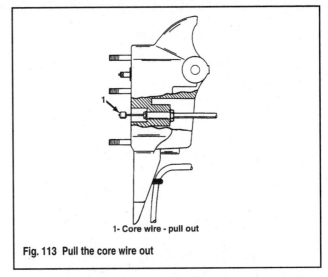

1- Core wire - pull out

Fig. 113 Pull the core wire out

■ The manufacturer strongly recommends that the small shift cable bellows clamp be the crimp type, not a worm type clamp and not a tie wrap. To crimp such a clamp, a special tool is required. However, a pair of common pliers may be easily and quickly modified to do the job. To customize a standard pair of pliers to crimp the bellows clamp, first tack weld a 3/4 in. nut to the pliers with the gripping surfaces of the pliers contacting two opposite sides of the nut. Clamp the pliers in a vice and drill out most of the threads using a 1/2 in. drill bit. With the pliers still in the vise, cut the nut in half, leaving an equal amount of the nut on each side of the pliers, as shown in the accompanying illustration.

10. Apply a coating of Perfect Seal to the locking nut threads. Secure the shift cable to the bell housing and tighten the nut securely using special cable tool. No more than two threads on the retainer should be visible.

11. Install the bellows clamp on the small end of the shift bellows with the end of the bellows 2 in. from the cable locking nut. Use the modified pliers, described in the previous Note, and squeeze the bellows clamp securely around the bellows. After the clamp is properly installed, water will not be able to enter the bellows and find its way into the boat.

12. Install the inner shift wire through the cable slide from the stern drive side. Install the Allen head screw securely, and then back it off about 1/8 turn. This adjustment will permit the inner wire to rotate freely. Install the safety wire, twist it tightly and cut off any excess.

13. Coat the inner shift wire with light weight oil, and then feed it into the shift cable until the cable slide enters the bell housing. Now, from inside the boat, carefully pull on the inner wire and be sure it is fully extended.

14. Install the protective wrapping around the cable.

15. Install the threaded tube on the shift cable end and tighten it until it just bottoms; now tighten the jam nut against the cable end.

16. Install a core wire locating tool (#91-17263).

17. Install the cable end guide over the core wire and then insert the core wire through the anchor. Tighten the screws securely.

18. Install the shift cable anchor adjustment tool (#91-17262) onto the end of the cable and make sure the end of the core ire is tight against the core wire locating tool.

19. Rotate the barrel on the cable threads until it lines up with the hole in the tool. Remove the tools and install the cable on the shift plate.

20. Adjust the cable.

ADJUSTMENT

◆ **See Figures 115a, 116, 117, 118 and 119**

■ The front propeller on Bravo III drives is always left hand rotation and the rear propeller is always right hand rotation, so the shift cable should always move as shown in the illustration. On Bravo I/II drives, if the cable moves in the direction of A when the control lever is moved to Forward, it is set up for RH propeller rotation. If it moves in the direction of B when the lever is moved to the Forward position, it is set up for LH propeller rotation.

1. Install the cable into the remote control.
2. Loosen the upper stud on the shift plate and move it within the slot

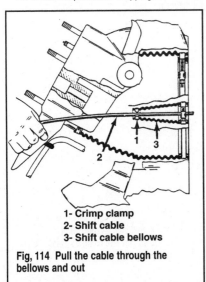

1- Crimp clamp
2- Shift cable
3- Shift cable bellows

Fig. 114 Pull the cable through the bellows and out

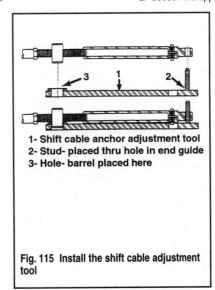

1- Shift cable anchor adjustment tool
2- Stud- placed thru hole in end guide
3- Hole- barrel placed here

Fig. 115 Install the shift cable adjustment tool

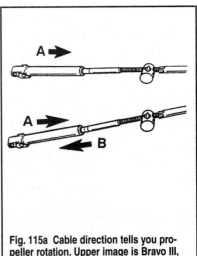

Fig. 115a Cable direction tells you propeller rotation. Upper image is Bravo III, lower image is Bravo I/II

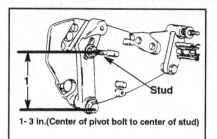

1- 3 in.(Center of pivot bolt to center of stud)

Fig. 116 Set the stud dimension...

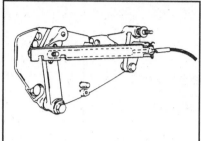

Fig. 117 ...and then install the tool

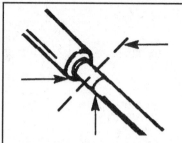

Fig. 118 Locate the center point of the cable backlash

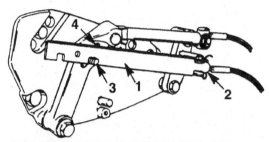

RH ROTATION BRAVO ONE AND TWO, AND ALL BRAVO THREE MODELS

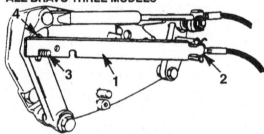

LH ROTATION BRAVO ONE AND TWO MODELS

1- Adjustment tool
2- Barrel retainer
3- Shift lever stud
4- Shift lever adjustment slot

Fig. 119 Fit the stud into the slot on the tool for final adjustment

until it is exactly 3 in. from the center of the pivot bolt and then install the shift cable. There should be a washer on each side of the guide, tighten the lock nut until it contacts the washer and then back it off 1 full turn.

3. Place the adjustment tool (#91-12427) over the shift cable as shown and then tape it in place over the barrel retainer.

4. Shift the remote control lever to Neutral, push in on the control cable end just enough to relieve the play and mark the position on the tube. Now pull out on the cable end until all play is removed and mark that position. Mark the center point of these two positions.

5. Install the control cable end guide into the shift lever and insert the anchor pin. Adjust the cable barrel so that the hole in the barrel is centered with the vertical centerline of the stud. Make sure that the center mark you made in the last step is in alignment with the edge of the cable end guide.

6. Remove the cable end guide from the shift lever by pulling out the pin and then install the barrel onto the stud. Tighten the barrel locknut until it bottoms out on the barrel. Now reinstall the other end into the shift lever.

7. Remove the adjustment tool.

8. Shift the control lever into the full Forward position—the rear slot in the tool should fit over the shift lever stud on Bravo I/II RH rotation models and all Bravo III models; on Bravo I/II LH rotation models it should be the forward slot. If the slot does not fit over the stud, loosen the stud and slide it up or down until it fits. Tighten the stud.

Exhaust Bellows

REMOVAL & INSTALLATION

◆ See Figures 121, 122, 123 and 124

■ If the stern drive has been removed, simply undo the bellows and replace them. The following procedure is with the drive still attached and may prove frustratingly difficult.

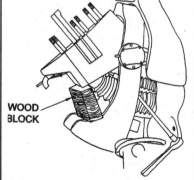

Fig. 121 With the stern drive removed from the boat, a 2 X 4 wood support under the bell housing will be helpful to disconnect the clamps on the exhaust bellows

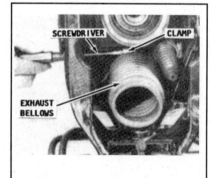

Fig. 122 An access hole in the housing provide a way to get your screwdriver on the clamp screw

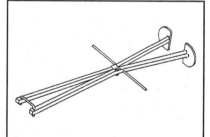

Fig. 123 A bellows expander is necessary to get the bellows back onto the flange

1. Tilt the stern drive slightly up and move it to the port side. Access to the aft clamp on the exhaust bellows is from the bottom of the bell housing. Remove the clamp. Move the stern drive to the center position. Access to the forward clamp on the exhaust bellows is through the access hole on the port side of the gimbal housing. Insert a screwdriver through the access hole and remove the forward exhaust bellows clamp.

2. Pull the exhaust bellows from the gimbal housing and bell housing exhaust flanges. It may be necessary to exert considerable force to pull the bellows loose because of the adhesive used during installation. Raising the stern drive to the full up position, will be helpful.

To install:

3. Clean the bellows mounting flanges of the upper bell housing with a wire brush or sandpaper, and then wipe the surface clean with lacquer thinner to remove any old glue.

4. If new bellows are being installed (and if you've removed them, you should install new one), be sure to clean the powder residue from the bellows surface with warm soapy water and dry thoroughly. The powdery substance is a mold release agent and the adhesive will not bond if this powder is left on the sealing surfaces.

5. Check the flanges and housing for cracks, nicks, or corrosion. Clean the bellows clamps thoroughly to ensure a good ground. New bellows have a clip on each end of the bellows to ground the clamp. Check the clamps for cracks or nicks. Replace the clamp if in doubt as to their condition. Always use stainless steel clamps for satisfactory service.

✳✳ WARNING

Bellows Adhesive must be used to ensure a satisfactory installation. This adhesive is extremely toxic and flammable. Therefore, make every effort to ensure adequate ventilation in the work area during its use. Work in the outdoors, if at all possible. Vapors from the adhesive may cause flash fire or ignite explosively.

Keep the adhesive away from heat, sparks, and open flame. Observe no smoking. Extinguish all flames and pilot lights in the area. Turn off stoves, heaters, electric motors, and all other possible sources of ignition while using the adhesive and until there is no doubt but what all vapors have left the area. Close the container immediately after use. The adhesive is harmful or fatal if swallowed. Avoid prolonged contact with the skin or breathing of the vapors. If swallowed, do not induce vomiting. Call a physician immediately. Keep the adhesive out-of-reach of children.

6. Apply a coating of Bellows Adhesive to the inside of each end of the bellow and around the mounting flanges. Allow the adhesive to dry approximately 10 minutes, or until the material is no longer tacky. Install the clamp on one end of the bellows, and then install it to the gimbal housing flange so that the clamp screw is accessible through the access hole. Tighten the clamp screw to 35 inch lbs. (4 Nm).

7. Now the hard part. Install the bellow clamp on the other end of the bellows.

■ A special bellows tool is almost a necessity to "pull" the bellows over the flange of the stern drive. Therefore, obtain expander tool (#91-45497A1) and place the tool into the first bellows convolution. Pull on the tool until the tool touches the flange on the bell housing—the hose starts to slip onto the flange—then release the tool.

The accompanying cross-section line drawing illustrates the position of the tool in the first convolution of the bellows.

8. Move the tool into the third bellows convolution and pull the bellows onto the bell housing flange. Tighten the hose clamp to a torque value of 35 in lbs (4Nm).

9. Start the engine and check for leaks.

U Joints Bellows

OEM ⑧ DIFFICULT

REMOVAL & INSTALLATION

◆ See Figures 125, 126, 127 asnd 128

■ Although not necessary, we recommend removing the stern drive. The bell housing does not have to be removed in order to replace the U-joint bellows. However, this is not an easy task. Therefore, work slowly and carefully. If the job proves too difficult, the decision may be made to remove the exhaust bellows and the water intake hose first.

1. Remove the single hose clamp securing the bellows to the gimbal housing using a long shank screwdriver. Work the bellows off the gimbal housing flange. Considerable force may be required to remove the bellows due to the adhesive used during installation.

2. There is no hose clamp securing the other end of the bellows to the bell housing. Instead, a sleeve fits tightly inside the bellows to hold it against the bell housing flange. The sleeve must be pried free of the bellows. Obtain a can of aerosol cleaner such as "Gunk" or WD-40. Spray the cleaner around the edge of the sleeve to help loosen it. Pry the sleeve free using a thin blade screwdriver. Push the bellows back from the bell housing flange until it is free. No adhesive is used at this end of the bellows, therefore, just push.

To install:

3. Remove all bellow adhesive from the inside diameter of the bellows, if the bellows are to be reused. Inspect the bellows for cracks, cuts, punctures and to be sure it is still flexible. If the least bit of doubt exists concerning the condition of the bellow, install new bellows.

4. Clean the bellow mounting flanges of the bell housing with a wire brush or sandpaper, and then wipe the surface clean with lacquer thinner.

5. Check the flanges and housing for cracks, nicks, or corrosion. Clean the bellow clamps thoroughly. Check the clamps for cracks or nicks. Replace the clamps if in doubt as to their condition. Always use stainless steel clamps as a replacement.

✳✳ WARNING

Bellows Adhesive must be used to ensure a satisfactory installation. This adhesive is extremely toxic and flammable. Therefore, make every effort to ensure adequate ventilation in the work area during its use. Work in the outdoors, if at all possible. Vapors from the adhesive may cause flash fire or ignite explosively.

Keep the adhesive away from heat, sparks, and open flame. Observe no smoking. Extinguish all flames and pilot lights in the area. Turn off

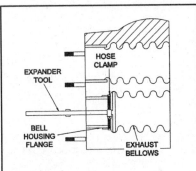

Fig. 124 Pulling the bellows onto the flange with the tool

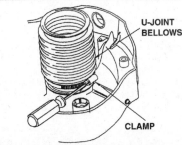

Fig. 125 The U-joint bellows clamp is placed in approximately the 3 o'clock position and tighten. A sleeve is wedged into the bell housing end as explained in the text

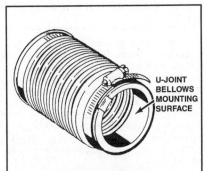

Fig. 126 Coat the gimbal housing end of the bellows with adhesive

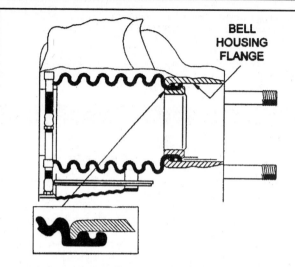

Fig. 127 Make sure that the bell housing flange rests in the second groove on the bellows

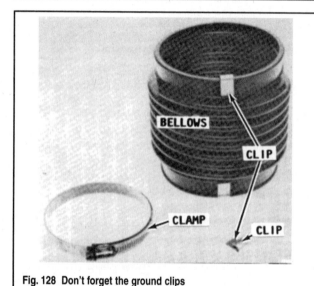

Fig. 128 Don't forget the ground clips

stoves, heaters, electric motors, and all other possible sources of ignition while using the adhesive and until there is no doubt but what all vapors have left the area. Close the container immediately after use. The adhesive is harmful or fatal if swallowed. Avoid prolonged contact with the skin or breathing of the vapors. If swallowed, do not induce vomiting. Call a physician immediately. Keep the adhesive out-of-reach of children.

6. Apply a coating of Bellows Adhesive to the inside diameter of the gimbal housing bellow end. Do not apply the adhesive to the bell housing end. Allow the adhesive to dry approximately 10 minutes, or until the material is no longer tacky. Install a grounding clip over the forward edge of the bellows with the shorter side of the clip on the inside of the bellows.

7. Find the word TOP embossed on the bellows. This word must face up after installation. Be sure to position the bead on the inner diameter of the bellows in the groove on the gimbal housing flange. Install the U-joint bellows over the flange and position the clamp tightening screw at the 3 o'clock position. Tighten the screw 35 inch lbs. (4 Nm). Don't forget the ground clip.

8. Apply Locquic Primer "T" to the internal threads of the bell housing. Apply the primer to the external hinge pin threads. Allow the primer at both locations to dry. Apply Loctite 271 to the threads of the bell housing, and then install the hinge pins. Tighten the hinge pins to a torque value of 95 ft. lbs (129 Nm).

9. Position the U-joint bellows on the bell housing. Ensure the bell housing flange is indexed in the second groove from the end of the bellows.

10. Obtain Sleeve Installation Tool (#91-818162). Lubricate the outside surface of the sleeve with soapy water or engine cleaner, and then install the sleeve tool. Use a suitable driving rod and move the sleeve in place on the bell housing.

11. If the water hose and the exhaust bellows were removed, install them to the gimbal housing.

12. Replace the stern drive.

13. Start the engine and check the completed work.

Bell Housing

OEM ③ DIFFICULT

REMOVAL & INSTALLATION

◆ **See Figures 129, 130, 131, 132 and 133**

1. Remove the stern drive as detailed previously.

2. Loosen the mounting screws, bend back the retainer clip and remove the trim position sender from the starboard side and the trim limit switch from the port side. Let them dangle temporarily.

3. Remove the shift cable.

4. Screw a tapered insert tool (#91-43579) into the water hose insert and turn out the insert (counterclockwise). Push the tube through the gimbal housing.

5. On models with a remote oil reservoir, remove the 90° hose barb fitting and the nut/washer from the thru-bulkhead fitting so that when you pull off the bell housing it will pull out of the gimbal housing.

6. Insert a sleeve removal tool (#91-818169) into the U-joint bellows hole, tighten the nut and remove the sleeve.

7. Remove the exhaust bellows.

8. Pop out the speedometer tubing clip at the bottom of the housing.

9. Attach a hinge pin tool (#91-78310) to a socket wrench. Swivel the housing to the port side and remove the starboard hinge pin. Now swivel it over to starboard and remove the port hinge pin. Carefully remove the housing.

10. If you need to replace the trim limit or position switches, loosen the screw and remove the retaining clamp. Disconnect the wire leads at the engine harness and pull them through the hole in the gimbal housing.

To install:

11. Install new fiber washers on the gimbal ring and secure them with Reiswald Sealer.

12. Bring the bell housing together with the gimbal housing. Push evenly on the perimeter of the bell housing until the U-joint bellows slides into the flange on the gimbal housing. Check to be sure the exhaust bellows aligns with the flange on the bell housing. Refer to the appropriate procedures in this section for pertinent details on final bellows connections.

13. Coat the hinge pin threads (inside the housing and on the bolts) with Loctite 271. Install and tighten the hinge pins to 140–150 ft. lbs. (190–203 Nm).

14. Install the speedometer hose clip.

15. Install the bellows to the gimbal housing.

16. Lubricate the end of the shift cable with 2-4-C lubricant and insert the shaft cable into the shift cable bellows.

17. Check final bellows installation against the individual procedures given earlier.

18. Install and adjust the shift cable.

19. Insert the water tube through the gimbal housing and is protruding through the bell housing by about 1/8 in. (3mm). Coat the threads of the tapered insert with 2-4-C lubricant and screw it back into the tube.

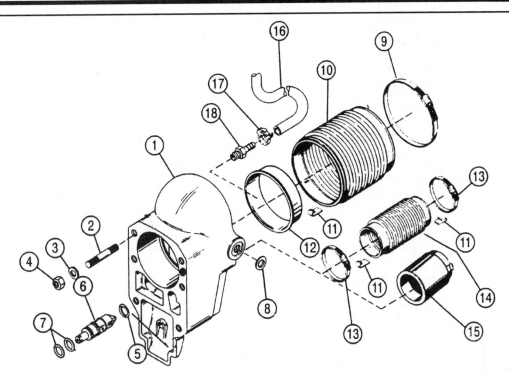

1 - Bell Housing
2 - Stud
3 - Washer
4 - Locknut
5 - O-ring
6 - Gear Lube Valve
7 - O-rings
8 - Hinge Pin Washer
9 - Bellows Clamp
10 - U-Joint Bellows
11 - Grounding Clip

12 - Sleeve
13 - Bellows Clamp
14 - Exhaust Bellows
15 - Exhaust Tube (Some Models)
16 - Lube Monitor Hose
17 - Hose Clamp
18 - Bayonet Fitting

Fig. 129 Exploded view of the bell housing

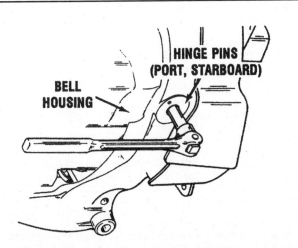

Fig. 130 The hinge pins are removed and installed with a special tool. Without this tool, removal of the hinge pins will almost be an impossible task

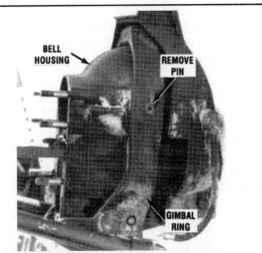

Fig. 131 Swivel the housing to one side in order to remove the opposite side's hinge pin

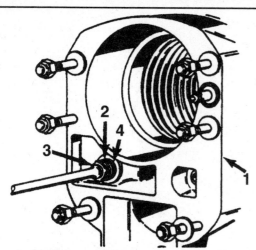

1- Bell housing
2- Tapered insert -turn clockwise
3- Tapered insert tool
4- Water hose

Fig. 132 Use the special tool to remove the tapered insert from the water tube

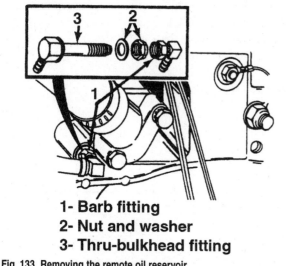

1- Barb fitting
2- Nut and washer
3- Thru-bulkhead fitting

Fig. 133 Removing the remote oil reservoir

20. If you disconnected the trim switches, route the trim limit switch and trim position sender leads, as indicated in the accompanying illustration. Install a sta-strap around both sets of leads approximately 5" (12cm) from the retaining cover.
21. Install the trim switches on either side of the gimbal ring
22. Install the stern drive.

Gimbal Ring

REMOVAL & INSTALLATION ③ DIFFICULT

This procedure can be completed with or without the engine and transom plate being removed.

Engine/Transom Assembly Installed

◆ See Figures 134, 135, 136, 137, 138, 139 and 140

■ Unless a previous owner has already drilled and access hole in the gimbal housing, you will need to drill (and plug) your own hole in order to remove the steering lever.

1. Remove the stern drive and the bell housing.
2. Pump a liberal amount of grease into the fitting at the top of the gimbal housing.
3. You'll need Access Plug Kit (#22-88847A-1) from your dealer. Use the template provided and position it on the housing using the dimple in the front of the housing—it will be under the decal.
4. Use a center punch and the template to mark the location of the holes on each side of the gimbal housing and then drill a 1/4 in. (6mm) hole through each side of the housing.
5. Using a 1-1/8 in. hole saw with a 1/4 in. (6mm) pilot rod, drill a hole in each side using the previous holes as a guide. Make sure the drill is perpendicular .to the housing and do not rush this.
6. Use a piece of tape and mark a 1 in. #180 pipe tap (hardware store) about 1-1/8 in. from the end. Coat it with grease and then thread it into the hole.
7. Insert a socket wrench into the port access hole and an open end into the starboard hole. Loosen and remove the steering lever clamping bolt. Now insert a punch through the starboard hole and unthread the elastic locknut. You may have to move the steering wheel a few time to jockey the lever and nuts into position.
8. Disconnect the trim cylinders at the gimbal ring and carefully set them aside—they should already have been disconnected on the other end when you removed the stern drive.
9. Disconnect the continuity wire and then remove the lower swivel cotter pin. Drive out the lower swivel pin and remove the anti-gauling washer.
10. Loosen the two upper gimbal ring bolts and then remove the upper swivel shaft and large washer. You may need a slide hammer if it is frozen. Lift out the gimbal ring. Look for the steering lever and its hardware and remove them along with the ring.
11. Remove the gimbal ring lower shaft bushing, using a suitable driver. Remove the gimbal ring upper swivel shaft bushing and single oil seal, using a slide hammer with a two jaw puller attachment.
12. Remove the gimbal ring hinge pins, using a Hinge Pin tool (#91-78310) to remove the pins, if they are threaded into the housing.
13. Obtain a suitable driver and drive out both hinge pin bushings.
To install:
14. Clean all metal parts in solvent and blow them dry with compressed air. Never, spin ball bearings with compressed air, because such action will ruin the bearings.
15. Remove all bellow adhesive from the inside diameter of the bellows, if the bellows are to be reused. Inspect the bellows for cracks, cuts, punctures and to be sure they are still flexible. If there is the least bit of doubt concerning their condition, do not hesitate to install new bellows. Clean the bellow mounting flanges of the bell housing with a wire brush or sandpaper and then wipe the surface clean with lacquer thinner.
16. Check the shift cables for cuts or damage caused by a cable being pinched or bent too short.
17. Inspect the water hose for cracks, cuts, punctures, or worn spots. Inspect the shift shaft oil seal for tears, wear, or any roughness. Check the shift shaft and shift shaft bushings for wear.
18. Inspect the O-ring and rubber gasket for cuts, nicks, hardness, or cracks.
19. Inspect the surface of the lower swivel pin in the area where the needle bearing rides. Any pitting, grooves, or uneven wear is cause to replace the bearing and swivel pin. Inspect the lip surface of the oil seals for wear, tears, and roughness.
20. Obtain an ohmmeter and test the trim limit switch. Set the switches of the ohmmeter for continuity reading and connect the leads of the ohmmeter to the leads of the trim limit switch. Rotate the outer case while holding the inner rotor and observe the ohmmeter. The meter should indicate continuity through most of the movement and then no continuity. If the meter indicates

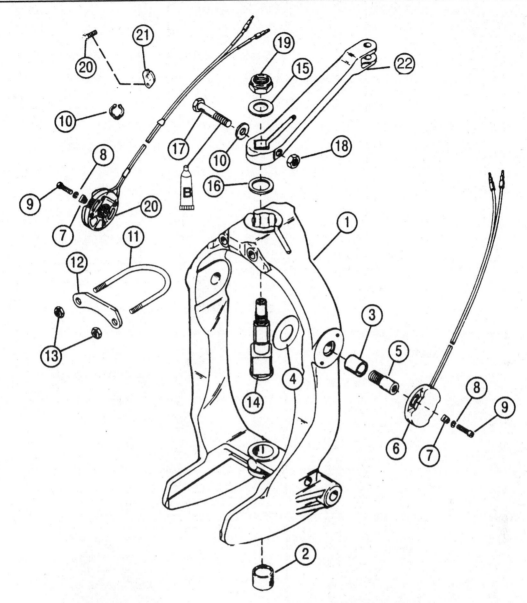

1 - Gimbal Ring
2 - Bushing
3 - Bushing
4 - Flat Washer
5 - Hinge Pin
6 - Trim Position Sender
7 - Clip
8 - Lockwasher
9 - Screw
10 - Clip
11 - U-Bolt
12 - Plate
13 - Locknuts
14 - Swivel Shaft
15 - Flat Washer (Smaller I.D.)

16 - Flat Washer (Larger I.D.)
17 - Clamp Screw
18 - Locknut
19 - Nut
20 - Screw
21 - Clamp Plate
22 - Steering Lever

Fig. 134 Exploded view of the gimbal ring

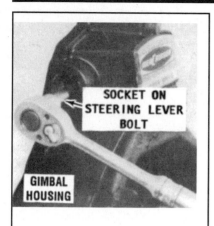

Fig. 135 Use a socket on the port side when removing the steering lever bolt

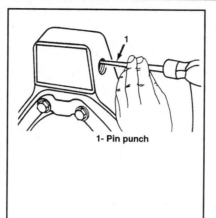

1- Pin punch

Fig. 136 Use a punch to drive off the elastic nut

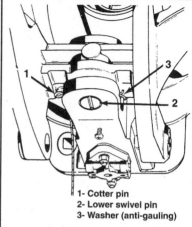

1- Cotter pin
2- Lower swivel pin
3- Washer (anti-gauling)

Fig. 137 Removing the lower swivel pin

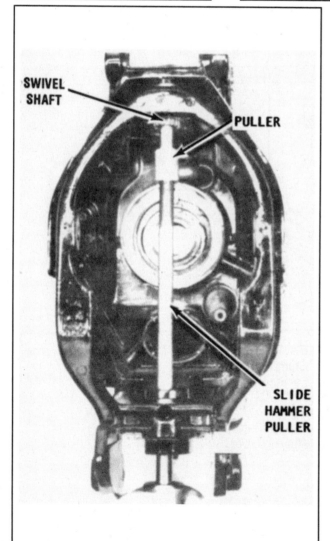

Fig. 138 You may need a slide hammer to remove the upper swivel shaft

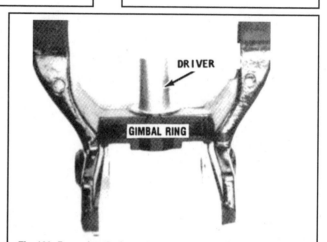

Fig. 139 Removing the lower bushing...

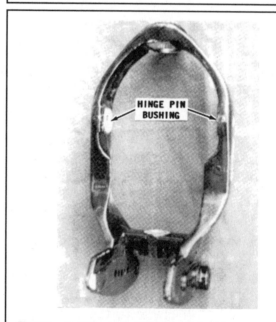

Fig. 140 ...and then remove the hinge pin bushings

continuity all the time or no continuity, the switch is defective and must be replaced.

21. Inspect the long steering lever retaining bolt. Any grooves found on the bolt are a result of friction against the upper swivel shaft. If grooves are discovered, both the steering lever and the bolt must be replaced.

22. Obtain Resiweld Sealer. Apply a coating of this sealer to the outer diameter of the lower bushing. Using a suitable driver, tap the bushing into place.

23. To install the small upper swivel shaft bushing, first obtain a Bearing and Seal Driver tool (#91-43578). Next, place the bushing on the tool, and then use a hammer and tap the bushing into place. Install the larger bushing in a similar manner.

24. Pack the single oil seal lip with Multi-Purpose lubricant and place the seal on the driver with the lip facing the smaller diameter of the tool. Install the oil seal into the gimbal ring until the seal seats against the larger bushing.

25. Check the condition of the synthane washers on the gimbal ring. If they are worn, or damaged, remove and discard the washers. Clean the washer mounting surface free of all adhesives, dirt, grease and oil. Peel the backing from a new washer and place it on the mounting surface, and then press the washer down firmly. The adhesive backing will hold the washer in place.

26. Apply a coating of Resiweld to the outer diameter of both hinge pin bushings. Using a suitable driver, tap the bushings into place on the port and starboard sides of the gimbal ring.

27. Apply a coat of 2-4-C Lubricant onto the surface of the lower swivel pin. Insert the switch harness, if equipped, through the gimbal housing, and then secure it in place with the screws. Place the gimbal ring in position in the gimbal housing. Place a washer between the gimbal ring and the housing. Secure the ring in place with the lower swivel pin and cotter pin.

28. Install the long bolt and nut into the steering lever. A washer is not used at this location. Tighten the nut until it is just "snug". If a new swivel shaft elastic nut is to be used, thread the nut onto the swivel shaft until it is tight, and then remove it. This action cuts the threads into the new nut (which should come with the access plug kit).

29. Check to be sure the gimbal ring rests squarely within the gimbal housing. Before installing the swivel shaft, check for a flat area machined onto the shaft splines. If a flat exists, the flat must face forward after installation. Place the larger washer on top of the ring, followed by the steering lever, then the smaller washer and finally, the large nut. Hold these components aligned together while the upper swivel shaft is passed through the gimbal ring and these parts. Start the large nut onto the shaft threads.

30. Tighten the bolt on the steering lever to 60 ft. lbs. (81 Nm). Use a punch through one of the side access holes to tighten the nut, until a clearance of 0.002–0.010 in. (0.05–0.25mm) exists between the lower swivel pin washer and the gimbal housing.

31. Use a rawhide mallet and strike the top of the gimbal ring flanges a couple times each to seat both pins. Recheck the clearance and adjust the large nut (using the punch and hammer method), as necessary to maintain the required clearance.

32. Tighten the two gimbal ring bolts to 55 ft. lbs. (74 Nm). Apply a coating of Perfect Seal to the threads or the sealing surface of the plugs and close the two large openings drilled earlier on the sides of the gimbal housing for access to the steering lever.

33. Install the trim cylinders by first attaching them in place with the forward anchor pin. Next, install the two hydraulic hoses to the connector. Finally, install the connector to the gimbal housing.

34. Connect the steering link rod to the steering lever. Tighten the castellated nut to 24 inch lbs. (3 Nm), and then back it off slightly and insert the cotter pin. If the assembly is secured with a Nylok nut instead of the cotter pin, tighten the nut to 5 ft. lbs. (7 Nm).

35. Install the bell housing and stern drive.

Engine/Transom Assembly Removed

◆ See Figures 134, 141 and 142

1. Remove the stern drive and the bell housing.
2. Loosen the two upper gimbal ring screws.
3. Reach in through the transom and loosen the steering lever clamp bolt.

4. Reach in through the transom and remove the elastic locknut on top of the steering lever with a 1-1/16 in. wrench.

5. Disconnect the trim cylinders at the gimbal ring and carefully set them aside—they should already have been disconnected on the other end when you removed the stern drive.

6. Disconnect the continuity wire and then remove the lower swivel cotter pin. Drive out the lower swivel pin and remove the anti-gauling washer.

7. Remove the upper swivel shaft and large washer. You may need a slide hammer if it is frozen. Lift out the gimbal ring. Look for the steering lever and its hardware and remove them along with the ring.

8. Remove the gimbal ring lower shaft bushing, using a suitable driver. Remove the gimbal ring upper swivel shaft bushing and single oil seal, using a slide hammer with a two jaw puller attachment.

9. Remove the gimbal ring hinge pins, using a Hinge Pin tool (#91-78310) to remove the pins, if they are threaded into the housing.

10. Obtain a suitable driver and drive out both hinge pin bushings.

To install:

11. Clean all metal parts in solvent and blow them dry with compressed air. Never, spin ball bearings with compressed air, because such action will ruin the bearings.

12. Remove all bellow adhesive from the inside diameter of the bellows, if the bellows are to be reused. Inspect the bellows for cracks, cuts, punctures and to be sure they are still flexible. If there is the least bit of doubt concerning their condition, do not hesitate to install new bellows. Clean the bellow mounting flanges of the bell housing with a wire brush or sandpaper and then wipe the surface clean with lacquer thinner.

13. Check the shift cables for cuts or damage caused by a cable being pinched or bent too short.

14. Inspect the water hose for cracks, cuts, punctures, or worn spots. Inspect the shift shaft oil seal for tears, wear, or any roughness. Check the shift shaft and shift shaft bushings for wear.

15. Inspect the O-ring and rubber gasket for cuts, nicks, hardness, or cracks.

16. Inspect the surface of the lower swivel pin in the area where the needle bearing rides. Any pitting, grooves, or uneven wear is cause to replace the bearing and swivel pin. Inspect the lip surface of the oil seals for wear, tears, and roughness.

17. Obtain an ohmmeter and test the trim limit switch. Set the switches of the ohmmeter for continuity reading and connect the leads of the ohmmeter to the leads of the trim limit switch. Rotate the outer case while holding the inner rotor and observe the ohmmeter. The meter should indicate continuity through most of the movement and then no continuity. If the meter indicates continuity all the time or no continuity, the switch is defective and must be replaced.

18. Inspect the long steering lever retaining bolt. Any grooves found on the bolt are a result of friction against the upper swivel shaft. If grooves are discovered, both the steering lever and the bolt must be replaced.

19. Obtain Resiweld Sealer. Apply a coating of this sealer to the outer diameter of the lower bushing. Using a suitable driver, tap the bushing into place.

20. To install the small upper swivel shaft bushing, first obtain a Bearing and Seal Driver tool (#91-43578). Next, place the bushing on the tool, and then use a hammer and tap the bushing into place. Install the larger bushing in a similar manner.

21. Pack the single oil seal lip with Multi-Purpose lubricant and place the seal on the driver with the lip facing the smaller diameter of the tool. Install the oil seal into the gimbal ring until the seal seats against the larger bushing.

22. Check the condition of the synthane washers on the gimbal ring. If they are worn, or damaged, remove and discard the washers. Clean the washer mounting surface free of all adhesives, dirt, grease and oil. Peel the backing from a new washer and place it on the mounting surface, and then press the washer down firmly. The adhesive backing will hold the washer in place.

23. Apply a coating of Resiweld to the outer diameter of both hinge pin bushings. Using a suitable driver, tap the bushings into place on the port and starboard sides of the gimbal ring.

24. Apply a coat of 2-4-C Lubricant onto the surface of the lower swivel pin. Insert the switch harness, if equipped, through the gimbal housing, and then secure it in place with the screws. Place the gimbal ring in position in the gimbal housing. Place a washer between the gimbal ring and the

housing. Secure the ring in place with the lower swivel pin and cotter pin.

25. Install the long bolt and nut into the steering lever. A washer is not used at this location. Tighten the nut until it is just "snug". If a new swivel shaft elastic nut is to be used, thread the nut onto the swivel shaft until it is tight, and then remove it. This action cuts the threads into the new nut (which should come with the access plug kit).

26. Check to be sure the gimbal ring rests squarely within the gimbal housing. Before installing the swivel shaft, check for a flat area machined onto the shaft splines. If a flat exists, the flat must face forward after installation. Place the larger washer on top of the ring, followed by the steering lever, then the smaller washer and finally, the large nut. Hold these components aligned together while the upper swivel shaft is passed through the gimbal ring and these parts. Start the large nut onto the shaft threads.

27. Tighten the bolt on the steering lever to 60 ft. lbs. (81 Nm). Now tighten it further until a clearance of 0.002–0.010 in. (0.05–0.25mm) exists between the lower swivel pin washer and the gimbal housing.

28. Use a rawhide mallet and strike the top of the gimbal ring flanges a couple times each to seat both pins. Recheck the clearance and adjust the large nut (using the punch and hammer method), as necessary to maintain the required clearance.

29. Tighten the two gimbal ring bolts to 55 ft. lbs. (74 Nm).

30. Install the trim cylinders by first attaching them in place with the forward anchor pin. Next, install the two hydraulic hoses to the connector. Finally, install the connector to the gimbal housing.

31. Connect the steering link rod to the steering lever. Tighten the castellated nut to 24 inch lbs. (3 Nm), and then back it off slightly and insert the cotter pin. If the assembly is secured with a Nylok nut instead of the cotter pin, tighten the nut to 5 ft. lbs. (7 Nm).

32. Install the bell housing and stern drive.

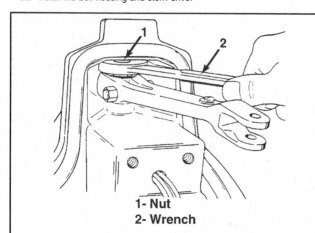

1- Nut
2- Wrench

Fig. 141 Reach through the transom to remove the steering lever

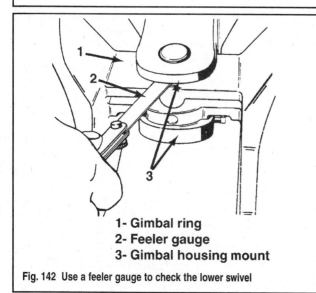

1- Gimbal ring
2- Feeler gauge
3- Gimbal housing mount

Fig. 142 Use a feeler gauge to check the lower swivel

REMOVAL & INSTALLATION  ◁DIFFICULT

◆ **See Figures 143 and 144**

1. Remove the stern drive, bell housing, gimbal ring and bushing.
2. Remove the engine.
3. If equipped with power steering, disconnect the clevis from the steering lever and then disconnect the cable at the clevis. Loosen the pivot bolts and then remove the power steering unit.
4. If equipped with manual steering, disconnect the steering cable at the lever. Loosen the pivot bolts and remove the steering swivel ring.
5. If equipped with a Gear Lube Monitor, disconnect the hose at the transom plate.
6. Disconnect the speedometer tube.
7. Tag and disconnect the MerCathode wires—see Section 3 for an illustration.
8. Tag and disconnect the trim switch wires at the engine and pull them through the assembly.
9. Disconnect the hydraulic lines at the plate and plug them securely. Carefully move them out of the way.
10. Remove the exhaust pipe at the plate.
11. Disconnect the ground wire at the steering lever.
12. Loosen the six nuts and two bolts that secure the inner plate to the transom. Support the gimbal housing and then remove the retainers and lift out the transom plate and gimbal housing.

To install:

13. Rotate the inside race of the gimbal housing bearing and check for rough or any sign of binding. Push and pull on the inner race to check the bearing for side play. The bearing should move only a slight amount. Any excessive movement is just cause to replace the bearing.
14. Check the upper swivel shaft roller bearing and bushing, by inspecting the area of the swivel shaft, where the bearing and bushing ride. Any pits, grooves, or uneven wear is cause to replace the bearing, bushing, or swivel shaft.
15. Check the transom plate for pitting or worn housing. Clean any old sealer from the back side of the plate. Clean the exhaust outlet. Use a tap, and "chase" the threads of the screw holes where the exhaust elbow is mounted.
16. Secure the water tube to the water hose and tighten the hose clamp securely. Install the water hose and tube through the gimbal housing. Route the trim limit switch and trim position sender leads down over the installed tube.
17. Secure both sets of leads with a sta-strap 5 in. (12cm) away from their retaining bracket. Tilt the bell housing up and secure the water hose over the flange on the bell housing. Tighten the hose clamp over the tube. Install the grommet and water tube cover and connect the engine water inlet hose to the water tube.
18. Insert the shift cable, the trim limit switch lead, and the hydraulic hose through the opening in the transom, and then move the gimbal housing into position. The hydraulic hose must be positioned on the starboard side.

■ **Do not hold onto the trim limit switch wires to support the gimbal housing while installing the unit. The trim limit switch or the switch leads will be damaged, if the gimbal housing is supported by the switch leads while fastening the inner transom plate.**

19. Hold the gimbal housing in position, and at the same time insert the shift cable and hydraulic hose through the large opening; insert the trim limit switch leads through the center hole of the three small holes in the inner transom plate; and then set the plate in position.
20. Thread the two short cap screws, with lockwashers, through the top two holes of the inner transom plate and into the gimbal housing. Insert the two special anode head bolt assemblies, with rubber seals between the bolt and gimbal housing, through the bottom two holes of the gimbal housing, the boat transom, and the inner transom plate.
21. Insert a flat washer and elastic stop nut on each bolt but do not tighten them at this time. Install a flat washer and elastic stop nut onto each of the two studs protruding through the inner transom plate.
22. Do not attempt to drive the cap screws through the transom, because

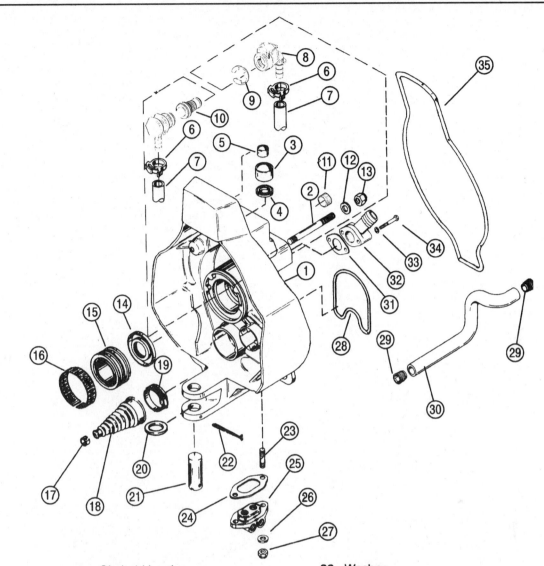

1 - Gimbal Housing
2 - Stud
3 - Lower Swivel Shaft Bushing
4 - Seal
5 - Upper Swivel Shaft Bushing
6 - Clamp
7 - Lube Monitor Hose
8 - Quick Disconnect Fitting
9 - E-Clip
10 - Gear Lube Fitting
11 - Water Bypass Plug
12 - Flat Washer
13 - Locknut
14 - Seal
15 - Gimbal Bearing
16 - Tolerance Ring
17 - Crimp Clamp
18 - Shift Cable Bellows
19 - Bellows Clamp

20 - Washer
21 - Lower Swivel Pin
22 - Cotter Pin
23 - Stud
24 - Gasket
25 - Hydraulic Manifold
26 - Washer
27 - Locknut
28 - Exhaust Passage Seal
29 - Water Hose Insert
30 - Water Hose
31 - Water Fitting Gasket
32 - Water Fitting
33 - Lockwasher
34 - Screw
35 - Gimbal Housing Seal

Fig. 143 Exploded view of the gimbal housing

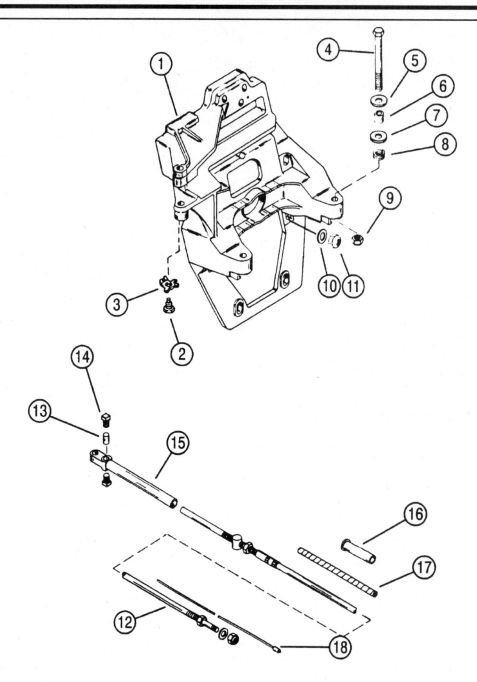

1 - Transom Plate Assembly
2 - Pivot Bolts
3 - Tab Washers
4 - Rear Engine Mounting Bolt
5 - Washer
6 - Spacer
7 - Washer - Fiber
8 - Lockwasher - Double Wound
9 - Locknut

10 - Washer
11 - Locknut
12 - Shift Cable Casing
13 - Core Wire Anchor
14 - Anchor Screws (2)
15 - End Guide
16 - Gimbal Housing Insert
17 - Core Wire

Fig. 144 Exploded view of the transom plate

the threads in the gimbal housing will be damaged. Install the two long cap screws; with a flat washer and lockwasher, through the remaining holes in the inner transom plate and into the gimbal housing. The square flat washer must be against the transom plate.

23. Now, tighten the transom plate cap screws and elastic stop nuts evenly. Work from the center up, and then down. Tighten the screws and nuts to 20–25 ft. lbs. (27–34 Nm).

24. Check to be sure the mating surfaces on the exhaust separator and the gimbal housing are clean. Place new O-ring seals into the grooves in the gimbal housing. Install the exhaust separator, with the exhaust elbows and exhaust bellows attached, to the gimbal housing. Thread the four hex-head cap screws and the one Allen-head cap screw, with lockwashers, into the gimbal housing. Tighten the screws to 20–25 ft. lbs. (27–34 Nm) and at the same time check to be sure the O-ring seals remain properly seated in the grooves.

■ Move quickly while working with the hydraulic hoses to prevent spilling any more hydraulic fluid than is necessary. Follow the sequence and take care to route the hoses properly, or the hoses may be damaged and the system become inoperative.

25. Connect the hydraulic lines and tighten the connections to 110 inch lbs. (12 Nm).

1. Install the steering gear and tighten the pivot bolts to 25 ft. lbs. (34 Nm). Bend the washer tabs up and against the bolt head.

26. Connect the Gear Lube line, speedometer tube and MerCathode wires.

27. Install the engine.

28. Install the bell housing and stern drive.

29. Bleed the hydraulic system. Adjust the shift cable.

TORQUE SPECIFICATIONS - BRAVO/BLACKHAWK

Component		ft. lbs.	inch lbs.	Nm
Shift cable bellows clamp			30-40	3.3-4.5
Steering cable coupler				
	Nut	35		48
	Locking plate screw		60-72	7-8
Steering lever clamp		45-55		61-74
Steering sys. Pivot bolts				
	Bravo	50		68
	Blackhawk ①	25		34
Stern drive-to-bell housing		50		68
Transom assembly		20-25		27-34
U-bolt nuts				
	1992-98	50		68
	1999-01	53		72
U-Joint bellows clamp			30-40	3.3-4.5
Upper unit				
	Back cover screws			
	Bravo	20		27
	Blackhawk	13-15		18-20
	Shift cam assembly		80	9
	Shift cam-to-shaft, screw		110	13
	Shift linkage-to-shaft, screw		110	13
	Top cover screws			
	Bravo	20		27
	Blackhawk	13-15		18-20
	U-joint retainer nut	200		271
	U-joint bearing preload			
	New		6-10	0.7-1.0
	Used		3-7.5	0.3-0.8
	Upper-to-lower unit	35		48

① Left hand thread

TORQUE SPECIFICATIONS - BRAVO/BLACKHAWK

Component		ft. lbs.	inch lbs.	Nm
Bellows hose clamps			35	4
Exhaust pipe-to-gimbal housing		23		31
Lower unit - Bravo I				
	Bearing carrier retainer	200		271
	Driveshaft bearing preload		3-5	0.3-0.8
	Lower-to-upper unit	35		48
	Pinion gear nut (1992-98)	100		135
	Pinion gear bolt (1999-01)	45		61
	Prop shaft bearing preload			
	New		8-12	0.9-1.4
	Used		5-8	0.6-0.9
	Trim tab	23		31
Lower unit - Bravo II				
	Driveshaft bearing preload		3-5	0.3-0.8
	Lower-to-upper unit	35		48
	Pinion gear nut	100		135
	Prop shaft bearing preload			
	New		8-12	0.9-1.4
	Used		5-8	0.6-0.9
	Trim tab	20		27
Lower unit - Bravo III/Blackhawk				
	Bearing carrier			
	Bravo III	150 ①		203 ①
	Blackhawk	250		339
	Bearing carrier retainer nut			
	Blackhawk	150		203
	Driveshaft bearing preload		3-10	0.3-1.1
	Lower-to-upper unit	35		48
	Oil fill/drain plug		30-50	3.4-5.6
	Oil vent plug		30-50	3.4-5.6
	Outer prop shaft bearing nut (Bravo	200 ①		271 ①
	Pinion gear bolt			
	1992-98	35		48
	1999-01	45		61
	Prop shaft bearing preload		8-18	0.9-2
	Rear housing plug screw	23		32
Gimbal ring		55		74
Gimbal ring-to-bell housing	Hinge pins	140-150		189-203
Propeller nut				
	Bravo I	55 MIN		75 MIN
	Bravo II			
	1992-98	55 MIN		75 MIN
	1999-01	60		82
	Bravo III/Blackhawk			
	Front	100		136
	Rear	60		81

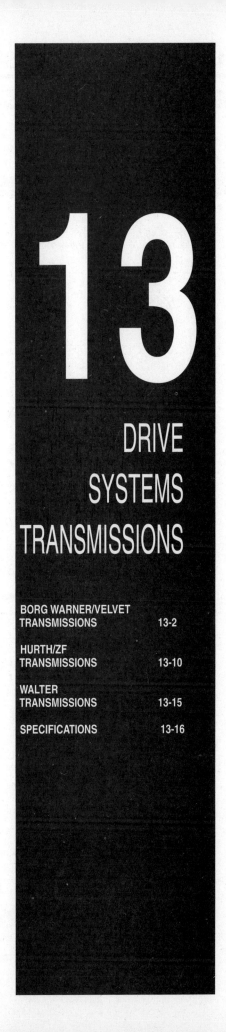

13

DRIVE SYSTEMS TRANSMISSIONS

BORG-WARNER/VELVET TRANSMISSIONS

This section contains service procedures for all Borg-Warner and Velvet in-line and V-drive transmissions. Disassembly and assembly procedures are not available, or recommended. Internal transmission problems should be referred to a local Borg-Warner or Velvet repair facility.

Identification

◆ See Figures 1 and 2

The transmission identification plate is located on the top left side of the transmission housing. All plates are color-coded to the transmission and ratio.

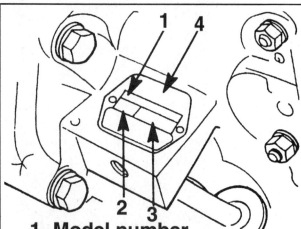

1- Model number
2- Ratio (in forward)
3- Serial number
4- ID plate model color code

Fig. 1 Transmission identification plate (exc. Velvet 5000 series)

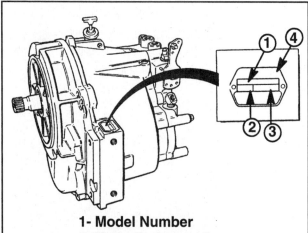

1- Model Number
2- Ratio (in forward)
3- Serial number
4- ID plate model color code

Fig. 2 Transmission identification plate (Velvet 5000 series)

B-W/Velvet In-Line Transmission Ratio 5.7L And 6.2L Engines

Ratio In Forward Gear	ID Plate Color Code
1:1	Red
1.52:1	Red
1.88:1 ①	Red
2.57:1	Red
2.91:1	Red

① Propeller shaft will turn opposite the engine on this transmission

B-W/Velvet In-Line Transmission Ratio 7.4L And 8.2L

Ratio In Forward Gear	ID Plate Color Code
1:1	Green
1.50:1	Green
2.50:1	Green
2.80:1	Green
2.91:1	Green

Borg-Warner V-Drive Transmission Ratio

Ratio In Forward Gear	ID Plate Color Code
1.51:1	Red
1.99:1	Red
2.49:1	Red

Velvet 5000A Down Angle Transmission Ratio

Ratio In Forward Gear	ID Plate Color Code
1.50:1	Black
2.00:1	Black
2.50:1	Black
2.80:1	Black

Velvet Borg-Warner V-Drive Transmission Ratio	
Ratio In Forward Gear	ID Plate Color Code
1.51:1	Red
1.99:1	Red
2.49:1	Red

Shift Cable

ADJUSTMENT

Please refer to the Maintenance section for complete details on lubrication points.

Except Velvet 5000 Series

◆ See Figures 3, 4, 5, 6, 7, 8, 9, 10, 11 and 12

1. Place the remote control lever in the Neutral position.
2. Disconnect the cable from the attaching studs on the shift plate.
3. Push in (toward the rest of the cable) on the cable end guide just enough to remove any play and then mark the inner end of the guide on the tube.
4. Now pull out on the cable end guide just enough to remove any play and mark the inner end of the guide once again.
5. Mark the halfway point between the first and second marks on the tube and move the end guide so its inner edge is now lined up with the mid-point mark on the tube.
6. Loosen the screw and adjust the cable barrel so that the mounting holes in it and the end guide line up with the mounting studs. Install the cable on the studs, but do not secure it.
7. Move the remote control shift lever to the full Forward position and check that the shift lever on the transmission is exactly as depicted in the illustration. The lever must be over the **F** stamped in the housing and the poppet must be in the last detent hole.
8. Move the remote control shift lever to the full Reverse position and check that the shift lever on the transmission is exactly as depicted in the illustration. The lever must be over the **R** stamped in the housing and the poppet must be in the first detent hole.
9. If the lever does not position correctly in either gear, move the shift lever stud from the top hole in the lever to the bottom hole and then recheck positioning. If it still won't position properly you'll need to replace the remote control. If the lever positions correctly in one position but not the other, recheck the adjustment made earlier in this procedure.
10. Install the nut and washer on the cable end guide stud, tighten it until its snug and then back it off one full turn on 1992-97 models; 1/2 turn on 1998-01 models.
11. Install the nut and washer to the barrel stud and tighten it until it just bottoms out—do not over-tighten it.

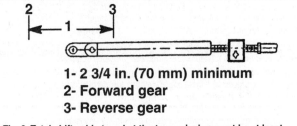

1- 2 3/4 in. (70 mm) minimum
2- Forward gear
3- Reverse gear

Fig. 3 Total shift cable travel at the transmission must be at least 2 ¾ in. (1992-97)

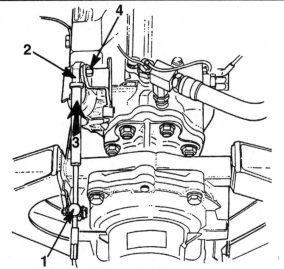

1- Cable barrel (does not move)
2- Cable end guide (does move, inner core wire extends)
3- Cable end fuide must move in this direction when remote control is shifted tp forward gear position
4- Transmission shift lever

Fig. 4 A view of the cable end guide

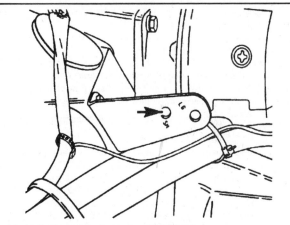

Fig. 5 Cable anchor stud hole – 1998-01 5.7L

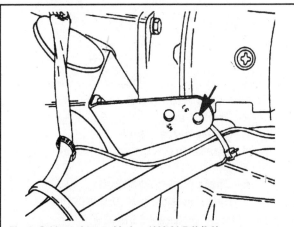

Fig. 6 Cable anchor stud hole – 1998-01 7.4L/8.2L

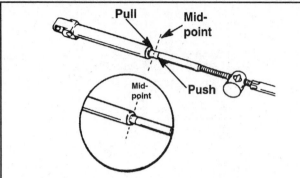

Fig. 7 When checking end-play, adjust the guide to the mid-point

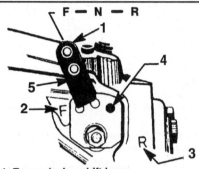

1- Transmission shift lever
2- Shift lever must be over this letter when propelling boat forward
3- Shift lever must over this letter when propelling boat in reverse
4- Proppet ball must be centered in detent hole for each F-N-R position (forward gear shown)
5- Install shift lever stud in this hole, if necessary, to center poppet ball in forward or reverse detent holes

Fig. 8 Shift lever positioning

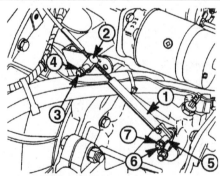

1- Cable end guide
2- Cable barrel
3- Cable barrel stud
4- Elastic stop nut and washer
5- Spacer
6- Cable end guide stud
7- Eleastic stop nut and washer

Fig. 9 Shift cable set-up—single cable, forward

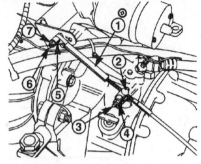

1- Cable end guide
2- Cable barrel
3- Cable barrel stud
4- Elastic stop nut and washer
5- Spacer
6- Cable end guide stud
7- Elastic stop nut and washer

Fig. 10 Shift cable set-up —single cable, rear

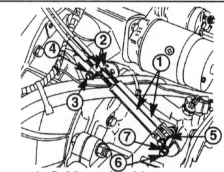

1- Cable end guide
2- Cable barrel
3- Cable barrel stud
4- Elastic stop nut and washer
5- Spacer
6- Cable end guide stud
7- Elastic stop nut and washer

Fig. 11 Shift cable set-up —dual cable, forward

Velvet 5000 Series

◆ See Figures 13, 14, 15, 16, 17, 18, 19, 20 and 21

Propeller rotation is determined by the cable installation on the transmission remote control. For standard left hand propeller rotation, the cable hook-up must cause the shift lever to move forward when the remote control lever is moved to the Forward position. For right hand propeller rotation, the cable hook-up must cause the shift lever to move backwards when the remote control lever is moved to the Forward position.

✳✳ CAUTION

All 5000 series transmissions are full power reversing and designed to work in conjunction with standard left hand rotation engines. Never install a 5000 transmission to an engine with right hand rotation.

1. Place the remote control lever in the Neutral position.
2. Confirm that the anchor stud is installed in the correct hole on the shift lever and bracket.

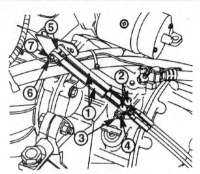

1- **Cable end guide**
2- **Cable barrel**
3- **Cable barrel stud**
4- **Elastic stop nut and washer**
5- **Spacer**
6- **Cable end guide stud**
7- **Elastic stop nut and washer**

Fig. 12 Shift cable set-up —dual cable, rear

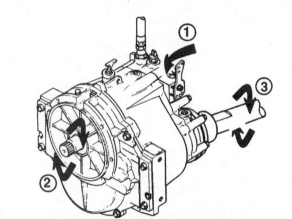

1- Direction of shift lever engagement (toward flywheel)
2- Engine/transmission input shaft rotation direction (LH)
3- Transmission Output/Propeller shaft rotation direction (LH)

Fig. 13 Shift lever engagement (LH rotation)

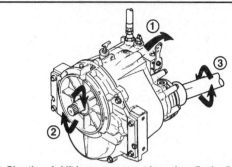

1- Direction of shift lever engagement (away from flywheel)
2- Engine/transmission input shaft rotation direction (LH)
3- Transmission Output/Propeller shaft rotation direction (RH)

Fig. 14 Shift lever engagement (RH rotation)

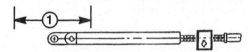

1- 2 3/4 inch (70 mm) minimum

Fig. 15 Total shift cable travel at the transmission must be at least 2 ¾ in.

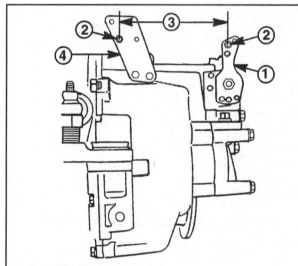

1- **Shift lever**
2- **Anchor stud**
3- **Dimension between stud (7 1/8 in. (181mm)**
4- **Shift cable bracket**

Fig. 16 The distance between studs on the bracket and the shift lever should be 7 1/8 in. (181mm)

   3. Confirm that the distance between the two studs (on the lever and the bracket) is 7-1/ in. (318mm). If not, loosen the clamping bolt at the bottom of the lever and move it slowly until it meets the dimension.
   4. Disconnect the cable from the attaching studs on the shift plate.
   5. Push in (toward the rest of the cable) on the cable end just enough to remove any play and then mark the inner end of the guide on the tube.
   6. Now pull out on the cable end just enough to remove any play and mark the inner end of the guide once again.
   7. Mark the halfway point between the first and second marks on the tube and move the end guide so its inner edge is now lined up with the mid-point mark on the tube.
   8. Loosen the screw and adjust the cable barrel so that the mounting holes in it and the end guide line up with the mounting studs. Install the cable on the studs, but do not secure it.
   9. Move the remote control shift lever to the full Forward position. Hold the lever in position and then confirm that the poppet ball is in the **B** detent hole.

   10. Move the remote control shift lever to the full Reverse and confirm that the poppet ball is in the **C** detent hole.
   11. If the lever does not position correctly in one or both gears, recheck the adjustment and travel. If it still won't position properly you'll need to replace the remote control.
   12. Install the nut and washer on the cable end guide stud, tighten it until its snug and then back it off one full turn.
   13. Install the nut and washer to the barrel stud and tighten it until it just bottoms out—do not over-tighten it.
   14. Refer to the accompanying illustrations for cable hardware layouts.

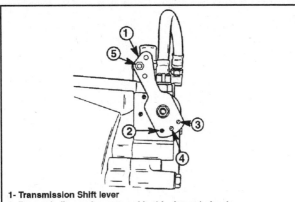

1- Transmission Shift lever
2- Poppet ball must be centered in this detent hole when left-handed propeller shaft rotation is desired
3- Poppet ball must be centered in this detent hole when right-handed propeller shaft rotation is desired
4- Poppet ball must be centered in this detent hole for Nuetral position
5- Install shift lever stud in this hole when using quicksilver shift cables

Fig. 17 Proper shift lever positioning

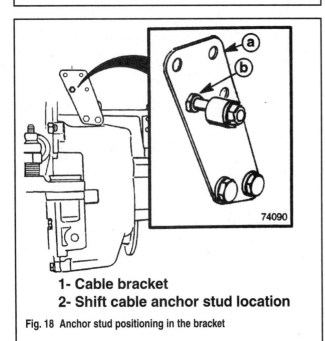

1- Cable bracket
2- Shift cable anchor stud location

Fig. 18 Anchor stud positioning in the bracket

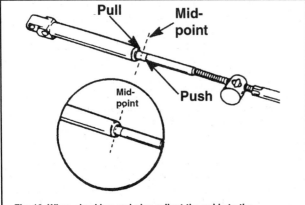

Fig. 19 When checking end-play, adjust the guide to the mid-point

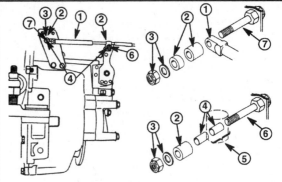

1- Cable end guide
2- Spacer (as required)
3- Elastic stop nut and washer
4- Bushing(s)
5- Cable barrel(s)
6- Cable barrel stud
7- Cable end guide stud

Fig. 20 Shift cable hardware—single cable, rear

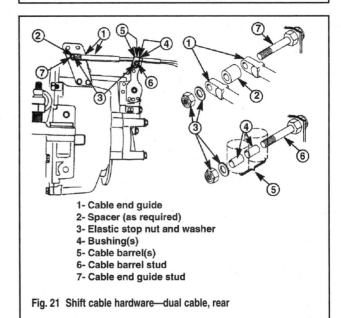

1- Cable end guide
2- Spacer (as required)
3- Elastic stop nut and washer
4- Bushing(s)
5- Cable barrel(s)
6- Cable barrel stud
7- Cable end guide stud

Fig. 21 Shift cable hardware—dual cable, rear

Transmission Fluid

Please refer to the Maintenance section for details on checking the transmission fluid level.

DRAINING FLUID

Except Velvet 5000 Series

◆ See Figures 22, 23, 24, 25 and 26

1. Clean the area around the cooler hose connection on the lower side of the housing.

2. Loosen the fitting nut and disconnect the hose at the elbow. Have some rags handy and be prepared to plug the hose end immediately.

3. Position a suitable container under the elbow and then remove the elbow from the bushing and allow the oil to drain on those 1992–97 in-line (5.7L engines) units with a 1:1 gear ratio and all V-drives.

On all other 1992–97 in-line units (7.4L/8.2L) and all 1998 and later units, remove the bushing and pull out the spring and plastic strainer tube.

■ **Do not remove the bushing, spring or strainer on 1992–97 1:1 units (5.7L) or you will have to replace each piece on installation. These units utilize a metal strainer that is deformed on installation.**

4. Unplug the cooler hose and drain the cooler into the container also.

5. Check the oil for large metal or rubber chips indicating a failing unit or damaged hose. Small metal particles are normal and should not be cause for alarm.

6. On transmissions except the 1992–97 non-1:1 in-line units, insert the plastic strainer into the bushing hole so the notch in the strainer is down and facing out toward the side of the housing. Position the spring and then screw in the bushing. On all transmissions, coat the threads with Perfect Seal and tighten the bushing to 25 ft. lbs. (34 Nm).

7. Coat the threads of the elbow with Perfect Seal and screw it into the bushing so it is secure.

8. Reconnect the cooler hose and tighten the fitting securely.

9. Remove the dipstick and fill the transmission with fluid until it registers FULL on the stick. For 1992–97 transmissions, use 10W hydraulic tractor fluid; about 2 qts. (1.9L) for the 1:1 in-line, about 3 qts. (2.8L) for all other in-lines, and 4.5 qts. (4.3L) for all V-drives. For 1998 and later models, use Dexron III® ATF; about 1.5 qts. (1.33L) for the 71C and 7.4L/8.2L 72C, 3 qts. (2.75L) for the 5.7L/6.2L 72 series and V-drive, and 2.5 qts. (2.4L) for the 72C Reduction models.

10. Install the dipstick, start the engine and allow it to run at 1500 rpm for about 2 min. Stop the engine and recheck the fluid level immediately. Add fluid as necessary until it meets the FULL mark.

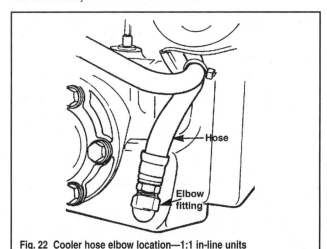

Fig. 22 Cooler hose elbow location—1:1 in-line units

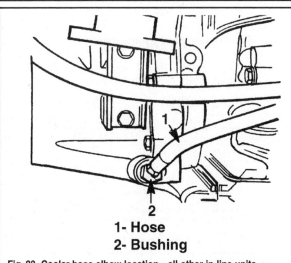

1- Hose
2- Bushing

Fig. 23 Cooler hose elbow location—all other in-line units

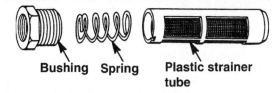

Bushing   Spring   Plastic strainer tube

Fig. 24 Cooler hose elbow location—V-drive units

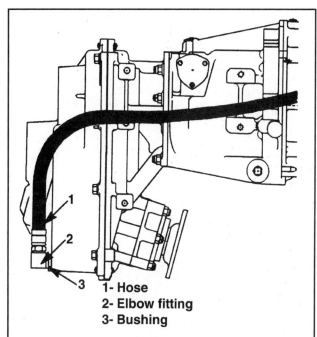

1- Hose
2- Elbow fitting
3- Bushing

Fig. 25 All units utilize a small strainer (note the notch used for positioning)

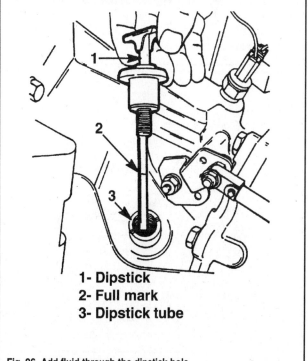

1- Dipstick
2- Full mark
3- Dipstick tube

Fig. 26 Add fluid through the dipstick hole

Velvet 5000 Series

◆ **See Figures 27 and 28**

1. Clean the area around the drain plug on the rear of the unit, below and starboard of the propeller shaft.
2. Position a suitable container under the plug and then remove the plug from the housing. Removing the dipstick will aid in draining.
3. Check the oil for large metal or rubber chips indicating a failing unit or damaged hose. Small metal particles are normal and should not be cause for alarm.
4. Coat the threads of the drain plug with Perfect Seal, thread it into the housing and tighten to 25 ft. lbs. (34 Nm).
5. Reconnect the cooler hose and tighten the fitting securely.
6. Remove the dipstick and fill the transmission with fluid (Dexron III ATF) until it registers FULL on the stick, about 3 qts. (2.75L) on all models except the 8.1L units. For 8.1L engines using the 5000A, about 2.75 qts. (2.6L), while its about 3.5 qts. (3.3L) for the 5000V.
7. Install the dipstick, start the engine and allow it to run at 1500 rpm for about 2 min. Stop the engine and recheck the fluid level immediately. Add fluid as necessary until it meets the FULL mark.
8. Reinstall the dipstick, making sure to tighten the T-handle securely.

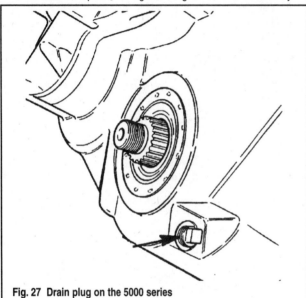

Fig. 27 Drain plug on the 5000 series

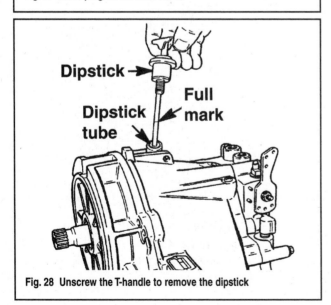

Fig. 28 Unscrew the T-handle to remove the dipstick

Dipstick →
Dipstick tube ↗
Full mark ↙

OEM ③ **DIFFICULT**

REMOVAL & INSTALLATION

Except Velvet 5000 Series

This procedure details removal of the transmission without removing the engine. Engine removal procedures can be found in the proper section.

1. Drain the transmission fluid as detailed previously.
2. Disconnect the fluid cooler hoses and the shift cable.
3. Tag and disconnect the electrical leads at the neutral safety and oil temperature switches. Move the wires out of the way.
4. Unscrew the flange bolts and disconnect the propeller/driveshaft at the transmission. Make sure you support the shaft(s) with a block of wood.
5. Loosen and remove the four rear mount-to-engine bed bolts.
6. Support the rear of the engine. You can wedge some 2 X 4s underneath the flywheel housing, but we recommend using an engine hoist. It's a bit more trouble, but it's certainly worth it from a safety standpoint!
7. Remove the two center transmission-to-flywheel housing bolts and insert two long non-threaded studs in their place to help support the transmission.
8. Remove the remaining bolts and then pull the transmission straight back and off of the engine, sliding on the studs. Once again, we recommend using a suitable lifting device; although it's not absolutely necessary.

To install:

9. On 1992–97 transmissions, check the transmission pump indexing for correct rotation as detailed following.
10. Check the output shaft rolling torque with a torque wrench and socket on the coupling nut. Refer to the chart.
11. Coat the input and drive plate shaft splines with Engine Coupler Spline grease.
12. Check the rear engine mount brackets on the transmission housing are tightened to 45 ft. lbs. (61 Nm).
13. Position the transmission so the input shaft splines are aligned with the drive plate and then slide it forward and into place. Tighten all mounting bolts to 50 ft. lbs. (68 Nm).
14. Lower the engine back into place with the hoist or remove the wooden blocks. Tighten the engine mounts-to-bed bolts securely.
15. Connect the electrical leads to the two switches. Connect the cooler hoses.
16. Connect and adjust the shift cable(s).
17. Check engine alignment as detailed in the Engine Removal & Installation section and then install the propeller shaft to the transmission output shaft. Install the washers, nuts and bolts and tighten to 50 ft. lbs. (68 Nm).
18. Refill the transmission with fluid.

Output Shaft Rolling Torque ①	
Ratio In Forward Gear	**Torque ft. lbs. (Nm)**
1:1	50 (68)
1.52:1	55 (75)
1.88:1	60 (81)
2.57:1	65 (88)
2.91:1	70 (95)

① Transmission not installed and no fluid

Velvet 5000 Series

This procedure details removal of the transmission without removing the engine. Engine removal procedures can be found in the proper section.

1. Drain the transmission fluid as detailed previously.
2. Disconnect the fluid cooler hoses and the shift cable.
3. Tag and disconnect the electrical leads at the neutral safety and oil temperature switches. Move the wires out of the way.
4. Loosen the trunion clamp fasteners on the port and starboard engine mounts.
5. Unscrew the coupling nuts and bolts at the output flange and disconnect the propeller/driveshaft at the transmission. Make sure you support the shaft(s) with a block of wood.
6. Loosen and remove the four rear mount-to-engine bed bolts.
7. Support the rear of the engine. You can wedge some 2 X 4s underneath the flywheel housing, but we recommend using an engine hoist. It's a bit more trouble, but it's certainly worth it from a safety standpoint!
8. Attach another hoist to the transmission and remove the rear mount brackets (base and trunion) from the transmission.
9. Remove the transmission-to-engine mounting bolts and then pull the transmission straight back and off of the engine. Once again, we recommend using a suitable lifting device; although it's not absolutely necessary.

To install:

10. Coat the input and drive plate shaft splines with Engine Coupler Spline grease.
11. Check the rear engine mount brackets on the transmission housing are tightened to 45 ft. lbs. (61 Nm).
12. Position the transmission so the input shaft splines are aligned with the drive plate and then slide it forward and into place. Tighten all mounting bolts to 55 ft. lbs. (75 Nm).
13. Install the rear engine mount brackets on the transmission housing and tighten to 45 ft. lbs. (61 Nm).
14. Lower the engine back into place with the hoist or remove the wooden blocks. Tighten the engine mounts-to-bed bolts securely.
15. Connect the electrical leads to the two switches.
16. Connect the cooler hoses and tighten the fittings to 25 ft. lbs. (34 Nm).
17. Connect and adjust the shift cable(s).
18. Check engine alignment as detailed in the Engine Removal & Installation section and then install the propeller shaft to the transmission output shaft. Install the washers, nuts and bolts and tighten to 50 ft. lbs. (68 Nm).
19. Refill the transmission with fluid.

PUMP INDEXING

1992–97 Only

◆ See Figures 29 and 30

The transmission pump must be properly indexed so as to match engine rotation. If improperly indexed, you will have insufficient oil pressure and the transmission will not shift properly. There are two arrows scribed into the pump housing and the pump must be positioned so that the arrow that points in the direction the input shaft and pump will be turned in by the engine is at the top side of the pump.

** CAUTION

Certain units may use the letters LH and RH in place of the arrows. These indicate rotation of transmission, NOT engine rotation.

1. Loosen and remove the four mounting bolts and then rap the fluid passage boss (at center of pump) softly with a rubber mallet to loosen the pump. DO not hit the bolt bosses with the mallet.
2. Make sure that the gasket is not sticking to the pump or transmission housing, and then slowly, and carefully, rotate the pump until the correct arrow (or letters) is at the top of the unit when the bolt holes align. It may not be exactly at 12:00, but get it as close as the bolt holes will allow.
3. Tighten the pump bolts to 204–264 inch lbs. (23–29 Nm).

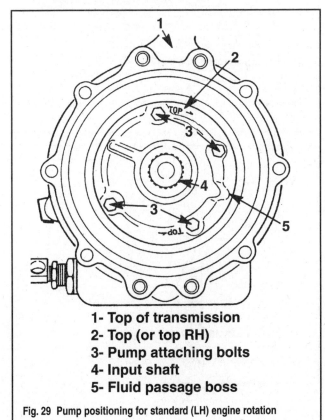

1- Top of transmission
2- Top (or top RH)
3- Pump attaching bolts
4- Input shaft
5- Fluid passage boss

Fig. 29 Pump positioning for standard (LH) engine rotation

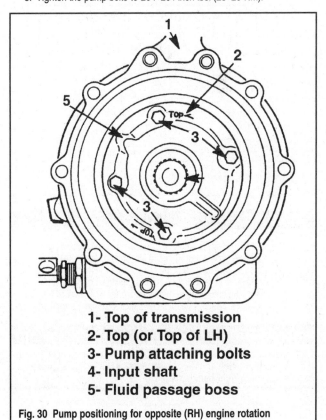

1- Top of transmission
2- Top (or Top of LH)
3- Pump attaching bolts
4- Input shaft
5- Fluid passage boss

Fig. 30 Pump positioning for opposite (RH) engine rotation

HURTH/ZF TRANSMISSIONS

This section contains service procedures for all Hurth and ZF down angle and V-drive transmissions. Disassembly and assembly procedures are not available, or recommended. Internal transmission problems should be referred to a local Hurth/ZF repair facility.

Identification

◆ See Figures 31 and 32

The transmission identification plate is located on the top rear side of the transmission housing. All plates contain the gear ratio, serial number and model.

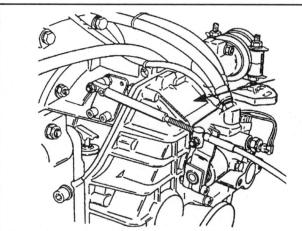

Fig. 31 The transmission identification plate is located here...

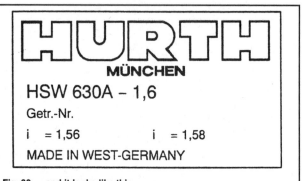

Fig. 32 ...and it looks like this

Shift Cable

Propeller rotation is determined by the cable installation on the transmission remote control. For standard left hand propeller rotation, the cable hook-up must cause the shift lever to move backwards (**B**) when the remote control lever is moved to the Forward position. For right hand propeller rotation, the cable hook-up must cause the shift lever to move forward (**A**) when the remote control lever is moved to the Forward position.

✳✳ CAUTION

All Hurth/ZF transmissions are built to work in conjunction with standard left hand rotation engines. Never install a Hurth/ZF transmission to an engine with right hand rotation.

ADJUSTMENT

◆ See Figures 33, 34, 35, 36, 37, 38, 39, 40 and 41

Please refer to the Maintenance section for complete details on lubrication points.

1. Place the remote control lever in the Neutral position.
2. Confirm that the anchor stud is installed in the correct hole for your transmission. The holes should be marked **630** and **800**.
3. Confirm that the shift lever on the transmission is positioned approximately 10° aft of vertical and that the distance between the two studs is 7-1/8 in. (318mm). If not, loosen the clamping bolt at the bottom of the lever and move it slowly until it meets the dimension.
4. Disconnect the cable from the attaching studs on the shift plate.
5. Push in (toward the rest of the cable) on the cable end just enough to remove any play and then mark the inner end of the guide on the tube.
6. Now pull out on the cable end just enough to remove any play and mark the inner end of the guide once again.
7. Mark the halfway point between the first and second marks on the tube and move the end guide so its inner edge is now lined up with the mid-point mark on the tube.
8. Loosen the screw and adjust the cable barrel so that the mounting holes in it and the end guide line up with the mounting studs. Install the cable on the studs, but do not secure it.
9. Move the remote control shift lever to the full Forward position. Hold the lever in position and then have a friend carefully slide the cable off the anchor points on the transmission. See if you can move the lever any farther forward.

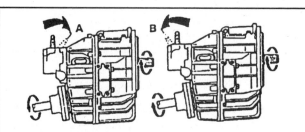

Fig. 33 Make sure the transmission shift lever moves in the right direction for intended propeller rotation

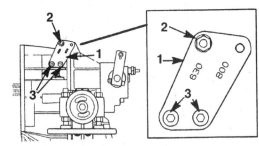

1- Shift cable bracket
2- Anchor stud location
3- Bracket mounting bolts

Fig. 34 The anchor stud must be in the correct hole

10. Reconnect the shift cable at the transmission.

11. Move the remote control shift lever to the full Reverse. Hold the lever in position and then have a friend carefully slide the cable off the anchor points on the transmission. See if you can move the lever any farther backward.

12. If the lever does not position correctly in either gear, move the shift lever stud from the top hole in the lever to the bottom hole and then recheck positioning. If it still won't position properly you'll need to replace the remote control. If the lever positions correctly in one position but not the other, recheck the adjustment made earlier in this procedure.

13. Install the nut and washer on the cable end guide stud, tighten it until its snug and then back it off one full turn.

14. Install the nut and washer to the barrel stud and tighten it until it just bottoms out—do not over-tighten it. On 1998 and later models, back the nut off 1/2 turn.

15. Refer to the accompanying illustrations for cable hardware layouts.

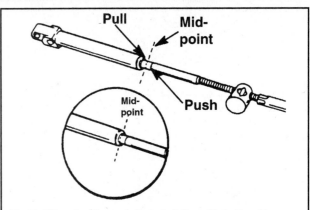

Fig. 36 When checking end-play, adjust the guide to the mid-point

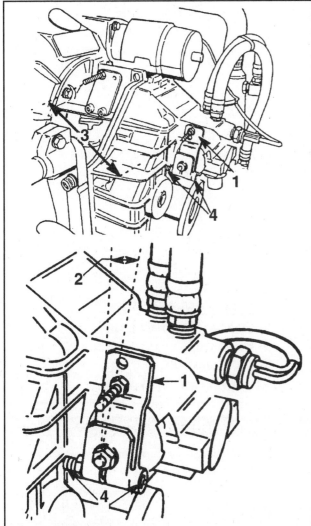

1- Shift lever

2- Lever, in neutral detent, must be approximately 10 degrees Aft

3- Dimension between studs - 7 1/8 in. (318mm)

4- Clamping bolt

Fig. 35 Shift lever positioning

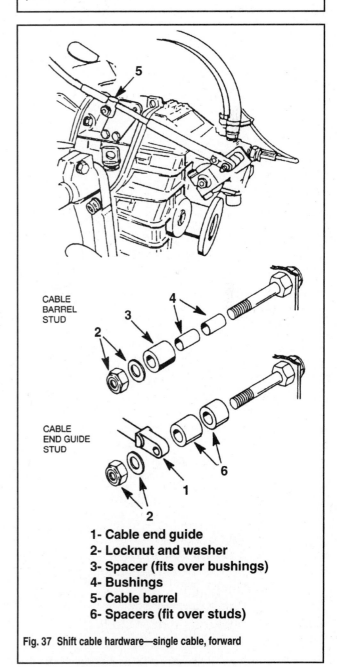

1- Cable end guide

2- Locknut and washer

3- Spacer (fits over bushings)

4- Bushings

5- Cable barrel

6- Spacers (fit over studs)

Fig. 37 Shift cable hardware—single cable, forward

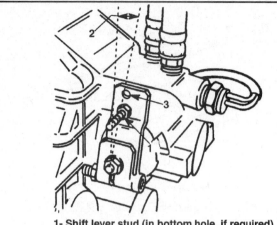

1- Shift lever stud (in bottom hole, if required)
2- Lever, in neutral detent, must be
   approximately 10 Aft of vertical
3- Shift lever top hole

Fig. 38 Moving the lever stud to the other hole

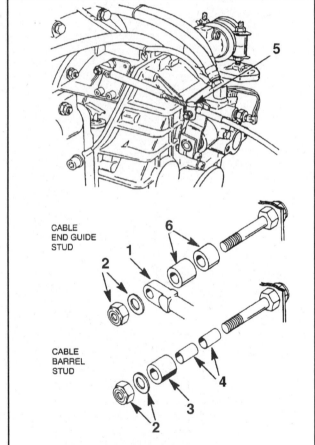

1- **Cable end guide**
2- **Locknut and washer**
3- **Spacer (fits over bushings)**
4- **Bushings**
5- **Cable barrel**
6- **Spacers (fit over stud)**

Fig. 39 Shift cable hardware—single cable, rear

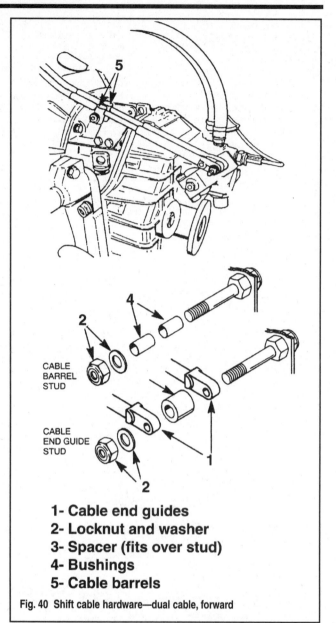

1- **Cable end guides**
2- **Locknut and washer**
3- **Spacer (fits over stud)**
4- **Bushings**
5- **Cable barrels**

Fig. 40 Shift cable hardware—dual cable, forward

Transmission Fluid

Please refer to the Maintenance section for details on checking the transmission fluid level.

DRAINING FLUID

◆ See Figures 42, 43, 44, 45 and 46

1. Clean the exterior of the transmission before proceeding.
2. Grab the oil filter top and turn it counterclockwise while pulling it up, remove the filter and set it aside, preferably in your oil drain container.

■ **Certain 1998-01 models may utilize a setscrew in the flange where the T-handle enters the filter top. Loosen the screw first.**

3. On models without a drain plug, insert a standard marine suction pump (available at almost any marine retailer) into the filter hole and pump out the fluid. Make sure you have plenty of rags available for those inevitable spills. On models with a drain plug (rear, starboard side of lower housing),

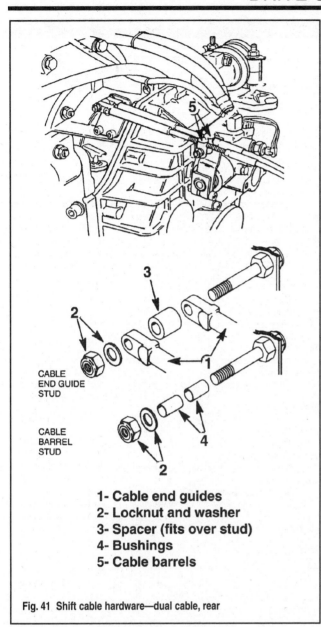

1- **Cable end guides**
2- **Locknut and washer**
3- **Spacer (fits over stud)**
4- **Bushings**
5- **Cable barrels**

Fig. 41 Shift cable hardware—dual cable, rear

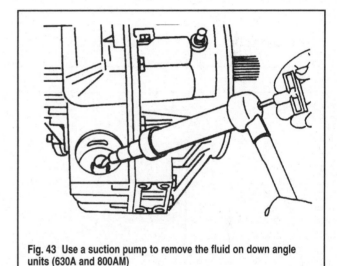

Fig. 42 Removing the oil filter

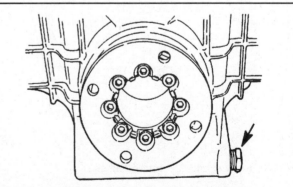

Fig. 43 Use a suction pump to remove the fluid on down angle units (630A and 800AM)

position a suitable container under the transmission, remove the plug and allow the fluid to drain.

4. Check the oil for large metal or rubber chips indicating a failing unit or damaged hose. Small metal particles are normal and should not be cause for alarm.

5. Fill the transmission with approximately 4.25 qts. (4L) of Dexron II-D automatic transmission fluid through the filter hole on 1992-97 630 models; 4.5 qts (4.2L) for 800 models; or 5 qts. (4.7L) for V-drives. On 1998 and later models, install the drain plug and tighten it securely; then fill the transmission through the filter hole with Dexron III ATF; 4.2 qts. (4L) for 630V, 3.2 qts. (3L) for the 630A and 5.8 qts. (5.5L) for the 800A.

6. Coat the O-ring on the filter assembly with a little ATF and then install the filter. Push down on the cover until it is seated on the housing and then turn the T-handle clockwise until it's tight. Don't forget to tighten the setscrew securely on later models.

7. Start the engine and allow it to run for about 2 min. Stop the engine and recheck the fluid level immediately with the dipstick on the other side of the housing. Add fluid as necessary until it comes between the MIN and MAX lines.

Fig. 44 Early V-drives (630V) and all 1998 and later models utilize a drain plug

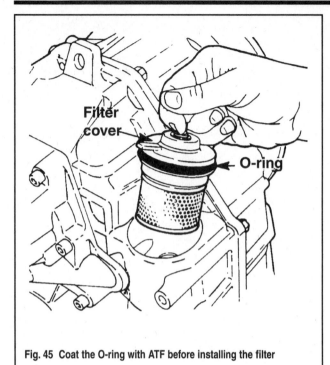

Fig. 45 Coat the O-ring with ATF before installing the filter

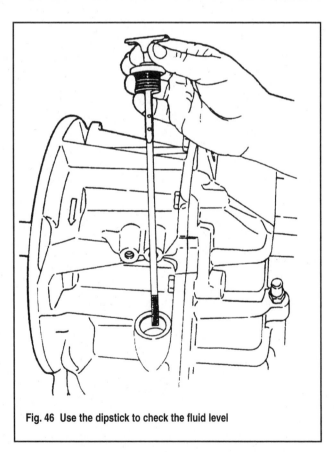

Fig. 46 Use the dipstick to check the fluid level

Transmission

OEM

REMOVAL & INSTALLATION ③ DIFFICULT

◆ **See Figure 147**

This procedure details removal of the transmission without removing the engine. Engine removal procedures can be found in the proper section.

1. Drain the transmission fluid as detailed previously.
2. Disconnect the seawater hoses from the transmission fluid cooler.
3. Disconnect the shift cable.
4. Tag and disconnect the electrical leads at the neutral safety and audio warning temperature switches. Move the wires out of the way.
5. Unscrew the flange bolts and disconnect the propeller/driveshaft at the transmission. Make sure you support the shaft(s) with a block of wood.
6. Support the rear of the engine. You can wedge some 2 X 4s underneath the flywheel housing, but we recommend using an engine hoist. It's a bit more trouble, but it's certainly worth it from a safety standpoint!
7. Loosen and remove the four rear mount-to-engine bed bolts.
8. Remove the two mounting bolts and four locknuts, and then pull the transmission straight back and off of the engine. Once again, we recommend using a suitable lifting device; although it's not absolutely necessary.

To install:

9. Coat the input and drive plate shaft splines with Engine Coupler Spline grease.
10. Position the transmission so the input shaft splines are aligned with the drive plate and then slide it forward and into place. Tighten all mounting bolts and nuts to 50 ft. lbs. (68 Nm).
11. Lower the engine back into place with the hoist or remove the wooden blocks. Tighten the engine mounts-to-bed bolts securely.
12. Connect the electrical leads to the two switches. Coat the connections with Liquid Neoprene.
13. Connect the seawater hoses to the cooler and tighten the hose clamps securely.
14. Connect and adjust the shift cable(s).
15. Check engine alignment as detailed in the Engine Removal & Installation section and then install the propeller shaft to the transmission output shaft. Install the washers, nuts and bolts and tighten to 50 ft. lbs. (68 Nm).
16. Refill the transmission with fluid.

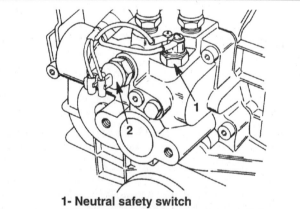

1- Neutral safety switch
2- Audio warning temperature switch

Fig. 47 Disconnect the two switches on the control block

WALTER TRANSMISSIONS

This section contains information on Walter RV26DV-drive transmissions. Disassembly and assembly procedures are not available, or recommended. Refer all internal transmission problems to an authorized Walter service and repair representative.

Transmission Fluid

DRAINING FLUID

② MODERATE

◆ **See Figures 48 and 49**

1. Position a suitable container under the transmission.
2. Loosen the magnetic drain plug on the lower aft, starboard side of the transmission housing. Keep an inward pressure on the plug as you're unscrewing it so as to minimize the initial oil splash until you get the plug out.
3. Wait until the oil stream from the drain hole has abated and then disconnect the oil hose at the strainer fitting elbow on the bottom case.
4. Loosen and remove the strainer fitting, but leave the elbow connected. Allow any remaining oil in the case to drain.

To install:

5. Clean the strainer and the magnetic plug and then install them both when the oil has drained completely. Tighten each securely.
6. Reconnect the oil hose to the elbow and tighten the clamp securely.
7. Pull up on the dipstick knob at the leading edge of the top cover and remove the dipstick.
8. Add 1 qt. (0.9L) of 30W heavy duty motor oil through the dipstick hole.
9. Replace the dipstick, start the engine and allow it to run at 1500 rpm for about 2 minutes. Turn the engine off and recheck the fluid level again. Add additional fluid until the level is up to the **H** mark on the dipstick.

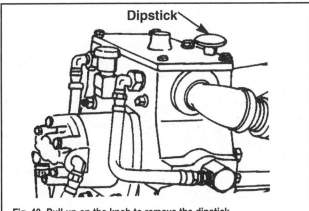

Dipstick

Fig. 48 Pull up on the knob to remove the dipstick

Transmission

REMOVAL & INSTALLATION

③ DIFFICULT

◆ **See Figures 50, 51, 52 and 53**

1. Drain the transmission fluid.
2. Drain the cooling system as detailed in the Cooling System section.
3. Remove the four nuts and disconnect the propeller shaft coupler from the transmission output shaft flange. Position 2X4s under the propeller shaft to support it while disconnected.

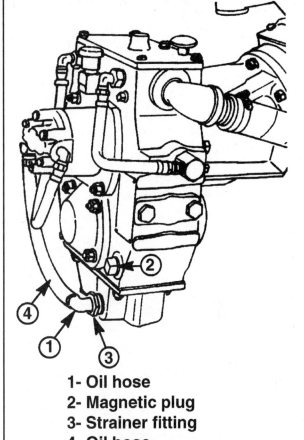

1- Oil hose
2- Magnetic plug
3- Strainer fitting
4- Oil hose

Fig. 49 Remove the drain plug and the strainer fitting

4. Disconnect the cooling hoses at the transmission top cover. Make sure the container is still under the unit as there will be some additional drainage from the hose.
5. Tag and disconnect the electrical leads at the oil pressure switch.
6. Support the transmission with a suitable lifting device. Remove the six nuts at the adapter housing and then carefully slide the unit straight back from the adapter housing and remove it

To install:

7. Move the transmission into position and slide it onto the adapter housing over the studs. You may have to rotate the flange a little to align the splines correctly. Install the nuts and tighten them to 50 ft. lbs. (68 Nm) in a crisscross pattern.
8. Connect the cooling hoses.
9. Connect the oil pressure switch leads.
10. Connect the propeller shaft coupling to the output shaft flange and tighten the nuts to 50 ft. lbs. (68 Nm) in a crisscross pattern. Now insert a 0.003 in. (0.07mm) feeler gauge between the coupler and the flange in at least four locations—there should be no instance of a gap greater then the gauge.
11. Fill the cooling system and then fill the transmission with fluid.
12. Start the engine and allow it to run for a few minutes. Turn it off and check for any leaks.

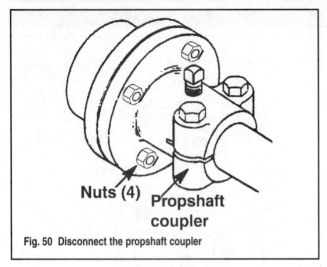

Fig. 50 Disconnect the propshaft coupler

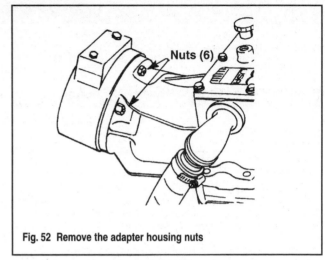

Fig. 52 Remove the adapter housing nuts

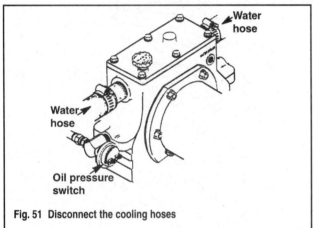

Fig. 51 Disconnect the cooling hoses

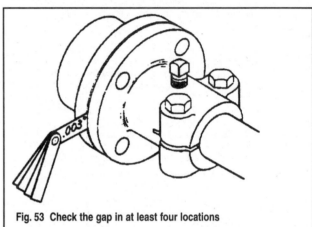

Fig. 53 Check the gap in at least four locations

TORQUE SPECIFICATIONS

Model	Component	ft. lbs.	inch lbs.	Nm
Borg-Warner/Velvet				
10-17, 10-18, 71-C, 72-C, V-drive, 5000V				
	Drain plug (bushing)	25		34
	Fluid hose-to-bushing	25		34
	Neutral start switch	8-11		11-14
	Prop coupler-to-output flange	50		68
	Pump-to-adaptor	17-22		23-29
	Rear mounts	45		61
	Shift lever		96-132	11-15
	Trans-to-flywheel	50		68
5000A	Drain plug	25		34
	Fluid hose-to-cooler	25		34
	Fluid hose-to-housing	25		34
	Neutral start switch		120	13
	Prop coupler-to-output flange	50		68
	Rear mounts	45		61

14

TRIM &
TILT

POWER TRIM AND TILT

Description

◆ **See Figures 1 and 2**

The power trim/tilt system consists of a valve body, hydraulic reservoir, electric motor, up and down solenoids, trim/tilt switch, trim sending unit, trim gauge, trim limit switch, two hydraulic trim cylinders, and the necessary hydraulic lines and electrical harnesses for the system to function efficiently.

The valve body, hydraulic reservoir, electric motor, and up down solenoids make up a single unit known as the pump assembly and is mounted inside the boat. Two hydraulic trim cylinders are mounted, one on each side of the stern drive unit. One end of the cylinder is connected to the gimbal housing and the opposite end is attached to the upper gear housing.

The trim sender unit is mounted on the starboard side of the gimbal housing and is connected to the trim gauge on the control panel through the wiring harness. When the stern drive is trimmed **IN** or **OUT** the gauge moves to indicate the stern drives position.

A trim limit switch is mounted on the port side of the gimbal ring and is connected to the **UP** solenoid on the trim/tilt hydraulic pump. This limit switch allows the stern drive to be trimmed out only to a preset position. If the stern drive is moved out past the support brackets under high throttle setting could impose high stress loads on the boat transom and possibly loss of steering control.

The relationship of the various units in the trim/tilt system are discussed in the following paragraphs.

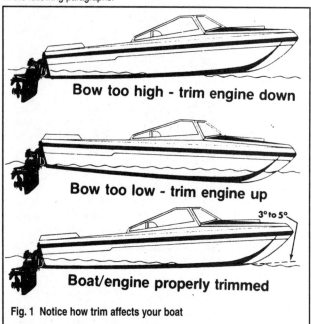

Bow too high - trim engine down

Bow too low - trim engine up

3° to 5°

Boat/engine properly trimmed

Fig. 1 Notice how trim affects your boat

HYDRAULIC UP CIRCUIT

When the **UP/UP** buttons on the control panel, or the **Trailer** switch on the remote control throttle, is pressed and held, the trim/tilt electric motor operates. The hydraulic pump is splined to the electric motor and the pump begins to create hydraulic pressure. This pressure created in the UP side of the system moves past the UP pressure relief valve, through lines and hoses to the trim/tilt cylinder.

At this point, the pressure pushes the dual hydraulic pistons outward. Fluid in the DOWN side of the system is forced out of the cylinder by movement of the piston. Fluid movement continues back through connecting hoses and lines to the DOWN pressure relief valve. The valve is pushed open and the fluid returns to the pump reservoir.

The UP pressure relief valve regulates the system pressure between 2200–2600 psi (15173–17932 kPa). A thermal relief valve prevents hydraulic lock-up in the full tilt-up position caused by the thermal expansion of the hydraulic fluid. A check valve in the end of the UP pressure relief valve maintains pressure in the UP side of the system when the pump is not

running and the unit is shifted into forward gear.

HYDRAULIC DOWN CIRCUIT

When the **IN** button is pushed on the control panel, or the throttle remote control lever, the electric motor and pump begin to operate. Hydraulic pressure is created in the DOWN circuit of the trim/tilt system. The pressure moves a slide valve in the valve body that unseats the poppet valve in the UP circuit. Hydraulic fluid moves past the DOWN pressure relief valve where the DOWN circuit pressure is regulated between 400–600 psi (2759–4183 kPa). Hydraulic fluid under pressure flows through lines, fittings and hoses to the DOWN side of the trim/tilt cylinder piston.

The hydraulic pressure moves the piston inward, forcing fluid out the UP side of the cylinder through hoses, fittings and check valves back to the reservoir. This fluid returning to the reservoir prevents the pump from operating under an out of fluid condition.

HYDRAULIC TRIM CIRCUITS

When the **OUT** or **IN** trim button on the control panel or the remote control throttle lever is pressed, the hydraulic trim/tilt motor begins to operate.

Trimming the stern drive OUT while at high engine power settings, could impose severe stress—possibly damaging the stern drive, boat transom and could cause loss of directional control. A trim limit switch is wired into the UP circuit to prevent such action. If the trim switch is held to the OUT position, the stern drive will raise or trim OUT. When the upper limit of the limit switch is reached, power to the UP solenoid on the pump is de-energized and the trim/tilt pump stops operating, preventing further trim movement of the stern drive. When the throttle lever is reduced to slightly above the idle position, the tilt **UP** switch is energized, allowing the stern drive to be raised further for beach and shallow water operation.

When the **IN** trim button on the control panel or the remote control throttle lever is pressed, the hydraulic trim/tilt motor begins to run and pressurizes the DOWN circuit. The stern drive moves inward until the gimbal housing contacts the transom stop pin.

A sliding check valve assembly and poppet valves mounted internally in the valve body seat to create hydraulic locks that prevent the stern drive from raising when the shift lever is moved to the **REVERSE** position. These same check valves create a hydraulic lock which holds the stern drive in a trim OUT position.

If the stern drive should strike an object under water while moving, the rapid return of fluid from the cylinders being extended will cause the check valve in the valve body and the cylinder pistons to open and release the hydraulic lock. This will allow the stern drive to kick-up and away from the underwater object.

TRAILERING OR LAUNCHING

Both the up buttons on the control panel must be pushed at the same time, or the **TRAILER** switch on the remote control throttle must be pressed, to raise the stern drive to the full UP position for trailering or launching the boat. By pushing the middle **UP/OUT** button, current is passed to the top **UP** switch. The current passing through the **UP** switch while the button is depressed, will by-pass the trim limit switch and permit the UP circuit to raise the stern drive to the full UP position. If the middle **UP/OUT** button is depressed during normal boat operation, a trim limit switch will prevent the stern drive unit from moving out beyond the gimbal ring support guides.

The **TRAILER** switch on the remote control throttle lever will raise the stern drive to the full UP position when approaching the beach or for trailering the boat. Never operate the engine above idle speed with the stern drive up past the limit switch settings. Such action would cause severe damage to the stern drive gimbal housing and possibly a loss of steering control.

When the hydraulic cylinders have reached their full extended travel, the pump motor will begin to "labor" if the control buttons are not released. Therefore, to prevent damage to the system, a bi-metal switch in the pump motor will open the circuit to stop the pump motor and prevent the motor from overheating. The switch will automatically close after the motor has cooled allowing the motor to again be operated.

The following is a short description of some additional major components in the trim/tilt system and their function.

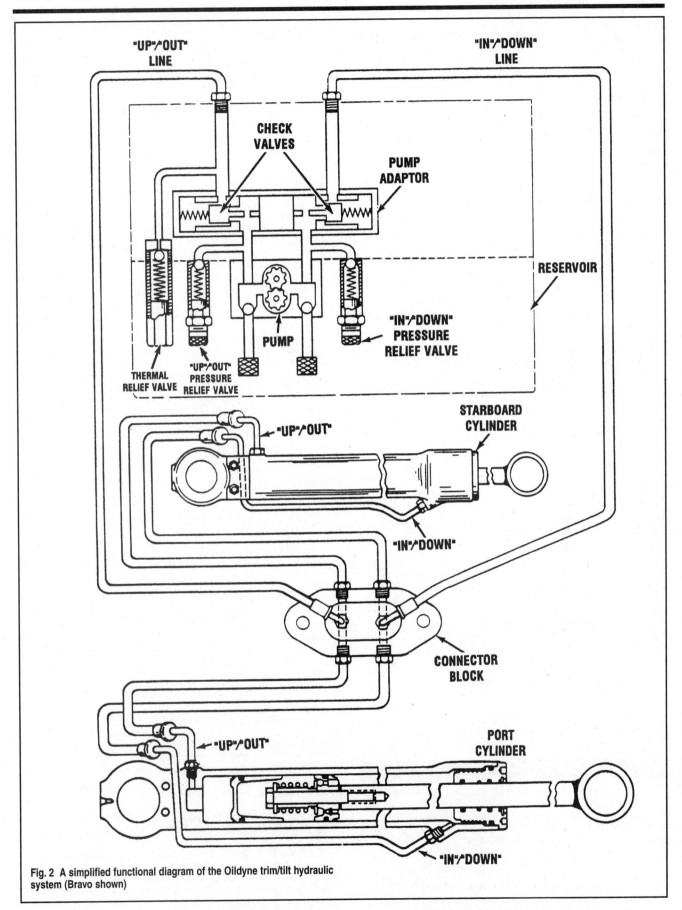

Fig. 2 A simplified functional diagram of the Oildyne trim/tilt hydraulic system (Bravo shown)

RESERVOIR

The reservoir is a translucent 1-1/2 quart reservoir that attaches to the pump valve body. A **MIN** and **MAX** mark cast into the reservoir makes it easy to check the fluid quantity. A large screw-on cap and opening provides for easy servicing of the reservoir. The cap vents the reservoir to atmosphere pressure when the fill neck seal is removed inside the cap.

HYDRAULIC PUMP

The hydraulic fluid pump is a gear- type pump, very similar to an engine oil pump. The pump is attached to the valve body with four fasteners and cannot be disassembled. If the pump fails, it must be replaced as an assembly. The reservoir must be drained and removed to gain access to the pump.

VALVE BODY

The valve body, or adapter, is the central mounting point for the pump, valves, reservoir and electric motor. The hydraulic pump is capable of creating pressure up to 3000 psi. Internal pressure relief valves set at the factory control the output pressure in the UP and DOWN circuits. A system thermal relief valve protects the pump, valves, lines, and cylinder packing from thermal expansion of the hydraulic fluid in the system when operating or sitting in the hot sun. The pump, pressure and thermal relief valves are all set at the factory and adjustments are not possible. If any of these components are found to be defective during troubleshooting or disassembly, they must be replaced with identical items. Each of the valves is color coded to ensure the correct valve is installed in the correct port. A replaceable filter is mounted on the inlet opening to the pump.

There are no provisions for a manual relief valve on the valve body. If the pump or motor fails, manually raising or lowering the stern drive would require the appropriate hydraulic line be loosened on the valve body external fitting.

Access to the valve body requires the reservoir to be drained and removed.

PRESSURE RELIEF VALVE—UP

The UP pressure relief valve is threaded onto the valve body. This valve regulates the maximum oil pressure in the hydraulic system during the UP cycle to between 2200–2600 psi (15173–17932 kPa). The valve is non-adjustable, and the replacement valve is color coded Blue.

PRESSURE RELIEF VALVE—DOWN

The DOWN pressure relief valve is threaded onto the valve body. The valve regulates the hydraulic pressure between 400–600 psi (2759–4138 kPa) in the DOWN circuit. The valve opens under pressure and allows hydraulic fluid to flow from the pump to the aft chamber of the trim/tilt cylinders. The aft chamber of the cylinder is identified as the chamber behind the piston. The valve prevents the stern drive from raising during Reverse operation, but does allow fluid to flow from the aft chamber of the hydraulic cylinders to the reservoir during the UP cycle. The valve is non-adjustable, and the replacement valve is color coded Green.

THERMAL RELIEF VALVE

The thermal relief valve prevents hydraulic lock-up in the full tilt-up position caused by the thermal expansion of the fluid. The pressure setting of this valve is non-adjustable, and the replacement valve is color coded Gold.

Troubleshooting

The electric pump motor must operate to drive the mechanical hydraulic pump. The hydraulic pump will convert the electrical energy into mechanical energy. The mechanical energy moves the hydraulic fluid to extend and retract the trim cylinders. Use the following troubleshooting procedures to determine if the problem is electrical or mechanical.

Determine If Problem Is Electrical Or Mechanical

1. Verify that the battery is fully charged before continuing the troubleshooting procedure. If the battery is not fully charged, the system could give a false indication of a problem when none actually exists.

2. Verify the fluid level in the reservoir is up to the **MAX** mark on the side of the reservoir.

3. Depress the **UP** or **DOWN** button on the control panel and listen for the pump motor to operate or the solenoid to click.

4. If the pump motor does not operate, and the solenoids do not "click", the problem is electrical. Refer to electrical troubleshooting.

5. If the pump motor operates, but the stern drive does not move, the problem is most likely a hydraulic or mechanical problem. Refer to mechanical troubleshooting.

ELECTRICAL

◆ **See Figures 3, 4, 5 and 6**

1. Obtain a DVOM and set the switches for 12-volt DC readings. Connect the meter Red lead to the 110 amp circuit breaker terminal. Connect the meter Black lead to the pump ground terminal. The meter should indicate 12-volts or battery voltage. If no voltage is indicated, check the battery cable connections to the pump 110 amp circuit breaker.

2. Set the DVOM switches for continuity reading. Connect the Red meter lead to the pump 110 amp circuit breaker terminal. Connect the Black meter lead to the solenoid buss bar. The meter should indicate no resistance and full continuity. If there is no continuity, or a high resistance is indicated, the 110 amp circuit breaker is defective and should be replaced with a new unit.

3. Connect a jumper wire (#10 AWG) to the Blue/White wire terminal on the UP solenoid. Connect the other end of the jumper wire to the pump 110 amp circuit breaker terminal. If the pump operates and the stern drive moves, the problem is in the control panel **UP** switch, the UP solenoid, the 20 amp in-line fuse, or the wiring between the control panel switch and the solenoids is defective. Check the in-line fuse and continue with the troubleshooting.

4. Connect a jumper wire (#10 AWG) to the Green/White wire terminal on the DOWN solenoid. Connect the other end of the jumper wire to the pump 110 amp circuit breaker terminal. If the pump operates and the stern drive moves, the problem is in the control panel **DOWN** switch, the Down solenoid, or the wiring between the control panel switch and the solenoid. Continue with the troubleshooting. If the pump fails to operate, the pump is frozen or the motor is defective. Separate the motor from the pump and repeat the test. If the motor still fails to operate, repair or replace the defective motor.

5. Depress and hold both **UP** switches on the control panel or the **TRAILER** switch on the remote control throttle. Listen for the UP solenoid to "click". If the solenoid fails to "click", connect the ohmmeter Red lead to the Blue/White wire terminal and the Black meter lead to the Black wire terminals on the UP solenoid. Again depress and hold both **UP** switches on the control panel or the **TRAILER** switch. If 12-volts is not indicated, the switch is defective or the 20 amp fuse is blown. Replace the blown fuse or replace the defective switch. If 12-volts is indicated on the meter, the UP solenoid is defective and must be replaced.

6. Repeat this test but hold in the switch for the **DOWN** position. Connect the ohmmeter to the Green/White wire on the DOWN solenoid terminal and the Black meter lead to the Black wire terminal on the DOWN solenoid terminals. Depress the **IN** or **DOWN** switch on the control panel. If 12-volts is not indicated, the switch or wiring is defective. Replace the defective switch or wiring. If 12-volts is indicated on the meter, the DOWN solenoid is defective and must be replaced.

SYMPTOMS AND PROBLEMS

Troubleshooting some problems can be done more easily by identifying the symptom first. Below is a list of the most common symptoms found on the trim/tilt system and the areas to check for defective components.

Trim Pump Motor Fails To Operate With OUT/UP or IN/DOWN Switches Pressed—Solenoids Fail To "Click"

1. Thermal breaker in pump open. Allow motor to cool. Replace breaker switch in motor end plate if it does not reset.

2. 20-Amp fuse blown. Check trim limit switch wires for broken insulation and shorted wires. Check pump wiring harness for damaged insulation, open or shorted wiring.

3. Check for 12-volts at pump 110 amp circuit breaker terminal.

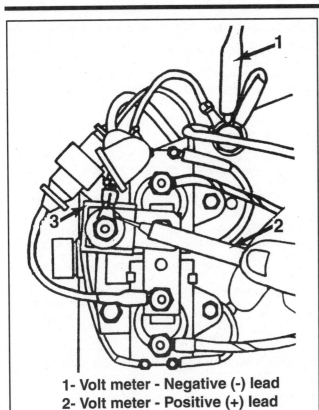

1- Volt meter - Negative (-) lead
2- Volt meter - Positive (+) lead
3- Fuse (Red)

Fig. 3 Check that there is battery voltage

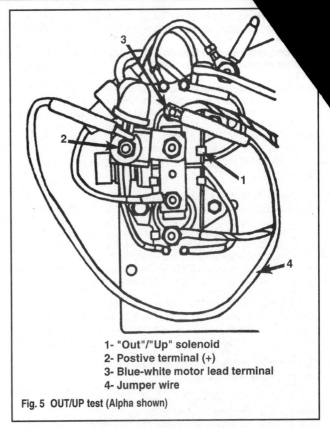

1- "Out"/"Up" solenoid
2- Postive terminal (+)
3- Blue-white motor lead terminal
4- Jumper wire

Fig. 5 OUT/UP test (Alpha shown)

1- 110 amp fuse (red)
2- Ohmmeter leads

Fig. 4 Checking the 110 amp fuse (breaker)

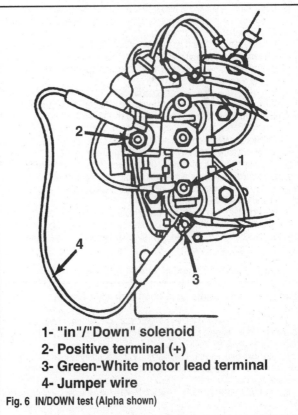

1- "in"/"Down" solenoid
2- Positive terminal (+)
3- Green-White motor lead terminal
4- Jumper wire

Fig. 6 IN/DOWN test (Alpha shown)

...p fuse. If no voltage is present, check
...ort or loose connection at the battery.

...With OUT/UP Or IN/DOWN
...k"

...corrosion or loose connection.
...out brushes, defective armature, field frame, or
...ed by water or oil. Repair or replace pump motor.
...ng harness has internal short, between OUT/UP and IN/DOWN
...cuits. Pump is trying to operate in both directions at same time. Check
wire harness for broken, damaged, melted insulation. Check control panel
switch assembly for damaged wiring and burned contacts in the switch
assembly.

Trim Pump Motor Will Not Operate In The OUT/UP Direction—Solenoid Does Not "Click"

1. Check UP solenoid wire connections for loose, dirty, or corroded
connections. Clean, tighten, repair or replace connector.
2. Check for 12-volts on Blue/White wire while holding OUT/UP switch
depressed. Voltage present—UP solenoid is defective and must be replaced.
Voltage not present—check for open Blue/White wire between solenoid and
control panel switch.
3. Possible bad field winding in pump motor. Connect 12-volt jumper
wire to Blue/White wire terminal on motor. If motor operates, solenoid is
defective and should be replaced. If motor does not run, the motor is
defective and must be replaced.

Trim Pump Motor Will Not Operate In The IN/DOWN Direction—Solenoid "Clicks"

1. Check DOWN solenoid wire connections for loose, dirty, or corroded
connections. Clean, tighten, repair or replace connector.
2. Check for 12-volts on Green/White wire while holding the IN/DOWN
switch depressed. If voltage is present—DOWN solenoid is defective and
must be replaced. Voltage not present—check for open Green/White wire
between solenoid and control panel switch.
3. Possible bad field winding in pump motor. Connect 12-volt jumper
wire to Green/White wire terminal on motor. If motor operates—solenoid is
defective and should be replaced. If motor does not operate—motor is
defective and must be replaced.

Trim Pump Motor Will Not Operate In The Out/Up Direction With TRAILER Switches Depressed—Solenoid "Clicks"

1. Check the UP solenoid Blue/White wire connections for loose, dirty, or
corroded connections. Clean, tighten, repair or replace connector.
2. Check for 12-volts on Blue/White wire while holding both TRAILER
switches depressed. Voltage present—UP solenoid is defective and must be
replaced. Voltage not present—check for open in Blue/White wire between
solenoid and control panel switch.
3. Possible bad field winding in pump motor. Connect 12-volt jumper
wire to Blue/White wire terminal on motor. If motor operates—solenoid is
defective and should be replaced. If motor does not operate, the motor is
defective and must be replaced.

HYDRAULIC AND MECHANICAL PROBLEMS

Stern Drive Trims Out/Up Or In/Down With Jerky Motion

1. Oil level in reservoir is low. Service reservoir to proper level. Cycle
stern drive Up and Down five times to remove any trapped air. Re-service
reservoir and check for leaks.
2. Trim cylinders sticking or binding. Disconnect trim cylinders from lower
unit end and cycle. Observe which cylinder is binding and replace the
defective cylinder.
3. Hydraulic hose(s) pinched in system. Examine the hydraulic hoses for

sharp bends, pinched hoses, or bent tubing. Replace any damaged hoses or
tubing.
4. Hose reversed on one cylinder at connection block. Reverse the hose
connection and test.
5. Pump pressure is low and one of the internal pressure relief valves
has malfunctioned. Repair or replace the pressure relief valve(s) on the
pump manifolds.

Stern Drive Cannot Be Lowered From Full Up Position Or Lowers With A Jerky Motion

1. Oil level low in reservoir or air in the system. Service reservoir to
proper level. Cycle stern drive UP and DOWN five times to remove any
trapped air. Re-service reservoir and check for leaks.
2. Hydraulic hose(s) pinched in system. Examine the hydraulic hoses for
sharp bends, pinched hoses, or bent tubing. Replace any damaged hoses or
tubing.
3. Trim cylinders sticking or binding. Disconnect trim cylinders from lower
unit end and cycle. Observe which cylinder is binding and replace the
defective cylinder.
4. Hose reversed on one cylinder at connection block. Reverse the hose
connection and test.
5. Stern drive unit is binding on gimbal ring housing. Check the hinge
pins and bushings in the gimbal ring for corrosion and binding. Check the
driveshaft housing for misalignment and/or damage. Check the upper gear
housing universal bearings for binding, damage and/or misalignment with
the engine. Replace any defective components.

Stern Drive Slowly Drops From Up Position During Extended Storage

1. Check entire hydraulic system for external leaking. Repair or replace
leaking components.
2. Pump thermal relief valve, pilot check valve or seal leaking internally.
Replace defective valve, seals or valve body.
3. Trim cylinder(s) leaking internally and the pump DOWN circuit has an
internal leak (both components must be defective for this condition to exist).
The manufacturer suggests installing a Trim Pump Rebuild Kit.

Stern Drive Will Not Stay In Out/Up Position While Underway

1. Check for air in the hydraulic system. Service the reservoir to the
proper level. Cycle the stern drive five times from full UP to full DOWN to
remove air from the system.
2. Check entire hydraulic system for external leaking. Repair or replace
leaking components.
3. Pump thermal relief valve, pilot check valve or seal leaking internally.
Replace defective valve, seals or valve body.
4. Trim cylinder(s) are leaking internally and the pump DOWN circuit has
an internal leak (both components must be defective for this condition to
happen). The manufacturer suggests installing a Trim Pump Rebuild Kit.

Stern Drive Trails Out/Up Upon De-acceleration Or When Shifting The Unit Into Reverse

1. Trim cylinder(s) are leaking internally. Rebuild or replace the trim
cylinders.
2. The pump DOWN circuit has an internal leak. The manufacturer
suggests installing a Trim Pump Rebuild Kit.

Oil Foams Out The Vent Fill Cap On The Reservoir

1. The hydraulic system oil is contaminated. Drain and flush the
hydraulic system with clean oil. Fill the reservoir and cycle the system four
or five times to remove any air.
2. Fluid level is low which caused the fluid to foam. Service reservoir to
the proper level. Cycle stern drive UP and DOWN five times to remove any
trapped air. Re-service reservoir and check for leaks.

Service Precautions

The following nine points should always be observed when installing, testing, or servicing any part of the power trim/tilt system.

1. Coat the threads of all fittings and O-rings with Quicksilver Power Trim and Steering Fluid, or an equivalent product.

2. Use clean Power Trim and Steering fluid when filling the hydraulic fluid reservoir. If the Power Trim and Steering fluid is not available, substitute with SAE 10W-30 or 10W-40 motor oil in an emergency and then drain the system as soon as possible..

3. The stern drive must be in the full DOWN position with the hydraulic cylinders collapsed when checking the fluid level or adding fluid to the reservoir. Normal fluid level should be maintained between the **MIN** and **MAX** marks on the side of the reservoir. If the quantity is low, add fluid until the level is up to the **MAX** mark on the side of the reservoir.

4. If the pump stops during long use, allow the pump motor to cool at least five minutes before starting again. An internal thermal circuit breaker protects the pump motor from overheating. If the pump still will not operate, the system 110 amp fuse may have blown.

5. Keep the work area clean when servicing or disassembling parts. The smallest amount of dirt or lint can cause failure of the pump to operate.

6. The valve body, pump assembly, pressure relief valves, and thermal relief valves must be replaced as individual assemblies. There are no piece parts available for these components except for the O-rings that are sold in a kit.

7. The motor is protected from internal fluid leaks and external moisture by seals, O-rings, and a grommet on the field winding harness. If the motor fails due to hydraulic fluid, clean the motor thoroughly and replace the pump internal seal and O-ring. If the motor fails from moisture as evidenced by internal corrosion, the wire harness grommet is damaged or deteriorated, and the pump motor should be replaced. Sometimes liquid neoprene can be used on the wire harness grommet to seal minor cracks. Use liquid neoprene with care to prevent contaminating the electric motor or hydraulic valves.

8. Keep the trim cylinders attached to the forward anchor pin during servicing and repairs. Do not allow the trim cylinders to hang by their hydraulic hoses. Such practice may kink and damage the hoses. Use care when handling trim cylinders during removal/installation of the stern drive unit. Rough treatment of the hoses could result in a weak-ended hose, partial separation at the fittings, or bending of the metal tubing, any one of which could restrict the flow of fluid to the trim cylinders.

9. Always hold both fittings with a wrench when tightening or loosening hydraulic hoses and fittings. When installing flex hoses, be sure there are no twists or sharp bends in the flex hose. The fitting on the cylinder end of the hose must point approximately 45º from the stern drive centerline towards the transom. A twisted hose will cause severe loads and result in stress that could bend the tubing ends. On the stern drive models covered in this manual the hoses are routed directly from the cylinder to a block under the gimbal housing. Take extra care when attaching the hoses to the block. If the fittings are cross threaded, the block would have to be replaced. This job would entail the stern drive being removed and possibly the engine.

System Testing

High pressure testing of hydraulic components requires expensive special gauges, tools and highly trained personnel to accurately interpret test results. A danger to the operator and others in the area always exists during the testing process. Therefore, it is highly recommended that the boat, with the trim/tilt system installed, be taken to a qualified repair facility having the necessary equipment to conduct the testing properly and safely.

Do not attempt to work on hydraulic hose connections without the proper tools designed for that specific purpose. The use of a common box end wrench will quite likely result in "rounding off" the flats on a hydraulic fitting. It is imperative that you have a set of "flare" or "line" wrenches that will almost guarantee the fitting will not to be damaged.

System Bleeding

◆ See Figures 7 and 8

■ **Please refer to the Maintenance section for procedures on filling the pump reservoir.**

The power trim/tilt system is designed to be self-purging with nothing to be done other than raising and lowering the stern drive unit several times. If, during service work, components were installed already filled with hydraulic fluid, or a line was disconnected and then reconnected, the self-purging feature of the system will adequately eliminate the small amount of unwanted air. However, if components were removed and installed "dry", the following procedures must be performed to adequately purge all air from the system.

✳✳ WARNING

Check to be sure the vent hole opening in the reservoir fill cap is free and clear. A small amount of air must be allowed to enter the reservoir tank as the fluid level drops or vacuum could develop in the tank making it difficult for the pump to draw the fluid from the reservoir.

1. Depress the **IN/DOWN** switch to operate the pump and lower the stern drive to the full DOWN position. Verify that the fluid level in the power trim reservoir is between the **MAX** and **MIN** marks cast into the side of the reservoir. If necessary, fill the reservoir with Quicksilver Power Trim and Steering fluid or an equivalent product.

2. Begin by bleeding the OUT/UP side of the trim cylinder. If one or both cylinders were replaced without filling the cylinders with fluid, be sure to check the reservoir fluid level often while performing this procedure. Disconnect the OUT/UP hose from the end of the trim cylinder. If one cylinder was rebuilt, then disconnect the one hose. If both trim cylinders were rebuilt, disconnect the hose from both trim cylinders. Place the ends of the hose(s) into a suitable container. Have an assistant press and hold the **UP** button on the control panel. When a solid stream of fluid, free of any air bubbles, is observed escaping from the end of the hoses, release the **UP** control button. Connect the hoses to the hydraulic cylinders, and service the reservoir, if needed.

3. Bleed the IN/DOWN side of the trim/tilt cylinder. Again, check the fluid level in the power trim pump reservoir and fill to the **MAX** mark cast into the side of the reservoir. Be sure to check the fluid level often while performing this procedure.

4. Disconnect the **IN/DOWN** hose from the hydraulic connector block (NOT the cylinder as before!). If one cylinder was rebuilt, then disconnect the one hose. If both trim cylinders were rebuilt, disconnect the hoses for both cylinders. Install a special pip thread plug P/N 22-38609, or equivalent into the opening in the hydraulic connector block.

5. Place the ends of the hose(s) into a suitable container. Stand clear of the stern drive because it will begin to move in this next step.

6. Have an assistant press and hold the **UP** button on the control panel. The cylinders will begin to extend raising the stern drive. At the same time, air and fluid will be forced out the end of the hoses. When the cylinders are fully extended—have your assistant release the control button.

✳✳ CAUTION

Be sure to check the power trim reservoir fluid level often during this procedure. If the reservoir fluid level drops to the MIN mark, stop and service the reservoir. Failing to maintain the proper fluid level will cause the pump to draw air into the system.

7. Remove the special pipe plug(s) from the connector block and just momentarily press the **DOWN** button on the control panel. This will remove any air from the DOWN circuit. Now, connect the trim cylinder hoses to the connector block and tighten them securely. Press and hold the **IN/DOWN** button on the control panel until the stern drive is in the full DOWN position.

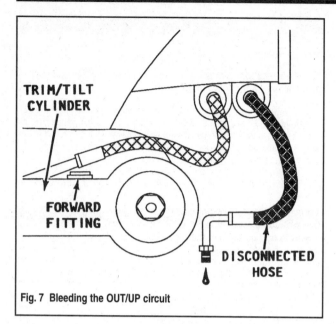

Fig. 7 Bleeding the OUT/UP circuit

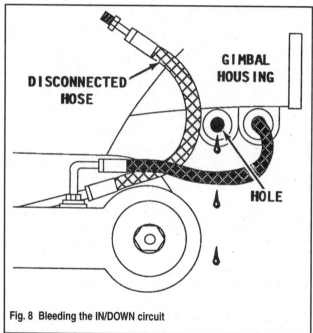

Fig. 8 Bleeding the IN/DOWN circuit

Check and service the reservoir fluid level, if needed. Raise and lower the stern drive four or five times to remove any trapped air in the trim system.

Trim Limit Switch/Trim Position Sender

The gauge mounted on the control panel indicates to the helmsperson the trim position of the stern drive at all times. An electrical sending unit on the starboard side of the stern drive is mechanically indexed to the hinge pin. When the stern drive is raised or lowered, the sending unit sends a variable voltage signal to the gauge unit on the control panel indicating the trim position of the stern drive. The sending unit and wire harness are sealed to prevent entry of water. Therefore, the unit is not internally serviceable. The gauge mounted on the control panel is also a sealed unit and is not internally serviced.

The trim sender is located on the starboard side of the stern drive and the trim limit switch is on the port side. Would you believe, one is a sending unit and the other is a switch, but their appearance on the outside is identical.

The trim limit switch on the other hand is only a switch. When the stern drive is trimmed OUT to a pre-determined limit, the switch will open, causing the OUT/UP solenoid to open. This action turns off the trim pump and the stern drive cannot be trimmed out past its pre-determined height. If the trim limit switch is adjusted incorrectly, it could cause the stern drive to trim out past the gimbal ring support flanges and damage the stern drive or boat transom.

ADJUSTMENT

Trim Position Sender

◆ See Figures 9 and 10

1. Lower the stern drive to the full DOWN position. Be sure the stern drive trim is fully IN and it is resting against the stop pin. Turn the stern drive hard-over to the port position exposing the trim sender on the gimbal housing hinge pin.
2. Pop the cap off the sender and check that the index mark on the center rotor is aligned with the index mark on the sender body.
3. Now, loosen the two adjusting screws on the slotted portion of the sender housing. Turn the ignition key switch to the RUN position without starting the engine. Rotate the sender housing in the required direction until the pointer on the gauge is at the bottom of the green arc. The trim position sender is now properly adjusted. Tighten the adjustment screws securely. Turn the ignition key switch to the OFF position.

Trim Limit Switch

◆ See Figures 9, 11, 12, 13 and 14

Turn the stern drive hard-over to starboard with the stern drive in the full DOWN position.

1. Pop the cap off the switch and check that the index mark on the center rotor is aligned with the index mark on the switch body.
2. Loosen the two adjusting screws on the slotted portion of the limit switch housing and turn it clockwise to the end of its slots.
3. Have an assistant press and hold the UP/OUT trim switch. Slowly rotate the trim limit switch housing counterclockwise until the trim cylinders

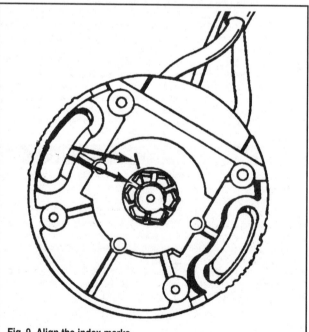

Fig. 9 Align the index marks...

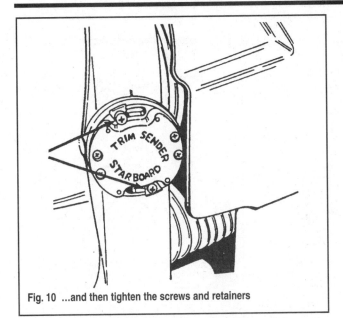

Fig. 10 ...and then tighten the screws and retainers

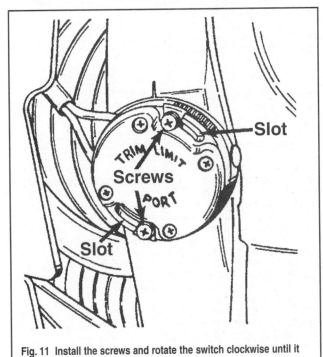

Fig. 11 Install the screws and rotate the switch clockwise until it stops...

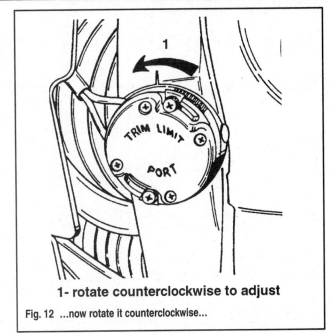

1- rotate counterclockwise to adjust

Fig. 12 ...now rotate it counterclockwise...

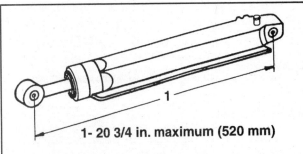

1- 20 3/4 in. maximum (520 mm)

Fig. 13 ...until the cylinder meets this dimension (Alpha shown)

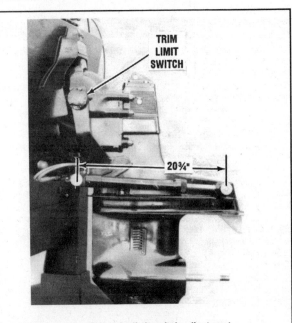

Fig. 14 A better view of the trim limit switch adjustment

have extended to a distance of 20-3/4 in. (520mm), measured between the forward and aft cylinder anchor pins on Alphas; On Bravos, the dimension should be 21-3/4 in. (552mm) maximum.

4. Tighten the screws on the switch housing securely.

REMOVAL & INSTALLATION

◆ **See Figures 15, 16 and 17**

1. Remove the stern drive as detailed in the Drive System section.
2. Remove the two mounting/adjustment screws with their washers and retainers and lift off the switch or sender unit, allowing it to dangle by the electrical leads.

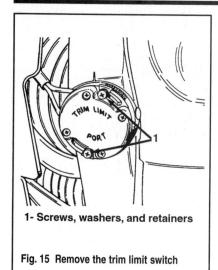

1- Screws, washers, and retainers

Fig. 15 Remove the trim limit switch

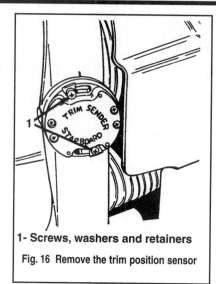

1- Screws, washers and retainers

Fig. 16 Remove the trim position sensor

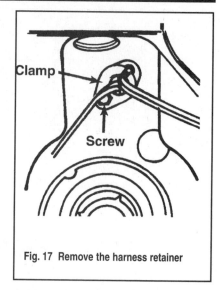

Fig. 17 Remove the harness retainer

3. Remove the U-joint bellows sleeve.

4. Remove the gimbal ring hinge pins.

5. Grab the bell housing and pull backwards on it while rotating it about 90°, until you have access to the wire harness retainer.

6. Remove the harness retainer, disconnect the wires at the engine and pull them through.

To install:

7. Feed the wires through the hole in the transom assembly and then slide the rubber grommet halves together in position so that the straight edges are vertical.

8. Install the retainer and tighten the screw to 90–100 inch lbs. (10–11 Nm).

9. Install the U-joint bellows, hinge pins and stern drive as detailed in the Drive System section.

10. With the drive unit in the full DOWN position, rotate the center rotor on the switch/sender until its index mark aligns with the index mark on the switch/sender body.

11. Install the cap and then install the switch/sender unit to the gimbal ring with the retainers and screws.

12. Adjust the units as detailed above.

Pump Assembly

The pump motor is a 12-volt DC bi-directional motor. The armature of the motor rests inside the field winding and is energized through brushes contacting the commutator on the end of the armature. When the up or down switches are closed, current is directed through the motor field in one of two directions causing the motor to run in one desired direction.

The opposite end of the armature is mechanically splined to the hydraulic pump. The gear rotor action of the pump causes the fluid to flow in either direction.

The motor case is sealed with O-rings and a shaft seal to prevent the entry of water or hydraulic fluid into the motor cavity.

Before going directly to the pump as a source of trouble, check the battery to be sure it is up to a full charge. Inspect the wiring for loose connections, corrosion and damaged wires. Take a good look at the control switches and connection for evidence of a problem.

The control switches may be quickly eliminated as a source of trouble by connecting the pump directly to the battery for testing purposes.

Testing the control panel switch must be done at the back of the switches because the wires are soldered to the switch terminals and cannot be disconnected. Unplug the trim harness from the trim pump.

Connect an ohmmeter, set for continuity reading, between the terminals on the back of the switch being tested. The meter must indicate an open circuit—no meter movement—with the button in the free position.

The meter must indicate continuity—full meter deflection—with the button(s) depressed. Remember to press both up buttons when checking the Trailering circuit on a 3-button control panel.

If the control panel switches are found to be satisfactory, the problem is with the solenoid(s), pump motor or the hydraulic pump. These components may be checked after the trim/tilt assembly is removed from the boat.

REMOVAL & INSTALLATION

◆ **See Figures 18 and 18a**

1. Using the down switch on the remote control throttle or the control panel, lower the stern drive to the full down position. If the stern drive will not lower electrically, place a suitable drain pan under the hydraulic lines. Now, loosen the UP/OUT line fittings on the hydraulic connector. Allow the fluid to flow into the drain pan until the stern drive is in the full down position and then tighten the fittings.

2. Disconnect the battery cables from the battery terminals. Disconnect the 12-volt positive cable from the 110 amp circuit breaker terminal.

3. Mark or tag the hydraulic hoses (right or left) before disconnecting them from the pump. The Black hydraulic hose on the left is for the UP circuit and the Gray hydraulic hose on the right is for the DOWN circuit. If possible, cap the hoses to prevent fluid leakage into the boat.

4. Pull to disconnect the three-prong trim harness connector from the trim pump.

5. Remove the bolt and nut on the side of the support bracket securing the engine wire harness, pump motor and solenoid Black ground wire terminals to the pump support bracket.

Remove the lag bolts securing the trim/tilt pump assembly to the boat. Lift the assembly out and free of the boat.

To install:

6. Position the trim/tilt pump assembly in the boat. Install the four lag bolts through the pump support bracket into the floor. Tighten the bolts securely.

7. Remove the cap from the hose and connect the Gray (DOWN) hose to the fitting marked DN on the adapter. Thread the hose fitting into the adapter several turns by hand to ensure it is not cross-threaded. Tighten the fitting to 70–150 inch lbs. (7.9–16.9 Nm) on the Alpha, or 110 inch lbs. (12 Nm) on the Bravo.

8. Remove the cap from the hose and connect the Black (UP) hose to the fitting marked up on the adapter. Thread the hose fitting into the adapter several turns by hand to ensure it is not cross-threaded. Tighten the fitting to 70–150 inch lbs. (7.9–16.9 Nm) on the Alpha, or 110 inch lbs. (12 Nm) on the Bravo.

9. Connect the three-wire trim/tilt harness to the pump bracket. Connect the battery (Red) positive 12-volt cable to the 110 amp circuit breaker. Apply a coating of liquid neoprene to the wire and terminal. Connect the battery (Black) negative to the ground terminal lug on the side of the support bracket. Secure the cable with the bolt and nut.

10. Remove the cap from the reservoir filler opening. Verify the hole in the cap is free and clear. If this hole becomes clogged, it could cause the hydraulic pump to cavitate, and the stern drive to raise or lower improperly.

11. Fill the reservoir with Quicksilver Power Trim and Steering Fluid to the **Max** line.

12. Cycle the stern drive Up and Down several times to bleed all air from the system. Check the reservoir level after each cycle and service, if needed.

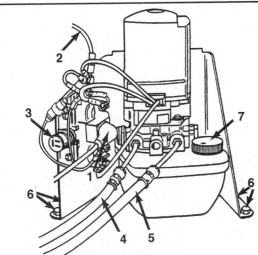

1- Positive battery lead
2- Negative battery lead
3- Harness connector
4- Black hydraulic hose ("Up" hose)
5- Gray hydraulic hose ("Down" hose)
6- Lag bolts and washers
7- Vent in fill gap

Fig. 18 Removing the trim pump

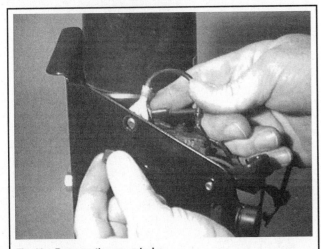

Fig. 18a Remove the ground wire

DISASSEMBLY & ASSEMBLY

◆ See Figure 19

Support Bracket

◆ See Figures 19a, 20, 21, 22, 22a, 22b, 22c

1. Remove the pump assembly.
2. Drain all the remaining fluid in the reservoir into a suitable container. Discard the fluid in accordance with any local hazardous waste ordinances.
3. Using a Phillips head screwdriver, loosen the screw securing the terminal cover on the solenoid buss. Slide the cover off the solenoid buss bar.
4. Remove the nuts and washers securing the Blue/White (Up) wire lead and Green/White (Down) wire leads from the pump motor to the solenoid terminals. Remember to tag the wires.
5. On the back of the support bracket, remove the two bolts and lockwashers securing the adapter with motor and reservoir to the support bracket. Slide the motor, adapter and reservoir out of the support bracket.
To assemble:
6. Align the back of the adapter to the support bracket and install the bolt and lock washer through the support bracket into the adapter. Tighten the bolts securely.
7. Connect the Green/White wire lead from the motor harness to the DOWN solenoid terminal. Secure the wire to the terminal with a flat washer and nut. Apply a coating of liquid neoprene to the wire and terminal.
8. Connect the Blue/White wire lead from the motor harness to the UP solenoid terminal. Secure the wire to the terminal with a flat washer and nut. Apply a coating of liquid neoprene to the wire and terminal.
9. Install the terminal cover over the solenoid terminals and secure it in place with a Phillips head screw in the center.
10. Connect the pump motor ground wire terminal to the support bracket and temporarily secure it in place with a bolt nut.

Solenoids

◆ See Figure 23

Replacement of the UP or DOWN solenoid is easily performed with the support bracket and pump assembly separated.
1. Remove the pump assembly and support bracket.
2. Mark and disconnect the wires from the defective solenoid terminal. The UP solenoid is on top and the DOWN solenoid is on the bottom.
3. Remove the two bolts securing the solenoid to the support bracket.
4. Install the new solenoid and secure it with two bolts. Connect the wires to the terminals as marked during disassembly.
Refer to the wiring schematic for the correct color code. Coat all wire terminals with liquid
neoprene after the system has been checked.

Pump Reservoir

◆ See Figures 24, 24a and 24b

1. Remove the pump assembly, support bracket and solenoids.
2. Place a scribe mark or a piece of tape on the motor case, adapter, and reservoir. This mark or tape will help orientate these components during assembling.
3. Using a 1/4 in. socket, remove the special shoulder bolt from the bottom of the reservoir. Slide the O-ring off the shoulder bolt and discard the O-ring.
4. Grasp the motor and adapter and lift them both straight up and free of the reservoir.
5. Slide the O-ring off the shoulder on the adapter and discard the O-ring.

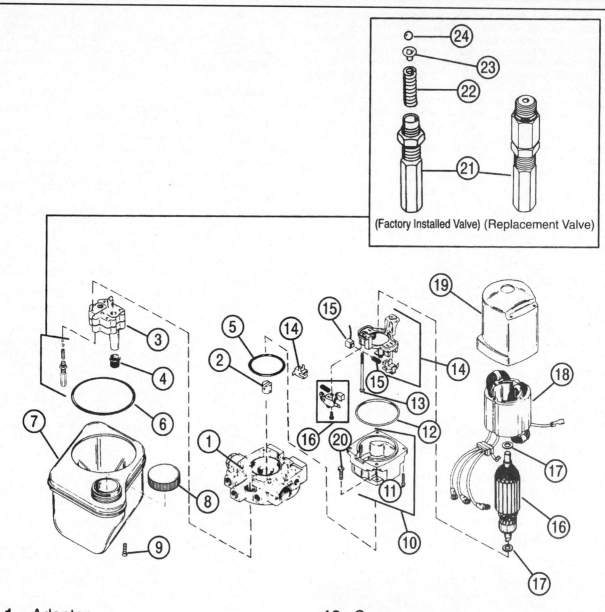

Fig. 19 Exploded view of the trim pump

1 - Adaptor
2 - Coupling
3 - Pump
4 - Filter
5 - O-ring-Motor End
6 - O-ring, Reservoir End
7 - Reservoir
8 - Cap
9 - Screw (Includes O-ring)
10 - End Cap w/Bearing
11 - Screw
12 - O-ring

13 - Screw
14 - Brush Holder Kit
15 - Brush set
16 - Armature
17 - Thrust Washer
18 - Field and Frame
19 - Housing
20 - Screw
21 - Relief Valve With:
22 - Spring
23 - Eyelet
24 - Check Ball

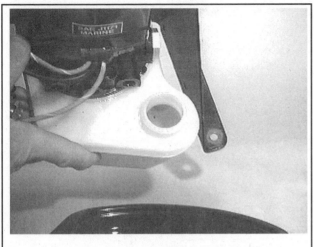

Fig. 19a Drain the fluid into a container

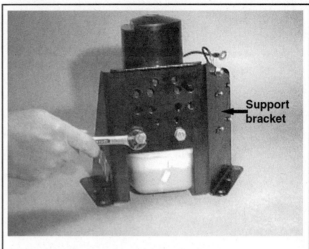

Support bracket

Fig. 22 Remove the mounting bolts...

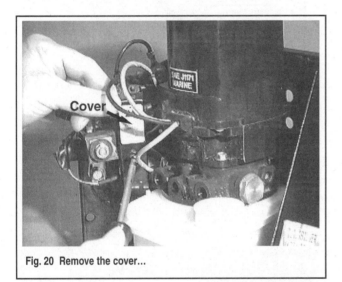

Cover

Fig. 20 Remove the cover...

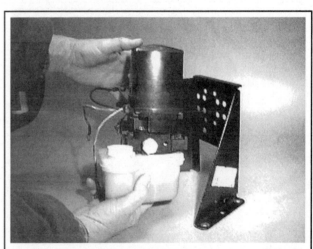

Fig. 22a ...and lift out the pump assembly

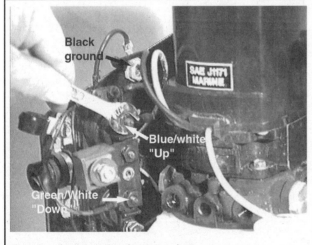

Black ground

Blue/white "Up"

Green/White "Down"

Fig. 21 ...disconnect the trim motor wires

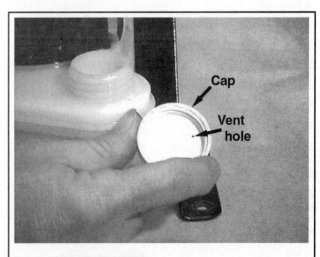

Cap

Vent hole

Fig. 22b Check the vent hole...

Fig. 22c ...and then fill the reservoir with fluid

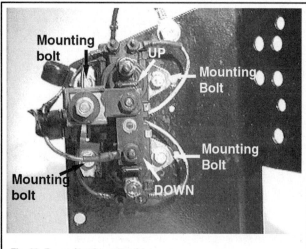

Fig. 23 Removing the solenoids

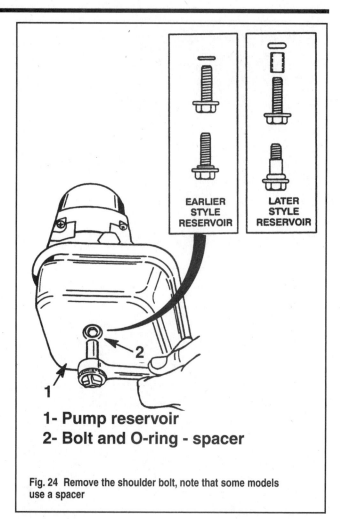

1- Pump reservoir
2- Bolt and O-ring - spacer

Fig. 24 Remove the shoulder bolt, note that some models use a spacer

To assemble:

6. Place a new O-ring onto the adapter and lightly lubricate it with Quicksilver Power Trim and Steering Fluid, or regular old motor oil.

7. Align the scribe marks or tape strips placed on the adapter and reservoir tank prior to disassembly.
Carefully lower the adapter onto the reservoir to avoid damaging the filters on the ends of the pickup tubes.

8. Place a new O-ring onto the shoulder bolt. Insert the bolt up through the bottom of the reservoir and into the adapter. Tighten the bolt securely.

Hydraulic Pump And Adaptor

◆ See Figures 25, 26, 27, 28, 29, 30, 31, 31a, 32, 33, 34 and 35

1. Remove or disassemble the support bracket, solenoids and reservoir.

2. Scribe mark or place a piece of tape across the motor case, adapter, and reservoir. This mark or tape will help orientate these components during assembling. Using a 1/4 in. socket, remove the special shoulder bolt from the bottom of the reservoir. Slide the O-ring off the shoulder bolt and discard the O-ring.

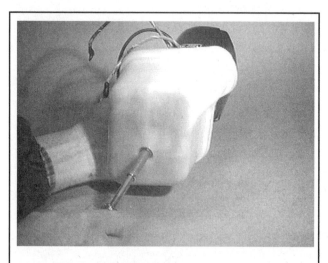

Fig. 24a Pull the bolt all the way out...

Fig. 24b   and lift out the pump

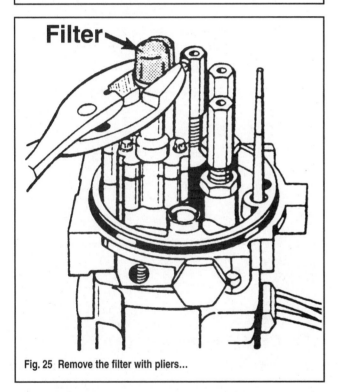

Fig. 25 Remove the filter with pliers...

3. Set the adapter and motor assembly back onto the reservoir temporarily. Remove the two bolts securing the electric motor to the adapter. Lift the motor straight up and free of the adapter. Remove and discard the O-ring on the shoulder of the adapter. Reach in and lift out the motor shaft coupler and set it aside for safe keeping.

■ The UP, DOWN and THERMAL pressure relief valves are installed and preset at the factory. A jam nut at the base of the valve holds this factory setting. Therefore, if the jam nut is loosened, as in removal, the factory setting of the valve will be changed and the valve must be replaced with a new replacement valve. Do not remove these valves unless you intend to replace them. The replacement valves are set at the factory and are secured to an adapter fitting. The adapter fitting is then threaded into the original pressure relief valve port and tightened. The spring, eyelet and ball are contained in the adapter for the replacement valve. Remove the factory valve only if it is known to have failed and then it must be replaced with a new replacement valve assembly.

All three valves look identical except the replacement valves are color-coded to prevent the wrong valve being installed. The colors for the replacement valves are as follows:

- UP Pressure Relief Valve—blue
- DOWN Pressure Relief Valve—green
- THERMAL Pressure Relief Valve—gold

Be sure the valve has failed before it is removed from the adapter. When installing a new replacement valve, make sure it is the correct color code and installed into the correct adapter position. Replacement valves may be installed and removed provided a new O-ring is used each time the adapter fitting is threaded into the adapter. Place the wrench only on the adapter fitting. Never tighten the valve by the hex flats on the valve or the jam nut.

4. If the filter is damaged or clogged, twist and pull the screen mesh and retaining ring off the end of the adapter inlet pipe. A new filter will be installed during assembly.

5. Using the correct size wrench, loosen the jam nut on one, two, or all three pressure relief valves. Un-thread the valves from the adapter and remove the ball, eyelet, and spring. Discard the valve and small parts.

■ The hydraulic pump is non-serviceable and is replaced only as a complete unit. The pump is secured to the adapter housing with two lobular head socket screws. If the special wrench for these lobular head screws is not available, use a standard 3/16 in. (5mm) six point socket.

6. Examine the base of the pump where it contacts the adapter. Two screws go through the pump and into the adapter. Two other screws only secure the pump halves together. Remove the two screws securing the pump to the adapter.

7. Lift the pump straight up and out of the adapter housing. Remove the two O-rings on the bottom of the pump and discard the O-rings. If the pump is defective, discard the pump.

8. Insert a flat blade screwdriver through the center of the pump shaft oil seal and pry the seal from the adapter. Discard the oil seal.

9. Place the adapter in a vise with soft face jaws with the large hex plugs exposed. Remove the large hex plug (2), O-ring(s) and spring(s).

10. Remove the adapter from the vise and remove the loose poppet valve(s) from the adapter.

11. A rebuild kit is required to remove and replace the check valve bodies and spool valve in the adapter. Insert the 1/8 in. diameter rod which comes with the rebuild kit against the check valve. Using a soft face mallet, tap the check valves, and spool valve out the other side of the adapter. Discard the check valve bodies. Clean the hex plugs, retainers and spool valve.

To assemble:

12. Clean all components in cleaning solvent and dry with low pressure compressed air. If compressed air is not available, make sure they air-dry completely. Replace all O-rings, seals and gaskets.

13. Inspect the check valve bodies and spool valve for signs of scoring and scratches. Check the poppet valve and seat for signs of grooves or wear on the poppet valve taper seat. If the seat or valve is worn, both the valve and valve body should be replaced.

14. Good shop practice dictates that a service rebuild kit for the hydraulic pump be purchased and all components in the kit be installed, any time the pump is disassembled.

15. Lubricate all components during assembly with Quicksilver Power Trim and Steering Fluid or motor oil.

16. Lubricate the check valve body, O-ring, and spool valve with Quicksilver Power Trim and Steering Fluid or motor oil. Slide the O-ring onto the check valve bodies and insert the spool valve into the center of the adapter. Slide a new check valve body onto each side of the spool valve. Work slowly and carefully so as not to force the check valve bodies into the adapter which could damage the O-rings. Center the spool and check valve bodies in the adapter.

17. Insert the poppet valves and place them into the check valve bodies on both sides.

18. Lubricate the new hex plug O-rings and slide the O-ring onto the hex plugs. Insert the spring into the center of the hex plug, and then thread the hex plug into the adapter so that it is just finger-tight. Working on both sides equally, tighten the hex plugs approximately 1/4 turn, and then back off 1/8 turn. Continue tightening both hex plugs equally in this manner until the plugs are tightened securely. This procedure will keep the spool valve centered and will prevent damage to the check valve O-rings.

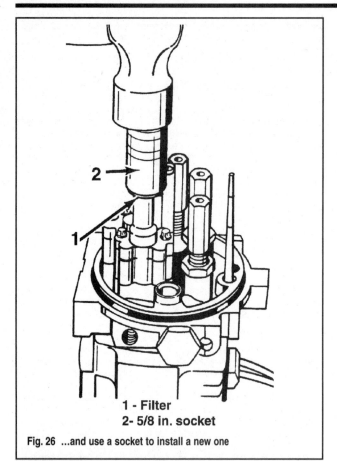

1 - Filter
2- 5/8 in. socket

Fig. 26 ...and use a socket to install a new one

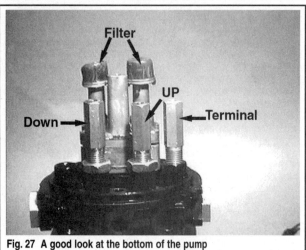

Fig. 27 A good look at the bottom of the pump

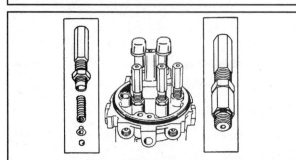

Fig. 28 Factory valve on left, replacement valve on right

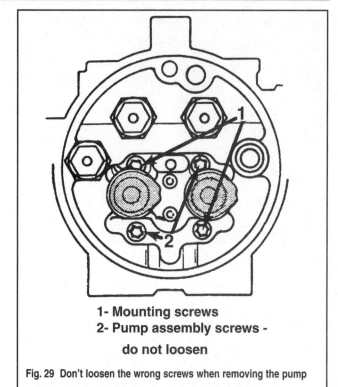

1- Mounting screws
2- Pump assembly screws -
do not loosen

Fig. 29 Don't loosen the wrong screws when removing the pump

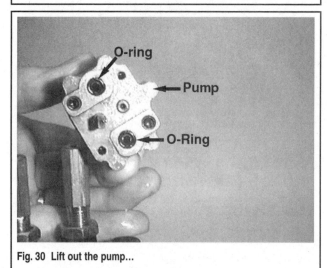

Fig. 30 Lift out the pump...

Fig. 30a...and pry out the adapter seal

Fig. 31 Don't forget to replace the O-ring on the adapter

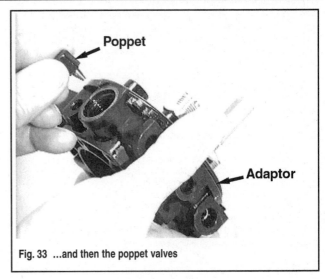

Poppet

Adaptor

Fig. 33 ...and then the poppet valves

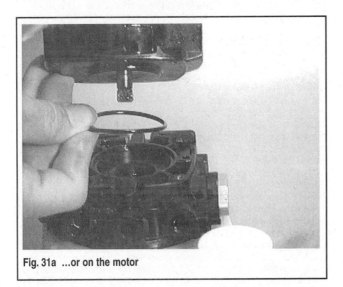

Fig. 31a ...or on the motor

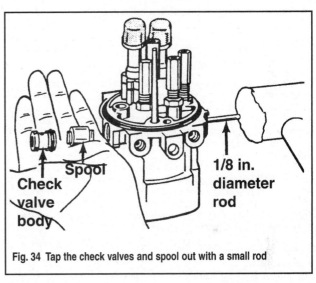

Check valve body

Spool

1/8 in. diameter rod

Fig. 34 Tap the check valves and spool out with a small rod

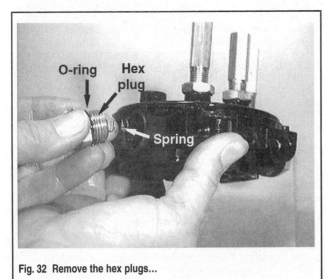

O-ring

Hex plug

Spring

Fig. 32 Remove the hex plugs...

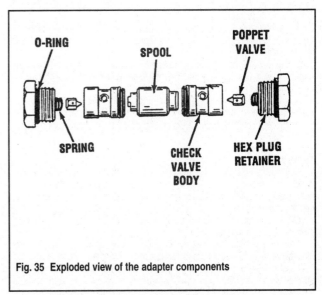

O-RING

SPOOL

POPPET VALVE

SPRING

CHECK VALVE BODY

HEX PLUG RETAINER

Fig. 35 Exploded view of the adapter components

19. Place a new pump shaft oil seal into the adapter with the lips of the seal pointing towards the pump. Press the seal into place with your thumb. Lubricate the lips of the seal with oil.

20. Place two new O-rings into the base of the hydraulic pump. Lower the pump into position on the adapter.

21. Using a 3/16 in. socket, tighten the two hex lobular screws alternately and evenly to 75 inch lbs. (8 Nm).

22. If you removed any of the relief valves, place a new O-ring onto the replacement valve fitting. Lubricate the O-ring with power steering fluid or motor oil and thread the color-coded valve into the port in the adapter. Tighten the new valve by the hex flats on the adapter fitting to 70 inch lbs. (7.9 Nm).

23. Place a new filter and sleeve over the pump inlet. Using a 5/8 in. deep socket, tap the end of the socket and drive the sleeve onto the end of the pump inlet. Repeat this step for the other filter.

24. Place a new O-ring onto the adapter and lightly lubricate it with Quicksilver Power Trim and Steering Fluid or motor oil.

25. Align the scribe marks or tape strips placed on the adapter and reservoir tank prior to disassembly. Carefully lower the adapter onto the reservoir to avoid damaging the filters on the ends of the pick-up tubes.

Pump Motor

◆ **See Figures 35a, 36, 36a, 37, 37a, 38, 38a, 39, 40, 41, 42, 43 and 44**

Be sure to keep the work area, tools and hands as clean as possible while working on the trim/tilt motor.

1. Place a scribe mark or a piece of tape on both halves of the motor case and the hydraulic adapter. These marks and/or tape will be very helpful during assembling of these components. Remove the two bolts securing the electric motor to the adapter. Lift the electric motor straight up and free of the adapter. Remove and discard the O-ring on the shoulder of the adapter.

2. Reach in and lift out the motor shaft coupler. Set the coupler aside for safekeeping.

3. Remove the four Phillips head screws securing the end cover to the motor. Lift off the end cover. It may be necessary to pry between the cover and motor because a tight seal may have formed around the wire harness grommet. Gently pull and twist on the grommet to break the tight bond. Once the grommet is free, lift off the cover. Be sure to remove the washer from inside the cover or from the armature shaft.

4. Remove the O-ring from the motor housing and discard the O-ring.

5. Loosen the screw and remove the tab securing the brush holder to the brush frame.

6. Grasp the brush holder and lift it from the brush frame. A spring behind the brush will push the brush out of the holder when it clears the end of the commutator on the armature.

7. Remove the Phillips head screw securing the thermal switch to the brush frame. Disconnect the Black wire terminal from the thermal switch and lift the switch free of the motor.

8. Remove the two Phillips head screws securing the brush frame to the motor housing. Gently move the wires aside while lifting the brush frame from the casing.

9. Grasp the armature and lift it out of the motor housing. If the thrust washer is on the end of the armature shaft, remove the thrust washer.

10. Place a scribe mark on the motor field frame and the motor housing. Lift out the field frame from the housing.

To assemble:

Any sign of oil in the pump motor indicates either the pump shaft oil seal is damaged or the vent hole in the reservoir fill cap is plugged. If the vent hole is plugged, air in the reservoir tank may not escape and oil is forced into the pump motor.

Clean the motor case with warm soap and water and blow-dry with low pressure compressed air.

Clean the armature, and field with a spray electrical contact cleaner and blow-dry with low pressure compressed air.

11. Check the armature on a growler for shorts, open windings, or shorted windings. If the commutator is worn, true it on a lathe, and undercut the mica. If a growler is not available check the armature with an ohmmeter.

Set the ohmmeter to the Rx1 scale. Connect the Black lead of the meter to the center of the armature shaft and connect the Red meter lead to each one of the commutator bars. If the meter indicates continuity between the commutator and the armature shaft, the armature is grounded and it must be replaced. If no continuity is indicated the armature is good.

12. Check the thermal switch for continuity. Obtain an ohmmeter and set the switches for Rx1 scale. Connect the Black meter lead to the thermal switch spade terminal, connect the Red meter lead to the brush lead. If continuity is indicated the switch is good. If no continuity is indicated the switch is defective and must be replaced. If the switch has high resistance, it must be replaced. Open the switch contacts and insert a piece of paper or other insulator. If continuity is indicated the switch is defective and must be replaced. If no continuity is indicated the switch is good.

13. Check the field frame for open circuits. Obtain an ohmmeter and set the switches for Rx1 scale. Connect the Red meter lead to the Blue/White wire lead on the field frame. Connect the Black meter Black lead to the brush lead. If zero ohms is indicated—full continuity—the field is good. If ohms are indicated or—no continuity—the field is open and must be replaced.

Move the Red meter Red lead over to the Green/White wire on the field. If zero ohms is indicated—full continuity—the field is good. If ohms are indicated or—no continuity—the field is open and must be replaced.

14. Check the field frame for a short. Obtain an ohmmeter and set the switches for Rx1 scale. Connect the Red meter lead to the brush lead. Connect the Black meter lead to the field metal frame. If zero ohms is indicated—full continuity—the field is defective, shorted out, and must be replaced. If zero ohms is not indicated or—no continuity—the field windings are good.

15. Check the positive brush lead on the field frame for damaged or broken insulation. Check the negative brush lead on the thermal switch for damage or frayed braided lead.

16. Check the amount of wear to the brushes. If they are worn to half their original length, approximately 3/8 in. or less, the brushes should be replaced. When replacing the brush connected to the field frame wires, cut the braided wire as close to the old brush as possible. Insert the new brush braided wire and the old field braided wire into a wire crimp provided in the brush kit. Squeeze the crimp tightly around both wires. The thermal cut-out switch and brush are replaced as an assembly.

17. Align the marks on the field frame with the marks on the motor case. These marks should have been made during disassembly. If no marks were made, the wire harness from the field must point towards the front of the motor case. The two screw holes in the field frame should also be aligned with the field winding. Slide the field frame down into the case.

18. Place the bronze thrust washer onto the end of the armature. Lower the armature through the center of the field frame and at the same time, guide the end of the armature into the motor case.

19. Align the brush frame with the field screw holes. Note the location of the thermal switch mounting pad. The pad must be directly in front of the Black spade wire or switch, if already installed. Gently pull back the wires on the field and lower the brush frame onto the end of the field frame. Align the screw holes in the brush frame, field, and motor case. Install the two Phillips head screws and tighten them securely.

20. Connect the Black wire terminal to the spade on the thermal switch. Place the thermal switch onto the mounting pad of the brush frame and secure it with the Phillips head screw.

21. Slide the brush holder over the spring and brush. Compress the spring while pushing the brush into the holder. Align the end of the brush with the commutator on the armature. Align the tabs on the bottom of the brush holder with the slots in the brush frame. Insert the brush holder into the brush frame and hold in place with finger pressure. Verify the brush holder is fully seated in the frame and the brush is contacting commutator end of the armature.

22. Place the lock tab over the brush holder and secure the brush holder in place with a Phillips head screw. Do the same for the other brush and brush holder.

23. Slide a new O-ring over the wire harness and grommet. Place the O-ring into the groove in the motor case.

24. Place the thrust washer over the end of the armature against the commutator. Align the cover with the wire harness grommet. Place a thin coating of liquid neoprene on the grommet to ensure a good seal.

25. Lower the cover over the end of the armature and align the tape marks and fastener holes. Check to be sure the O-ring remains in the groove and the grommet fits into place. Apply a drop of Loctite to the screw threads and secure the cover in place with the four Phillips head screws. Tighten the screws alternately and evenly in a cross-sequence. Do not over tighten the screws.

26. Apply a small amount of Quicksilver 2-4-C marine lubricant to the pump and motor coupler. Position the coupler with the narrow slot facing down towards the pump and the larger slot facing up. Place the coupler onto the end of the hydraulic pump shaft. Be sure the coupler is fully engaged with the slot on the end of the pump shaft.

27. Place a new O-ring into the groove in the adapter. Align the scribe marks or tape strip on the motor case with the adapter. The wire harness should be facing forward.

28. Slowly lower the motor onto the adapter. Be sure the flats on the end of the motor shaft engage the slot in the coupler. Secure the motor to the adapter with two hex head bolts. Tighten the bolts to 25 inch lbs. (2.9 Nm).

Install the pump and motor assembly into the support bracket.

Fig. 35a Loosen the mounting bolts...

Fig. 36 ...and lift the motor off the adapter

Fig. 37 Remove the end cover screws...

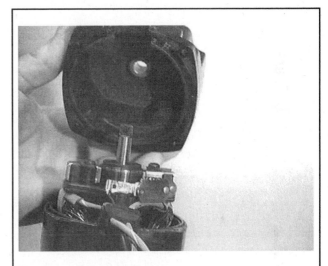

Fig. 37a ...and lift off the cover

Fig. 36a Lift out the motor shaft coupler

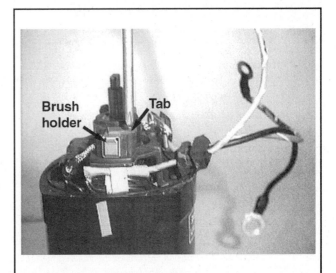

Fig. 38 Remove the brush hold down arms

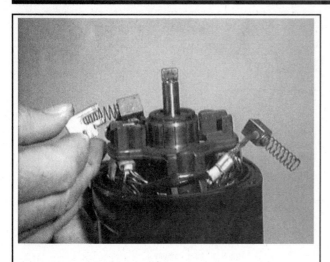

Fig. 38a Remove the brush holder, spring and brush

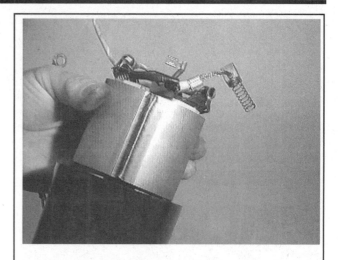

Fig. 40 ...and then the field frame

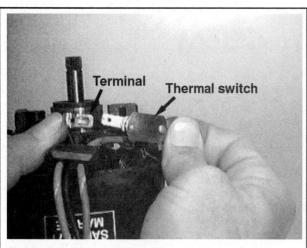

Fig. 38b Remove the terminal switch

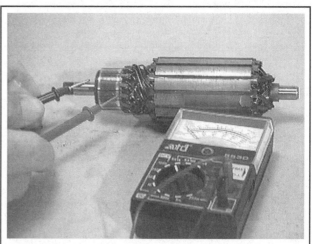

Fig. 41 Testing the armature

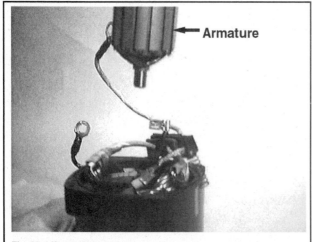

Fig. 39 Lift out the armature...

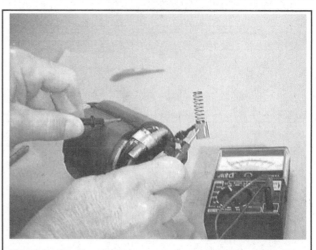

Fig. 42 Testing for a short

Fig. 43 Testing the thermal switch

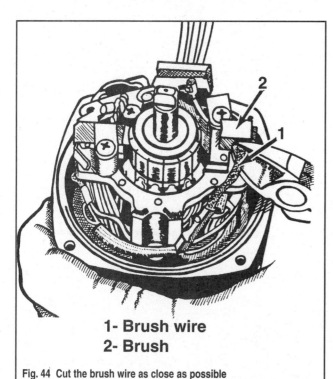

1- Brush wire
2- Brush

Fig. 44 Cut the brush wire as close as possible

Trim Cylinder

◆ See Figure 45

■ If troubleshooting procedures have isolated a problem to the trim cylinders, for example: a leaking oil scraper seal around the rod or a defective impact valve, it is strongly recommended that the cylinders be removed and replaced with new ones, rather than attempting to disassemble the cylinders.

This recommendation is based on the fact considerable difficulty may be encountered in removing the end cap from the cylinder. Even with the aid of the special tool for this purpose, the task is most difficult, sometimes impossible.

■ Removing the end cap without the use of the special tool, could be termed "impossible". The risk of damaging the small holes in the end cap used to hold the tool is very high. Should these holes be even slightly damaged, not even the special tool will remove the end cap. The cylinder would be completely unserviceable.

Other than the end cap removal, the rest of the components are relatively easily serviced.

The following procedures cover the removal and installation of a single cylinder. Simply repeat the procedures for the other side unless specific instructions are included.

■ 1998–01 Bravo units contain a Trim-In Limit insert in the drive unit to keep certain vessels from rolling onto their sides under extreme operating conditions. This system is detailed extensively in the Drive System section, but make sure you observe the position of the insert before attempting any work on these drive units—forward position on Bravo I and II, Aft position on Bravo III

REMOVAL & INSTALLATION

◆ See Figures 46, 47, 48, 49, 50, 51 and 52

1. Move the stern drive to the full DOWN position. Place a suitable drain pan under the hydraulic hoses

2. Disconnect the UP hydraulic hose from the end of the cylinder using the correct size "Flare Nut" or "Line Wrench". These wrenches will prevent damaging the hex flats on the line or hose fittings, if they are extremely tight and/or a slight amount of corrosion has built up on the fitting. Most standard open-end wrenches will flex under high torque loads, causing the wrench to slip, damaging the hex fitting on the hydraulic line.

3. Disconnect the DOWN hydraulic hose fitting at the hydraulic connector on the transom assembly. Install hydraulic plug P/N 22-38609, or equivalent plugs into the hoses and/or fittings to prevent draining the trim/tilt hydraulic system any more than necessary.

4. Pry off the plastic cover on the end of the forward anchor pin. Pull the E-Clip off the end of the anchor pin (Alpha) or remove the locknut (Bravo) and then slide off the flat washer and then the bushing from the anchor pin. Repeat this step for the opposite end of the trim cylinder.

5. Grasp the cylinder on the inside of each anchor pin and pull both ends of the cylinder off the anchor pins. Remove the flat washer and bushing on the inside surface cylinder ends or the anchor pins. There is also a snap-ring on Bravo drives

6. Repeat the above steps for the opposite cylinder if both cylinders are to be repaired or replaced.

To install:

7. Position the port and starboard trim cylinders so the offset of the piston rod eyelets face the stern drive. The UP port for the cylinder hose connections should also be facing up.

8. Install a bushing onto the forward and aft anchor pin for the cylinder. Place the ends of the cylinder over the bushings and push the cylinder ends onto the bushing and the anchor pins.

9. Install another set of bushings onto the forward and aft anchor pins for the cylinder. Push the bushing onto the anchor pin and into the end of the cylinder. Slide a flat washer onto each anchor stud and secure the cylinder to the anchor stud with an E-ring on the Alpha. On the Bravo, tighten all four locknuts finger-tight and then tighten each one until the nut and washer just bottom on the shoulder of the anchor pin.

■ On 1998–01 Bravo units, ensure that the trim-in insert is in the same position it was on cylinder removal—forward for the Bravo I and II units and aft for Bravo III units.

10. Install a new plastic end cap over the ends of the anchor pins. Repeat the last three steps for the other trim cylinder if it was also removed.

11. Remove the plugs from the end of the hose fittings and cylinder port fittings. Bleed the system as detailed previously and then connect the hydraulic hoses to the ports. Carefully tighten the fittings to 70–150 inch lbs. (8–17 Nm) on the Alpha, or 110 inch lbs. (12 Nm) on the Bravo.

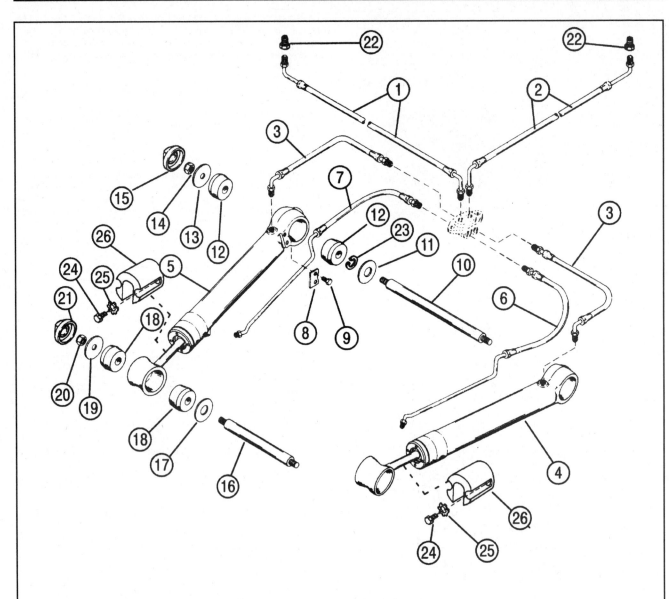

Fig. 45 Bravo trim system components

1 - IN/DOWN Hose to Trim Pump (Gray)
2 - UP/OUT Hose to Trim Pump (Black)
3 - Hose to Trim Cylinder
4 - Starboard Trim Cylinder
5 - Port Trim Cylinder
6 - Starboard Trim Cylinder Hose
7 - Port Trim Cylinder Hose
8 - Plate
9 - Screw
10 - Front Pin
11 - Washer
12 - Bushing
13 - Washer
14 - Nut
15 - Cap
16 - Rear Pin
17 - Washer
18 - Bushing
19 - Washer
20 - Nut
21 - Cap
22 - Connector (Trim Pump)
23 - Retainer
24 - Screw
25 - Continuity Washer
26 - Trim Cylinder Anode

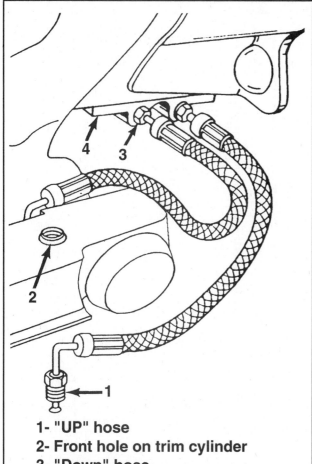

1- "UP" hose
2- Front hole on trim cylinder
3- "Down" hose
4- Hydraulic connector

Fig. 46 Disconnect the hydraulic hoses

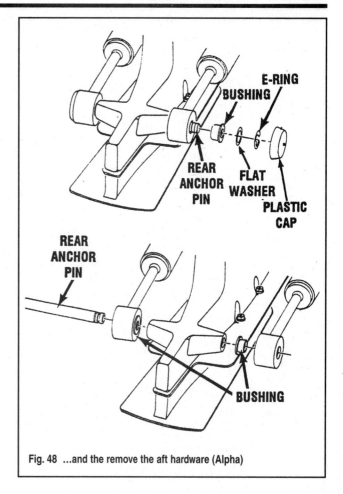

Fig. 48 ...and the remove the aft hardware (Alpha)

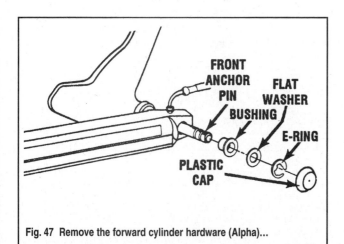

Fig. 47 Remove the forward cylinder hardware (Alpha)...

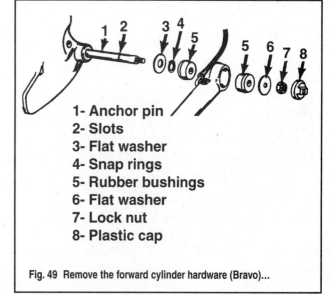

1- Anchor pin
2- Slots
3- Flat washer
4- Snap rings
5- Rubber bushings
6- Flat washer
7- Lock nut
8- Plastic cap

Fig. 49 Remove the forward cylinder hardware (Bravo)...

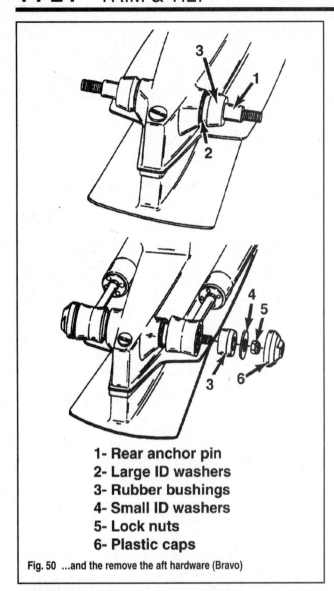

1- Rear anchor pin
2- Large ID washers
3- Rubber bushings
4- Small ID washers
5- Lock nuts
6- Plastic caps

Fig. 50 ...and the remove the aft hardware (Bravo)

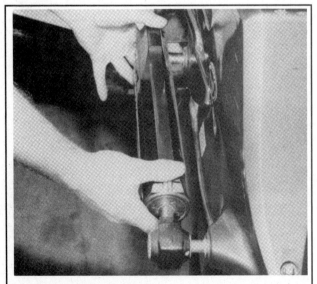

Fig. 51 Removing the cylinder

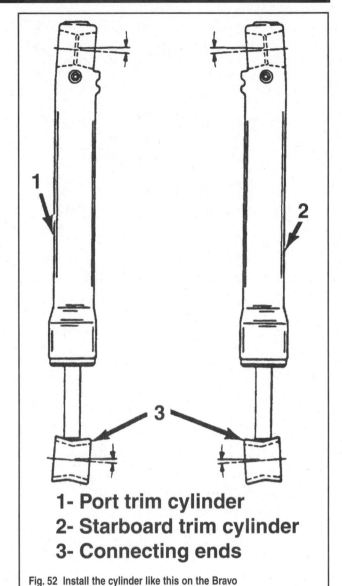

1- Port trim cylinder
2- Starboard trim cylinder
3- Connecting ends

Fig. 52 Install the cylinder like this on the Bravo

DISASSEMBLY & ASSEMBLY   OEM   ② MODERATE

◆ See Figure 53, 54, 55, 56, 57, 58, 59, 60, 61 and 62

1. Hold the cylinder over the drain pan; extend, and then retract the cylinder 2–3 times to remove all fluid from the cylinder.

2. Remove the two bolts securing the retaining plate on the DOWN hydraulic hose. Using the correct size line or flare nut wrench, disconnect the hose from the cylinder. Place the hose into a plastic bag to prevent contamination from entering the hose.

3. Remove the trim cylinder anode from the end cap (two screws) on 1998–01 units.

4. Place the front mounting flange the cylinder in a vise equipped with soft-face jaws and carefully tighten it.

5. Obtain a spanner wrench (#91-821709) or an equivalent tool. Removal of the end cap is difficult using the special tool: and could be termed "impossible" without the tool. Insert the tangs of the special tool into the holes in the end cap. If needed, slide a long breaker bar onto the tool so the bar is in the same plane as the tool. This position will provide maximum mechanical advantage. Remove the end cap.

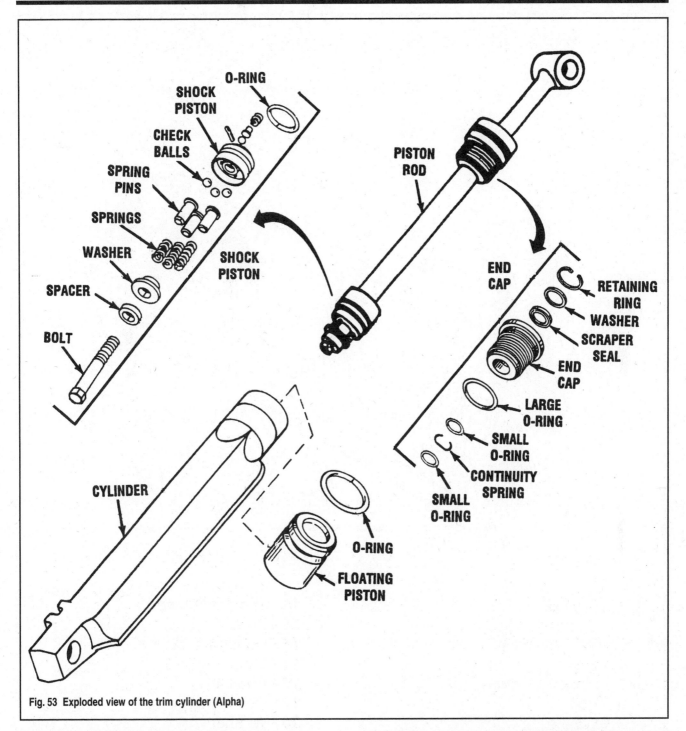

Fig. 53 Exploded view of the trim cylinder (Alpha)

Tap the breaker bar, if necessary, but bear in mind—if the holes in the end of the cap become damaged (elongated), the cylinder might as well be given the "deep six". If the attempt to remove the cap is successful, congratulations! Continue to unscrew the end cap until it is held by a single thread. Extend the rod, and then continue to remove the end cap and piston assembly.

6. Pop the tilt limit insert off of the piston rod on 1998–01 units

7. Remove the cylinder from the vise and invert it to remove the floating piston from the cylinder. Note the orientation of the piston as it is removed from the cylinder. Remove and discard the O-ring from the floating piston.

■ The factory, or boat builder, may have installed a spacer on the end of the piston rod to prevent full IN trim on Alpha drives. This spacer—if equipped—is located between the head of the piston bolt and the flat washer on the piston assembly. If this spacer is installed, it must be re-installed during assembling. Failure to ensure these parts are correctly installed could cause the bow to plow during full IN trim and possibly induce unwanted or dangerous steering changes. Replacement cylinders will usually have the capability for increased trim-in range (approx. 1-1/2° with spacer removed) which could improve acceleration on some boats, but we recommend discussing this with an authorized Mercury facility prior to deciding to remove the spacer.

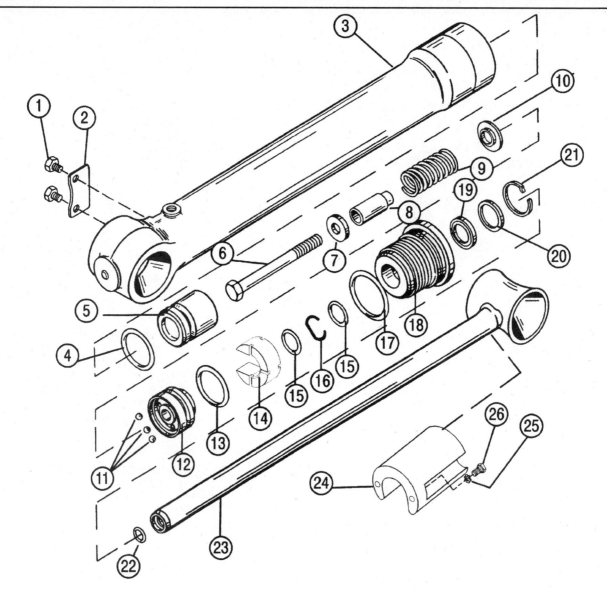

Fig. 54 Exploded view of the trim cylinder (Bravo)

1 - Screws
2 - Clamping Plate
3 - Trim Cylinder
4 - O-Ring
5 - Floating Piston
6 - Bolt
7 - Washer
8 - Spring Guide
9 - Spring
10 - Spring Guide Washer
11 - Check Balls
12 - Shock Piston
13 - O-Ring

14 - Trim Limit Insert
15 - Small O-Ring
16 - Continuity Spring
17 - Large O-Ring
18 - End Cap
19 - Rod Scraper
20 - Washer
21 - Retaining Ring
22 - Small O-Ring
23 - Piston Rod
24 - Anode
25 - Star Washer
26 - Screw

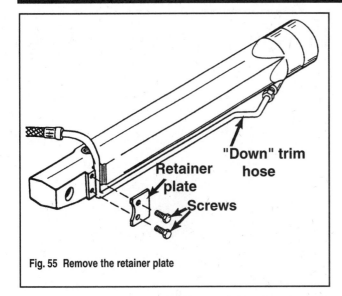

Fig. 55 Remove the retainer plate

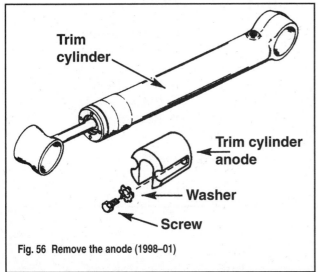

Fig. 56 Remove the anode (1998–01)

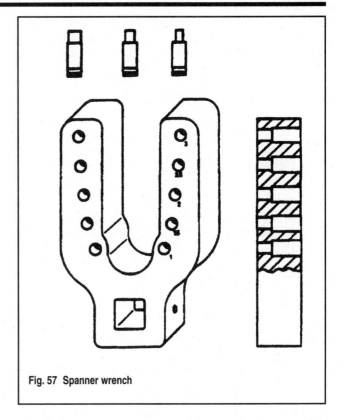

Fig. 57 Spanner wrench

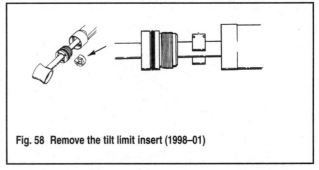

Fig. 58 Remove the tilt limit insert (1998–01)

✳✳ CAUTION

Never attempt to substitute a trim cylinder from another style drive unit or application.

8. Place the eyelet end of the piston into a vise and clamp it tight. Slowly remove the bolt from the end of the piston. As the bolt loosens, the tension on the shock piston and springs will lessen. When all tension has been removed, unscrew the bolt and slide off the spacer, washer, springs, spring pins, check balls, and shock piston assembly from the end of the piston rod.

9. Grasp the end cap and slide it off the piston rod. Remove the retainer ring, washer, scraper, O-rings, and continuity spring from the end cap.

10. Remove the small O-ring from inside the end of the piston rod.

To assemble:

11. Make an effort to keep the work area clean, because any contamination on the piston could lead to a malfunction. Inspect the interior of the cylinder for any signs of scoring or roughness. Clean all surfaces with safety solvent and blow them dry with compressed air.

12. The internal check balls in the shock piston are an integral part of the piston, and should not be removed. If the shock piston has any signs of corrosion or damage it must be replaced with a new unit.

13. Clean the end cap threads with a wire brush to remove all traces of the old sealant.

14. Lubricate all O-rings and interior components with Power Trim and Steering Fluid.

15. Insert a small O-ring, continuity spring, and another small O-ring into the rod cap.

16. Insert a new scraper seal flat washer and retaining ring into the opposite end of the rod cap. Slide the large O-ring over the threads of the rod cap and into the groove.

17. Lubricate the piston rod and rod cap with Quicksilver Power Trim and Steering Fluid. Now, slide the rod cap onto the piston rod.

18. Place a new small O-ring into the end of the piston rod.

19. Slide the large O-ring over the shock piston assembly. Position the shock piston on the end of the piston rod. Place the three check balls, spring pins, and springs into the shock piston. Place the large washer over the three springs and attempt to hold all these pieces in position. Apply Loctite 271 to the threads of the retaining bolt. Now, attempt to insert the retaining bolt with spacer—if equipped—through the washer and thread it into the end of the piston rod. If success was gained on the first attempt, congratulations, you're good!

20. If the parts slipped out from the shock piston, reposition all the parts again and attempt to thread the bolt into the end of the cylinder rod.

21. Tighten the shock piston retaining bolt to 15–20 ft. lbs. (20–27 Nm); 17 ft. lbs. (23 Nm) on 1998–01 Bravo units.

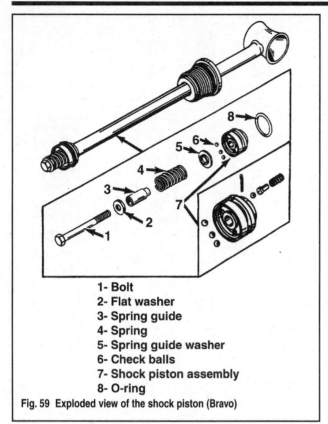

1- Bolt
2- Flat washer
3- Spring guide
4- Spring
5- Spring guide washer
6- Check balls
7- Shock piston assembly
8- O-ring

Fig. 59 Exploded view of the shock piston (Bravo)

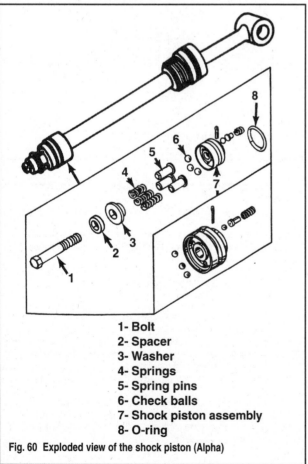

1- Bolt
2- Spacer
3- Washer
4- Springs
5- Spring pins
6- Check balls
7- Shock piston assembly
8- O-ring

Fig. 60 Exploded view of the shock piston (Alpha)

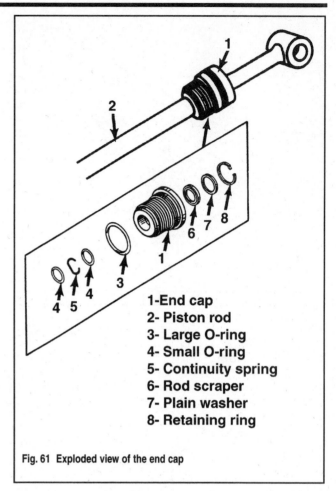

1-End cap
2- Piston rod
3- Large O-ring
4- Small O-ring
5- Continuity spring
6- Rod scraper
7- Plain washer
8- Retaining ring

Fig. 61 Exploded view of the end cap

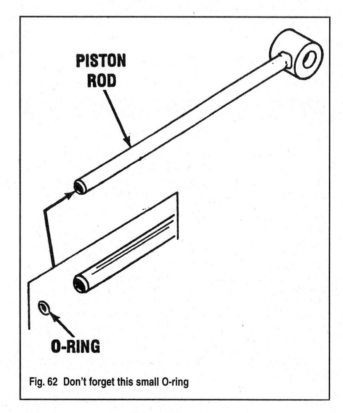

PISTON ROD

O-RING

Fig. 62 Don't forget this small O-ring

22. Install the tilt limit insert onto the piston rod if equipped.
23. Clamp the cylinder into a vise equipped with soft-jaws as detailed in the third step. Apply a coating of Quicksilver Power Trim and Steering Fluid to the interior surfaces of the cylinder and piston assembly.
24. Place a new O-ring onto the floating piston. Insert the floating piston into the cylinder in the correct position, as noted during disassemby.
25. Slowly lower the piston rod into the cylinder. Work the loose end of the piston rod in a circular motion to help start the shock piston O-ring into the cylinder. Once the piston starts down into the cylinder, press gently until the cap contacts the cylinder.

■ **Special Lubricant 101 is called out in the next step. This lubricant provides both a moisture seal and electrical bond to the trim cylinder. If this lubricant is not applied, corrosion build up on the trim cylinder, piston and rod could occur.**

26. Apply a coating of Special Lubricant 101 to the threads of the end cap. Thread the end cap into the cylinder. Tighten the end cap using spanner wrench to 40–50 ft. lbs. 55–68 Nm); 45 ft. lbs. (61 Nm) on 1998–01 Bravo units.
27. Install the anode to the end cap and tighten the two screws to 30 inch lbs. (3.4 Nm).
28. Push the rod into the cylinder until it bottoms out for installation.
29. Position the DOWN hydraulic line onto the cylinder. Start the line fitting into the cylinder finger-tight. Tighten the line fitting using the correct size flare or line wrench to 70–150 inch lbs. (8–17 Nm) on the Alpha, or 110 inch lbs. (12 Nm) on the Bravo. Place the clamping plate onto the cylinder and secure with the two bolts.

Dual Power Trim System

RELAY REPLACEMENT   DIFFICULT

◆ **See Figure 63 and 64**

1. Disconect the negative battery cables.
2. Remove the eight screws and lift off the control box cover.
3. Carefully cut off the relay electrical leads from the terminal block as close to the relay terminals as possible.

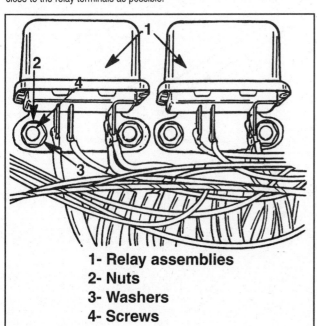

1- **Relay assemblies**
2- **Nuts**
3- **Washers**
4- **Screws**

Fig. 63 Relay removal

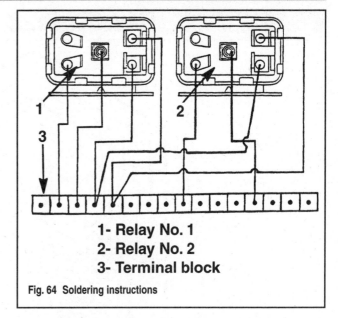

1- **Relay No. 1**
2- **Relay No. 2**
3- **Terminal block**

Fig. 64 Soldering instructions

4. Remove the two mounting screws and nuts and lift out the relay.
To install:
5. Strip some insulation odd the ends of each electrical lead and carefully re-solder each to its respective relay terminal.

■ **Use only 63/67 (tin/lead) alloy solder when re-attaching the leads. Never use acid core solder or you will damage the relay. Always coat the terminal connections with Liquid Neoprene.**

6. Reinstall the relay and tighten the mounting screws securely.
7. Install the control box cover and tighten the screws securely.
8. Connect the battery cable.

DIODE MODULE REPLACEMENT   MODERATE

◆ **See Figure 65**

1. Disconnect the negative battery cables.
2. Remove the eight screws and lift off the control box cover.
3. Tag and disconnect the electrical leads from the terminal block.
4. Remove the two mounting screws and nuts and lift out the module.
To install:
5. Connect the electrical leads to their respective terminals on the module.
6. Install the module and tighten the mounting screws securely.
7. Install the control box cover and tighten the screws securely.
8. Connect the battery cable.

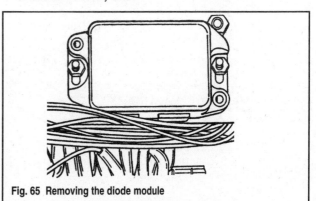

Fig. 65 Removing the diode module

WIRING SCHEMATICS

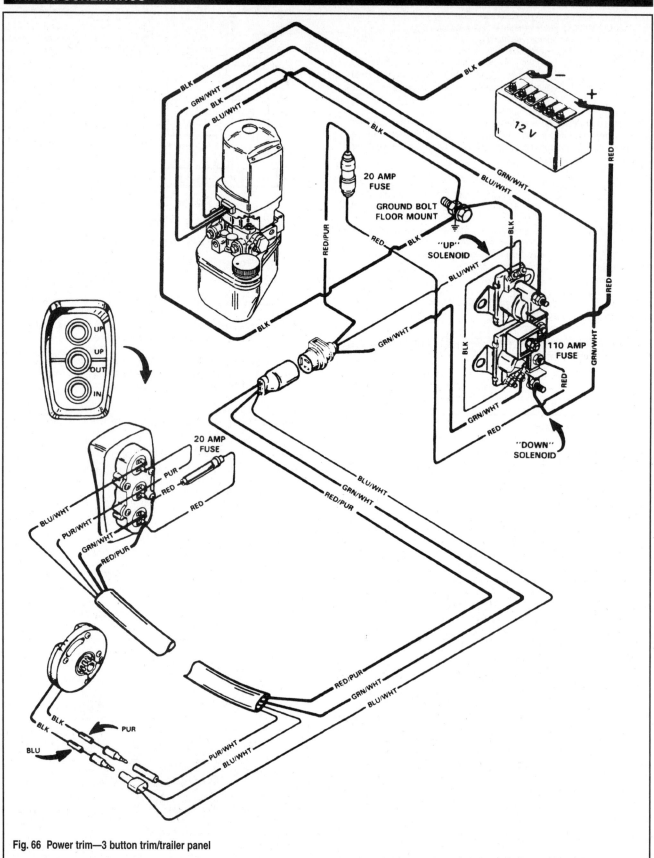

Fig. 66 Power trim—3 button trim/trailer panel

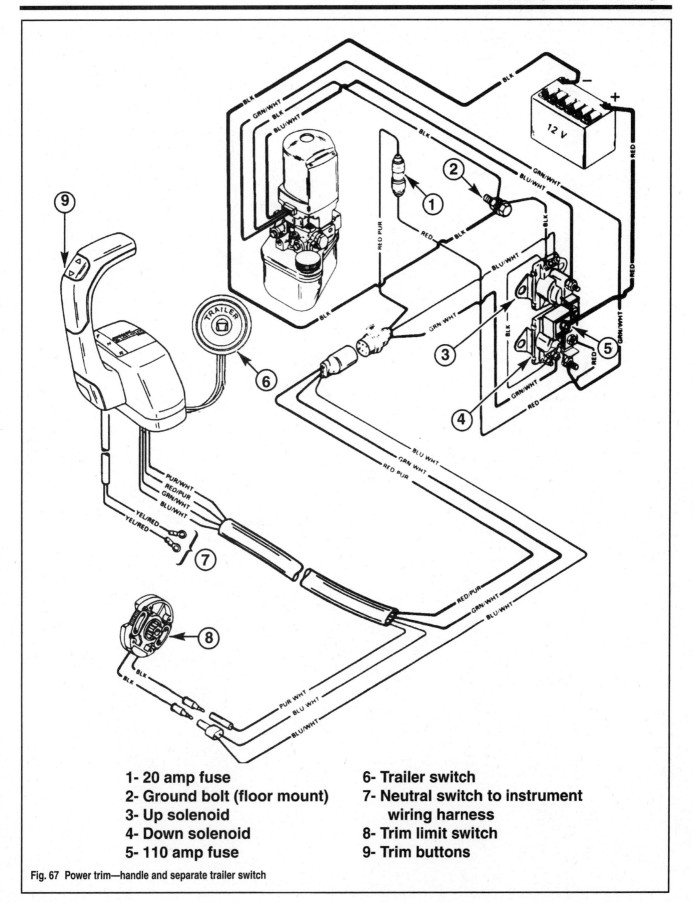

1- 20 amp fuse
2- Ground bolt (floor mount)
3- Up solenoid
4- Down solenoid
5- 110 amp fuse
6- Trailer switch
7- Neutral switch to instrument wiring harness
8- Trim limit switch
9- Trim buttons

Fig. 67 Power trim—handle and separate trailer switch

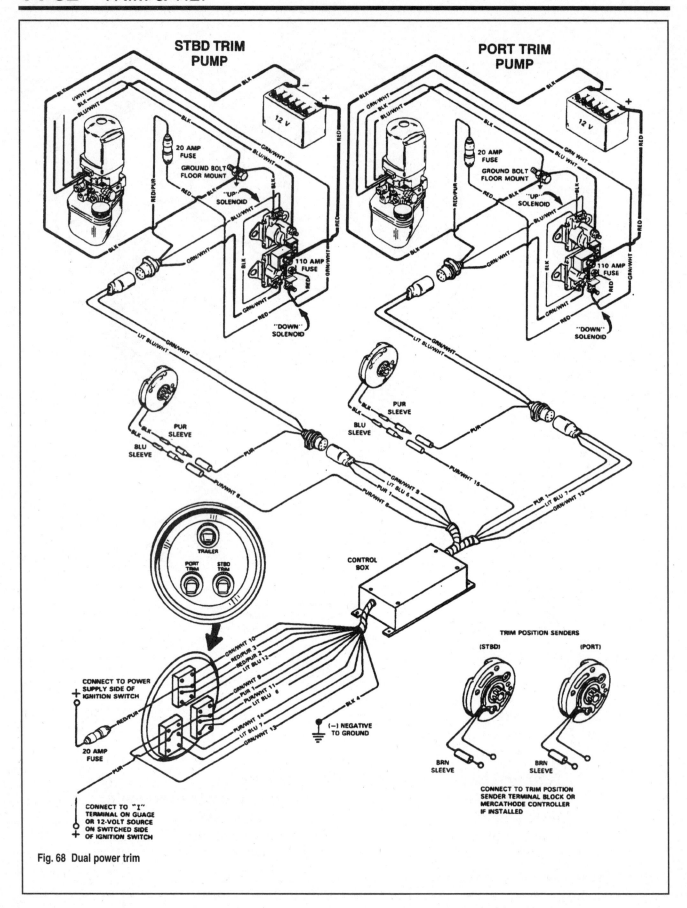

Fig. 68 Dual power trim

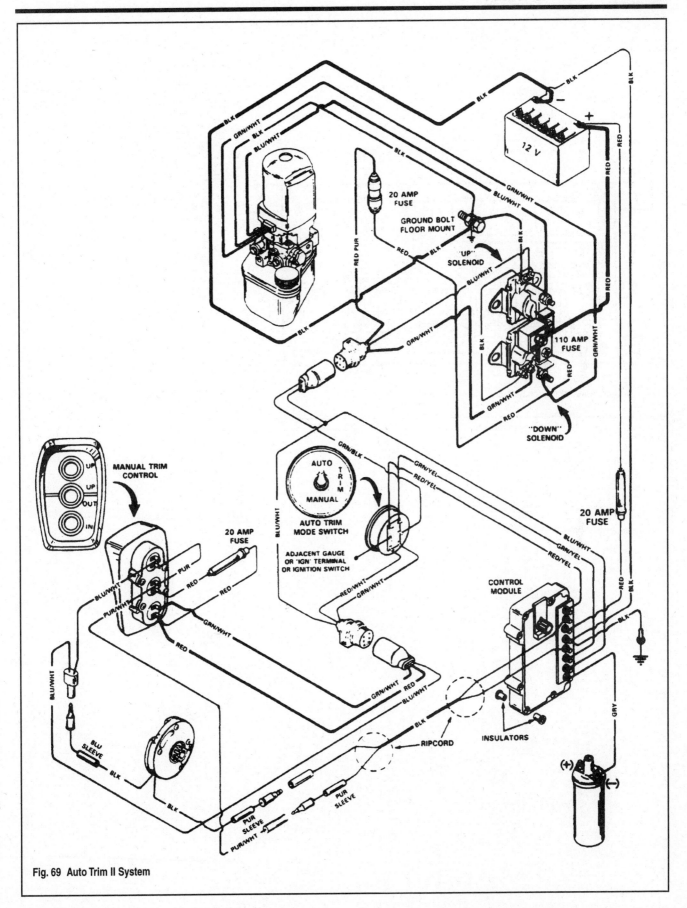

Fig. 69 Auto Trim II System

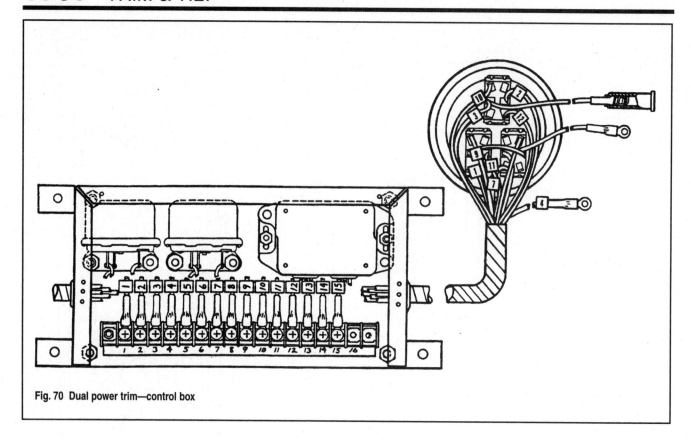

Fig. 70 Dual power trim—control box

15

STEERING

MANUAL STEERING SYSTEM

The manual steering system on your boat is very simple—a steering wheel, a steering cable and a steering lever in the stern drive unit. Due to the variety of steering systems available, the following procedures are general in their scope. Please refer to the Maintenance section for Lubrication procedures and Drive Systems for Steering Lever procedures.

Steering Cable

REMOVAL & INSTALLATION

◆ **See Figures 1. 2, 3, 4 and 5**

1. Locate the end of the steering cable at the transom. Pull out the cotter pin securing the clevis pin to the cable and the steering lever. Pull out the clevis and disconnect the cable from the steering lever. On dual engine installations, support the tie bar and then repeat the procedure on the opposite side. You will also need to disconnect the clevis assembly from the cable.

■ **All Quicksilver guides come standard with a self-locking coupler between the cable and the guide tube. On applications not coming with a self-locking coupler, and external locking device must be in place.**

2. On self-locking applications, loosen the coupler. On others, remove the cotter pin and slide the locking sleeve off of the coupler nut, and then loosen the nut. On either application, carefully pull the cable through the guide tube and swivel ring after loosening the nut.

3. Carefully remove the steering cable and then disconnect it at the helm station.

To install:

4. Clean the guide tube and steering lever thoroughly and inspect carefully for cracks, wear or other damage.

5. Connect the new cable to the helm as per the cable manufacturer's instructions and run it back through the boat.

6. Coat the end of the steering cable liberally with Special Lubricant 101 and feed it through the coupler and guide tube.

7. Attach the clevis assembly to the end of the cable (if equipped) and then align everything with the hole in the steering lever and insert the cotter pin.

8. Follow the manufacturer's instructions for final installation and adjustment. As a general rule of thumb though, turn the steering wheel all the way to starboard and check that the guide tube protrudes through the ring 3/4 in. (19mm). Thread the coupler nut onto the tube and tighten to 35 ft. lbs. (48 Nm). Check that the distance from the inner end of the nut to the centerline of the cable end hole is 21-3/8 in. 543mm). Mid-point of the cable travel should be 16-7/8 in. (429mm) with no more than 4-1/2 in. (114mm) travel in either direction.

9. Center the steering wheel and confirm that the steering lever is centered in the drive unit.

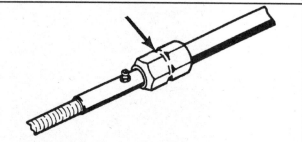

Fig. 1 Most units will have a self-locking coupler...

1- Locking sleeve kit (#71669A1)

Fig. 2 ...otherwise a locking sleeve must be installed

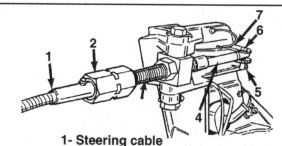

1- Steering cable
2- Cable coupler nut
3- Steering cable guide tube
4- Steering cable end
5- Cotter pin
6- Clevis pin
7- Steering lever

Fig. 3 A good look at the steering cable hardware

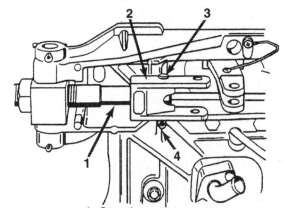

1- Steering cable end
2- Clevis assembly
3- Clevis pin
4- Cotter pin

Fig. 4 Certain applications will have a clevis assembly attached to the end of the cable

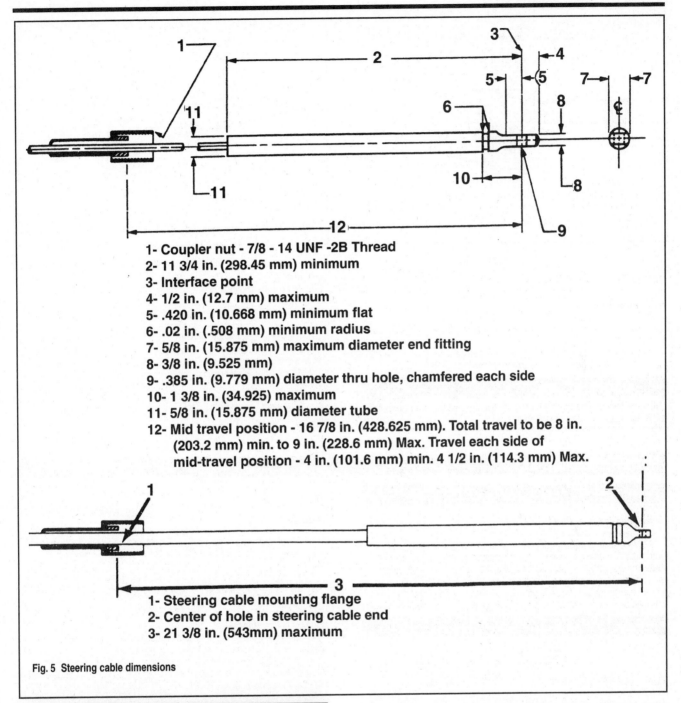

1- Coupler nut - 7/8 - 14 UNF -2B Thread
2- 11 3/4 in. (298.45 mm) minimum
3- Interface point
4- 1/2 in. (12.7 mm) maximum
5- .420 in. (10.668 mm) minimum flat
6- .02 in. (.508 mm) minimum radius
7- 5/8 in. (15.875 mm) maximum diameter end fitting
8- 3/8 in. (9.525 mm)
9- .385 in. (9.779 mm) diameter thru hole, chamfered each side
10- 1 3/8 in. (34.925) maximum
11- 5/8 in. (15.875 mm) diameter tube
12- Mid travel position - 16 7/8 in. (428.625 mm). Total travel to be 8 in.
   (203.2 mm) min. to 9 in. (228.6 mm) Max. Travel each side of
   mid-travel position - 4 in. (101.6 mm) min. 4 1/2 in. (114.3 mm) Max.

1- Steering cable mounting flange
2- Center of hole in steering cable end
3- 21 3/8 in. (543mm) maximum

Fig. 5 Steering cable dimensions

Swivel Ring

REMOVAL & INSTALLATION

② MODERATE

◆ See Figure 6

1. Remove the steering cable.
2. Use a screwdriver or pair of pliers and bend the locking tabs away from the two upper and lower pivot bolts. One side will be away from the bolt and the other will be away from the ridge in the transom assembly.
3. Remove the two pivot bolts and tab washers from the transom assembly and then pull out the swivel ring and guide tube assembly.
4. Lubricate the bushings inside the swivel ring with a good amount of Special Lubricant 101. Do the same to each pivot bolt.
5. Position the assembly into the transom assembly and then screw the bolts (don't forget the washers) in all the way by hand—no wrenches yet.
6. Make sure that the tab washers tangs are straddling the ridge in the transom plate on one side and then tighten them to 25 ft. lbs. (35 Nm)—you can use a wrench this time! Hopefully when you hit the correct torque figure, the remaining tab will align with a flat on the bolt. If so, bend it up and against the bolt head and then bend the other tab down over the transom assembly ridge. If it doesn't align, unscrew it and start over.
7. Move the swivel ring back and forth to ensure that it is pivoting correctly and then install the steering cable.

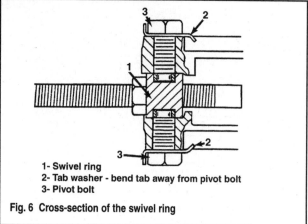

1- Swivel ring
2- Tab washer - bend tab away from pivot bolt
3- Pivot bolt

Fig. 6 Cross-section of the swivel ring

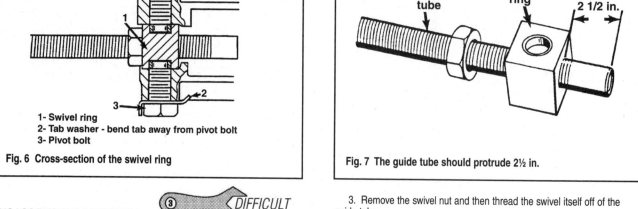

Fig. 7 The guide tube should protrude 2½ in.

DISASSEMBLY & ASSEMBLY DIFFICULT

◆ See Figure 7

1. Disconnect the steering cable and remove the swivel ring/guide tube assembly.
2. Mount the unthreaded end of the guide tube in a soft-jawed vise and then carefully apply heat to the locknut to loosen the Loctite.

✳✳ CAUTION

Be very careful when applying heat to the nut that you don't damage the bushings inside the swivel ring.

3. Remove the swivel nut and then thread the swivel itself off of the guide tube.
4. Pry out the two bushings.

To assemble:

5. Inspect the bushings for cracks, wear or other damage. It's not a bad idea to replace them anyway while they're out.
6. Clean the tube threads and the inside of the nut with a wire brush to remove any remaining Loctite and then reapply a good amount of Loctite 271 to the threads on the area of the tube where the swivel and nut will go.
7. Thread the swivel ring onto the guide tube until the unthreaded end of the tube is protruding 2-1/2 in. (64mm). Thread the nut onto the tube until it seats against the swivel and then tighten it to 35–40 ft. lbs. (47–54 Nm).
8. Install the assembly into the transom assembly and connect the steering cable.

POWER STEERING SYSTEM

Description

◆ See Figures 8, 9, 10, 11, 12 and 13

All MerCruiser power steering systems utilize an engine-driven, vane-type hydraulic pump supplying fluid and pressure, via hoses, to a control valve. The control valve controls flow and pressure to a booster cylinder, the two of which make up the power steering assembly. There are three basic power steering modes: Neutral, left turn and right turn. The control valve, activated by the steering wheel via a steering cable controls all three modes.

Testing

PUMP LUG TEST

Alpha And 1992–97 Bravo Only

◆ See Figure 14

This test will require the use of a power steering system pressure gauge kit.

1. Install a pressure gauge between the control valve and the pressure line from the power steering pump—this is the line on the left as you are looking at the power steering unit from inside the boat.

2. Ensure the the power steering fluid is at the correct level in the reservoir(s).
3. Open the valve on the test gauge, start the engine and allow it to idle.
4. Turn the steering wheel to the full left position and check the reading on the gauge. If higher than 300 psi, turn the engine OFF and check the following:

- Any obstruction between the gimbal ring and gimbal housing, and all moving parts of the steering system.
- The steering lever is contacting the cut-out in the transom. If so, modify or enlarge the cut-out.
- The steering cable guide tube is protruding 3/4 in. (19mm) from the adapter. If not, adjust as detailed later in this section.

5. Restart the engine and turn the wheel to the full right position and check the reading on the gauge. If higher than 300 psi, turn the engine OFF and check the following:

- Any obstruction between the gimbal ring and gimbal housing, and all moving parts of the steering system.
- The steering lever is contacting the cut-out in the transom. If so, modify or enlarge the cut-out.
- The steering cable dimension between the inner edge of the mounting flange and the center of the hole in the end of the cable is 21-3/8 in. (543mm) when the cable is fully extended as deatailed in the Steering Cable section.
- The steering cable guide tube is protruding 3/4 in. (19mm) from the adapter. If not, adjust as detailed later in this section.

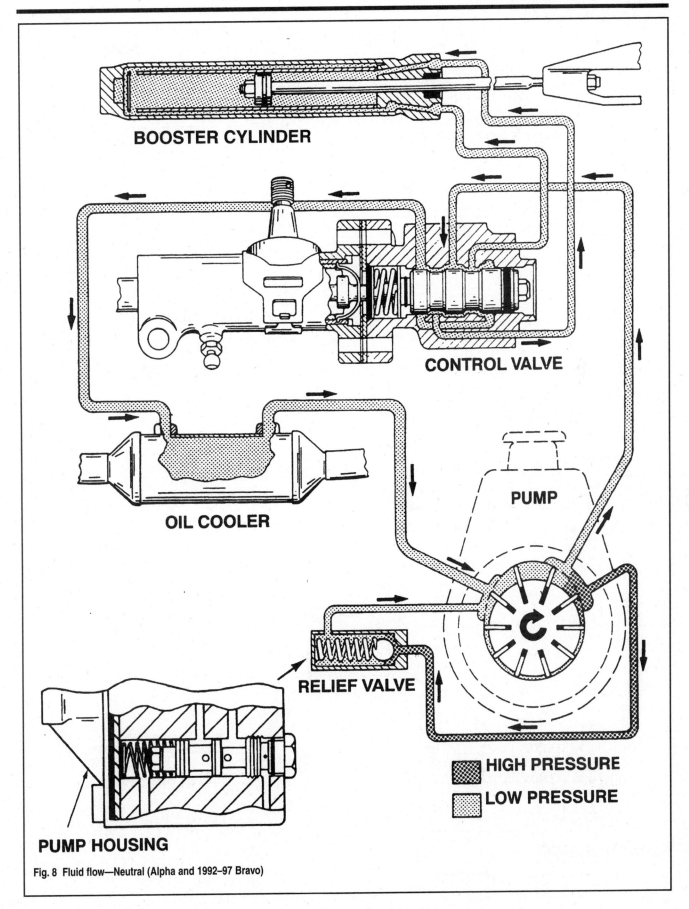

BOOSTER CYLINDER

CONTROL VALVE

OIL COOLER

PUMP

RELIEF VALVE

PUMP HOUSING

HIGH PRESSURE

LOW PRESSURE

Fig. 8 Fluid flow—Neutral (Alpha and 1992–97 Bravo)

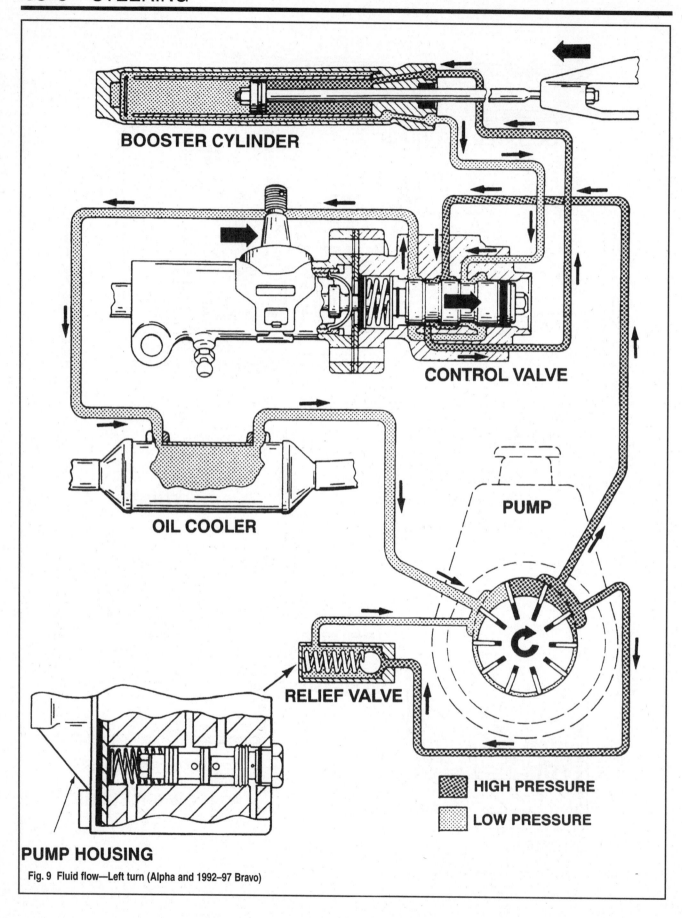

BOOSTER CYLINDER

CONTROL VALVE

OIL COOLER

PUMP

RELIEF VALVE

PUMP HOUSING

HIGH PRESSURE

LOW PRESSURE

Fig. 9 Fluid flow—Left turn (Alpha and 1992–97 Bravo)

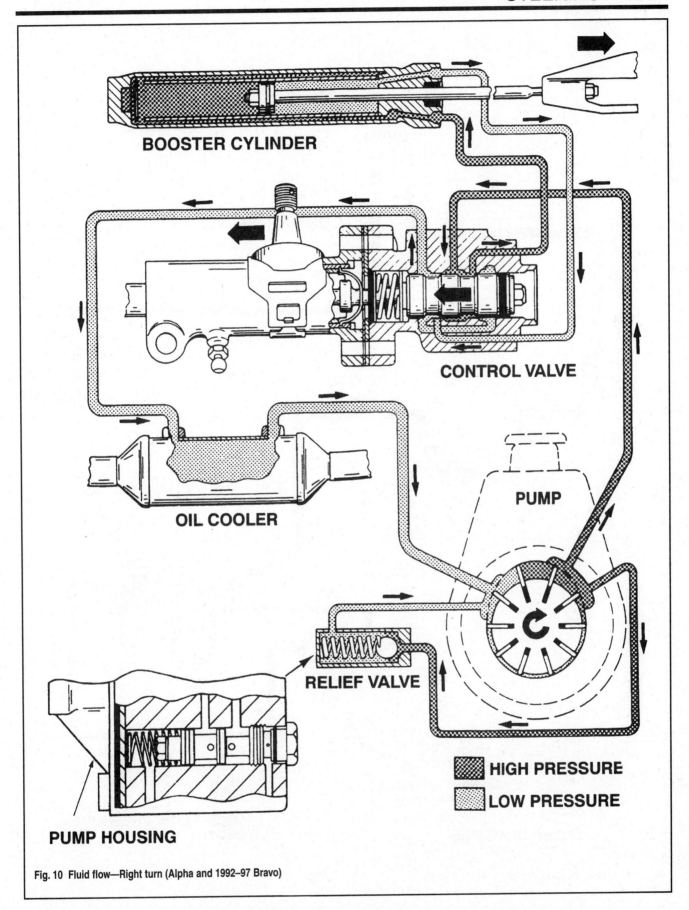

BOOSTER CYLINDER

CONTROL VALVE

OIL COOLER

PUMP

RELIEF VALVE

PUMP HOUSING

HIGH PRESSURE

LOW PRESSURE

Fig. 10 Fluid flow—Right turn (Alpha and 1992–97 Bravo)

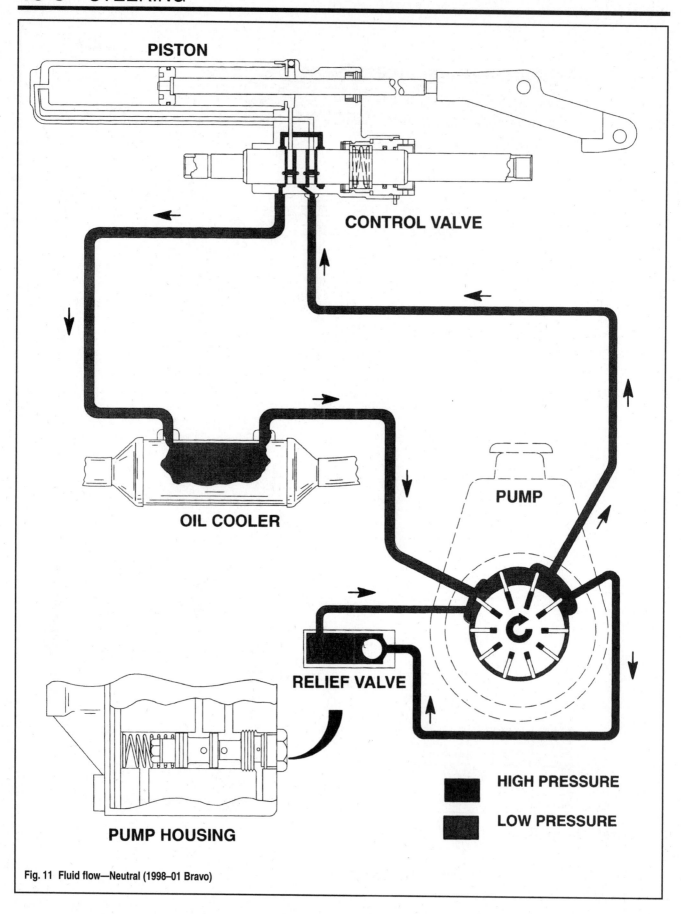

PISTON

CONTROL VALVE

OIL COOLER

PUMP

RELIEF VALVE

PUMP HOUSING

HIGH PRESSURE

LOW PRESSURE

Fig. 11 Fluid flow—Neutral (1998–01 Bravo)

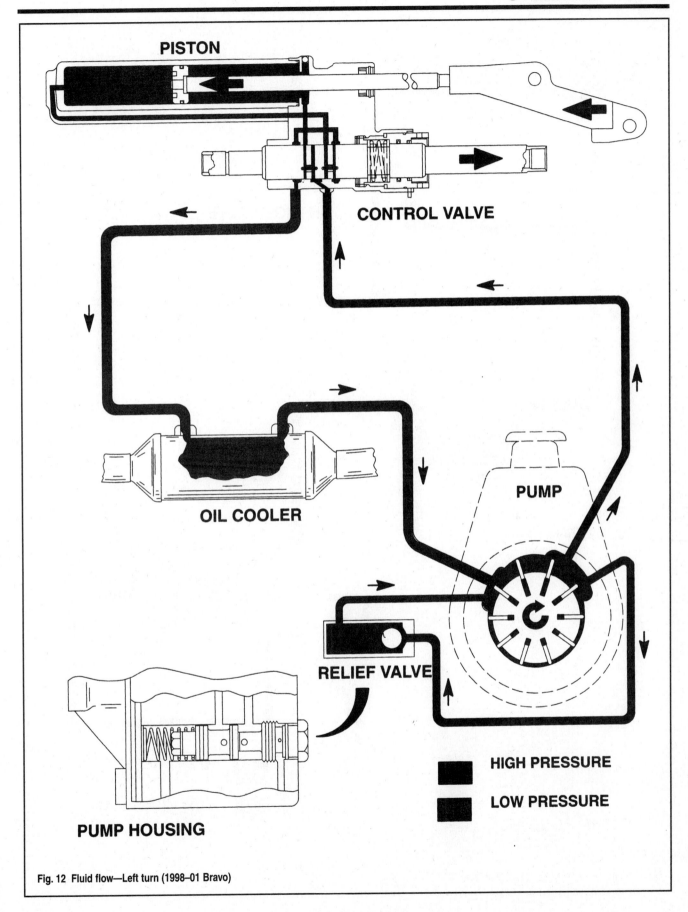

PISTON

CONTROL VALVE

OIL COOLER

PUMP

RELIEF VALVE

PUMP HOUSING

HIGH PRESSURE

LOW PRESSURE

Fig. 12 Fluid flow—Left turn (1998–01 Bravo)

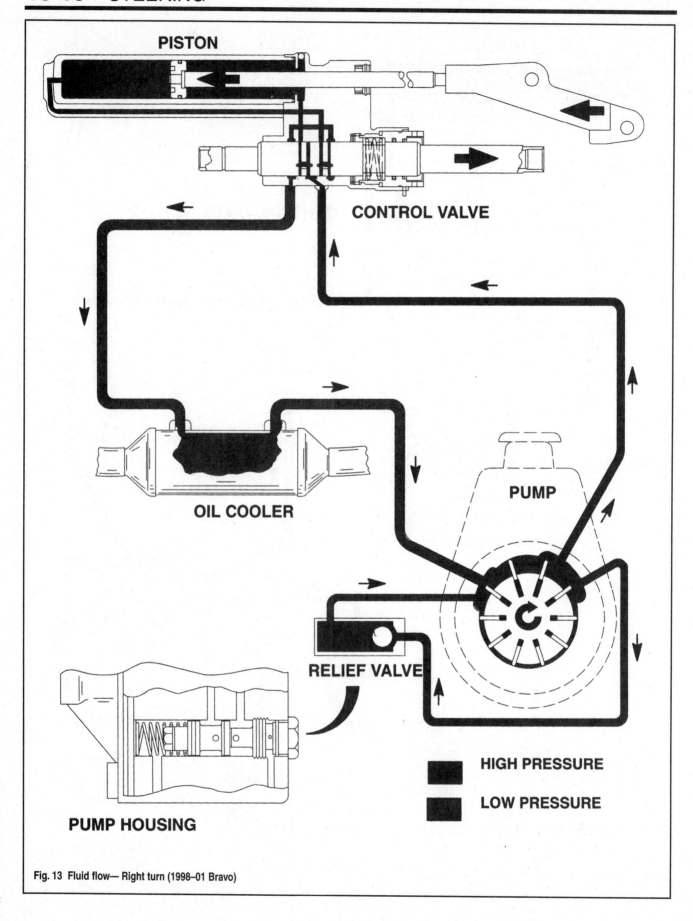

Fig. 13 Fluid flow— Right turn (1998–01 Bravo)

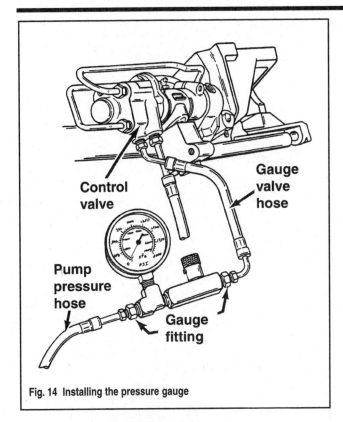

Fig. 14 Installing the pressure gauge

Labels on figure: Control valve, Gauge valve hose, Pump pressure hose, Gauge fitting

SYSTEM PRESSURE TEST

◆ See Figure 14

This test will require the use of a power steering system pressure gauge kit.

1. Disconnect and remove the steering cable. Disconnect the clevis assembly from the steering lever.
2. Install a pressure gauge between the control valve and the pressure line from the power steering pump—this is the line on the left as you are looking at the power steering unit from inside the boat.
3. Ensure the power steering fluid is at the correct level in the reservoir(s).
4. Open the valve on the test gauge.
5. Connect a flushing device between a water source and the flush port on the starboard side of the drive. Turn on the water to about half of its maximum flow and fill the cooling system. Do not use full tap pressure.
6. Start the engine and run it at 1100–1500 rpm until it reaches normal operating temperature.
7. Allow the engine to drop down to idle and check the reading on the gauge. If lower than 70 psi, move on to the Pump Pressure Test. If higher than 125 psi, check for restrictions in the hydraulic lines. Any pressure in between the above readings, move to the next step.
8. Push the control valve adapter block MOMENTARILY to the left and then to the right—do not hold the block over for more than 5 seconds. The gauge should show an instant increase in pressure when pushing the block in either direction.
9. Push the adapter block to the right until the booster cylinder piston rod is fully retracted into the cylinder. Once retracted, MOMENTARILY push the block to the right again until the highest pressure reading is obtained. If above 1000 psi, the system pressure is good. If below 1000 psi, move to the Pump Pressure Test.

PUMP PRESSURE TEST

◆ See Figure 14

This test will require the use of a power steering system pressure gauge kit.

1. Install a pressure gauge between the control valve and the pressure line from the power steering pump—this is the line on the left as you are looking at the power steering unit from inside the boat.
2. Ensure the power steering fluid is at the correct level in the reservoir(s).
3. Connect a flushing device between a water source and the flush port on the starboard side of the drive. Turn on the water to about half of its maximum flow and fill the cooling system. Do not use full tap pressure.
4. Start the engine and run it at 1100–1500 rpm until it reaches normal operating temperature.
5. Close the valve on the gauge just long enough to observe a maximum pressure reading.
6. Close and open the valve three times in a row; observing the maximum pressure reading each time. Record each readings.
7. If the three readings were between 1150 psi and 1250 psi, and were within 50 psi of each other, the pump is OK, you're done. If the pump is OK, but you got low readings in the System Pressure Test, proceed to the Booster Cylinder Test later in this section.
8. If the three readings were between 1150 psi and 1250 psi, but were not within 50 psi of each other, the flow control valve in the power steering pump is sticking or you have clogged hydraulic lines.
9. If all three readings were constant, but under 1000 psi, replace the power steering pump.

Steering Cable

REMOVAL & INSTALLATION

◆ See Figures 15, 16, 17, 18, 19, 20, 21 and 22

1. Locate the end of the steering cable at the transom. Pull out the cotter pin securing the clevis pin to the cable and the clevis assembly. Pull out the clevis and disconnect the cable from the assembly. On dual engine installations, support the tie bar and then repeat the procedure on the opposite side.

■ All Quicksilver guides except the 1992–97 Bravo come standard with a self-locking coupler between the cable and the guide tube. On applications not coming with a self-locking coupler, and external locking device must be in place. 1992–97 Bravos utilize a locking plate instead of the self-locking coupler.

2. On self-locking applications, loosen the coupler. On early model Bravos, remove the lock plate bolt and lift off the plate. On others, remove the cotter pin and slide the locking sleeve off of the coupler nut, and then loosen the nut. On all applications, carefully pull the cable through the guide tube after loosening the nut.
3. Carefully remove the steering cable and then disconnect it at the helm station.
To install:
4. Clean the guide tube and clevis assembly thoroughly and inspect carefully for cracks, wear or other damage.
5. Connect the new cable to the helm as per the cable manufacturer's instructions and run it back through the boat.
6. Coat the end of the steering cable liberally with Special Lubricant 101 and feed it through the coupler and guide tube.
7. Align the hole in the end of the cable with the holes in the clevis assembly, insert the clevis pin and then insert the cotter pin.

8. Follow the manufacturer's instructions for final installation and adjustment. As a general rule of thumb though, turn the steering wheel all the way to starboard and check that the guide tube protrudes through the ring 3/4 in. (19mm). On later model Bravos, position an open end wrench over the flats on the tube and rotate it so they are vertical. On all models, thread the coupler nut onto the tube and tighten to 35 ft. lbs. (48 Nm). If not equipped with a self-locking coupler, install the sleeve and cotter pin, or install the locking plate and tighten the bolt securely. Check that the distance from the inner end of the nut to the centerline of the cable end hole is 21- 3/8 in. 543mm). Mid-point of the cable travel should be 16-7/ in. (429mm) with no more than 4-1/2 in. (114mm) travel in either direction.

9. Center the steering wheel and confirm that the steering lever is centered in the drive unit.

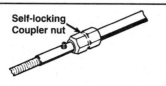

Fig. 15 Most units will have a self-locking coupler

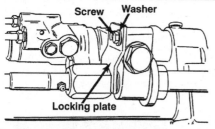

Fig. 16 Some models will use a locking plate...

1- Locking sleeve kit (# 71669A1)

Fig. 17 ...otherwise a locking sleeve must be installed

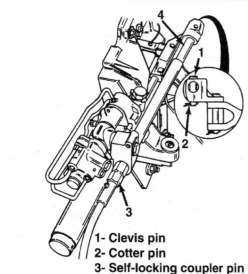

1- Clevis pin
2- Cotter pin
3- Self-locking coupler pin
4- Steering cable

Fig. 18 A good look at the steering cable hardware—Alpha

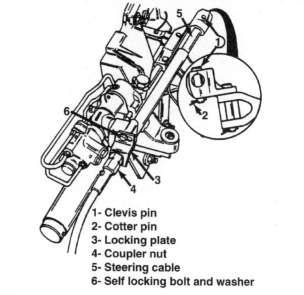

1- Clevis pin
2- Cotter pin
3- Locking plate
4- Coupler nut
5- Steering cable
6- Self locking bolt and washer

Fig. 19 Bravo—early models

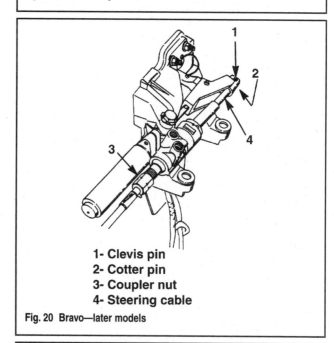

1- Clevis pin
2- Cotter pin
3- Coupler nut
4- Steering cable

Fig. 20 Bravo—later models

1- Flat (hold verticle)
2- Suitable wrench

Fig. 21 On late model Bravos, the flats on the tube must be vertical

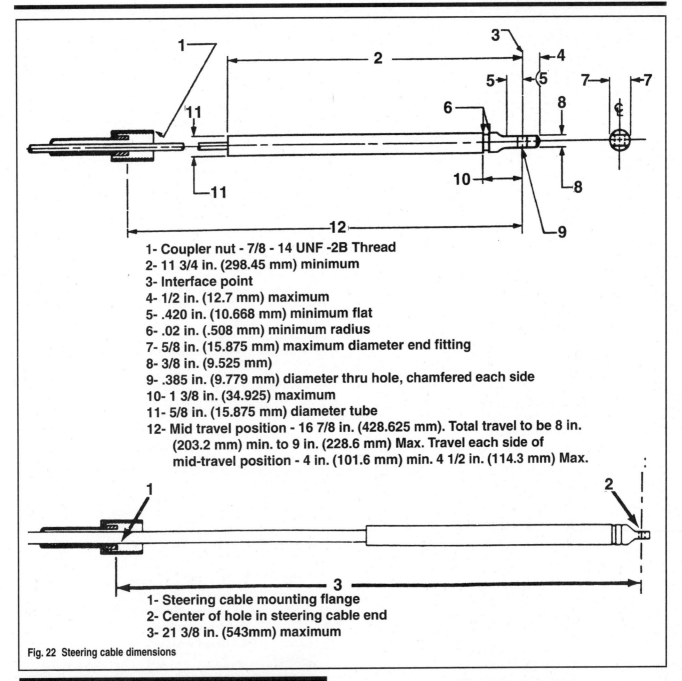

1- Coupler nut - 7/8 - 14 UNF -2B Thread
2- 11 3/4 in. (298.45 mm) minimum
3- Interface point
4- 1/2 in. (12.7 mm) maximum
5- .420 in. (10.668 mm) minimum flat
6- .02 in. (.508 mm) minimum radius
7- 5/8 in. (15.875 mm) maximum diameter end fitting
8- 3/8 in. (9.525 mm)
9- .385 in. (9.779 mm) diameter thru hole, chamfered each side
10- 1 3/8 in. (34.925) maximum
11- 5/8 in. (15.875 mm) diameter tube
12- Mid travel position - 16 7/8 in. (428.625 mm). Total travel to be 8 in. (203.2 mm) min. to 9 in. (228.6 mm) Max. Travel each side of mid-travel position - 4 in. (101.6 mm) min. 4 1/2 in. (114.3 mm) Max.

1- Steering cable mounting flange
2- Center of hole in steering cable end
3- 21 3/8 in. (543mm) maximum

Fig. 22 Steering cable dimensions

Steering Cable Guide Tube

REMOVAL & INSTALLATION

③ DIFFICULT

Alpha And 1992–97 Bravo Only

◆ See Figures 23, 24 and 25

1. Disconnect the hydraulic lines at the control valve assembly. Be sure to plug both of the lines and the inlets on the valve. Have some rags handy because you're going to spill some power steering fluid.
2. Disconnect and remove the steering cable.
3. Loosen the mounting bolt on the adapter block and then tap the block lightly with a rubber mallet to break it loose from the assembly. Remove the bolt and then lift off the adapter block and guide tube.
4. Carefully mount the adaptor body in a soft-jawed vise. Apply heat,

very carefully, to the inner tube nut to loosen the Loctite and then loosen the nut. Pull out the guide tube, nut, guide and bushing.
To install:
5. Check the tube, bushing and block for any cracks, wear or other signs of damage. Clean them all thoroughly.
6. Clean the threads of the guide tube with a wire brush to remove any residual Loctite and then coat the area of the threads where the adapter block and nut will rest with Loctite 271.
7. Thread the nut onto the tube and slide on the guide and bushing. Install the tube into the adapter block until the threaded end protrudes 3/4 in. (19mm) through the block. Tighten the nut to 40 ft. lbs. (54 Nm) on Alphas or 15–20 ft. lbs. (20–27 Nm) on 1992–97 Bravos.
8. Install the adapter block assembly and tighten the mounting bolt to 30–40 ft. lbs. (41–54 Nm).
9. Install and connect the steering cable.
10. Unplug and reconnect the two hydraulic lines. Tighten the large fitting nut to 20–25 ft. lbs. (27–34 Nm) and the small fitting to 96–108 inch lbs. (11–12 Nm).

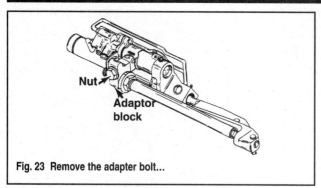

Fig. 23 Remove the adapter bolt...

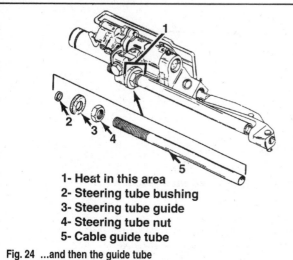

1- Heat in this area
2- Steering tube bushing
3- Steering tube guide
4- Steering tube nut
5- Cable guide tube

Fig. 24 ...and then the guide tube

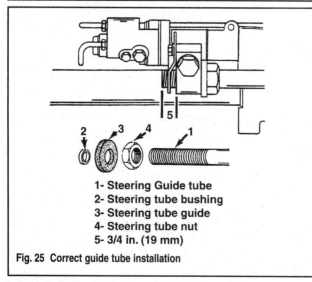

1- Steering Guide tube
2- Steering tube bushing
3- Steering tube guide
4- Steering tube nut
5- 3/4 in. (19 mm)

Fig. 25 Correct guide tube installation

1998–01 Bravo

The power steering unit on these models is not serviceable and must be replaced as a unit.

Power Steering Unit

The power steering unit is made up of the control valve assembly, adapter and booster cylinder. It is attached to the inner transom assembly

REMOVAL & INSTALLATION

♦ See Figures 26, 27 and 28

1. Disconnect the hydraulic lines at the control valve assembly. Be sure to plug both of the lines and the inlets on the valve. Have some rags handy because you're going to spill some power steering fluid.
2. Disconnect and remove the steering cable.
3. Remove the lower cotter pin and pull out the clevis connecting the clevis assembly and booster piston rod to the steering lever.
4. Use a screwdriver or pair of pliers and bend the locking tabs away from the two, upper and lower, pivot bolts. One side will be away from the bolt and the other will be away from the ridge in the transom assembly.
5. Remove the two pivot bolts and tab washers from the transom assembly and then pull out the power steering unit.

To install:

6. Lubricate the bushings inside the swivel ring (at the inner end of the booster) with a good amount of Special Lubricant 101. Do the same to each pivot bolt.
7. Position the assembly onto the transom assembly and then screw the bolts (don't forget the washers) in all the way by hand—no wrenches yet.
8. Make sure that the tab washers tangs are straddling the ridge in the transom plate on one side and then tighten them to 25 ft. lbs. (34 Nm)—you can use a wrench this time! Hopefully when you hit the correct torque figure, the remaining tab will align with a flat on the bolt. If so, bend it up and against the bolt head and then bend the other tab down over the transom assembly ridge. If it doesn't align, unscrew it and start over. If you need to tighten the bolt slightly to get the flat and tab to line up, it's Ok; do not though, loosen the bolt to line things up.
9. Connect the piston rod to the steering lever and slide in the clevis pin. Install the cotter pin.
10. Grab the booster cylinder and move it back and forth to ensure that it is pivoting correctly and then install the steering cable.
11. Unplug and reconnect the two hydraulic lines. Tighten the large fitting nut to 20–25 ft. lbs. (27–34 Nm) and the small fitting to 96–108 inch lbs. (11–12 Nm).

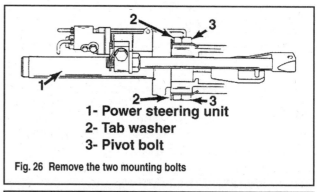

1- Power steering unit
2- Tab washer
3- Pivot bolt

Fig. 26 Remove the two mounting bolts

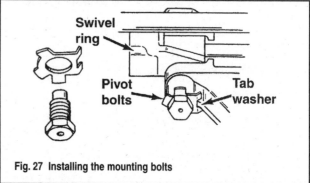

Fig. 27 Installing the mounting bolts

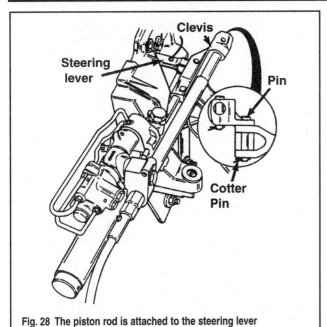

Fig. 28 The piston rod is attached to the steering lever

Control Valve

REMOVAL & INSTALLATION

Alpha And 1992–97 Bravo Only

◆ **See Figure 29**

1. Disconnect the hydraulic lines at the control valve assembly. Be sure to plug both of the lines and the inlets on the valve. Have some rags handy because you're going to spill some power steering fluid.
2. Disconnect and remove the steering cable.
3. Loosen the two line fittings on the side of the valve and disconnect the lines. Be sure to plug both of the lines and the inlets on the valve. Have some rags handy because you're going to spill some power steering fluid.
4. Remove the guide tube assembly.
5. Loosen the mounting bolt and remove the valve from the adapter assembly.
To install:
6. Clean the control valve thoroughly and check it carefully for cracks, wear or any other damage.
7. Mount the valve on the adapter assembly and tighten the bolt to 25–35 ft. lbs. (34–47 Nm).
8. Unplug the two hydraulic lines from the booster and thread them into the control valve. Tighten the fitting nuts securely.
9. Install and connect the steering cable.
10. Unplug and reconnect the two hydraulic lines. Tighten the large fitting nut to 20–25 ft. lbs. (27–34 Nm) and the small fitting to 96–108 inch lbs. (11–12 Nm).

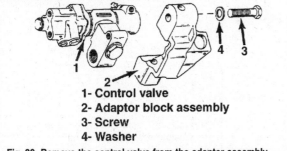

1- Control valve
2- Adaptor block assembly
3- Screw
4- Washer

Fig. 29 Remove the control valve from the adapter assembly

1998–01 Bravo

The power steering unit on these models is not serviceable and must be replaced as a unit.

DISASSEMBLY & ASSEMBLY

Alpha And 1992–97 Bravo Only

◆ **See Figures 30, 31, 32, 33, 34, 35, 36 and 37**

1. Remove the control valve and remove the adapter block if not already done.
2. Pop the dust cap from the starboard side of the valve and remove the adjusting nut.
3. Remove the two screws and washers and separate the control valve housing from the adapter.
4. Carefully remove all the internal valve components as shown in the accompanying illustration.
5. Being very careful not to nick the top surface of the adapter housing, turn the valve shaft counterclockwise to remove the adjuster plug, spring and upper ball seat.
6. Pull out the lower ball seat and bearing sleeve. Pull back on the rubber boot and remove the ball stud.
To assemble:
7. Clean all parts thoroughly and inspect fro cracks, wear or other damage.
8. Press the lower seat into the bearing sleeve and then insert them into the bore. Press the ball stud through the boot and into the sleeve. Install the upper seat.
9. Install the ball seat spring with the small coil facing downward. Insert the valve shaft into the plug and thread it in until it's tight. Back off the plug until the slot lines up with the notches in the bearing sleeve. Install the key over the shaft so that the tangs in the key fit into the notches.
10. Slide the small washer over the shaft and then position the gasket and spacer. Slide on the large washer.
11. Install a new O-ring into the adapter side of the valve housing.
12. Install a V-block seal onto the valve spool so that the lip on the seal faces the lands on the spool. Insert the assembly into the adjusting nut side of the housing.

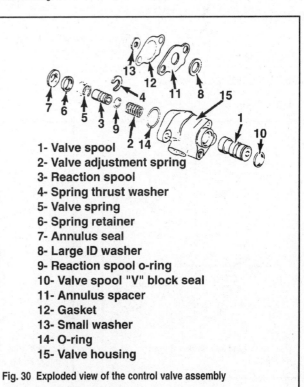

1- Valve spool
2- Valve adjustment spring
3- Reaction spool
4- Spring thrust washer
5- Valve spring
6- Spring retainer
7- Annulus seal
8- Large ID washer
9- Reaction spool o-ring
10- Valve spool "V" block seal
11- Annulus spacer
12- Gasket
13- Small washer
14- O-ring
15- Valve housing

Fig. 30 Exploded view of the control valve assembly

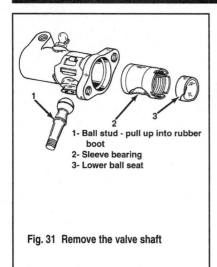

Fig. 31 Remove the valve shaft

1- Ball stud - pull up into rubber boot
2- Sleeve bearing
3- Lower ball seat

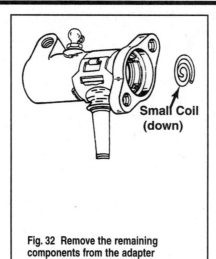

Fig. 32 Remove the remaining components from the adapter

Small Coil (down)

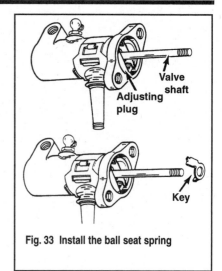

Fig. 33 Install the ball seat spring

Valve shaft
Adjusting plug
Key

13. Assemble the reaction spool as shown in the illustration. Insert the valve adjustment spring into the inner opening on the valve and then press in the reaction spool.

14. Line up the gasket and spacer holes with those on the adapter and then position the valve housing onto the adapter. Tighten the mounting bolts to 20–30 ft. lbs. (27–41 Nm). Don't forget the lock washers.

15. Press down the valve spool and thread the adjusting nut onto the shaft about 4 rotations. Press in the dust cover.

16. Install the control valve to the adapter assembly and tighten the bolt to 25–35 ft. lbs. (34–47 Nm).

17. Connect the hydraulic lines from the booster and tighten the fitting nuts securely. Do not over-tighten.

18. Install the cable guide and steering cable.

19. Unplug and reconnect the two hydraulic lines. Tighten the large fitting nut to 20–25 ft. lbs. (27–34 Nm) and the small fitting to 96–108 inch lbs. (11–12 Nm).

1998–01 Bravo

The power steering unit on these models is not serviceable and must be replaced as a unit.

BALANCING

③ ⟨DIFFICULT

Alpha And 1992–97 Bravo Only

◆ See Figure 38

All control valves are balanced when delivered from the factory and should not, in regular operation, require further adjustment. However, occasionally a drive unit will develop a creeping problem—engine running, in Neutral, hands off the steering wheel, yet the drive moves slightly in one direction or another. If you are experiencing this, the control valve must be balanced.

1. With the engine off, disconnect the steering cable at the clevis assembly.

2. Disconnect the clevis assembly on the end of the piston rod from the steering lever.

3. Pry off the dust cover on the starboard side of the control valve assembly to expose the adjustment nut.

4. Connect a flushing device between a water source and the flush port on the starboard side of the drive. Turn on the water to about half of its maximum flow. Do not use full tap pressure.

5. Start the engine (make sure you're in Neutral) and observe what the booster cylinder piston rod does.

6. If the booster cylinder piston rod moves to Starboard, turn the adjusting nut clockwise until the rod just begins to move to Port and note the position of the nut. Turn the nut counterclockwise slowly until the rod starts to move to Starboard again and note this position. Now, turn the nut clockwise to the midpoint between the two positions you noted.

7. If the booster cylinder piston rod moves to Port, turn the adjusting nut counterclockwise until the rod just begins to move to Starboard and note the position of the nut. Turn the nut clockwise slowly until the rod starts to move to Port again and note this position. Now, turn the nut counterclockwise to the midpoint between the two positions you noted.

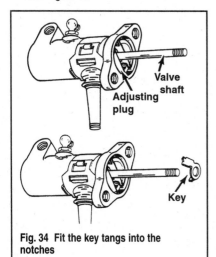

Fig. 34 Fit the key tangs into the notches

Valve shaft
Adjusting plug
Key

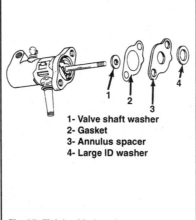

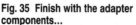

Fig. 35 Finish with the adapter components...

1- Valve shaft washer
2- Gasket
3- Annulus spacer
4- Large ID washer

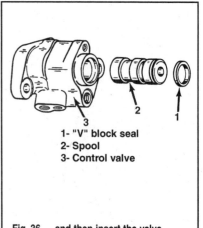

Fig. 36 ...and then insert the valve spool in the valve housing

1- "V" block seal
2- Spool
3- Control valve

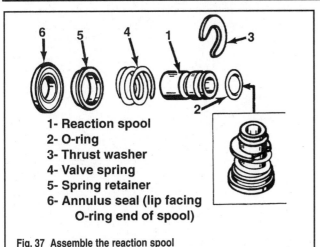

1- Reaction spool
2- O-ring
3- Thrust washer
4- Valve spring
5- Spring retainer
6- Annulus seal (lip facing
      O-ring end of spool)

Fig. 37 Assemble the reaction spool

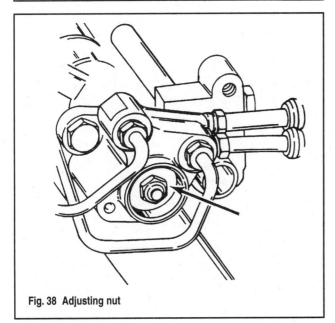

Fig. 38 Adjusting nut

8. Turn off the engine and disconnect the flushing device.
9. Fill the adjustment nut cavity in the control valve with 2-4-C Marine Lubricant and then install the dust cover.
10. Connect the clevis assembly and the steering cable.
11. Start the engine and confirm that there is no longer any drive unit creeping. If there is and you are confident of your adjustments, check the steering cable adjustment.

1998–01 Bravo

The power steering unit on these models is not serviceable and must be replaced as a unit.

Booster Cylinder

REMOVAL & INSTALLATION

 DIFFICULT

Alpha And 1992–97 Bravo Only

◆ See Figures 39 and 40

■ Although not absolutely necessary, we recommend removing the power steering unit before performing this procedure.

1. Remove the power steering unit.
2. Remove the control valve.
3. Loosen the nut on the end of the booster piston and pull off the clevis assembly.
4. Remove the cotter pin and then thread a small screw into each of the two retaining pins and remove the pins from the adapter block assembly.
5. Disconnect the booster cylinder from the adapter assembly.
6. If necessary, remove the two hydraulic lines at the end of the booster. Mark which line went to which fitting.
7. Remove the snap-ring and then pry out the large and small oil seals.
To install:
8. Clean the assembly thoroughly and inspect for cracks, wear or any other damage.
9. Install the small oil seal over the piston rod with the lip facing toward the cylinder.
10. Install the large oil seal over the rod with the lip facing away from the cylinder. Install the snap-ring.
11. Reconnect the hydraulic lines to their appropriate fittings and tighten securely. Do not over-tighten.
12. Install the booster cylinder into the adapter assembly and press in the two retaining pins. Install the cotter pin.
13. Install the clevis assembly over the piston rod and tighten the end nut securely.
14. Install the control valve.
15. Install the power steering assembly.

1998–01 Bravo

The power steering unit on these models is not serviceable and must be replaced as a unit.

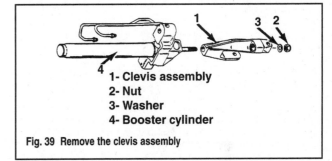

1- Clevis assembly
2- Nut
3- Washer
4- Booster cylinder

Fig. 39 Remove the clevis assembly

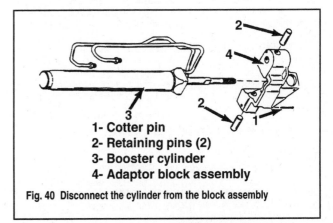

1- Cotter pin
2- Retaining pins (2)
3- Booster cylinder
4- Adaptor block assembly

Fig. 40 Disconnect the cylinder from the block assembly

TESTING

DIFFICULT

1. Connect a flushing device between a water source and the flush port on the starboard side of the drive. Turn on the water to about half of its maximum flow. Do not use full tap pressure.
2. Start the engine and reach in and push the control valve adapter block to the right until the booster cylinder piston rod is fully retracted into the cylinder.

3. Turn the engine OFF and disconnect the upper hydraulic line (the metal one to the booster, not the line from the power steering pump) on the control valve. Have some rags handy to take care of the inevitable spilled fluid and make sure you plug both the valve fitting and the line end.

4. Start the engine again and MOMENTARILY push the valve adapter block to the right and observe what happens.

✳✳ WARNING

Do not lug the pump at maximum pressure for more than 5 seconds in the above step or you risk damaging the pump.

5. If the piston rod extends while pushing the adapter block, the booster cylinder has a leak and will require replacement. Repeat this test again after replacement; if pressure is still low (the rods moves) you will need to replace the control valve also.

6. If the rod does not move, but you are still experiencing low system pressure, replace the control valve.

7. Turn the engine OFF and connect the hydraulic line, clevis assembly and steering cable.

Power Steering Pump

■ **The power steering pump utilizes Metric fasteners.**

 DIFFICULT

REMOVAL & INSTALLATION

1992–97 V6/V8 And All 4 Cylinder Engines

◆ **See Figures 41 and 42**

1. Relieve drive belt (or serpentine belt) tension as detailed in the Maintenance section and remove the belt.

■ **On certain engines applications you may have to remove the pump before disconnecting the hydraulic lines.**

2. Loosen the hose clamp on the return line (lower) and pull the line off the fitting on the pump housing. Make sure to have a suitable container and some rags available. Drain the fluid reservoir. Plug the line and secure it somewhere with the plugged end facing up.

3. Loosen the pressure line fitting on the rear of the pump housing and remove the line. Plug it and tie it up as you did with the return line.

4. Loosen the pump to brace bolt (if you haven't already) and disconnect the pump from the engine brace. If necessary, remove the brace from the engine.

5. Remove the two mounting bracket bolts and lift out the pump assembly.

■ **On dual engine installations there may be a remote reservoir connected by hoses to the top of each pump reservoir. Remove the hose if necessary.**

6. Remove the pump pulley and mounting bracket if necessary.

To install:

7. Install the mounting bracket and pulley if removed.

■ **On certain engines applications you may have to connect the hydraulic lines prior to installing the pump.**

8. Install the pump and bracket assembly onto the brace and tighten the bolts securely.

9. If removed, attach the pump brace to the engine and tighten it to 30 ft. lbs. (41 Nm).

10. Connect the pump to the engine brace and tighten the bolt securely.

11. Connect the pressure line to the flow control fitting and tighten it securely; be sure to use a new O-ring. Slide the return line over the pump fitting and tighten the hose clamp.

12. Install and adjust the drive/serpentine belt.

13. Fill the reservoir with power steering fluid and bleed the system.

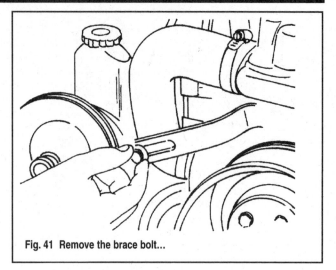
Fig. 41 Remove the brace bolt...

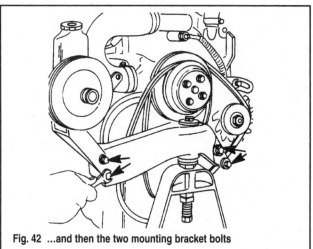
Fig. 42 ...and then the two mounting bracket bolts

1998–01 V6/V8 Engines

EXCEPT 8.1L

◆ **See Figure 43**

1. Loosen the locknut on the idler pulley adjusting stud, turn the stud to release belt tension and remove the serpentine belt from around the pump.

2. Loosen the hose clamp on the return line (low pressure) and pull the line off the fitting on the pump housing. Make sure to have a suitable container and some rags available. Drain the fluid reservoir. Plug the line and secure it somewhere with the plugged end facing up.

3. Loosen the pressure line fitting on the rear of the pump housing and remove the line. Plug it and tie it up as you did with the return line.

4. Loosen the pump to brace/bracket fasteners (bolts, nuts or both and lift out the pump assembly.

■ **On dual engine installations there may be a remote reservoir connected by hoses to the top of each pump reservoir. Remove the hose if necessary.**

5. Remove the pump pulley and mounting bracket if necessary.

To install:

6. Install the mounting bracket and pulley if removed.

7. Connect the pump to the engine brace/bracket and tighten the nuts and/or bolts to 30 ft. lbs. (41 Nm).

8. Connect the pressure line to the flow control fitting and tighten it securely; be sure to use a new O-ring. Slide the return line over the pump fitting and tighten the hose clamp.

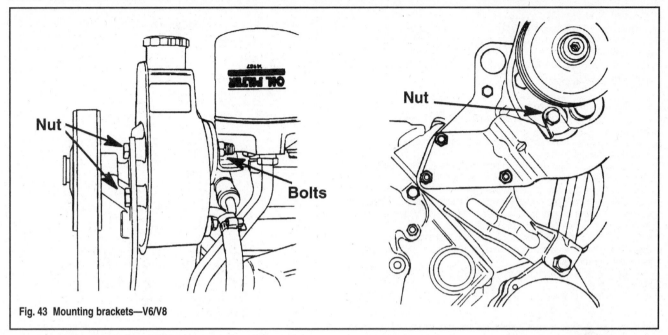

Fig. 43 Mounting brackets—V6/V8

9. Install and adjust the serpentine belt.
10. Fill the reservoir with power steering fluid and bleed the system.

8.1L ENGINES

◆ See Figure 44

These engines utilize a remote power steering fluid reservoir.

1. Loosen the serpentine belt idler pulley and ease belt tension. Remove the serpentine belt.
2. Loosen the hose clamp on the return line (low pressure) and pull the line off the fitting on the pump housing. Make sure to have a suitable container and some rags available. Drain the fluid reservoir. Plug the line and secure it somewhere with the plugged end facing up.
3. Loosen the pressure line fitting on the front of the pump housing and remove the line. Plug it and tie it up as you did with the return line.
4. Loosen the three pump to brace/bracket bolts and lift out the pump assembly.

To install:
5. Connect the pump to the engine brace/bracket and tighten the bolts to 30 ft. lbs. (41 Nm).
6. Connect the pressure line to the larger fitting and tighten it to 23 ft. lbs. (31 Nm); be sure to use a new O-ring. Slide the return line over the pump fitting and tighten the hose clamp.
7. Install and adjust the serpentine belt.
8. Fill the reservoir with power steering fluid and bleed the system.

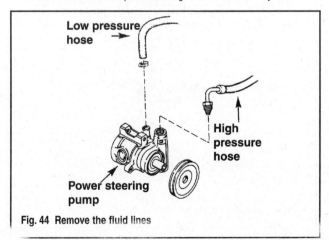

Fig. 44 Remove the fluid lines

FLOW CONTROL VALVE SERVICE

Except 8.1L Engines

◆ See Figure 45

■ Although removal of the flow control valve may be possible without removing the power steering pump first on certain installations, we recommend removing the pump.

1. Loosen and remove the fitting nut. The outer O-ring will probably be attached; discard it.
2. Pull out the control valve assembly and spring. Pry out the inner O-ring and discard it.
3. Clean all components thoroughly. Inspect them for cracks, wear or other damage.
4. Install as they were removed, using new O-rings. Tighten the fitting nut to 35 ft. lbs. (47 Nm).

8.1L Engines

The power steering pump on these engines is not serviceable.

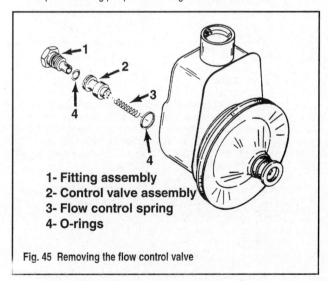

1- Fitting assembly
2- Control valve assembly
3- Flow control spring
4- O-rings

Fig. 45 Removing the flow control valve

PUMP PULLEY

◆ **See Figures 46 and 47**

1. Remove the power steering pump.
2. Carefully lay the pump on a clean surface with the pulley facing upward.
3. Install a pulley removal tool (#J-25034) or equivilent onto the end of the pulley and pump shaft.
4. Hold the flats of the tool with an open end wrench while turning the tool bolt clockwise until the pulley breaks loose from the shaft.
5. Press the pulley onto the shaft with a pulley installation tool (#91-93656A1). Thread the stud all the way into the shaft, place the bearing over the stud—don't use the spacer supplied with the kit. Thread the nut onto the shaft and then thread the shaft all the way onto the stud.
6. Install the pump (with the tool still attached). Install the drive belt and place a long straight edge across the pulley on the other side of the belt (crankshaft or circulating pump).
7. Turn the large pusher nut on the tool until the drive belt is parallel with the straightedge.
8. Remove the tool and adjust the belt tension.
9. Fill and bleed the system.

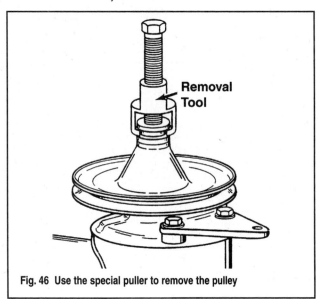

Fig. 46 Use the special puller to remove the pulley

OIL SEAL REPLACEMENT

Except 8.1L Engines

◆ **See Figures 48 and 49**

1. Remove the power steering pump and then remove the pulley.
2. Wrap a small piece of shim stock (0.005 in.) around the shaft and feed inside the seal until it bottoms in the body of the pump; about 2-1/2 in.
3. Use a small chisel to cut the seal and tear the metal body about 1 in.
4. Insert a small pry bar, or awl, between the seal edge and the housing and then pry out the seal. Remove the shim stock.
5. Install a new seal over the shaft with the metal side facing up.
6. Support the pump housing/reservoir on two pieces of wood, position a 1 in. socket over the seal and lightly tap it into place in the housing.
7. Install the pump and pulley.

8.1L Engines

The power steering pump on these engines is not serviceable.

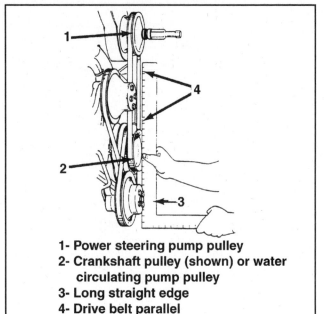

1- Power steering pump pulley
2- Crankshaft pulley (shown) or water circulating pump pulley
3- Long straight edge
4- Drive belt parallel

Fig. 47 Checking pulley alignment

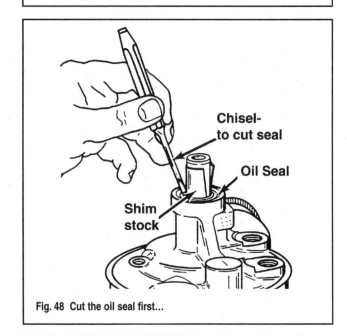

Fig. 48 Cut the oil seal first...

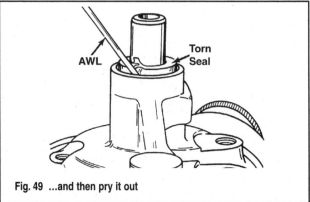

Fig. 49 ...and then pry it out

DISASSEMBLY & ASSEMBLY

 ③ DIFFICULT

Except 8.1L Engines

◆ **See Figures 50, 51, 52, 53, 54 and 55**

1. Remove the pump and pulley.
2. Remove the flow control valve assembly.
3. Loosen the studs and separate the pump housing from the reservoir. You may have to persuade the separation a little, but don't hit it too hard and use a rubber mallet.
4. Insert a small awl into the hole in the end plate and maneuver the retaining ring around in its recess until the end of the ring is approximately 1 in. (25mm) past the hole. Press in on the ring with the awl while prying up on it with a small screwdriver until it pops out of the recess.
5. Lift out the internal pump components as shown in the illustration. Be careful not to lose any of the pump vanes.
6. Remove the two dowel pins. Pry the two O-rings out of the housing and discard them.
7. Remove the small C-clip on the end of the pump shaft and pull off the rotor and thrust plate.
8. Remove the magnet from inside the housing.

To assemble:

9. Clean all metal components thoroughly. Inspect all parts for wear or other damage.

■ **It's a good idea to purchase a seal kit (#5688044) from your local General Motors dealer.**

10. Install a new pressure plate O-ring in the third groove of the housing and then insert the two dowel pins into their recesses. Make sure to coat the O-ring with power steering fluid.
11. Slide the thrust plate onto the pump shaft and then install the rotor with the countersunk side facing the thrust plate. Install the C-clip.
12. Insert the shaft assembly into the housing and then position the pump ring over it so the dowel pins fit into the two holes in the ring.
13. Insert all pump vanes into the slots in the rotor so that rounded edges face outward toward the pump ring. Make sure that each vane moves freely in its slot.
14. Install the pressure plate so the spring groove is facing up.
15. Coat a new large O-ring with power steering fluid and install it into the second groove in the housing.

16. Position the spring on the pressure plate and install the endplate. Squeeze the retaining ring together and install it over the plate and into the groove. An arbor press will make this easier, but you should be able to get it in without it.
17. Coat the remaining large O-ring and the two small ones with power steering fluid and install them.
18. Position the magnet into the housing.
19. Slide the pump housing into the reservoir and install the studs. Tighten to 35 ft. lbs. (47 Nm).
20. Install the flow control valve and the pulley and then install the pump.

8.1L Engines

The power steering pump on these engines is not serviceable.

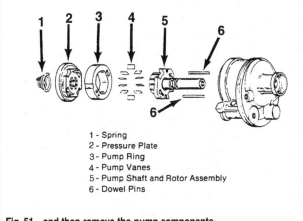

1 - Spring
2 - Pressure Plate
3 - Pump Ring
4 - Pump Vanes
5 - Pump Shaft and Rotor Assembly
6 - Dowel Pins

Fig. 51...and then remove the pump components

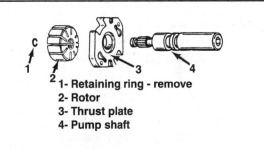

1- Retaining ring - remove
2- Rotor
3- Thrust plate
4- Pump shaft

Fig. 52 Separate the rotor and thrust plate

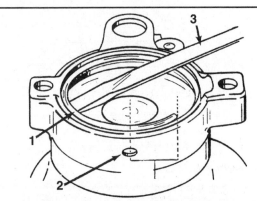

1- Retaining ring - position so that ring end is 1 in. (25mm) from end of hole in housing
2- Hole- insert into Awl in housing to push ring from recess
3- Screwdriver - Pry ring from pump body

Fig. 50 Pry out the retaining ring...

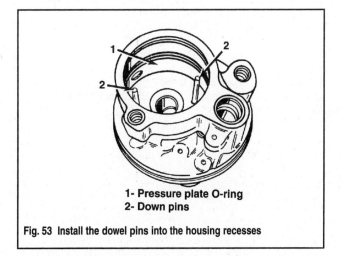

1- Pressure plate O-ring
2- Down pins

Fig. 53 Install the dowel pins into the housing recesses

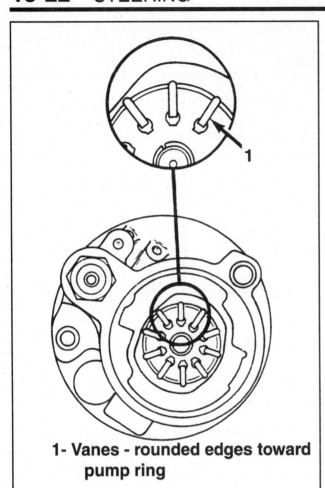

1- Vanes - rounded edges toward pump ring

Fig. 54 Correct pump vane positioning

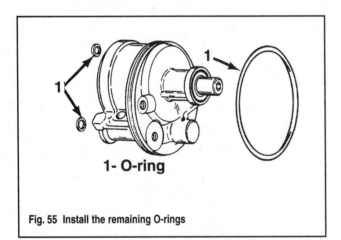

1- O-ring

Fig. 55 Install the remaining O-rings

Power Steering Cooler

All engines covered here utilize a raw-water cooled steering cooler on the low pressure (return) side of the power steering system; located in-line between the control valve and the power steering pump.

REMOVAL & INSTALLATION

◆ See Figure 56

1. Locate the cooler. It can generally be found at the rear of the engine, although certain engines will have it tucked in at the front of the engine. If it's not readily visible, just follow the lines from the pump.
2. Drain the cooling system as detailed in the Maintenance section.
3. Loosen the hose clamps and wiggle off the cooling hoses on each end of the unit. Have some rags and a container handy as there will still be some residual water in the cooler and lines. Tie the hoses up and out of the way.

■ **Although not absolutely necessary, we recommend draining the power steering system before attempting this procedure.**

4. Loosen the hydraulic lines running into and out of the cooler unit. Most will simply use a hose clamp, but a few application will use a screw-in fitting. Have some rags and a container handy as there will still be some residual fluid in the cooler and lines. Tie the hoses up and out of the way. Plug the openings to prevent contamination.
6. Loosen and remove the mounting brackets—usually just hose clamps, and lift out the cooler.
To install:
7. Clean the cooler thoroughly and inspect it for cracks, wear or other damage.
8. Position the cooler and tighten the mounting hardware.
9. Slide the cooling hoses over the end flanges and tighten the hose clamps securely. Check the hose ends for cracks, crimping or bulges.
10. Install the hydraulic lines onto their original fitting and tighten the hose clamps or nuts securely.
11. Refill the cooling and power steering systems.

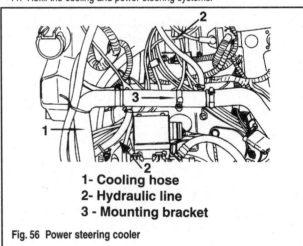

1- Cooling hose
2- Hydraulic line
3 - Mounting bracket

Fig. 56 Power steering cooler

COMPACT HYDRAULIC SYSTEM

This system uses no conventional pump, cable or control valve. Steering is controlled hydraulically via the helm.

Hydraulic Fluid

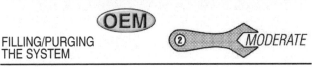

FILLING/PURGING
THE SYSTEM

◆ See Figures 57 and 58

To complete successful purging of this system you will need two people.

■ Hydraulic fluid must be visible in the filler tube during the entire procedure. Never allow the fluid container used for filling to empty, causing the filler tube to be empty and allowing air to enter the system.

1. Remove the vent/fill plug from above the helm.
2. Obtain a Filler Kit from your local Mercury representative and screw the filler tube into the vent/fill hole. Hand tighten the tube and then screw in a container of hydraulic fluid to the top of the tube.
3. Turn the container upside down, support it and then carefully punch a small hole in its bottom.
4. With fluid still visible in the fill tube and in the container, remove the container and install another one. Punch a hole in this one also…and you thought this was going to be neat!
5. When no air bubbles are visible in the filler tube any longer, the helm is full. Leave the fluid container connected, and securely supported.
6. Pop off the caps on the bleeder valves in the T-fittings on the steering cylinder. Attach a sufficient length of clear plastic tubing from each bleeder and into a suitable container.
7. Have an assistant start to turn the steering wheel SLOWLY clockwise while you open the Starboard bleeder. Continue turning the wheel until you observe an air-free (no bubbles) stream of fluid coming through the tube and then close the bleeder. Continue turning the wheel until the piston rod is extended fully from the cylinder.
8. Now, have your assistant start to turn the steering wheel SLOWLY counterclockwise while you open the Port bleeder. Continue turning the wheel until you observe an air-free stream of fluid coming through the tube and then close the bleeder. Continue turning the wheel until the piston rod is fully retracted into the cylinder and the wheel comes to a stop.
9. Open the Starboard bleeder again. Hold the piston rod in the retracted position while continuing to turn the steering wheel counterclockwise until an air-free stream of fluid is visible. Close the bleeder while continuing to turn the wheel.
10. Check the fluid level.

CHECKING THE LEVEL

1. Turn the steering wheel until the steering cylinder piston rod is fully retracted into the cylinder.
2. Obtain a Filler Kit from your local Mercury representative and screw the filler tube into the vent/fill hole above the helm.
3. Pour a container of hydraulic fluid into the tube very slowly until it is about 1/2 full. Make sure there are no air bubbles in the tube.
4. Pop off the cap on the Starboard bleeder valve in the T-fitting on the steering cylinder. Attach a sufficient length of clear plastic tubing from each bleeder and into a suitable container.
5. Have an assistant start to turn the steering wheel SLOWLY clockwise while you open the Starboard bleeder. Continue turning the wheel until the fluid level in the tube drops to the top of the plastic fitting. Continue turning the wheel for another 1/4 of a turn and then stop. Close the bleeder.
6. Remove the filler tube and install the vent/fill plug. The fluid level should be at the bottom of the filler hole. Never allow the fluid level to drop more than 1/4 in. (6mm) below the hole.

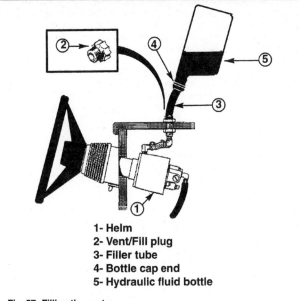

1- Helm
2- Vent/Fill plug
3- Filler tube
4- Bottle cap end
5- Hydraulic fluid bottle

Fig. 57 Filling the system

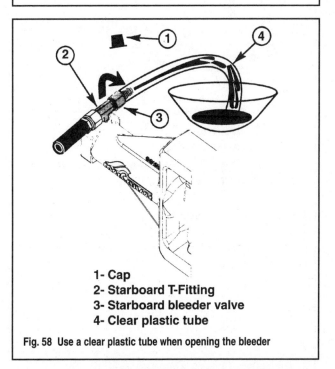

1- Cap
2- Starboard T-Fitting
3- Starboard bleeder valve
4- Clear plastic tube

Fig. 58 Use a clear plastic tube when opening the bleeder

Steering Unit

REMOVAL & INSTALLATION

◆ See Figures 59, 60, 61 and 62

1. Disconnect the hydraulic lines at the steering cylinder assembly T-fittings. Be sure to plug both of the lines and the inlets on the cylinder. Have some rags handy because you're going to spill some power steering fluid. Plug the lines as quickly as possible and tie them up and out of the way.

2. Remove the lower cotter pin and pull out the clevis connecting the clevis assembly and piston rod to the steering lever.

3. Use a screwdriver or pair of pliers and bend the locking tabs away from the upper and lower pivot bolts. One side will be away from the bolt and the other will be away from the ridge in the transom assembly.

4. Remove the two pivot bolts and tab washers from the transom assembly and then pull out the steering cylinder.

To install:

5. Lubricate the upper and lower assembly with a good amount of Special Lubricant 101. Do the same to each pivot bolt.

6. Position the assembly onto the transom assembly and then screw the bolts (don't forget the washers) in all the way by hand—no wrenches yet.

7. Make sure that the tab washers tangs are straddling the ridge in the transom plate on one side and then tighten them to 25 ft. lbs. (34 Nm)—you can use a wrench this time! Hopefully when you hit the correct torque figure, the remaining tab will align with a flat on the bolt. If so, bend it up and against the bolt head and then bend the other tab down over the transom assembly ridge. If it doesn't align, unscrew it and start over. If you need to tighten the bolt slightly to get the flat and tab to line up, it's Ok; but do not though, loosen the bolt to line things up.

8. Connect the cylinder piston rod to the steering lever and slide in the clevis pin. Install the cotter pin. Make sure that the angled notch in the clevis is facing aft as shown.

9. Unplug and reconnect the two hydraulic lines. Lubricate a new O-ring with hydraulic fluid and press the hoses all the way into their fitting. Tighten the fitting nuts to 130 inch lbs. (15 Nm).

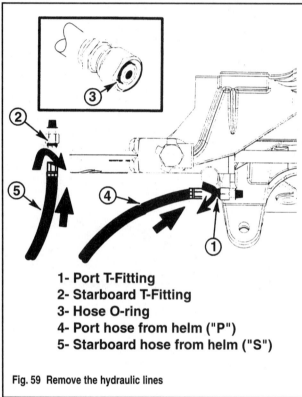

1- Port T-Fitting
2- Starboard T-Fitting
3- Hose O-ring
4- Port hose from helm ("P")
5- Starboard hose from helm ("S")

Fig. 59 Remove the hydraulic lines

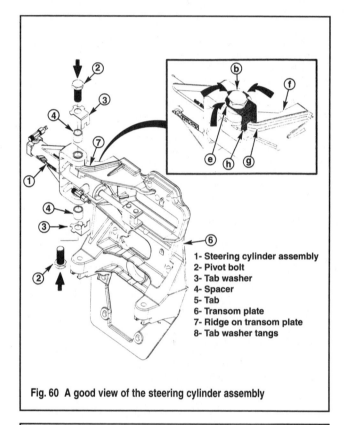

1- Steering cylinder assembly
2- Pivot bolt
3- Tab washer
4- Spacer
5- Tab
6- Transom plate
7- Ridge on transom plate
8- Tab washer tangs

Fig. 60 A good view of the steering cylinder assembly

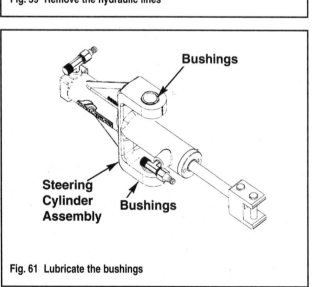

Bushings

Steering Cylinder Assembly

Bushings

Fig. 61 Lubricate the bushings

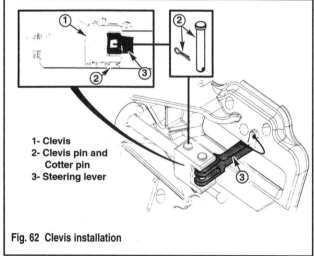

1- Clevis
2- Clevis pin and Cotter pin
3- Steering lever

Fig. 62 Clevis installation

GLOSSARY

Accumulator An air-filled tank used to smooth out pressure in a freshwater system; also a tank used in a refrigeration system to trap liquid refrigerant that might otherwise damage the compressor.

Aerial: See antenna.

Aft: Toward the rear of your vessel.

Alternating Current (AC): Electrical current reversing its direction at regular intervals. Each repetition of these changes is a cycle and the number of cycles that take place in 1 second is the frequency.

Alternator: A machine for generating electricity by spinning a magnet inside a series of coils. The resulting power output is alternating current (AC).

Ambient Conditions: The surrounding temperature or pressure, or both.

Ammeter: An instrument for measuring current flow.

Ampere: A measure of the rate of electric current.

Anode: A sacrificial alloy (zinc>) that attracts electrolysis before the metal (usually aluminum) in your drive system

Anti-Siphon Valve: A valve that admits air to a line and prevents siphonic action.

Armature: The rotating windings in a generator (AC or DC).

ATDC: After top dead center. The point after the piston reaches the top of its travel on the compression stroke.

BTDC: Before top dead center. The point just before the piston reaches the top of its travel on the compression stroke.

Backlash: The clearance, or play, between two parts, such as gears.

Ball Valve: Either a valve with a spring-loaded ball or one with a ball rotating in a spherical seat.

Beam: A vessel's dimension, measured at its widest point.

Bearing: A device for supporting a rotating shaft with minimum friction. It may take the form of a metal sleeve (a bushing), a set of ball bearings (a roller bearing), or a set of pins around the shaft (a needle bearing). Also, heading, or position on a compass.

Bearing Race: The outer cage within which a set of balls rotates in a roller bearing.

Binnacle: A housing for a compass.

Bleeding: The process of purging air from a fuel or hydraulic system.

Bow: The front of your vessel.

Breaker Points: A set of points inside the distributor, operated by a cam, which make and break the ignition circuit.

Bridge Rectifier: An arrangement of diodes for converting alternating current (AC) to direct current (DC).

Cable Clamp: A U-shaped bolt with a saddle used to join or to make loops in wire rope.

Camshaft: A shaft in the engine on which are the lobes (cams) which operate the valves. The camshaft is driven by the crankshaft, via a belt, chain or gears, at one half the crankshaft speed.

Carburetor: A device, usually mounted on the intake manifold of an engine, which mixes the air and fuel in the proper proportion to allow even combustion.

Cavitation: The rapid formation and collapse of water vapor bubbles in a low pressure area on the leading edge of the propeller.

Check Valve: An electrical or mechanical valve that allows flow in only one direction.

Chine: The intersection of the bottom and sides on a flat or V-bottomed boat.

Circlip: A split steel snapring that fits into a groove to hold various parts in place.

Circuit: The path of electric current. A closed circuit has a complete path. An open circuit has a broken or disconnected path. A short circuit has an unintentional direct path bypassing the equipment (appliance, resistance) in the circuit.

Circuit Breaker: A load-sensitive switch that trips (opens a circuit) if a threshold-exceeding current flows through it.

Cleat: Hardware used to tie up line. Usually on the deck or dock.

Clevis Pin: A metal pin with a flattened head at one end and a hole for a cotter pin (split pin) at the other. It is used to fasten rigging together.

Clew: The lower, aft corner of a sail.

Clutch: A device used to couple and uncouple a power source from a piece of equipment. It may be manually, hydraulically, or electro-magnetically operated. A cone clutch forces a tapered seat onto a tapered friction pad. A brake-band clutch tightens a friction band around a smooth face on a gear. A disc clutch holds alternating metal and friction plates together.

Compression Ratio: The volume of a combustion chamber with the piston at the top of its stroke as a proportion of the total volume of the cylinder with the piston at the bottom of its stroke.

Connecting Rod: The connecting link between the piston and the crankshaft.

Corrosion: A process that leads to the destruction of two metals. Galvanic corrosion arises when dissimilar, electrically connected metals are immersed in an electrolyte (e.g., salt water). A current is generated, leading to the destruction of the anode (less noble metal) and the protection of the cathode (more noble metal). Pinhole and crevice corrosion are the results of galvanic corrosion occurring in just one piece of metal due to differences in the composition of the metal. Stray-current corrosion is the result of external current leakage through metal fittings in contact with an electrolyte, such as salt water. Massive corrosion can occur where the current leaves a fitting. The term electrolysis refers to the passage of electricity through the electrolyte.

Cotter Pin: A pin with two legs. With legs together the pin is placed through the hole in a clevis pin. The legs are then opened outward to prevent the cotter pin from backing out of the hole. The cotter pin, in turn, prevents the load-bearing clevis pin from backing out of its retaining hole.

Crankshaft: The engine component that converts the reciprocating (up and down or back and forth) motion of the pistons into the rotary motion used to turn the driveshaft.

CV-Joint (constant velocity joint): A type of propeller shaft coupling that permits considerable engine misalignment.

Displacement Hull: A type of hull (usually round) that displaces water with forward motion. Common in sailboats.

Distributor: A mechanically driven device on an engine which is responsible for electrically firing the spark plug at a pre-determined point of the piston stroke.

Dowel: A round metal or wooden pin.

Drift: Any suitable sized round metal bar used to knock out bushings, clevis pins and the like.

DVOM: Digital volt ohmmeter.

Electrolysis: A chemical change or breakdown of an electrolyte caused by an electric current.

Electrolyte: The solution in a battery. A liquid conductor of electricity. Also, an electrically conductive solution (saltwater) with a high mineral content.

Feeler Gauge: Thin strips of metal machined to precise thicknesses or round metal wires of precise diameters that are used for measuring small gaps.

Firing Order: The order in which combustion occurs in the cylinders of an engine.

Flap Valve: A simple rubber flap, sometimes weighted. Fluid pressure opens it in one direction and closes it in the other.

Fuel Filter: A filter used to prevent impurities from entering the engine via the fuel system.

Fuel/Water Separator: In line filter used to remove water particles from the fuel prior to entering the engine.

Fuse: A protective device in a circuit which prevents circuit overload by breaking the circuit when a specific amperage is present. The device is constructed around a strip or wire of a lower amperage rating than the circuit it is designed to protect. When an amperage higher than that stamped on the fuse is present in the circuit, the strip or wire melts, opening the circuit.

Fusible Link: A piece of wire in a wiring harness that performs the same job as a fuse. If overloaded, the fusible link will melt and interrupt the circuit.

Gear Ratio: The relative size of two gears. If the gears are in contact, their relative speed of rotation will be given by the gear ratio. Example: If the gear ratio is 8:1, the smaller gear will rotate eight times faster than the larger gear.

Generator: A machine for generating electricity by winding a series of coils inside a magnet. The resulting power output is alternating current. In DC systems, this output is rectified via a commutator and brushes.

Glow Plug: A heating element installed in diesel engine precombustion chambers to aid in cold-starting.

Governor: A device for maintaining an engine or electric motor at a constant speed, regardless of load.

Ground: A connection between an electric circuit and the earth, or some conducting body serving in place of the earth.

Gudgeon: One-half of a rudder hinge, the other half being a pintle.

Gunwale: The upper edge of the sides of a vessel.

Header Tank: A small tank set above an engine on heat exchanger-cooled systems. The header tank serves as an expansion chamber, coolant reservior, and pressure regulator (via a pressure cap).

Heat Exchanger: A vessel containing a number of small tubes through which cooling water is passed, while raw water is circulated around the outside of the tubes to carry off heat from the cooling water.

Heat Sink: A mounting for an electronic component designed to dissipate heat to the atmosphere.

Helm: The driver's control center

Hose Clamp: An adjustable stainless steel or plastic band used primarily for securing hose ends to their fittings.

Hull: The bottom of a vessel, structurally.

Hydrometer: A float-type instrument that measures specific gravity. Most often used to determine the state of charge of a battery by measuring the specific gravity of the electrolyte.

Impeller: The rotating fitting that imparts motion to a fluid in a rotating pump. Commonly found in marine water pumps and jet drives.

Inboard: An engine located inside the hull of a vessel

Injector: A device for atomizing fuel and spraying it into a cylinder or fuel delivery chamber. Used on all diesel and gasoline injected engines.

Jet Drive: Propulsion system where water is drawn into a drive unit via an impeller and expelled at high pressure through an outlet directed away from the vessel.

Keel: The centerline of the bottom of the hull.

Leeward: The side of a landform or object sheltered from the wind.

Mast: A vertical pole, usually wood or aluminum, that a sail is attached to.

Noble Metal: A metal high on the galvanic table. Noble metals are likely to form a cathode in many cases of galvanic corrosion and therefore are unlikely to corrode.

OEM: Original equipment manufacturer.

Ohm: A unit of electrical resistance

Ohmmeter: An instrument for measuring resistance. Usually incorporated as one function of a multi-meter.

Outboard: An engine/drive system connected to the transom of a vessel.

Outdrive: The lower half of a stern drive unit.

PFD: Personal flotation device

Pinion: A small gear designed to mesh with a large gear (for example, a starter-motor drive gear).

Pintle: One-half of a rudder hinge, the other half being a gudgeon.

Pitch: The total distance a propeller would travel in one revolution, as determined by the amount of deflection of its blades, if there were no losses as it turned.

Planetary Gears: An arrangement of small gears around a central drive gear, with a large ring gear around the outside of the small gears.

Planing Hull: A type of hull that lifts out of the water at a certain speed in order to reduce drag.

Port: The left side of the boat as you face the bow.

Potentiometer: A variable resistor used for adjusting some voltage regulators.

Propeller: A multi-bladed device at the end of the vessel's drive system used to propel the vessel through the water.

PWC: Personal watercraft.

RPM: Revolutions per minute (usually indicating engine speed).

Raw Water: The water the boat sits in, which is used to cool the engine and components. May be used in conjunction with a closed system that uses conventional anti freeze.

Reed Valve: Essentially a one-way check valve(s) allowing the air/fuel mixture from the carburetor to enter the crankcase, but not exit it. Usually found on 2-stroke engines.

Resistance: The opposition an appliance or wire offers to the flow of electric current, measured in ohms.

Rudder: A flat, vertical structure usually attached to the stern of a boat for steering purposes.

Shim Stock: Very thin, accurately machined pieces of metal.

Skeg: A small keel aft, used to support a rudder.

Spade Rudder: A rudder with no support beneath the hull.

Specitic Gravity: A measure of the density of a liquid when compared with water. Most often used to measure the density of electrolyte in a battery, i.e., the strength of the acid and therefore the battery's state of charge.

Split Pin: See cotter pin

Spreader: A strut on a mast to improve the angle of shrouds and stiffen the mast panels.

Starboard: The right side of the boat as you face the bow.

Starter: A high-torque electric motor used for the purpose of starting the engine, typically through a high ratio geared drive connected to the flywheel ring gear.

Stern: The rear of your vessel.

Stern Drive: A drive unit attached to the stern of a vessel and connected to an inboard engine.

Stroke: The distance that the piston travels from bottom dead center to top dead center.

TDC: Top dead center. The point at which the piston reaches the top of its travel on the compression stroke.

Thermistor: A resistor that changes in value with changes in temperature.

Thermostat: A heat-sensitive device used to control the flow of coolant through an engine; or a heat sensitive switch used to turn a water-heating element off and on.

Thickness Gauge: See feeler gauge.

Tiller: The lever used to turn the rudder from side-to-side.

Transom: The stern of a square-ended boat

Trim Tab: A small rudder hinged to the trailing edge of a main or auxiliary rudder, or the drive unit. Also, hydraulic devices mounted to the transom on many planning hull vessels, used to adjust the vessel's attitude (fore-to-aft or port-to-starboard)while on plane.

Turbocharger: A blower driven by engine exhaust gas and used to pressurize the inlet air.

Turnbuckle: An adjustable fitting used to tension standing rigging.

Valve Clearance: The gap between a valve stem and its rocker arm when the valve is fully closed.

Valve Cover: The housing of an engine bolted over the valve mechanism. Also, cylinder head cover.

Viscosity: The ability of a fluid to flow. The lower the viscosity rating, the easier the fluid will flow. 10 weight motor oil will flow much easier than 40 weight motor oil.
Volt: Unit used to measure the force or pressure of electricity.

Voltage Regulator: A device that controls the current output of the alternator or generator.

Voltmeter: An instrument used for measuring electrical force in units called volts. Voltmeters are always connected parallel with the circuit being tested.

Windlass: Any of various mechanisms used to hoist or haul and anchor.

Windward: The side of a landform or object that the wind in blowing on.

Zinc: See Anode.

SELOC PUBLISHING'S FULL-LINE MASTER LIST

OUTBOARDS

ISBN	PART NO.	TITLE/DESCRIPTION	YEARS
089330018-7	1000	Chrysler Outboards, All Engines	1962-84
089330055-1	1100	Force Outboards, All Engines	1984-99
089330048-9	1200	Honda Outboards, All Engines	1978-01
089330007-1	1300	Johnson/Evinrude Outboards, 1-2 Cyl	1956-70
089330008-X	1302	Johnson/Evinrude Outboards, 1-2 Cyl	1971-89
089330009-8	1306	Johnson/Evinrude Outboards, 3-4 Cyl	1958-72
089330010-1	1308	Johnson/Evinrude Outboards, 3, 4 & 6 Cyl	1973-91
089330063-2	1311	Johnson/Evinrude Outboards - All V Engines	1992-01
089330052-7	1312	Johnson/Evinrude Outboards, All In-line engines/2 & 4 Stroke	1990-01
089330015-2	1400	Mariner Outboards, 1-2 Cyl	1977-89
089330016-0	1402	Mariner Outboards, 3, 4 & 6 Cyl	1977-89
089330012-8	1404	Mercury Outboards, 1-2 Cyl	1965-91
089330013-6	1406	Mercury Outboards, 3-4 Cyl	1965-89
089330014-4	1408	Mercury Outboards, 6 Cyl	1965-89
089330051-9	1416	Mercury/Mariner Outboards, All 2-Stroke Engines	1990-00
089330067-5	1418	Mercury Outboards, All 2-Stroke Models	2001-05
089330050-0	1600	Suzuki Outboards, All 2-Stroke Engines	1988-99
089330064-0	1701	Yamaha Outboards, All Engines	1984-96
089330065-9	1703	Yamaha Outboards, All Engines, All 2-Stroke Engines	1997-03
089330066-7	1705	Yamaha/Mercury & Mariner - All 4-Stroke Engines	1995-04

STERN DRIVES

ISBN	PART NO.	TITLE/DESCRIPTION	YEARS
089330005-5	3200	Mercruiser Stern Drives	1964-91
089330053-5	3206	Mercruiser Stern Drives	1992-00
089330068-3	3208	Mercruiser Stern Drives	2001-06
089330004-7	3400	OMC Stern Drives	1964-86
089330056-X	3404	OMC Stern Drives	1986-98
089330011-X	3600	Volvo/Penta Stern Drives, All Gas Engines	1968-91
089330038-1	3602	Volvo/Penta Stern Drives, Volvo Engines	1992-93
089330057-8	3606	Volvo/Penta Stern Drives, All Gas Engines	1992-03

INBOARDS

ISBN	PART NO.	TITLE/DESCRIPTION	YEARS
089330049-7	7400	Yanmar Inboards	1975-98

PERSONAL WATERCRAFT

ISBN	PART NO.	TITLE/DESCRIPTION	YEARS
089330032-2	9200	Kawasaki	1973-91
089330042-X	9202	Kawasaki	1992-97
089330045-4	9400	Polaris	1992-97
089330033-0	9000	Sea-Doo/Bombardier	1988-91
089330043-8	9002	Sea-Doo/Bombardier	1992-97
089330034-9	9600	Yamaha	1987-91
089330044-6	9602	Yamaha	1992-97

Seloc-on-Line (Internet Access)

089330075-6	5000	One Mfg/Model - Subscription per user 3 years	1965-06

ENGINE FINDER

The following listings contain all engines covered in this manual

Model	Engine/Cylinder	Year Coverage
3.0	181, 4 cyl	1992 - 2001
4.3	262, V6	1992 - 2001
4.3	262 Mag, V6	1997
5.0	305, V8 (GM)	1992 - 2001
5.7	350, V8 (GM)	1992 - 2001
5.7	350 Mag, V8	1992 - 2001
6.2	377, V8	1998 - 2001
7.4	454, V8	1992 - 2001
7.4	454 Mag, V8	1992 - 2001
8.1	496, V8	1998 - 2001
8.1	496 Mag, V8	1998 - 2001
8.2	502, V8	1992 - 2001
8.2	502 Mag, V8	1992 - 2001
Alpha	Sterndrive	1992 - 2001
Blackhawk	Sterndrive	1992 - 2001
Borg Warner	Sterndrive	1992 - 1995
Bravo 1/11/111	Sterndrive	1992 - 2001
Hurth	Sterndrive	1992 - 2001
Velvet	Sterndrive	1996 - 2001
Walter	Sterndrive	1992 - 2001
ZF	Sterndrive	1996 - 2001

SelocOnLine

SelocOnLine is a maintenance and repair database accessed via the Internet.
Always up-to-date, skill level and special tool icons, quick access buttons to wiring diagrams, specification charts, maintenance charts, and a parts database.
Contact your local marine dealer or see our demo at www.seloconline.com

SelocOnLine

Step 1

Select your manufacturer
Year / Model
Engine

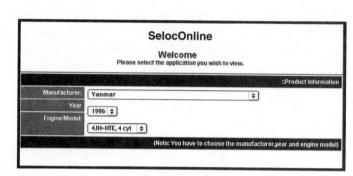

Step 2

Select an Engine System

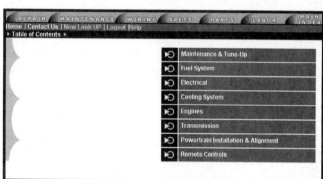

Step 3

Select the repair

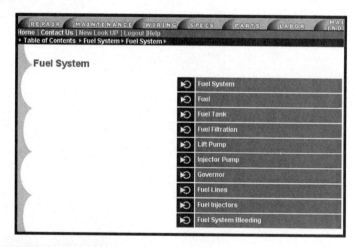

Step 4

Print the repair procedure

Home | Contact Us | New Look UP | Logout |Help
Table of Contents ▶ Fuel System ▶ Fuel System ▶ Lift Pump ▶ Removal & Installation ▶

JH Series Engines

Fig. 12 On JH engines, the fuel lift pump mounts to the injector pump ▶ see image

1. Turn the fuel tank petcock to the **OFF** position.
2. Place a small receptacle under the lift pump to contain any excess fuel.
3. Remove the fuel line fittings attached to the lift pump. Be careful not to lose the copper washers.

   ➤ Be sure to mark the fuel lines before they are removed from the pump. If the lines are reversed, the lift pump will not supply fuel to the injector pump.
4. Remove the two bolts that attach the lift pump to the engine block.
5. Draw the lift pump from the engine block.

   To install:

6. Carefully scrape any gasket residue from the mating surfaces of the lift pump and injector pump.
7. Lightly grease the end of the lift pump piston where it rides on the cam.
8. Using a new gasket, install the lift pump to the injector pump.
9. Install the fuel lines to the lift pump in the proper direction.
10. Once the lines are attached, turn the fuel petcock to the **ON** position.
11. Loosen the air bleed screw on the secondary (between the lift pump and injector pump) fuel filter.
12. Operate the priming lever on the lift pump until fuel runs out of the air bleed screw opening, then tighten the screw.

Seloc PRO

Real World Solutions In Real Time

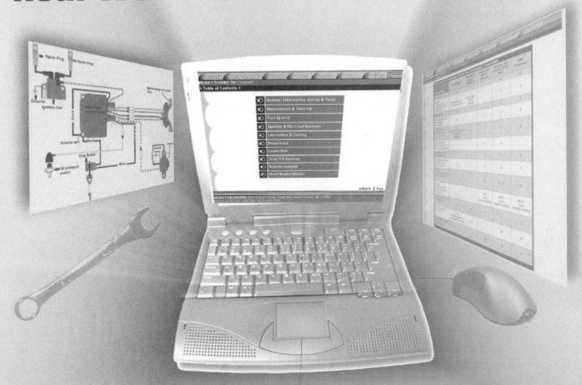

Seloc Pro is a mechanical repair database that is accessed via the internet. There is no other tool in your shop that will save you as much time or provide you with as much productivity as Seloc Pro. Unlike manufacturer information you may presently use, Seloc Pro allows you to navigate the same way thru our database regardless of the manufacturer. Our database is written so that each procedure is 1-3 pages in length. New content or changes to content can be added in minutes.

Seloc Pro is always up-to-date.

Features Include:

Quick access buttons to databses for Wiring Diagrams, Specifications, Parts and Labor Times

Seloc Labor Times - Real world freshwater and saltwater times for hundreds of operations for Johnson, Evinrude, Yamaha, and Mercury from 1980 thru 2000

Hyper-linked index for quick access to unit repair sections

Mfgs covered include Force, Honda, Johnson, Evinrude, Mercruiser, Mercury, Suzuki, Yamaha and Yanmar. Coming during the 4th quarter are Volvo Penta and OMC Stern Drive

Contact your local distributor for a demo or Seloc at 866-735-6255